PROGRESS IN BOTANICAL RESEARCH

Progress in Botanical Research

Proceedings of the 1st Balkan Botanical Congress

Edited by

Ioannes Tsekos
and
Michael Moustakas

Aristotle University of Thessaloniki, Greece

SPRINGER SCIENCE+BUSINESS MEDIA, B.V.

A C.I.P. Catalogue record for this book is available from the Library of Congress.

ISBN 978-94-010-6219-0 ISBN 978-94-011-5274-7 (eBook)
DOI 10.1007/978-94-011-5274-7

Printed on acid-free paper

Contents

Biochemistry, Metabolism and Bioenergenetics

Ecology and Ecophysiology

Structure and its Dynamics

Genetics, Plant Breeding and Biotechnology

Growth, Development and Differentiation

Preface

This volume is the final document of the 1st Balkan Botanical Congress and comprises after reviewing the full texts of the Congress Lectures submitted by their authors. The articles refer to all branches of plant sciences in the field of pure and applied research. The subjects dealt with in the Congress, and each representing a separate section in this book covered the following areas of interest:

I. Taxonomy, geobotany and evolution
II. Biochemistry, metabolism and bioenergetics
III. Ecology and ecophysiology
IV. Structure and its dynamics
V. Genetics, plant breeding and biotechnology
VI. Growth, development and differentiation.

The Congress was organized by the Department of Botany, Aristotle University of Thessaloniki and the Hellenic Botanical Society. The city of Thessaloniki was chosen by the Organizers since it enjoyed the accolade of the "cultural capital of Europe" for the year 1997.

The Thessaloniki Congress has taken on the character of an International Congress since 320 scientists, mainly from the Balkan countries and the rest of Europe (26 countries in all) took part in it. The 11 invited speakers who shared their experience with us, were well-known specialists from all the European countries. The aim of the Congress was to bring together scientists from a broad range of disciplines related to Plant Science. It provided an opportunity for discussions of all branches of Plant Science, and particularly of interdisciplinary subjects, and of aspects of wider significance in order to exchange new knowledge, ideas, experiences and techniques. The major objectives were a) to obtain a comprehensive perspective of the current knowledge of Botany, b) to present and discuss new experimental data, and c) to promote new research and collaboration between research teams in Balkan countries.

It has been a number of months since the Congress (September 19-22 1997) took place and it is still a pleasure to recall its main events and recapitulate its happenings. Both the quantity and quality of scientific interactions, within and outside the Congress, have exceeded our most optimistic expectations. Hopefully it will be remembered as an active forum of interdisciplinary exchange and as a pleasant moment. On behalf of the organizing committee we want to thank all participants again for making the 1st Balkan Botanical Congress very successful. We have received a number of very nice e-mails and letters after the Congress, and we are pleased to hear that many of the participants made new scientific contacts, were inspired with new ideas, and had a good time during it. We could not have wished for more! One of its goals has been the continuation of this Congress. Therefore upon its closure the Congress has been decided to be organized every three years. Istanbul (Turkey) has been chosen as the place where the 2nd Balkan Botanical Congress is to be held, the year 2000.

The Editors

Taxonomy, Geobotany and Evolution

PATTERNS OF EVOLUTION AND THE EFFECTS OF TOXIC METALS

A. D. BRADSHAW
School of Biological Sciences, University of Liverpool
Liverpool L69 3BX, U. K.

1. Abstract

We have often thought that evolution is a slow and gentle process. Yet natural conditions are always severe; as a result natural selection is likely also to be severe. Simple models show that severe selection can have profound effects on the genetic characteristics of a population. This is born out by many evolutionary studies on plant species growing in normal habitats. Species occurring in metal contaminated habitats provide particularly good evidence. The metal tolerance that evolves allows species to occupy habitats from which they would otherwise be excluded. The evolutionary change can be very rapid, taking only a few generations. It can also be highly localised; a patch of ground only a few metres across can maintain a genetically distinct population, despite gene flow.

But all this can happen only if the appropriate variation is present. Many species never evolve metal tolerance. This is because they do not possess the necessary variation, a condition of *genostasis*. We should respect the power of evolution, but not expect that it can always achieve adaptation.

2. Introduction - the power of the environment

At the present time, when we tend increasingly to respect nature and natural processes, it is all too easy to fall into the trap of seeing the natural world as something rather benign and kindly. As a result we often see evolution as a slow and gentle process. Certainly it can be a very slow process. Many of the species that are present today have existed in an identical form for many millions of years.

Yet ecologists know how severe, even violent, the environment can be. Death and elimination is a normal feature. Species can suffer set backs to their growth in the habitats they have occupied for thousands of years, and be completely eliminated from habitats to which they are not ecologically 'suited' but into which they regularly attempt to migrate. The degree of restriction of species to particular habitats is a fundamental aspect of all species with which every ecologist is familiar.

This has important implications in evolution. It indicates that natural selection can be very powerful. The condition of exclusion means that while the relative fitness of one species in a habitat can be 1.0, that of another can be 0. So the coefficient of selection against the unadapted species can be up to 1.0. This carries the implication that coefficients of selection against genotypes within populations can be similar - not the 0.001 or 0.01 assumed by Haldane and others in early work on evolutionary processes (Haldane 1932). I am sure that no ecologist, familiar with the power of the environment, would be surprised at this.

3. The consequences of powerful natural selection

Simple mathematical models of the effects of natural selection on the genetic composition of populations show that such high levels of selection can have profound effects on the genetic characteristics of a population, in two respects.

Firstly they can cause rapid change - measured in terms of a few generations. A coefficient of selection of 0.5 can cause the decline in frequency of a disfavoured gene of intermediate dominance from 0.9 to 0.25 in 10 generations; a coefficient of selection of 0.9 can cause the same change in less than 4 generations (Shorrocks 1978). So if there is genetic variability available for selection, we should be able to observe genetic changes occurring in populations within a few years.

Secondly they can cause localised evolutionary change - even in situations where there is considerable exchange of genes between populations. Any exchange of genes will tend to keep the gene frequencies of two populations being selected in different directions similar. As a result many people have not expected to find genetic differences between adjacent populations. Yet models show that if coefficients of selection operating on two populations differ by 50%, these populations can show substantial evolutionary divergence, even if they are no more than 4 metres apart and the pollen flow occurring between them is at the level found in wind pollinated grasses (Jain & Bradshaw 1966)

Theory is always a good guide to what may happen. But do we find such consequences of powerful natural selection in practice? Some years ago the famous evolutionary zoologist, A.J.Cain (1977), wrote that "the golden rule is always to ask questions of the animals, not of the pundits". So what sort of evolution do plants growing in ordinary habitats reveal? There are many examples. One of the most revealing is that of Snaydon and his collaborators on the populations of the grass *Anthoxanthum odoratum* growing in the Park Grass fertiliser and lime experiment originally started in 1852 at Rothamsted Experimental Station near London. The liming treatments were started in 1903 and have been continued up to the present. The populations of *Anthoxanthum* on the limed and unlimed plots have become completely different genetically in their nutrient responses, and in a number of other features, within 60 years (Snaydon 1970).

Agrostis stolonifera is a creeping grass found in pastures and on exposed cliffs in different parts on Britain. On cliffs it has evolved a dense tufted growth form, quite in contrast to its normal very stoloniferous pasture form. These difference are genetic and are maintained in cultivation. A detailed study of the populations growing on a single sea cliff in North Wales (Aston & Bradshaw 1966) has shown that the changes in growth form can occur over a distance of 5 metres, despite the fact that the species is outbreeding and wind pollinated and that there is good evidence of gene exchange between the populations.

Similar localised differentiation is widespread in *Agrostis capillaris (tenuis)* growing in pastures in different parts of mid-Wales (Bradshaw 1959). Even sharper differentiation occurs in *Anthoxanthum odoratum* in the plots in the Park Grass experiment. Soil conditions and vegetation change over a distance of 10 cm and the populations of *Anthoxanthum* differ over distances less than 150mm (Snaydon & Davies 1976). Evolutionary differentiation, therefore, means that most species consist of a graded patchwork of distinct, localised populations, matching the variations of the environment.

But these differences are not just in morphology, but in attributes of fitness. The cliff populations of *Agrostis stolonifera*, for instance, are four times more resistant to wind exposure than the pasture populations. In *Anthoxanthum odoratum* reciprocal transplants experiments of the species within pairs of contrasting Park Grass plots show that plants originating from the contrasting plot have a fitness on average only half that of the plants native to the plot (Davies & Snaydon 1976). Evolutionary differentiation therefore makes an important contribution to both total and local fitness.

4. Evolution in heavy metal contaminated sites

In my early years as a scientist most plant ecologists did not study metal contaminated sites of any sort. They were considered to be too unnatural and therefore of little relevance to normal situations. Yet in 1934, Prat, in an overlooked paper, had shown that localised evolution must have been occurring. Plant material of *Silene dioica* from a copper mine had an ability to grow on copper contaminated soils not possessed by normal populations of the same species. Later studies on a variety of species have shown that populations from metal contaminated habitats can indeed provide us with extensive and elegant evidence about evolutionary processes, bearing out and extending what has already been shown to occur in more normal habitats.

As many ecologists at this Conference will know, the first feature of species growing in metal contaminated habitats is the evolution of specific populations with metal tolerance (see review by Shaw 1989). It manifests itself most clearly as the ability of metal populations to develop roots in contaminated soils not possessed by normal populations of the same species (Bradshaw 1952). This can be examined most easily in solution culture, from which a simple rooting test has been developed (Wilkins 1957). Although some cross tolerance can occur, tolerance is essentially specific to individual metals (Jowett 1958), although these may be combined where the metals occur together (Wong 1982). The ability to root is closely connected with survival. As Prat originally showed, populations without tolerance do not survive on metal contaminated soils.

The speed of this evolution is particularly impressive. It was clear from early work that tolerance could be found in populations in recent mining areas and might therefore have evolved within 100 years. The situation explored by Wu in the areas around the copper refinery at Prescot, Merseyside shows that evolution can be far more rapid than this. In a series of polluted grasslands of different ages, significant evolution towards metal tolerance could be found in populations of *Agrostis stolonifera* exposed to copper pollution for only four years, and substantial evolution in populations only 25 years old (Wu et al. 1975).

In parallel with this, it has been shown that almost full metal tolerance can be achieved experimentally in a single cycle of selection, if a large number of individuals from a normal population are screened for tolerance. Typically 5-10 fully tolerant individuals can be found in a total of 1000 individuals screened (Walley et al. 1974). Although this represents a very high level of selection, it is exactly what can occur in nature.

Localisation of evolution is also evident in metal contaminated areas. Evolution of metal tolerant populations can occur on small areas of metal mine waste not more than 30 metres across, such as on Trelogan lead/zinc mine in North Wales. A transect across the sharp boundary of this mine shows that the evolutionary differentiation is, in fact, highly localised. An almost complete change from tolerance to non-tolerance occurs over a distance of eight metres in the zinc tolerance found in the grass *Anthoxanthum odoratum*. Strong selection for and against tolerance must be occurring on the two sides of the environmental boundary, sufficient to overcome the effects of gene flow (Putwain in Jain & Bradshaw 1966). Other characters in the same plants along the same transect, however, change much more gradually, suggesting that weaker selection is acting upon them (Antonovics & Bradshaw 1970). Such evolutionary differentiation means that species form a graded patchwork of some complexity, as each character is adjusted independently to the interplay of selection and gene migration affecting it.

5. Evolution under electricity pylons

Against such a wealth of evidence of patterns of evolution provided by metal contaminated sites, one more example seems hardly necessary. Yet the fact that the zinc contamination occurring under electricity pylons (transmission towers) has its own evolutionary effects deserves our attention, just because it seems so unexpected. From a personal standpoint it is certainly something that, when I started my own evolutionary work, I would never have imagined could happen, or be worth looking for. Yet most electricity pylons are made of steel which has been galvanised, zinc-coated, to prevent the steel rusting. The zinc forms a sacrificial coat, so that underneath each pylon quite high levels of zinc can accumulate, sufficient to be toxic to vegetation especially in the vicinity of each pylon leg where the rain water collects. This metal contamination is high enough, even in these small (10x10 m) and isolated areas (pylons are about 300 m apart), to promote the evolution of zinc tolerance in a number of plant species (Al Hiyaly et al 1988). The toxicity takes sometime to build up, but evolution of tolerance occurs within 15 - 30 years.

6. Evolutionary failure

From all this it is clear that, despite its apparent slowness, the power of evolution to cause rapid local change is remarkable. Its effects should not to be underestimated. But despite this, we do not see evolutionary change taking place everywhere. In particular, evolution may occur in some species, but not in others. There seems to be a contradiction.

This very clear where metal tolerance is involved. Only a relatively few species seem to have evolved tolerance. This might be because they have not had the opportunity, by not being present in metal contaminated situations, and so having not been subject to natural selection. At the Prescot copper refinery only five species occur in the contaminated areas and have tolerant populations; other species do not seem to have evolved tolerance Pollution only began, however, when the refinery was constructed at the beginning of the century. Before that there would have been at least 100 species present in the area, all of which would have become exposed to the copper pollution. The species which have disappeared must have failed to evolve tolerance. Yet they would all have been exposed to copper pollution and therefore the power of natural selection. Since evolution depends on the impact of a) natural selection on b) random genetic variation, it follows that any failure of evolution must be because of lack of variation .

It is possible to examine populations for the occurrence of variation for metal tolerance using the bead screening technique, in which seed is some on a raft of poly-ethylene beads floating on a solution containing the relevant heavy metal. Individual seedlings possessing metal tolerance can be quickly recognised by their root growth. Using this technique Ingram showed that species which fail to evolve tolerance do so because they do not possess the necessary variability in their normal populations. (Bradshaw 1984).

The absence of the necessary variability may affect populations within a species also. It seem quite common for variability for metal tolerance to be present only in certain populations of a species (Gartside & McNeilly 1974, Symeonidis et al. 1985). Electricity pylons uncover the evolutionary significance of this because they provide an opportunity to study evolution in a replicated situation. Evolution of zinc tolerance does not take place under every pylon. The few tolerant individuals occurring in normal populations of *Agrostis capillaris* must the starting point for this evolution. At pylons where the evolution does not occur, zinc tolerant individuals are absent from the surrounding normal populations (Al-Hiyaly et al. 1990).

7. Conclusions

Evolution is clearly a powerful process, able to change the characteristics of species at the population level and contributing significantly to ecological fitness. The changes can occur within a few generations and on a very local scale. But evolution is not always successful. Indeed, in most situations, the evidence is that species are not evolving. This cannot be because selection has failed - it is always present. It must be because the appropriate variation is not present. Evolution is held back by lack of variability - a condition of *genostasis* (Bradshaw 1993)

Natural habitats are therefore not the only ones to study. The nasty ones which we have too often overlooked may have interesting and important stories to tell.

8. References

Al-Hiyaly, S.A., McNeilly,T. and Bradshaw, A.D. (1988) The effects of zinc contamination from electricity pylons - evolution in a replicated situation. *New Phytologist* **110**, 571-580.

Al-Hiyaly, S.A.K., McNeilly,T., Bradshaw, A.D. and Mortimer, A.M. (1990) The effects of zinc contamination from electricity pylons. Genetic constraints to the evolution of zinc tolerance. *Heredity, London* **70**, 22-32.

Antonovics, J. & Bradshaw, A.D. (1970) Evolution in closely adjacent plant populations. VIII Clinal patterns at a mine boundary. *Heredity, London* **25**, 349-362.

Aston , J. L. & Bradshaw, A.D. (1966) Evolution in closely adjacent plant populations. II *Agrostis stolonifera* in maritime habitats. *Heredity, London* **21**, 649-664.

Bradshaw, A.D. (1952) Populations of *Agrostis tenuis* resistant to lead and zinc poisoning. *Nature, London* **169**, 1098.

Bradshaw, A.D. (1959) Population differentiation in *Agrostis tenuis* Sibth. I Morphological differentiation. *New Phytologist* **58**, 208-227.

Bradshaw, A.D. (1984) The importance of evolutionary ideas in ecology - and vice versa, in B.Shorrocks (ed), *Evolutionary Ecology,* Blackwell, Oxford, pp 1-25.

Bradshaw, A.D. (1993) Genostasis and the limits to evolution. *Philosophical Transactions of the Royal Society, London,* **B 333**, 289-305.

Cain, A.J. (1977) The efficacy of natural selection in wild populations. In *The Changing scenes in Natural Sciences 1776-1976.* Academy of Natural Sciences, Special Publication **12**, pp. 111-133.

Davies, M.S. & Snaydon, R.L. (1976) Rapid population differentiation in a mosaic environment. III Coefficients of selection. *Heredity, London* **36**, 56-66.

Gartside, D.W. and McNeilly, T. (1974) The potential for evolution of heavy metal tolerance in plants II Copper tolerance in normal populations of different plant species. *Heredity, London* **33**, 303-308.

Haldane, J.B.S. (1932) *The Causes of Evolution.* Harper, London.

Jain, S.K. & Bradshaw, A.D. (1966) Evolutionary divergence among adjacent plant populations. I The evidence and its theoretical analysis. *Heredity, London* **21**, 407-441.

Jowett, D. (1958) Populations of *Agrostis* spp. tolerant of heavy metals. *Nature, London* **182**, 816.

Prat, S. (1934) Die Erblichkeit der Resistenz gegen Kupfer. *Bering Deutches Botanisches Gesellschaft* **102**, 65-67.

Shaw, A.J. ed. (1989) *Heavy Metal Tolerance in Plants: Evolutionary Aspects.* CRC Press, Boca Raton, Florida.

Shorrocks, B. (1978) *The Genesis of Diversity.* Hodder and Stoughton, London.

Snaydon, R.W. (1970) Rapid population differentiation in a mosaic environment I The response of *Anthoxanthum odoratum* populations to soils. *Evolution* **24**, 257-269.

Snaydon, R.W. & Davies, M.S. (1976) Rapid population differentiation in a mosaic environment IV Populations of *Anthoxanthum odoratum* at sharp boundaries. *Heredity, London* **37**, 9-25

Symeonidis, L., McNeilly, T. and Bradshaw, A.D. (1985) Interpopulation variation in tolerance to cadmium, copper, lead, nickel and zinc in nine populations of *Agrostis capillaris* L. *New Phytologist* **101**, 317-324.

Walley, K.A., Khan, M.S.I. and Bradshaw, A.D. (1974) The potential for evolution of heavy metal tolerance in plants I Copper and zinc tolerance in *Agrostis tenuis* . *Heredity, London* **32**, 309-319.

Wilkins, D.A. (1957) A technique for the measurement of lead tolerance in plants. *Nature, London* **130**, 37.

Wong, M.H. (1982) Metal cotolerance to copper, lead, and zinc in *Festuca rubra*. *Environmental Research* **29**, 42-47.

Wu, L., Bradshaw, A.D. and Thurman, D.A. (1975) The potential for evolution of heavy metal tolerance in plants III The rapid evolution of copper tolerance in *Agrostis stolonifera* . *Heredity, London* **34**, 165-187.

THE EARLY BOTANICAL EXPLORATION OF GREECE

W. GREUTER
Botanischer Garten & Botanisches Museum Berlin-Dahlem der Freien Universität
Königin-Luise-Str. 6-8, D-14191 Berlin, Germany

Abstract

An account is given of botanical exploration and explorers of Greece in its present-day boundaries, from antiquity to the end of the 19th Century. Our admittedly fragmentary knowledge of the herb lore and botanical skills of the ancients (Theophrastos, Dioskourides) shows Greece to be the cradle of modern botanical science. For early European botanists, visiting Greece became a pilgrimage to the pristine sources of their science. Modern investigation, starting in the 16th Century, concentrated almost exclusively on the few areas still under Venetian control, in particular Crete. When Crete fell to the Turks, exploration came to a halt. Few travellers ventured into the Ottoman Empire, including its Greek parts: the French Tournefort (1700-1702), Olivier and Bruguière (1792-1798), and Dumont d'Urville (1819) who, at the end of the Greek liberation war, were followed by Bory (1829); and the Englishman Sibthorp (1786-1787, 1794-1795). After 1832, the independent Greek Kingdom was wide open to visiting botanists, often of German origin, and hosted resident ones among whom Heldreich is prominent alongside with the first modern Greek plant scientist, Orphanides. Meanwhile some islands and northern Greeke (Epirus, Macedonia, and Thrace), still under the Turkish yoke and therefore of difficult access, were being explored by botanical pioneers such as Grisebach (1839), Formánek (1889-1899), and a few others. To close the period, Halácsy between 1900 and 1904 published the first Greek national Flora, just about to be relayed by Phitos & al.'s incipient *Flora hellenica*.

Classical antiquity: the fathers of scientific botany

Greece is well known as the cradle of scientific botany, fathered by the Greek nature philosophers of classical antiquity. Not for nothing the word "botany", a comparative neologism, is Greek in origin. It is particularly timely to consider the roots and early days of botanical exploration of this country at a meeting held in the Greek city of Thessaloniki.

This paper must by needs begin with Theophrastos, born in Eresos on the island of Lesbos 372 years before Jesus Christ and who ended his long and fruitful life in Athens at the venerable age of 85. He may not have been the first human to look at plants with a scientific eye, but he is certainly the first whose botanical writings *(De historia plantarum,* or *Περι φυτων ιστοριας)* have, through the copying zeal of ancient scholars and erudite monks, survived to the present day. As a pupil of the famous Aristotle he was, so to say, on collegial terms with that other famous pupil of the same master, the Great Alexander of Macedonia.

Theophrastus knew best the plants of his home island and of the Troad peninsula facing it, of which he must have climbed Mt Ida (Kaz dağ) repeatedly. The plants of Mt Ida play a choice role in his writings, and since he rarely if ever took care to explicit which Mt Ida he was referring to, he was to induce later itinerant botanists such as Tournefort into a vain search of some plants on the even more famous Mt Ida of Crete where they never grew, such as "*Vaccinium vitis-idaea*" (Theophrastus's *ιδαιον αμπελος,* which in reality must rather be *V. arctostaphylos*), or the raspberry, named *ιδαιος βατος* by Theophrastus and even nowadays known as *Rubus idaeus.*

In the Renaissance, when the learning of Antiquity was rediscovered, interpreting the old writings became a crucial problem. Most famous among the ancient physicians knowledgeable of herb lore and healing powers was the Cilician Dioskourides, who lived in the 1st century of our era. He is known to have built on a rich and manifold tradition stemming from his peers and predecessors, such as Krataivas, the physician of Mithridates in Pontus, whose writings were however lost, so that Dioskourides is the only one whose memory has survived. Of his "Materia medica" *(Περι υλης ιατρικης)* illustrated handwritten copies exist, which apparently integrate a pictured herbal of Krataivas and draw on contemporary sources. The most ancient and most valuable of them, the *Codex Aniciae Julianae,* made about the year 512 for presentation to a noble lady of Constantinople, is now in the National Library in Vienna, having been acquired in 1576 for 100 Louisdor on behalf of Emperor Maximilian II (see Mazal 1981). This codex, also known as *Codex Vindobonensis,* includes colour representations of almost 400 plant species, painted on parchment, many of which can be identified reliably while others are highly stylised and hardy recognisable with any degree of confidence.

Renaissance: Venice and Italian botanists

Adepts of the plant sciences, perhaps rather herb sciences, as they developed forcefully in the mid 16th century of Europe, were hard put to interpret correctly what they could glean out of the writings of old. They tried to match as best they could the imperfect and partly corrupted texts, copied and recopied by generations of monastic scribes, and the few, often rudimentary images that had survived, with the plants they found on their own door-sills. It was then unheard of that plants differ in different regions, and that floras change when you travel, but even if they had wanted to, Renaissance naturalists could not easily have gone and verified what Dioskourides might have had at hand in his eastern Mediterranean home country. Just bear in mind: in 1453, about 90 years before the first Botanical Garden was founded in Pisa (in 1543 or 1544), Constantinople had fallen to the Turks, and the former Eastern Roman Empire thus became virtually closed to occidental travellers except a few who were on diplomatic mission. It is an obvious and by no means amazing fact that exploration of the territories under Turkish administration lagged far behind that of the areas governed by Christian powers.

A look at the maps of those times shows that soon after the fall of Constantinople the only readily accessible parts of Greece were those still under the control of Venice, being basically the Ionian Islands and Crete. Crete and Venice therefore played a key role in the early exploration of the Greek flora, starting in the mid-16th Century (see Lack 1996 for a detailed presentation). Disregarding some very early general naturalists, the first such explorer with mainly botanical interests appears to have been Luigi Squalermo, better know as Anguillara (c. 1512-1570), the first Prefect of the Botanical Garden of the University of Padua (founded 1545). While nothing much is known of his travels, it is evident from his only published work, the now very rare *Semplici* (Anguillara 1561), that he did indeed visit Crete and other points along the coast of the Ionian and Adriatic Sea in the early 1550s. Other names of fame associated with early Cretan botany, all of them in the 2nd half of that same Century, are Prospero Alpino (1553-1616), 5th Prefect of the Padua Garden, who explored Crete on his travel to and back from Egypt in the years 1581 and 1584 (Alpino 1627, Turland 1995); Jodocus de Goethuysen or Goedenhuizen, or with his Italianised name, Giuseppe Casabona or Benincasa (1535-1595), Prefect, first of the Botanical Garden in Florence (founded as "Giardino delle Stalle" in 1545 by Cosimo I Grand Duke of Tuscany; Moggi 1975) then of the Botanical Garden of Pisa University, who visited the Island in 1590; and physician Onorio Belli (1550-1603), of Vicenza, who resided in Hania from 1583 until 1599 and sent seeds and dried plants to correspondents throughout Italy and other parts of Europe. These were high times for Cretan botany, particularly when local governors, such as duke Girolamo Cappello repeatedly mentioned in the literature, were fostering the natural sciences and the exploration of the island's natural patrimony.

A remarkable fact that remained for a long time unnoticed is that it was Casabona who, by hiring a botanical artist to illustrate his plants, initiated a tradition that was to be upheld by the most prominent among his successors. The name of that artist, and what remains of his unpublished work, were recently brought to light through the study of the old manuscript collections at the University library in Pisa (Tongiorgi Tomasi 1984, Garbari & al. 1991: 43, 134): he was a Dutch or German soldier named Georg Dyckmann, and his 35 surviving water-colour paintings (see Greuter 1996; colour reproductions in Garbari & al. 1991: 39-41, 278-279) are most remarkable early examples of the art of botanical illustration.

Greece under the Turkish yoke: some French and two other explorers

In the 17th Century, bellicose events, culminating in the fall of Iraklio and the whole of Crete to the Turks in 1669, brought botanical activities in Greece to an abrupt halt. Venice, and Italy in general, were henceforth out of business. France appeared on the scene, with King Louis XIV sending a team of naturalists to explore the Levante down to Transcaucasia (Armenia and Georgia) with respect to its natural history and ancient and modern geography. The heroes of that extraordinary expedition were French botanist and academy member Joseph Pitton de Tournefort (1656-1708), German physician Andreas Gundelsheimer (1668-1715) as well as painter and illustrator Claude Aubriet to tread in Dyckmann's footsteps. Crete was the first foreign land to be explored by the Tournefort party, who stayed there for three full months, from 2nd May to 1st of August 1700, visiting many parts of the island and all the main mountain massifs. Having left Crete, they practised "island hopping" mainly in the Cyclades and East Aegean Islands, leaving Greece for Constantinople in mid-March 1701. Tournefort and Aubriet again crossed the Aegean on their swift return cruise from Smyrna to Livorno, 13 April to 23 May 1702 (Tournefort 1717).

Tournefort and Gundelsheimer brought home a rich botanical harvest in the form of dried plants, a list of which was published soon after in Tournefort's (1703) *Corollarium.* Numerous duplicates were distributed to a number of European botanists. Unfortunately the full scientific account that Tournefort planned to produce was never written, and Aubriet's uniquely beautiful vellum paintings, produced from this partly water-coloured travel sketches, ended up well-guarded in the famous vellum collections of the Paris Natural History Museum, only few of them ever to be published (e.g. by Desfontaines 1808). In spite of its enormous importance and high fame, the results of Tournefort's expedition did not reflect themselves adequately in later, Linnaean and post-Linnaean botanical writings.

Probably due to the constraints of international politics, France was to remain for well over a century the leading nation in exploring the Ottoman Empire. Two French biologists, Guillaume Antoine Olivier (1756-1814) and Jean Guillaume Bruguière (1750-1798), were sent by the French Government on a mission to the shah of Persia, which journey was to last six years (1792-1798) and cost the latter's life (he died of contracted illness in Ancona, almost back home, in 1798). A full account of their travels was published by Olivier (1801-1807), but the botanical aspects, having been mainly Bruguière's concern, were rather poorly treated, and the collections were inadequately labelled, so that the expedition's importance for E. Mediterranean and S.W. Asian botany was not what it should have been. Olivier and Bruguière were among the first to study the autumn and winter flora, having spent three full months (August to October 1794) in Crete; they also visited various other Aegean islands until they left in 1795 for Lebanon and the interior of S.W. Asia. On their way back, in 1798, they travelled from Constantinople to Athens and hence to the Ionian Islands of Ithaki, Kefallinia, and Kerkira.

Keeping to France, next comes Jules Sébastien César Dumont d'Urville (1790-1842), a surveyor and hydrographer with pronounced botanical skills and interests. As a young navy officer he joined captain Gautier on the vessel *Chevrette,* whose task it was to survey the coasts and islands of the Mediterranean, on two of its annual cruises in 1819 and 1820. The former was entirely devoted to exploring the Aegean islands, mainly the Cyclades and Eastern Sporades. An assiduous collector, Dumont d'Urville (1822) published a full account of his harvest. His most famous booty, however, was a flower of a different kind: an ancient Greek marble torso, bought from a peasant of Milos on behalf of the French ambassador in Constantinople, de Rivière, that was soon to become famous as the Venus of Milo and up to the present day is felt to incorporate the beauty ideal of classical Greece.

The last of the famous French expeditions to involve Greece stands in a different context. It was one of those war-time expeditions which, in the tradition established by Napoleon in Egypt, followed the victorious French armies to explore the conquered territories. Only this time the French were not conquerors but liberators. In August 1828 general Maison landed with an army corps of 14,000 in "Morea" and swept the Turkish occupants out of the country. This was part of a concerted action of the powers (France, Britain, and Russia) in support of Greek independence that finally tilted fortunes in the Greek Liberation War. No sooner was this accomplished than a governmental expedition, with three separate parties, disembarked in Pilos (then Navarino). The team in charge of natural sciences was headed by colonel Jean Baptiste Geneviève Marcellin Bory de Saint-Vincent (1778-1846) and included the botanist J. M. Despréaux who, however, appears to have gone his own ways later on during the expedition. Bory and his companions explored most of the southern and central Peloponnese, coastlands and interior alike, between early March and mid-July of 1829, when the whole team except Bory himself was taken down with fever (obviously malaria) contracted in the Eurotas swamps. They were barely saved, in their retreat in Monemvasia (Naples de Malvoisie), thanks to the lucky intervention of a Bavarian physicist and renowned botanist, Joseph Gerhard Zuccarini (1797-1848) who happened to be in Nafplio (Naples de Roumanie). After which Bory continued on his own, visiting several Aegean islands, in particular of the Cyclades, to return to his native France on new-year's day of the following year (see Lasègue 1845: 112-115). Bory was a naturalist of many interests, yet his botanical harvest was impressive, and fully accounted for in the botanical part of the expedition account (Fauché & al. 1832).

Back to the 18th Century and the days of the Ottoman Empire, the single prominent example of an explorer-botanist from Britain must now be mentioned: John Sibthorp (1758-1796) of Oxford, one of the many representatives of the young gentry to travel around Europe (see Lack 1998). In fact, there was nothing truly exceptional in what he did: there were several of his like to travel the Balkans and the Near East, even with some naturalistic interest; some have published travel accounts, mentioning the odd plant. Some may have collected dried specimens or, more likely, bulbs and seeds for the homely rock garden. What singles out Sibthorp, then, is the seriousness of his endeavour, the fact that he did plant collecting on a large scale and, most importantly, that by the terms of his will he posthumously contracted the professional skill of botanists to bring his harvest to fruition.

Sibthorp was well prepared for his travel by having studied beforehand the Paris collections with Tournefort's herbarium and Aubriet's paintings. Also, he passed through Vienna at the onset and got hold of a set of engravings made from the already mentioned *Codex Aniciae Julianae*. Furthermore, he went to Feldsberg (now Valtice) where he hired as a travel companion a man who was to become the best famed botanical artist of his time, Ferdinand Bauer. Sibthorp travelled to Greece twice, in journeys also extending to Asia Minor and Cyprus: first from June 1786 to September 1787, when after a stopover at Milos and Kimolos he started in Tournefort's footsteps by visiting Crete in July and August, then went island-hopping through the Central and Eastern Aegean (Samos) to Izmir; having spent the winter in Constantinople and the following spring on Cyprus, he resumed his Aegean cruise and later in the summer ascended Mt Parnassos, Mt Dirfis in Evvia, and Mt Athos. When he undertook his second journey (1794-1795) Sibthorp was already a sick man, and his travelling record is less impressive. Yet, he visited Limnos, then again the Holy Mountain and several other places, on his way from Constantinople to Zakinthos where he was to spend some winter months with his friend Hawkins (who got around Greece a lot more, and for longer time, than Sibthorp and, while not really a botanist himself, would ever and again harvest some peculiar plant to please him). The following year they both made an extensive two-month trip through the Peloponnese in March and April before Sibthorp embarked on his way home, an arduous and adventurous one which, through adverse circumstances, finally ruined his health and caused his premature death, shortly after his return to England.

This is not the place to dwell on the history of the publication, by Smith (1806-1816), of the *Prodromus* and the monumental *Flora graeca,* the latter of which took 34 years (Sibthorp 1806-1840) before all of its ten volumes were achieved, to be one of the most costly and most impressive monuments of botanical illustration ever published. Nor is it the place to delve into the inadequacies of Sibthorp's legacy, the chaotic state of the documents he left and complete absence of labelling of any of his plants. Also, I must forbear telling the full story of how Bauer worked in the field, using a newly invented colour chart to annotate his sketches with the appropriate colour codes, eventually to produce the most gorgeous yet faithful plant paintings ever made (Lack & Ibáñez 1997).

We cannot leave the older, Turkish times without mentioning Franz Wilhelm Sieber (1789-1844), the universal naturalist, traveller and plant collector from Prague who spent almost a full year in Crete, landing on the quarantine islet of Dia, off Iraklio, on 7 January 1817 and departing the island on 26 November, to continue for Egypt (where he was also to collect mummies – Sieber 1820) and Palestine. Not only did he explore Crete to an unprecedented degree, visiting all its parts and climbing all its mountain tops; not only did he, quite naturally under these circumstances, discover a large number of species new to science, including some of the rarest endemics; but he also left us, known but to a lucky few, one of the most colourful and merrily idiosyncratic travel accounts ever written (Sieber 1823). You will find in it treasures unknown and unsuspected, including a dozen of Sieber's own, original poems, hints at a recipe for healing rabies, and an erudite and convincing demonstration that what the travellers of old (including Tournefort) had accepted to be Minos's mythic labyrinth was in fact a Roman quarry used to build the temples and houses of Gortyna. Sieber's is a truly enchanting personality, and he managed to do a lot of further travelling in many parts of the world before sadly finishing his days in a lunatics' asylum.

The Greek Kingdom and the German influence

Space limitations preclude giving a full account of the botany of the Greek Kingdom, which became part of what we consider the civilised world as soon as it was freed of the Turkish yoke. It may be interesting, however, to rehearse how late this kingdom came into being, and how gradually it reached its present size and boundaries. When it was created in 1832, the Kingdom of Greece under the Bavarian King Othonos merely comprised the Peloponnese, Sterea Ellas, and the islands of the western and central Aegean. The Ionian Islands (previously English) were added in 1864, Thessaly in 1881, Epirus, Macedonia and Crete (the latter autonomous since 1898) after the Balkan Wars of 1912/14, Thrace and the Eastern Sporades in 1920, and lastly, the Dodekanisos just barely 50 years ago, in 1947.

Reflecting perhaps the Bavarian regency, a number of German or Austrian botanists were active in Greece in the times of the young kingdom. Names like Spruner, Berger, Friedrichsthal, Fiedler, Fraas, and Unger come to mind, some of them resident in the country, other itinerant visitors. By far the most important of the former was of course Theodor von Heldreich (1822-1902), and although his life and travels are fairly generally known (see, e.g., Halácsy 1902, and archival information at Geneva), there is no way of bypassing him. Born in Dresden in Germany, Heldreich learnt his job of a botanist in Montpellier (1837) and Geneva (1838-42) where he was befriended and employed by the Candolles. He made his first travel to Sicily for Candolle, but then got in a way kidnapped by Boissier (who had himself visited Attica and Boeotia in May 1842 on his Oriental trip). Boissier contracted Heldreich, then settling down in Athens, to collect for him; and indeed the early Heldreich collections of 1844 (Peloponnese), 1845 (W. and S. Anatolia) and 1846 (Crete) were bulk-shipped to Geneva and distributed from there with the familiar small yellow or bluish labels with Boissier's tiny handwriting reproduced by lithography. Boissier held Heldreich in high esteem for all his life, and associated both their names in authorship for the many new species Heldreich was to discover. Yet the intimate partnership of these early years was not upheld in the long run. Perhaps Boissier, a typical Swiss capitalist, was not sufficiently generous after all in his financial arrangements. It is also possible that Heldreich, when forced to leave Greece in 1849, having lost everything to some fraudulent business partner, was disappointed at the lack of support by his Geneva friends. This critical period of his life was not fortunately to last, nor did it cause him to emigrate to the New World as he is said to have planned. In 1851 he was back in Athens, where he was to stay for the rest of his life.

Among his numerous important expeditions and collecting trips, let me mention some which exceeded the Greek boundaries of the time: in 1851 he visited Salonica and Mt Hortiatis then went on to Mt Olimbos which he was the first botanist to explore; in 1870 he went to Crete for a second time; in 1882 and 1883 he travelled to Thessaly, just liberated, climbing Mts Ossa and Pilio; deeper still into Thessaly he penetrated in 1885, in the company of Heinrich Carl Haussknecht (1838-1903), when he pioneered in exploring the Pindos range. To round off the picture of this unique man and scientist, let me hint at his passion for collecting shells and other zoological items – which of course fitted in perfectly with his 25-year assignment (1858-1883) as a keeper of the Natural History Museum in Athens: I will add that he was a remarkable polyglot, having published papers not only in Latin but also German, Italian, French, and Greek all of which he wrote and spoke fluently.

Theodoros Orphanides (1817-1886), the first Greek botanist since antiquity to acquire fame, I will just mention in passing – not because he would not deserve attention, which he does, but because I feel that I know less of him than many of those in attendance. He was the first Professor of Botany of Athens University, where his extensive and invaluable herbarium is still housed, and was a keen plant collector and mountaineer, who climbed most of the Greek summits from the Peloponnese to far-off Mt Athos. Famous he was, among other things, for the discovery, in 1851, of *Biebersteinia orphanidis* on Mt Killini. Many after him have looked in vain for this showy and unmistakable plant, which he was suspected to have eradicated completely in his zeal of amassing a full centuria (he was a greedy collector if there ever was, and must have decimated many stands of rare and threatened plants during his lifetime). In 1994, almost one and a half centuries after its discovery, two young botanists from Patras have just recently managed to find it again (Vassiliades & Constantinidis in Phitos & al. 1995: 92-93).

Northern Greece in the 19th Century: adventures galore

I shall finish this account by mentioning two botanists who pioneered by exploring the wildest and remotest parts of what is now Greece: the plains and mountains of Macedonia. As I have mentioned, this remained Turkish territory until 1912, and was not secure ground by any modern standard.

August Heinrich Rudolph Grisebach (1814-1879) was aged 24, a lecturer at Göttingen University soon to become tenured professor there, when out of the blue he decided to spend his next summer (of 1839) in the mountains of northern Albania. He bought the basic equipment (to include a barometer and thermometer, a microscope, two plant presses with 1000 sheets of blotting paper, plus two pistols and ammunition) in Vienna and Pest, where he was told that his original plan to travel to Albania directly and seek protection from the local chieftains was unworkable since he would need the sultan's firman and an interpreter, both to be obtained only in Constantinople – so there he went by boat via the Danube River and the Black Sea. Just figure out for yourself the small detour he made! To think that at the end of four months of travelling, probably somewhat worn out, he crossed his target area in just four days on horseback, from Prizren to Shkodr (Scutari).

Before leaving Constantinople for his long travel through unheard-of, virtually unexplored country he had to find his dragoman (interpreter). Having learnt that all interpreters were (of course) Greek and therefore (sorry to say) unreliable, he decided to try out the honesty of the man of his choice on an fortnight's extra trip, that led him to Bursa and Mt Olympus of Bithynia (Ulu dağ) – after which he was well content with his dragoman but was hard put to convince him to stay on and come along.

The Greek portion of Grisebach's jouney lasted from 30 May till 29 June. It took him from the port of Enos (reached by land) to the coast of Thrace near Makri, the island of Thasos and the Athos Peninsula which he circled, climbing the Holy Mountain to the top, then on to Salonica, Jannitsa, Edessa (Vodena), and Arnissa (Ostrovo), from where he climbed Mt Piperitsa (Nidgé) S.W. of Mt Kaimakčalan (he established its height at 5544 feet by measuring the boiling temperature of water as compared to air temperature). Upon his descent he crossed today's national border, heading for Bitola from where he climbed Mt Pelister (Peristeri). His main stations thereafter were Prilep, Skopje, Tetovo, the Šar planina, Prizren, Shkodr, and the southernmost Austrian harbour in Dalmatia, Budva ("Lastua"). Grisebach (1841) was to publish a lively and most readable account of his travel, and soon after, a complete revision of his, and a few others', botanical booty (Grisebach 1843-1845).

Our second hero is Eduard Formánek (1845-1900), college teacher at Brno in present-day Czechia (Brünn in Moravia). From 1887 to 1899, the year before his demise at age 55, he spent each summer holiday, regularly from early July to mid-September, in the mountains of the central and southern Balkan Peninsula, bringing back no less than 40,000 plant specimens – often appallingly pressed and scrappily labelled. By 1887, two important changes had occurred as compared to Grisebach's time. First, between 1878 and 1881, the Ottoman Empire had lost several of its Balkan territories: Bosnia being under Austro-Hungarian administration, Serbia (now independent) having annexed the Niš area, Bulgaria as well as Rumania having acquired autonomy, and Thessaly being united to Greece – which still left to Turkey the whole of Albania, Macedonia, and Thrace. Second, the Balkan railway had just been built, linking Salonica (and Athens) directly with Austria and Central Europe. While this was essential in allowing a quick approach to the areas to be explored, it does in no way detract from Formánek's exploits in the field.

On 6 of his 13 travels (1889, 1894-1896, 1898-1899) Formánek visited present-day Greece. The areas he explored include Epirus (1894), the Pindus range of Thessaly (1889.1895), Mt Hasia (1896), Mts Vermio and Voras (1898-1899), Mt Belles (1899), the Salonica area with Mt Hortiatis (1889, 1898-1899), Mts Kato Olimbos and Ossa, Timbi valley, Volos and Mt Pilio (1889, 1895), and Mt Othrys (1896). He also explored many areas just outside the northern Greek border. On two further travels (1891, 1893) and in 1898 he was partly based in Bitola, ascending Mts Pelister, Baba, and Galičica, and investigating the area to the west as far as the shores of Lakes Prespa and Ohrid. In 1895 and 1898 he botanised on the northern slopes of Mt Kožuf and in the area west of Gevgelija.

Formánek published the results of his expeditions annually at great speed, which led to many misidentifications and to most of his new species and subspecies having since been sunk (see Vandas 1909). His results remain valuable, however, thanks to the fact that all his collections are preserved at the Moravian Museum in Brno. Formánek gave but little details on the ways and means of his travelling, but in accounting for his last (1899) trip he complains of many adversities, including being flooded at night and having a bad fall with his horse. Whether there might have been any connection between these mishaps and his premature death I was unable to find out.

The 1890s brought almost an inflation of botanical visitors to northern Greece, including the Germans Paul Ernst Emil Sintenis (1847-1907) and Joseph Bornmüller (1862-1948) visiting Macedonia, the Athos Peninsula and the island of Thasos in 1891; Austrian plant collector Ignaz Dörfler (1866-1950) exploring the Mt Voras area in 1893 (Degen & Dörfler 1895); and Antonio Baldacci of Bologna (1867-1950), the famous explorer of Albania, who also collected, i.a., in Epirus (1893-1896) and Crete (1893, 1899).

I have left for the end Jenő de Halácsy (1842-1913), Hungarian botanist living in Vienna. He must be mentioned, not so much because of his travel to the northern Peloponnese then from Etolo-Acarnania through Epirus to Thessaly in June and July 1893, but because he synthesised the results of all his predecessors and contemporaries in that marvellously accurate and complete 3-volume Flora (Halácsy 1900-1904) which up to the present day remains the principal if not the only national Flora of this country. Being almost a century old, and leaving out large portions of today's national territory, it is of course badly out of date.

Epilogue: a memorable date

It is not for nothing that I just specified: "up to the present day": this [20 September 1997] is indeed a memorable day for Greek botany. I brought with me, as a special carrier from Sven Koeltz the publisher, the first copy of the first volume of Greece's new and future national Flora, *Flora hellenica* (Phitos & al. 1997), to be presented to Dimitrios Phitos, teacher and mentor of so many of those present at the Symposium, who as the chairman of the Flora's team will forever remain the spiritual father of this whole great and ambitious undertaking.

Acknowledgements

I owe sincere thanks to H. Walter Lack for helpful advice, in particular for providing data from his still partly unpublished work on John Sibthorp, and for critically reading the present manuscript.

References

Alpino, P. 1627: De plantis exoticis, **1-2.** – Venezia.

Anguillara L. 1561: Semplici, liquali in più pareri a diversi nobili huomini scritti appaiono. – Venezia.

Degen, Á. von & Dörfler, I. 1895: Beitrag zur Flora Albaniens und Mazedoniens. Ergebnisse einer von I. Dörfler im Jahre 1893 unternommenen Reise. – Denkschr. Kaiserl. Akad. Wiss. Math.-Naturwiss. Kl. **64:** 701-748.

Desfontaines, R. L. 1808: Choix des plantes du Corollaire des Instituts de Tournefort, publiées d'après son herbier et gravées sur les dessins originaux d'Aubriet. – Paris.

Dumont d'Urville, J. S. C. 1822: Enumeratio plantarum quas in insulis Archipelagi aut littoribus Ponti-Euxini, annis 1819 et 1820, collegit atque detexit. – Mém. Soc. Linn. Paris **1:** 255-387.

Fauché, M., Brongnart, A. T., Bory de Saint-Vincent, J. B. G. M. & Chaubard, L. A. 1832[-1836]: Expédition scientifique de Morée. Section des sciences physiques, **3(2)**, Botanique. – Paris & Strasbourg.

Garbari, F., Tongiorgi Tomasi, L. & Tosi, A. 1991: Giardino dei semplici. L'Orto botanico di Pisa dal XVI al XX secolo. – Pisa.

Greuter, W. 1996: [Review of] Fabio Garbari, Lucia Tongiorgi Tomasi & Alessandro Tosi – Giardino dei semplici. – OPTIMA Newslett. **30:** (47)-(48).

Grisebach, H. R. A. 1841: Reise durch Rumelien und nach Brussa im Jahre 1839. – Göttingen.

– 1843-1845: Spicilegium florae rumelicae et bithynicae exhibens synopsin plantarum, quas aest. 1839 legit. – Braunschweig.

Halácsy, E. von, 1900-1904: Conspectus florae graecae. – Leipzig.

– 1902: Theodor von Heldreich. Ein Nachruf. – Magyar Bot. Lapok **1:** 325-336.

Lack, H. W. 1996: Die frühe botanische Erforschung der Insel Kreta. – Ann. Naturhist. Mus. Wien **98B,** Suppl.: 183-236.

– 1998: To die for Greek flowers. Sibthorp, Bauer and Hawkins in the Levant. – Oxford.

– & Ibáñez, V. 1997: Recording colour in late 18th Century botanical drawings – Sydney Parkinson, Ferdinand Bauer and Thaddäus Haenke. – Curtis's Bot. Mag. Kew Mag. **14:** 87-100.

Lasègue, A. 1845: Musée botanique de M. Benjamin Delessert. – Paris.

Mazal, O. 1981: Pflanzen, Wurzeln, Säfte, Samen. Antike Heilkunst in Miniaturen des Wiener Dioskurides. – Graz.

Moggi, G. 1975: Il museo botanico ed il giardino dei semplici dell'Università di Firenze. – Boll. Soc. Ital. Iris **1975:** 26-39.

Olivier, G. A. 1801-1807: Voyage dans l'Empire Othoman, l'Egypte et la Perse, fait par ordre du gouvernement pendant les six premières années de la République. – Paris.

Phitos, D., Strid, A. & Snogerup, S. (ed.) 1997: Flora hellenica, **1.** – Königstein/Taunus.

– , –, – & Greuter, W. (ed.) 1995: The red data book of rare and threatened plants of Greece. – Athina.

Sibthorp, J. 1806-1840: Flora graeca. – London.

Sieber, F. W. 1820: Beschreibendes Verzeichnis der in den Jahren 1817 und 1818, auf einer Reise durch Creta, Aegypten und Palästina gesammelten Alterthümer und anderen Kunst- und Natur-Produkte nebst einer Abhandlung über ägyptische Mumien. – Wien.

– 1823: Reise nach der Insel Kreta im griechischen Archipelagus im Jahre 1817. – Leipzig & Sorau.

Smith, J. E. 1806-1816: Florae graecae prodromus: sive plantarum omnium enumeratio, quas in provinciis aut insulis Graeciae invenit Johannes Sibthorp, M.D. – London.

Tongiorgi Tomasi, L. 1984: L'isola dei semplici. – Kos **1:** 61-84.

Tournefort, J. P. de, 1703: Corollarium Institutionum rei herbariae, in quo plantae 1356 ... in orientalibus regionibus observatae recensentur et ad genera sua revocantur. – Paris.

– 1717: Relation d'un voyage du Levant, fait par ordre du roi. – Paris.

Turland, N. J. 1995: Linnaeus's interpretation of Prospero Alpino's De plantis exoticis, with special emphasis on the flora of Crete. – Bull. Nat. Hist. Mus. London (Bot.) **25:** 127-159.

Vandas, K. 1909: Reliquiae Formánekianae. – Brno.

A TAXONOMIC SURVEY ON LICHENS OF FOÇA (FUKIA)

B. AKDEMIR, G. ÇOBANOGLU
Department of Biology, Atatürk Education Faculty, Marmara University, 81040 Istanbul, Turkey

1. Introduction

Lichen samples were collected from Foça (Fukia)- Izmir district, located on the west part of Turkey and close to the Aegean Sea, in 1995. The collecting region is a treeless slope of 10 m height from the sea level. Dominant plant cover is richly widespread maquia. There is no residental area in and around the collecting site, and no air pollution problem. Specimens were morphologically observed under binocular dissecting microscope and taken pictures; further examinations were made with light microscope. They were determined by using several floras and keys [1,2,3,4,5,6,7,8,9, 10,11,12]. Twenty one species belong to seventeen genera were identified.

2. Materials and Methods

Lichens were collected with their substrates. They were let air-dried in room conditions and deposited in herbarium within 21x13 cm sized paper bags and 16x11 cm sized envelopes with their localities, altitudes, dates and substrate types noted on.

The specimens were morphologically investigated under a dissecting microscope (Olympus x40) in order to observe the types of fruiting bodies and vegetative organs, and chemically tested with the reagents P, I, K, C, KC and N for identification especially in species level. 10 % HCl was used for defining types of rocky substrates.

Anatomical observations were made by hand-sectioned materials mounted in water and by using standard microscopic techniques.

3. List of the Identified Species

Aspicilia caesiocinerea (Nyl. ex Malbr.) Arnold s.l. (*Fig. 1*)
Aspicilia cinerea (L.) Körber
Buellia subdisciformis (Leighton) Vainio (*Fig. 2*)
Caloplaca ceracea Laundon *(Fig. 3)*
Caloplaca flavovirescens (Wulfen) DT. & Sarnth.

Cladonia foliacea (Hudson) Willd. (*Fig.4*)
Collema undulatum Laurer ex Flotow (*Fig.5*)
Lasallia pustulata (L.) Merat (*Fig.6*)
Lecanora atrosulphurea (Wahlenb.) Ach. (*Fig.7*)
Lecanora cenisia Ach.
Lecidea fuscoatra (L.) Ach.
Parmelia pulla Ach.
Parmelia tinctina Maheu & A.Gillet (*Fig.8*)
Pertusaria monogona (*Fig.9*)
Polyblastia cupularis Massal. (*Fig.10*)
Protoparmelia badia (Hoffm.) Haf.
Ramalina siliquosa (Hudson) A. L. Sm.
Rhizocarpon geographicum (L.) DC.
Rhizoplaca chrysoleuca (Sm.) Zopf (*Fig.11*)
Umblicaria crustulosa (Ach.) Frey (*Fig.12*)
Xanthoria parietina (L.) Th. Fr.

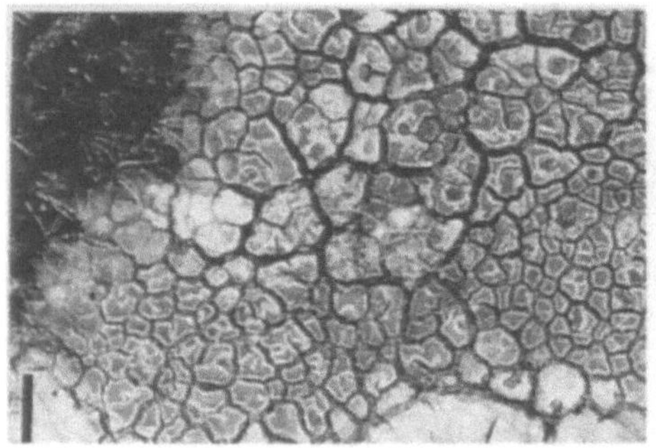

Figure 1. A. caesiocinerea

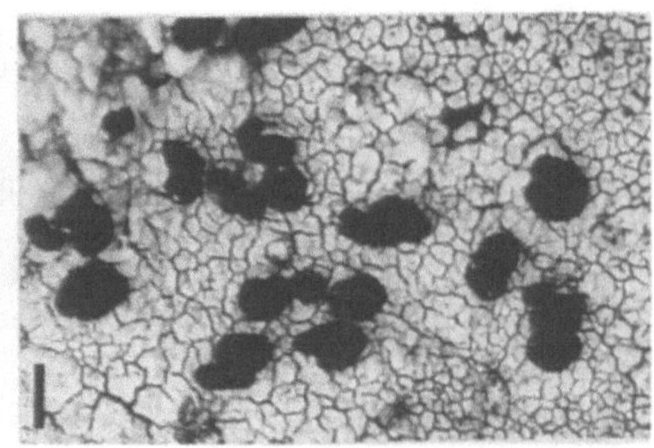

Figure 2. . B. subdisciformis

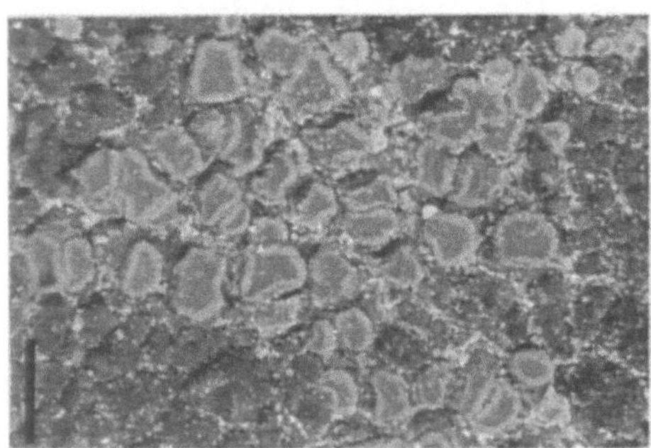

Figure 3. C. ceracea

Figure 4. C. foliacea

In the figures, the bar — indicates 0.5 cm.

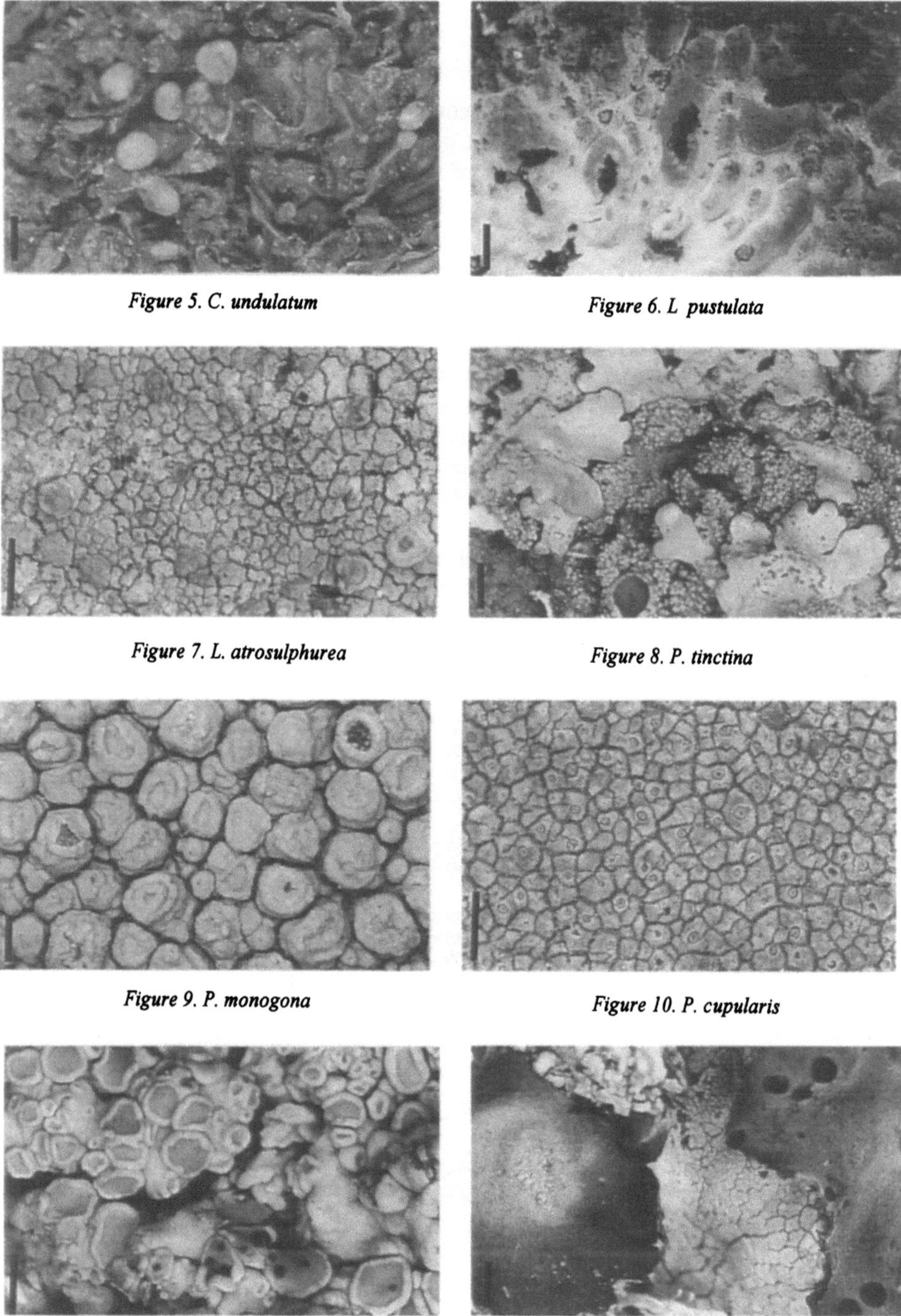

Figure 5. C. undulatum

Figure 6. L pustulata

Figure 7. L. atrosulphurea

Figure 8. P. tinctina

Figure 9. P. monogona

Figure 10. P. cupularis

Figure 11. R. chrysoleuca

Figure 12. U. crustulosa

4. Results

Twenty one species belong to seventeen genera were identified. Photobiont included in their thalli is from Chlorophyta. Only *''Collema undulatum''* has a blue-green algae '*Nostoc*' as a photobiont partner. Most of the specimens are saxicolous growing on silicious rocks and stones, except one terricolous species *''Cladonia foliacea''*. No epiphytic lichens observed because the collecting site is lack of trees. In Foça, 50 % of the identified species are crustose, the others are foliose except *Ramalina siliquosa*; and most of them are well-developed. At least locally no obstructive impact of sulfur dioxide (SO_2) concentration on the development of lichens is observed.

Three species collected from this region are new records for Turkish lichen flora: *Lecanora atrosulphurea, Pertusaria monogona* and *Polyblastia cupularis.*

5. Acknowledgements

We are very grateful to *Dr.Süleyman Akdemir* who assisted in collecting lichens.

6. References

1. Çobanoğlu, G. Istanbul-Kinali, Burgaz, Heybeli ve Büyükada Likenleri Üzerinde Taksonomik Bir Çalisma (Thesis). Marmara Üniversitesi, Fen Bilimleri Enstitüsü, Istanbul, 1996.
2. Dobson, F. *Lichens an Illustrated Guide*. The Richmond Publishing Co. Ltd. , Surrey, 1981.
3. Jahns, H.M. *Collins Guide to the Ferns, Mosses, Lichens of Britain and Nothern and Central Europe*, Collins Grafton Street, London, 1987.
4. John, V. *Atlas der Flechten in Rheinland-Pfalz*, 13/1, Oppenheim, 1990.
5. Moberg, R., Holmasen I. *Flechten von Nordund Mitteleuropa Ein Bestimmungsbuch*. Gustav Fischer Verlag, Stuttgart. Jena. New York, 1992.
6. Öztürk, S. Uludag Liken Türleri Üzerinde Taksonomik Arastirmalar. (Thesis). Uludag Üniversitesi, Fen Bilimleri Enstitüsü, Biyoloji Ana Bilim Dali, Bursa, 1989.
7. Pisut, I. Intressante flechtenfunde aus der Turkei. Preslia. *Praha: 42* (1970), 379-383.
8. Poelt, J., Vezda, A. Bestimmungsschlüssel Europaischer flechten. Erganzungsheft II. *Bibliotheca Lichenologica* (1981), Bd. 16., Cramer, Vaduz.390.
9. Purvis, O.W., Coppins, B.J., Hawksworth, D.L., James, P.W. and Moore, D.M.*The Lichen Flora of Great Britain and Ireland*. Natural History Museum Publications in association with the British Lichen Society, London, (1992).
10. Vitt, D.H., Marsh, J.E., Bovey, R.B. *Mosses Lichens & Ferns of Northwest North America*. University of Washington Press-Seattle, (1988).
11. Wirth, V. *Flechtenflora*. Verlag Eugen Ulmer Stuttgart, (1980).
12. Wirth, V. *Die Flechten Baden-Wurttembergs*. Verlag Eugen Ulmer Stuttgart, 1987.

MARINOPALYNOLOGICAL EVIDENCE FOR THE LATE QUATERNARY VEGETATION AND CLIMATIC CHANGES ALONG THE BULGARIAN BLACK SEA COAST

J.R. ATANASSOVA AND E.D. BOZILOVA
University of Sofia, Faculty of Biology, Department of Botany, Bld Dragan Tzankov 8, Sofia-1421, Bulgaria

1. Introduction

The marinopalynological investigations on sediments from the Bulgarian Black Sea shelf and continental slope, combined with the determination of dinoflagellate cysts and C-14 dating are of great importance for the reconstruction of the palaeoecological conditions along the sea coast as well as within the Black Sea basin for the Late Glacial and the Holocene (Filipova et al., 1989; Shopov et al., 1992; Atanassova 1995).

The results from the palynological investigations of 2 cores - one from the bulgarian shelf (C-4a) and one from the continental slope (XK-71) are presented and the changes of the vegetation and climate during the last 11000 years are discussed.

Due to the specificity of the bental streems and processes of sedimentation on the different parts of the bulgarian shelf, the thiknes of the Late Quaternary sediments is quite different - from several centimetres to 20-30 metres.

2. Material and methods

Location of the cores (*Figure 1.*):

- core C-4a, from the central part of the northern bulgarian shelf at water depth about 50 m.

- core XK-71, from the middle part of the northern bulgarian continental slope at water depth 520 m.

The samples for pollen analysis are prepared according the standard method of Faegri and Iversen (1975) treated in advance with HF acid.

The frequency of each pollen type in the pollen diagrams is presented in percentage of the pollen sum (P), which includes all arboreal (AP) and nonarboreal (NAP) pollen types: P = AP + NAP = 100 %. The frequency of the dinoflagellate cysts is presented in percentage on the base of the total pollen sum. The pollen diagrams are presented in simplified form containing the major pollen taxa and the dinoflagellate cysts.

3. Results and discussion

The predominance of *Chenopodiaceae* and *Artemisia* pollen in the pelitic muds on the shelf between 35-21 m (core C-4a) and in the clay muds between 140 - 110 cm on the continental slope (core XK-71) (*Figure 2.*) show that this sediments were deposited during the conditions of dry and cold climate characteristic for the Late Glacial (probably the Younger Dryas - 11000 BP). Xerophytic herb communities were widely distributed along the bulgarian coast. Most probably in refugial places were presented groups of deciduous trees (*Quercus, Corylus, Ulmus e ct.*). The predominance of the dinoflagellate cysts *Spiniferites cruciformis* and *Tectatodinium psilatum* living in low water salinity (Wall and Dale, 1974) show a regressive phase of the sea and low sea level.

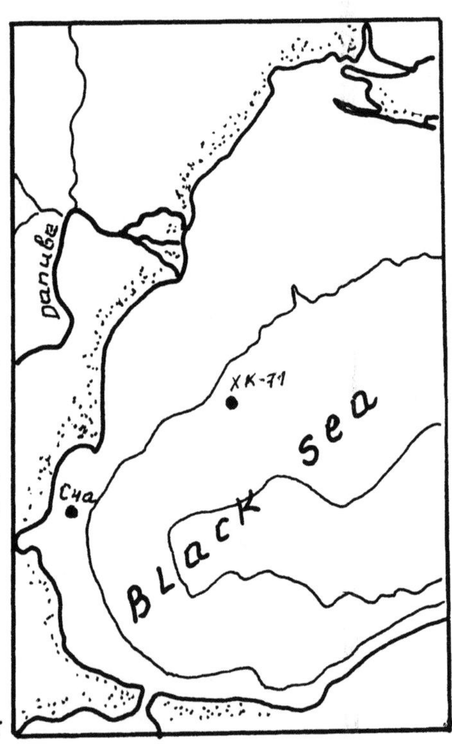

Figure 1. Location of the cores

The transition Late Glacial/Holocene (about 10000 BP) is characterised in the pollen diagrams by the increase in the quantitative participation of deciduous trees - first of all *Quercus, Corylus* and *Ulmus*.

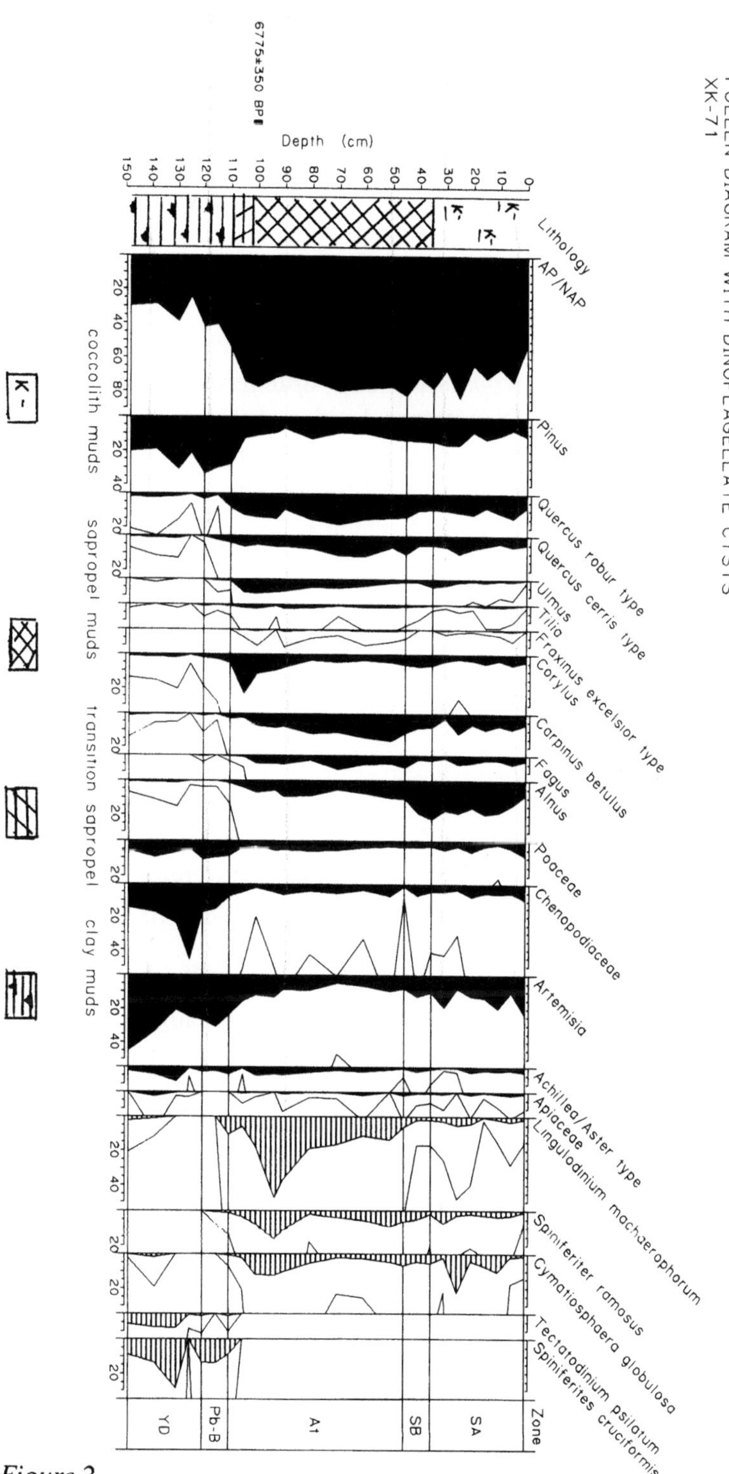
POLLEN DIAGRAM WITH DINOFLAGELLATE CYSTS
XK-71
6775±350 BP
Depth (cm)
Lithology
AP/NAP
Pinus
Quercus robur type
Quercus cerris type
Ulmus
Tilia
Fraxinus excelsior type
Corylus
Carpinus betulus
Fagus
Alnus
Poaceae
Chenopodiaceae
Artemisia
Achillea/Aster type
Apiaceae
Lingulodinium machaerophorum
Spiniferites ramosus
Cymatiosphaera globulosa
Tectatodinium psilatum
Spiniferites cruciformis
Zone
YD
Pb-B
A1
SB
SA
coccolith muds
sapropel muds
transition sapropel
clay muds

Figure 2.

The quick expansion of the mixed oak forest was caused by the rise of the temperatures and the increase of the humidity at the beginning of the Holocene (Preboreal and Boreal periods).

The sapropel muds on the continental slope (core XK-71, 102-36 cm) (*Figure 1.*) are characterised by maximal participation of deciduous pollen (*Quercus, Corylus, Ulmus, Tilia, Fraxinus* and *Carpinus betulus*). This reflects the vast distribution of the mixed oak forests along the coast during the climatic optimum of the Holocene (Atlantic period, 8000 - 5000 BP). The radiocarbon age of the sapropel muds at 99-102 cm - 6775±350 BP (Lb. Hv. 15165) confirms this conclusion. These sediments are rich in dinoflagellate cysts - *Lingulodinium machaerophorum* and *Spiniferites ramosus* which indicate an increased salinity of the water and a rise of the sea level.

The coccolith muds at the western continental slope (core XK-71, 36-0 cm) (*Figure 2.*) are deposited during the last 3000 years. Characteristic is the rise of the *Alnus* curve in the pollen diagram which reflects the formation of the flooded longoz forests along the coast.

4. References

Atanassova, J. (1995) Palynological data of tree deep water cores from the western part of the Black Sea, in E.D.Bozilova and S.B.Tonkov (eds), *Advances in Holocene palaeoecology in Bulgaria*, Sofia, 68-84

Faegri K.J. and J.Iversen (1975) *Textbook of pollen analysis*, Black Well Sciences Publishers Oxford

Filipova M., Bozilova E. and Dimitrov P. (1989) Palynological investigation of the Late Quaternary deep-water sediments from the Southwestern part of the Black Sea, *Bulletin Musee National Varna*, 25, 40, 177-187

Shopov V, Bozilova E. and Atanassova J. (1992) Biostratigraphy and radiocarbon data of upper Quaternary sediments from western part of Black Sea, *Geologica Balcanica*, 22, 2, 59-69

Wall D. and Dale B. (1974) Dinoflagellates in Late Quaternary deep water sediments of the Black Sea, *The Black Sea - geology, chamistry and biology*, 20, 364-375

PHENOLOGICAL PATTERNS OF ORO-MEDITERRANEAN VEGETATION IN THE WHITE MOUNTAINS (WESTERN CRETE)

E. BERGMEIER
Universität Freiburg, Institut für Biologie II (Geobotanik)
Schänzlestraße 1, D-79104 Freiburg, Germany

1. Introduction

The White Mountains (Lefka Ori) form the western of the three high mountain massifs of Crete. It is the most extensive high-altitude area of the island: nearly 4.000 ha reach elevations of more than 2.000 m. The major part consists of crystalline limestones and dolomites. The terrain is rich in dolines, clefts and caves shaped by karstic erosion. There are almost no springs, and permanent streams are missing.

The vegetation of the slopes is extremely scattered and locally almost without plants, at least on first glance. This is why the area above tree-line has been termed „High Desert", a designation which is figurative for the lunar appearance but incorrect both for climatic and botanical reasons. Precipitation may be estimated to be well above 2.000 mm per year, falling almost exclusively during the winter months and as snow. The flora above tree-line comprises more than 200 taxa some 35 % of which are regional or local endemics (Strid, 1993). Perennial herbs and subshrubs are well represented, geophytes are scattered while annuals and shrubby phanerophytes occur in small numbers and decrease with altitude (Bergmeier, 1995).

Except for some phenological observations reported by Hager (1985) this is the first contribution to phenological knowledge on Greek oro-mediterranean vegetation. In landscape ecology, phenological patterns in vegetation not only reflect climatic conditions on various scales (Bergmeier, in press) but may also be a tool in studies concerning ecology and dynamics in vegetation. Both aspects will be addressed here.

2. Methods

At various levels in the southern part of the Lefka Ori, above the settlements of Chora Sfakion and Anopoli, metal fences were established to keep sheep and goats out. The present account concerns the highest fenced site, at 1.620 m a.s.l., just above tree-line (35°16'N/24°05'E). Both doline and slope habitats were included. Study period was from 1993 to 1995 throughout the vegetation periods. Quantitative phenological data were sampled in eight plots of 2 m^2 each, two plots in a grazed doline, two in non-grazed doline, two on a grazed slope, and two on a non-grazed slope. Flowering structures such as heads, umbels, spikes, single flowers, etc. according to the species morphology, were recorded in each plot at intervals of c. 5-8 days.

3. Results and Discussion

Maximum flowering intensity of most species is between May and July. Some taxa (*Anchusa cespitosa, Thesium bergeri, Petrorhagia illyrica* subsp. *taygetea, Bufonia stricta, Crepis sibthorpiana*) show an extended flowering period comprising several months with a more or less distinct phenological optimum in the beginning of the flowering period. *Taraxacum* re-flowers in autumn after flowering maximum in spring and a summer break. *Colchicum cretense* is a specialized late autumn flowering geophyte. Shrubs and subshrubs show maximum flowering intensity between mid-May and beginning of August (Fig. 1). Development of annuals in April/May is forced in between snow break and drying-out of upper soil-layer. Late snow break as in 1993 prevents annuals from completing life-cycle. This is why annuals are almost missing in Crete at levels above 1800 m. From June onwards the phenological rhythm in '93 is synchronous with that in subsequent years (Fig. 2).

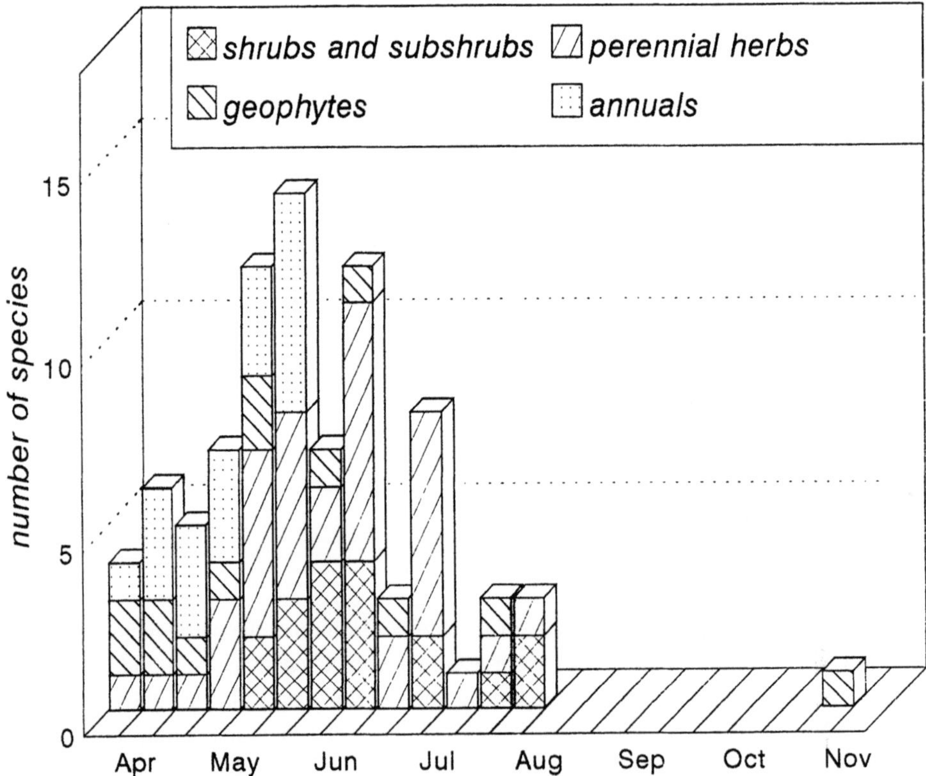

Figure 1. Distribution of maximum flowering activities of species grouped by life-form, at 1.620 m a.s.l., during the vegetation period.

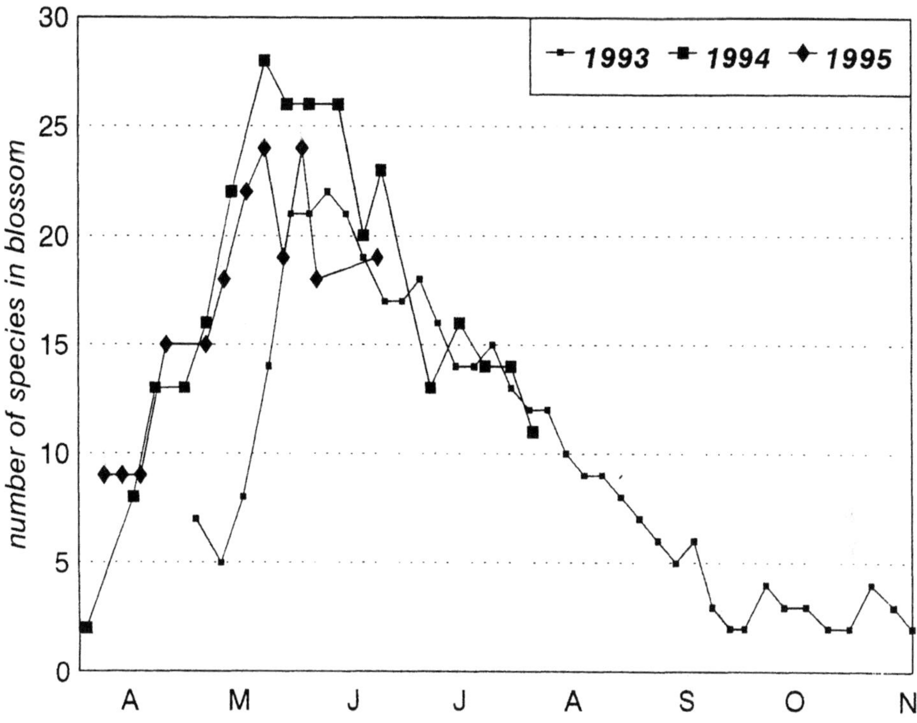

Figure 2. Community phenology at 1.620 m in different years.

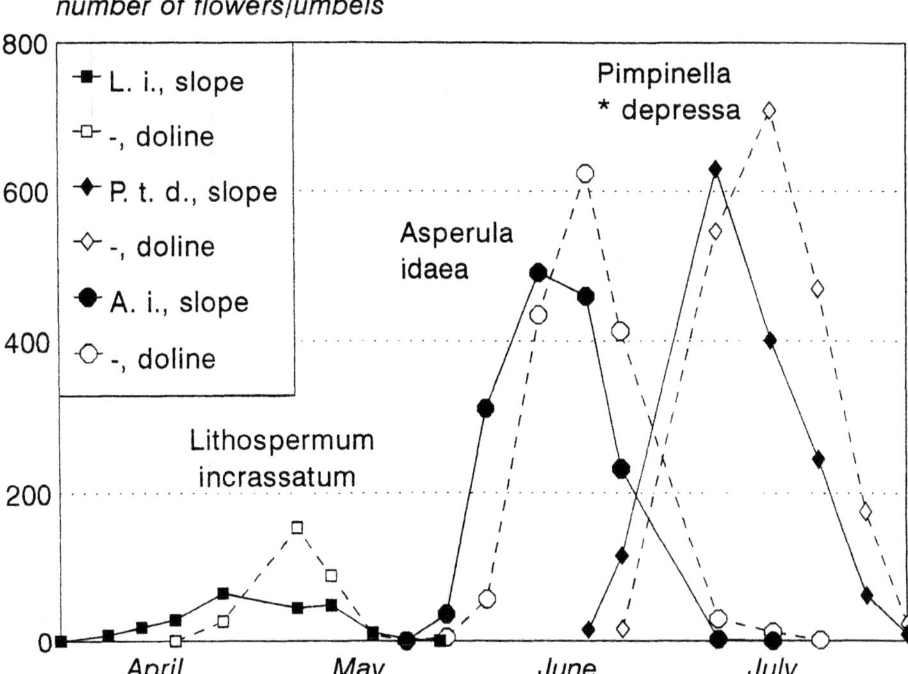

Phenological stages of plants depend on elevation. Taxa with wide altitudinal distribution range, *e.g.*, *Euphorbia acanthothamnos*, *Allium rubrovittatum* and *Coridothymus capitatus*, show 43-50 days of phenological difference between 180 m and 1.620 m a.s.l. The ratio of phenological delay is 3-3.5 days delay per 100 m altitudinal difference, or in other words, with mean temperature decline of 0.6°/100m, the species show 5-5.8 days phenological delay per 1° adiabatic temperature decline.

Phenological analysis of plants growing on doline grounds revealed flowering periods being some 5-8 days later than in plants of the same species growing on the adjacent north-exposed slope (Fig. 3). Cold air is flowing downhill into dolines which form frost pockets. Since phenological delay occurs in early as well as late flowering species, doline grounds represent habitats with less favourable microclimatic conditions than slopes of whatever aspect throughout the vegetation period.

High endemism rates in the Cretan high mountains make the regions particularly relevant for nature conservation efforts. The recently published Red Data Book of Greece (Phitos *et al.,* 1996) includes several high-mountain endemics of Crete which are considered threatened due to extensive grazing by sheep and goats. However, the assessments are not always sufficiently substantiated by facts from vegetation ecology and population biology. Can the fencing experiment and quantitative phenological studies contribute to a better understanding of vulnerability of endemics in the Lefka Ori? Present data do not support the hypothesis of browsing being currently a critical factor for the survival of the endemics in the study site nor seems vulnerability status appropriate for *Alyssum fragillimum* (considered vulnerable by Turland in Phitos *et al.,* 1996), *Anchusa cespitosa* (rare, Strid in Phitos *et al.,* 1996), and *Crepis sibthorpiana* (vulnerable, Kamari in Phitos *et al.,* 1996). Several species, including various endemics show increasing flowering activity in the non-grazed plots but are frequent on grazed pastures as well. Prostrate growth and appressed habit of many oro-mediterranean plants provides protection against browsing pressure. For possible exceptions among extremely rare species such as *Thlaspi zaffranii*, *Ranunculus radinotrichus* and *Nepeta sphaciotica* more thorough investigations on habitat preferences and population development are needed.

4. References

Bergmeier, E. (1995) Die Höhenstufung der Vegetation in Südwest-Kreta (Griechenland) entlang eines 2450 m-Transektes, *Phytocoenologia* **25**, 317-361.

Bergmeier, E. (in press) The phenological approach in Mediterranean landscape analysis, *Phytocoenologia*.

Hager, J. (1985) *Pflanzenökologische Untersuchungen in den subalpinen Dornpolsterfluren Kretas.* Diss. Bot. **89**, Cramer, Vaduz.

Phitos, D., Strid, A., Snogerup, S. and Greuter, W. (eds.)(1996, „1995") *The Red Data Book of rare and thereatened plants of Greece*, WWF, Athens.

Strid, A. (1993) Phytogeographical aspects of the Greek mountain flora, *Fragm. Flor. Geobot. Suppl.* **2**(2), 411-433.

LATHYRUS BINATUS PANČ. IN SERBIA

A species of global importance in biodiversity conservation

B. BUTORAC [1], P. BOŽA[2], R. IGIĆ [2], I. PANIĆ[1] AND D. SAVIĆ[3]
[1] *Institute on Protection of Nature of Serbia, 21000 Novi Sad, Radnička 20*
[2] *Institute of Biology, PMF, 21000 Novi Sad, Trg Dositeja Obradovića 2*
[3] *National Park "Fruška Gora", Znajev trg 1, 21208 Sremska Kamenica*

1. Abstract

A West Serbian mountain Mokra Gora is a "locus classicus" of the eastern Illyrian endemic species *Lathyrus binatus*. This mountain was the only known locality of this species until 1992 when it was found in the National Park " Tara" (Čemerište)[8]. In 1994 we found this species at three new localities (Zvijezda, Crvene Stene, and Crveni Potok). Its survival points out its importance and also global interest in both regional and species protection to preserve ecosystem, specific, and genetic diversity.

2. Introduction

The species *Lathyrus binatus* was found in the National Park "Tara". We recorded this east-Illyrian endemic plant, described for the first time by Pančić[9], from a number of localities and reservations (Zvijezda, Karaula, Štula, Crvene Stene and Crveni Potok) of the investigated area. The presented data on the distribution of this Pančić's species and its global importance for the plant genetic resource conservation provide the basis for future action.

3. Results and discussion

Mokra Gora (western Serbia) is "locus classicus" of the Pančić's endemic species *Lathyrus binatus*. The first finding of this plant supplemented with certain species characteristics was published by Pančić in 1874[9]. Hayek [6] described it as a form of the species *Lathyrus pratensis* L. Beck-Mannagetta [2] quoted *Lathyrus pratensis* L. v. *binatus* (Pančić, J., Fl.knez.Serb.:256 (1874) pro.spec) for a number of localities in Bosnia and the Prenj mountain in Hercegovina. In the same source also Maly.K. was mentioned who, in 1908, cited this taxon as *Orobus pratensis* v. *binatus* Asch.,Kanitz. Ball [1] in Flora Europaea, in his study of the genus *Lathyrus* also mentioned *Lathyrus binatus* Pančić as an intermediary taxon between *Lathyrus pratensis* L. and *L. halersteinii* Baumg. for the region of central Yugoslavia. Diklić [4] described it as a valid species. Pančić's data on the occurrence of *Lathyrus binatus* at Mokra Gora is cited as an unique record of this species for the Serbian flora. In volumes IX [7] and X [10] no additional data on the range of this rare and endemic plant in the Serbian flora were presented. According to Gajić [5], *L. binatus* belongs to the group of east-Illyrian endemic species. This statement is in agreement with the *Lathyrus binatus* distribution data in certain regions of Bosnia and certain microhabitats of westernmost Serbia. These localities refer to Mokra Gora and Zvijezda and Tara mountains. They are in most part within the boundaries of the National Park "Tara" representing a center of floristic-vegetation diversity not only in Serbia, but also, of the whole Balkan peninsula. Šilić [11] in his survey into the endemic species also mentioned the species *Lathyrus binatus* with its ranges in Serbia and Bosnia and Hercegovina, but without details on its localities. On the basis of these range data he classified it into the group of the endemic plants of Yugoslavia. Taking into consideration the fact that the endemic character is primarily related to the phytogeographical aspects of a region, and that *Lathyrus binatus* belongs to the eastern parts of the Illyrian floristic province (certain localities of Bosnia and western Serbia), we agree with the Gajić's data [5] on its nature. Obratov et al. [8] in their supplement to flora of the National Park "Tara" quoted for the first time the species *Lathyrus binatus* emphasizing the locality Čemeriste.

Butorac et al. [3] in their survey into plant cover of this Park, recorded this endemic species from certain new localities reffering to the refuge habitats of officially recognized reservations of Zvijezda, Crvene Stene, Crveni Potok, and Karaula Štula. These new data are a valuable contribution to the knowledge of the distribution of this Pančić's species within the boundaries of its endemic range. Knowledge of vegetation and habitats types throws more light upon the synecological characteristics of the analyzed species enabling adequate conservation steps in our attempts to reduce the risk of its extinction.

According to Pančić [9]and Flora of Serbia [4], *Lathyrus binatus* ocċurs in stony habitats. Obratov et al. quoted it for limestone substratum. Our investigation results reffering to the reservation Zvijezda (Topli Do locality) are in agreement with the above data. In addition to the species *Lathyrus binatus*, permanent members of the rocky vegetation are *Globularia cordiifolia, Genista lydia, Athamantha haynaldii, Helianthemum nummularium, Pyrus nivalis, Dianthus petraeus, Sempervivum tectorum,* and *Rosa spinosissima* f. *spraeta*. According to Šilic [11], this endemic species occurs also in open habitats at forest margins and forest clearings. Our finding at Crvene Stene locality reffers to the transition area between the forest stands of beech, fir, and spruce community and mesophilous meadows of the alliance *Arrhenatherion elatioris*. These stands are dominated by *Rhinanthus rumelicus, Danthonia calycina, Sanguisorba officinalis* and *Filipendula hexapetala*. The habitat in the reservation Crveni Potok is a refuge in character due to an unique forest peat bog in this country. *Lathyrus binatus* also endures at the margin of stands of the association *Alno-Piceetum omorikae* Čolić et Gigov. It is cold, marshy relict habitat where soil and air temperatures are lowered due to permanent water impact therefore imposing central European character upon the recorded phytocenosis. The occurrence of this species also under such conditions point out its broad ecological range and its adaptability. Also worthy of note is the most recent record of the species *Lathyrus binatus* from the National Park 'Tara" reffering to the locality Zaovine (according to V.Stevanović report in 1997).

4. Conclusion

A review of all the available data, presented in this paper, shows that in addition to the first record of *Lathyrus binatus* made by Pančić [9], those of Obratov et al. [8], Butorac et al. [3], and most recently by Stevanović are the only findings furnishing the evidence on the distribution of this rare plant species in Serbia. Its localities restricted to a narrow zone and endemic character of this east- Illyrian taxon have ensured its position in the Red List of Serbia. In addition, this Pančić's taxon requires not only improvement and protection of its habitats, but, also, conservation of the species itself. This step will much contribute to conservation of ecosystem, species, and genetic diversity.

Acknowledgement

We thank Dr Vladimir Stevanović for ceding his most recent data on the distribution of the species *Lathyrus binatus*

5. References

1. Ball, P. W. in Tutin, T. G. ed. 1968 Genus *Lathyrus L.* in *Flora Europaea* 2: 140, Cambridge University Press.
2. Beck-Mannagetta, G. 1927 Genus *Lathyrus L.* in *Flora Bosnae, Hercegovinae et regionis Novi Pazar*, III Choripetalae, Državna štamparija u Sarajevu, Beograd - Sarajevo.
3. Butorac, B., Panić, I., Savić, D. 1995 Novi prilozi za vaskularnu floru Nacionalnog parka "Tara". *Zaštita prirode* **46-47**, 133-143.
4. Diklić, N. in Josifović, M. ed. 1972 Rod *Lathyrus L. Flora SR Srbije*, tom IV, Srpska Akademija Nauka i Umetnosti, Beograd.
5. Gajić, M. 1980 Pregled vrsta flore SR Srbije sa biljnogeografskim oznakama. *Glasnik šumarskog fakulteta*, ser A, **54**, 111-141.
6. Hayek, A. 1927 Genus *Lathyrus L.* in *Prodromus Florae Peninsulae Balcanicae*, 1 Band: 816, Dahlem bei Berlin.
7. Josifović, M., ed. 1977 *Flora SR Srbije*, tom IX Srpska Akademija Nauka i Umetnosti, Beograd.
8. Obratov, D. et al. in Gajić, M. et al. 1992 *Drugi dodatak Flori i Vegetaciji nacionalnog parka "Tara" in Vegetacija nacionalnog parka "Tara"*, Šumarski fakultet - Beograd, Nacionalni park "Tara" - Bajina Bašta.
9. Pančić, J. 1874 *Flora Kneževine Srbije (Flora Principatus Serbiae)*, Beograd.
10. Sarić, M. ed 1986: *Flora Srbije*, tom X, Srpska Akademij Nauka i Umetnosti, Beograd.
11 Šilić, Č. 1988 *Endemične biljke*, Svjetlost, Oour za udžbenike i nastavna sredstva - Sarajevo, Zavod za udžbenike i nastavna sredstva - Beograd.

CONTRIBUTION TO THE STUDY OF VEGETATION OF MOUNT VERMION (NC GREECE)

S. CHOCHLIOUROS & TH. GEORGIADIS
Botanical Institute - Plant Ecology, Division of Plant Biology, Department of Biology, University of Patras, GR 265 00, Patras-Greece

1. Abstract

Mount Vermion is located in NC Greece and it is a mainly calcareous mountain. The Selliotis stream divides the mountain into N and S orientations where its highest peaks are located. The N, NE, S, SA orientations are colonized by abundant wood vegetation with *Fagus, Pinus, Quercus* and *Abies* species dominating. On the contrary, in the western areas, there is no wood vegetation. From the bioclimatic point of view, Mount Vermion belongs to the sub-humid bioclimatic level where the winters are harsh and cold. According to the meteorological data, the studied area is characterized by the type of Mediterranean climate of the Para-Mediterranean regions, but it is also intensively effected by the Meso-European climate.

2. Materials and methods

250 releves were done according to the Braun - Blanquet phytosociological method (NETSHOFF & VAN DER MAAREL 1978) in the studied area and from represantative biotopes of various vegetation types. All species of vascular plants as well as some environmental factors (elevation, aspect, slope inclination and geological substratum) were recorded at each releve site.
The initiation of the total phytosociological material was carried out with the assistance of TURBOVEG software.
The phytosociological units were defined according to the Braun-Blanquet method (1964) and with the assistance of multi-factor analysis (Taxonomy, Classification).
The altitudinal limits of our studied area are defined by the level ≥500m.

3. Discussion - Conclusions

The analysis of releves which is done by the multivariant method of classification and ordination, has shown the destinction of the main vegetation types which are occurred on the mountain: Forests: The Alliances *Fagion moesiaceae hellenicum* (Dafis 1973) Ellenberg & Glavac 1974, *Pino-Chamaecytision* Nov, *Quercion frainetto* Horvat 1954, *Tilio - Acerion* Klika 1955, *Ostryo - Carpinion* Horvat 1956-1958, Mixed stands with *Fagus* and *Pinus* species, mixed stands with *Fagus* and *Abies* species, *Prunetalia* (Figure 1), and Pastures: Communities in wet places with *Blysmus, Juncus* and *Scirpus* species and All. *Ramondion nathaliae* Horvat 1935 in calcareous rocks, communities with *Astragalus angustifolius* species, formations with *Juniperus communis* and *Stipa* species, *Pinus heldrachii* stands, communities with *Festuca* species, formations with *Globularia* species, formations with *Stipa* species.

In some plant communities, several of the existing references are verified as: *Sesleria robustae - Pinetum pallasianae* Barbero et Quezel 1976 car. Bergineier, *Tilio tomentosae-Castanetum* Dafis 1966, *Fagus sylvatica* et *Physospermum aquilegifolium* (Quezel, 1967), *Ramonda nathaliae - Campanula fonuameriana* Qezel 1967, while is shown the existence of new ones is also shown.

Considering the location of the mountain, as well as the high percentage of species of Balkan origin and also our data analysis, we can conclude that there are new, mainly Balkan, syntaxa in the area which are occurred in the forest ecosystems as well as in the pasture ecosystems.

Further analysis of our data, in combination with a common analysis of all the already existed data related to our study area and neighboring areas as well as the comparison to mountains of the Balkan Peninsula, will bring out a lot of new data on the vegetation of N. Greece and surrounding areas.

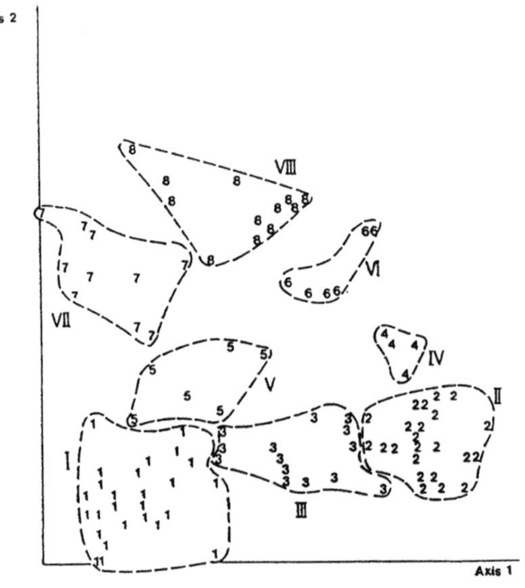

Figure 1. Ordination A

A Forests

I (1) - : Querco - Fagetea Br.-Bl. et Vliegeger in Vlieger 1937
Fagetalia sylvatica Pawlovski 1928
Fagion moesiacae hellenicum (Dafis 1973) Ellenberg & Glavac 1974
II (2) - : Quercetea pubescentis (Oberd. 1948) Doing Kraft 1955
Quercetalia pubescentis Br.-Bl. 1932
Pino - Chamaecytision Quezel 1976
Seslerio robustae - Pinetum pallasianae Barbero et Quezel 1976 corr. Bergmeier 1990
III (3) - : Mixed stands with *Fagus sylvatica* and *Pinus nigra* subsp. *pallasiana*
IV (4) - : Mixed stands with *Fagus sylvatica* and *Abies borisii-regis*
V (5) - : Communities with *Quercus pubescens*, *Quercus frainetto* species
VI (6) - : Quercetea pubescentis (Oberd. 1948) Doing Kraft 1955
Quercetalia pubescentis Br.-Bl. 1932
Tilio - Acerion Klika 1955
Tilio tomentosae - Castanetum Dafis 1966
VII (7) - : Prunetalia
VIII (8) -: Quercetea pubescentis (Oberd. 1948) Doing Kraft 1955
Quercetalia pubescentis Br.-Bl. 1932
Ostryo - Carpinion (Horvat, 1956-1958)

Pastures

Rocky pastures with low scrubs (*Juniperus*, *Astragalus*) and stands with *Pinus heldreichii*:
Communities with *Astragalus angustifolius* species
Formations with *Juniperus communis* species
Formations with *Juniperus communis* and *Stipa* species
Pinus heldreichii stands

Stepped and grassy pastures:
Communities with *Festuca* species
Formations with *Globularia* species
Formations with *Stipa* species

Communities in wet places with *Blysmus, Juncus and Scirpus* species

In calcareous rocks:
Asplenietea - Trichomanis (Br.-Bl. 1934) corr. Oberdorfer 1977
Potentilletalia speciosae Quezel 1964
Ramondion nathaliae Horvat 1935
Ramonda nathaliae - Campanula formanekiana Quezel 1967

4. References

Athanasiadis, N. Hp. (1978, 1986) *Dasiki Fitokinoniologia*, I. Fokianaki and Jiachoudi-Jiapouli, Thessaloniki.
Barbero, M. & Quezel, P. (1976) Les groupements forestiers de Grce Centro Meridionale, *Ecologia Mediterranea* **2**, 3-86.
Horvat, I., Glavac, V., Ellenberg, H. (1974) *Vegetation Sudosteuropas*. Gustav Fisher Verlag, Stuttgart.
Dafis, S. (1973) Taxinomisis tis dasikis blastisis tis Ellados, *Epist. Epet. Geop. and Das. School. of Univ. of Thessaloniki* **15/2**, 75-91.
Quezel, P. (1964) Vegetation des hautes montagnes de la Grece meridionale, *Vegetatio* **12** (5/6), 289-386.
Quezel, P. (1967) La vegetation des hautes sommets du Pinde et d' Olympe de Thessalie, *Vegetatio* **14**, 127-228.

PHYTOGEOGRAPHICAL ANALYSIS OF THE FLORA OF MOUNT VERMION (NC GREECE)

S. CHOCHLIOUROS & TH. GEORGIADIS
Botanical Institute - Plant Ecology, Division of Plant Biology, Department of Biology, University of Patras, GR 265 00, Patras-Greece

1. Abstract

Mount Vermion is located in NC Greece. Its orientation is N-S. It is a large mountain with calcareous substrate. The altitude of its highest peaks varies from 1650 to 2052 m. Due to its location within the boundaries of the Mediterranean and Middle-European regions and its geology, Vermion is a significant mountain from both a floristic and phytogeographical point of view. The total of the Greek endemic, Balkan endemic and Sub-Balkan taxa which occur on Vermion are analyzed as well as the chorological and life-form spectrum of the total flora which is composed of 1042 taxa (859 species, 183 subspecies). This analysis is particularly useful because it helps us to determine the phytogeographical connections existing between Vermion and other massives or phytogeographical regions of Greece and the Balkan Peninsula.

2. Results - Discussion - Conclusions

Considering the exclusive presentations among Vermion' s neighbouring NC phytogeographical regions, of the total Greek endemic taxa (32 in total) which occur in Vermion' s flora, we have reached the conclusion that the relations between these neighbouring NC phytogeographical regions and more eastern regions (e.g N.E) are weak. On the contrary, the relations with those regions to the west (e.g. N.P.) are stronger, since 4 exclusive presentations are found between NC and NP. Correspondingly, the comparison to southern regions shows that there are greater phytogeographical connections to St.E than to Peloponnisos.
From the analysis of the common Greek endemic taxa (Table 1) in the same phytogeographical region (NC), we conclude that more grow on Olympos (11 taxa) and ther the mountains: Vourinos, Voras, Pieria, Varnous and Pinovo in ascending order. Similarly, the percentages of common taxa among the other phytogeographical regions of Greece are greater in the northern regions (NP, SP, NE) and smallest in southern regions (St.E, PEL.).

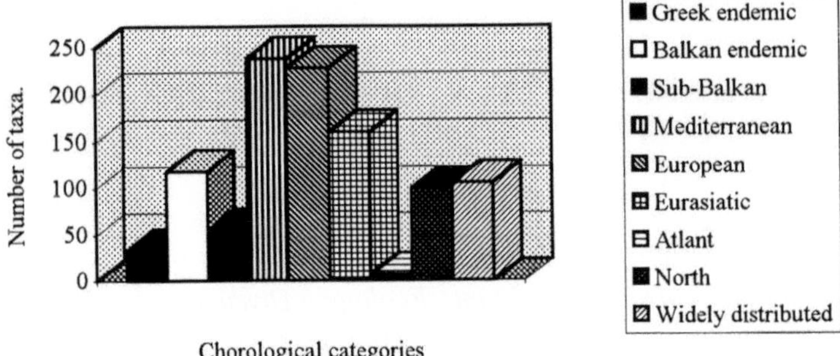

Figure 1. Schematic representation of the chorological spectrum of Vermion flora

TABLE 1. Geographical distribution of Greek taxa of Mount Vermion in the same geographical region and in wider geographical regions

NC							NE						NP	SP	St.E	PE	EC	AE	Cr
Ve	Ol	Vu	Vo	Pi	Va	Pn	At	Ch	Pa	Me	Be	Ro							
32	11	5	3	3	1	1	3	1	1	1	1	1	11	8	17	7	2	3	3

If we consider the common Balkan endemic taxa (Table 2) within the NC region we conclude that there is also a closer relation with Olympos (51 taxa) and ther with the mountains: Vourinos, Voras, Pinovo, Pieria, Tzena and Varnous. The Balkan endemics predominate in the NC region and afterwards in NP, NE, SP, St.E, PEL, EC, AE, IO, CR .

TABLE 2. Numerical distribution of Balkan endemic taxa of Mount Vermion relative to other mountains and wider geographical regions

NC								NE						NP	SP	EC	StE	PE	AE	Io	Cr
Ve	Ol	Vu	Vo	Pn	Pi	Tz	Va	Pa	Me	Ch	At	Ro	Be								
118	51	44	40	27	24	24	14	31	28	28	16	10	7	65	56	30	54	41	12	8	2

These consitute the largest element of the extension towards Northern and Central mainland parts. Their significance is reduced from St. E. towards Pel. and they have scanty presence in Cr.

Considering the percentages of common endemic taxa (Greek and Balkan), the intimate phytogeographical connection of Mount Vermion with Mount Olympus confirms Strid' s ascertainment (1993) that Mounts Vermion and Olympus are connected by a relatively high level of similarity in the endemics as well as the common taxa.

In comparison to the Balkan Peninsula (Table 3) it is shown that the NC region has close relations with the Former Yugoslavia. The eldest section of the Balkan taxa which are widely distributed in Balkan Peninsula, participates in the studied flora.

TABLE 3. Analysis of Balkan taxa of Mount Vermion

Categories	Number of taxa	%
Gr, Ju, Al, Bu	31	26.27
Gr, Ju, Al	30	25.42
Gr, Ju	21	17.79
Gr, Al	15	12.71
Gr, Ju, Bu	14	11.86
Gr, Al, Bu	1	0.84
Gr, Bu	6	5.08
Total	118	100.00

The percentage of the Balkan-An taxa in Vermion' s flora appears to have a primary position if we compare it to other categories of Sub-Balkan taxa (Table 4).
The Balkan-An taxa present a higher percentage of representation in the NE phytogeographical region with an extension towards NC (Strid 1993). Besides, it is well-known that the Balkan Peninsula was connected during the Miocene with NW Anatolia through the mountain chain of Rodhopi (Kronberg et al., 1970).

TABLE 4. Analysis of Sub-Balkan taxa of Mount Vermion

Balkan + It	Balkan + Rm	Balkan + An
14	7	34

Future floristic study of other massives will constitute the key for better phytogeographical approaches between the massives and the phytogeographical regions.

3. References

Chochliouros, S. & Georgiadis, Th. (1997) Additions to the Flora of Mount Vermion (N.C. Greece), *Willdenowia* **27** (1/2).

Ganiatsas, K. (1929) *Phytogeographia*, Thessaloniki (in Greek).

Kronberg, P., Meyer, W. & Pilger, A. (1970) Geologie der Rilla-Rhodope Masse Zwissen Strimon und Nestos, *Beih. Geol. Jb.* **88**, Hannower.

Strid, A. (1986) *Mountain Flora of Greece* **1**, Cambridge.

Strid, A. & Tan, K. (1991) *Mountain Flora of Greece* **2**, Edinburgh.

Strid, A. & Tan, K. (1993) Phytogeographical aspects of the Greek mountain flora, *Fragm. Flor. Geobot. Suppl.* **2** (2), pp. 411-433.

Turril, W. B. (1929) *The Plant - life of the Balkan Peninsula*, Oxford.

EVALUATION SYSTEM OF THE RELATIVE IMPORTANCE OF VEGETATION SYNTAXA FOR NATURE CONSERVATION IN GREECE

P.[1] DIMOPOULOS, E[2]. BERGMEIER, E[3]. PAPASTERGIADOU, K.[4] SYKORA, S[3] DAFIS.

1:Pindarou 25, GR- 13562 Agioi Anargyroi, Athens, Greece.

2:Univ. of Freiburg, Institute of Biology II Schaenzlestrasse 1, D-79104 Freiburg, Germany.

3:Greek Biotope-Wetland Centre, 14th km Thessaloniki-Mihaniona, GR57001 ThermiGreece.

4:Department of Terrestrial Ecology and Nature Conservation, Agricultural University of Wageningen, Bornsesteeg 69, NL-6708 PD Wageningen, The Netherlands.

1. Introduction

Basic knowledge of floristic composition, structure, dynamics and ecology of vegetation is of fundamental importance for the conservation of areas of biological interest (Dierschke 1995). A thorough classification and description of plant syntaxa is of a great importance for nature conservation. Currently, a critically revised syntaxonomic system is in progress in Greece. A syntaxonomic typology of Greek Habitats has recently been established (Papastergiadou et al. 1997). It is mainly based on the phytosociological classification of European syntaxa (Commission of European Communities 1991). A detailed ecological evaluation and mapping of the sites eligible for the NATURA 2000 Ecological Network is a dire neceessity and should be based on the syntaxa described from Greece. The main reasons for the use of vegetation syntaxa: i) Plant communities integrate relevant environmental variables and summarize the total floristic diversity (Gehu 1992), ii) They strongly indicate the possibilities of restoration of Potential Natural Vegetation.

2. Materials and Methods

The procedure employed by Seibert (1980), Loidi (1995) and partly Rodwell & Cooch (1996) for Germany, Spain and United Kingdom, respectively, is followed. The deviced evaluation system combines several ecological criteria in an index. The criteria of the Nature Conservation Review, the examples from Germany (Seibert 1980), Spain (Loidi 1995), France (Gehu 1992) and United Kingdom (Rodwell & Cooch 1996), have been taken into account. For our evaluation system, the following criteria were considered: Rarity (R), Naturalness (N), Threat (T), Replaceability (P) and Floristic-Phytocoenotic value (F). For each criterion, a 0-10 point-scale is given. Some other parameters not intrinsic to the syntaxa, but relevant for conservation have also been taken into consideration. Numeric values according to the scales can be assessed for each criterion and related to each of the cartographic units. By assigning a numerical value to each of the mapped syntaxa, a vegetation map can be transformed into a map of ecological values. However, we must point out with emphasis, that high-level syntaxa, such as orders or classes, may differ much concerning the respective criterion and may thus be attributed to 3 or even 5 scale numbers.

3. The Criteria and the Scoring System

3.1 NATURALNESS (N)

It expresses the degree of human interference comprising two aspects: i) actualistic approach: extent of transformation by man or dependence of plant communities on either traditional or intensive management (Sukopp 1972), ii) historical approach: successional proximity to the historical natural vegetation (Kowarik 1988).

0: Polluted, intensively urbanized areas, almost completely occupied by buildings, roads, etc., **1:** Peri-urban and industrial areas, harbours, garbage heaps, **2:** Parks, gardens, intensively cultivated fields, roadsides, walls, abandoned cropfields, **3:** Exotic tree plantations for intensive exploitation, overgrazed pastures, **4:** Intensively grazed grasslands and meadows, degraded phrygana, traditionally cultivated fields with rich weed vegetation, **5:** Heavily degraded scrub and woodlands, mature phrygana, severely polluted aquatic, riparian and coastal habitats, extensively grazed pastures, **6:** Abandoned pastures and phrygana, grazed scrub, moderately polluted aquatic, riparian and coastal habitats, **7:** Grazed and/or coppicced woodlands, naturally regenerating post-fire scrub and woodland, disturbed aquatic, riparian and coastal habitats, initial woodland succession areas, oro-mediterranean pastures, **8:** Young natural woodland mixed with mantle and other seral communities linked to the forest system, little to moderately grazed woodlands, moderately exploited forests, little disturbed wetland and coastal habitats, **9:** Potential natural and permanent vegetation subject to light exploitation, regenerated woodlands, coastal dunes, salt marshes without direct human impact, **10:** Mature non-exploited forests, inaccessible rocks, cliffs, screes.

3.2 THREATS (T)

The evaluation of threats is is correlated to the response of each plant community to habitat changes (fragility), the degree of accessibility of the syntaxon, etc. Criteria of primary concern to express whether a community is threatened or not are: trend of stand numbers (decreasing, constant, increasing); trend of stand quality (species combination currently constant, changing); likeliness of human impact (constant or changing) and species-in-community trends (one, few to many species ubiquitous, widespread, regionally rare, regionally vulnerable, absolutely vulnerable).

0: No vegetation, **1:** Aggressive communities with negative effect on surrounding vegetation, **2:** Depauperate communities of ubiquitous species with increasing stand numbers, **3:** Vegetation with ubiquitous or widespread species with constant stand numbers, **4:** Stand numbers and species combination constant, no rare or vulnerable species, **5:** Stand numbers and species combination almost constant, with regionally rare but no vulnerable species, **6:** Stand number constant, species combination changing, no or few regionally rare species, **7:** Stand number decreasing, species combination changing, few regionally rare species, **8:** Stand number decreasing, species combination changing, many regionally rare species, **9:** Stand number rapidly decreasing, species combination changing, several regionally rare and few vulnerable species, **10:** Stand number rapidly decreasing, species combination changing, few to many vulnerable species.

3.3 REPLACEABILITY OR RESILIENCE AFTER DISTURBANCE (P)

It is defined as the capability of a vegetation type to recover after destruction or strong transformation. An inverse scale is proposed, since the less replaceable plant communities are evidently more demanding of protection.

0: No vegetation, **1:** Fragmented annual weed communities, **2:** Nitrophytic annual and fragmented perennial pioneer communities, **3:** Degraded macchia and phrygana vegetation, pine forests, non-fragmented weed and other annual communities, pioneer scrub, reeds, shortlived aquatic communities, wall vegetation, **4:** High macchia and mature phrygana, pre-forest, fragmented chasmophytic, dune, aquatic, littoral and inland saline perennial vegetation, fragmented mesophytic meadows and dry grasslands, **5:** Non-fragmented meadows and salt marshes, coastal white dunes, aquatic, littoral and inland saline vegetation, fragmented vegetation in rock crevices and rocky perennial grasslands, **6:** Mediterranean oak and riparian woodlands, hedges, scrub along wood-margins, coastal grey dunes, mature grasslands, rocky pastures, **7:** Species-poor deciduous forests, isolated wetland and grassland stands, **8:** Oro-mediterranean vegetation (p.p.), montane coniferous forests, species-rich broadleaved forests, **9:** Mires, rocks and cliffs, oro-mediterranean vegetation (p.p.), **10:** Isolated rocks and cliffs with relict plants.

3.4 FLORISTIC-PHYTOCOENOTIC VALUE (F)

The objective assessment of this criterion is difficult, though it is considered, together with naturalness, as the main factor in order to estimate the ecological value of a vegetation type. After Rodwell & Cooch 1996, this criterion includes the following elements: *i) floristic value:* specific diversity (species richness, e.g. total number of plants per vegetation type), *ii) phytosociological value*: phytosociological diversity (richness of syntaxa at different levels), *iii) phytogeographical value*: participation of endemic or rare taxa in each syntaxon, *iv) structural complexity.*

0: no vegetation, 1: Low species and community diversity, simple structure, widespread to ubiquitous species, 2: Low to moderate species and community diversity, simple structure, mostly widespread species, 3: Moderate species and community diversity, stand structure rather simple, mostly widespread flora, 4: Moderate species and community diversity, stand structure of two layers, widespread flora, geographically fairly variable, 5: Fairly rich species and community diversity, stand structure of two or more layers, rare but widespread species, geographically variable, 6: Fairly rich species and community diversity, stand structure of two or more layers, high geographic variability, Balkan endemics present, 7: Fairly rich species and community diversity, stand structure of two or more layers, high geographic variability, Greek endemics present, 8: Rich species and community diversity, stand structure of two or more layers, geographically very variable, few regional endemics present, 9: Rich species and community diversity, stand structure of two or more layers, geographically very variable, several regional endemics present, 10: Rich species and community diversity, stand structure of two or more layers, geographically very variable, many regional or local endemics present.

3.5 RARITY (R)

Since data about the distribution of syntaxa in Greece are scarce, assessment of rarity of phytocoenoses can only be preliminary and tentative. Under this criterion we summarize the following subcriteria: i) general distribution pattern at national level, very widespread (occurrence in 8-13 phytogeographical regions); fairly widespread (in 5-7 regions); regional (in 3-4 regions); restricted (in 1-2 regions). ii) regional dispersal pattern, although such information is usually incomplete and variable in quality, may be described as common, occurring abundantly in a given region; scattered, occurring throughout the region, but in stands which are scarce and widely dispersed;–local, within a region confined to a small area where it is frequent; rare, within a region confined to a small area with one or few stands only, iii) total syntaxon area at national level (although accurate and detailed data are mostly unavailable). Areas can be preliminary categorized as: large (> 50.000 ha) medium (500-50.000 ha), small (5-500 ha), punctual (< 5 ha); iv) ratio of actual and potential area (area that would have been covered without human impact): < 10%, 10-50%, 50-100%, >100%.

1: Very widespread, common, medium to large, 50 ->100%, 2: Widespread, common, small to medium, 1->100 %, 3: Widespread, scattered, small to large, 10 - 100 %, 4: Widespread, regional to fairly common medium to large, 10 - >100 %, 5: Regional, common, small to medium, < 10 - 100 %, 6: Restricted, common, small to large, <10 - 100 %, 7: Regional, scattered, small to large, 10 - >100 %, 8: Restricted, scattered or local, small to medium, 50 - 100 %, 9: Regional, rare to local, small to medium, <10 - 100 %, 10: Restricted, local or rare, one to few, small, < 10 - 100 %.

4. Conservation Interest (CI)

The Conservation Interest of a site, taking into consideration only its floristic and vegetation characteristics (botanical value of the site) can be expressed in the formula:

CI= N+P+T+F+R

N=Naturalness, P= Replaceability, T= Threat, F= Floristic-Phytocoenotic value, R= Rarity.

The values for N, P, T and F can be linked to cartographic units, which means that each colour on a vegetation map is associated with a concrete value for each of these parameters.

5. Concluding remarks

Vegetation maps using phytosociological units are of great significance for application of management plans and for accurate evaluations. The criteria used for the evaluation, should be represented on the vegetation map (corresponding to each of the cartographic units) and the Total Ecological Value of one site (Total Conservation Interest) will be the sum of the Conservation Interests of each of the mapped units. Our knowledge on distribution, variability, ecology and diversity of most syntaxa in Greece is still far from sufficient. Having this in mind, the present attempt can only be a preliminary one which s in need of re-evaluation, discussion and continuous improvement.

6. References

Commission of European Communities (1991) *CORINE biotopes manual. Habitats of the European community.* Office for Publications of the European Communities. Luxembourg.

Dierschke, H. (1995) The Braun-Blanquet approach to phytosociology as a basis for nature conservation, exemplified by montane grassland areas, in Song, Y., Dierschke, H., Wang, X. (eds.), Applied Vegetation Ecology. Proceed. 35th Symposium IAVS, East China Normal University Press., Shangai, pp. 1-9.

Gehu, J.-M. (1992) Sur la fragilite structurale de la frange terrestre des bords de mer et son analyse par les methods de la phytosociologie moderne dans un souci conservatoire. *Coll. Phytosociologiques* **19,** 37-44. Berlin.

Kowarik, I. (1988) Zum menschlichen Einfluss auf Flora und Vegetation. *Landschaftsentwicklung und Umweltforschung (Berlin)* **56,** 280 pp.

Loidi J. (1995) Phytosociology applied to nature conservation and land management In Song, Y., Dierschke, H., Wang, X. (eds.) *Appl. Vegetation Ecology,* Proc. 35th Symposium IAVS, Univ. Press., China, p. 17-30.

Papastergiadou, E., Dafis, S., Dimopoulos, P. and Lazaridou Th. (1997) Syntaxonomic typology of Greek Habitats. *Folia Geobotanica Phytotax.* **32,** 335-341.

Rodwell, J. and Cooch, S. (1996*) Red Data Book of British Plant Communities. Rationale and Methodology.* WWF, Unit of Vegetation Science, Lancaster University, United Kingdom, pp. 12.

Seibert, P. (1980) Oekologische Bewertung von homogenen landschaftsteilen, Oekosystemen und Pflanzen gesellschaften. *Ber. ANL.* **4,** 10-23.

Sukopp, H. (1972) Wandel von Flora und Vegetation in Mitteleuropa unter dem Einfluss des Menschen. *Berichte ueber Landwirtschaft (Bonn)* **50,** 112-139.

INTRASPECIFIC VARIATION OF *AMYGDALUS ORIENTALIS* MILLER (ROSACEAE) IN TURKEY

A. A. DÖNMEZ & Ş. YILDIRIMLI
Hacettepe University, Faculty of Science, Department of Biology
06532-Beytepe/Ankara TURKEY

1- Abstract

Amygdalus orientalis Miller has many varieties in Turkey. This variation is mainly caused by genetical and ecological factors. The species could be divided into small geographical groups, based on the diversity of habitus and other morphological features varying even in small population. However, some morphological features of the groups overlap. Therefore, the species should not be divided into species or lower taxonomic categories.

2- Introduction

The genus *Amygdalus* is represented by 12 species in Turkey (Browicz, 1972). While some species of the genus have a small distribution area, other species are distributed over a wider area.

Amygdalus orientalis is found nearly throughout Turkey. Bioclimatic diversity and habitat richness in Turkey have some effects on the species. As a result, many local populations of the species could be seen.

The genus *Amygdalus* and other genera of the tribe Pruneae (*Padus*, *Laurocerasus*, *Prunus*, *Cerasus*, *Armeniaca* and *Persica*) in Turkey have been revised by the former author in his ongoing PhD study. This poster has been extracted from the thesis.

3- Materials and method

Many flowered and fruited samples of *A. orientalis* were collected from various parts of Turkey during field trips. These samples were examined both morphologically and anatomically. Leaf and wood sections of the samples were prepared by using paraffin method. All sections were stained with safranine and fast green. Pollen samples were also studied by light microscope and some selected samples by scanning electron microscope. While anatomical and palynological works were carried out on 5 different local populations, morphological work was done on a large number of samples collected from many localities in Turkey.

4- Conclusion

Our study has revealed that *A. orientalis* population in Turkey has variations in terms of ecological preference, habitus and morphology such as leaf shape and indumentum, colour of flowers, fruit shape and indumentum, stone shape and surface ornamentation. On the basis of these characters, the species could be divided into 3 groups or subpopulations:

Figure 1. A sample of the first group *A. orientalis*.

The first group includes the plants which are spinescent and 1-2 m high (Figure 1). Their leaves are obovate, oblanceolate; indumentum white canescens to pubescens, rarely pilose. Petals are pale to deep pink or creamy white in colour. Fruit shape is ovate-oblong; apex obtuse or acute. Stone shape is similar to that of fruit and its surface is smooth, pitted or sometimes furrowed. The plants are widely distributed. The plants in the second group are nearly spineless and up to 3 m in height (Figure 2).

Their leaf shape ranges from elliptic to obovate; leaves are generally white canescens to pubescens. Petals are pale pink to creamy white in colour. Fruit is ovate, rarely oblong; apex obtuse or acute. Stone shape is similar to that of fruit; its surface ornamentation is smooth or pitted, rarely furrowed. The plants grow among gardens and in mixed scrub and open areas. They prefer deep soils. Population can be seen only in south-east Turkey.

Figure 2. A sample of the second group *A. orientalis*.

The last group is characterized by the plants which are spiny and 0.5-1 m high (Figure 3). Their leaves are obovate to oblanceolate, pubescens to pilose. Petals are pink in colour. Fruits are generally smaller than those of other groups, being ovate. Stone shape is similar to that of fruit and stone surface is smooth or pitted. The plants thrive on calcareous rock cracks. They are restricted to the Abanoz Pasture of Anamur at the Taurus Range in southern Turkey.

Browicz (1993) described the population of *A. orientalis* located at the Abanoz Pasture of Anamur as a new species, namely *A. zielinskii*. However, in the light of our observations on various populations of the species distributed all over

Figure 3. A sample of the third group *A. orientalis*.

Turkey, we disagree with the idea of Browicz. The Abanoz Pasture samples collected for this study seem not to have any distinguishable characters from other *A. orientalis* samples in other parts of Turkey. It should also be stated that Browicz and Zohary (1996) have not considered *A. zielinskii* in the latest revision of the genus *Amygdalus*.

Our both anatomical and palynological works have showed that there are no remarkable differences between *A. orientalis* populations in Turkey.

Although the species has been tentatively divided into 3 groups by the present authors, the groups share some morphological characters in common. Because of the presence of intraspecific variation encountered in *A. orientalis*, this complex species should not be divided into new species or lower categories.

5- References

Browicz, K. (1972) ***Amygdalus*** L., in P. H. Davis (ed.) *Flora of Turkey and The East Aegean Islands*, **4**, 6-28, Edinburgh University Press, Edinburgh.

Browicz, K. (1992) A new species of ***Amygdalus*** from south Anatolia, *The Karaca Arboretum Magazine* **1**, 125-126.

Browicz, K. & Zohary, D. (1996) The genus ***Amygdalus*** L. (Rosaceae): Species relationships, distribution and evolution under domestication, *Genetic Resources and Crop Evolution* **43**, 229-247.

FLORA, VEGETATION AND ECOLOGICAL MANAGEMENT OF KALAMAS RIVER DELTA

E. ECONOMIDOU
9 Vrioullon st, Byron 16232, Athens, Greece.

1.Introduction

The Kalamas river delta lies in N.W. Greece, very close to the Greek-Albanian border. The delta area occupies 7,000 ha and consists of two estuaries which were formed after the initial bed was partially deviated for irrigation purposes. A large area of the delta floodplain was converted into agricultural land, but also new coastal habitats developed.

As a consequence a great habitat variety of lagoons, salt and brackish water marshes, wet meadows and reed-thickets exist today along the coast. Further habitats of the island including remains of riparian forest and the island-like hills between the two deltas, are covered with dense phryganic vegetation.

On the contrary, large portions of the wider delta plain have been drained and converted to farmland. Irrigation canals and draining trenches have also been constructed.

2.Methods

To make a description of the vegetation many data were collected. More than one hundred samples were made. They are mainly made to make a description of the communities at the salt marshes and in accordance to the methods of Braun-Blanquet (1951) and were elaborated with methods of cluster analysis in computer programmes. They were classified mainly by Twinspan and Decorana (Hill 1979 a, b.).

3.Results and discussion

3.1.FLORA

During the floristic research 346 plant taxa have been registered but no species is rare, threatened etc. However, some species occur locally in a small number of habitats, which are often degraded or regressed.

In the previous group belong some hygrophilous species such as *Alisma plantago-aquatica* and *Cladium mariscus*, certain relicts of once developed riparian forests, f. ex. *Salix alba, Alnus glutinosa* and *Ulmus minor*. Generally the psammophilous species occur in small numbers in few sites. *Pseudorlaya pumila* and *Calystegia soldanella* should be considered as locally threatened.

From the systematic spectrum, it appears that the proportions of *Graminae* (16,9%) and *Compositae* (11,1%) are high, as well as the families of *Leguminosae* and *Cyperaceae* (6,9% for each family).

3.2. VEGETATION

Nowadays, Kalamas wetland comprises the following coastal habitats and plant communities:

1. Sandy beaches with ammonitrophilous vegetation of *Cakiletea maritimae* enriched by many *Ammophiletea* species (*Echinophora spinosa, Elymus farctus, Eryngium maritimum*). This community has many common species with the community of *Salsola Kali* and *Xanthium strumarium* Oberdorfer et Tuexen, 1956 and also with the community *Salsolo- Cakiletum aegyptiacae* Costa et Manzarrot, 1981.
2. Lagoons which lie in the old estuary region and are characterised by the presence of halophytic vegetation areas either along the banks or on islets in the lagoons. The main species are *Sarcocornia perennis* and *Arthrocnemum macrostachyum*. On the periphery of the lagoons and at sites of fresh water supply, plant communities of *Juncetum* and *Phragmitetum* are developing while in the lagoon interior the *Ruppietum* community is present.
3. The new river estuary which was created 40 years ago has a) thicket of various species of *Tamarix*. This thicket pre-existing in the area expanded and took over a large part of the new estuary b) Salt marshes with *Halocnemo-Arthrocnemetum* and *Halocnemetum strobilacei* c) Newest deposits with plant communities of *Sarcocornia perennis, Arthrocnemum macrostachyum* and *Salicornia europaea*. These three halophytic communities interchange with eachother as well as with the *Tamarix* community, depending on soil salinity and water flooding frequency d) Reedbeds of *Phragmites australis*.
4. Salt and brakish marshes, where many types are expanded a) salt marshes with *Arthrocnemum macrostacyum* and *Limonium narbonense* b) brakish marshes with communities of *Juncus acutus* or *Juncus effusus (Juncetum)* in zones close to irrigation channels c) brakish marshes with communities of *Scirpus maritimus* and *Juncus subulatus (Junceto-Scirpetum maritimi)* d) reed thickets of *Phragmites australis* and *Scirpetum litorali* in areas where fresh water is present almost throughout the year.
5. Wet meadows with hygro-nitrophilous vegetation dominated by *Lolium multiflorum* and *Rumex conglomeratus*.

6. Remains of riparian forests with willow-, elm,-alder,-and plane-trees, which have usually the form of hedges concentrated in zones between the cultivations and the hills.
Phrygana and maquis on the hills, which were islands in earliest times. Phrygana are dominated by *Phlomis fruticosa* and are affected by intense grazing. In other sites the vegetation is thick and high, e.g. a type of maquis. The main species of this vegetation are *Quercus coccifera, Olea europaea* and *Pistacia lentiscus*. On some hills *Querues ithaburensis ssp. macrolepis* occures in small or large populations. Finally, on certain hot and dry slopes, communities of *Ephorbia dendroides* are located.

3.3.ECOLOGICAL IMPACTS

The impacts of human activities on the studied area and certain enviromental problems are presented here:
1. The land drainage engineering works some decades ago in the deltaic plain for agricultural purposes caused problems in the region's hydrology. The impacts of the intensive agriculture are the pollution's increase, the degradation of water quality and the modification of habitats and plant communities. The impacts are catastrophic in the wet meadows, the fresh water marshes and the riparian vegetation; strong impacts have been observed in brakish marshes of *Juncus* spp. etc. The diversity and the density of the nesting aquatic birds and the habitats of many mammals, amphibians and reptiles have been affected. The formation of new coastal habitats has been considered as positive influence by the river's diversion.
2. Grazing in the salt and brakish marshes and wet meadows resulted in the reduction of certain plant communities, the animal grasses and other herbs and caused the introduction of many nitrophilous and thorny species. The impacts are more intense in riverside vegetation, reed beds and phrygana. Negative impacts by the grazing have been provoked on the animals and mainly on the avifauna but it could have positive influences by favouring animal species of open habitats.
3. Fishing has not affected the area seriously. Impacts on the birds are caused when the fishermen are at the same time hunters.
4. Tourism is nowadays an activity very much restricted in the studied area. Strong, negative impacts have been observed in the large sandy beach of Drepanon and the beach of the old river estuary.
5. Hunting has considerable impacts on birds, in spite of the many game refuges in the studied area.
6. Enviromental problems arise from illegal solid waste disposal, particularly in salt marshes. The discharge of urban wastewater from Ioannina in the river of Kalamas and the deliberate vegetation fires are key threats.

3.4. ECOLOGICAL MANAGEMENT

The preliminary management proposals are summarised as follows:

A: General

1. Statutory provisions for the protection of the wetland.
2. Designation of the Delta as protected area (art. 19 of Law 1650/86)
3. Incorporation of the Delta in networks of protected areas (Ramsar Convention, EU directives 79/409.92/43).
4. Listing of the Delta in the Enviromentally Sensitive Areas (Regulation 2078/92).
5. Establishment and operation of a Management Authority for the whole delta.

B: Specific

1. Reestablishment of riparian forests and restoration of sandy beaches, creation of hedgerows, development and creation of fresh water marshes, restoration of vegetation of wet meadows as well as the former estuary and the area of Valtos, restoration of vegetation of the island-like hills and sustainable management of range ecosystems.
2. Arresting pollution and developing ecotourism.

4. References

Braun-Blanquet, J. (1951). *Pflanzensoziologie.* 2ste Auflage. Springer Verlag. Wien.

Economidou, E., Vassilakis K., Bousbouras, D. & Witte R. (1993). *Assessment of the ecological value of the Kalamas river delta and preliminary proposals for a management plan.* Vol. A. Phase 1: 1992. Greek Biotope/Wetland Center and University of Patras. Department of Biology. Greece. 78p. (in Greek, English summary)

Economidou, E., Vassilakis, K., (1995). *Assessment of the ecological value of the Kalamas river delta and preliminary proposals for a management plan.* Vol. B. Phase 2: !993. Greek Biotope/Wetland Centre. Greece. (In Greek , English summary.)

Gilberston, D.D., Kent, M. & Pyatt, F.B. (1985). *Practical Ecology* pp. 274 - 297. Hutchinson. London, Melbourne, Sydney, Auckland, Johannesburg.

Hill, M.O. (1979a). *Twinspan. A. Fortran Programme for arranging by classification of the Individuals and Attributes.* Ecology & Systematics. Cornell University. Ithaca, New York 14850

Hill, M.O (1979b) *Decorana. A Fortran Programme for Detrended Correspondence Analysis and Reciprocal Averaging.* Ecology & Systematics. Cornell University. Ithaca, New York 14850.

GEOGRAPHICAL DISTRIBUTION OF *ARABIS PROCURRENS* WALDST. & KIT, *LILIUM RHODOPAEUM* DELIP. AND *LATHRAEA RHODOPEA* DINGLER IN GREECE

E. ELEFTHERIADOU, N. ATHANASIADIS,
K. THEODOROPOULOS AND I. TSIRIPIDIS
Institute of Forest Botany - Geobotany,
Department of Forestry and Natural
Environment, Aristotle University,
GR 540 06 Thessaloniki, Greece.

1. Abstract

The first records of *Arabis procurrens* Waldst. & Kit. as well as new localities of *Lilium rhodopaeum* Delip. and *Lathraea rhodopaea* Dingler in Greece are given. Furthermore, the geographical distribution of the above mentioned species in the Greek mainland and their chorology are cited.

2. Introduction

This is a contribution to the knowledge of the Greek flora by giving the up to now known geographical distribution (fig. 1) in the Greek mainland of three rare plants, as well as information about their ecology and their populations.

3. Arabis procurrens Waldst. & Kit., *Pl. Rar. Hung. 2: 154 (1803)*

It is an endemic species of the Balkan peninsula and the Carpathians. According to Greuter & al. (1986) and Akeroyd (1993) there is no record of this species for the Greek area. An old record by Zaganiaris (1940) for the mt. Karlik (Papikio) needs confirmation, because it is not clear if the collection is from the Greek or Bulgarian area (the herbarium speciment was not found in TAU herbarium).

It has been found in the Greek part of Rodhopi, on moist granite or limestone substrates, either on mossy rocks or directly on the soil with a thin layer of leaf litter. It usually grows in *Fagus* forests, small openings and rarely in *Quercus* spp., *Pinus nigra* and *Pinus sylvestris* forests, at altitudes (380)800 -1600 m.

Nomos Dramas: Frakto, 41° 30′ 26′′N, 24° 29′ 53′′E, alt. 1390 m, 31.07.96, *Fagus* forest; Idem, 41° 30′ 17′′N, 24° 29′ 49′′E, alt. 1320 m, 28.07.95, *Fagus* forest; Idem, 41° 30′ 25′′N, 24° 27′ 24′′E, alt. 1120 m, 31.07.96, *Fagus* forest; Idem, 41° 30′ 08′′N, 24° 27′ 07.5′′E, alt. 1160 m, 19.06.97, rocky slopes of

Pinus nigra forest; Frakto, Zarkadia, 41° 24′ 17.1′′N, 24° 28′ 23.1′′E, alt. 380 m, 19.06.97, roadsides of *Carpinus orientalis - Quercus pubescens* forest; Stamna, 41° 26′ 41′′N, 24° 34′ 26′′E, alt. 1220 m, 15.08.96, *Fagus* forest; Idem, 41° 24′ 57′′N, 24° 33′ 01′′E, alt. 1090 m, 16.08.96, *Fagus* forest; Idem, 41° 24′ 22.2′′N, 24° 32′ 34,0′′E, alt. 1070 m, 30.07.97, *Fagus* forest; Idem, 41° 25′ 04.9′′N, 24° 32′ 06.6′′E, alt. 1100 m, 19.06.97, roadsides of *Fagus* forest (obs.); Idem, 41° 26′ 36.7′′N, 24° 34′ 31.8′′E, alt.1250 m, 19.06.97, *Fagus* forest; Idem, 41° 26′ 24.4′′N, 24° 34′ 34.6′′E, alt. 1250 m, 19.06.97, *Fagus* forest; Silli - Tholos, 41° 23′ 48,4′′N, 24° 32′ 34,0′′E, alt. 1060 m, 30.07.97, roadsides of *Fagus* forest; Trachoni-Dipotama, 41° 22′ 33.6′′N, 24° 43′ 26.8′′E, alt.1550 m, 20.06.97, opening; Idem, 41° 22′ 59.4′′N, 24° 37′ 09.2′′E, alt. 940 m, 21.06.97, roadsides of *Quercus dalechampii* forest; Trachoni-Dipotama, 41° 23′ 32′′N, 24° 37′ 41′′E, alt. 970 m, 24.07.96, *Fagus* forest. **Nomos Xanthis**: Drymos, 41° 19′ 56′′N, 24° 40′ 56′′E, alt. 1100 m, 20.08.96, *Fagus* forest; Idem, 41° 20′ 45′′N, 24° 41′ 39′′E, alt. 1250 m, 20.08.96, *Fagus* forest; Idem, 41° 20′ 46′′N, 24° 42′ 10′′E, alt. 1260 m, 20.08.96, *Fagus* forest; Idem, 41° 19′ 50′′N, 24° 43′ 37′′E, alt. 1460 m, 04.08.95, *Fagus* forest; Idem, 41° 20′ 37,5′′N, 24° 42′ 42,6′′E, alt. 1240 m, 10.08.95, *Fagus* forest; Idem, 41° 20′ 37′′N, 24° 42′ 57′′E, alt. 1250 m, 10.08.95, *Fagus* forest; Idem, 41° 20′ 37.5′′N, 24° 42′ 42.6′′E, alt. 1240 m, 28.05.97, roadsides of *Fagus* forest; Filakio Drymos, 1440 m, 01.06.90, *Fagus* forest; Crossing Mesoplagia - Kerasos, alt. 1190 m, 31.05.90, *Fagus* forest; Ergotaxio Paleochoriou, alt. 1300 m, 28.05.90, damp places in *Pinus sylvestris* forest; Pagokristali, alt. 1500 m, 29.05.90, on rocks in opening of *Pinus sylvestris* forest; Erymanthos - Kerasos, alt. 1240 m, 03.07.90, *Fagus* forest; Before crossing Mesoplagia - Kerasos, alt.1210 m, 31.05.90, *Fagus* forest; After crossing Mesoplagia, alt. 1120 m, 05.08.92, roadsides of *Fagus* forest; Meteora, alt. 1480 m, 08.05.92, rocky roadsides in opening of *Fagus* forest; Oreo, 41° 19′ 02′′N, 24° 45′ 11′′E, alt. 1270 m, 25.08.96, *Fagus* forest; Idem, 41° 19′ 05′′N, 24° 45′ 25′′E, alt. 1280 m, 25.08.96, *Fagus* forest; Idem, 41° 19′ 05′′N, 24° 45′ 25′′E, alt. 1280 m, 29.05.97, *Fagus* forest; Idem, 41° 18′ 15′′N, 24° 46′ 01′′E, alt. 850 m, 19.08.96, *Fagus* forest; Idem, 41° 17′ 51′′N, 24° 46′ 14′′E, alt. 810 m, 19.08.96, *Fagus* forest; Idem, 41° 19′ 24,5′′N, 24° 45′ 45,5′′E, alt. 1120 m, 03.08.95, *Fagus* forest.

4. *Lilium rhodopaeum* Delip., *Naučhi Trud. Agron. Fak. Plovdiv 1:129 (1952)*

It is a rare species endemic of C. Rodhopi (Matthews 1980, Strid & Tan 1991). It occurs at c. 1300 m in open alpine meadows, within the mixed conifer belt of *Picea abies, Abies alba* and *Pinus sylvestris* (Lucas & Synge 1978). In the Greek mainland, it occurs in scattered groups on rocky schistose slopes at the edge of forests with *Picea abies* and *Fagus sylvatica* dominant, and on alpine meadows in the Zagradenia area, at altitudes of 1600-1800 m, also near Livaditis village in the northern part of Nomos Xanthis (Tan 1995). The up to now known records are from Frakto and Livaditis (Strid & Franzén 1982, Sfikas 1985, Strid & Tan 1991, Tan 1995, Eleftheriadou & Raus 1996).

According to our findings, it grows in subalpine meadows with other species such as *Pteridium aquilinum, Genista lydia, Calamagrostis arundinacea/varia, Anthoxanthum odoratum, Inula hirta, Geranium sanguineum, Vincetoxicum hirundinaria* etc and scattered shrubs such as *Pinus sylvestris, Juniperus communis, Corylus avellana, Crataegus* sp., *Rosa* sp. etc, at altitudes of 1200-1800 m. The greatest population was found in the area of Trachoni-Dipotama. It occurs in an extended area and counts over 500 individuals. The new localities are cited below.

Nomos Dramas: Trachoni - Dipotama, 41° 22′ 41.6′′N, 24° 39′ 29.5′′E, alt. 1260 m, 20.06.97, (obs.); Idem, 41° 22′ 42.5′′N, 24° 39′ 35.1′′E, alt. 1260 m, 20.06.97; Idem, 41° 23′ 56.6′′N, 24° 41′ 29.4′′E, alt. 1340 m, 20.06.97, (obs.); Idem, 41° 24′ 07.2′′N, 24° 42′ 04.8′′E, alt. 1390 m, 20.06.97, (obs.); Idem, 41° 22′ 33.6′′N, 24° 43′ 26.8′′E, alt. 1550 m, 20.06.97, (obs.). **Nomos Xanthis**: Drymos, peak Erymanthos, 1480 m, 04.10.90, opening of *Pinus sylvestris* forest; Idem, 1480-1567 m, 05.08.92, rocky ground in opening of *Pinus sylvestris* forest; Idem, 1550 m, 31.05.90, opening on the summit; Idem, 1450-1560 m,

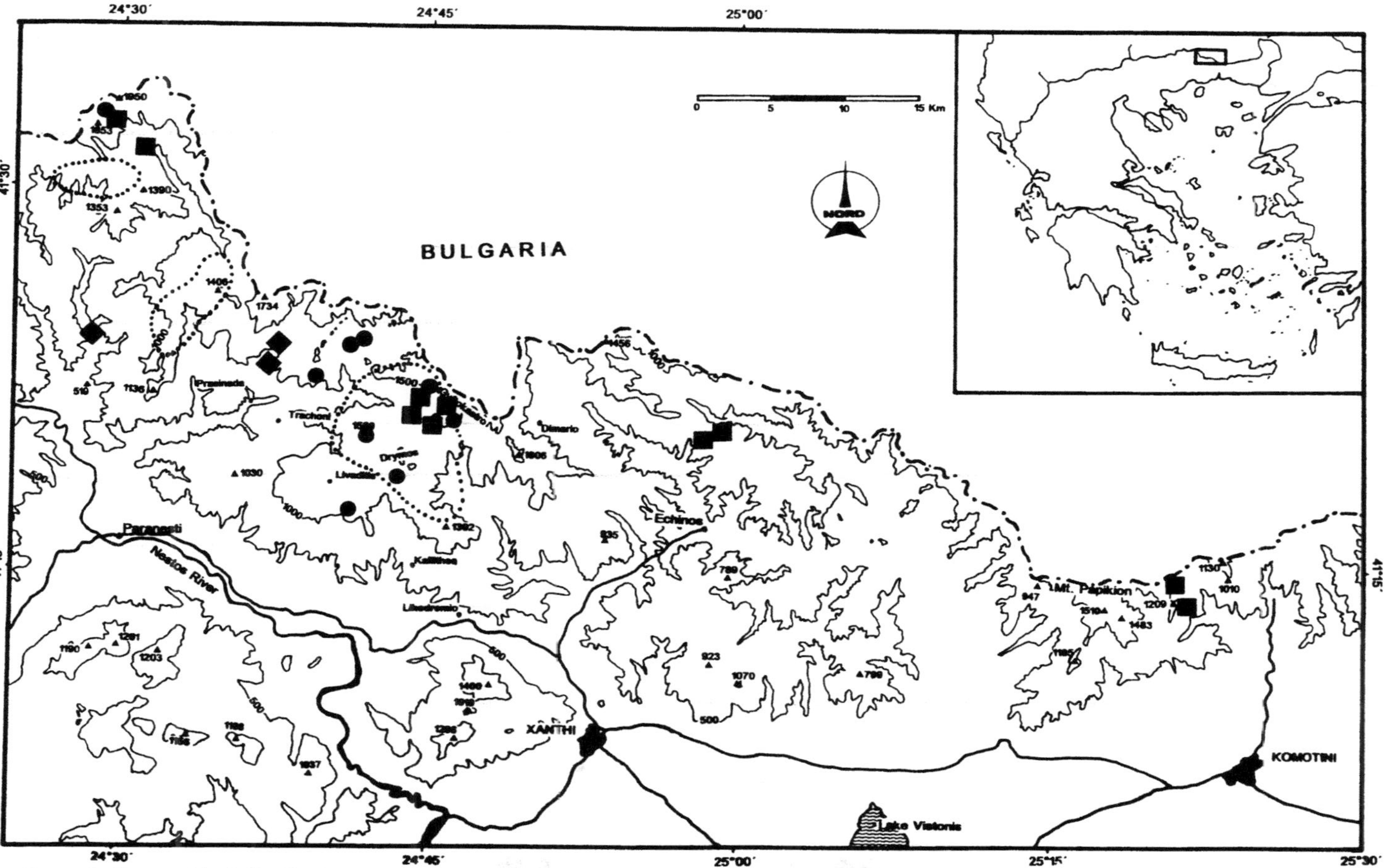

Figure 1. Geographical distribution of *Arabis procurrens* Waldst. & Kit. (····, ◆), *Lilium rhodopaeum* Delip. (●) and *Lathraea rhodopea* Dingler (■) in Greece.

04.07.90, opening and damp places; Drymos, before Livaditis village, 1350 m, 06.08.92, damp places in opening of *Pinus sylvestris* forest; Drymos, Meteora, 1470 m, 03.07.90, dry rocky places; Drymos, Mnimio Analogidi, alt. 1350 m, 06.07.90, opening of *Fagus* forest; Oreo, from Meteora towards Gyftokastro, 1500 m (obs.).

5. Lathraea rhodopaea Dingler, *Bot. Zeit. 35: 74 (1877)*

It is an endemic species of the Balkan peninsula (S Bulgaria, NE Greece), parasitic on the roots of various trees and shrubs (Webb 1972). The only record for Greece is from Frakto (Strid 1995, Eleftheriadou & Raus 1996). The record by Zaganiaris (1940) for mt Karlik (Papikio) needs confirmation, because it is not clear if the collection is from the Greek or the Bulgarian area (the herbarium speciment was not found in TAU herbarium). It has been mainly found in *Fagus* forests, parasitic on the roots of various trees and shrubs such as *Fagus* sp., *Corylus avellana* and *Salix caprea*. It grows in restricted areas in small populations of about 10 individuals, usually along rivulet banks, as well as in small depressions in *Fagus* forest with a thick layer of leaf litter. The new localities are cited below.

Nomos Dramas: Frakto, 41° 31′ 35′′N, 24° 30′ 53.8′′E, alt. 1010 m, 18.06.97, *Fagus* forest with *Corylus avellana*, (obs.). **Nomos Xanthis**: Drymos, 41° 21′ 30.8′′N, 24° 45′ 45.0′′E, alt. 1300 m, 28.05.97, *Fagus* forest; Idem, 41° 20′ 44.2′′N, 24° 44′ 39.8′′E, alt. 1250 m, 28.05.97, *Fagus* forest and *Salix caprea* group; Idem, 41° 21′ 26.9′′N, 24° 44′ 42.6′′E, alt. 1110 m, 30.05.97, *Fagus* forest; Drymos, Lakkes, alt. 1360 m, 29.05.90, *Fagus* forest;Thermes, 41° 20′ 12.5′′N, 24° 58′ 32.3′′E, alt. 710 m, 26.05.97, *Fagus* forest; Idem, 41° 20′ 18.9′′N, 24° 59′ 08.8′′E, alt. 680 m, 26.05.97, *Fagus* forest. **Nomos Komotinis**: Pandrosos, 41° 13′ 52.6′′N, 25° 20′ 45′′E, alt. 1010 m, 23.05.97, *Fagus* forest; Idem, 41° 14′ 40.4′′N, 25° 20′ 41.3′′E, alt. 910 m, 23.05.97, *Fagus* forest.

6. References

Akeroyd, J. R. (1993) *Arabis*, in Tutin,T. G. & al. (ed.), *Flora Europaea, vol. 1: Psilotaceae to Platanaceae*, Cambridge University Press, Cambridge, pp. 352-356.

Eleftheriadou, E. & Raus, Th. (1996) The vascular flora of the nature reserve Frakto Virgin Forest of Nomos Dramas (E Makedonia, Greece), *Willdenowia* **25**, 455-485.

Greuter, W., Burdet, H. M. & Long, G. (1986) *Med - Checklist, vol. 3 (Dicotyledones: Convolvulaceae - Labiatae)*, Conservatoire et Jardin Botanique, Med - checklist trust of OPTIMA, Geneve.

Lucas, G. & Synge, H. (1978) *The IUCN Plant Red Data Book*, International Union for Conservation of Nature and Natural Resources (IUCN).

Matthews, V. A. (1980) *Lilium*, in Tutin, T. G. & al. (ed.), *Flora Europaea, vol. 5: Alismataceae to Orchidaceae (Monocotyledones)*, Cambridge University Press, Cambridge, pp. 34-35.

Sfikas, G. (1985) *I chlorida tis ditikis Rodopis. Simvoli sti meleti tis*, Ekdosi E.O.S. Dramas, Drama.

Strid, A. (1995) *Lathraea rhodopaea Dingler*, in Phitos, D., Strid, A., Snogerup, S. & Greuter, W. *The red data Book of rare and threatened plants of Greece*, World Wide Fund for Nature 1995, pp. 328-329.

Strid, A. & Franzén, R. (1982) New floristic records from the mountains of northern Greece (Materials for the Mountain Flora of Greece, 12), *Willdenowia* 12, 9-28.

Strid, A. & Tan, K. (1991) *Mountain flora of Greece, vol. 2*, Edinburgh University Press, Edinburgh.

Tan, K. (1995) *Lilium rhodopeum Delip.*, in Phitos, D., Strid, A., Snogerup, S. & Greuter, W. *The red data Book of rare and threatened plants of Greece*, World Wide Fund for Nature 1995, pp. 334-335.

Webb, D. A. (1972) *Lathraea*, in Tutin, T. G. & al. (ed.), *Flora Europaea, vol. 3: Diapensiaceae to Myoporaceae*, Cambridge University Press, Cambridge, pp. 281.

Zaganiaris, D. N. (1940) Herbarium macedonicum. Tertium et quartum mille, *Sci. Ann. Fac. Phys. Math. Aristotelian Univ. Thessaloniki* 6, 38-141.

MONITORING OF MODERN POLLEN DEPOSITION BY MEANS OF POLLEN TRAPS IN THE PIERIA MOUNTAINS, GREECE: PRELIMINARY RESULTS

A. GERASIMIDIS, N. ATHANASIADIS, A. DRIVALIARI,
K. THEODOROPOULOS, E. ELEFTHERIADOU
Institute of Forest Botany-Geobotany, Faculty of Forestry and Natural Environment, Aristotle University of Thessaloniki GR 540 06 Thessaloniki, Greece.

1. Introduction

A modern pollen deposition monitoring program has been initiated in Greece for the first time. For this purpose six modified Tauber traps (Hicks and Hyvarinen, 1986), were placed on a transect from black pine (*Pinus nigra*) forest to the tree-line of Scots pine (*Pinus sylvestris*) in the Pieria mountains.

2. Investigation area

The pollen traps transect area is altitudinaly extended at a length of about 9 km on the western part of the Pieria mountains (N. Greece). Its altitude extends from about 1200 m to over the tree-line (*c.* 2000 m). The climate is humid continental with short warm summers and harsh winters (Gerasimidis, 1985). The wind direction and wind speed are dominant factors of pollen dispersal and pollen deposition. Following the analysis of the Velvendos meteorological data according to the method of Hicks *et al.* (1996), it was found that northern winds were prevailing.

In Pieria mountains the following vegetation zones occur with increasing altitude: The Quercetalia ilicis zone with evergreen shrubs, the Quercetalia pubescentis of mesophytic oak forests and *Pinus nigra*, the Fagetalia zone with *Fagus,* mainly on northern exposures, occasionally *Abies* and *Pinus nigra,* the Vaccinio-Piceetalia zone with *Pinus sylvestris* forest, extending to the western part of the mountain crest and the Astragalo-Acantholimonetalia, which appears above timberline with dwarf shrubs, such as *Juniperus nana* and *Daphne oleoides*.

3. Pollen traps location and the surrounding vegetation

The geographic location of the traps, as well as information about the surrounding vegetation in a circular catchment area of 600 m diameter are presented in Table 1.

TABLE 1.Location of the pollen traps and description of the surrounding vegetation.

Trap	Coordinates	Altitude	Vegetation type	Detailed vegetation description and coverage (%)
1	22°10'13" E 40°14'19" N	1950 m	Unforested Grassland	Herbs 70%. Dwarf shrubs 80% (*Vaccinium myrtillus, Juniperus nana, Bruckenthalia spiculifolia* in proportion 5:4:1). *Pinus sylvestris* (single trees and shrubs) 5%.
2	22°10'01" E 40°14'27" N	1870 m	Unforested Grassland 80% *P. sylvestris* forest 20%	Herbs 50%. Dwarf shrubs 40% (*Juniperus nana, Vaccinium myrtillus* in proportion 8:2).Trees (or shrubs) *Pinus sylvestris* 20%.
3	22°10'15" E 40°15'34" N	1645 m	Unforested Grassland 50% *P. sylvestris* forest 50%	Herbs 40%. Dwarf shrubs (*Juniperus nana*) 40% Shrubs (*Juniperus communis, Pinus sylvestris*) 10%. Trees (*Pinus sylvestris*) 50%.
4	22°09'55" E 40°16'27" N	1500 m	*P. sylvestris* forest	Herbs 40%. Trees (*Pinus sylvestris*) 70%.
5	22°09'15" E 40°17'06" N	1450 m	Grassland 20% *P.sylvestris* forest 25%. *P. sylvestris* /*P. nigra* mixed forest 55%	Herbs 40%. Shrubs (*Juniperus communis*) 5%. Trees (*Pinus sylvestris* & *Pinus nigra*) 75%.
6	22°10'11" E 40°18'20" N	1260	*P. nigra* forest	Herbs 60%.Shrubs(*Juniperus communis*) 5%. Trees (*Pinus nigra*) 80%.

4. Pollen data collection and processing

The late placement of traps (May 1996), after the annual flowering starting point of many forest taxa in Greece (Athanasiadis, 1986), resulted in dissimilarities of pollen losses (qualitatively, quantitatively) from the annual pollen deposition of 1996.

The laboratory preparation of the pollen traps content has followed the methodology described by the outlines of the E.P.M.P.- Project Description and Guidelines (1996). At least 500 AP grains were counted in each pollen slide. For the total sum all pollen types were taken into account except for Pteridophyta spores and grains of hydrophytes. Pollen percentages and pollen influx values are given in Figure 1.

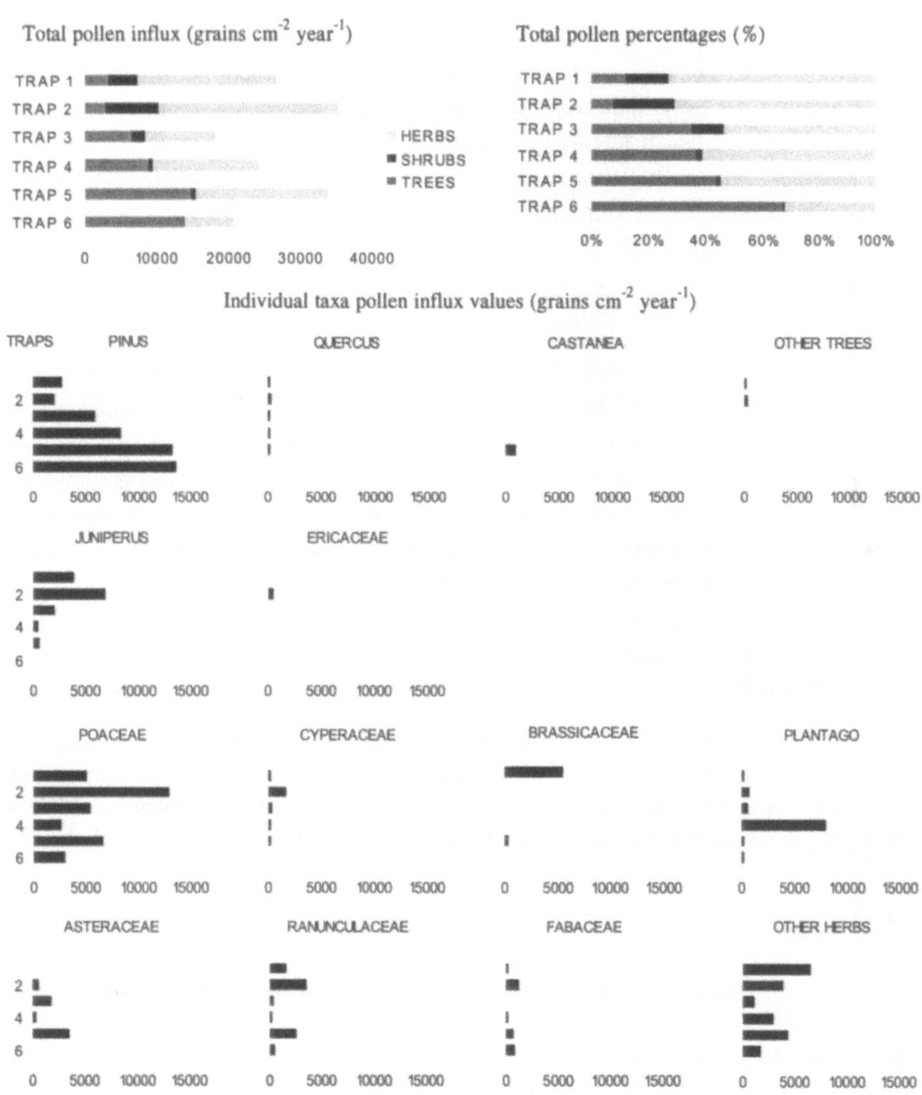

Figure 1. Pollen trap results. May - October 1996.

5. Results - Discussion

The correlation between the pollen traps results and the surrounding vegetation, as well as time differences in annual flowering season timing of different species were taken into account for interpretation of results. The Arboreal Pollen/Non Arboreal Pollen (AP/NAP) ratio decreases with increasing altitude, with a slight deviation of trap 2

(Fig. 1). The pollen values show a correspondence between AP/NAP ratio and forested/unforested areas (Tab. 1).

Pinus is dominant among AP, though its flowering season almost ended when the traps were placed. The time lag between the collection period and the flowering season further affected the usual underrepresentation of *Abies* and *Fagus* in pollen diagrams, though fir stands or beech forests are extended in the Pieria mountains. On the other hand, *Quercus,* whose flowering season lasts until June, was constantly represented. The significant presence of *Castanea* in trap 5 may be explained by the dominance of northern winds transporting pollen from northern Chestnut stands or woods, as well as by its late flowering starting point. *Juniperus* is more frequent in traps 1,2,3, which are closer to the tree-line (Fig. 1). *Juniperus* pollen may be mainly provided by *Juniperus nana*. Other dwarf shrubs, frequent in the surrounding vegetation, such as Ericaceae (Tab. 1), are not well represented in the pollen traps content (Fig. 1).

Poaceae are naturally dominant among NAP, due to their abundance in surrounding vegetation and their flowering season timing. Cyperaceae seem underrepresented in pollen diagrams, event related probably to their shorter flowering season ending before the flowering season of Poaceae. Finally, it is worth mentioning the variety of the NAP types from trap to trap. This fact emphasizes the local character of the different vegetation units around the traps.

6. Acknowledgments

This research was connected with the FOREST project (No ENV4 CT95-0063). The pollen traps were placed with the participation of Dr. S. Hicks whose guidance and help were determinant for the initiation of the project. Dr. A. Tsiontsis from the Greek Forest Research Institute of Thessaloniki is gratefully acknowledged for providing the Meteorological Data.

7. References

Athanasiadis, N. (1986) *Forest Botany II. Trees and Shrubs of the Greek Forests* (in Greek), Thessaloniki.

Gerasimidis, A. (1985) Standortskundliche Verhältnisse und postglaziale Vegetationsentwicklung der Wälder Lailias, Bez. Serres und Katafygi im Pieria-Gebirge (Nordgriechenland), *Wiss. Jahrb. Fak. Forstwiss. nat. Umwelt Univ. Thessaloniki* **26**(7), 1-133.

Hicks, S. and Hyvarinen, V.- P. (1986) Sampling modern pollen deposition by means of "Tauber traps" : some considerations, *Pollen et Spores* **28**, 219-242.

Hicks, S., Ammann, B., Latatowa, M., Pardoe, H., and Tinsley, H. (1996) *European Pollen Monitoring Programme. Project Description and Guidelines*, Oulu Univ. Press, Oulu.

Hicks, S., Pellikka, P., and Eeronheimo, H. (1996) The relationship of modern pollen deposition to local and regional vegetation in the Pallas area using high accuracy numerical vegetation mapping, Pallas - Symposium 1996.

THE SAGE PLANTS IN GREECE: MORPHOLOGICAL VARIATION AND ITS TAXONOMIC IMPLICATIONS

E. HANLIDOU, R. KAROUSOU, S. KOKKINI.
Laboratory of Systematic Botany and Phytogeography, School of Biology, Aristotle University, 540 06 Thessaloniki, Greece

1. Introduction

The commercial name "sage" is attributed to aromatic species of the genus *Salvia,* which produce essential oils rich in 1,8-cineole, α- and β-thujone, and camphor. Among the 22 *Salvia* species occurring in Greece, three species are characterised by these features, *viz. S. fruticosa* Miller, *S. pomifera* L. and *S. officinalis* L. In the frame of a wider research project concerning the Labiatae plants in Greece, the distribution and morphological variation of the three wild growing "sage" species is studied.

2. Material and Methods

The study of the distribution of the three *Salvia* species, is based on our own collections deposited in TAU, on material examined from the herbaria ATH and B (abbreviations according to 1) and literature information [2, 3, 4, 5]. The study of the morphological variation is based on herbarium material and field observations.

3. Results and Discussion

The distribution of the three species in Greece is presented in Figs 1, 2 and 3. *S. fruticosa* (commercially known as "Greek sage") is an east mediterranean species reaching westwards S Italy and Sicily [6]. It mainly occurs in the maquis and phrygana of S Greece and the Ionian and Aegean islands (Fig. 1), at altitudes between 0 and 1000(-1350)m. *S. pomifera* (Cretan sage) occurs in similar habitats of SE Greece and Aegean islands (Fig. 2), at altitudes between 0 and 1350m, as well as in SE Anatolia [7]. Finally, we have found *S. officinalis* (Dalmatian sage), a species restricted in W Balkans and Italy, in shrublands of the mixed deciduous zone in NW Greece (Mts Voras, Pinovo, Vermion, Smolikas, Mitsikeli, Timfi) (Fig. 3), at altitudes between 600 and 950m.

The overall distribution pattern of the three *Salvia* species seems to be related to the climatic gradient of the country. *S. fruticosa* and *S. pomifera* occur only in the

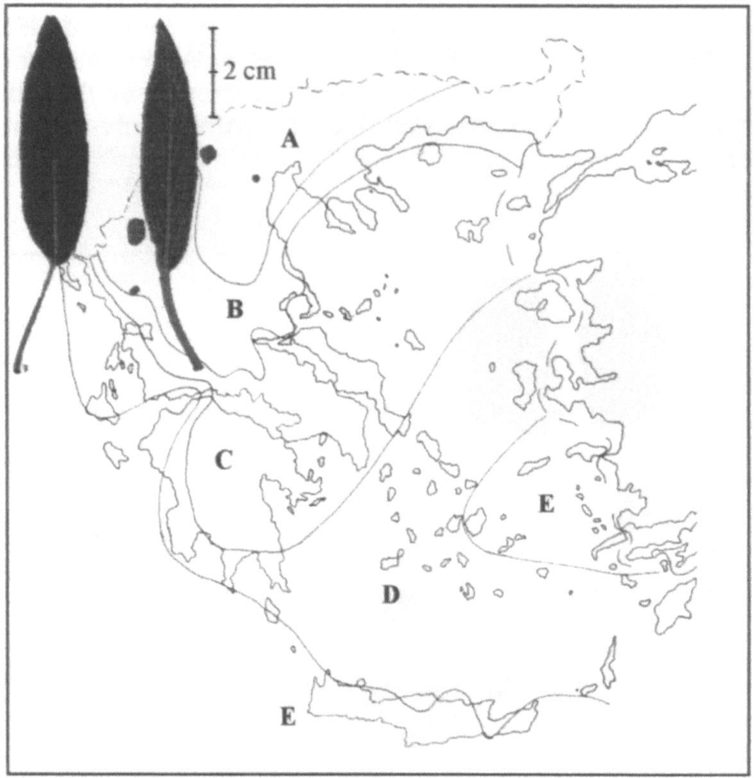

Figure 3. Representative leaves and distribution of *Salvia officinalis* in Greece. A-E: climatic zones, after [8].

5. Boratynsky, A., Browicz, K., and Zielinski, J.: *Chorology of trees and shrubs in Greece*, Poznanska Drukarnia Naukowa, Poznan, 1992.
6. Greuter W., Burdet, H.M., and Long, G.: *Med-Checklist*, Vol. 3, Conservatoire et Jardin botaniques, Geneve, 1986.
7. Hedge, I.C.: *Salvia* L., in P.H. Davis (ed.), *Flora of Turkey and the East Aegean islands*, Vol. 7, Edinburgh University Press, Edinburgh, pp. 188-192, 1982.
8. Kotini-Zambaka, S: Symvoli stin kata mina meleti tou klimatos tis Elladas, PhD thesis, Aristotle University, Thessaloniki, 1983.
9. Hayek, A.: *Prodromus Florae Peninsulae Balcanicae*, Vol. 2, Verlag des Repertorius, Berlin, 1931
10. Hedge, I.C.: *Salvia* L., in T.G. Tutin, V.H. Heywood, N.A. Burges, D.M. Moore, D.H. Valentine, S.M. Walters and D.A. Webb (eds), *Flora Europaea*, Vol. 3, University Press, Cambridge, pp. 188-192, 1972.
11. Boissier, E.: *Flora Orientalis*, Vol. 4, H. Georg, Geneve and Bale, 1879.
12. Szwarcbaum, S.: Influence of leaf morphology and optical properties on leaf temperature and survival in three mediterranean shrubs, *Plant Science Letters* **26** (1982), 47-56.
13. Lewis, M.C.: The physiological significance of variation in leaf structure, *Sci. Progr. Oxf.* **60** (1972) 25-51.
14. Fitter, A.H. and Hay, R.K.M.: *Environmental Physiology of Plants*, ed. 2, Academic Press, London, 1987.
15. Bentham, G.: Labiatae, in A. de Candolle (ed.), *Prodromus Systematis Naturalis Regni Vegetabilis*, Vol. 12, Victor Masson, Paris, 1848.

"FLORA" A DATABASE AND SOFTWARE FOR FLORISTIC AND VEGETATION ANALYZES

B. KARADZIC, V. SASO-JOVANOVIC, Z. JOVANOVIC AND R. POPOVIC
Institute for Biological Research "Sinisa Stankovic", 29 Novembra 142, Belgrade, Yugoslavia

1. Introduction

Data banks of the biological resources are a desirable prerequisite for studies in both applied and theoretical ecology. Data that are collected from biological surveys, and stored in a suitable database may help in management plans, land use decisions and a development of basic strategies of the conservation reserve systems.

Moreover databases may help in theoretical studies. Numerous problems in biogeography and ecology can be resolved using powerful data banks. Stored data enable accurate definitions of biogeographic chorions (floristic provinces, regions etc.) as well as borders between them. Ecology of each species can be defined using stored information of its distribution. Monitoring of population and community dynamics also requires suitable databases.

Serbia in particular, and Balkan peninsula in general are characterized by extremely great biodiversity. More than 3000 different taxa of vascular plants were described in Serbia. They form heterogeneous and very diverse plant communities. So far there were described 633 basic syntaxonomic units (480 associations and 153 subassociations) which are included in 138 alliances, 73 orders and 40 vegetation classes. The great complexity and heterogeneity of vegetation in Serbia is caused by various factors. The most prominent among them are: specific geographic position of Serbia, different climate and soil conditions, variable orography and specific phytogeographic relations in this part of the Balkan peninsula.

Such floristic diversity and vegetation heterogeneity requires a suitable software for both storing and processing data sets. In this article we described a Windows software named "FLORA" which may serve for both aims. The software is written in Microsoft VisualBasic.

2. Description of the software

The "FLORA" is divided into three functionally liked parts: 1. Data input, 2. Data Processing and 3. Output of results.

Data input is designed for two types of data sets. The dialog window for species-oriented input (Fig 1) wait for the user to enter information on species name, locality, angular coordinates of the locality(latitude and longitude), altitude , exposition and inclination of the site where species occurred, as well as some additional information like date of the found, reference (if data is found in a literature source), degree of threatens. Entered data are automatically stored in a Data Access database.

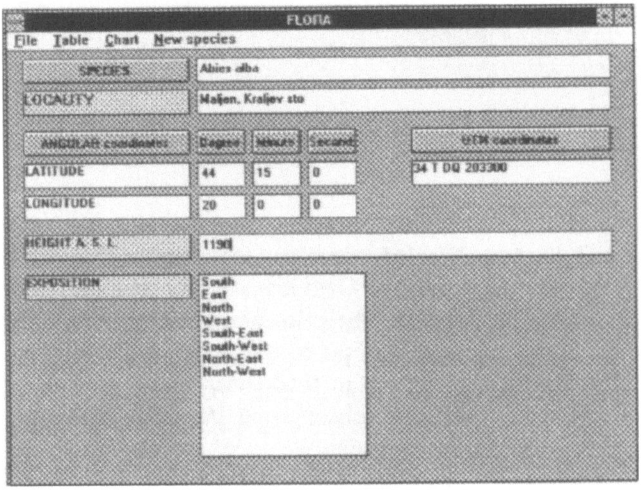

Fig. 1. The window for species-oriented input.

The community oriented input is divided in two steps. The first step requires the information on characteristics of the habitat (locality, longitude, latitude, altitude, inclination, exposure) to be entered. The window for the second step (Fig. 2) allows the creation of a phytocoenotic table.

The importance of a species in analyzed community should be assessed using the combined abundance-cover scale that was proposed by Westhoff & van der Maarel (1973). In contrary to the combined cover-abundance scale that was proposed by Braun-Blanquet (1928), this scale is completely numeric. The formed phytocoenotic tables are stored in a database together with information of biotopes where communities occur. The stored phytocoenotic tables can be retrieved and then used for different types of data analyzes. FLORA has in-built routines for table rearrangement, classification, ordination and calibration. Several table rearrangement algorithms have been proposed in ecological literature. The procedure for the table rearrangement in the FLORA is based on the two-way sorting of sites and species according to their scores of correspondence analysis. The FLORA offer numerous classification methods such as single linkage, complete linkage, flexible clustering, weighted and unweighted variants of both arithmetic and centroid clustering (Sneath & Sokal, 1973), optimal clustering (Orloci, 1967; Karadžić 1990; Karadžić & Popović, 1993) and TWINSPAN (Hill, 1979a).

FLORA

File Edit Table rearrangement Classification Ordination Calibration

	Species\Quadrats	1	2	3	4	5	6	7	8	9	10	11	12
1	Abies alba	5	5	9	3	5	3	3	7	1	7	3	2
2	Fagus sylvatica	5	3	3	7	5	7	7	2	5	3	5	5
3	Acer pseudoplatanus	1	0	1	2	3							
4													
5													
6													
7													
8													
9													
10													
11													
12													
13													

Fig 2. The window for community-oriented input.

Available ordination methods in FLORA are: polar ordination (Gauch & Scruggs, 1979), Principal Components Analysis, Correspondence Analysis and its detrended form (Hill 1979b). Original data sets can be standardized using the Generalized Standardization Procedure (Karadžić & Popović, 1994).

Environmental conditions within each community may be assessed using the calibration procedure (ter Braak, 1987). Calculation of the weighted average of an environmental factor is based on indicator values of vascular plants that are defined by Ellenberg (1986) as well by Kojić et al. (1994, 1997).

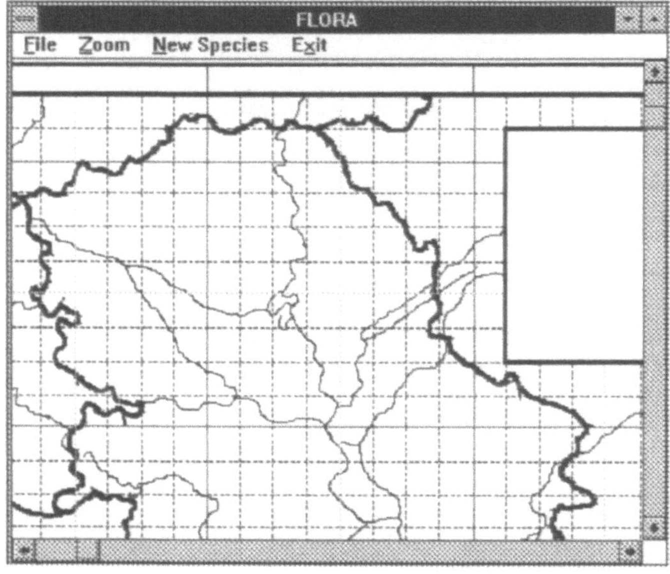

Fig. 3 The window with UTM map. Distribution of species and plant communities may be plotted in 10x10 km or 5x5 km squares, using stored information in a data base.

Outputs of "FLORA" are designed to provide information on both ecology and biogeography of either species or plant communities. A user of the software can select desirable species and view its distribution in different environmental conditions. Geographic distribution of species (or community-types) can be plotted directly on an UTM (Universal Transverse Mercator) map which is divided by a grid of either 10 x 10 km or 5x5 km squares (Fig. 3).

3. Conclusion

In this article we described main characteristics of the "FLORA", a powerful database and software that was created at the Institute for Biological Research "Siniša Stanković", in Belgrade. The program enables the storing and processing of large data sets. The software "FLORA" is useful for both applied and theoretical studies which involve investigations of ecology as well as biogeography of plant taxa and plant communities. The UTM maps of the first version of the program are designed for users in Serbia. Subsequent versions will be extrapolated for users in adjacent countries, and finally in all Balkan area.

4. References

Braun-Blanquet,J. (1928) *Phlanzensoziologie. Grundzuge der Vegetatinskunde.* 1. Ed. Berlin..

Ellenberg, H. (1986) Vegetation Mitteleuropas mit den Alpen. Ulmer, Stuttgart..

Gauch, H. G. & Scruggs. W. M. (1979) Variants of polar ordination. Vegetatio 40: 147-153.

Hill, M. O. (1979a) *TWINSPAN- a FORTRAN program for arranging multivariate data in an ordered two-way table by classification of the individuals and attributes.* Cornel University, Ithaca, N. Y.

Hill, M. O. (1979b) *DECORANA- a FORTRAN program for detrended correspondence analysis and reciprocal averaging.* Cornel University, Ithaca, N. Y.

Karadžić, B. (1990) On heterogeneity levels in numerical classification. I. Reversals generated by application of some agglomerative methods. *Arch. Biol. Sci.* 42, 69-82.

Karadžić, B. & Popović, R. (1993) On the incompatibility of the chord distance with some classification and ordination algorithms. *Zh. obsch. biol. Russ. Acad. Sci.* 54: 430-437.

Karadžić, B. & R. Popović, (1994) A generalized standardization in ecological ordination: tests with principal components analysis. *Journal of Vegetation Science* 5: 259-262.

Kojić, M., Popović, R. & Karadžić, B. (1994) *Fitoindikatori i njihova uloga u proceni ekoloških uslova u staništu.* Nauka, Beograd.

Kojić, M., Popović, R. & Karadžić, B. (1997) *Vaskularne biljke Srbije kao indikatori staništa.* Institut za istraživanja u poljoprivredi "Srbija", Institut za biološka istraživanja "Siniša Stanković", Beograd.

Orloci, L. (1967) An agglomerative method for classification of plant communities. J. Ecol. 53: 193-205.

Sneath, P. H. A. & Sokal, R. R. (1973) *Numerical taxonomy-The principles and practice of numerical classification.* W. H. Freeman, San Francisco.

ter Braak, C. J. F. (1987) Calibration. In: Jongman, R. H. G., ter Braak C. J. F. & van Tongeren, O. R. F. (eds.) *Data analysis in community and landscape ecology.* 78-90 pp. Pudoc, Wageningen..

Westhoff, V. & van der Maarel, E. 1973: The Braun-Blanquet approach. In: R. H. Whittaker (ed.) *Handbook of vegetation science V. Ordination and classification of communities.* 617-726 pp. Junk, The Hague.

Origanum vulgare L. IN THE BALKAN PENINSULA AND ANATOLIA: DISTRIBUTION AND MORPHOLOGICAL VARIATION

R. KAROUSOU, A. DARDIOTI, S. KOKKINI
Laboratory of Systematic Botany & Phytogeography, Department of Botany, School of Biology, Aristotle University, 540 06 Thessaloniki, Greece

1. Introduction

Origanum vulgare L. is a polymorphic species, growing wild throughout the Mediterranean, the Irano-Turanean and the Euro-Siberian region [1]. It is distinguished into six subspecies, four of which *viz.* subsp. *vulgare*, subsp. *hirtum* (Link) Ietswaart, subsp. *gracile* (Koch) Ietswaart and subsp. *viridulum* (Martrin-Donos) Nyman are found in the Balkan Peninsula and Anatolia [1]. However, due to overlap of the diagnostic characters, some confusion still exists at the subspecies level.

In the present work, a detailed study of the distribution and morphological characters of *O. vulgare* is attempted, in order to reveal any variation pattern along the species range in the Balkans and Anatolia.

2. Material and Methods

Voucher *O. vulgare* specimens from the herbaria TAU, B, G, ATH, K, LD and W (abbreviations according to 2) have been used for the study of the distribution. The specimen identification has been based on contemporary dichotomous keys [1,3].

TABLE 1. Characters used for the morphometric analysis of *Origanum vulgare* (Plants scored were fully flowering)

Characters	Score
type of sessile glands	1: inconspicuous; 2: semi-conspicuous; 3: conspicuous
bract colour	1: green; 2: green with purple margins; 3: purple
leaf length (mm)	
leaf width (mm)	
No of sessile glands on the lower leaf face (per mm^2)	
bract length(mm)	
bract width (mm)	
calyx length	
bract to calyx length ratio	

In total 401 specimens have been used for the morphometric analysis. The specimens were classified on basis of their geographic occurrence into 8 entities (Fig. 2). The characters studied are presented in Table 1. Discriminant analysis was applied based on Wilk's lambda stepwise method, by using the appropriate programs of SPSS [4].

3. Results and Discussion

The distribution of the four *O. vulgare* subspecies is presented in Fig. 1. Subspecies *hirtum* and subsp. *gracile* are found in the most hot and arid areas of *O. vulgare* range, whereas subsp. *viridulum* and subsp. *vulgare* occur in areas characterised by lower temperatures and higher humidity. In particular, subsp. *hirtum* is found on the islands and in coastal areas of the Mediterranean sea, whereas subsp. *gracile* is restricted in inner Anatolia. On the other hand, subsp. *viridulum* and subsp. *vulgare* occur in inland mountainous areas, as well as in northern Anatolia.

Discriminant analysis revealed that two main groups can be distinguished within *O. vulgare* (Fig. 2). The characters mainly responsible for this distinction are presented in Table 2.

TABLE 2. Main discriminant characters between the two *O. vulgare* groups provided by discriminant analysis

Group A	Group B
Conspicuously glandular leaves, bracts & calyces	Inconspicuously glandular leaves, bracts & calyces
Leaves 7-23 x 5-16 mm	Leaves (11)15-42 x 6-25 mm
No of sessile glands on the lower leaf face : 2-26/mm^2	No of sessile glands on the lower leaf face: 0-10/mm^2
Bract length: 2-5 mm	Bract length: 2.5-6.5 mm

The presence of numerous, conspicuous sessile glands is associated to the amount of the produced essential oil. Group A (subsp. *hirtum* and subsp. *gracile*) comprises plants having a strong oregano-like odor, whereas plants belonging in Group B (subsp. *viridulum* and subsp. *vulgare*) are non- or slightly aromatic. Published data suggest that besides the morphological differences, the *O. vulgare* subspecies may also distinguished by their essential oils and leaf anatomy [5,6]. Thus, subsp. *hirtum*, mainly found in areas where a Mediterranean climate occurs, has a high content of phenol-rich essential oils. Due to their strong aromatic odor, subsp. *hirtum* plants are widely used as a spice under the commercial name "Greek oregano". On the other hand subsp. *vulgare* plants, grown in areas with a Continental type of climate contain very low amounts of essential oils and devoid the characteristic oregano odor [7,8].

Our results indicate that *O. vulgare* is largely differentiated along its range in the Balkan Peninsula and Anatolia. The two distinct groups formed suggest that *O. vulgare* should be considered as a complex including two closely related species *viz.* *O. vulgare* L. (include. subsp. *vulgare* and subsp. *viridulum*) and *O. hirtum* Link

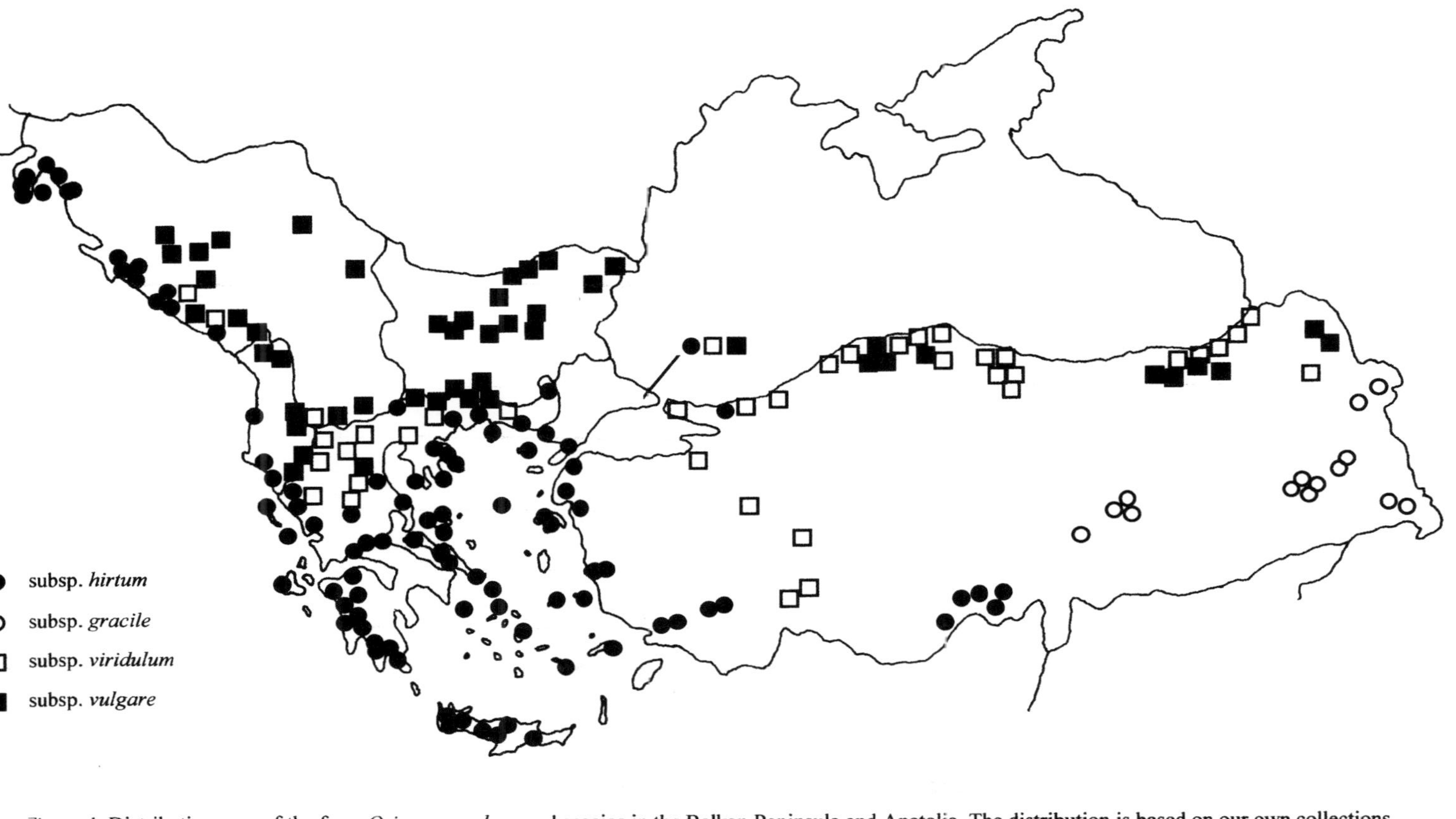

Figure 1. Distribution map of the four *Origanum vulgare* subspecies in the Balkan Peninsula and Anatolia. The distribution is based on our own collections and specimens examined.

(include. subsp. *hirtum* and subsp. *gracile*). Further study of material from the rest *O. vulgare* s. lat. range will lead to a final taxonomic conclusion.

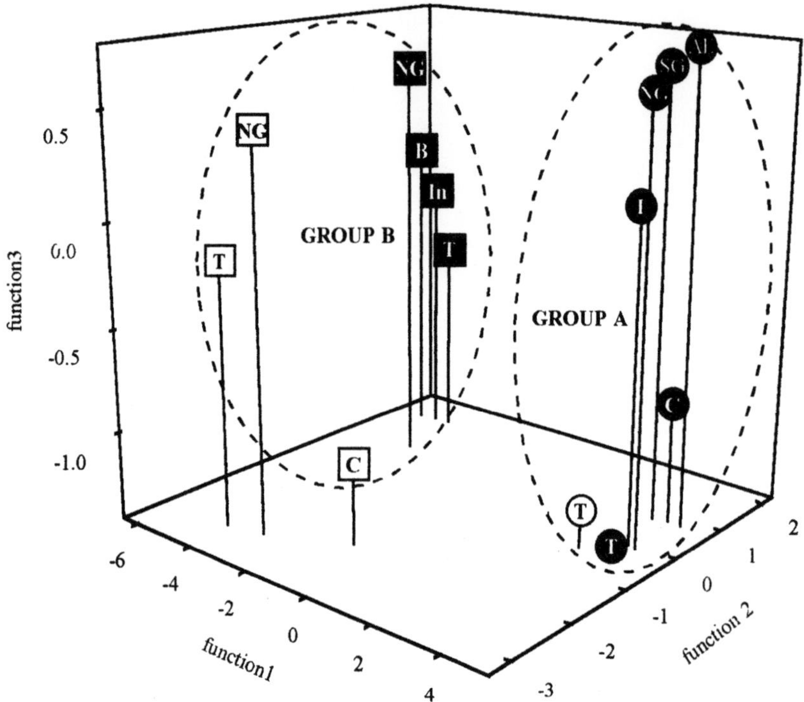

Figure 2. Discriminant score means of the four *Origanum vulgare* subspecies from 8 geographical entities of the Balkan Peninsula and Anatolia. ●: subsp. *hirtum*; O: subsp. *gracile*; ■: subsp. *vulgare*; □: subsp. *viridulum*. T: Turkey & Anatolia; B: Bulgaria; AE: Aegean islands; SG: S Greece; NG: N Greece; I: Ionian islands; C: Coastal areas of Albania & former Jugoslavia; In: Inland mountainous areas of Albania & former Jugoslavia.

4. References

1. Ietswaart, J.H.: *A Taxonomic Revision of the Genus Origanum (Labiatae)*, Leiden Botanical Series, Vol. 4, Leiden University Press, The Hague, 1980.
2. Holmgren, P.K., Holmgren, H,N., and Barnett, L.C.: Index Herbariorum, Vol. 1, ed. 8, *Regnum Veg.* **120** (1990).
3. Kokkini, S., Vokou, D., and Karousou, R.: Morphological and chemical variation of *Origanum vulgare* L. in Greece, *Botanika Chronika* **10** (1991), 337-346.
4. Norussis, M.:*SPSS/PC+*, *Statistical Package for the Social Sciences*, ed. 4, SPSS, UK Ltd. 1991.
5. Bosabalidis, A.M. and Kokkini, S.: Infraspecific variation of leaf anatomy in *Origanum vulgare* grown wild in Greece, *Bot. J Linn. Soc.* **123** (1997), 353-362.
6. Kokkini, S. (1997): Taxonomy, diversity and distribution of *Origanum* species, in S. Padulosi (ed.), *Oregano*, IPGRI, Rome, pp. 2-12.
7. Kokkini, S. and Vokou, D.: Carvacrol-rich plants in Greece, *Flav. Fragr. J* **4** (1989), 1-7.
8. Kokkini, S., Karousou, R. and Vokou, D.: Pattern of geographic variation of *Origanum vulgare* trichomes and essential oil content in Greece, *Biochem. Syst. Ecol.* **22** (1994), 517-528.

FLORAL MECHANISMS AND BREEDING SYSTEMS OF *GENTIANA* SPECIES PRESENTED IN BULGARIAN FLORA

E. KOZUHAROVA
Institute of Botany, Bulg. Acad. Sci., "Acad G. Bonchev" str. Bl. 23 Sofia 1113, Bulgaria

1. Introduction

Some data of the floral mechanisms and the breeding systems of *Gentiana lutea* ssp. *lutea* L., *G. punctata* L., *G. asclepiadea* L., *G. pneumonanthe* L., *G. cruciata* L., *G. verna* L., *G. utriculosa* L. and *G. nivalis* L. are available (Müller, 1881, Petanidou, and den Nijs 1991, Petanidou, et al. 1995, Petanidou, et al. 1995). The different populations of one and the same plant species living in different territories may have differences in its pollination ecology (Grant and Grant, 1965, Stebins, 1974). The aim of this paper is to investigate the floral mechanisms and the breeding systems of the *Gentiana* species presented in the Bulgarian flora.

2. Material and methods

Observations and experiments are conducted in 1990, 1991, 1992 and 1994 in Bulgarian Mts. (Dafni 1992). The development of flowers and inflorescences was observed periodically on labelled plants. The spontaneous self-pollination ability was tested on flowers covered with paper bags at bud stage. Self-compatibility was tested by hand self-pollination of previously covered flowers.

3. Results and discussion

In Bulgaria occurs only *Gentiana lutea* ssp. *symphyandra* (Murb.) Hayek. Its flowers are self-compatible (all hand-selfed flowers set fruits, n=10 and their seed set was 66% of the seed set of free pollinated flowers). Müller describes mechanism of spontaneous self-pollination in the typical subspecies (Müller, 1881). The spontaneous self-pollination of *Gentiana lutea* ssp. *symphyandra* is avoided by fusion of the anthers (all covered flowers did not set seeds, n=319).

The life span of the flower is 4-5 days. This subspecies have protandry of one day, while the typical subspecies have homogamous flowers according to Müller (Müller, 1881).

Gentiana punctata flowers have anthers, which are fused in a ring below the stigma. Thus spontaneous self-pollination is avoided (all covered flowers did not set seeds, n=19). The life span of the flower is 3-4 days with protogyny of one day.

The anthers of *Gentiana asclepiadea* are fused in a ring below the stigma. The flowers are not spontaneously self-pollinated (all covered flowers did not set seeds, n=95). Their self-compatibility decreases during the life span (77 % of the flowers hand-selfed at early female stage set fruits, n=22, while only 40% the flowers hand-selfed at late female stage set fruits, n=25; their seed set was respectively 62 % and 50% of the seed set of free pollinated flowers). The life span of the flower is 6-11 days with protandry of 2-4 days.

Gentiana pneumonanthe have self-compatible flowers (Petanidou, and den Nijs 1991). Its anthers are fused in a ring below the stigma in most of the flowers. The stigma of some flowers is situated into the ring of anthers. Such flowers are partially spontaneously self-pollinated (45 % of the covered flowers set fruits, n=11; their seed set was 14% of the seed set of free pollinated flowers but the number of seeds varied significantly per capsule, min=2; max=118). The flower life span is 7-10 days with 3-4 days protandry.

Gentiana cruciata have self-compatible flowers (Petanidou, et al. 1995), Its anthers are free. The flowers are partially spontaneously self-pollinated (48% of the covered flowers set fruits, n=208, but their seed set was 35% of the seed set of free pollinated flowers). The life span of the flower is 4-9 days with 4-5 days protandry.

Gentiana pyrenaica have flowers with protandry of 3-4 days and partial ability for spontaneous self-pollination according to their morphological characters.

Gentiana verna have self-compatible flowers (all hand-selfed flowers set normal fruits with fertile seeds, n=10). The flowers are not spontaneously self-pollinated (all covered flowers did not set fruits, n=15). The flower is homogamous. The labeled flowers lived 7-10 days.

Gentiana utriculosa have self-compatible self-pollinated flowers (all covered flowers set fruits, n=30, with seed set equal to that of free pollinated flowers). The flowers are homogamous. The anthesis have duration of 6-8 days. Spontaneous self-pollination takes place a day after opening of the flower and its mechanism is described by Kozuharova and Anchev (in press).

Gentiana nivalis have self-compatible, self-pollinated flowers (all covered flowers set fruits, n=30, with seed set equal to that of free pollinated flowers). They are homogamous and the anthesis is 7-9 days.

The adaptation to long proboscis pollinators goes in two directions in the Bulgarian *Gentiana* species. The lowest level of adaptation is presented by *Gentiana lutea* and *G. punctata* with more or less exposed nectar. One line of adaptation is directed to narrower and deeper nectar pockets (formed by fusion of the stamen filaments to the base of the corolla). It is presented by *G. asclepiadea, G. pneumonanthe, G. cruciata,* and *G. pyrenaica* (with nectar pockets between 9 and 15 mm). The other line is towards deep and narrow corolla tube in addition to the deep nectar pockets presented in *G. verna, G. utriculosa* and *G. nivalis*.

The flowers of some *Gentiana* species wilt immediately after the pollination (Webb and Littleton, 1989). The flowers of all investigated species are still open 3-4 days after the free pollination have been completed.

Sect. *Gentiana* (*Coelanthe*) is primitive (Kuznetzov, 1984 and Scharfetter, 1953). Homogamy and all types of dichogamy are presented in this section (Fig. 1). The other sections contain protandrous species (Fig. 1). Sect. *Coelanthe* (*Cyclostigma*) is highlyspecialized (Kuznetzov, 1984 and Scharfetter, 1953). The species in this section have homogamous flowers (Fig. 1).

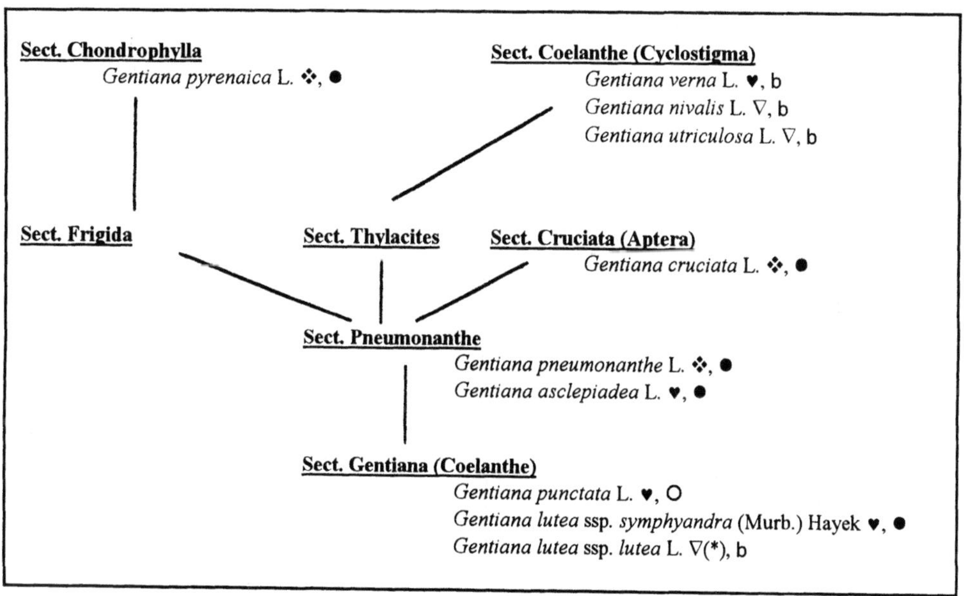

Figure 1. Phylogenetic relationship between the sections according to Kuznetzovand Scharfetter. b Homogamy, O Protogyny, ● Protandry, ∇ Spontaneous self-pollination, (* According to Müller, 1881), ♥ Lack of spontaneous self-pollination, ❖ Partial spontaneous self-pollination

One of the reasons for the reduced self-compatibility of the species is a decreased pollen viability during the anthesis. The pollen had been exposed 2 days in the hand-selfed flowers of *Gentiana lutea*, 4 and 10 days in the flowers of *G. asclepiadea* hand-selfed respectively at early and late female stages. The fruit set was respectively 100%, 77% and 40%.

The lack of pollinators is considered as one of the main reasons for the shift from allogamy to autogamy, even in plants with highly co-adapted flowers both, at annual and perennial level (Grant and Grant, 1965, Stebins, 1974, Richards, 1990). There is a tendency to avoid the autogamy in the Bulgarian *Gentiana* species. Its shift to autogamy in the annuals *G. utriculosa* and *G. nivalis* is secondary as well as the partial ability for spontaneous self-pollination of *G. pneumonanthe*. The origin of the partial ability for spontaneous self-pollination is not evolutionary clear.

The development of the inflorescences as a rule is complicated. Thus male stage flowers and female stage flowers are not separated in the space. Geytonogamy is probable.

The breeding mechanisms and the pollinator activity reported by Kozuharova indicates that xenogamy dominates in the populations of *Gentiana lutea, G. punctata, G. asclepiadea, G. cruciata, G. pneumonanthe, G. verna* and *G. pyrenaica*, while autogamy is a main reproductive mechanism for *G. utriculosa* and *G. nivalis* (Kozuharova in press). Partial spontaneous autogamy depending on many factors may occur in *G. cruciata, G. pneumonanthe* and *G. pyrenaica*. Geytonogamy is more probable in *G. verna* and *G. pyrenaica* as well as in *G. lutea* and *G. punctata.*

4. References

1. Dafni, A. (1992) *Pollination ecology*. IRL Press. 270 pp.
2. Grant, V. and Grant, K. (1965) *Flower pollination in the Phlox family* Columbia University Press 180p.
3. Kozuharova E. (in press) Pollination ecology of *Gentiana* species presented in Bulgarian flora. *Comp. Acad. Sci. Bul.*
4. Kozuharova E., M. Anchev. (in press) Floral biology, pollination ecology and breeding systems in *Gentiana verna, G. utriculosa* and *G. nivalis* (sect. *Calatianae, Gentianaceae). Annuire de l'Universite de Sofia "st K. Ohridski" Faculte de Biologie Livre 2 Botanique*
5. Kuznetzov, N. (1894) Subgenus Eugentiana, genus *Gentiana* Tournefort, Systematical, morphological and geographical working. (Russ.) - *Travaux de la societe des Naturalistes de st. Petersbourg vol. XXIV Section de Botanique*, 1-540.
6. Scharfetter, R. (1953) *Biographien von Pflanzensippen*. pp. 312-333. - Wienn.
7. Muller, H. (1881) *Alpenblumen, ihre Befruchtung durch Insecten und ihre Anpassungen an dieselben.* Engelman Verlag, Leipzig, pp. 329-349.
8. Petanidou, Th. and den Nijs, H. C. M. (1991) Comparative pollination ecology of two rare Dutch *Gentiana* species, in relation to population size. *Acta Norticulturae* 288 6th Pollination Symposium, pp. 308-312
9. Petanidou, Th., den Nijs, J. C. M., Ostermeijer, J. G. B., and Ellis-Adam, A. C.. (1995) Pollination ecology and patch dependent reproductive success of the rare perennial *Gentiana pneumonanthe L. New Phitol.* 129, 155-163.
10. Petanidou, Th., den Nijs, H. C. M. and Oostermeijer, J. G. B. (1995) Pollination ecology and constrains on seed set of *Gentiana cruciata* L. in The Netherlands. *Acta. Bot. Neerl.*, 44 (1), 55-74.
11. Stebins, G. L., 1974: *Flowering plants - Evolution above species level.* - Edward Arnold Ltd., London, 399 pp.
12. Richards, A. (1990) *Plant breeding systems.* - Unwin Human London, University Press, Cambridge, 529 pp.
13. Webb, C., Littleton, J. (1983) Flowering longevity and protandry in two species of *Gentiana* (Gentianaceae). - *Annals of the Missouri Botanical Garden* 74 (1), 51-57.

THE VASCULAR FLORA OF THE WALLS OF THESSALONIKI (N GREECE)

E. LAGIOU, N. KRIGAS, E. HANLIDOU, S. KOKKINI
Laboratory of Systematic Botany and Phytogeography, School of Biology, Aristotle University, 540 06 Thessaloniki, Greece

1. Introduction

Wall, one of the most distinct man-made constructions, is an extreme environment in many respects, such as the available room for settlement, the hardness and the alkalinity of the substratum, the scarcity of soil and humus, the inclination, the temperature and the humidity (Segal 1969). Thus, beeing a peculiar habitat, which is suitable for only a limited number of taxa, walls are the subject of an increasing number of publications.

In the frame of a wider research project concerning the urban and suburban flora of Thessaloniki, the old fortification Walls of the city were studied (Fig. 1). The Walls were firstly built during the Hellenistic period and partly reconstructed several times between the 4th and 10th c AD. They are constructed out of blocks of greenschist, bricks and soft lime mortar. Their length is c. 4 km, their height varies from 2 to 12 m, and their width is between 1.4 and 2.0 m. Nowadays the Walls lie in the city centre; they are surrounded by narrow green areas (their width ranging from 5 to 25 m) planted with few ornamental trees and shrubs or small surfaces of vacant land covered with spontaneous vegetation. Because of their age, the construction of their masonry, and the low human impact on their top, the Walls, unlike the modern buildings, are covered by a well developed vegetation. Thus, they represent a unique habitat in the city.

2. Results and Discussion

Our floristic survey, which started in 1995 and still continues, includes the spontaneously growing taxa both on the Walls and, for comparison, in the surrounding open area. Specimens are kept in the Herbarium of the Laboratory of Systematic Botany and Phtytogeography, Aristotle University of Thessaloniki (TAU). A total number of 217 vascular taxa have been collected and identified (taxonomy and nomenclature follows Tutin *et al.* 1968-1980, 1993). Among them a percentage of 15% is found exclusively on the Wall top, 40% is found only in the surrounding area, while 45% is found in both of them. Thus, the Walls' flora consists of 119 taxa.

On the basis of their abundance, the most characteristic taxa of the Walls are *Parietaria judaica* L., *Mercurialis annua* L., *Hyoscyamus albus* L., *Verbascum undulatum* Lam., *Capparis spinosa* L. subsp. *spinosa, Micromeria juliana* (L.) Bentham ex Reichenb., *Diplotaxis tenuifolia* (L.)DC., *Sedum album* L., *Lactuca serriola* L., *Aspodelus fistulosus* L. and *Bromus madritensis* L..

Figure 1. A view of the northern part of the Walls of Thessaloniki

From the total number of taxa recorded on the Wall, 7 are cultivation escapes. Two of them, *Erysimum cheiri* (L.) Crantz and *Antirrhinum majus* L., are among the most frequently naturalised on the walls taxa all over Europe (Segal 1969). Between the spontaneously growing taxa, 101 are native in Greece, while 11 are indroduced and naturalised. Most of them are of American origin (neophytes), e.g. *Solanum eleaegnifolium* Cav. and *Amaranthus* spp.

Concerning the spectrum of the total areas of distribution of the taxa growing on the Walls (Fig. 2), a prevelance of the Mediterranean element is observed (c. 47%). Examples that can be given are *Ephedra foeminea* Forskål, *Capparis spinosa* subsp. *spinosa, Chrysanthemum coronarium* L., *Micromeria juliana, Diplotaxis tenuifolia, Sisybrium orientale* L., *Ficus carica* L., *Hyoscyamus albus, Parietaria judaica, Asphodelus fistulosus, Allium guttatum* Steven subsp. *sardoum* (Moris) Stearn and *Peganum harmala* L. Widespread taxa, i.e. those having a wide European and/or Eurasiatic (e.g. *Artemisia campestris* L., *Sonchus oleraceous* L., *Mercurialis annua,*

Trifolium repens L. subsp. *repens* or (sub-) cosmopolitan distribution (e.g. *Arenaria serpyllifolia* L., *Stellaria media* (L.) Vil. subsp. *media, Cichorium indybus* L., *Tribulus terrestis* L., also constitute a high percentage (in total 33%). The Balkan element is represented by a small percentage, comprising of only five taxa, i.e. *Ptilostemon afer* (Jacq.) W. Greuter, *Sedum urvillei* DC., *Aurinia corymbosa* Griseb, *Thymus sibthorpii* Bentham and *Verbascum undulatum*. The Endemic element is represented by *Silene flavescens* Waldst. & Kit. subsp. *thessalonica* (Boiss. & Heldr.) Nyman, a species mainly occurring in NE Greece.

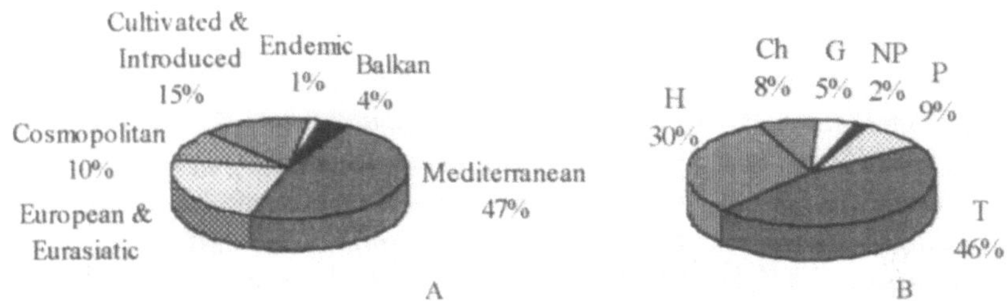

Figure 2. Spectra of the total areas of distribution (A) and life forms (B) of the taxa growing on the Walls of Thessaloniki

Among the different life forms (Fig. 2), therophytes prevail (c. 46% of the total spectrum), followed by hemicryptophytes (30%). Phanerophytes show a relatively high contribution, although the taxa included, with the exception of *Ficus carica.* and *Ailanthus altissima* (Miller) Swingle, are represented only by few individuals (e.g. *Opuntia ficus-indica* (L.) Miller, *Robinia pseudacacia* L., *Nerium oleander* L. The prevalence of both therophytes and Mediterranean taxa may be attributed to the habitat characteristics, as well as the micro- and macroclimatic conditions.

Among the taxa found growing on the Walls, 34 were found only on their top and not in the surrounding area. Several of these taxa, e.g. *Sedum acre* L., *S. sediforme* (Jacq.) Pau, *S. urvillei, Ephedra foeminea, Allium guttatum* subsp. *sardoum, Lactuca viminea* and *Silene flavescens* subsp. *thessalonica,* are inhabitants of rocky places, to which the wall, as a habitat, shows a close resemblance (Segal 1969, Hruska 1987). Another group of the exclusive wall-dwellers includes species which are rare or not found throughout the city, e.g. *Muscari comosum* (L.) Miller, *Scilla autumnalis* L., *Allium paniculatum* L. subsp. *paniculatum, Micromeria juliana, Thymus sibthorpii, Trigonella monspeliaca* L., *Carlina corymbosa* L. subsp. *corymbosa*. These species may be either relicts of the old natural vegetation existing before in the area (both the Walls' age and the low human impact on their top support this suggestion) or may have more recently invaded from the surroundings of Thessaloniki.

Both on the Walls and the surrounding area were found 85 taxa. Our research all over Thessaloniki, as well as the literature information concerning other European cities (cf. Brandes 1995, Wittig *et al.* 1985) shows that the majority of these taxa are common in urban habitats (e.g. *Amaranthus albus* L., *Capsella bursa-pastoris* (L.) Medicus, *Senecio vulgaris* L., *Lactuca serriola, Tribulus terrestis, Parietaria judaica, Ballota nigra* L. subsp. *nigra*), thus characterised as urbanophilous.

A comparison between the flora of the Walls of Thessaloniki and other European and Mediterranean wall floras (cf. Brandes 1988, Brandes 1992, Brandes 1995, Guggenheim 1992, Hruska 1987, Lötschert 1984, Segel 1969, Werner *et al.* 1989) reveals both similarities and differences. The similarities are exemplified by the presence of common wall-dwelling taxa, which are found:

a) all over Europe, such as *Parietaria judaica*, *Diplotaxis tenuifolia*, *Sonchus oleraceus* and *Sedum album*

b) mainly in the Mediterranean area, e.g. *Bromus madritensis*, *Hyoscyamus albus*, *Mercurialis annua*, *Capparis spinosa* subsp. *spinosa*, *Veronica cymbalaria* Bodard, *Calendula arvensis* L. and *Ficus carica*.

The differences are exemplified by the:

a) dominant occurrence, not recorded elsewhere, of several taxa, e.g the Balkan *Verbascum undulatum* and the Mediterranean *Peganum harmala*, *Allium guttatum* subsp. *sardoum*, *Asphodelus fistulosus*, *Salsola kali* L. subsp. *tragus* (L.) Nyman, *Clypeola jonthlaspi* L., *Sisymbrium orientale*. However, it should be noted that only few Mediterranean walls have been studied. Further information may reveal that these taxa have a wider than the known todate occurrence in the wall floras of the Mediterranean area.

b) restricted occurrence of some typical European wall-dwellers, such as *Cymbalaria muralis* P. Gaertner, B. Meyer & Scherb. and *Erysimum cheiri*.

c) absence of pteridophytes, which are frequently found in wall floras all over Europe.

3. References

Brandes, D. (1988) Zur Kenntnis der Ruderalvegetation von Mallorca. 1. Die Vegetation der Mauern und Mauerfüsse, *Documents phytosociologiques* **11**, 111-123.

Brandes, D. (1992) Flora und Vegetation von Stadtmauern, *Tuexenia* **12**, 315-339.

Brandes, D. (1995) The flora of old town centres in Europe, in H. Sukopp, N. Numata and A. Huber (eds), *Urban ecology as the basis of Urban planning*, SPB Academic Publishing bv, Amsterdam, pp. 49-58.

Guggenheim, E. (1992) Mauervegetation in der Stadt Zurich, *Ber. Geobot. Inst. ETH, Stiftung Rubel, Zürich* **58**, 164-191.

Hruska, K. (1987) Syntaxonomical study of Italian wall vegetation, *Vegetatio* **73**, 13-20.

Lötschert, W. (1984) Mauerfugen-Gesellschaften im Hohen Westerwald. Ein synökologischer Beitrag, *Tuexenia* **4**, 39-44.

Segal S (1969) *Ecological notes on wall vegetation*, Dr. W. Junk N.V. Publishers, The Hauge.

Tutin, T.G. *et al.* (1968-1980) *Flora Europaea* vols 2-5, Cambridge University Press, Cambridge.

Tutin, T.G. *et al.* (1993) *Flora Europaea* vol. 1 ed. 2, Cambridge University Press, Cambridge

Werner, W., Gödde, M. & Grimbach, N. (1989) Vegetation der Mauerfugen am Niederrhein und ihre Standortverhältnisse, *Tuexenia* **9**, 57-73

Wittig, R., Diesing D. & Gödde, M. (1985) Urbanophob-Urbanoneutral-Urbanophil. Das Verhalten der Arten gegenüber dem Lebensraum Stadt, *Flora* **177**, 265-282.

MICROMORPHOLOGY, ANATOMY AND MYXOCARPY OF NUTLETS OF SELECTED *MICROMERIA* SPECIES (LAMIACEAE)

P.D. MARIN, S. DULETIĆ-LAUŠEVIĆ, M. RISTIĆ, P. JANAĆKOVIĆ
Faculty of Biology, Inst. of Botany, Takovska 43, 11000 Belgrade, Yugoslavia

1. Introduction

Lamiaceae is a large family comprising 220 genera and nearly 4.000 species, widely distributed. One of the area of major morphological diversity of this family is Mediterranean region. A number of species is recognized as Balkan endemics.

Morphology, shape, colour and size of the nutlets were used as diagnostic characters in Lamiaceae classification, but more precise characters were found to be the pericarp structure and especially the micromorphological characteristics. In addition, several authors have reported the presence of myxocarpy in the family Lamiaceae, the phenomenon of producing mucilage when the nutlets becoming wet.

In this study the nutlets of *Micromeria* species were analysed in order to provide a detailed description of the pericarp structure and data about mucilage production on the nutlets, as well as the micromorphological characters.

2. Material and methods

This study is based on the nutlets taken from herbarium material of the Botanical Garden "Jevremovac" Inst. Botany, Fac. Biology, Belgrade, where the voucher specimens are deposited.

For the micromorphological analysis, the specimens were mounted on stubs and coated with 30nm gold in a Polaron E 5150 sputter coater and examined with a JEOL JSM T. 20.

Studies of shape, surface structure and measurements of the nutlets were made using Bausch & Lomb 1350 UP microscope.

Cross sections of the nutlets were studied and photographed from the permanent slides by Diastar^{TM} microscope.

For the investigations of mucilage production, at least five nutlets were treated with distilled water, examined and mucous layer measured under the light microscope. The measurements were taken every 15 minutes in the first hour, and then after each hour during 8 hours. No reaction after 8 hours was taken to indicate the absence of mucilage reaction.

3. Results

Micromorphology. The nutlet sculpturing of *M. juliana* and *M. parviflora*, from the section *Eumicromeria*, showed high prominent ridges with deep furrows or tracts. On the nutlets of these two species trichomes were not found.

The nutlet sculpturing of analyzed species from section *Pseudomelissa* (*M. dalmatica, M. albanica* and *M. thymifolia)* showed shallow discoid patterns with undulating ridges. On the posterior part of nutlets unicellular, small to medium size trichomes were found.

Pericarp structure. Data considering nutlet size, thickness of pericarp and sclerenchyma region, mucilage reaction and presence of trichomes are given in Table 1.

Five species from two sections of the genus *Micromeria* were analyzed. Trichomes were present in species from Sect. *Pseudomelissa. M. thymifolia* and *M. dalmatica* have trichomes at the posterior part. The pericarp of *Micromeria* species is very thin, 17-45 µm, especially in *M. juliana* and *M. parviflora*. Exocarp and mesocarp (one layer) occupy 35-40% of the pericarp thickness, dark coloured. Sclerenchyma region is one-layered, with vertically arranged bone cells.

Myxocarpy. The cells which produce mucilaginous material upon wetting are placed in the exocarp. By measuring the reaction of the nutlets after wetting in distilled water the following results have been obtained:

- Moderately strong (0.1-0.5 mm): *M. parviflora, M. thymifolia*
- Weak, but distinct mucilaginous reaction (less than 0.1 mm): *M. dalmatica*
- Very weak or only swolling reaction (less than 0.05 mm): *M. albanica, M. juliana*

TABLE 1. The investigated characteristics of nutlets in the studied *Micromeria* species
1. Nutlet dimensions (mm); 2. Pericarp thickness (µm); 3. Ratio of nutlet lenght to pericarp thickness (%); 4. Sclerenchyma region thickness (µm); 5. Ratio of pericarp to sclerenchyma region thickness (%); 6. Mucilage production on the nutlets upon wetting; 7. Presence of trichomes at the nutlet surface.

Section Species	1.	2.	3.	4.	5.	6.	7.
Pseudomelissa							
M. albanica	1.1x0.5	45	4.1	23	51	+/-	+
M.thymifolia	1.0x0.4	30	3.0	15	50	++	+
M. dalmatica	0.9x0.4	30	3.3	19	63	+	+
Eumicromeria							
M. juliana	1.0x0.3	17	1.7	9	53	+/-	-
M. parviflora	0.9x0.4	20	2.2	10	50	++	-

4. Discussion

Macromorphology, shape and dimensions of nutlets are characters of limited taxonomic significance, but nutlet ornamentation is much more important in taxonomic studies of Lamiaceae genera. The surface characters of nutlets have been studied by many authors and it is proven that they could be useful in a study of phylogeny, particularly within genera. However, similarities in surface structure do not always reflect homology in anatomical structure.

Micromorphology. *Micromeria* was for the first time described as a genus by Bentham (1853). In recent literature authors mainly do not divide this genus into the sections. However, our results suggest the possibility of separating the five species into two well-defined groups. These results are correlate with previous works based on micromorphology (Husain et al. 1990) and flavonoid profile (Marin 1989; Tomás-Barberán et al. 1991) and fits with Boissier's (1879) infrageneric classification of this genus. According to given results of nutlet microcharacters, *Micromeria* can be distinguished from other genera of *Satureja* complex.

Pericarp structure. The pericarp of labiates in transverse section is divided into three regions: exocarp (epicarp, i.e. nutlet epidermis), mesocarp usually thin and soft, below which is layer of vertically aranged bone cells and/or layers of stone cells. The inner epidermis is often referred to as the endocarp which is usually thin, as well as the seed coat.

The pericarp thickness is mostly correlated with nutlet dimensions. The pericarp thickness of *Micromeria* species from the sect. *Pseudomelissa* is significantly higher than the one of the species from the sect. *Eumicromeria*. The strucure is basically similar among species within genera.

In *Micromeria* pericarp characteristics are well correlated with infrageneric classification, which is in agreement with previous results based on micromorphology and chemotaxonomy (Husain et al. 1989, 1990; Marin 1989; Tomás-Barberán et al. 1991).

Myxocarpy. According to Ryding (1992) presence of myxocarpy is one of the features that distinguish the subfamily Nepetoideae from all the other subfamilies of Lamiaceae. This phenomenon occurs in about 70-75% of the Nepetoideae genera. In the tribe Mentheae 70% species have been found to have mucilaginous reaction (Ryding, 1992). In our investigation *Micromeria* species, devided in two sections, showed moderately strong, weak or very weak reaction, with no significant difference between sections.

Hedge (1970) and Ryding (1992) suggested that absence of myxocarpy is correlated with certain characteristics of nutlets, for instance hairy or glandular nutlets were found to lack distinct mucilage production. The nutlets of *Micromeria* spp. which belong to sect. *Pseudomelissa* are hairy at the posterior part, while the nutlets of the species from sect. *Eumicromeria* are hairless and narrower than the previous ones. However, species of both sections showed similar mucilage reaction, i.e. the mucilage reaction could not be correlated with the sections.

The thickness of pericarp was not found to be correlated with the mucilaginous reaction.

5. Conclusions

Our results showed clear differences among *Micromeria* species belonging to different sections. Species from the sect. *Pseudomelissa* were characterized by trichomes on nutlet surface, different surface patterns, significantly thicker pericarp, sclerenchyma thickness and ratio between nutlet length and pericarp thickness. This study confirms usefulness of analyzing of nutlet microcharacters and anatomical features as aditional taxonomic characters in taxonomical studies of the Lamiaceae family.

Taxonomical implications of the ability for mucilage production are rather limited. Probably, differences of mucilage production even among related taxa could be explained by different strategy of dispersal of the nutlets during evolution of certain taxa.

6. References

Bentham, G. (1853) *Labiatarum genera et species*, London.

Boissier, E. (1879) *Flora Orientalis 4*, Genevae, Basileae, Georg.

Hedge I.C. (1970) Observations on the mucilage of *Salvia* fruits, *Notes Roy. Bot. Gard. Edinburgh* **30**, 79-95.

Husain, S.Z., Marin, P.D., Diklić, N., Petković, B. (1989) Micromorphological and phytochemical studies in two new endemic *Nepeta* (Lamiaceae) species in Yugoslavia, *Pak. J. Bot.* **21(2)**, 210-217.

Husain, S.Z., Marin, P.D., Šilić, Č., Qaiser, M., Petković, B. (1990) A micromorphological study of some representative genera in the tribe Saturejeae (Lamiaceae), *Bot. J. Linn. Soc.* **103**, 59-80.

Marin, P.D. (1989) Chemotaxonomical and micromorphological study of *Micromeria* Benth. and related genera from the tribe Saturejeae (Lamiaceae), Ph.D., Univ. Belgrade.

Ryding, O. (1992) The distribution and evolution of myxocarpy in Lamiaceae, in R. Harley and T. Reynolds (eds.), *Advances in Labiate Science*, Royal Botanic Garden, Kew, pp. 85-96.

Tomás-Barberán, F.A., Gil, M.I., Marin, P.D., Tomás-Lorente, F. (1991) Flavonoids from someYugoslavian *Micromeria* species: Chemotaxonomical aspects, *Biochem. Syst. Ecol.* **19(8)**, 697-698.

DISTRIBUTION OF *DACTYLIS GLOMERATA* SUBSP. *SLOVENICA* (DOM.) DOM. (POACEAE) IN EUROPE

M. MIZIANTY
Department of Plant Systematics, Laboratory of Plant Variability, W. Szafer Institute of Botany, Polish Academy of Sciences Lubicz 46. PL-31-512 Kraków, Poland,

Dactylis glomerata subsp. *slovenica* (Dom.) Dom., first recorded by Domin in 1923 from Western Slovakia, is a tetraploid ($2n = 28$) representative of the polyploid complex of the genus *Dactylis* L. It is characterized by its great height (up to 1.5 m); culm thickened at base (to a diameter of ca 5 mm or more); long (20-25 cm) and broad (6 mm or more) dark-green leaves; long, relatively diffuse panicle (up to 15 cm or longer) with markedly spread branches, elongate-lanceolate glumcs; non-ciliate lemma, smooth on the surface, with keel glabrous or ciliate.

In his monograph on the genus *Dactylis* Domin (1943) recorded the occurrence of *D. glomerata* subsp. *slovenica* mainly from the West Carpathian Mts., but also from the East Carpathians, and from the Bohemian-Moravian Highlands. Tutin (1980) listed the taxon in *Flora Europaea* as being present only in the West Carpathians.

From long-term detailed population and experimental studies of morphological features (Mizianty 1988a), cytological, karyological (Mizianty 1991a), and chemotaxonomic analyses (Mizianty 1993) we may conclude that *Dactylis glomerata* subsp. *slovenica*, is a genetically fixed taxon.

Based on an analysis of material from 59 main European herbaria, as well as the author's own field studies, 49 localities for *Dactylis glomerata* subsp. *slovenica* (Dom.) Dom. have been recorded (Fig. 1).

The distribution of *D. g.* subsp. *slovenica* in Europe was given for the first time by Mizianty 1997 (in press). The main distribution areas for this taxon are the Carpathian Mountains (East, West, and South), the Alps (East and West as well as southern alpine belts: Venetian, Apuan, and Ligurian, which are the transition areas between the Alps and the Apennines), the Foreland of the Alps and the Jura. The presence of subsp. *slovenica* was also recorded from one locality in the Northern Apennine Mountains, which are also obviously a part of the same alpine system.

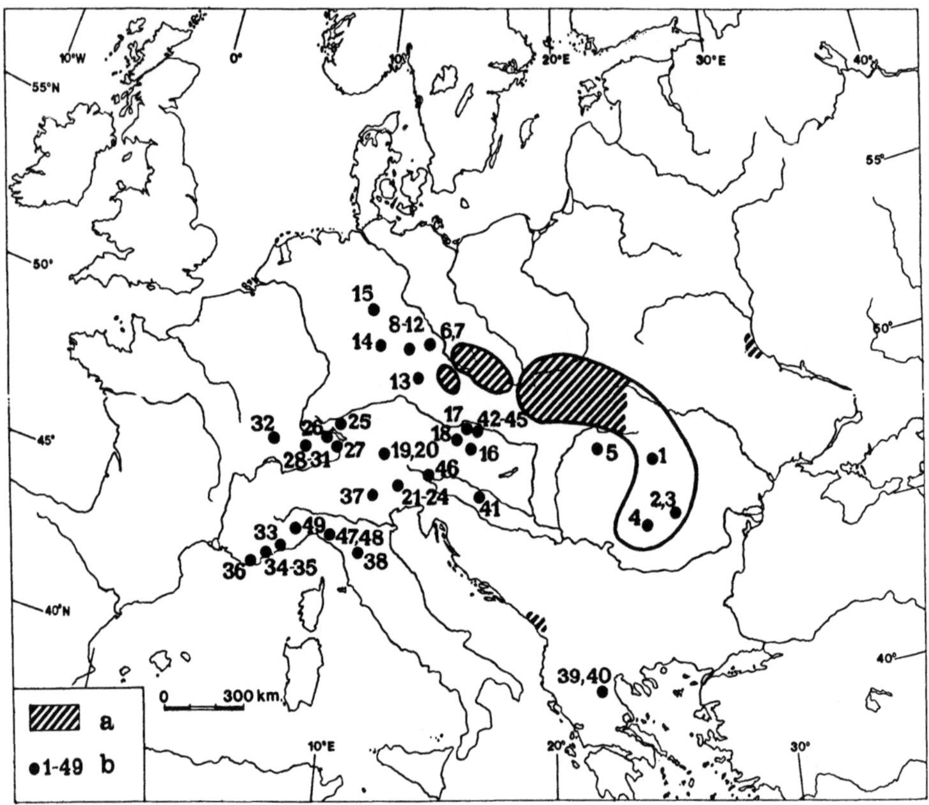

Figure 1. Distribution of *Dactylis glomerata* subsp. *slovenica* (Dom.) Dom. in Europe; a - areas of distribution according to Mizianty 1988b: West Carpathians, Czech Massif (Sudeten Mts., Karkonosze Mts., Jesioniki Mts., Czech Valley, Roztocze Upland, Dnieper Upland, Kiev Polesie, Dinaric Alps, b - numbers assigned to localities according to Mizianty 1997 (in press): East Carpathians (1), South Carpathians (2 - 4), Apuseni Mts. (5), Erzgebirge Mts./Krušné Hory (6 - 12), Bayerischer Wald (13), Frankenwald (14), Harz Mts. (15), Foreland of Alps (25 - 27), Jura (28 - 32), Eastern Alps (16 - 24, 42 - 46), Western Alps (33 - 36), Venetian Alps (37), Apuan Alps (47 - 48), Ligurian Alps (49), North Apennines (38), Kalnik Mts. (41), Olympus (39 - 40)

Apart from the above-mentioned mountain ranges, *Dactylis glomerata* subsp. *slovenica* also occurs in the Czech Massif (extending across the Sudeten Mts., Bohemian-Moravian Highlands, and Erzgebirge Mts.) as well as in such mountain ranges as the Bayerischer Wald, Frankenwald , Harz, Apuseni (Romania), and Kalnik (Croatia). Apart from this compact distribution area, single localities for *Dactylis glomerata* subsp. *slovenica* were recorded from the Dinaric Alps, Olympus Massif, and in the environs of Kiev: the Dnieper Upland and Kiev Polesie (Mizianty 1988b).

All localities for *Dactylis glomerata* subsp. *slovenica* occurring outside the main distribution area need to be thoroughly recorded. Possibly, more intensive studies may fill the gap in information between the main distribution area of this subspecies and the locality in the Dinaric Alps, which are an offshoot of the Alps and part of the same alpine system. Specimens from the Olympus Massif are very similar to *D. g.* subsp. *slovenica,* but their hybrid origin cannot be ruled out (e.g. *D. g.*subsp. *slovenica* × *D. g.* subsp. *hispanica)*.

Subsp. *slovenica* grows on limestone or dolomite, in the montane or subalpine belts, occasionally descending to lower elevations (Mizianty 1991b). Most frequently, it has been recorded between altitudes of 600 and 1300 m.

The distribution area of tetraploid *Dactylis glomerata* subsp. *slovenica* is medium-sized when compared with other known taxa of the polyploid complex of *Dactylis* and is completely enclosed within the distribution area of tetraploid *D. g.* subsp. *glomerata*, and, partially, within the distribution area of diploid *D. g.* subsp. *aschersoniana* (Mizianty 1990). Apparently, these two subspecies contributed to the creation of *D. g.* subsp. *slovenica* (Mizianty 1991b).

It may be conjectured that *Dactylis glomerata* subsp. *slovenica* originated in lower mountain elevations where both *D. g.* subsp. *glomerata* and *D. g.* subsp. *aschersoniana* occur. Subsequently, *D. g.* subsp. *slovenica* dominated higher mountain elevations, where its parental species do not occur. *Dactylis glomerata* subsp. *slovenica* most likely came into existence before the glacial period and subsequently its distribution area was fragmented during glaciation.

References

Domin, K. (1923) Problémy a metody rostlinné sociologie a jejich použiti pro výzkum lučnich a pastviných porostů Republiky Ceskoslovenské ["Problems and methods of plant sociology and their application for the analysis of meadows and pastures"], *Publ. Min. Zem.* **39**, 1-375 (in Czech).

Domin, K. (1943) Monografická studie o rodu *Dactylis* L. [Notulae criticae ad cognitionem et dispositionem stirpium generis *Dactylis* L.] , *Acta Bot. Bohem.* **14**, 3 -147 (in Czech).

Mizianty, M. (1988a) Biosystematic studies on *Dactylis* L. 2. Original research. 2.1. Morphological differentiation and occurrence of representatives of the genus *Dactylis* in

Poland. 2.1.1. Field studies and experimental cultures, *Acta Soc. Bot. Pol.* **57**, 589-621.

Mizianty, M. (1988b) Biosystematic studies on *Dactylis* L. 2. Original research. 2.1. Morphological differentiation and occurence of representatives of the genus *Dactylis* in Poland. 2.1.2. Distribution of *Dactylis glomerata* subsp. *slovenica* (Dom.) Dom. in Poland and adjacent regions, *Acta Soc. Bot. Pol.* **57**, 623-636.

Mizianty, M. (1990) Biosystematic studies on *Dactylis* L. 1. Review of the previous studies. 1.2. Cytology, genetics experimental studies, and evolution, *Acta Soc. Bot. Pol.* **59**, 105-118.

Mizianty, M. (1991a) Biosystematic studies on *Dactylis* L. (Poaceae). 2. Original research. 2.2. Cytological differentiation of the genus in Poland, *Fragm. Flor. Geobot.* **36,** 301-320.

Mizianty, M. (1991b) Biosystematic studies on *Dactylis* (Poaceae). 3. Conclusions, *Fragm. Flor. Geobot.* **36**, 321-338.

Mizianty, M. (1993) Biosystematic studies on *Dactylis* (Poaceae). 4. Chemotaxonomy of the genus in Poland., *Fragm. Flor. Geobot.* **38**, 61-68.

Mizianty, M. (1997 in press). Distribution of *Dactylis glomerata* subsp. *slovenica* (Poaceae) in Europe, *Fragm. Flor. Geobot.* **42,**

Tutin, T.G. (1980) Dactylis L., in: T.G. Tutin, V.H. Heywood, N.A. Burges, D.M. Moore, D.H. Valentine, S.M. Walters and D.A. Webb (eds.), *Flora Europaea: Alismataceae to Orchidaceae, 5: (Monocotyledones)*, Cambridge University Press, Cambridge, pp. 170-171.

BIOSYSTEMATIC INVESTIGATION OF GENUS *ACHILLEA* L., SECT. *FILIPENDULINAE* (DC.) AFAN. (*ASTERACEAE*) IN BULGARIA

A.M. NEDELCHEVA
Sofia University, Biological Faculty, Department of Botany
Bld. Dragan Tzankov 8, 1421 Sofia, Bulgaria

1. Introduction

Section *Filipendulinae* (DC.) Afan. includes endemic taxa with indistinct origin and high infraspecific variability. These species are insufficiently investigated and contradictions are found in the Flora and monographical research for Bulgaria (Velenovsky, 1891,1898; Stojanov and Stefanov, 1924, 1933, 1948; Stojanov, Stefanov and Kitanov, 1967; Kuzmanov, 1984).

2. Materials and Methods

Living plants and herbar specimens are investigated (SO, SOM. SOA, BP, W, WU, G). The pollen morphology is studied by the acetolysis method (Erdtman, 1969). The pollen fertility is established by the method of Alexander (1969) modificated for the present study. A dendrogram for karyotaxonomical relationships with euclidian distances is also constructed (Chemeris and Androshchuk, 1982).

3. Results and Discussion

3.1. MORPHOLOGICAL INVESTIGATION

The analysis of variability presents a parallelism in the fenotype changes of sect. *Filipendulinae* species and in g. *Achillea* as a whole. This tendency is a well expressed by the change of the folia rachis and of the drastic increase in

the size of capitula and relate to decrease of its number in corymbs. Many authors make reference for parallel variability in different sections in g. *Achillea* as well as the same between g. *Achillea* and g. *Anthemis*. In this way the data confirm the law of Vavilov (1920). As a result of the comparative morphological analysis the following groups among the investigated species could be differentiated:
a *A. thracica* Vel., *A. clypeolata* Sm., *A. chrysocoma* Friv., *A. coarctata* Poir.; **b** *A. leptophylla* Bieb.; **c** *A. depressa* Jka.

3.2. KARYOLOGICAL INVESTIGATION

The karyotypes of the diploid species (*A. thracica, A. clypeolata, A. coarctata, A. leptophylla, A. depressa*; 2n=18) are very similar in chromosome morphology - symmetrical karyotype, with predominance of M- and SM -chromosomes.

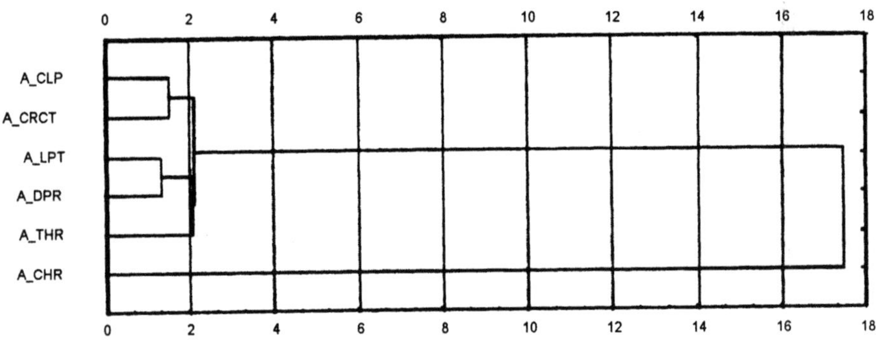

Figure 2. Dendrogram showing the level of similarity between karyotypes of species *A. clypeolata* (A_CLP), *A. coarctata* (A _CRCT), *A. leptophylla* (A_LPT), *A. depressa* (A_DPR), *A. thracica* (A_THR), *A. chrysocoma* (A_CHR).

The dendrogram (Fig.2) presents the level of similarity. Insignificant karyotaxonomical difference is observed to morphologically similar species *A. clypeolata* and *A. coarctata* (1.55%). The highest difference between *A. chrysocoma* (hexaploid) and the other diploids (17.71%) is in connection with its origin - infraspecific hybridization with subsiquent polyploidy. *A. thracica* is a well distinguished from the other diploids. These data together with the morphological evidence show that *A. thracica* is a well distinguished species, no conspecific with *A. clypeolata* as the opinion of Richardson (1976).

3.3 POLLEN INVESTIGATION

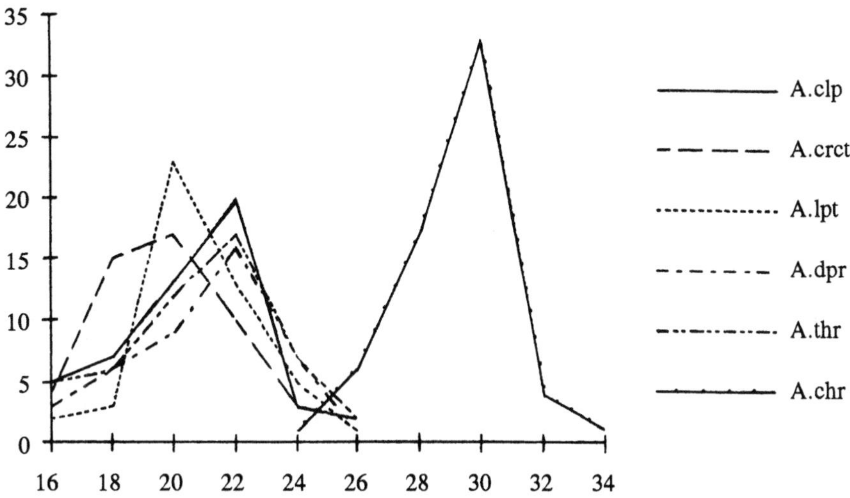

Figure 3. Variant curved lines for the size of pollen grains.

The morphology of the pollen grains is typical for *Achillea* species. In some populations of diploids 4-colporate type pollen is observed. The correlation between the chromosome number and the pollen size in g. *Achillea* is confirmed and for the species of sect. *Filipendulinae* as well. Folllowing values for polyar dimeter are estublished: *A. clypeolata*, *A. coarctata* , *A. leptophylla*, *A. depressa* and *A. thracica* (diploids) - 18-22 μm, *A. chrysocoma* (hexaploid) - 28-30 μm. Fig.3 shows a high amplitude of the variability in the size of the pollen grains. They are similar for the diploids. This sign is appropriate for differentiation between diploids and hexaploid but can not be used as taxonomical criteria among the diploids. A high percent of aborted pollen grains is established. In 44% of investigated populations the pollen sterility ranged around 10-30%. This high pollen sterility is probably determinated by the xerophytic habitats and the possibilities for vegetative propagation.

3.4 TAXONOMY AND EVOLUTION

The biosystematic investigations of the bulgarian species from sect. *Filipendulinae* showed that it is heterogeneous group. The comparative morphological analysis, the karyological and phytochemical data give a possibility to differentiate of these species in three groups. They corespond with the taxonomic system of Klokov and Krickaja (1984): Sect. *Filipendulinae* (DC.) Afan.: *A. thracica* Vel., *A. clypeolata* Sm., *A. chrysocoma* Friv., *A. coarctata* Poir.; Sect.*Micrantha* Klok. et Krytzka: *A. leptophylla* Bieb; Sect.*Ochroleucae* Klok. et Krytzka: *A. depressa*Jka. The main evolutionary mechanisms for the investigated section are the chromosome abberations and translocations. The polyploidy is less important in the evolution in comparison with sect. *Milledolium* .(Mill.) Koch..The intraspecific hybridization is present to a small degree, as compared to the taxonomic groups with expressed polyploidy and species without great genetic incompatibility. The herbar specimen pointed by Rechinger (1930) (N 8250/3, G-!) is incorrect by determinated as *A. biebersteinii*. On the basis of this conclusion together with the floristic study of the marked localities *A. biebersteinii* should be excluded from the bulgarian flora.

4. References

Alexander M.P. (1969) Differential staining of aborted and nonaborted pollen, *Stain Techn.* **44** **(3)**, 117-122.

Chemeris, E. and Androshchuk, A. (1982) Primenenie evklidovogo rasstojanija v kariotaksonomicheskom analize, *Dokl. AN Ukrainy* **4**, 73-77.

Erdtman G. (1969) .*Handbook of Palinology,* Hafner Publ. Co., New York.

Klokov, M. and Krickaja, L. (1984) Sistematika i geografitcheskoe rasprostranenie pri rodov *Ptarmica* Mill. and *Achillea* L. v flory Ukrainy, in K. Sytnik (ed.), *Tysjatchelistniki*, Naukova dumka, Kiev, pp. 195-255.

Kuzmanov, B. (1984) *Taksonomija i Evoljutsija na Balgarskite Slozhnotsvetni Rastenija*, Disertatsija, BAN, Sofija.

Richardson I. (1976) *Achillea* L., in Tutin T (ed.), *Flora Europaea*, Cambridge Univ. Press., Cambridge.

Stojanov, N. and Stefanov, B. (1924) Flora na Balgarija,Nauka i izkustvo, Sofija.

Stojanov, N. and Stefanov, B. (1933) Flora na Balgarija,Nauka i izkustvo, Sofija.

Stojanov, N. and Stefanov, B. (1948) Flora na Balgarija,Nauka i izkustvo, Sofija.

Stojanov, N., Stefanov, B. and Kitanov, B. (1967) Flora na Balgarija,Nauka i izkustvo, Sofija.

Velenovsky J. (1891) *Flora Bulgarica*, Prostat apud Fr. Rivnac, Pragae.

Velenovsky J. (1898) *Flora Bulgarica. Suplementum I*, Pragae.

PLANT COMMUNITIES WITH ENDEMIC SPECIES *PANCICIA SERBICA* VIS. (*APIACEAE, ARALIALES*) IN YUGOSLAVIA

D. OBRATOV and M. DJUKIC
Faculty of Forestry
Kneza Višeslava 1, 11030 Belgrade, Yugoslavia

1. Introduction

Some parts of the Balkan Peninsula, during the period of glaciation, were refuges for a great number of European Pliocene species. Some Tertiary species which reduced their range in Europe in that period, remained as endemorelics or relics, and one of such species in the region of Dinarides is *Pancicia serbica* Vis. As for horology, endemorelic species *Pancicia serbica* ranges in the central and West parts of Yugoslavia and North and East parts of Albania.

2. Material and Methods

Phytocoenological research has been carried out according to the principles and methods of the Braun - Blanquet's French-Swiss school (1965).

3. Results and Discussion

3.1. ASSOCIATION PANCICIO-NARDETUM STRICTAE

After Horvat (1946), plant communities of matgrass occur on almost all mountains from the Alps to Pelister, and they are described under the general name *Nardetum stricatae* Greb. 1950. At some localities in southwest Serbia, the community *Pancicio-Nardetum strictae* spreads over limestone bedrock. In this community, *Pancicia serbica* is absolutely dominant, as it was created by the degradation of spruce forests, at the altitude of 1290-1325 m. The community is composed of 47 species (Table 1). The dominant species in the community are hemi-cryptophytes, and the greatest effect is that of Middle European floral elements.

3.2. ASSOCIATION PANCICIO-PICEETUM ABIETIS

At some sites in southwest Serbia, such as Jadovnik, Zlatar, Ozren, Koritnik, etc.,

TABLE 1: Association: *Pancicio - Nardetum strictae*

Locality	Gradina				Brojila				Osoje		Degree of presence
Altitude	1310	1315	1300	1290	1300	1310	1295	1325	1340	1300	
Exposition	NE	N	NE	N	NE	N	E	NE	NE	NE	
Slope	25°	15°	20°	30°	25°	15°	15°	5°	3°	30°	
Parent rock	Limestone										
Area m2	100 x 100										
Number of sample	1	2	3	4	5	6	7	8	9	10	
Ground layer											
H *Nardus stricta*	2.2	2.2	2.3	3.3	2.3	2.3	2.3	2.3	2.2	2.3	V
H *Pancicia serbica*	2.2	1.2	1.1	1.1	1.1	+.1	1.2	1.1	1.1	1.1	V
H *Agrostis capillaris*	+.1	+.1	+.1	+.1	1.1	1.1	1.1	1.1	1.1	1.1	V
H *Hieracium hoppeanum*	+.1	1.1	+.1	+.1	+.1	+.1	1.1	1.1	1.1	1.1	V
Ch *Chamaespartium sagittatus*	+.1	1.1	1.1	+.1	+.1	1.1	.	1.1	+.1	+.1	V
H *Carlina acaulis*	+.1	+.1	+.1	+.1	+	+.1	+	1.1	+	+	V
H *Alchemilla flabellata*	+.1	+.1	+.1	+	+	+.1	+.1	+.1	+	+	V
Ch *Thymus glabrescens*	+.1	+.1	+.1	+.1	+.1	+.1	+	+	+	+	V
H *Veratrum album* subsp. *lobelianum*	+.1	+	+	+.1	+.1	+.1	+	+.1	+	+.1	V
Ch *Vaccinium myrtillus*	+	+.1	+.1	+	+	+.1	+	+	+.1	+.1	V
H *Scabiosa dubia*	+.1	+	+	+.1	+	.	+	.	+.1	+	IV
H *Pedicularis commosa*	+	+.1	.	+	+	+	+.1	+.1	+	.	IV
H *Dianthus barbatus* subsp. *liburnicus*	+	+.1	+	.	+	.	+	+.1	+	+	IV
H *Poa trivialis*	+	+	+.1	.	.	1.2	.	+	2.2	+	IV
H *Briza media*	+.1	1.1	+	.	+	+.1	.	+.1	.	+.1	IV
H *Leucanthemum vulgare*	+.1	1.1	+	+	.	+	.	+	.	+	IV
H *Carex caryophyllea*	+.2	+.2	.	.	+	+	.	.	+.2	+	IV
H *Lathyrus pratensis*	.	+.2	+.2	+.1	+.1	.	+.1	.	+.1	.	IV
H *Danaa cornubiensis*	.	1.1	+.1	+	.	.	+	1.1	.	+	IV
H *Pimpinella saxifraga*	.	+.1	+.1	.	+.1	+	.	+	+.1	.	III
H *Gentiana verna*	+.1	+.1	.	+	.	.	+	+.1	.	+.1	III
H *Fragaria vesca*	+	+.1	.	+	.	+	+.1	+	.	.	III
H *Stachys officinalis*	+.1	+.1	+	+	+	.	.	.	+	.	III

Number of sample	1	2	3	4	5	6	7	8	9	10	D.of p.
H *Saxifraga aizoon*	+.1	+	+	.	+.1	+	.	+	.	.	III
H *Galium vernum*	+.2	.	.	+	.	+	.	+.2	.	+.2	III
H *Plantago lanceolata*	.	+.1	.	+.1	.	+.1	.	+.1	+	.	III
H *Tragopogon pratensis*	.	.	+.1	.	+.1	.	+.1	+.1	.	+	III
P *Rosa pendulina*	.	.	.	+	+	+.1	.	+.1	.	.	II
H *Stellaria graminea*	.	.	+	.	+	.	.	.	+.1	+.1	II
H *Sanguisorba minor*	.	.	.	+.1	.	+.1	+	+	.	.	II
H *Hieracium bauhini*	+.1	.	.	.	+	.	+	.	+.1	.	II
H *Digitalis ferruginea*	.	.	.	.	+	.	+	.	+	+	II
H *Anthylis vulneraria*	.	.	+.1	.	+.1	.	.	.	.	+	II
G *Anemone ranunculoides*	.	.	.	+.1	.	.	.	.	+	+	II
H *Geum montanum*	.	.	.	.	.	.	+	.	+	+.1	II
H *Filipendula hexapetala*	.	+	.	+	.	+	.	.	.	.	II
H *Euphorbia cyparissias*	.	+	.	+	.	.	+	.	.	.	II
G *Dactylorhiza maculata*	.	.	+	.	.	+	.	+	.	.	II
T *Capsella bursa-pastoris*	.	+		.	+	.	.	.	.	+	II
H *Viola silvestris*	.	.	+	+	.	.	.	+	.	.	II
H *Marrubium peregrinum*	+	.	.	.	+	.	.	.	.	+	II
H *Bromus erectus*	.	.	+	.	.	.	.	+	+	.	II
H *Potentilla argentea*	+	.	.	+	.	+	.	.	.	.	II
H *Phyteuma spicatum*	.	.	.	.	+.1	.	.	+.1	.	.	I
H *Veronica officinalis*	.	.	.	.	.	+	+.1	.	.	.	I
H *Galium album*	.	.	+.1	.	.	+	.	.	.	.	I
H *Lychnis flos-cuculi*	.	.	.	.	+	.	+	.	.	.	I

In one sample there are : *Leontodon autumnalis* (+), *Creprs conysifolia* (+), *Carex leporina* (+), *Centaurea jacea* (+), *Stellaria nemorum* (+), *Cerastium arvense* (+), *Deschampsia flexuosa* (+), *Primula elatior* (+), *Campanula cervicaria* (+), *Poa compressa* (+), *Centaurea phrygia* (+), *Poa bulbosa* (+), *Gentiana punctata* (+).

Pancicia serbica forms plant communities with spruce. The communities occur mainly on limestone, at various exposures, on the terrain with mild (10°), semi-steep and steep slopes (up to 45°), at the altitude of 1365-1400 m. Rendzina, as the soil type, occurs very nonuniformly. Phytocoenological records were made at the localities of the mountain Zlatar. The community is composed of 59 species. The spectre of life forms determines the community as hemi-cryptophyte, and from plant-geographical aspect, the greatest effect is that of Eurasian floral elements.

3.3 ASSOCIATION PANCICIO-CORYLETUM AVELLANAE

On the mountain Zlatar, in the former sites of beech forests, hazel thickets are formed. They grow on limestone bedrock, on hillsides and hollows, predominantly on the north and northeast exposures.

In other parts of Yugoslavia, in Monte Negro, there occur the communities *Pancicio-Lilietum bosniacae* Bjelč. et Lak. 1969: *Ranunculo-Pancicietum serbicae* Lak. 1969 and *Festuco-Pancicietum serbicae* Tom. 1970.

4. Conclusion

The communities with endemorelic species *Pancicia serbica* Vis. occur on limestone parent rock and at higher altitudes. The community *Pancicio-Piceetum abietis* grows in the zone of spruce forest in southwest Serbia, and especially on the mountain Zlatar. The community *Pancicio-Nardetum strictae* ranges in the zone of degraded spruce forest, and in the zone of degraded beech forest there occurs the community *Pancicio-Coryletum avellanae.* The communities *Pancicio-Lilietum bosniacae, Ranunculo-Pancicietum serbicae* and *Festuco-Pancicietum serbicae* have been described in the region of Monte Negro.

5. References

Braun-Blanquet,J.(1965)Plant sociology - *The study of plant communities*, Hofner Publisching Company, New York.

Grebenščikov, O. (1950) *O vegetaciji centralnog dela Stare planine.* SANU. Beograd.

Horvat, I. (1946) *Šumske zadruge Jugoslavije.* Šumarski priručnik. HANU. Zagreb.

Obratov, D., Matović,M. (1978) Two new forms of Pancicia serbica Vis., *Biosistematics*, 13, 1-6. Beograd.

Obratov, D., Matović,M., Djukić, M. (1993) Chorological, Ecomorphologic and phytocoenological characteristics of the species Pancicia serbica Vis. *Abstracts VII OPTIMA Meeting*, 89, Bulgaria.

AQUATIC PLANTS OF SULTAN MARSHES WETLAND

M. ÖZTEKİN AND Ş. YILDIRIMLI
Hacettepe University, Faculty of Science, Department of Biology, 06532 Beytepe/ Ankara-Turkey

1. Abstract

Sultan Marshes, one of the five wetlands in Turkey, which is candidate for the Ramsar Agreement from the point of the Nature Protection Area is 17.200 hectare. This area consists of the ecosystem of fresh water and salty water which surround them with the parts of salty area and marshes. Sultan Marshes, spreading on the wetlands of Europe and Balcan, which consists of the typical aquatic plants is an aquatic ecosystem. Sultan Marshes, which is found on the center of Anatolia Peninsula that has a class criteria from the point of different species of birds is an aquatic area. Sultan Marshes is rather fertile from the point of aquatic flora and this flora is divided into three parts: These are; The plants that floats on the surface of the water, The plants whose root is found in the water attaching to the soil and The plants which is near the coast. The plants, which are the most widespread and which call more attention than the others, are *Lemna gibba* L. (Lemnaceae), *Sparganium erectum* L. subsp.*erectum* (Sparganiaceae), *Phragmites australis* (Cav.) Trin ex Steudel (Poaceae), *Typha latifolia* L. (Typhaceae). Sultan Marshes with its aquatic flora, is also important for birds for their accomodation and a shelter place.

2. Introduction

Sultan Marshes" which is one of the 5 candidate places for **Ramsar Contract** within the frame of Protecting Nature, is one of the most important wetlands in Turkey (Erdem, 1994). The universal protection system that was developed in the "**Meeting of International Wetlands Ecology**" held in Turkey in 1967, formed a background for the preparation of an international contract in the subject of protecting the wetlands. Four years after this meeting, in 1971, "**The Contract of the Protection of Wetlands**" focusing on the international habitation of water birds was opened signature in Ramsar, Iran. The contract was put into practise with the participation of Greece as a seventh supporting country. These countries contributed to the incorporation of total 641 wetlands in the Ramsar list.Turkey took side with this contract which is utterly important for the protection of wetlands on 30th December 1993 and put forward the ecosystems of Kuş Lake, Burdur Lake, Seyfe Lake, Sultan Marshes, and Göksu Delta as candidates for the contract list (Erdem, 1994).

Except for the Group of Independent States, Turkey has the richest wetlands of Europe and the Middle East in terms of the variety of species. Within International criteria, Sultan Marshes is an **A class** wetland.The **A class** wetlands are the wetlands that contain 25000 or more water birds a day. The fundamental reason that the wetlands of Turkey have an international importance stems from the geographical situation of this country. The two of the most important ways of bird migration in the Western Palearctic region pass over Turkey (Erdem, 1994). 301 species of bird were found as a result of studies in the region (Özçelik, 1995). According to (Kiziroğlu vd., 1992), 426 species of bird were found in Turkey. These numbers indicate how important the region is in terms of birds.

Sultan Marshes" is a wetland in the middle of Anatolia to 70 km south of Kayseri city, 1071m. altitude and occuping 17200 hectare field. The area is surrounded by the mount Erciyes (3916m.) from the north and by the Aladağs of Taurus (3756m.) from the south. The altitude of the mountains of its east and west is 2000 meters. These are very different ecosystems in "Sultan Mashes". The pure water swamp and marshy ecosystems cover an area of 5200 hectares. In this area, 1090 hectare-lenght Kepir Marshes and 3300 hectare-lenght Örtülüakar Marshes are located. The Yay Lake, which takes place between these two systems, is a salty water ecosystem covering an area of 3650 hectares. The Çöl Lake is located to the North-Western corner of the area. The water of this lake is salty, unplanted and a completely dry ecosystem in summers. We can add a new ecosystem in which local swamps and floods were seen in the earlier years. But later the area was dried and grassed, becoming suitable for the types of bird in terms of nutrition and protection. The new ecosystem covers an area of 8350 hectare.

3. Materials and Methods

In Sultan Marshes which is the field of investigation, 61 water plant taxons were found as a result of field research which lasted two days and which was realised regularly once in every fifteen days starting from March 1993 to December 1994. The samples of plants were dried, labeled and stuck on the herbarium card. Later they were put in the Herbarium of Biology Departmant, Faculty of Science, Hacettepe University (Seçmen vd., 1992 ; Erik vd., 1995). In the description of samples, mainly Flora of Turkey and the East Aegean Islands (Davis, 1965-1988) were used. In the description of water plants Fassett, (1966); Mühlberg, (1982); Güner, (1985) ;Altınayar, (1988) were especially reffered to. All of the taxons in the findings section were given according to the evolutionary order (Davis, 1965-1988) in the Flora of Turkey. On the list, the taxons not collected from the research field, but collected and published by some other researchers can be found (Davis, 1965-1988 ; Kasparek, 1985). The information concerning the phytogeographical regions of the species of water plants were taken from (Davis, 1965-1988).

4. Results

The water plants found in the research field can be divided categories according to our observations:
a) The floating plants
b) The plants which are attached to the soil under water
c) The plants which are near the coast
The major part of the pure water swamp to the southern parts of the area is covered with *Phragmites australis.* Except for the open water surfaces of this species, there is nearly a coverage of 90 % at every place. *Phragmites* is also very common along with the channels in the area. In Sultan Marshes the region containing *Phragmites* is an important source of earning. It's fresh parts are used as animal food.*Phragmites* is exported abroad.
In addition to the *Phragmites,* the kinds of secondary dominance on Eğrigöl Island, Kalaycı Island and Çorak Island which are located in Örtülüakar Marshes to the south are as folows: *Cyperus fuscus* is met very often in the open areas near water. When you go to the areas where *Phragmites* is found openly, *Schoenoplectus lacustris* subsp. *tabernaemontanii* and *Bolboschoenus maritimus var. maritimus* are seen frequently. In the transitional areas of the swamps surrounding Islands, *Cirsium alatum* subsp. *alatum* can be found. Between *Phragmites* species, *Lysimachia vulgaris, Calystegia sepium* subsp. *sepium* and *Althaea officinalis* add to the color and variety of the marshes.
The unplanted parts of the water surfaces between *Phragmites* can be mainly seen in Eğri Lake, Sarp Lake,Üç lakes and Ağyatağı behind lakes. The under water types of the unplanted parts of the water surfaces between *Phragmites* are as follows in the density order : *Myriophyllum spicatum, Utricularia australis, Nymphaea alba, Ranunculus sphaerospermus, Sparganium erectum* subsb. *erectum, Butomus umbellatus, Polygonum amphibium, Potamogeton pectinatus.* On the water coasts of the unplanted parts of the water surfaces between *Phragmites,* the types of *Lythrum salicaria, Epilobium hirsutum, Solanum dulcamara* are dominantly seen reaching 70 %. For example in the channels of Tuzla in the south, the kinds of *Nymphaea alba* and *Myriophyllum spicatum* are seen in the average 90 % dominance. The groups of the kinds of *Sparganium erectum* subsp. *erectum* are found in the Devetepesi channels and Kepir Marshes to the north of the field. However, the kinds of *Polygonum amphibium* and *Hydrocharis morsus-ranae* which are seen very rarely between the kinds of *Nymphaea alba* can also be seen. The kints of *Bidens tripartita, Ranunculus sceleratus* and *Berula erecta* which are found places of water in *Örtülüakar Marshes.* In the swamp areas of the channels around the pomp stations of Camız Lake, the kinds of *Spergularia media and Potentilla reptans.* The kind of *Lemna gibba* around Soysallı springs and Çayırözü channel in the north forms approximately a 100 % coverage on the water surfaces. Although *Lemna minor* species does not form a very close coverage, it can still be found in Yerköy channel in the south. *Spirodela polyrhiza* is also found in the Kepir Marshes.
Ranunculus sphaerospermus forms a white surface on the water and indicates a domination of nearly 100 % in the marshy and swamp places in the regions around the

pomp station of Camız Lake. *Typha latifolia* is much more common in the channels of the southern regions. It has a domination of 60 % together with the species of *Phragmites australis* and *Cyperus fuscus* especially in Gözbaşı region of the south. *Calamagrostis pseudophragmites* is also found near the water of the second channel.

5. Discussion and Conclusion

As a result of field investigations some total of 61 taxon types were found. 7 of these are the plants which float on the water, 16 of these are the plants having root under the water and 38 of these are the ones which are found along the water. 36 of these are dicotly, 25 of these are monocotly. 36 dicotly taxons are found in 22 families while 25 monocotly taxons are found 10 families (Davis, 1965-1988). According to these findings, the dicotly water plants are greater than others. As it was explained in the findings, although the number of species of monocotly water plants is little, the area they cover is wide.

The family of Cyperaceae is the one which contains 5 different genera in the research field. Generally, this family is a typical species of the wetlands. Later the family of *Poaceae* is notable for containing 3 different genera. The genus of *Potamogeton* is the one which contains the most dominant species in the research field. There are 4 species belonging to this genus. Apart from this, although *Rumex* genus of Polygonaceae family contains 4 species, it is not very common in the field. Following this, *Typha, Epilobium, Mentha, Plantago* genera contain 3 species respectively. 8 of found taxons are Euro-Siberian phytogeographical elements, 3 of them are Irano-Turanian phytogeographical elements. However, the other taxons are scattered or widespread elements.

6. References

Altınayar, G., 1988, Su Yabancı Otları, DSİ Genel Müdürlüğü, İşletme ve Bakım Dairesi Başkanlığı Yayını, Ankara.

Davis, P.H., (ed.) 1965- 1988, Flora of Turkey and The East Aegean Island: 1 (1965), 2 (1967), 3 (1970), 4 (1972), 5 (1975), 6 (1978), 7 (1982), 8 (1984), 9 (1985), 10 (1988), Edinb. Univ. Press.

Dugan, P.J., 1990, Sulak Alanların Korunması, çeviri: DHKD, İstanbul, IUCN- International Union for Conservation of Union.

Erdem, O., 1994, Türkiye'nin Kuş Cennetleri, T.C. Çevre Bakanlığı, Yeşil Seri: 4, Ankara.

Erik, S., Güner, A., Yıldırımlı, Ş., Sümbül, H., 1995, Tohumlu Bitkiler Sistematiği Laboratuvar Kılavuzu, Beytepe, Ankara.

Fasset, N.C., 1966, A Manual of Aquatic Plants, The University of Wisconsin Press, London.

Güner, H., 1985, "Hidrobotanik" Su bitkileri, Ege Üniversitesi Fen Fakültesi Kitapları Serisi, No. 91, Bornova- İzmir.

Kasparek, M., 1985, Die Sultanssümpfe, Naturgeschicte Eines Vogelparadieses in Anatolien , Heidelberg.

Kiziroğlu, İ., Turan, L., Erdoğan, A., 1992, Sultansazlığı'nın Ekoornitolojisi ve son durumu, H.Ü. Eğitim Fakültesi Dergisi, 217-227.

Mühlberg, H., 1982, The Complete Guide to Water Plants, EP Publishing Limited. German Democratic Republic.

Özçelik, M.F. ve Sezer, G.A., 1995, Ornito- Turizm kavramı, Türkiye'nin Ornito- Turistik potansiyeli ve Sultan Sazlığı modeli, Sulak Alanların Korunması Uluslararası Toplantısı, Kapadokya- Türkiye, 27 Eylül- 1 Ekim 1995, T.C. Çevre Bakanlığı, Çevre Koruma Genel Müdürlüğü, Ankara.

Seçmen, Ö., Gemici, Y., Leblebici, E. Görk, G., Bekat, L., 1992, Tohumlu Bitkiler Sistematiği, Bornova-İzmir.

POLLEN MORPHOLOGY AND ITS CONTRIBUTION TO THE TAXONOMY OF SOME NATIVE GREEK *AEGILOPS* SPECIES .

I. *Diploid and polyploid species of the wider group of genomes C, U.*

S. PANAJIOTIDIS[1], N. ATHANASIADIS[2],
L. SYMEONIDIS[1] and S. KARATAGLIS[1]
[1]Department of Botany , [2]Institute of Forest Botany-Geobotany.
Aristotle University of Thessaloniki ,
GR-540 06 , Thessaloniki , Greece.

1. Introduction

Pollen morphological characters (mainly those of the exine sculpture) have been used as a means in studying the phylogeny and taxonomy of certain genera of the *Gramineae* family(Grant 1972, Rajendra et al. 1978, Sun and Liang 1991). In the present study some native Greek *Aegilops* species of the wider group of genomes C,U which comprises the diploids *Ae. umbellulata* U and *Ae. caudata* var. *typica* and *polyathera* C, and tetraploid hybrids *Ae. triuncialis* UC̲ and *Ae. cylindrica* DC (genome symbols after Kimber and Tsunewaki 1988) are investigated for their pollen quantitative characters. The results are compared to those of different taxonomical discriminations based on morphological (Zhukovsky 1928, Eig 1929) and cytogenetical data (Kihara 1954, Chennaveeraiah 1960).

2. Materials and methods

Spikes of the investigated species were collected at a period slightly before anther dehiscence. The selected pollen samples were acetolysed using Erdtman's method (1960). 30 intact pollen grains from each taxon (with their pore in center position) were scored under LM for the following parameters: Polar axis , Equatorial axis, Pore diam., Anulus diam., Exine width (fig.4). Data were subjected to stepwise discriminant analysis and a UPGMA phenogram was produced based on the Mahalanobis distances calculated between the taxa.

3. Results and discussion

Discriminant analysis yielded five functions while the first two accounted for the 98,5% of the total variation (each function is a linear combination of the pollen variables measured). Fig.1 represents the projection of the taxa onto those functions. The first function separates the taxa into three clusters: a major one comprising the species *Ae. cylindrica* and *Ae. caudata* with its varieties *typica* and *polyathera* and two single-species clusters comprising *Ae. umbellulata* and *Aetriuncialis* respectively. The second function slightly separates the two species of the major cluster.

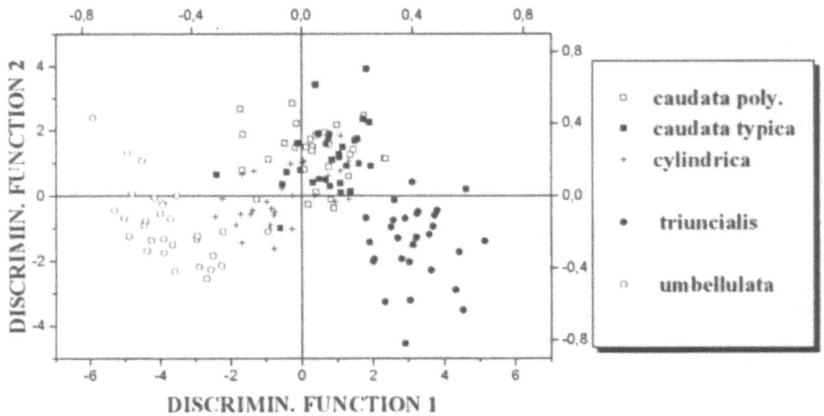

Fig.1 Projection of the investigated taxa onto the first two discriminant functions. *Ae. caudata* and *Ae. cylindrica* form a cluster which is clearly dinstinguished from the rest of the species (*Ae. umbellulata* and *Ae. triuncialis*)

The squared Mahalanobis distances D^2 calculated among the species centroids and the UPGMA phenogram based on them (tab.1, fig.2) supplement the results of discriminant analysis (smaller values of D^2 imply larger degree of morphological affinity and the opposite).The distance between the two varieties *typica* and *polyathera* is rather small and statistically insignificant. Both varieties of *Ae. caudata* are almost identical regarding their pollen quantitative characters. Moreover the two varieties and especially the latter are closely related to *Ae. cylindrica* (tab.2). *Ae. umbellulata* and *Ae. triuncialis* are quite remote from each other and from the above mentioned species.

	Ae. caudata var. *polyathera*	*Ae. caudata* var. *typica*	*Ae. cylindrica*	*Ae. triuncialis*	*Ae. umbellulata*
Ae. caudata var. *polyathera*	0,000				
Ae. caudata var. *typica*	0,355	0,000			
Ae. cylindrica	2,661	3,770	0,000		
Ae. triuncialis	16,189	13,628	16,639	0,000	
Ae. umbellulata	21,189	24,062	11,313	48,730	0,000

Tab.1 Squared Mahalanobis distances calculated between the investigated taxa based on five pollen parameters.

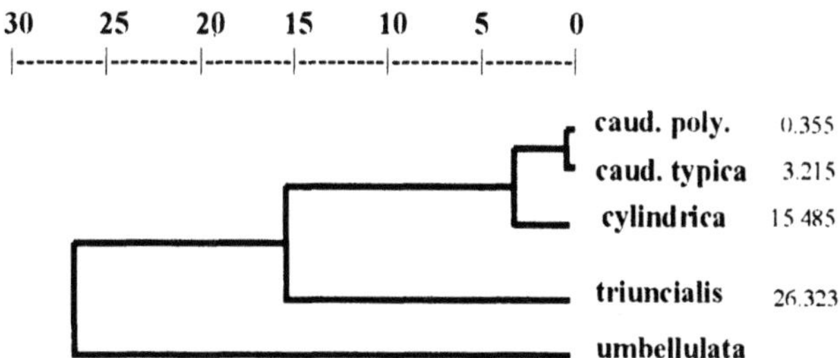

Fig.2 UPGMA phenogram of the species investigated based on the Mahalanobis distances calculated between their centroids.

Comparing the clustering of the species on basis of their pollen quantitative characters with the different taxonomical discriminations cited in tab.2 we conclude that it is more in agreement with those based on cytogenetical studies. Kihara studying the species genome affinity and Chennaveeraiah their chromosome morphology concluded in placing *Ae. cylindrica* and *Ae. caudata* together in section *Cylindropyrum* and in agreement with Eig they placed also *Ae. umbellulata* and *Ae. triuncialis* in the same section. The fact that *Ae. triuncialis* is quite remote from *Ae. umbellulata* (which confirms partly Zhukovsky's classification) is due to the very large values of the pollen parameters of the former.

Kihara (1954) Chennaneeraiah (1960)	**Eig (1929)**	**Zhukovsky (1928)**
sect. ***Polyeides***	sect. ***Pleionathera***	sect. ***Polyeides***
Ae. umbellulata ZHUK.	subsect. ***Libera***	*Ae. umbellulata* ZHUK.
Ae. triuncialis L.	*Ae. umbellulata* ZHUK.	sect. ***Surculosa***
sect. ***Cylindropyrum***	*Ae. triuncialis* L.	*Ae. triuncialis* L.
Ae. cylindrica HOST.	sect. ***Monoleptathera***	sect. ***Cylindropyrum***
Ae. caudata L.	*Ae. cylindrica* HOST.	*Ae. cylindrica* HOST.
	sect. ***Macrathera***	sect. ***Comopyrum***
	Ae. caudata L.	*Ae. caudata* L.

Tab.2 Different taxonomical discriminations in sections of the investigated taxa based on cytogenetical (Kihara , Chennaveeraiah) and morphological (Eig , Zhukovsky) studies.

4. References

Chennaveeraiah, M.S. (1960) Karyomorphologic and cytotaxonomic studies in Aegilops, *Acta Horti. Gotoburgensis* **23**, 85-178

Eig, A. (1929) Monographisch-kritische uebersicht der gattung Aegilops, *Feddes Repertorium Beih.* **55**, 1-228

Erdtman , G. (1952) *Pollen morphology and plant taxonomy - Angiosperms*, Almqvist and Wiksell, Stockholm.

Grant, C.A. (1972) A scanning electron microscopy survey of some maydeae pollen, *Grana* **12**, 177-184

Kihara, H. (1954) Considerations on the evolution and distribution of Aegilops species based on the analyser-method, *Cytologia* **19**, 336-357

Kimber, G. and Tsunewaki, K. (1988) Genome symbols and plasma types in the wheat group, in T.E. Miller, and R.M.D. Koebner, (eds.), *Proc. 7th Int. Wheat Genet. Symp.*, Cambridge, pp. 1209-1211.

Sun, Y. and Liang, G.H. (1991) Pollen surface sculpture of wheat, rice, maize and five Sorghum species, *Cytologia* **56**, 659-664

Zhukovsky, P.M. (1928) A critical-systematical survey of the species of the genus Aegilops L., *Bull. Appl. Bot. Gen. Pl. Br.* **18**, 417-609

POLLEN MORPHOLOGY AND ITS CONTRIBUTION TO THE TAXONOMY OF SOME NATIVE GREEK *AEGILOPS* SPECIES.

II. *Diploid and polyploid species of the wider group of genomes U,M,N.*

S. PANAJIOTIDIS [1], N. ATHANASIADIS [2],
L. SYMEONIDIS [1] and S. KARATAGLIS [1]
[1]*Department of Botany*, [2]*Institute of Forest Botany-Geobotany.*
Aristotle University of Thessaloniki,
GR-540 06, Thessaloniki, Greece.

1. Introduction

Native Greek polyploid *Aegilops* species sharing two common genomes (*Ae. ovata* MU, *Ae. triaristata* UM, *Ae. triaristata* UMN) and the parental diploid species (*Ae. umbellulata* U, *Ae. comosa* ssp. *comosa* var. *thessalica*, ssp. *heldreichii* var. *biaristata* and *subventricosa* M, *Ae. uniaristata* N) are investigated for their pollen quantitative characters (genome symbols according to Kimber and Tsunewaki 1988).As in a previous paper (Panajiotidis et al. 1998) results are compared to those of different taxonomical discriminations based on morphological (Zhukovsky 1928, Eig 1929) and cytogenetical data (Kihara 1954, Chennaveeraiah 1960).

2. Materials and methods

Spikes of the investigated species were collected at a period slightly before anther dehiscence. Selected pollen samples were treated as in previous paper (Panajiotidis et al. 1998) and the same statistical analysis was applied.

3. Results and discussion

Discriminant analysis yielded five functions while the first two accounted for the 97%

of the total variation. Fig.1 represents the projection of the taxa centroids onto those functions. Three major clusters can be distinguished: the first comprising the species *Ae. umbellulata* , *Ae. ovata* and *Ae. triaristata* 4x the second comprising the species *Ae. uniaristata* and *Ae. comosa* with its varieties and the third the species *Ae. triaristata* 6x. Moreover a slight separation in the second cluster can be recognized as both varieties of ssp. *heldreichii* (*biaristata* , *subventricosa*) form a group that is equally distant to the third variety *thessalica* and *Ae. uniaristata*. Hexaploid *Ae. triaristata* is wide distant to both clusters.

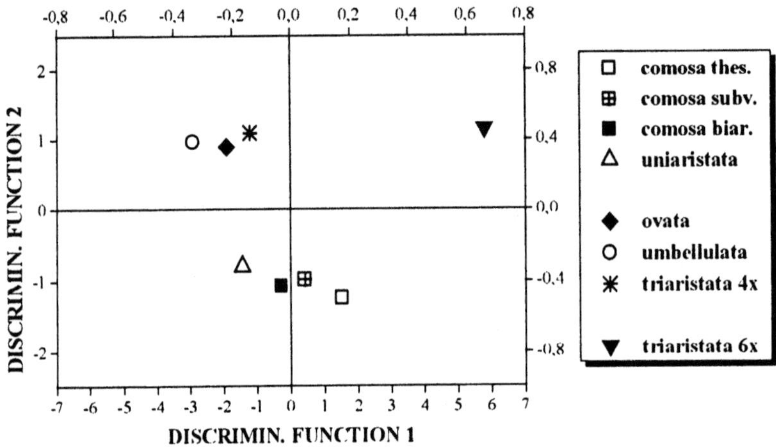

Fig.1 Projection of the centroids of the investigated taxa onto the first two discriminant functions. The species of the section *Polyeides*(*Ae. umbellulata*, *Ae. ovata* and *Ae. triaristata* 4x, 6x) are separated from those of the section *Comopyrum*(*Ae. comosa* with its varieties and *Ae. uniaristata*)

Squared Mahalanobis distances D^2 calculated between their centroids are cited in tab.2 (smaller values of D^2 imply closer morphological relationship) and the UPGMA phenogram based on them in fig.2 . Phenogram shows clearly that :

1. *Ae. triaristata* 6x is widely distant to the rest of the species
2. *Ae. uniaristata* is closely related to the varieties of *Ae. comosa* and especially those of ssp. *heldreichii*.
3. *Ae. ovata* and *Ae. triaristata* 4x are closely related to their parental species *Ae. umbellulata* . *Ae. ovata* shows greater resemblance than *Ae. triaristata* 4x.

Comparing clustering of the species on the basis of their pollen morphological affinity with the different taxonomical discriminations cited in tab.2 we can point out that:

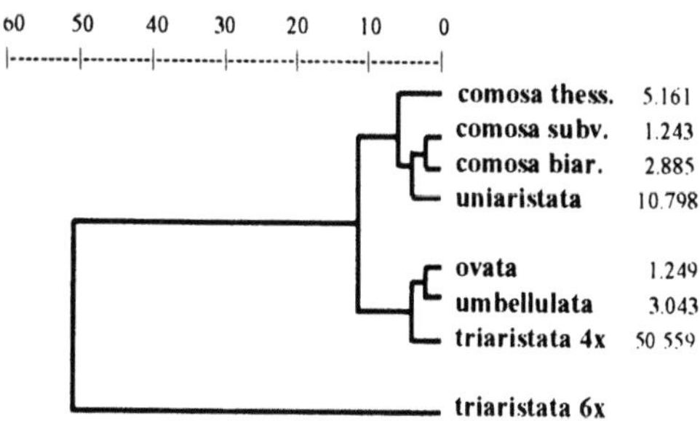

Fig.2 UPGMA phenogram of the species investigated based on the Mahalanobis distances calculated between their centroids.

	comosa thessalica	*comosa subventricosa*	*comosa biaristata*	*uniaristata*	*umbellulata*	*ovata*	*triaristata 4x*	*triaristata 6x*
comosa thessalica	0,000							
comosa subventricosa	2.071	0.000						
comosa biaristata	3,831	1.243	0,000					
uniaristata	9.580	4,031	1.739	0,000				
umbellulata	25.811	15.664	11.808	5.744	0,000			
ovata	17.210	9.315	7.480	3.568	1,249	0,000		
triaristata 4x	13.730	8.724	6.392	4.130	4.212	1,874	0,000	
triaristata 6x	25.342	34.940	43.332	58.321	78,682	61.699	51.599	0,000

Tab.2 Squared Mahalanobis distances between the investigated taxa of *Aegilops* based on the five pollen parameters.

1. All researchers with the exception of Chennaveeraiah have placed *Ae. uniaristata*and *Ae. comosa* in the same section recognizing a close relatioship between them. The present study seems to confirm that relationship.
2. The rest of the species were placed in the same section with no exception. Moreover Chennaveeraiah and Kihara included both tetraploid and hexaploid *Ae. triaristata*. Our data confirm placing of these species in the same section and furthermore suggest the possibility of discriminating between tetraploid and hexaploid *Ae. triaristata* on the basis of their pollen quantitative characters.

Kihara (1954)	**Chennaveeraiah (1960)**	**Zhukovsky (1928)**	**Eig (1929)**
sect. ***Polyeides***	sect. ***Polyeides***	sect. ***Polyeides***	sect. ***Pleionathera***
Ae. ovata L.	*Ae. ovata* L.		subsect. ***Libera***
Ae. umbellulata Zhuk.	*Ae. umbellulata* Zhuk.	*Ae. ovata* L.	*Ae. ovata* L.
Ae. triaristata (4x) Willd.	*Ae. triaristata* Willd.	*Ae. umbellulata* Zhuk.	*Ae. umbellulata* Zhuk.
Ae. triaristata (6x) Willd.	*Ae. recta* *(Zhuk.) Chen.	*Ae. triaristata* Willd.	*Ae. triaristata* Willd.
sect. ***Comopyrum***	sect. ***Comopyrum***	sect. ***Comopyrum***	sect. ***Macrathera***
Ae. comosa Sibth. et Sm.	*Ae. comosa* Sibth. et Sm.	*Ae. comosa* Sibth. et Sm.	*Ae. comosa* Sibth. et Sm.
Ae. uniaristata Vis.	sect. ***Uniaristatopyrum***	*Ae. uniaristata* Vis.	*Ae. uniaristata* Vis.
	Ae. uniaristata Vis.	*Ae. heldreichii* (Holmz.) Halac.	

Tab.2 Discrimination of the investigated taxa in sections based on cytogenetical (Kihara, Chennaveeraiah) and morphological (Eig , Zhukovsky) data.
* Synonym of *Ae. triaristata* (6x) Willd.

4. References

Chennaveeraiah, M.S. (1960) Karyomorphologic and cytotaxonomic studies in Aegilops, *Acta Horti. Gotoburgensis* **23**, 85-178

Eig, A. (1929) Monographisch-kritische uebersicht der gattung Aegilops, *Feddes Repertorium Beih.* **55**, 1-228

Kihara, H. (1954) Considerations on the evolution and distribution of Aegilops species based on the analyser-method, *Cytologia* **19**, 336-357

Kimber, G. and Tsunewaki, K. (1988) Genome symbols and plasma types in the wheat group, in T.E. Miller, and R.M.D. Koebner, (eds.), *Proc. 7th Int. Wheat Genet. Symp.*, Cambridge, pp. 1209-1211.

Panajiotidis, S., Athanasiadis, N., Symeonidis, L. and Karataglis, S. (1998) Pollen morphology and its contribution to the taxonomy of some native Greek Aegilops species **I.** *Diploid and polyploid species of the wider group of genomes* C,U. (this volume)

Zhukovsky, P.M. (1928) A critical-systematical survey of the species of the genus Aegilops L., *Bull. Appl. Bot. Gen. Pl. Br.* **18**, 417-609

THE FLORISTIC DIVERSITY OF THE GREEK FLORA AS EXEMPLIFIED BY THE EXPLORATION OF SOME EAST AEGEAN ISLETS.

M. PANITSA & D. TZANOUDAKIS
Division of Plant Biology, Department of Biology, University of Patras, GR-26500, Patras, Greece

1. Introduction

Greek flora is well known for its richness and diversity. Although the East Aegean area is one of the biogeographical subdivisions of Greece which don't have a large number of rare or local endemic species, our explorations showed an impressing diversity.
Subject of our explorations was the inventory of the plant species and the vegetation units of the isolated small islands of the East Aegean area and generally the study of the floristic diversity in these islet groups.
The majority of the 75 small islands studied, belonging to the prefectures of Lesbos, Chios and Dodekanisos, lie North-Easternwards of the main East Aegean islands between the geographical longitudes 27° 08′ 50′′ and 26 ° 14′ 00′′ east and latitudes 39° 19′ 08′′ and 36 ° 59′ 43′′ north (Map 1).
Most of the islets have an individual area which is not exceeding 25 ha. The larger ones are Lipsos, Enousa, Agathonisi and Arki from which the island groups take their names. The total land surface is about 66.5km^2 and the highest elevation is on Lipsos, at 277m.
The islets studied were connected with Asia Minor during the most extensive Pleistocene glacial periods, when the sea level dropped by c.120m. The geological substrate is mainly schist, crystalline and semi-crystalline limestone and limestone schist.
The climate of the studied area is typically mediterranean with mild winters and long dry periods of five and a half to seven months. The summer winds are dry and contribute heavily to the aridity of the area during summer months. From the bioclimatic point of view, the areas where the meteorological stations are situated (Lesvos, Chios, Samos and Leros), belong to the subhumid bioclimatological stage with mild winters.

2. Materials and Methods

The islet groups studied have been visited many times in different seasons from 1990 to 1996 for collection of plant material as well as for field observations. Voucher specimens have been deposited in Patras (UPA) and floristic lists for the whole area and/or each individual islet are given by Panitsa (1997). In a total of 24 islets, comparison of the floristic composition of each one registered at different periods was possible since Prof. Runemark kindly gave us his data from his visits (with R. Bothmer) on them in 1974. In order to count the floristic independence of the islets studied, Sorensen's similarity coefficient (S=0, for floristic independence and S=1 for maximum floristic similarity) is used.

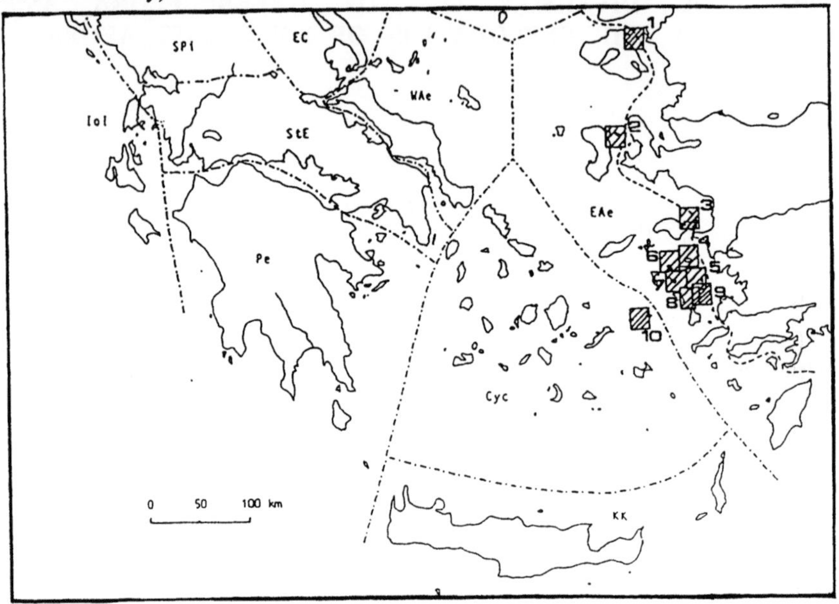

Map 1. Geographical position of the islets studied.
1.Tokmakia (I1-I4). 2. Enousses (I5-I10). 3. Diaporia (I11-I13). 4. Agathonisi (I14-I21). 5. Farmakonisi (I22). 6. Arki (I23-I34). 7. Lipsi (I35-I39). 8. Islets eastwards of Leros (I60-I68). 9. Kalolimnos-Imia islet group (I69-I73). 10. I[s]. Mavres (I74-I75).

3. Results and Discussion

In the 75 small islets studied, 725 plant taxa were found. Six of them are pteridophytes and 719 spermatophytes belonging to 337 genera and 71 families. According the results of the analysis of the total flora, it was found that therophytes dominate (58.6%), while the percentage of hemicryptophytes is 15.3%, of geophytes 14.5%, chamaephytes 6.5% and phanerophytes 5.1%. This biological spectrum is in accordance with the bioclimatological position of the area at the subhumid stage and with its intense thermo-mediterranean character. The percentages of the biological types vary from islet to islet and sometimes have extreme values.

From the analysis of the data, different levels of diversity can be distinguished concerning the floristic richness (a-diversity), the floristic differences between the islets (b-diversity) and the floristic dynamics.

The number of species (725) per surface unit (66.5 km^2) is very impressive (10.9 species/km^2). It gives an idea of the role of these small islands to the conservation of the diversity of the Aegean area if is it compared with 0.38-0.4 species/km^2 for the whole East Aegean area, with 0.54 species/km^2 for Cyclades and with 0.2 species/km^2

for Crete and Karpathos (estimated number of plant taxa on these biogeographical regions in: Phitos & al. 1995). It is noteworthy that the number of species is not augmented as the surface augments, but the above comparative numbers give an idea of the richness in species of such small areas as the islets.

None of the 725 taxa comprising the flora of the small islands studied occur in all the 75 of them. *Sedum litoreum* and *Pistacia lentiscus* are the most frequently encountered taxa and occur in 51 and 48 islets respectively, while 211 out of 725 taxa were not present on more than one islet.

Comparisons of the floras of the islets were carried out using Sorensen's similarity coefficient and significant differences from islet to islet were noted (Panitsa 1997). Even the comparison between islets of the same group has given a very low value for the similarity coefficient; this indicates the floristic independence of each of them. The following three groups have been recognized according to size and also, the impacts of human influence: a) Islands with a very small size (islets). b) Small or medium-sized islands, with a well-established phrygana vegetation. c) Larger islands with more similar, constant and balanced floras.

The islands of the first group show a significant floristic independence. The floristic composition varies from islet to islet depending mainly on random factors and on the geomorphology of the islets. The flora and the vegetation cover of these islets is characterized by the presence of some species that prefer mainly such biotopes (Islet specialists). Such taxa are: *Convolvulus oleifolius* subsp. *scopulorum, Allium commutatum, Lavatera arborea, Anthemis scopulorum, Silene holzmannii, Salsola aegaea, Helichrysum orientale, Scariola acanthifolia και Scorzonera cretica.* The last three species also occur on the larger islands or in continental areas, on isolated cliffs. It is noteworthy that these species do not occur on the larger islands of the study area.

The islands of the second group present a low to moderate floristic similarity. Human influences (grazing, fires etc.) seem to play a role in their floristic composition. Among the factors influencing the floristic differences seem to be the different stages in vegetation evolution after fire, the grazing intensity; this depends on the distance between the islet and the larger inhabited one etc.

In the floristic comparison among islands of the third group the higher Sorensen's coefficient values (0.32-0.69) have been observed and their floristic similarity seems to be due to the more stabilized environmental conditions and the more intense human influences which contribute to the increase of the same anthropochorous elements. Except the floristic differences from islet to islet, significant differences were observed concerning the floristic composition of even the same islet in time space. When our data were compared with those of Runemark and Bothmer, who visited 24 of the islets in the study area in 1974, an important quantitative and qualitative fluctuation in flora composition was discovered. An analysis of the data showed that the percentage of the constant flora (i.e., plant species found by Runemark and Bothmer at 1974 and from the current investigation) is 42-66%. With respect to the temporary flora (i.e., plant species found only by one group of workers), the percentages are 8-48% (Fig.1). These percentages underline that the flora of the small islands is not constant but at a very dynamic state. The frequency of the biological types of plant species comprising constant as well as temporary floras is not differing very much. That means that plant species comprising temporary flora cover more or less all the biological types (Fig.2).

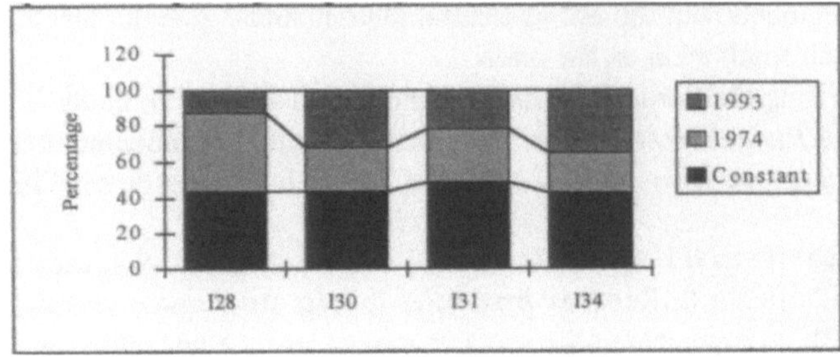

Figure 1.Percentages of the constant and the temporary flora of Arki islets group (I28: Minaronisi, I30: Zouca, I31: small Zouca, I34: Spartonisi).

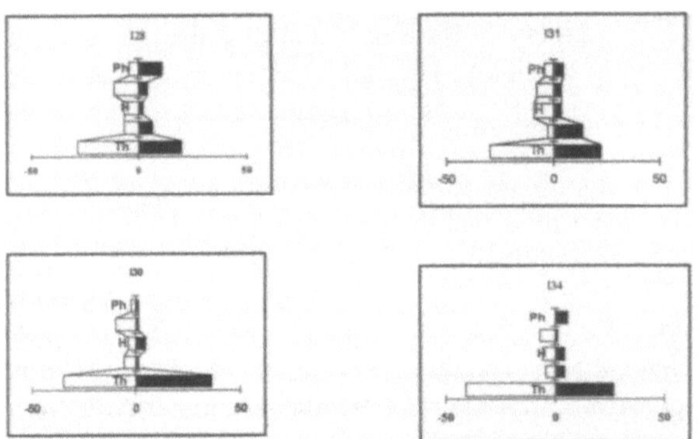

Figure 2. Percentages of the biological types of plant species comprising the temporary (white bars) and the constant (black bars) flora of islets belonging to Arki group (Ph: phanerophytes, Ch: chamaephytes, H: hemicryptophytes, G: geophytes, Th: therophytes).

It should be noticed that islets are very fragile ecosystems on which random factors, intense environmental fluctuations (storms, wave action) or human interference of even a low intensity can affect the flora and vegetation very greatly. The small plant species populations are consequently threatened and risk being eliminated. Any protection or management plan concerning the area and in general, vulnerable ecosystems, as those of small islands and islets should be based on available information concerning their diversity and their restoration possibilities.

4. Acknowledgement Financial support by the General Secretariat of Research and Technology and the Ministry of Environment is kindly acknowledged. The authors also thank Prof. Runemark for his help and hospitality.

5. References

Panitsa, M. (1997). *Contribution to the knowledge of the flora and vegetation of the East Aegean islets.* Ph.D. thesis, University of Patras. Patras, Greece.

Phitos, D., Strid, A., Snogerup, S. & Greuter, W. (Eds.) (1995): *The Red Data Book of rare and threatened plants of Greece.* World Wide Fund for Nature (WWF) 1995.

PHYTOSOCIOLOGICAL RESEARCH OF THE ABOVE THE TIMBERLINE GRASSLANDS OF MOUNT PANGEON (NE GREECE)

K. PAPADEMETRIOU [1)] V. KARAGIANNAKIDOU [1)] M. KONSTANTINOU [1)] & P. DIMOPOULOS [2)]
1) Department of Biology, Division of Botany, Institute of Systematic Botany & Phytogeography, Aristotle University of Thessaloniki, 540 06 Thessaloniki, Greece.
2) Pindarou 25, 13562, Agioi Anargyroi, Athens, Greece

1. Introduction

The grasslands of Greece, have remarkably been degraded, mostly due to overgrazing and fires. In order that an efficient conservation and management of the high altitude grasslands is acheived, the detailed phytosociological and ecological research is absolutely necessary for the estimation of their natural potential and contributes to a more rational management.

Phytosociological research of the grassland vegetation of Greece is still poor. Detailed phytosociological studies exist only for a few sites in certain areas (QUEZEL 1964, 1967, 1969, 1989, RAUS 1987, ZAFFRAN 1990, DIMOPOULOS 1993, GEORGIADIS & DIMOPOULOS 1993, KARAGIANNAKIDOU 1994) of Greece. Especially for the grasslands of NE Greece, which present high floristic-phytogeographical and ecological significance, even fewer phytosociological data are available (QUEZEL 1989, KARAGIANNAKIDOU 1994). The coexistence of European and Balkan geoelements in their floristic composition, due to their geographical position, often leads to a complex vegetation structure, which makes difficult and at the same time very interesting their syntaxonomic status.

Mount Pangeon is sufficiently well studied from a floristic point of view (STRID & PAPANICOLAOU 1981, PAPANICOLAOU 1985), whilst it is totally neglected from a phytosociological point of view. The objectives of the present study were to analyze the floristic features of the syntaxonomy and the synecology of the above the timberline grasslands of Mount Pangeon, on the basis of the numerical treatment of the releves sampled over their entire distribution area. An extensive study of the grassland vegetation of neighbouring mountains (Menikion, Falakron, Orvilos etc.) of Northern Greece is absolutely necessary.

2. Study Area

Mount Pangeon is located in East Macedonia (North East Greece) on the boundaries between the prefectures of Kavala and Serres. It lies in a NE-SW direction between 24°

01' to 24° 09' N and 40° 52' to 40° 57'E. Its highest peak is called Mati (Pilaf Tepe) and it is located in the central part of the mountain reaching the altitude of 1956 m.
The geological substrates of Mount Pangeon belong to the Rhodope massif, which consists of 4 series of rocks (gneisses, marbles, mica-schists, marbles alternating with amphibolites and mica-schists).
All the study sites covering an area of ca. 18 Km^2 , are located in the areas above the timberline of Mount Pangeon, at altitudes ranging from 1300m in the northern slopes (lowest limits of the grasslands of Pangeon) to 1440m and 1300m respectively on the western and south-eastern slopes and 1100 m in the south-western parts of the mountain. In the southern parts of the mountain, the limits which follow the line of 1100m and 1300m ascend to 1500 m in the south-eastern parts, whereas in the eastern parts, the limits ascend to 1660m and 1800m in the area Dikorfo.
From a bioclimatic point of view, the study area belongs to the humid zone with cold winters presenting sub-mediterranean bioclimatic character (MAVROMMATIS 1980).

3. Material and Methods

The study is based on a total of 259 releves carried out on the grasslands of Mount Pangeon, which share 145 species and subspecies. These were sampled and classified following the Braun-Blanquet phytosociological approach (BRAUN-BLANQUET 1964, WESTHOFF & VAN DER MAAREL 1978).
TURBO(VEG) (HENNEKENS 1996a) was used for input and handling of vegetation releves. In order to obtain an effective description of community types, both classification (Two-Way INdicator SPecies Analysis: TWINSPAN, HILL 1979b) and ordination techniques (Detrended Correspondence Analysis with the DECORANA package, HILL 1979a) were employed.
As a rule, the nomenclature of the taxa follows Mountain Flora of Greece (STRID 1986, STRID & KIT TAN 1991) and Flora Europaea (TUTIN et al. 1964-1980). Syntaxonomic nomenclature follows the references cited, when not otherwise specified. The form of names follows the Code of Phytosociological Nomenclature (BARKMAN et al. 1986). The formal description of new syntaxa and their typification will be presented elsewhere, together with clarification of syntaxonomic nomenclature on the association level.

4. Results

4.1 ORDINATION

From the successive analyses on the basis of the Detrended Correspondence Analysis,three ordination plots gave the best results, in order to interpret the community variation in terms of ecological gradients. In the first ordination diagram of the 16 syntaxa (259 releves) from Mt. Pangeon, the *Daphno-Festucetea* and the *Festuco-Brometea* are more or less clearly separated along the first axis. All the *Daphno-Festucetalia* communities have been plotted on the left side of the diagram, whereas communities of *Festucetalia valesiacae* (clusters 13,14,15,16) occur on the right side. In the middle part of the diagram, the associations and subassociations of the *Daphno-Festucetalia* overlap. After removing the overlapping releves of the *Daphno-*

Festucetalia included in the *Astragalo-Seslerion*, the remaining 11 clusters were reanalysed. In the resulting ordination diagram, the grasslands of the two above mentioned high-rank syntaxa are now clearly separated from one another. The discrimination observed between the communities of *Astragalo-Seslerion* and *Festucion valesiacae* is due to their floristic differences and become very evident on the synoptic phytosociological table which is not included here. From the ordination diagram originated from the reanalysis of the remaining 5 clusters (6, 7, 8, 9, 10), a better separation between the clusters has been acheived, though many releves are characterized by transitional floristic compositions, which makes very difficult to draw the clear boundaries between the syntaxa.

4.2 PHYTOSOCIOLOGICAL INTERPRETATION

In the above the timberline grassland vegetation of Mount Pangeon, sixteen vegetation syntaxa have been distinguished and listed below, accompanied by their altitudinal distribution and the group of diagnostic taxa:

***DAPHNO-FESTUCETEA* Quezel 1964**

Daphno-Festucetalia Quezel 1964

Astragalo-Seslerion Quezel 1964 **[clusters 1-12]**

Cerastio decalvani-Astragaletum angustifolii **ad interim [cl. 1]**: Diagnostic taxa: *Cerastium decalvans, Astragalus angustifolius, Bellardiochloa violacea.* Altitudinal distribution: 1200-1300m.

Festuco pauciniae-Astragaletum angustifolii **ad interim [cl. 2]**: Diagnostic taxa: *Festuca rupicola ssp.paucinia, Astragalus angustifolius.* Altitudinal distribution: 1200-1720m.

Globulario meridionalis-Festucopsietum sanctae **ad interim [cl. 3-6]**: Diagnostic taxa: *Festucopsis sancta, Globularia meridionalis.* Altitudinal distribution: 1200-1950m.

erysimetosum drenowskii **ad interim [cl. 3]**: Diagnostic taxa: *Erysimum drenowski, Campanula orphanidea.* Altitudinal distribution: 1200-1650m.

asperuletosum purpureae **ad interim [cl. 4]**: Diagnostic taxa: *Asperula purpurea ssp. purpurea, Dianthus drenowskianus, Edraianthus graminifolius.* Altitudinal distribution: 1250-1700m.

seslerietosum latifoliae **ad interim [cl. 5]**: Diagnostic taxa: *Sesleria latifolia.* Altitudinal distribution: 1250-1800m.

galietosum rhodopei **ad interim [cl. 6]**: Diagnostic taxa: *Galium rhodopeum.* Altitudinal distribution: 1200-1300m.

Asperulo nestiae-Festucetum cyllenicae **ad interim [cl. 7]**: Diagnostic taxa: *Asperula aristata ssp. nestia, Festuca cyllenica.* Altitudinal distribution: 1150-1800m.

Paronychio rechingeri-Linetum elegantis **ad interim [cl. 8]**: Diagnostic taxa: *Linum elegans, Paronychia rechingeri, Juniperus communis ssp. alpina.* Altitudinal distribution: 1560-1700m.

Onobrychido montanae-Thymetum thracici **ad interim [cl. 9-10]**: Diagnostic taxa: *Onobrychis montana, Thymus thracicus.* Altitudinal distribution: 1400-1950m.

centaureetosum pindicolae **ad interim [cl. 9]**: Diagnostic taxa: *Centaurea pindicola.* Altitudinal distribution: 1500-1550m.

typicum **[cl. 10]** : Altitudinal distribution: 1500-1700m.

Fragmentary community of the *Asperulo nestiae-Festucetum cyllenicae* **[cl. 11]**:

Altitudinal distribution: 1500-1550m.

Genisto lydiae-Festucetum macedonicae **ad interim [cl. 12]**: Diagnostic taxa: *Genista lydia, Genista carinalis, Vaccinium myrtillus, Centaurea affinis ssp affinis, Luzula campestris, Botrychium lunaria.* Altitudinal distribution: 1750-1800m.

FESTUCO-BROMETEA Br.-Bl. et R.Tx. in Br.-Bl. 1949

Festucetalia valesiacae Br.-Bl. et R. Tx. 1943

Festucion valesiacae Br.-Bl. et R. Tx. 1943 **[cl. 13-16]**

Festuco amethystinae-valesiacae **ad interim [cl. 13]**: Diagnostic taxa: *Festuca amethystina, Avenula pubescens.* Altitudinal distribution: 1650-1700m.

Stachyo alpinae-Nepetetum nudae **ad interim [cl. 14]**: Diagnostic taxa: *Stachys alpina, Calamintha alpina, Seseli rigidum, Silene ciliata, Trifolium medium.* Altitudinal distribution: 1300-1450m.

Trifolio alpestri-Festucetum valesiacae **ad interim [cl. 15-16]**: Diagnostic taxa: *Trifolium alpestre, Festuca valesiaca.* Altitudinal distribution: 1400-1700m.

bellardiochloetosum violaceae **ad interim [cl. 15]**: Diagnostic taxa: *Bellardiochloa violacea, Luzula sylvatica.* Altitudinal distribution: 1400-1700m.

scabiosetosum ochroleucae **ad interim [cl. 16]**: Diagnostic taxa: *Scabiosa ochroleuca, Rosa pimpinellifolia.* Altitudinal distribution: 1400-1500m.

5. References*

Dimopoulos, P. (1993) *Floristic and phytosociological research of mountain Killini - An ecological approach.* Ph. D. Thesis, University of Patras (In Greek).

Hennekens, S. M. (1996b) *TURBO(VEG). Software package for input, processing, and presentation of phytosociological data, User's guide. Version July 1996.* IBN-DLO, Wageningen & Lancaster University, Lancaster.

Hill, M.O. (1979a) *DECORANA - a FORTRAN program for detrended correspondence analysis and reciprocal averaging.* Ecology & Systematics, Cornell University, Ithaca.

Hill, M.O. (1979b) *TWINSPAN - a FORTRAN program for arranging multivariate data in an ordered two way table by classification of the individuals and the attributes.* Ecology & Systematics, Cornell University, Ithaca.

Karagiannakidou, V. (1994) Contribution to the study of mountain-subalpine grassland vegetation of Mount Menikion, northeastern Greece , *Ecol. Medit.* **XX 3/4,** 73-84.

Papanicolaou, K. (1985) Contribution to the flora of mount Pangaion (Pangeon), North East Greece, *Ann. Mus. Goulandris* **7,** 67-156.

Quezel, P. (1964) Vegetation des hautes montagnes de la Grece meridionale. *Vegetatio* **12 (5-6),** 289-386.

Quezel, P. (1967) La vegetation des hauts sommets du Pinde et de l' Olympe de Thessalie. *Vegetatio* **14(1-4),** 127-228.

Quezel, P. (1989) Contribution a lT etude phytosociologique des pelouses ecorchees culminales du massif du Falakron, *Bios (Thessaloniki)* **1,** 187-193.

Raus, Th. (1987) Gipfel Vegetation des Ossa Gebirges im Vergleich zum Olymp (Thessalien, Griechenland), *Munestersche Geor. Arbeiten* **27,** 199-206.

Strid, A. and Papanicolaou, K. (1981) Floristic notes from the mountains of Northern Greece, *Nord J. Bot.* **1,** 66-82.

* Selected references are only included

GROWTH CHARACTERISTICS AND DISTRIBUTION PATTERN OF C_4 GRASSES *DICHANTHIUM ISCHAEMUM* AND *CHRYSOPOGON GRYLLUS* IN N. GREECE

V. P. PAPANASTASIS
Laboratory of Range Science, Aristotle University
54006 Thessaloniki, Greece

1. Introduction

It is well known that vegetation has a contagious distribution which contains one or many scales of pattern. Such a pattern may be related to the morphology of the individual species, to environmental factors or to the interaction between species and the environment (Kershaw, 1973). Its detection largely depends on the size of quadrat used for its measurement (Greig-Smith, 1983; Kershaw, 1973). Also, the shape of quadrat will affect the pattern of a particular species, particularly if it is a perennial that forms tussocks or rosettes.

Dichanthium ischaemum (L.) Roberty and *Chrysopogon gryllus* (L.) Trin. are two perennial C_4 grasses widely distributed in low to middle elevation (0-800 m a.s.l.) grasslands of northern Greece. They usually grow together on a variety of soil types but *D. ischaemum* is more drought resistant and more productive than *C. gryllus* (Papanastasis, 1981; Noitsakis, 1988).

The aim of this paper was to investigate the growth characteristics and the distribution pattern of each of the two species in relation to shape and size of quadrat as well as the association between them.

2. Materials and Methods

The experiment was carried out in a foothill grassland located 25 km east of the city of Thessaloniki, fenced against grazing for 20 years. Besides *D. ischaemum* and *C. gryllus*, several other plants, mainly annuals were grown (Papanastasis, 1977). The study area has a semi-arid Mediterranean climate with 430 mm annual rainfall and 15°C mean air temperature.

Three quadrat shapes: square, circular and rectangular, and five quadrat sizes, 0.0625, 0.125, 0.250, 0.500 and 1m^2 were tested (for details on sizes see Papanastasis, 1977). From here on, the five quadrat sizes will be also referred to as 1, 2, 3, 4 and 5 from the smallest to the largest size respectively. Quadrats were made of iron rods of 8 mm diameter.

For each size and shape combination five samples were randomly taken from each of five blocks, 10X10 m each. In every sample, the number of individual plants from each of the two species *D. ischaemum* and *C. gryllus* was counted and their basal area was measured by taking two cross dimensions. In case that a specific plant partly fell in the sampling quadrat, it was measured for its basal area but counted only if one half or more of its area included in the quadrat. In addition, the aboveground biomass of each of the two species was estimated by clipping to the ground level, taking the samples to the laboratory and weighing them after ovendrying at 65°C. The determination of biomass was done only in three blocks, randomly selected out of the five. All the measurements were carried out at the end of the growing season (in June).

Density, basal area and weight values were subjected to analysis of variance; if significant differences were found, the Duncan test was used. To investigate the degree of association of the two species, their correlation coefficient was calculated. Finally, the detection of their distribution pattern was made by applying the variance/mean ratio (Kershaw, 1973).

3. Results and Discussion

3.1. GROWTH CHARACTERISTICS

Square quadrats tended to have the highest number of plants of *D. ischaemum* and rectangular the least while in *C. gryllus* the trend was the opposite. Differences however were not statistically significant. On the other hand, the number of individuals was increased as the quadrat size increased but only the last two large sizes (4 and 5) produced significant results in both species (Table1). In general, *D. ischaemum* had more individuals (about 50%) per unit area than *C. gryllus*.

TABLE 1. Mean number of individual plants of the two species in relation to quadrat size.

Species	Size of quadrat (m^2)				
	0.0625	0.125	0.250	0.500	1.000
Dichanthium ischaemum	1.5a[1]	2.7a	4.0a	8.0b	15.7c
Chrysopogon gryllus	1.0a	2.5a	3.8a	6.4b	10.8c
Total	2.5a	5.2a	7.8a	14.4b	26.5c

[1/] Means of each species followed by the same letter are not significantly different at $P \leq 0.05$

As in density, the basal area of *D. ischaemum* was largest in the square quadrat and of *C. gryllus* in the rectangular but the differences were statistically insignificant. On the contrary, quadrat size produced significant results (Table 2). However, the two smaller sizes (1 and 2) were not statistically significant between each other as well as 2 and 3 in both species.

TABLE 2. Basal area means (cm^2) of the two species in relation to quadrat size.

Species	Size of quadrat (m^2)				
	0.0625	0.125	0.250	0.500	1.00
Dichanthium ischaemum	35.29a[1]	109.74ab	202.10b	402.87c	702.03d
Chrysopogon gryllus	41.24a	83.76ab	293.49b	596.39c	1190.00d
Total	76.54a	195.01a	495.58d	999.25c	1892.63d

[1/] Means of each species followed by the same letter are not significantly different at P≤0.05

The failure of the shapes as well as of the smaller sizes to produce significant differences should be attributed to the high variation of vegetation caused by the sparseness of the two species sampled. Indeed, *D. ischaemum* occupied no more than 7% per m^2 while *C. gryllus* was a little bit more abundant (12%). By comparing the density and basal area data it comes out that individual *D. ischaemum* plants had an average basal area of 45 cm^2 while in *C. gryllus* it was twice as much larger (90 cm^2).

Weight tended to be higher in the square than in the rectangular shape in *D. ischaemum* but the differnces were not significant. On the contrary, the rectangular shape produced significantly higher weight than the square shape in *C. gryllus*. On the other hand, weight was increased as the quadrat size increased in both species but only the larger quadrats 3, 4 and 5 produced significant results (Table 3).

TABLE 3. Mean weight (g) of the two species in relation to quadrat size.

Species	Size of quadrat (m^2)				
	0.0625	0.125	0.250	0.500	1.000
Dichanthium ischaemum	8.61a[1]	23.78ab	45.59b	86.10c	166.85d
Chrysopogon gryllus	2.48a	5.01ab	13.47b	33.10c	47.00d
Total	11.09a	28.79ab	59.06b	119.2c	213.85d

[1/] Means of each species followed by the same letter are not significantly different at P≤0.05

By comparing weight with the basal area and density it comes out that the average ratio weight/basal area was about 0.23 in *D. ischaemum* and only 0.04 (g/cm^2) *in C. gryllus*; while the ratio weight/individual plant was about 10 and 4 g respectively for the two species. These results indicate that individual plants of *D. ischaemum* had smaller basal area but more weight and they were more abundant than plants of *C. gryllus*; also, the former species is more flexible and a more effective colonizer than the latter species. On the other hand, the biomass that both species produced represented about 67% of the total biomass recorded in the study area (Papanastasis, 1977).

3.2. CORRELATION

The correlation coefficients of the basal area of the two species were found negative while the ones of the last three largest quadrats were also statistically significant, almost regardless of the shape of quadrat (Table 4). These results suggest that as the presence of one species increases the other decreases apparently because they tend to occupy the same niches and they are, therefore, competitive.

TABLE 4. Correlation coefficients(r) of the basal area of the two species

Shape of quadrat	Size of quadrat (m^2)				
	0.0625	0.125	0.250	0.500	1.000
Square	- 0.26	- 0.35	- 0.50*	-0.48*	- 0.55*
Circular	- 0.21	- 0.25	- 0.54*	- 0.48*	- 0.31
Rectangular	- 0.26	- 0.26	- 0.47*	- 0.55*	- 0.59*

* Indicates significance at P≤0.05.

3.3. DISTRIBUTION PATTERN

Small quadrat sizes produced insignificant variance/mean ratios in both species regardless of the quadrat shape thus indicating random or regular distribution (Table 5). Significantly different ratios were found only in the two largest sizes. In *D. ischaemum*, the largest size 5 gave significant ratios in all shapes suggesting a contagious distribution of square shape since size 4 also produced a ratio significantly higher than unity. In *C. gryllus*, only the rectangular shape of the largest size gave significantly higher than unity ratio suggesting a rectangular contagious distribution.

TABLE 5. Variance/mean ratios of the two species

Species	Shape of quadrat	Size of quadrat (m^2)				
		0.0625	0.125	0.250	0.500	1.000
Dichanthium ischaemum	Square	0.88	0.80	0.64	1.71*	1.89*
	Circular	1.13	0.89	0.78	0.65	1.81*
	Rectangular	0.75	0.75	0.68	1.18	1.51*
Crysopogon gryllus	Square	1.25	1.13	1.23	2.73*	1.07
	Circular	0.91	0.95	1.17	0.98	1.08
	Rectangular	0.96	1.01	1.34	1.14	1.73*

*Indicates significance at P≤0.05.

The discrepancy of small and large sizes in showing the distribution of the two species may be attributed to the fact that pattern in plant species has several scales (Kershaw, 1973). It seems that the particular species tend to aggregate in large areas but the shape of their pattern is probably affected by their competition between each other as well as by other environmental factors.

5. References

Greig-Smith, P. (1983) *Quantitative Plant Ecology*, 3rd Edition, Butterworths, London.

Kershaw, K. A. (1973) *Quantitative and Dynamic Ecology*, 2nd Edition, Edward Arnold Publishers, London.

Noitsakis,B. (1988) Dynamique de croissance en relation avec le potentiel hydrique folaire chez deux especes fourrageres, *Agronomie* **8**(2):115-120.

Papanastasis, V. P. (1981) Species structure and productivity in grasslands of northern Greece, in N. S. Margaris and H. A. Mooney (eds.), *Components of Productivity of Mediterranean-climate Regions- Basic and Applied Aspects*, Dr W. Junk Publishers, The Hague, pp. 205-217.

Papanastasis, V. P. (1977) Optimum size and shape of quadrat for sampling herbage weight in grasslands of northern Greece, *J. Range Management* **30**(6):446-449.

IMPORTANT PLANT AREAS OF THE NATURA 2000 NETWORK OF GREECE.

E. PAPASTERGIADOU
The Goulandris Natural History Museum, Greek Biotope Wetland Centre, 14th Km Thessaloniki- Mihaniona, GR 57001 Thermi, Greece.

1. Introduction

NATURA 2000 is a European network of protected areas, and is the core of EU nature conservation policy. This planned "ecological" network of key sites is due to be fully in place by 2004. The EU has set minimum nature conservation standards through two wildlife Directives which apply across all Member states. The 1979 "Birds Directive" and the 1992 Directive on the Conservation of Habitats and Wild Fauna and Flora ("Habitats Directive") seek to counter a dramatic loss of wild habitats, flora, and fauna in Europe. The emphasis of the "Habitats Directive" (92/43/EEC) is to protect habitats, both in their own right and as locations of threatened species. Member states must be designate Special Areas of Conservation (SACs) for all Annex I habitats and Annex II species. The resulting network of SACs, together with Special Protection Areas designated under the Birds Directive, will comprise the NATURA 2000 network. The Directive's list of 410 plant species whose habitats are to be protected (on Annex II) fails to include many species which should have been there on scientific grounds but were excluded for political expediency. In Greece a total number of 296 sites were proposed to be included in this network. This list included all the protected areas of Greece such as National Parks, Biogenetic reserves, Ramsar wetlands, etc.

Wild plants are severely threatened all over Europe. Yet, despite widespread concern for environmental issues, far too little attention has been paid to the immense dangers they face. To this end, the first European Conference on wild plant conservation created the "PLANTA EUROPA" network which is a network of organisations whose mission would be to conserve the wild plants, both higher and lower, of Europe and their habitats. The "flagship" project of the above network is the identification and protection Europe's Important Plant Areas (IPAs). The number of known Greek plants (estimated at 6,300 species and subspecies of vascular plants), is one of the greatest within all the other European countries of similar or even larger areas (Strid & Kit-Tan, 1992). The most impressive is that the Greek Flora has almost the half number of the total number of species of the European land (IUCN-WWF 1988). However the Greek Flora has not only a great number of plant taxa, but many of these species are endemics and are found in smaller or bigger areas of Greece. The total number of Greek endemics is not thoroughly known. It is estimated that about 1300 taxa are endemics (20%) (Iatrou, 1986). Thirty nine plant taxa listed on Annex II to Directive 92/43/EEC, have been recorded in Greece, 23 of them are included in the Red Data list of rare and threatened plants of Greece (Phytos et.al. 1995). Thirty one of 39 taxa of the Directive are Greek endemic plants, 3 are subendemic and the other 5 are very rare in Greece. Furthermore, 1852 plant taxa have also been recorded in the relative database of the NATURA 2000 network as Other and Greek Important species.

The present work, is a preliminary approach to recognise the richest areas of important plant species among the 296 sites of the NATURA 2000 network in Greece. In the present study the most Important Plant Areas were determined mainly based on the total number of threatened, endemic, and/or protected species present in each site. The ranking of the NATURA sites of Greece according to the number of the threatened and endemic species as well as of those that are protected under International Conventions, has not taken into account the total number of plant species present in every site. It is possible that in this case a different ranking would be obtained.

2. Materials and Methods

Data on the plant species of the NATURA 2000 sites of Greece were obtained from the relevant Database "BioGreece '95" (Tsantas, 1995). For the purposes of the present study the following adaptations were made: The list of Other Important Plant Species with motivation A was updated following the publication of the Red Data List of rare and threatened plants of Greece (Phytos et.al. 1995). The total number of Other and Greek Important Plant species (1852) was adjusted to include the number of plant species listed in Annex II to Directive. After that, the total number of the Important Plant Species reached 1891 taxa. The motivation categories according to which each taxon was included in the field Important Plant Species are as follows:

A: Red Data List of rare and threatened plants of Greece.
B: Greek endemic taxa
C: Taxa included in the lists of International Conventions (Bern, Biodiversity, Cites '92), as well as plants listed in Annex II to Directive 92/43/EEC.
D: Important plants for other reasons i.e.: included in the WCMC database as threatened, European Red data list (1992), Greek Presidential Decree (67/81), Balkan endemics and subendemics, and rare in Greece.

3. Results and Discussion

Site assessment based on the total number of Important Plant Species is given. As it was expected, the mountaineous areas proved to be the most interesting sites from the floristic point of view. Nine sites host more than 150 plant taxa with motivation A, B, C or D. Thirteen sites host a number of taxa ranging between 101 and 105, 26 site between

51 and 100, 97 sites between 11 and 50, whereas 118 sites don't have more than 10 other important plant taxa. The 9 higher in ranking sites of this assessment have been identified as the most Important Plant Areas (IPAs) for Greece and are the following:

GR2550006 OROS TAYGETOS
GR2530001 KORYFES OROUS KYLLINI
GR1250005 ETHINIKOS DRYMOS OLYMPOU
GR1250001 OROS OLYMPOS
GR2450002 OROS GIONA
GR2130002 KORYFES OROUS SMOLIKAS
GR2320002 OROS CHELMOS KAI YDATA STYGOS
GR4340008 LEFKA ORI
GR2450001 ORI VARDOUSIA

The number of Important Plant Species per motivation category for the 9 richer sites are given in Fig. 1. The majority of species with motivation A are present in the site "LEFKA ORI" in Crete, whereas the majority of endemic species (motivation B) are present in "OROS TAYGETOS"," KORYFES OROUS KYLLINI", and "LEFKA ORI". Olympus National Park host the majority of species with motivation C whereas "OROS CHELMOS KAI YDATA STYGOS" host the greater number of species with motivation D.

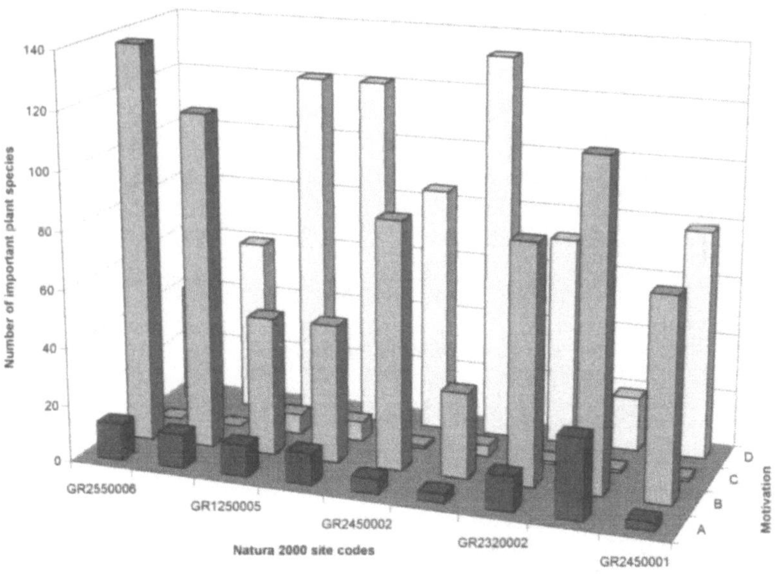

Fig. 1. Number of Important Plant Species per motivation category in the richest sites of the NATURA 2000 Network in Greece. (GR 2550006 OROS TAYGETOS, GR2530001 KORYFES OROUS KYLLINI, GR1250005 ETHINIKOS DRYMOS OLYMPOU, GR1250001 OROS OLYMPOS, GR2450002 OROS GIONA, GR2130002 KORYFES OROUS SMOLIKAS, GR2320002 OROS CHELMOS KAI YDATA STYGOS,GR4340008 LEFKA ORI, GR2450001 ORI VARDOUSIA).

If the assessment criterion of the "number of species included in the Red Data List" (motivation A), is alternatively applied, site ranking differentiate and the richest site become "LEFKA ORI" (28 taxa), followed by "IDI OROS (15 taxa), "OROS TAYGETOS" (13 taxa), "OROS CHELMOS KAI YDATA STYGOS" (12 taxa), "CHARADRA FLAMBOURITSA" (12 taxa), and "ETHNIKOS DRYMOS OLYMPOU" (11 taxa). Between all the NATURA 2000 sites, 7 host a number of Important Plant Species with motivation A ranging between 11 and 50, while 112 sites have no more than 10 of these species.

If the assessment criterion changes to the "number of endemic taxa" (motivation B), then the highest degree of endemism is observed on mount Taygetos (138 taxa), followed by the sites "CHARADRA FLAMBOURITSA" (116 taxa), and "LEFKA ORI" (113 taxa). Other 10 sites host between 51 to 100 endemic taxa, whereas in 76 and other 104 sites this number ranges between 11 to 50 and 1 to10 respectively.

The ranking of the sites according to the last two evaluation criteria, motivation C (sites which hosts species included in International Conventions), and D (sites hosting threatened or rare species), are as follows:

The sites with more than 10 taxa (motivation C) are "OROS OLYMPOS" (28), "KORYFES OROUS GRAMMOS" (15), and "KORYFES OROUS VORAS" (11).

The sites "KORYFES OROUS SMOLIKAS" (133 taxa), "KORYFES OROUS VORAS" (124) and "ETHNIKOS DRYMOS OLYMPOU"(121) have the majority of species with motivation D

Conclusively, it is suggested all the NATURA 2000 sites to be considered as IPA for Greece. Priority should be given to the following sites: GR255006 to GR4210007 (more than 10 important plant taxa) (BioGreece '95). However, it needs to be stressed out that for all these sites, and possibly more, further study need to be conducted. The characterisation of the above sites as the most important from a floristic point of view affects in no way the ecological value of other sites, with high biodiversity of habitat and species, where cosmopolitan flora grows.

Acknowledgements
The research and collection of data have been performed in the framework of the Life-Habitat project.

4. References

European Commission for Europe. (1992). European Red List of globally threatened animals and plants.
Iatrou, G. (1986) Contribution to the study of endemism of the flora of Peloponnissos. Diss. Univ Patras 310p
IUCN-WWF (1988). Biodiversity. The key role of plants. Plant Conservation Office, UK.
Official Journal of the European Communities No 206/7/1992: Council Directive 92/43/EEC of the 21 may 1992 on the conservation of natural habitats and wild fauna and flora.
Phytos, D., Strid,A., Snogerup, S., & Greuter, W. (1995). The Red Data Book of rare and threatened plants of Greece. WWF, 527 p.
Strid, A., & Kit Tan (1992). Flora Hellenica and the threatened plants of Greece. Opera Bot. 113: 55-67.
Tsantas, N. (1995) "BioGreece '95". NATURA 2000 Network Database manual. Commission of the European Communities DG XI, The Goulandris Natural History Museum- Greek Biotope Wetland Centre, 28p. + Annex 60 p.
WCMC (1993). Greece. Conservation status listing of plants. Compiled from the World Conservation Monitoring Centre Plants Database. Report.

BRYOPHYTES AND THEIR ENVIRONMENTAL CONDITIONS AT ENIPEAS (Mt. OLYMPOS) AND LYKORREMA (Mt. OSSA) STREAMS (GREECE)

B. PAPP[1], E. TSAKIRI[2] and D. BABALONAS[2]
[1] *Botanical Department of the Hungarian Natural History Museum, Budapest, Hungary.*
[2] *Department of Botany, School of Biology, Aristotle University, 540 06 Thessaloniki, Greece.*

1. Introduction

Greece is a bryologically scarcely explored area. Even in the more investigated parts of the country plenty of new records can be explored. The present study deals with the aquatic and riparian bryophyte vegetation of two mountain streams in Greece in correlation with water quality. Studies on the bioindication role of bryophytes in aquatic environment started in several countries in the mid-seventies. Several publications contain data on water quality indications (saprobity) of bryophytes (Frahm 1974, Empain 1973, 1978, Peñuelas & Sabater 1987, Vrhovšek et al. 1984, 1985).

2. Materials and Methods

The investigated streams are situated in Northern Greece (Fig. 1). Mt. Olympos belongs to the northcentral chorological district of Greece, while Mt. Ossa to the eastcentral. The stream valleys face to the east. The areas of the streams of the two mountains are different from geological point of view. At Enipeas valley (Mt. Olympos) crystalline limestones and dolomites dominate. At Lycorrema (Mt. Ossa) the substrate consists mainly of blue schist and gneiss.

The bryophyte vegetation influenced by the water was investigated during 1995/96. The frequency of each bryophyte species was estimated on a 1-4 scale along sections of the streams about 100 m long, at 4 zones according to the water level [zone a: (-10),(-5)-0 cm, b: 0-10 cm, c: 10-20 cm, d: 20-30 cm].

Chemical, physical parameters such as pH, conductivity, NO_3-N, NO_2-N, NH_4-N, PO_4-P, HCO_3^- were measured in April 1996. The methods of A.P.H.A. (1976), Alexiades (1977) (HCO_3^-) and Grasshoff (1976) (NH_4-N) were followed.

The sampling points were 4 on Mt. Olympos - Enipeas stream: OlPr - at Prionia waterfall (alt. 1100m), OlAD - at Agios Dionisios monastery (alt. 800 m), OlLb - before Litochoro village (alt. 275 m) and OlLa - after Litochoro (alt. 260 m). On Mt. Ossa - Lycorrema stream there were 3 sampling points: Os70 (alt. 700 m), Os40 (alt. 400 m) and OsKa - after Karitsa village (alt. 270 m).

Canonical correspondence analysis (CCoA) (Ter Braak 1988) was carried out to get information on the effect of environmental variables on the bryophyte vegetation.

3. Results and Discussion

3.1. FLORISTICAL RESULTS

Five (5) liverworts and fifty-eight (58) mosses were found along the streams. In the list of table 1 are included all the species that were collected along the two streams and not only those of the sampling sites.

One taxon, *Orthotrichum cupulatum var. riparium*, found in Enipeas valley, is new to Greece. Eleven (11) species were new in the eastcentral region of Greece where Mt. Ossa belongs to. One species was new to the northcentral region according to Düll (1995). They are marked with asterisk in the list.

Also *Bryum cellulare* is a vulnerable species in Europe according to the Red Data Book of the European Bryophytes (1995).

3.2. WATER CHEMISTRY

The different bedrock chemistry of the two streams is reflected in the HCO_3^- concentration. At the sites of the calcareous Mt. Olympos the concentration was 5-7 times higher (146-195 mg/l) than in Mt. Ossa (12-30 mg/l).

The NH_4-N concentration was about 10 times higher (0.122 mg/l) at the sampling point after Litochoro village (OlLa) than at the other sampling points of the stream (0.005-0.014 mg/l). The PO_4-P concentration was more than 7 times higher (0.052 mg/l) than in the other sampling points at Mt. Olympos (0.007 mg/l) and 3-4 times higher than at the sampling points of Mt. Ossa (0.013-0.015). Such changes of the parameters indicate pollution of the water due to human activities of Litochoro village.

On Mt. Ossa such clear connection between the changes of chemical parameters and the effects of Karitsa village was not found. But the NH_4-N concentration was 2-4 times higher (0.014 mg/l) after the village (OsKa) than in the upper sections of the stream (0.003-0.005 mg/l).

The NO_3-N concentration was usually 7-10 times higher (0.040-0.047mg/l) at the sampling points of Mt. Olympos than at Mt. Ossa. But at the sampling point after Karitsa village (OsKa), on Mt. Ossa, a higher value (0.037 mg/l) was measured than at the other sampling points of the mountain (0.004-0.007 mg/l).

3.3. BRYOPHYTE VEGETATION AND WATER QUALITY

Canonical correspondence analysis was carried out for the comparison of bryophyte vegetation and the environmental variables of the water (Fig. 2, 3). Samples of b and c level were used, as the species number in level a was too low and level d was not affected directly by the water.

The bryophyte vegetation of the sampling sites of the upstream sections of Enipeas valley at Mt. Olympos (OlPr, OlAD) correlates well with the altitude. On both levels *Cratoneuron commutatum* is predominant. *Bryum pseudotriquetrum* and *Cinclidotus fontinaloides* are also important species. Besides them some colonists are also frequent

TABLE 1. List of the species found along the streams.

	Hepaticae	Olympos	Ossa
1	Jungermannia sp.	+	
2	Lunularia cruciata (L.) Dum. ex Lindb.	+	+
3	Pellia endiviifolia (Dicks.) Dum	+	+
4	Porella cordaeana (Hueb.) Moore		+
5	Pressia quadrata (Scop.) Nees	+	
	Musci	**Olympos**	**Ossa**
6	Amblystegium humile (P. Beauv.) Crundwell	+	
7	Amblystegium varium (Hedw.) Lindb.	+	
8	Brachythecium rivulare B., S. & G.	+	+
9	Brachythecium rutabulum (Hedw.) B., S. & G.	+	+
10	Bryum alpinum With.	+	+
11	Bryum barnesii Wood	+	
12	Bryum caespiticium Hedw.	+	
13	Bryum capillare Hedw.		+
14	Bryum cellulare Hook. ex Schwaegr.		+
15	Bryum pallens (Brid.) Sw. ex Roehl.	+	
16	Bryum pseudotriquetrum (Hedw.) Gaertn., Meyer & Scherb.	+	+
17	Bryum sp.		+
18	Cinclidotus aquaticus (Hedw.) B. & S.	+	
19	Cinclidotus fontinaloides (Hedw.) P. Beauv.	+	
20	Cinclidotus riparius (Brid.) Arnott		+
21	Cirriphyllum crassinervium (Tayl.) Loeske & Fleisch.	+	+
22	Cratoneuron commutatum (Hedw.) Roth var. commutatum	+	
23	Cratoneuron filicinum (Hedw.) Spruce	+	+
24	Dicranum sp.	+	
25	Didymodon fallax (Hedw.) Zander	+	+
26	Didymodon luridus Hornsch.	+	
27	Didymodon rigidulus Hedw.	+	
28	Didymodon sinuosus (Mitt.) Delogn.	+	
29	Didymodon tophaceus (Brid.) Lisa	+	
30	Didymodon vinealis (Brid.) Zander var. flaccida (Bruch & Schimp.) Zander	+	+
31	Eucladium verticillatum (Brid.) B., S. & G.	+	
32	Eurhynchium hians (Hedw.) Sande Lac. var. hians		+
33	Eurhynchium hians (Hedw.) Sande Lac. var. rigidum (Boul) Duell		
34	Eurhynchium pumilum (Wils.) Schimp.	+	+
35	Eurhynchium speciosum (Brid.) Jur	+	
36	Fissidens cristatus Wils. ex Mitt.	+	
37	Fissidens rivularis (Spruce) B., S. & G.		+
38	Funaria hygrometrica Hedw.	+	
39	Grimmia pulvinata (Hedw.) Sm.	+	
40	Grimmia trichophylla Grev.	+	
41	Gymnostomum aeruginosum Sm.		+
42	Gyroweisia reflexa (Brid.) Schimp.	+	
43	Hygroamblystegium tenax (Hedw.) Jenn	+	+
44	Hygrohypnum luridum (Hedw.) Jenn.	+	
45	Isopterygium pulchellum (Hedw.) Jaeg. & Sauerb.	+	
46	Isothecium alopecuroides (Dubois) Isov.		+
47	Orthotrichum cupulatum Brid. var. riparium Hueb.	+	
48	Philonotis calcarea (B. & S.) Schimp.	+	
49	Plagiomnium undulatum (Hedw.) T. Kop.		+
50	Pohlia melanodon (Brid.) J. Shaw	+	
51	Racomitrium aciculare (Hedw.) Brid.		+
52	Rhynchostegiella curviseta (Brid.) Limpr.		+
53	Rhynchostegium confertum (Dicks.) B., S. & G.	+	+
54	Rhynchostegium megapolitanum (Web. & Mohr) B., S. & G.	+	+
55	Rhynchostegium murale (Hedw.) B., S. & G.	+	
56	Rhynchostegium riparioides (Hedw.) C Jens	+	+
57	Schistidium apocarpum (Hedw.) B. & S.	+	+
58	Scleropodium touretii (Brid.) L. Koch	+	
59	Scorpiurium circinatum (Brid.) Fleisch. & Loeske	+	+
60	Thamnobryum alopecurum (Hedw.) Gang. var. alopecurum		+
61	Tortula muralis Hedw.	+	
62	Weisia sp. 1	+	
63	Weisia sp. 2	+	

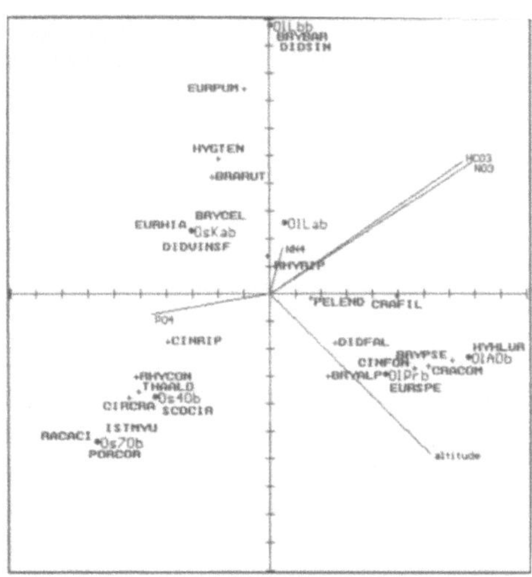

Fig. 2. Triplot of the CCoA ordination of level **b**.

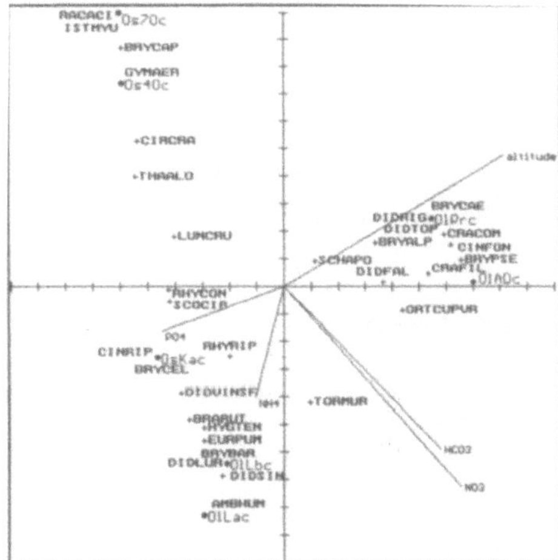

Fig. 3. Triplot of the CCoA ordination of level **c**.

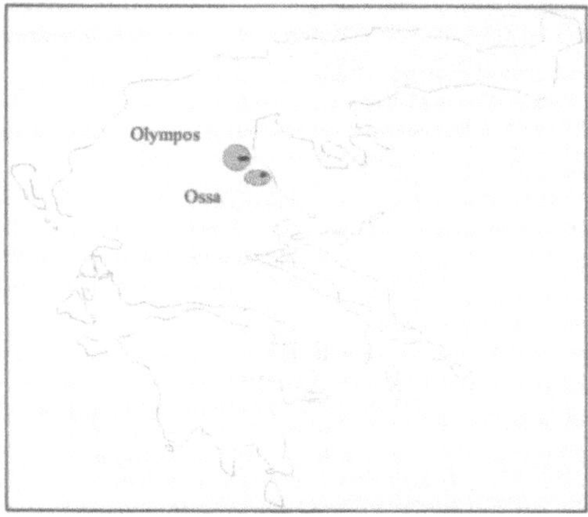

Fig. 1. Location of the study area.

on level b (e.g. *Didymodon fallax, Bryum alpinum*). On level c *Cratoneuron filicinum* and another colonist *Didymodon tophaceus* became important.

The lower concentrations of HCO_3^- and NO_3-N are characteristic features of the sampling sites at Mt. Ossa (Os70, Os40). Here *Thamnobryum alopecurum*, *Cirriphyllum crassinervium*, *Racomitrium aciculare* are important elements of the bryophyte vegetation. The species mentioned above are not really aquatic mosses. This fact indicates the seasonal supply of water. There was no water during summer at these upstream sections. Despite of this some aquatic bryophytes are also frequent here e.g. *Cinclidotus riparius*.

The bryophyte vegetation of the sampling sites near villages (OlLa, OsKa, OlLb) is under anthropogenic influence, pollution in the widest sense. It is indicated by higher NH_4-N and sometimes higher PO_4-P concentrations. *Rhynchostegium riparioides* is predominant on both levels. *Hygroamblystegium tenax*, *Brachythecium rutabulum*, *Eurhynchium pumilum* are also important constituents at some sites. Colonist species also appear (e.g. *Bryum barnesii, Didymodon sinuosus*).

The water pollution is manifested by the decrease of species number and the changes in the species composition of bryophytes. Therefore, they can be used as indicator organisms of the changes in water ecosystems. The sources of pollution and disturbance can be traced by bryophytes.

Acknowledgement - Many thanks to Dr. Miklós Rajczy for the computation of the data. The study was supported by IKY (Idrima Kratikon Ypotrofion, Greece) and MÖB (Hungarian Scholarship Committee).

4. References

Alexiades, K. (1977) Physical and chemical analysis of the soil. Aristotle University of Thessaloniki, Thessaloniki.

American Public Health Association (1976) Standard methods for the examination of water and wastewater, 14th ed., New York.

Düll, R. (1995) Bryophytes of Greece. *Bryologische Beiträge* **10**, IDH-Verlag.

Empain, A. (1973) La végétation bryophytique aquatique et subaquatique de la Sambre belge, son déterminisme écologique et ses relations avec la pollution des eaux. *Lejeunia N.S.* **69**: 1-58.

Empain, A. (1978) Relations quantitatives entre les populations de bryophytes aquatiques et la pollution des eaux courrantes. Définition d'un indice de qualité des eaux. *Hydrobiologia* **60**: 49-74.

Frahm, J.P. (1974) Wassermoose als Indikatoren für die Gewässerverschmutzung am Beispiel des Niederrheins. *Gewässer und Abwasser* **53/54**: 91-106.

Grasshoff, K. (1976) Methods of seawater analysis. Verlag & Chemie, New York.

Peñuelas, J. & Sabater, F. (1987) Distribution of Macrophytes in relation to environmental factors in the Ter river, N.E. Spain. *Int. Revue ges. Hydrobiol.* **72**: 41-58.

Ter Braak, C.J.F. (1988) CANOCO. A FORTRAN program for canonical community ordination by partial detrended canonical correspondence analysis. GLW. Wageningen. The Netherlands.

Vrhovšek, D., Martincic, A. & Kralj, M. (1984) The applicability of some numerical methods and the evaluation of Bryophyta indicator species for the comparison of the degree of pollution between two rivers. *Arch.Hydrobiol.* **100**: 431-444.

Vrhovšek, D., Martincic, A., Kralj, M. & Štremfelj, M. (1985) Pollution degree of the two alpine rivers evaluated with Bryophyta species. *Biološki vestnik* **33**: 95-106.

European Committee for Conservation of Bryophytes (1995) Red Data Book of the European Bryophytes. ECCB, Trondheim.

THE SERPENTINE FLORA IN THE EASTERN RHODOPES MOUNTAINS (BULGARIA)

D.K. PAVLOVA[1], E.K. KOZUHAROVA[2], D.S. DIMITROV[1]
[1] *University of Sofia, Department of Botany, Faculty of Biology*
bul. DraganTzankov 8, 1421 Sofia
[2] *Institute of Botany, Bulgarian Academy of Sciences*
bul. Acad. G. Boncev bl. 23, 1113 Sofia

1. Introduction

The object of the present investigation is the serpentine flora in the Eastern Rhodopes Mts. (Cherna planina). Until now no special attention was paid to the serpentine flora in Bulgaria. An exception is the find of *Festuca oviniformis* Vett., in the serpentines in the Eastern Rhodopes Mts. (Kozuharov, 1985). The interest in the serpentine flora arouses by the fact that the bulgarian serpentine bodies are older with a greater chemical variety of the rocks, the soils and the plants respectively (Kozuharova, 1984) compared to those found in Albania, Greece and former Yugoslavia. Our pilot results do not claim to comprehensiveness.

2. The Study Area and Methods

The serpentine complexes in the Cherna planina were selected as being the largest and the most typical ones. The serpentine bodies with their specific ring-shaped distribution are incorporated into the pre-Cambrian crystalline complex and the dimensions vary from several meters up to 10-12km^2 (Kozuharova, 1984).

Four transects for our pilot study were selected in May-July 1996: I- village Avren; II- village Kazak; III- village Zulti chal; IV- village G. Kamenjane (Fig. 1). The displacement is between 300 and 650 m a.s.l.

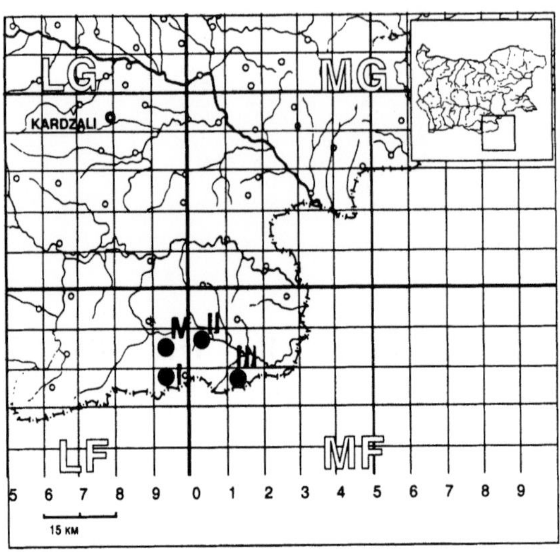

Figure 1. Map of Bulgaria and the study area

The transacts pass terrains which have an outlet to the basic rock. The material collected is deposited in the Herbarium of the Sofia University (SO). The floristic analysis was performed using the approach of Meusel (1965). The calculation of the percentage values was done on basis of the data summarized by Kozuharov et al. (1992).

3. Results

3.1. GENERAL CHARACTERISTICS OF THE VEGETATION

A large part of the serpentines is covered by coppice oak woods. On southern slopes they change into shiblyak patent in some places. Another part of the serpentines mainly on southern and western slopes completely lacks tree vegetation and the herb flora is represented by small dispersed or mosaic populations.The indigeneous vegetation is of south-eastern european or asiatic type dominated by *Quercus frainetto* Ten., *Q. pubescens* Willd. and *Q. cerris* L.. In the shiblyak communities are also found *Jniperus oxycedrus* L., *Fraxinus ornus* L., *Acer campestre* L., *Pistacia terebinthus* L., *Paliurus spina-cristi* Mill., *Cystus incanus* L., *Dichantium ischaemum* L. etc. The herb layer of the coppice woods is composed of mesophilous plants most of them with middle-european and transitional-mediterranean character.

3.2. TAXONOMIC STRUCTURE OF THE SERPENTINE FLORA

In the study area are established 281 species, subspecies and varieties belonging to 167 genera and 53 families (annuals- 62, perennial herbs - 186, semi-shrubs - 4, trees and shrubs - 29). The families with the highest number of species are *Poaceae* (34), *Fabaceae* (31), *Asteraceae* (21), *Caryophyllaceae* (21), *Scrophulariaceae* (13), *Rosaceae* (12), *Cyperaceae* (12). The grass family is represented by 22 species followed by *Asteraceae* (15), *Fabaceae* (13), *Lamiaceae* (12), *Caryophyllaceae* (11), *Scrophulariaceae* (9), *Apiaceae* (8) *Brassicaceae* (5) , *Liliaceae* (5). The richest genera in species are *Carex* (10), *Trifolium* (8), *Bromus, Potentilla, Galium* (5). Altogether 19 taxa appeared as new ones for the flora of the Eastern Rhodopes Mts.

The phytogeographical specificity of the serpentine flora is determined by the following basic groups: circumboreal (8.54%), middle-european (13.16%), mediterranean (32.38%), euroasiatic (10.67%), pontic-middleasiatic (16.72%), oriental-turanian (5.69%), euroafrican (2.84%), cosmopolitans (1.77%) elements.

The Balkan endemics *Armeria rumelica* Boiss., *Digitalis viridiflora* Lindl., *Dianthus moesiacus* Vis. et Panc., *Fritillaria pontica* Wahl., *Linum tauricum* Willd. *ssp. bulgaricum* (Podp.) Petrova, *Verbascum adrianopolitanum* Podp., *Galium mirum* Rchb., *Dianthus gracilis* S. et S., *Moenchia graeca* Boiss. et Heldr., *Saponaria stranjensis* Jord., *Thymus bracteosus* Vis. ex Bentham and the bulgarian endemic *Oenanthe millefolia* Jka. are established.

The taxa - *Oenanthe millefolia* Jka, *Convolvulus boissieri* Stend. *ssp. parnassicus* (Boiss. et Orph.) Kuzm., *Fritillaria pontica* Wahl., *Linum tauricum* Willd. *ssp. bulgaricum* (Podp.) Petrova, *Verbascum adrianopolitanum* Podp., *Saponaria stranjensis* Jord., *Bupleurum gerardii* All., *Limodorum abortivum* (L.) Shwartz with a rank "rare" as started in the Red Data Book for Bulgaria are present in the composition of the serpentine flora.

The species *Apera interrupta* (L.) Beauv. and *Bromus intermedius* Guss. dubious for the bulgarian flora are also confirmed. The species *Thymus bracteosus* Vis. ex Bentham is a new one for the country.

4. Discussion

The present-day serpentine flora in the Eastern Rhodopes Mts. has not undergone long time for its development. According to Stefanov (1943) 500-1000 years ago had existed oak and black pine forests. These forests did not enable the penetration of heliophilous species. This has happened after the vast deforestation by Man. The first herb immigrants have irradiated from the Northern Greece (Stefanov,

1943) where the serpentine terrains had been earlier deforested (Turrill, 1929). All this information leads to the conclusion about the young age of the active speciation processes on the serpentines. The short biological history of the serpentine terrains and the local climate ensure the prerequisites for the refugial character of the processes rather than for an intensive fireplace. Nevertheless even at this altitude in the xerothermic oak belt the speciation processes are active and result in the formation of a specific flora.

The basic influence in the formation of the serpentine flora is exerted by the mediterranean, submediterranean, oriental-turanian and middle-european floristic elements. These are mainly xeromorphic species which find suitable phytoclimatical conditions for their growth provided by the serpentine rocks and the soils above them. The relatively high participation of endemics is synchronous with the analyses performed by Brooks (1987).

The preliminary observation on the serpentine flora shows the mosaic character of the plant populations. Beyond doubt the future investigations will enrich our knowledge on the serpentine flora in Bulgaria.

5. Acknowledgements

The authors are greateful to the deceased Prof. Dr. St. Kozuharov for his idea and assistance to start this research.

This paper is a contribution to Project B-511 financed by the National Science Fund in Sofia.

6. References

Brooks, R. (1987) *Serpentine and its vegetation*, Groom Helm, London.

Kozuharov, St. (1985) *Zhitnite rasteniya (sem. Poaceae) v Balgariya - Genofond, razprostranenie i evolyutsionni strategii*, Dissertation. Inst. Bot., Bulg. Acad. Sci., Sofiya.

Kozuharov, St., Andreev, N., Anchev, M., Markova, M., Peev, D. and Petrova, A. (1992) *Opredelitel na vishite rasteniya v Balgaria*, Nauka i izkustvo, Sofiya.

Kozuharova, E. (1984) Proizkhozhdenie i struktornoe polozhenieserpentinizirovanykh ultrabazitov dokembriiskoi ofiolitovoi asotsiatsii v Rodopskom massive. I Geologicheskoe polozhenie i sostav ofiolitovoi asotsiatsii, *Geologika Balkanica* **14 (4)**, 9-36.

Meusel, H., Jager, E., and Weinert E. (1965) *Verleichende Chorologie der Zentraleuropaischen Flora*, **1,** VEB Gustav Fischer Verlag, Jena.

Stefanov, B. (1943) Fitogeografski elementi v Balgaria, *Sbornik na BAN* **39**, 1-509.

Turrill, W. B. (1929) *The Plant-life of the Balkan Peninsula,* Clarendon Press., Oxford.

MORPHOLOGICAL AND PALLINOLOGICAL RESEARCHES ON CORYLUS COLURNA L.(BALKAN HAZELNUT)SPREADING NATURALLY IN TURKEY

M. SARIBAŞ
Zonguldak Karaelmas University Faculty of Forestry, 74100, Bartın, Turkey

1. Introduction

Hazelnut tree grows prevalently in Anatolia, especially in Black Sea region. İn places close to sea cultural hazelnuts; in forest areas natural hazelnuts are generally seen. People of Blacksea region establish hazelnut gardens by usually opening the areas from forest. Hazelnut, walnut and olive are among nourishing sources of Anatolian people. These three fruit trees are also common to the Mediterranean countries. Likewise, hazelnut is one of the most important export products of Turkey.

Corylus colurna L. which is the subject of our research, is defined as tree. Being appropriate to the name of " The Botany Congres", this species is also named as "Balkan hazelnut". The names Bear Hazelnut, Gökbulatan Hazelnut and Turkish Hazelnut are sometimes used. Although cultural hazelnuts usually grow to 5-6 m, the tree ones can grow up to 25 m. on average (Güner, 1993).

This taxon belongs to Betulaceae family's, Coryleae Sub-family, Corylus L. species of Angiospermaes.The synonyms of them are C.arborescens Mönch; C. byzantina Clus. As mentioned above, it is a tree which has a smooth stem, with diametre of 1,5 m. and lives about 80-100 years.

The wood properties of *C. colurna L.* had been investigated by Shucla-Sharma (1990) and results were compared with *Tecdonia grandis*. Besides, there are some publications about this species as ornamental tree (Koltsov,1988; Kosenko, 1991). The publications about *C. colurna L.* pollens could not been found, but Aytuğ (1971) and Sarıbaş (1993) carried out research involving a similar species, *C. colurna L.*

2. Material and Methods

Polen and wood materials had been taken from *C. colurna L.* trees which exist in 1200 m. elevation and within the borders of west Blacksea. Bolu province Central Forest exploitation Kale region. Pollen materials had been observed starting from January 1997 and the pollens were collected in February (Metcalfe and Chalk, 1957). The measurement of the pollens were made in preparations prepared with Woodehouse method (Woodehouse, 1954) According to this method, wood pieces in size of match stick were put in seperate crucibles and heated lightly after putting pure nitric acid. Then, the measurement were been realised by making preparation of fibres.

3. Results

3.1. NATURAL DISTRIBUTION OF *C. colurna L.* IN TURKEY

This taxon is found widely in the Nortwest region of Anatolia. İt is possible to see them individually or in small groups both in deciduous forests like Oak, Beech, Maple and in mixed forests like Beech, Fir. Their vertical distribution is between about 800-1700 height.

3.2. EXTERNAL MORPHOLOGICAL PROPERTIES OF *Corylus colurna L.*

This type of Corylus is a tree which can grouw up about 15-25 m. height. There is a monumental Corylus tree with 26,2 m height and 2,90 m. diameter in Bolu (Güner, 1993). So, this district has been protected as.

"Biogenetic Reserve Field" by the Ministry of Forest. Tall columnar tree to 25 m, with corky, deeply furrowed dark, grey bark. Young (annotinous) schoots tawny, hairy (pubescent and glandular-setose). Leaves ovate to broadly ovate, rarely obovate, 7-14x5x11 cm, with 7-10 pairs of veins, slightly lobulate, biserrate or crenate-serrate, acute to cuspidate at apex, cordate to obliquely cordate at base, dark green and glabrous above, pale green or pale glacous-green and glabrous beneath except for pilose primary veins and hairs in axils. Petioles 2,5-4 cm, glandular-hairy and puberulent. Stipules lanceplate, acuminate. Staminate catkins to 12 cm, with ovate - acuminate bracts. Involucre divided almost to base into many long acuminate serrate lobes, pubescent or glandular-setose, 2-3x as long as nut. Nuts sub-globose to ovoid, flattened laterally, 15-20x10-18 mm, clusters of 2-6, very thick. Valed . Pericarp is thicker. than other hazelnut variety. The most part of nut that is connected to cupula is called "flue part". This part reaches aproximethly to half of nut. This property is caracteristic *for. C. colurna L.*

3.3. THE WOOD PROPERTİES OF *C. colurna L.*

Its wood are sound, durable and heartwood is red. It is used in furniture manufacture preparation of some tools and carpentry.

It is investigated only wood fiberfrom between inside morphologic properties. D. mensions of fibres orman given in table 1. Wood fiber lenghts are minumum 680 micron, maximum 1700 micron and average 1190 micron as noted in it.

TABLE 1: Dimension of Fibres *C. colurna L.*

Properties of Fibers	Fiber Length (mic)			Fiber width (mic.)			Lumen width(mic)			Cell-wall thicknes (mic)		
	min	mean	max	min	mean	max	min	mean	max	min	mean	max
R1 (1)	700	1178	1650	15,2	27,3	35,4	7,6	16,9	22,9	2,6	5,2	7,8
R1 (2)	740	1115	1650	17,7	24,8	30,4	10,1	15,9	20,2	2,5	4,7	6,3
R1 (3)	700	1263	1420	20,2	25,9	35,4	10,1	15,5	22,8	2,5	5,2	8,8
R2 (1)	940	1290	1900	20,3	31,4	32,9	10,1	17,0	22,8	2,5	4,8	6,8
R2 (2)	970	1364	1700	19,2	27,2	38,0	7,6	14,3	25,3	2,5	5,1	11,4
R2 (3)	860	1307	1770	15,2	27,9	38,0	10,1	18,4	22,8	2,6	4,8	9,8
R3 (1)	680	1236	1450	20,2	26,2	35,4	7,6	15,9	22,8	2,4	5,0	8,5
R3 (2)	740	1067	1400	22,8	28,0	32,9	15,2	18,4	22,8	3,8	4,6	6,4
R3 (3)	680	1236	1450	20,2	26,2	35,4	7,6	15,2	22,8	2,4	5,0	8,9

3.4. POLLEN PROPERTIES OF *Corylus colurna L.*

The measuring of the fresh pollen was done using Woodehouse Method after one month. The results of the measuring and also some microphotographs *of C._colurna L.* have been given below (Fig. 1).

<u>*Corylus colurna L.*</u>
Origin of Sample= A4; Bolu, Kale around, 1200 m, M.Sarıbaş and Z. Kaya
Collection Date= 20.2.1997
Pollen Type= Triporatae
Pollen Shape= Suboblata, P/E = 0.87 (W).
Exine= 1.19 μ (W).
Aperture= Pori; small and not clear.
Structure= Tectatae
Sculpture= Granule;
İntin= Thin, Exlint ≅ 2/1
<u>Fresh Pollen</u>

	M	&
P:	25.60	± 1.99 μ
E:	29.10	± 1.25 μ

plg:could not be measured
plt: could not be measured
I:6.49 ± 0.45 μ
i:Thin, could not be measured
$\sqrt{x}$: 1.19 ± 0.27 μ

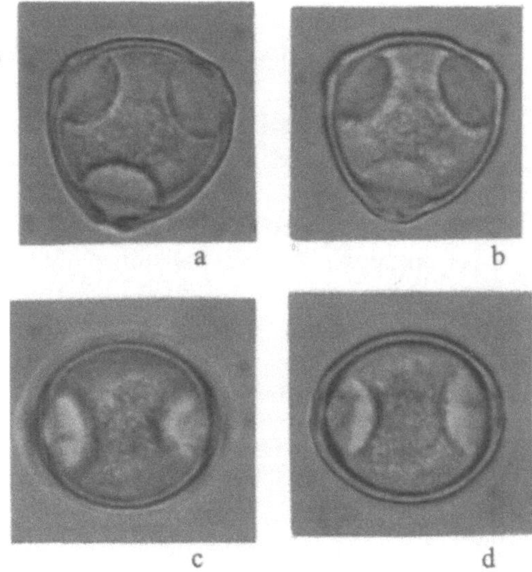

Figure 1. The pollen of Corylus colurme L. (a,b; polor wiew, c,d; profile) x 800.

4. Conclusions

C.colurna L. is present widely in the northwest region of Anatolia, mostly, in soliter. It has precious wood for making furniture and can be used for many kinds of works. For this reason, it has recently been consumed excessivelly. Therefore, The Ministry of Forestry has preserved a region that is 347 ha, as 'Biogenetic reserve' in Bolu (Anşin, 1989).

İn this research, it has been obtained some information about *C.colurna L.* pollen for the first time. There is not any other research abaut this species in Turkey. Nevertheless, it has been compared with *C.avellana L.* pollen from the same category. There are some differences between their dimensions. For example; while p is 25,60 ± micron in *C.colurna L.*, p is 23.49 ± 1.56 micron in *C.avellana L.* (Aytuğ, 1971)

Also the dimensions of the fibres were measured for the first time. Statistical results belonging to the fibers are compared with *Betula L.* and *Alnus L.* İt has been seen that they had similar data. For exemple the length of fibre of *Betula papirifera* is 1350 micron; and *Alnus rubra* is 1190 micron; of *C.colurna L.* 1190 micron (Bozkurt, 1992).

What apitythat we dont know the worth of it. Also it has been cut by forest villagers continiously in Turkey. Furthermore, they aren't used for Landscape planning. Their fruits are used to produce nut-sugar, a product of Bolu province. It has both esthetical and touristical value. Suggestions:

The natural expansion regions of this plant should be determined in the world and Turkey.

- The important natural groups of *C. colurna L.* schould be preserved.
- It shold be used in plantatıon projects and landscape planning and produced vegetatively.
- The genetic properties of *C. colurna L.* shoud be investigated.
- The properties of seeds and pollen of this plant and its geographical variations should be studied.

5. References

Aytuğ, B. (1971). Atlas Des Pollens Des Environs D' İstanbul. İ.Ü. Yayın No: 1650. İstanbul.

Anşin, R. (1989). Türkiye'de Ormansızlaşma ile birlikte Yaşamlarını Giderek Yitirmekte Olan Önemli Odunsu Bitkilerimiz Çevre ve Orman Dergisi, Cilt 5, Sayı 6.5. 17-19, Ankara

Bozkurt, Y. (1992). Odun Anatomisi, İ.Ü. Yayını, İstanbul.

Davis, P.H. (1982). Flora of Turkey and the East Aegean İslands. Vol VII, P. 686, Edinburgh.

Güner, A. (1993). Monumental Trees of Turkey, Karaca Ar. Mag. Volum:2, pp. 22-24, Yalova.

Koltsov, A.F. (1988). Rooting of Turkısh tilbert Summer cutting usıng auxin stimulants of rhizogenesis. Vliyanie Regulyatorov Rosta na Razvitie produktivnostr Rastenii., pp. 121-134, Stavropolr, USSR.

Kosenko, I.S. (1991). Rooting of Turkısh tilbert Summer cutting usıng auxin stimulants of rhizogenesis. Vliyanie Regulyatorov Rosta na Razvitie produktivnostr Rastenii., 121-134, Stavropolr, USSR.

Metcalfe, C.R. - Chalk, L., (1957). Anatomy of Dicotyledons. Volum 1, pp.542-549, Oxford.

Sarıbaş, M., (1993). Recherches Morphologiques sur les Pollens des Peupliers. Doğa-Tr. J. of Agricultural and Forestry 17, 707-714.

Shukla, N.K. Sharma, R.R. (1990). Physical and mechanical properties of Corylus columa from Jubbal Forest Division. FRI, Dehra Dun, India.

Wodehouse, R.(1954) Pollen Grains, Their Structure, İdentifacation and Significance and Medicine, Newyork.

COMMITTEE FOR EXPLORATION OF THE FLORA AND VEGETATION OF SERBIA - THREE DECADES OF WORK

M. R. SARIĆ[1], N. DIKLIĆ, O. VASIĆ
[1]*Serbian Academy of Sciences and Arts, Belgrade, Yugoslavia*
Natural History Museum, Belgrade, Yugoslavia

1. Introduction

The Committee for Exploration of the Flora and Vegetation of Serbia of the Serbian Academy of Sciences and Arts was founded in 1966. The object of the Committee is to organise and supervise both fundamental and applied studies of flora and vegetation on the territory of Serbia.

What moved the Serbian Academy to establish the Committee and steer a part of its activities towards exploring the plant world in Serbia?

Perhaps the origin of the Committee for Exploration of the Flora and Vegetation should be sought in the remote past in the spark briefly kindled over the hitherto unknown flora of the present territory of Serbia.

The year was 1580. On his journey through Serbia, S. von Hauzen from Nürnberg found in the vicinity of Belgrade a plant which he entrusted to Clusius who described the plant as "*Crocus vernus latifolius flavo flore*" (Jurišić 1905) According to the modern classification and nomenclature that was *Crocus flavus* Weston 1771 (Mathew 1980).

The first and the oldest floristic record from Serbia, for a long time it was the only one. A series of untoward circumstances directly or indirectly postponed a detailed exploration of Serbian flora until the middle of XIX century.

At the beginning of XIX century, however, Serbian flora was still unexplored and virtually unknown to the scientific community. But at that time, European naturalists developed an interest for the Balkan Peninsula, the most part of which was still under Ottoman rule.

Thus the naturalists such as Friwaldsky (1835, 1836, 1832-38) and Boué (1840) cited the plants they observed on their journeys through Serbia. Unfortunately, these data are sparse and random.

Grisebach (1843-1845) was only slightly more attentive to the flora of this region. He listed the meagre 140 species, 134 of which were flowering plants, and also described several species from Serbia which were rare, or new to science. However, he visited only a part of Serbia, and inferring from the length of his list, he obviously collected plants *an passe*. What casts a shadow over his explorations of plant world of Serbia is his observation relating to the character and scientific importance of Serbian flora, put into the following words: "Serbia ... shows little promise to botanists".

Today we can only speculate about the reasons for such a confident, but certainly a hasty and superficial judgement. At any rate, he paid no attention to geological, pedological, geographical, and climatic features of Serbia. Had he done that, he would have realised that that area promised far more interesting flora.

In the years and decades following Griesbach's decisively negative assessment, numerous investigations (Pančić 1856, 1874, Visiani 1860, Petrović 1882) demonstrated that he had judged both wrongly and unfairly. Serbian flora, being exceptionally rich and complex, as well as one of the oldest floras, occupies an important place not only in Europe, but on a global scale.

Josif Pančić was the first to recognise it: "The flora of the Principality of Serbia is unusually rich, not so much in the number of species, of which nothing can be said with accuracy, as in its monotypes or plants belonging to it only, and a multitude of species reaching Serbia from east and south" (Pančić 1874).

The middle of XIX century is the turning point in the studies of Serbian flora. The age when people recognised and used plants from experience, when scientific results were scant, was succeeded by an age of intensive scientific investigations.

That new age was marked by the name, person, and scientific results of the father of Serbian botany - Josif Pančić (1814-1888).

His explorations, pioneer though they were, marked the golden age of Serbia's botany. Pančić did not only increase the number of species known to occur in Serbia. More importantly, he demonstrated that Serbia's flora comprises numerous endemites and relicts, envoys from the bygone days. Likewise, that it is very complex taxonomically, chorologically, and evolutionary. Pančić discovered a total of 47 valid species new to science (Mayer & Diklić 1967), which he described from the material mainly from Serbia.

The crowning achievement of Pančić's floristic studies is the Flora of the Principality of Serbia published in 1874, while its Supplement appeared ten years later (Pančić 1884).

The Flora of the Principality of Serbia, being the first monograph of that kind, lays claim to more than historical importance in Serbian botany. Its value, and the influence it exerted on the future development of botany are manifold. With the Flora, Pančić firmly established Serbian botany among European sciences. He ascertained that Serbia's flora was rich, and worthy of further studies.

Pančić was the first and, for a long time, the only man to meticulously and comprehensively investigate Serbian flora. With his death in 1888, the remarkable progress in botanical science was suddenly arrested.

It was naturally to be expected that Pančić's disciple, follower and associate Sava Petrović (1839-1889) would continue floristic investigations in Serbia. It was not to be; although younger than Pančić, he outlived his master only by a year. Still, he published the Flora of the Neighbourhood of Niš and the Supplement (Petrović 1882, 1885). Petrović's work is in fact an addition to the Flora of the Principality of Serbia, because at the time when Pančić was engrossed in plant collecting, the area

of Niš was not yet a part of the Principality of Serbia.
After the sudden, almost concomitant departure of these two eminent botanists from the scientific scene, floristic explorations in Serbia went into decline.
The beginning of XX century witnessed the revival of botanical explorations in Serbia. That period was marked by the efforts of two distinguished scientists, Lujo Adamović (1864-1935) and Nedeljko Košanin (1874-1934). Both were Botany professors at the Belgrade University, and were, at different times, directors of the Botanical Garden in Belgrade.
Adamović's (1909) abiding interest in the vegetation of the Balkan Peninsula resulted in his master work. However, he also explored exhaustively the flora of south-eastern Serbia (Adamović 1908-1912), and deserves special credit for the discovery of species new to science.
Košanin pursued a wide range of subjects. He studied flowering plants, mosses, ferns, algae, and explored the flora and vegetation of Serbia and what is now the F R Yu of Macedonia, and described a few species new to science (Košanin 1913, 1914, 1921, 1923, 1926).
After Košanin, around 1934, floristic explorations in Serbia came to a standstill.
Still, even during that period, there was a group of curators at the Natural History Museum in Belgrade who had not neglected floristic explorations. Their efforts, however, were directed to the floras of Macedonia, Greece, Bulgaria, and Albania, rather than to the flora of Serbia (Vasić 1993).
In the forties and fifties, the genera *Thymus*, *Evonymus*, *Trapa*, *Fagus*, *Vicia*, were represented monographically, and so were, partly, *Quercus*, *Salix*, *Populus*.
However, the only work treating the flora of Serbia was still Pančić's Flora from 1874. Being relevant only to the much smaller area of the Principality of Serbia, it could not be fully of use for the territory of the modern-day Serbia. Furthermore, while the floristic and chorological data remained invaluable, notable changes had taken place in the areas of taxonomy, nomenclature, and synonymy.
Although the intensity of floristic explorations after Pančić waned, the knowledge of flora gradually increased considerably. It became clear that the study of Serbian flora was profoundly important, not only on Balkan but on European scale, for the understanding of complex processes of history and evolution of plant cover in this part of the world. In effect, it proved necessary to critically analyse and systematise all data and present them to the public in accordance with scientific principles and criteria.

2. Committee for Exploration of the Flora and Vegetation of Serbia

Although the explorations since Pančić differed in intensity and results, a noteworthy level of knowledge of flora of Serbia's entire territory was acquired. During the sixties, Serbian botanists were examining the conditions, possibilities, and most advantageous methods for the composition of a modern Flora.
A compilation of a national Flora, is a huge, complex, and responsible task

involving many years of hard work. The best guarantee of success for such a venture is to undertake it under the auspices of the leading scientific institution.
Thus, in 1966 the Serbian Academy of Sciences and Arts founded the Committee for Exploration of the Flora of Serbia. The necessity for establishing a Committee responsible for a compilation of Flora was first recognised by Serbian botanists and the academician Mladen Josifović (1897-1981) was fully responsive to it. On the basis of his rationale about the scientific, cultural, and economic significance of compiling national Flora, Serbian Academy of Sciences and Arts, founding the Committee, undertook to publish an edition entitled "Flora SR Srbije" [Flora of the Social Republic of Serbia] as one of its priorities.
M. Josifović was elected president of the Committee, and its first members were chosen among the eminent Serbian botanists of that time.
The Committee activities were soon extended to the studies of vegetation. In consequence, the Committee received more members, and in 1977 it was renamed the Committee for Exploration of Flora and Vegetation of Serbia.
M. Josifović conducted the Committee activities with notable success until his death. He was succeeded by the academician Miloje Sarić under whose judicious guidance the Committee has been thriving for 16 years.
Today, the Committee has 20 members. Although they belong to different generations, and accordingly differ in the degrees of scientific experience, all are united in their eagerness to explore every corner of this country and become thoroughly acquainted with its flora and vegetation.
Therefore, the Committee meetings, which take place in the Serbian Academy, are the least part of its activities. The best part of the work is carried out in the field, in herbaria, in scientific institutions. In addition to the Committee members, repersentatives of certain botanical disciplines, an army of flora and vegetation investigators participate, indirectly, through the results of their investigations, in the work of the Committee.
Thirty years ago, when the Committee was founded, its principal object was the compilation and publication of the Flora of SR Serbia. At the outset, a comprehensive, responsible, and complex task lay before it.
Ninety-two years have elapsed from the publication of the Pančić's Flora to the establishment of the Committee. Nevertheless, Pančić may be described as the prae-founder of the Committee, since his work served as the basis for organising work on the modern Flora, which is another proof of its enduring value.
It was estimated on the basis of Pančić's Flora of the Principality of Serbia, as well as papers, studies, and monographs published in the following decades that the Flora of SR Serbia should be composed of eight volumes, which were to be published in the period 1970-1976. The final volume of the Flora was intended to mark the 10th anniversary of the Committee's foundation. It was certainly an ambitious plan, and one may well wonder how it was possible to carry it out.
The organisation of activities was exceptionally good, the credit for which should

be given to M. Josifović, the then Committee director and editor of the Flora. Distinguished for his superlative scientific achievements, M. Josifović was also an excellent professor and pedagogue, sportsman, and was esteemed for his culture, broad views, and delicate sense of beauty. With his gently exercised authority and understanding attitude, he was ideally suited to preside over the Committee, as well as to supervise the work on the Flora.

The scientific staff from fundamental and applied faculties of Belgrade and Novi Sad Universities, and from Natural History Museum in Belgrade was numerous and had sufficient scientific experience to commence work on flora.

Elaborate preparations were made, including: (1) research into the flora of Serbia from the days of Pančić to the present time, accompanied with a critical analysis of the already existing literature data from XIX and the beginning of XX century, which required validation of some published taxa in the field as well; (2) taxonomical, classificational, nomenclatural, and chorological revision of certain taxa, the data for which were incomplete, doubtful, or entirely missing.

The Herbarium of the Balkan Peninsula housed in the Natural History Museum in Belgrade was of huge importance during the work on the Flora (Vasić 1993). Being, so to speak, a regular database, the Herbarium was instrumental in compiling the basic inventory of flora, completing data on accurate distribution for the majority of plant species, as well as for infraspecific taxa, analysing variability of species, and particularly in verifying the already available taxonomic interpretations of the polymorphic nature of certain taxa.

Based on an analysis of the data obtained from literature and Herbarium, plans were made and an exhaustive field research was undertaken with the aim of acquiring more complete, and more recent data on taxa and their frequency and distribution in Serbia.

The modern Flora of Serbia was planned to contain vascular flora from the area of the present-day Serbia, from the earliest to the most recent records. The analysis and the presentation of data were to be in keeping with the latest methods for compiling national Floras.

The Flora delineates basic characteristics of higher taxonomic categories, but the focus of attention are exhaustive data for species. Entries for species include general data, such as synonymy, total range, descriptions, and, if necessary, information about specific morphological features for the given area, and actual data for the territory of Serbia.

Thus, an entry for a species comprises: (1) scientific, Latin name, together with literature sources, (2) in cases when the actual name is a result of revision, the bazionim is also provided and other synonymy, (3) exhaustive description of morphological characteristics, (4) data on habitat features, (5) general distribution, *i.e.* the total range of the species, (6) data on distribution on the territory of Serbia, together with the character of distribution; widespread, rare, restricted to certain habitat types or areas. For polymorphous species infraspecific structure is also analysed, and distribution is given for each taxa.

The largest number of species is illustrated, with their characteristic features enlarged. The complete first edition, and the first volume of the second edition were illustrated by Milovan Gajić (1926-1994), a Botany professor and also a gifted painter, mainly from the material from the Herbarium of the Natural History Museum in Belgrade.

The first two volumes were published in 1970. The six remaining volumes were issued in subsequent years, one per year, so that 1976 saw the publication of the eighth and final volume of the Flora of SR Serbia (Josifović 1970-1976). It was exactly 102 years since the appearance of the celebrated Pančić's Flora of the Principality of Serbia. Concurrently with the publication of the eight volume of the modern Flora, Serbian Academy of Sciences and Arts photoprinted Pančić's Flora. Thus the past and the present met.

The Flora of SR Serbia is not merely a catalogue of all available, critically organised data. It is also a guide and stimulus for new investigations. Thus the ninth volume was published as the first supplement to the Flora (Josifović 1977).

During the following decade the Committee continued with undiminished energy to encourage, organise and direct investigations in Serbia. The results were numerous scientific papers, studies, monographs, and, to crown these efforts, the tenth volume (second supplement) of the Flora (Sarić & Diklić 1986).

The first edition of the Flora of SR Serbia encompasses 7799 autochthonous taxa, namely 3428 species and 4371 subspecies, varieties and forms.

The first volumes were sold out while the tenth volume was being printed. But even more than that, new data which came to the surface in the meantime (Vasić & Diklić 1996), showed that the investigation of flora and the Committee activities did not terminate with the publication of the first edition. In response to the need which soon arose for a new edition, the first volume of the second, amended edition, entitled the Flora of Serbia, appeared in 1992 (Sarić 1992). The manuscripts for volumes II and III have likewise been prepared.

In recent years, the Committee has intensified work on the edition of Flora of the Algae of Serbia. The first volume treating *Cyanophyta* was published in 1996 (Blaženčić 1996).

The work on the edition of Vegetation of Serbia began at a later date. Its first volume (Janković *et al.* 1984) presents the basic features and geohistorical development of vegetation on the territory of Serbia. Furthermore, it offers a survey and analysis of floristic elements and living forms occurring in Serbia. The second volume (Jovanović *et al.* 1997), relating to forest vegetation deals with formation principles, geographic and altitudinal distribution, composition and structure, classification, the degree of endangerment, and protection measures.

The editions of Flora and Vegetation proved to be an invaluable source of information not only for further fundamental investigations, but also in the field of the so-called applied sciences and disciplines. It also appeared expedient to compile specific floras from the basic flora. The specific floras were planned to encompass the plants directly useful to men. That venture was the next step in employing theoretical knowledge in solving experimental and practical problems of exploiting properly natural resources. The results of fundamental investigations together with pertinent data from certain fields would be highly valuable in planning sustainable development.

Thus, at the proposal from the present director of the Committee, the academician Sarić, a new edition entitled Specific Floras was undertaken. The titles of the volumes planned clearly indicate which plant characteristics would determine the choice of the subject and for what kind of activities they are intended. The volumes include Wild Allies of the Cultivated Plants, Weed Plants, Mellifluous Plants, Poisonous Plants, Ornamental Plants, Wild Edible Plants.

The Medicinal Plants of Serbia (Sarić 1989) as well as the Aromatic Plants (Jančić *et al.* 1995) have already been published.

The Committee not only pursues goals more or less directly related to the compilation and publication of the aforesaid editions, it also supervises, stimulates, and supports all aspects of fundamental explorations of flora on the territory of Serbia. The least contribution is a priceless stone leading to the completion of the picture of the flora of Serbia, as well as of the Balkans and European regions.

3. Comments

Although the working conditions have not always been of the best kind, particularly during the last few years, it is not with small pleasure that we look back over the thirty years of activities of the Committee for Exploration of the Flora and Vegetation of Serbia of the Serbian Academy of Sciences and Arts.

The editions of Flora, Vegetation, Flora of Algae, and Specific Floras, entrusted to the Committee, are not merely thousands of printed pages. They are the fruit of vast knowledge amassed by generations of botanists. The recognition of the value of the knowledge of plants and the unwavering spirit of dedicated research into flora which links Pančić with the founders of the Committee has affected the youngest generation of botanists as well. In spring, when the snow has not melted yet, they scatter throughout Serbia spurred and led by the desire to advance the present knowledge.

From that enormous body of data we would like to single out for special mention that in Serbia, since Pančić, 90 species, 20 subspecies, 87 varieties, and 187 forms new to science have been discovered.

3. References

Adamović, L. (1909) *Die Vegetationsverhältnisse der Balkanländer*, Wilhelm Engelmann, Leipzig.

Adamović, L. (1908-1912) Flora jugoistočne Srbije, *Rad Jugoslovenske Akademije znanosti i umjetnosti*, **175**, 153-214; **177**, 193-240; **179**, 125-176; **181**, 325-336, **183**, 83-176; **185**, 194-242; **188**, 84-140.

Blaženčić, J. (ed.) (1996) *Flora algi Srbije* **1:** *Cyanophyta*, Naučna knjiga, Beograd.

Boué, A. (1840) *La Turquie d'Europe.*

Friwaldsky, I. (1832-1838) Mitteilungen über die im Balkan Gebite gemachte naturwissenschaftliche Reise.

Friwaldsky, I. (1835, 1836) Botanische Zeitschrift "Flora" in Regensburg.

Grisebach, A. H. R. (1843-1845) *Spicilegium Florae Rumelicae et Bythinicae, exhibens Synopsin Plantarum quas Aest. 1839 legit* **1**, **2**, Brunsvigae [Brunswick].

Jančić, R., Stošić, D., Mimica Dukić, N. and Lakušić, B. (1995) *Aromatične biljke Srbije*, Dečje novine, Beograd - Gornji Milanovac.

Janković, M., Pantić, N., Mišić, V., Diklić, N. and Gajić, M. (1984) *Vegetacija SR Srbije* **1**, SANU, Beograd.

Josifović, M. (ed.) (1970-1976) *Flora SR Srbije* **1-8**, SANU, Beograd.

Josifović, M. (ed.) (1977) *Flora SR Srbije* **9**, SANU, Beograd.

Jovanović, B., Mišić, V., Vukićević, E., Diklić, N. and Dinić, A. (1997) *Vegetacija Srbije* **2**(1), SANU, Beograd.

Jurišić, Ž. J. (1905) *Flora Kraljevine Srbije za proteklih sto godina*, Beograd.

Košanin, N. (1913) *Narthecium scardicum* spec. nova, *Öster. Botan. Zeitschrift*, 1-3.

Košanin, N. (1914) *Dioscorea balcanica* Koš. n. sp., *Öster. Botan. Zeitschrift*, 1-3.

Košanin, N. (1921) Geografija balkanskih Ramondija, *Glas. Srp. kr. akad.* **101**, 34-49.

Košanin, N. (1923) Život tercijernih biljaka u današnjoj flori, *Glas. Srp. kr. akad.* **107**, 1-13.

Košanin, N. (1926) Nove vrste u flori Južne Srbije, *Glas. Srp. kr. akad.* **119**, 19-29.

Mathew, B. F. (1980): *Crocus* L., in Tutin, T. G. *et al.* (eds.): *Flora Europaea* **5**, 92-99, University Press, Cambridge.

Mayer, E. and Diklić, N. (1967) Nomenclator Pančićianus, *Pančićev zbornik*, SANU, Beograd.

Pančić, J. (1856) Verzeichniss der in Serbien wildwachsenden Phanerogamen, nebst den Diagnosen einiger neuer Arten, *Verhandlungen des Zoologisch-botanischen Vereins in Wien* **6**, 475-598.

Pančić, J.(1874) *Flora Kneževine Srbije*, Beograd.

Pančić, J.(1884) *Dodatak Flori Kneževine Srbije*, Beograd.

Petrović, S. (1882) *Flora okoline Niša*, Beograd.

Petrović, S. (1885) *Dodatak Flori okoline Niša*, Beograd.

Sarić, M. R. (ed.) (1989) *Lekovite biljke SR Srbije*, SANU, Beograd.

Sarić, M. R. (ed.) (1992) *Flora Srbije* **1**, SANU, Beograd.

Sarić, M. R. and Diklić, N. (eds.) (1986) Flora SR Srbije **10**, SANU, Beograd.

Vasić, O. (1993) Herbarium of the Natural History Museum in Belgrade as a basis for the shaping and publishing of the Flora of Serbia I-X (Ed. 1, 1970-1986), *Webbia* **48**, 259-265.

Vasić, O. and Diklić, N. (1996) Flora of Serbia - a review of phytotaxonomic studies published during the period 1981-1993, *Bocconea* **5**(1), 45-52.

Visiani, R. de (1860) *Pemptas plantarum Serbiae novarum*, Venezia.

AN OVERVIEW OF THE COASTAL VEGETATION OF GREECE BASED ON MULTIVARIATE ANALYSIS: DUNES.

K.V.[1], SYKORA, D.[2], BABALONAS, E.[3], PAPASTERGIADOU
[1]*Department of Terrestrial Ecology and Nature Conservation, Agricultural University of Wageningen*
[2]*Department of Botany, Aristotle University of Thessaloniki, Greece*
[3]*Greek Biotope Wetland Centre, 14 Km Thessaloniki Mihaniona, Greece.*

1. Introduction

For the conservation of Greek nature an overview of its plant communities is indispensable. Most classes have insufficiently been surveyed. The extent of gaps in knowledge is very often unknown. In order to detect the gaps and to give an overview of the existing knowledge on coastal vegetation of Greece, the releves available have been collected and a Greek Phytosociological Database has been constructed. In this scope a review of the plant communities occuring along the Greek coast is in preparation. This stu-dy presents an overview of plant communities of dunes occuring along the Grcck coasts.

2. Materials and methods

From literature 2565 coastal releves have been selected including 744 plant species. These releves have been obtained from all the available published tables describing the vegetation on Greek dunes, which were made according to the Braun -Blanquet phytosociological method. Species names are according to Flora Europaea. These releves were put into a database by means of TURBOVEG. MEGATAB was used to structure the entire raw table and to divide it into clusters representing the different classes. The main cluster representing the dunes consisted of 825 releves. These releves comprise the classes, Crithmo-Staticetea, Cakiletea, Ammophiletea, and Juncetea maritimi. After further clustering, checking the homogeneity and removal of 140 heterogeneous releves 13 associations could be described.

3. Results

The following communities have been found in Greek coastal dunes. The synecology of each association and in some cases of subassociations has also given.

Ammophiletea BR.BL. et TX. 1943.
Pioneer communities from upper sandy or gravely beaches or from coastal dunes.The communities mainly consist of geophytes and hemicryptophytes. Sporadically inundated by sea water during storms.

Ammophiletalia BR.BL. et TX. 1943.
* *Community of Otanthus maritimus*
Synecology: Seaward side of sand dunes.
* *Community of Eryngium maritimum*
Synecology: *Eryngium maritimum* is stimulated by the deposition of organic matter, i.e. strand line or anthropogenic waste.
* *Agropyro juncei-Sporoboletum pungentis* (Arènes 1929) nom. mut. Mayer 1995.
Synecology: Vegetation growing on the higher part of gravely or sandy beaches and on small embryo dunes on places that are temporarily inundated by the sea. It is the most halophytic of the dune communities. It grows on salt encrusted sand in contact with the Arthrocnemetea or on little cushion like sand dunes in contact with the Cypero mucronati-Agropyretum juncei.
- *cakiletosum maritimae* / subass.nova

Synecology: *Cakile maritima* and *Euphorbia peplis* are halonitrophile species of strand lines and coastal jetsam and indicate the eutrophication by organic matter.
* *Cypero mucronati-Agropyretum juncei* Fernández-Prieto & Díaz-Gonzáles 1991.
Synecology: Vegetation of small, embryo dunes formed at the highest part of the beach.
- *typicum:* On embryo dunes still influenced by mobile sand.
- *cakiletosum maritimae* / subass.nova

Synecology: The high frequency of *Cakile maritima* indicates the deposition of mineralizing organic matter originating from the tide mark.
- *otanthetosum maritimi* / subass.nova

Synecology: On stabilized embryo dunes, often occuring more inland than the other subassociations. In the successional sequence the othanthetosum is followed by initial stages of the Ammophiletum.
- *echinophoretosum spinosae* / subass.nova

* *Elymetum sabulosi* Babalonas 1979
Synecology: This nitrophylous community grows on embryonic dunes along the fringes of beaches where the salt content is to high for *Ammophila* communities.
- *cynodontetosum dactylii* / subass.nova
- *elymetosum farcti:* Elymus farctus, Pancratium maritimum and some species with low frequency only.

* *Medicagini marinae-Ammophiletum australis* Fernández-Prieto Diaz Gonzales 1991.
Synecology: Vegetation of yellow, mobile dunes, landwards of the Agropyretum juncei. Usually more than 1 meter high.
- *sporoboletosum* / subass.nova
- *scabiosetosum maritimae* Mayer 1995. On stabilized sands.
- *centaureetosum pumilionis* Mayer 1995.

Synecology:This subassociation is found on yellow dunes, which in general are moderately steep. Occasionally these dunes are also moderately flat. They are generally more stabilized and show a lower degree of influx of sand than in the case of the Ammophiletum typicum.

- *leymetosum racemosi* / subass.nova

Synecology: This community has been described by Géhu and Uslu (1989) from the coast of the Black sea where it occurs on extensive dunes.

Helichryso-Crucianelletalia maritimae Géhu, Riv.-Mart., & R. Tx. 1975
Crucianellion maritimae Rivas-God. et Rivas-Mart. 1963
* *Euphorbio- Silenetum nicaeensis* Lavrentiades 1964

Synecology: On stabilized dunes, growing more inland than the Ammophiletum. According to Lavrentiades (1964) this association grows aproximately 200 m from the sea on dunes of about 5 m high. Stabilization of the dunes can be natural or anthropogenic i.e. by plantation of an artificial hedgerows of pine trees.

* *Ephedra distachya-Silene subconica*-ass. Oberd. 1952

Synecology: This association occurs on the inland side of the Ammophiletum dunes on fixed sandy soil containing some humus. The pH-value ranges between 6.5 and 6.6. It appears as a progressive stage of development of the psammosere upon grey, stabilized sand dunes, in which the heigth varies between 1,5 and 5 meter above sea level. It can grow at a distance of 10 to several hundreds of meters from the sea-line. It grows on flat to slopy soils. Vegetation cover between 10 and 75%

- *allysetosum minus* / subass.nova

Synecology: This is transitional to the Artemisietum campestris, on old stabilized dunes.

- *sideritosum montanae* / subass.nova
- *hypericetosum olympicae* / subass.nova
- *nigelletosum arvensis* / subass.nova
- *ammophiletosum arenariae* / subass.nova

Synecology: The environmental conditions are transitional to that of the Ammophiletum. This subassociation indicates dunes with less or recently fixed and thus more mobile sand.

- *pancratietosum maritimi* / subass.nova

Synecology: This subassociation occurs landwards from the Ammophiletum on well drained, grey and humose sandy dunes stabilized for many years. In the soil composing these old dunes, abundant organic material is built up and the soil is more compact and has a better water-retention because of the presence of clay particles.

Medicagini marinae-Triplachnion nitentis Mayer 1995.

Synecology: This association grows on flat sandy or gravel beaches along the SE and E coast of Crete, a region characterized by low precipitation.

* *Limonio graeci-Triplachnetum nitentis* Mayer 1995.

Synecology: The association occurs on flat cobble and gravel banks mixed with varying quantities of sand. The drainage is higher than in standard sandy beaches, while the precipitation in the area of its occurrance is recorded to be the lowest in Crete.

- *centaureetosum aegialophilae* Mayer 1995.

Synecology: On excessively drained, coarse sands, gravel or rounded cobble stones.

- *atriplicetosum halimi* Mayer 1995.

Synecology: This subassociation is characterized by a higher supply of nutrients through plant material washed into the cobble stone beach. It grows closer to the waterline than the subassociation centaureetosum aegialophilae.

- *typicum* This subassociation is characterized by the absence of differential species.
- *silenetosum pinetori* / subass.nova

Synecology: This subassociation grows farther away from the sea than the other subassociations. It occurs on the flat crest of very dry, excessively drained gravel or cobble stone beaches.

* *Corydothymo capitatae-Centaureetum spinosae* Géhu 1991

Synecology: This plant community forms a garrigue dominated by small cushionlike shrubs on sand deposited on gravel or river stones.

coridothymetosum capitati / subass.nova

Synecology: In this subassociation the coastal sand dune character is most replaced by the garigue (phrygana) character.

- *Pseudorlayetosum pumili* / subass.nova

Synecology: It grows inland from the upper part of beaches. The presence of Cakiletea species indicates nutrient enrichment caused by accumulation of organic matter.

* *Medicagini marinae-Stachyetum spinosae* Géhu et al 1987.

Synecology: It can be found on inner dune systems. *Stachys spinosa* is a species of dry rocky and stoney places and cliffs and also grows in garigue-vegetation.

eryngietosum maritimae Géhu et al 1987.

Synecology: This subassociation grows on inner dunes and in depression between dunes.

- *coridothymetosum capitati* Géhu et al 1987.

Synecology: This subassociation occurs relatively far from the beach, on extensive level areas with sandy-gravely soils .

* *Artemisietum campestris* Babalonas 1979

Synecology: The Artemisietum campestris grows on stabilized inner dunes. It extends where the Ammophiletea disappear because of the fixation of the sand. In the zonation the Artemisierum campestris can be found to follow after the Elymetum sabulosi in the inland direction. The habitat conditions are comparable to that of the *Ephedera distachya-Silene subconica*-ass. occuring on other localities along the northern part of the Aegean sea.

4.References (Selected References).

Babalonas, D. (1984). Chorologie und soziologie von *Leymus racemosus* susp *sabulosus* (Poaceae). Phyton 24: 165-171.

Babalonas, D., Sykora, K.V. and Papastergiadou, E. (1995). Review of plant communities from Greek dunes and salt-marshes, a preliminary summarizing list. *Ann. Bot. (Roma)* 53, 107-117.

Lavrentiades, G. (1964). The ammophilous vegetation of the Western Peloponnesos coasts. Vegetatio 12: 223-287.

Lavrentiades, G. (1993). Dry Coastal Ecosystems of Greece. Pp429-441. In : E.van der Maarel (ed) Ecosystems of the World 2A. Dry coastal ecosystems, Polar Regions and Europe, Elsevier publ.

Papastergiadou, E.& D. Babalonas (1996). Salt marshes and sand dune ecosystems in the North Aegean coastal areas. In: Coastal Mnagement and Habitat Conservation, 65-74p.

TWO RARE TAXA IN THE SOUTHERN PELOPONNESE

K. TAN* AND G. IATROU
Botanical Institute, University of Copenhagen, 140 Gothersgade, Copenhagen 1123-K, Denmark
Division of Plant Biology, Department of Biology, University of Patras, GR-265 00 Patras, Greece

At the 5th Scientific Conference of the Hellenic Botanical Society held at the European Cultural Centre in Delphi on 21-23 October 1994, we presented a paper on the supposedly 'extinct' status of a taxon, *Onobrychis aliacmonia* Rech. fil. in the Leguminosae. It was gratifying to note the results of our work, to find that the species was no longer extinct and that we had discovered another endemic taxon for the Peloponnese --- a new subspecies of *O. aliacmonia* which we have named *O. aliacmonia* subsp. *peloponnesiaca* Iatrou & Kit Tan. This was subsequently published in the Festschrift volume of the Annales Nat. Hist. Museum, Wien in 1996, in honour of Professor Karl Heinz Rechinger's 90th birthday.

This kind of research forms the basis for a good Red Data Book of Greece, mainly depending on fieldwork and original conclusions, never the acceptance nor following of another's information and opinions without verification of fact.

We continue our research in the study area which is situated in the south Peloponnese on the road from Skala to Molai near the village of Vlachiotis.

In the locality where *O. aliacmonia* subsp. *peloponnesiaca* was discovered, was a puzzling *Astragalus* which Walter Strasser (a Swiss biology teacher) had collected in 1985 and had named *Astragalus thracicus* Griseb. As *A. thracicus* is a mountain steppe species, its occurrence at such low altitudes in the southern Peloponnese is rather interesting and the subject deemed worthy by us of further investigation. The plant proved to be an undescribed new taxon which we now name *Astragalus laconicus* Iatrou & Kit Tan. It has affinities to *A. creticus* Lam., the type subspecies of which is probably endemic to central and eastern Crete.

This Astragalus also occurs in the Malea Peninsula, the easternmost prong of the Peloponnese. Here, it was collected by Professors Hans Runemark and Sven Snogerup in May 1964, more than 20 years before Strasser was aware of its existence in the Peloponnese, and more than 30 years before we now, at this meeting, present its real identity.

A. laconicus is a dwarf shrub forming dense, spiny tussocks 10-20 cm tall. The leaves are paripinnate, with the rachis protracted into a yellowish spine; the leaflets greyish-green and adpressed-pubescent. The stipules are conspicuous. The flowers are sessile in the leaf axils and crowded in inflorescences 2-3 cm below the tip of the branches. The calyx is completely hidden by the dense long white hairs; the teeth are lanceolate-setaceous. The corolla is pale pink drying cream, and persistent; the standard with lamina contracted into a narrow claw without a distinct constriction between limb and claw, and longer than the wings and keel.

We now come to a second unusual taxon in our study area. Kit Tan together with Gert Vold from the Copenhagen Botanic Garden, explored the study area further and found a very unusual species of *Thymus*. This thyme is distinct from all thymes known in Greece and at first could not easily be keyed out in the relevant volume of *Flora Europaea*, leading them to believe that it could possibly be a taxon new for Greece. The plant is none other than *Thymus laconicus* Jalas which was originally described as prostrate in habit. Our plant was a much-branched dwarf shrub, strongly woody at the base and with erect, leafy branches. The plant does not have a creeping, prostrate habit common in many thymes. Another unusual feature is the presence of large, conspicuous, broadly ovate floral bracts which are strongly veined and differ from the cauline leaves both in shape and texture, the whole forming a compact, swollen inflorescence. The corolla is purplish to rose-pink and much exserted from the calyx.

This thyme has an interesting botanical history. Dr Constantinos Goulimis, the well-known and keen amateur naturalist who, by the way, was also the co-discoverer with Professor Karl Heinz Rechinger of *Onobrychis aliacmonia,* found this thyme in the Malea Peninsula in May 1955. It was not yet in full flower. He realised it was distinct from other thymes in Greece and in a letter dated 27 June 1955 to Dr Turrill then director of the herbarium at the Royal Botanic Gardens, Kew, asked for an identification, enclosing some material. Back came Dr Turrill's swift reply on 1st July. Turrill stated in his letter that he was unable to match the *Thymus* with any taxon from the Balkan Peninsula. He asked Dr Goulimis to try to obtain more material, stating that it be worth studying in detail and offering to describe it if it be new. On 11 July of the same year, Goulimis sent material in flower to Turrill stating in his letter that the bracts of the plant recall those in *T. membranacea* Boiss. from Spain. The credit for obtaining this new material must go to Goulimis' mule driver who collected part of the plant in June and it was this material sent to Kew that formed the type of the new species which was formally described by Turrill as *T. pastoralis*. Turrill thought it appropriate to name the plant *T. pastoralis* in recognition of the shepherds who have pastured their flocks on the hills of Greece for many centuries. The plant had also been highly praised by local shepherds as being good fodder and with an attractive odour. However, we doubt that *T. pastoralis* was indeed the plant thus designated by the shepherds as it does not occur in sufficiently large quantities for easy browsing by the flocks, and another thyme, perhaps *T. atticus* elak., could have been the candidate.

When Professor Jalas at the Botanical Museum, University of Helsinki prepared the account of *Thymus* for the relevant volume of *Flora Europaea*, he only had the material at Kew at hand and this consisted of only a few branches. However, he was aware that the name *T. pastoralis* was illegitimate being a later homonym of a Russian taxon, and thus he renamed the plant *T. laconicus*.

When the plant collected in our study area did not match the description in *Flora Europaea,* Kit Tan, suspecting it could possibly be *T. laconicus*, asked for the loan of the type of *T. laconicus* from Kew. When it arrived it matched our plant in all respects. The only difference had been the misleading statement in the description which stated that the plant was prostrate. The branches are sometimes arcuate-ascending but the plant, at least not in our material, is not creeping nor prostrate. The strongly woody habit and the conspicuously bracteate inflorescences are characters found in a group of woody thymes from southern and eastern Spain and not in the Balkans. *T. laconicus* is rarely collected and even Professor Jalas, the expert on European thymes, admitted he was not aware that the species could also have such an erect woody habit (pers. comm to KT in 1996). Its distribution is restricted to a few scattered localities in the southern Peloponnese and it is certainly rare. It has a superficial resemblance to *Coridothymus capitatus* (L.) Reichenb. fil. when in the sterile state and thus may also have been overlooked. However, in late bud and in full flower, the crowded bracteate inflorescences are unmistakeable.

T. laconicus and *Astragalus laconicus* both occur in the south of the Malea Peninsula where they had first been discovered, one in 1955, the other in 1964. They are now collected in our study area north of the Malea Peninsula, and taxonomically resurrected, one as regards to its proper description of habit, the other given the status of a new species. Both are rare taxa with small populations. Certainly the existence of *A. laconicus* in our study area is under threat. During a visit we noticed a "Caterpillar" steadily approaching. By the destruction of the habitat for land use and development, it would only be a matter of time before this interesting plant disappears from our locality. As marked and threatening as the jaws of the vehicle relentlessly grinding on the hill slopes above the *locus classicus*, thus as marked and threatened is the fate of our poor Astragalus ....

What of a taxon already listed as 'perhaps extinct' in Europe?

Bongardia chrysogonum (L.) Griseb. in the Berberidaceae was only known from historical records in the Aegean area (from 1796--1836) until unexpectedly discovered in 1987 by Professor Arne Strid and co-workers in a small field in the northern Peloponnese. There are no other published records from mainland Europe but the species has a large distribution further east, from central Anatolia to Pakistan where it occurs in semi-natural habitats. The population consisted of 30--40 flowering plants when first discovered but has now disappeared because of a change in agricultural practice. *Bongardia chrysogonum* is a tuberous weed of cultivation and depends on annual tilling by old-fashioned methods, not deep ploughing and extensive use of herbicides. Once the field is abandoned and reverts to garigue or woodland it cannot survive as it cannot withstand competition. It is the opinion of Professor Strid that the species must now be classified as extinct in its only European locality.

However, the species was known in 1984 from a small, semi-cultivated cereal field in east central Peloponnese, c. 100 km away. The significance of the discovery unfortunately was not known to the discoverer, G. Iatrou and the population had not since been monitored. A search in 1994 by Kit Tan and Gert Vold revealed two plants (one a voucher kept in C). In 1995 the field was burned and deeply ploughed and the plants were not seen. In 1996, *Gladiolus italicus* Miller still edged the field margins and *Leontice leontopetalum* L. was abundant, but the *Bongardia* plants had disappeared, a consequence once again of deep ploughing. On a more hopeful note, a few kilometres to the south are several fields abundant with *Tulipa undulatifolia* Boiss., another bulbous weed of cultivation, and it is possible that the *Bongardia* may still have a refuge there.

STUDY ON THE VEGETATION AND THE DYNAMICS OF *EUPHORBIA DENDROIDES* AND *JUNIPERUS PHOENICEA* PLANT COMMUNITIES IN S. GREECE (N. KORINTHIAKOS GULF)

M. THEOCHAROPOULOS, G. DIMITRELLOS & TH. GEORGIADIS
Botanical Institute - Plant Ecology, Division of Plant Biology, Department of Biology, University of Patras, GR 265 00, Patras-Greece

1. Indroduction - Climate - Geology

The communities of *Juniperus phoenicea* and *Euphorbia dendroides* species constitute a significant part of Thermo-Mediterranean ecosystems in Greece. Locally, *Juniperus phoenicea* enters into the Meso-Mediterranean level. In the littoral and sublittoral zone of Nafpaktos, Itea, Antikira in the North Korinthiakos Gulf, they present a significant development which is particularly interesting from both an ecological and syntaxonomic point of view. Their syntaxonomic position, until now, appears to vary according to different authors (Braun-Blanquet, Barbero & Quezel and Rivas-Martinez).

From a bioclimatic point of view, the studied area is characterized by the succession of two bioclimatic zones: the littoral zone with a semi-arid bioclimate, and the sublittoral zone with a sub-humid bioclimate. These two zones show variants to temperate and cool bioclimates respectively, and also to altitude and relief. The dominate geological substratum is mesozoic limestones, while mixed flysch and alluvial deposits occur locally.

2. Methods - Results

During 1996-1997, 80 releves were made in the studied area, according to Braun-Blanquet's method (Guinochet, 1973) and their elaboration was carried out with the assistance of the multi-variate analysis (STATOS, Roux M. 1994 and TWINSPAN, Hill 1979b). Thirty-three soil profiles were also made using the yardstick of the geological substrate and the condition of natural vegetation. Corresponding soil analyses were then made aiming to extract definate conclusions on the connections between soil and plants.

The clusters of releves for the *Juniperus phoenicea* and *Euphorbia dendroides* communities obtained from the classification were compared to the ordination diagram (Figure 1).

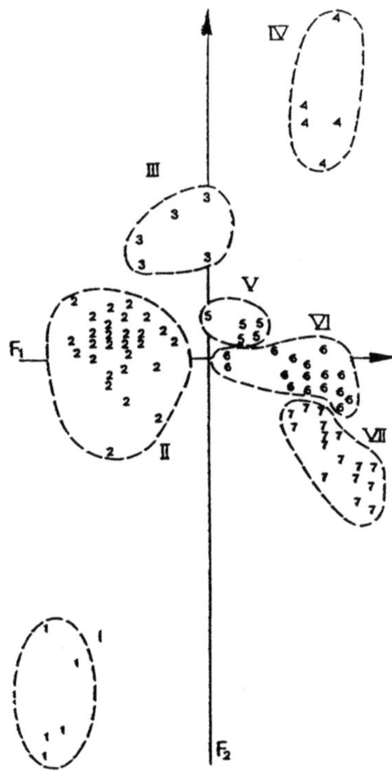

Figure 1. Ordination diagram of the eighty (80) releves made in plant communities with *Juniperus phoenicea* and *Euphorbia dendroides* (Pre-forest and Shrub formations): I: Inulo parnassicae - Ptilostemetum chamaepeuce, II: Phlomido fruticosae - Euphorbietum dendroidis, III: *Coridothymus capitatus* community, IV: Pistacio lentisci-Juniperetum phoeniceae Pinetosum halepensis, V: Pistacio lentisci-Juniperetum phoeniceae, VI: Anthylli hermanniae-Sarcopoterietum spinosae, VII: Phillyreo latifoliae-Querco cocciferetum Juniperetosum phoeniceae.

The phytosociological elaboration of 80 releves shows that they form preforest ecosystems (Pistacio-Rhamnetalia) followed by degraded phryganic formations (Cisto-Micromerietea) and Therophyte ranges (Thero-Brachypodietea). On the vertical calcareous cliffs we described new associations with *Inula parnassica* and *Ptilostemon chamaepeuce* which form the transition towards *Euphorbia dendroides* formations. In total 4 associations, 2 sub-associations and 1 community have been found, of which 3 associations and 2 sub-associations are new. The complete syntaxonomic classification of syntaxa present in the study area is displayed in Table 1.

TABLE 1. Syntaxonomic scheme

Preforest vegetation
Quercetea ilicis Br.-Bl. 1947
Pistacio lentisci - Rhamnetalia alaterni Rivas - Martinez 1974.
Ceratonio siliquae - Rhamnion oleoidis Barbero & Quezel 1979
Pistacio lentisci - Juniperetum phoeniceae Ass. nova
Pistacio lentisci - Juniperetum phoeniceae Pinetosum halepensis sub. ass. nova
Pistacio lentisci - Rhamnion alaterni Quezel & Barbero 1985
Phillyreo latifoliae - Querco cocciferetum Barbero & Quezel 1976 Juniperetosum phoeniceae sub. ass.nova

Phryganic vegetation
Cisto - Micromerietea Oberdorfer 1954
Cisto - Micromerietalia Oberdorfer 1954
Hyperico empetrifolii - Micromerion graecae Barbero & Quezel 1989
Phlomido fruticosae - Euphorbietum dendroidis Ass. nova
Anthylli hermanniae- Sarcopoterietum spinosae Zohary & Orshan 1966.
Coridothymus capitatus communities

Calcareous cliff vegetation
Asplenietea trichomanis Br.-Bl. 1934
Onosmetalia frutescentis Quezel 1964.
Campanulion versicoloris Quezel 1964.
Inulo parnassicae - Ptilostemetum chamaepeuce Ass. nova

The study of their dynamics in combination with human impacts has enabled the definition of the relations between the above phytosociological units as shown in Table 2.

TABLE 2. Dynamics of *Juniperus phoenicea* and *Euphorbia dendroides* communities

Belt		Semi-arid		Sub-humid	
	Calcareous cliffs	Ca	Fl	Ca	Fl
MesoMediterranean	*Sideritis roeseri* *Alkana graeca*	Phillyreo-Cocciferetum Juniperetosum phoeniceae *Cistus creticus* communities	Anthylli-Sarcopoterietum spinosae	Phillyreo-Cocciferetum Juniperetosum phoeniceae *Cistus creticus* communities	Anthylli-Sarcopoterietum spinosae
ThermoMediterranean	Inulo-Ptilostemetum	Pistacio- Juniperetum phoeniceae Pinetosum halepensis Phlomido-Euphorbietum dendroidis *Coridothymus capitatus* communities	Anthylli-Sarcopoterietum spinosae	Pistacio-Juniperetum phoeniceae *Phlomis fruticosa* communities	

From the correlation of the plant communities to particular soil characteristics, it was shown that the soil types present in the study area, in combination with relief, can be distinguished into: a) plain soils, where alluvial formations dominate and which are suitable for agriculture and b) mountainous soils, where the studied plant communities occur. The soils of the second category are located on: a) hard limestone where they have a clay and clay-loam texture, are usually red in colour, slightly alkaline with presence of active calcium and satisfactorily provided with bases of K, Na, Ca and Mg, or b) mixed flysch where they have a sandy-loam texture, are red or brown in colour, slightly alkaline or acid without the presence of active calcium and satisfactorily or fairly provided with bases of K, Na, Ca and Mg.

3. Conclusions

The *Juniperus phoenicea* communities constitute stable ecosystems in the Thermo-Mediterranean level in the frame of edaphoclimatic conditions. The evolution towards forest ecosystems is very difficult and slow. The stands of *Pinus halepensis* with *Juniperus phoenicea* form the sub. ass. of Pistacio-Juniperetum Pinetosum halepensis. The degradation of *Juniperus phoenicea* and *Quercus macrolepis* stands leads to the dominance of *Euphorbia dendroides, Phlomis fruticosa* and other phryganic species. Finally, the extension of *Juniperus phoenicea* in the Meso-Mediterranean level forms the sub. ass. of Phillyreo-Cocciferetum Juniperetosum phoeniceae (Table 2).

4. References

Barbero, M. & Quezel, P. (1976) Les groupements foréstièrs de Grèce Centro-Méridionale, *Ecologia Mediterranea* **2**, 3-86.

Barbero, M. & Quezel, P. (1979) Le problème des manteaux forestiers des Pistacio-Rhamnetalia alaterni en Méditerranée orientale, Essai syntaxonomique, *Doc. Phytosociologiques* **8**, 9-21.

Barbero, M. & Quezel, P. (1980) La végétation forestière de Crète, *Ecologia Mediterranea* **5**, 175-209.

Barbero, M. & Quezel, P. (1989) Contribution à l' étude phytosociologique des matorrals de Méditerranée orientale, *Lazaroa* **11**, 37-60.

Braun-Blanquet, J. (1964) *Pflanzensoziologie*, **3**, Springer Verlag, Auflage Wien.

Guinochet, M. (1973) *Phytosociologie,* Masson et Cie, Paris.

Roux, M. (1994) *Le statoscope*, Méthodes stadistiques pour la Biolologie, Version 1.5, Faculté des Sciences de St. Jerôme, 13397 Marseille Cedex 13.

Hill, M. O. (1979b) *TWINSPAN - a FORTRAN program for arranging multivariate data in ordered two way table by classification of the individuals and the attributes*, Cornell University, Department of Ecology and Systematics, Ithaca.

Oberdorfer, E. (1954) Nordagaische Kraut-und Zwergstrnchfluren im Vergleich mit den entsprechenden vegetationseinheiten des westlichen Mittel meergebietes, *Vegetatio* **5-6**, 88-96.

Quezel P., (1964) Végétation des hautes montagnes de la Grèce meridionale, *Vegetatio* **12**, 289-385.

Quezel, P., Barbero, M., (1990) Les forêts méditerranéennes, problemes poses par leur signification historique écologique et leur conservation, *Acta Botanica Malacitana* **15**, 145-178.

Theocharopoulos, M. & Georgiadis, Th. (1984) Contribution a l' étude de la végétation de l' Attique orientale (Nea Makri) en Grèce (prise en compte des impacts urbains et touristiques), *Ecologia mediterranea* **5** (3-4), 133-157.

CONTRIBUTION TO THE BIODIVERSITY (BOTANICAL DIVERSITY) AND PHYTOGEOGRAPHY OF THE ISLAND OF EUBOEA.

P. TRIGAS AND G. IATROU

University of Patras, Department of Biology, Division of Plant Biology, GR-265 00, Patras, Greece.

1. Introduction

The island of Euboea lies along the eastern coasts of Sterea Ellas. With a total area of 3,654 km^2 is the second island of Greece in size, after Crete. Its length is 175 km and its width ranges from 7 to 45 km. The complicate geological history of the island in combination with the varying climatological conditions have created a high diversity of habitats, which shelter an impressive diversity of floristic elements.

The flora of Euboea has been studied by the authors for the last three years and during this period several collections have been made at different seasons from all over the island. For the needs of our study we separate the island into three distinct regions, according to phytogeographical, geological, geomorphological and climatic criteria (Fig. 1).

2. Materials and methods

This study is based on a survey of 1,635 vascular plant taxa (species and subspecies), resulted from our own field collections and the known literature (e.g. Rechinger 1961, Krause & al. 1963, Boratynski & al. 1988). The nomenclature is after Tutin & al. (1964-1980, 1993) and, where possible, Greuter, & al. (1984-1989). The life-form categories follow the system by Raunkiær (1934). For the chorological types, the classification by Pignatti (1982), and Davis (1965-1985), has been used, adapted in our area. For comparative chorological observations within Greece, the phytogeographical subdivision of Greece according to the Flora Hellenica Project is followed.

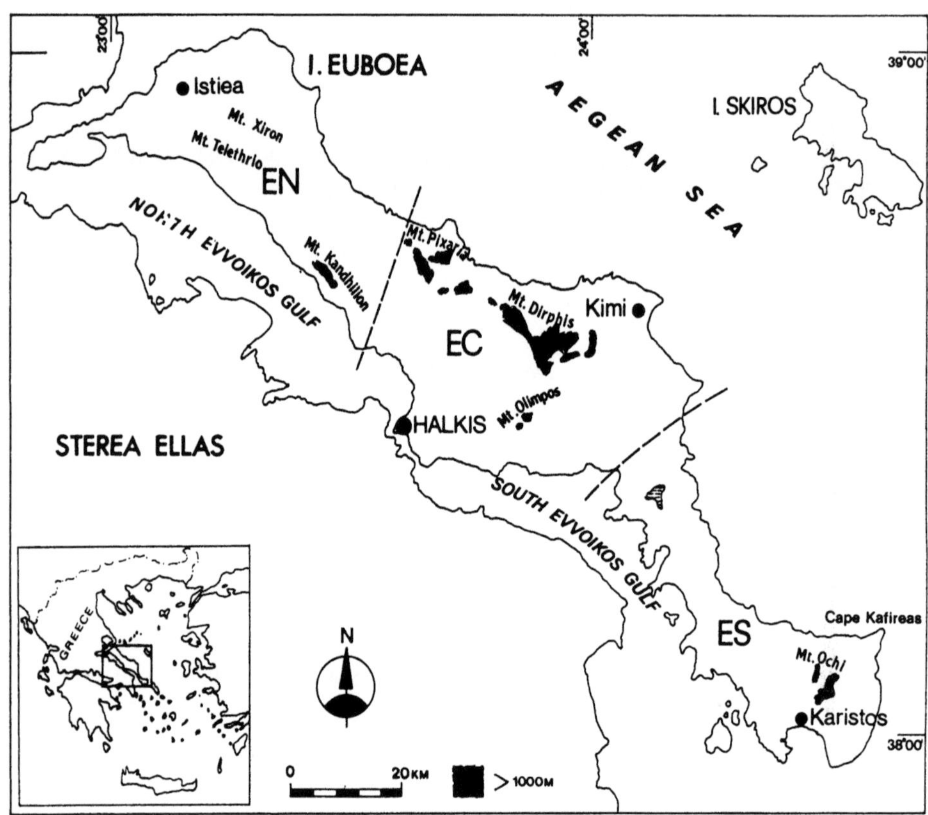

Fig. 1: Geographical position of the island of Euboea within Greece and detailed map of the area studied.

3. Results and discussion

According to the present status the flora of Euboea consists of 1,635 vascular plant taxa included in 582 genera and 112 families. In addition, 15 hybrids prosper in the island, but these are not numbered and have not been taken into account in the floristic analysis.

From the standpoint of richness of species, *Compositae*, *Fabaceae* and *Poaceae* are represented by 187, 160 and 121 taxa, respectively. These three families comprise 28.6% of the taxa found in the island.

The richness and the composition of the species differ into the three regions of Euboea. Thus, the flora of North, Central and Southern parts of the island consist of 1048, 945 and 635 taxa, respectively.

Regarding the life-form spectrum of the flora of Euboea, it appears that the percentages of Therophytes (33.88%) and Hemicryptophytes (33.03%) are higher than those of the other life-forms, which occur in much smaller percentages (Geophytes 13.40%, Chamaephytes 9.11%, Phanerophytes 10.22% and Hydrophytes 0.37%).

The distribution of the life-forms given for the island of Euboea, is not uniform all over the island, but differs from area to area, depending on the environmental and especially on the varying climatological conditions. Comparing the life-form spectra of the three parts of Euboea, the highest percentage of Therophytes appears in N Euboea (37.31%), followed by S Euboea (33.91%), while C Euboea appears in much lower percentage (24.34%), and differentiates itself from the other areas, because of its more humid bioclimate and mountainous bas-relief.

In contrast , the percentage of Hemicryptophytes is the highest in the flora of C Euboea (38.14%), followed by S Euboea (31.02%), and N Euboea (29.77%). The geological composition of S Euboea which mainly consists of crystalline schist justifies the higher percentage of Hemicryptophytes in the flora of the semi-arid S Euboea in relation to the more humid N Euboea. This substrate absorbs great quantities of water, which is available to the plants during the arid period of the year.

The chorological spectrum of the flora of the island could be divided into 14 main groups; 637 (38.96%) of the taxa can be regarded as widely distributed. The Sub- and Cosmopolitan group comprises 115 taxa, while the European, Mediterranean-Turanian and Paleotemperate groups are represented by 93, 86 and 73 taxa, respectively. The rest chorological groups of this category are represented by fewer taxa. Taxa within this category, whose centre of distribution lies to the north of the island of Euboea, make up a significant part of the flora. Concerning also the Balkan elements, we estimate that at least 21.2% of the taxa of Euboea belong to these northern taxa which probably migrated southwards into the island during wetter and cooler periods. In addition, at least 13.4% of the taxa of Euboea seem to be of eastern origin. These facts, corroborate the hypothesis that the island of Euboea has been acted as a crossroads to the migration ways of the plants from North→South as well as East→West directions.

The Mediterranean elements are represented in the flora of Euboea with a percentage of 43.79% (716 taxa). From these, the E Mediterranean element comprises 209 taxa (12.78%).

The Balkan elements are represented in the flora of Euboea by 112 taxa (6.85%), half of these are Balkan endemic and the other half are Balkan subendemic.

The endemic flora of Euboea consists of 170 taxa (10.4%) from which 123 taxa are endemic to Greece, while 47 taxa (2.9%) are considered endemic to the island. The taxonomic position of some of the endemic taxa of the island and especially some of the taxa growing on ultramafic substrates, is in dispute. Noteworthy is the fact that only 12 from the local endemic taxa of Euboea are distributed in more than one regions of the island. The rest 35 local endemic taxa are distributed only in one of the regions of the island, and most of them are single area endemic (17 taxa are exclusively distributed in N Euboea, 12 taxa in C Euboea and 6 taxa in S Euboea).

The highest proportion 59.6% (or 28 taxa) of local endemic taxa are found on a variety of calcareous habitats, mostly on rocky places and in crevices of limestone cliffs. The ultramafic substrates shelter nine local endemic taxa (19.1%), while four taxa grow on calcareous as well as on ultramafic substrates.

From the local endemic taxa found in S Euboea, three taxa (*Allium karistanum*, *Armeria johnsenii*, *Limonium runemarkii*), grow on crystalline schist, while two more taxa (*Campanula celsii* subsp. *carystea*, *Viola euboea*) grow also on calcareous substrates. One more taxon (*Ammi topalii*) grows in coastal sandy habitats around the city of Halkis, but it has not been found for many years, and probably has become extinct because of the constructive and touristic development of the area.

The presence of 123 Greek endemic taxa in the flora of Euboea, give us the opportunity to make some assumptions for the phytogeographical connections of the flora of the island, with those of the other phytogeographical regions of Greece.

The close phytogeographical connection of Euboea with Sterea Ellas is supported by the presence of 99 Greek endemic taxa distributed in both regions, while 15 of them are distributed exclusively in these regions. Their presence strengthen the paleogeographical hypothesis, that the definite isolation of the island from Sterea Ellas happened only 18,000 years B.P.

From the other phytogeographical regions of Greece, Kyklades, East Central Greece, Peloponnisos and North Sporades are the most close to Euboea phytogeographically.

Noteworthy is the example of *Fritillaria euboeica*, which prospers in C and N Euboea, and on Mt. Athos. This is a rather "interesting" phytogeographical connection (we believe that plants with horticultural interesting should not be used for phytogeographical connections).

References

Boratynski, A., Browicz, K. & Zielinski, J. (1988) Woody flora of Euboea (Evvoia), *Arbor. Kornickie* **33**: 13-74.

Davis, P. H. (ed.) (1965-1985) *Flora of Turkey and the East Aegean Islands, Vol. 1-9*, Edinburgh Univ. Press.

Greuter, W., Burdet, H. M. & Long, G. (1984-1989) *Med-Checklist. A critical inventory of vascular plants of the circummediterranean countries,Vol. 1,3,4,* Genève.

Krause, W., Ludvig, W. & Seidel, F. (1963) Zur Kenntnis der Flora und Vegetation auf Serpentinstandorten des Balkans. 6. Vegetationsstudien in der Umgebung von Mantoudi (Euböa), *Bot. Jahrb.* **82(4)**: 337-403.

Pignatti, S. (ed.) (1982) *Flora d' Italia, Vol. 1-3*, Bologna.

Raunkiær, C. (1934) *The life-forms of plants and statistical plant geography*, Clarendon Press, Oxford.

Rechinger, K. H. (1961) Die Flora von Euboea, *Bot. Jahrb.* **80**: 383-465.

Tutin ,T. G. et al. (eds.) (1964-1980) *Flora Europaea, Vol. 1-5*„ Cambridge Univ. Press, Cambridge.

Tutin, T. G. et al. (eds.) (1993) *Flora Europaea Vol. 1 (second edition)*, Cambridge Univ. Press, Cambridge.

GENERAL OVERVIEW AND CHOROLOGICAL ANALYSIS OF BRYOPHYTES IN NORTHERN GREECE (MACEDONIA AND THRACE)

E. TSAKIRI, V. KARAGIANNAKIDOU and D. BABALONAS
Department of Botany, School of Biology, Aristotle University, 540 06 Thessaloniki, Greece.

1. Introduction

In this work is presented the bryophytic flora of Northern Greece (Macedonia and Thrace). Unfortunately Greece belongs to the countries that have not widely been exploited, as far as Bryophytes are concerned. From the investigation achieved until now, is concluded that there are plenty of these species. The number of species and a chorological analysis is presented, based on the published inventories and some unpublished records.

2. Study area and Methods

Macedonia and Thrace extend between 39° 50' and 41° 44' 50'' N latitude and 20° 48' 30'' and 26° 38' E longitude, and with the islands of Samothrace and Thasos cover an area of c. 43.000 Km^2. Macedonia and Thrace belong to the Northern Greece and are areas mainly mountainous. Mountain ranges are interrupted in some places by the valleys that form the rivers Aliakmon, Axios, Strimon, Nestos and Evros.

Limestones and marble form the substrate on some mountains of north-east Greece (Pangeon, Falakro, etc.) and igneous rocks occur particularly in Northern Greece. Granite is found in some places in Northern Greece, with the most important occurrences being on Varnous and Vrondous mountains. Gneisses and crystalline schists are widespread in north-east (e.g. on Belles and Rodopi). Of great botanical interest is also a small area of ophiolite (serpentine) (Strid 1986).

According to the average temperature and the precipitation, the climate of Northern Greece can be classified as of mainly Mediterranean type but it widely varies depending on the altitude and on the orientation of the slopes. Divergence of this type occurs on high altitudes above 700 m. Generally it can be said that in the areas that are away from the coastal zone the climatic type can be characterized as transitional between the Mediterranean and the Continental type of climate (Balafoutis 1977).

For this work the last published inventory of Düll (1995) was mainly used and also some additional unpublished records (Papp and Babalonas, Papp et al., in press). The chorological analysis was based on Düll (1983, 1992, 1995).

3. Results and Discussion

For the investigated territory of Macedonia and Thrace, the total number of taxa is 505. This includes 99 liverworts (20%) and 406 (80%) mosses. The liverworts are represented by five orders with the most important of them the order of Jungermanniales (67%), though there are fifteen orders of mosses with Hypnobryales (26%) and Pottiales (21%) to have the prevalence. There is a total of 32 families with 123 genera of mosses, and 31 families with 45 genera of the liverworts. In the class of mosses the genus Bryum (Bryaceae) includes 26 species, and the genus Orthotrichum (Orthotrichaceae) follows with 23 species. The liverworts genera with the majority of species are Scapania (9 sp.), Lophozia (7 sp.), Cephaloziella and Riccia (6 sp.) (Table 1).

Table 2 provides a comparison of the number of taxa of the area with those of other Mediterranean areas from which exist bryophytes inventories. It is still obvious that Macedonia and Thrace bryoflora is comparatively rich, although is still a lot of work to be done. The only area from which a bigger number of bryophyte species is known is Turkey, which has however a surface which is larger than Macedonia and Thrace. From table 2 it can also be seen the slightly bigger number of liverworts in Sicily (Dia et al. 1991).

From the chorological analysis three (3) basic chorological groups were defined: the Circumboreal, the Mediterranean and the Atlantic element. The Circumboreal element is best represented in Northern Greece by 61%, followed by the Mediterranean with 30% and the Atlantic element with 9% (Fig. 1, Table 3). The high percentage of the Circumboreal and Mediterranean element in Northern Greece is attributed to the geographical position and to the mountainous character of the studied area. This explains why the group of taxa that have "a montane distribution" constitute a considerable big group (c. 170 taxa) (Table 3).

The results presented here is certain that they will be modified since the bryological exploration of Greece is in progress. It is in process this data to be also classified in ecological groups, based on some of the main factors that affect their distribution like water, pH of the substratum, the type of substratum and the light.

TABLE 1. Species numbers of the most important genera of Northern Greece bryoflora (Macedonia and Thrace).

MOSSES		LIVERWORTS	
genera	number of species	genera	number of species
Bryum	26	Scapania	9
Orthotrichum	23	Lophozia	7
Eurhynchium	14	Cephaloziella	6
Brachythecium	12	Riccia	6
Fissidens	12	Marchantia	4
Syntrichia	12	Porella	4
Didymodon	11	Barbilophozia	3
Grimmia	10	Fossobronia	3
Hypnum	8	Frullania	3
Polytrichum	8	Jungermannia	3
Tortula	8	Lophocolea	3

TABLE 2. A numerical comparison of the bryoflora of Northern Greece (Macedonia and Thrace) with other Mediterranean areas (Data from Dia & Not, 1991. The * indicates that the number of Crete species was modified by us, according to Düll, 1995).

	MOSSES			LIVERWORTS		
area	families	genera	species	families	genera	species
Turkey	38	154	549	35	56	135
Sicily	34	121	367	33	47	111
Sardinia	30	105	336	26	33	62
Israel	20	71	202	16	16	33
Crete *	24	84	255	28	35	76
North Greece (Macedonia-Thrace)	32	123	406	31	45	99

TABLE 3. The chorological groups of Bryophytes in Northern Greece (Macedonia and Thrace). The number in the parenthesis indicates the Bryophyte species that have a "montane distribution".

	Liverworts	Mosses	Total
Circumboreal	54 (26)	251 (82)	305 (108)
Mediterranean	32 (7)	121 (32)	153 (39)
Atlantic	13 (9)	34 (16)	47 (25)
Total	99 (42)	406 (130)	505 (172)

4. References

Balafoutis, C. (1977) Contribution to the study of the climate of Macedonia and West Thrace. Ph.D. Thesis, University of Thessaloniki, Thessaloniki.

Dia, M.G. and Not, R. (1991) Chorological and ecological analysis of the bryophyte flora in Sicily, *Flora Mediterranea* **1**, 143-156.

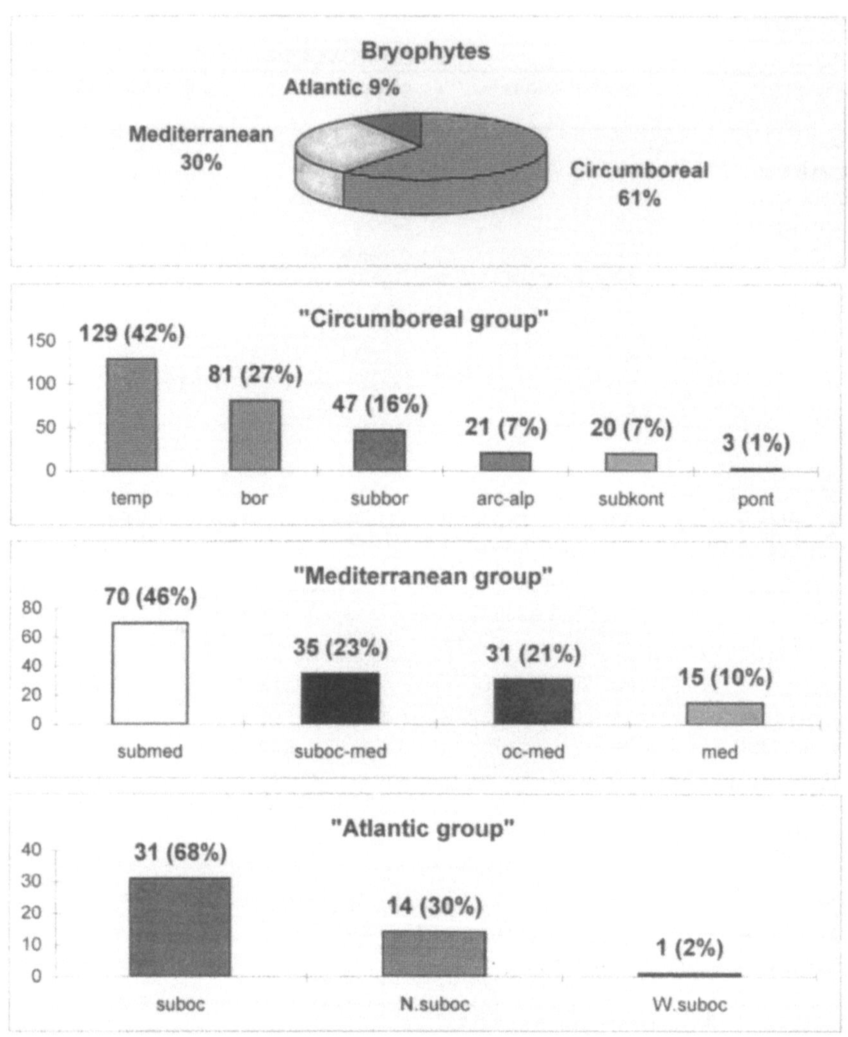

Fig. 1. Chorological spectra of the bryoflora in Northern Greece (Macedonia and Thrace).

Düll, R. (1983) Distribution of the European and Macaronesian liverworts (Hepaticophytina). *Bryologische Beiträge* **2**, IDH-Verlag.

Düll, R. (1992) Distribution of the European and Macaronesian mosses (Bryophytina). Annotations and progress. *Bryologische Beiträge* **8/9**, IDH-Verlag.

Düll, R. (1995) Bryophytes of Greece, *Bryologische Beiträge* **10**, IDH-Verlag.

Papp, B. and Babalonas, D. (in press) Bryophytes of some maquis and phrygana stands in Greece, *BIOS* (Macedonia, Greece).

Papp, B., Tsakiri, E. and Babalonas, D. (in press) Bryophytes and their environmental conditions at Enipeas (Mt. Olymp) and Lykorrema (Mt. Ossa) streams. Proc. 1st Balkan Botanical Congress.

Strid, A. (1986) *Mountain flora of Greece*, Vol. 1, Cambridge University Press.

PRELIMINARY RESULTS ON THE WOODY PLANTS COMPOSITION OF CRETAN PHRYGANIC ECOSYSTEMS

G. M. TSIOURLIS, Z. TSEKOURA, P. KASAPIDIS, M. CHATZAKI and M. MAROUKLI
University of Crete, Biology Department
P.O.Box 1470, 71110 Iraklion Greece
**present address: NAGREF, Forest Research Institute*
57006 Vasilika, Thes/niki Greece

1. Introduction

The island of Crete, as one of the first places of civilization, has been subject to intensive human activities for the last 6.000 years. During this long period, fire, grazing and wood cutting have degradated and drastically changed the natural climatic sclerophyllous-evergreen forests (Kermes Oak -*Quercus coccifera*- and Pine -*Pinus brutia*) and high matorrals (Wild Olive -*Olea europaea* var. *sylvestris*- and Carob -*Ceratonia siliqua*-) to lower vegetal formations. Nowadays, the additional impact of the dry Mediterranean climate and the eroded soils, make phrygana, a low matorral of hemishperical dwarf-shrubs, the most common natural ecosytem in Crete with a cover of about 25% of the island (Kypriotakis *et al.*, 1996).

In this work we present some preliminary results on the structure of the phryganic ecosystems. We record and classify the different types that exist in Crete, according to the composition and diversity of the woody plant species and the plant coverage.

2. Materials and methods

Crete is one of the largest islands of the Mediterranean Basin. Having a sharp relief with altitudes up to 2400 m, all the Mediterranean climatic zones are present (thermo, meso, oro and supra-Mediterranean). Rainfall and temperature, the first rises and the latter falls from east to west, from south to north and from the coasts to the top of the mountains. In the present work we selected 32 stations of different altitudes from the sea level up to 1400m and on the main geological substrates. The stations are large, homogeneous areas (as far as the species composition and plant coverage are concerned) and scattered throughout the island. We used the line-intercept method to estimate the coverage of the woody plant species. In each station 4 transects (50m long) were selected and in every 0.50m the plant species were recorded as well as their height and projection of the canopy on the transect. The diversity of the woody plant species was calculated using the Shannon-Wiener index H'.

3. Results

The total number of the recorded woody species is 49 (Table 1). The main families represented are *Lamiaceae* (12), *Fabaceae* (7), *Cistaceae* (4), *Euphorbiaceae* (3), *Ericaceae* (3), and *Compositaea* (3). There are three endemics of Crete (*Ebenus cretica*, *Phlomis lanata* and *Verbascum spinosum*) and three endemics of the South Aegean area (*Phlomis cretica*, *Ononis spinosa* and *Stachys spinosa*). The commonest and most abundant species, present at the two third of the stations each, are the Thorny Burnet -*Sarcopoterium spinosum*-, which represents 25% of the relative coverage (RC), the Thyme -*Thymus capitatus*- (RC 14%) and the Spiny Broom -*Calicotome villosa*- (RC 8%).

The mean plant cover is 41% and is greatly variant among the stations ranging between 23% and 80% (Figure 1). According to the plant cover, we distinguish sparse (0-30%), medium (low medium 30-45% and high medium 45-60%), and dense (>60%) phrygana. Most stations are in the medium class (25) and particularly in the low medium (20).

The species diversity index H' also varies greatly, between 0.12 and 0.86 (Figure 1). We distinguish phrygana with a low diversity (< 0,33), with a moderate diversity (0,33<H'<0,66) and with a high diversity (>0,66). In phrygana with a low diversity one species dominates the plant cover: *S. spinosum*, *Phlomis fruticosa* or *Euphorbia dendroides*. In phrygana with a moderate diversity two species represent the 80% of the total cover: *T. capitatus* with *C. villosa* or *S. spinosum*; or three species *C. villosa*, *Euphorbia acanthothamnos* along with *T. capitatus*. In phrygana with a high diversity, five or more species are codominant.

Sarcopoterium's relative cover declines with the raise of diversity while other important species as *T. capitatus*, *C. villosa*, *Genista acanthoclada* or *Salvia fruticosa* appear almost exclusively at Shannon index values over 0,5 (Figure 2).

At altitudes above 1000m mid-elevated phrygana, characterized by *Berberis cretica*, *E. acanthothamnos* and *Astragalus angustifolius,* replace the typical phrygana.

TABLE 1: Woody plant species and their mean coverage in 32 phryganic ecosystems through the island of Crete (Greece). (+ = < 0,3%; CE = Crete endemic; SAE = South Aegean Endemic)

WOODY PLANT SPECIES: *Sarcopoterium spinosum* (10,5%), *Thymus capitatus* (5,9%), *Calicotome villosa* (3,4%), *Phlomis fruticosa* (2,%), *Cistus creticus*, *C. parviflorus* & *C. salvifolius* (2,8%), *Euphorbia acanthothamnos* (2,0%), *Genista acanthoclada* (1,7%), *Phlomis lanata*[EC] (1,7%), *Salvia fruticosa* (1,4%), *Euphorbia dendroides* (1,3%), *Satureja thymbra* (1,3%), *Berberis cretica* (0,9%), *Pistacia lentiscus* (0,8%), *Ebenus cretica*[EC] (0,5%), *Phlomis cretica*[SAE] (0,4%), *Olea europaea* var. *sylvestris* (0,4%), *Rhamnus oleoides (0,4%)*, *Astragalus angustifolius* (0,3%), *Ballota pseudodictamnus*(0,3%), *Erica manipuliflora* (0,3%), *Ononis spinosa*[SAE] (0,3%), *Arbutus unedo* (0,2%), *Asparagus aphyllus* (0,2%), *Erica arborea* (0,2%), *Helichrysum microphyllum* (0,2%), *Origanum onites*, *Helichrysum* sp., *Osyris alba*, *Euphorbia characias*, *Quercus coccifera*, *Acer creticus*, *Ceratonia siliqua*, *Hypericum empetrifolium*, *Pyrus spinosa*, *Teucrium divaricum*, *Teucrium* sp., *Thymelaia hirsuta*, *Thymelaia hirsuta*, *Verbascum spinosum*[EC], *Anthyllis hermanniae*, *Astragalus* sp., *Cupressus sempervirens*, *Fumana arabica*, *Phagnalon graecum*, *Alysum fallacinum*, *Ruta halepensis* and *Stachys spinosa*[sae] (+)

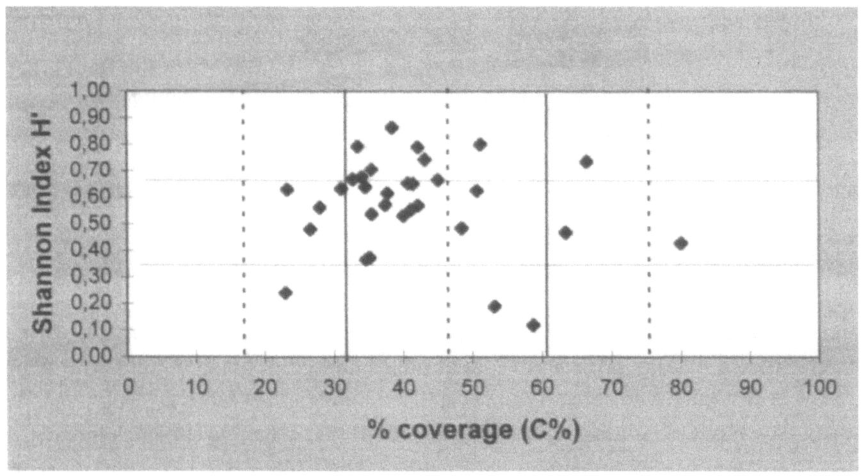

FIGURE 1: Classes of phrygana according to the woody plant cover and the species diversity of 32 stations in Crete (Greece).

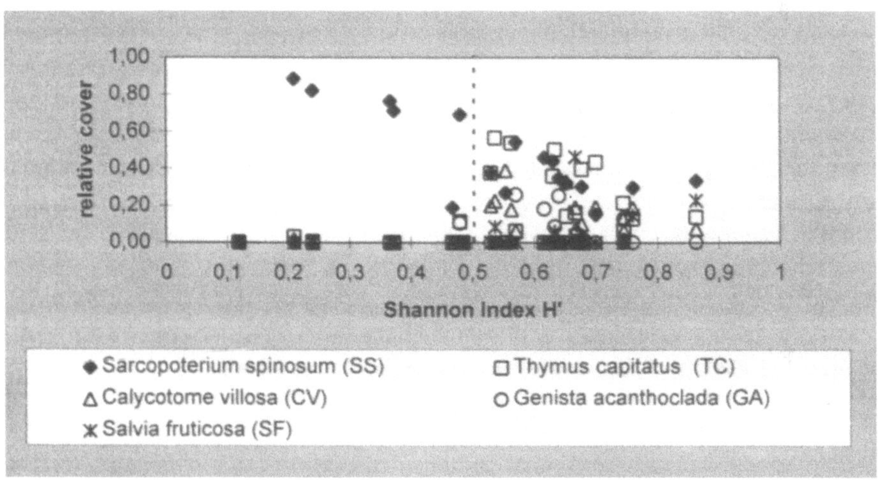

FIGURE 2: Woody species' relative cover in relation to the diversity of 32 Cretan phryganic ecosystems (Greece).

4. Discussion

Cretan phrygana ecosystems are particularly rich in woody species in comparison to the rest of Greece. The number of the typically phryganic species recorded in the 32 selected stations is about thirty five. The characteristic species are *S. spinosum*, *T. capitatus*, *C. villosa*, *Phlomis* spp., *Cistus* spp., *Euphorbia* spp., *Genista acanthoclada*, *Salvia fruticosa*, *Satureja thymbra*, *B. cretica*, and *E. cretica*. The other recorded woody species (e.g. *Q. coccifera*, *C. siliqua*) are relicts of higher vegetation that existed at the same place before degradation. Most species have an East-Mediterranean distribution, while *E. cretica*, *P. lanata* and *V. spinosum* are endemics of Crete. The Thorny Burnet is the commonest species and is present in the 2/3 of the stations. It presents a strong dynamics in post fire succession and in colonizing abandoned fields. Because of these reasons, along with its low palatability, *S. spinosum* is the most abundant phryganic species not only in Crete, but in the whole of Greece (Economidou, 1976). The Thyme and the Spiny Broom are also widely dispersed among sites but they never dominate. Other species such as *S. fruticosa*, *G. acanthoclada*, *E. acanthothamnos*, *P. fruticosa*, *P. lanata* and *Cistus* spp. are recorded in less than the 1/4 of the stations and they usually represent less than 50% of the total plant cover of each station, with the exception of *P. fruticosa*, *Cistus* spp. and *E. dendroides* that dominate each in one station. The dominance of *E. dendroides* is also reported by Diamantopoulos *et al.* (1994) in certain greek phrygana.

In sites with a low species diversity *S. spinosum* often dominates. Most of these sites can be considered as highly disordered due mainly to heavy grazing but also to frequent fires. Sites with a medium diversity are usually dominated by thorny species (such as *S. spinosum*, *C. villosa*, *G. acanthoclada*) which are more resistant to grazing. In sites with a high diversity, that can be considerd as less disordered, there is a significant increase of non thorny species (*Phlomis* spp., *Cistus* spp. and *S. fruticosa*) that as more palatable they are less resistant to the grazing pressure.

The most representative type of phrygana is that with a low medium plant cover (about 40%) and a medium diversity (H' about 0,5).

Similar types of phrygana can be found from the sea level up to 1000m in the whole island. The climatic gradients do not imply differences in the woody plant species composition unlike the anthropogenic management practices. In this first approach the effect of the geological substrate is not clear, but, as Bergmeier (1996) and Kasapidis and Tsiourlis (in press) demonstrate, some species prefer limestone substrate (e.g, *T. capitatus, E. acanthothamnos*) where there is also a higher diversity of woody plants. Above this altitude, there is the Cretan endemic habitat of mid-elevated phrygana, which is characterized by the spiny *B. cretica*, *E. acanthothamnos* and *A. angustifolius*. The dominance of mid-elevated species follows the altitudinal gradient. *S. spinosum* gradually falls off up to 1.100m and *E. acanthothamnos* steps up above 500m.

5. References

Bergmeier E., 1996. Zur Bedeutung der Beweidung fuer die griechische Phrygana. *Ber.d.Reinh. - Tuxen-Ges.* 8, 221-236.

Diamantopoulos J., Pirintsos S.A., Margaris N.S. and Stamou G.P., 1994. Variation in Greek phrygana vegetation in relation to soil and climate. *Journ. Vegetation Science*, 5, 355-360.

Economidou E., 1976. La répartition des phrygana en Grèce et ses rapports avec le climat et l'influence anthropique. *Doc. phytosocio. Lille*, Fasc. 15-18, 45-56.

Kasapidis P. and Tsiourlis G.M. (in press). Preliminary results on the structure of ecosytems and the desertification processes in relation to overgrazing in Asterousia mountain (Crete, Greece). *Proceedings of the Ént. Conf. on Mediterranean Desertification*, Crete, 1996, 9 p.

Kypriotakis Z., Tzanoudakis D. and Tsiourlis G.M., 1996. Vegetation map of Crete. *Proceedings of the 6th Botanical Scientific Conference*, 5-11 April 1996, Paralimni - Cyprus, 301-306. (In greek).

Acknowlodgements: This study is supported by the University of Crete.

THE GENUS *PAEONIA* IN GREECE: A SUBSPECIES AND A CYTOGEOGRAPHICAL DISTRIBUTION PATTERN NEW TO THE GREEK AREA.

D. TZANOUDAKIS [1] and TH. ARAMPATZIS.[2]

1. *Division of Plant Biology, Dept. of Biology, University of Patras, 26500 Patras, Greece.*

2. *Dept. of Forestry, Technological Education Institute of Kavala, Proastio,66100 Drama, Greece.*

1. Introduction

Paeonia is one of the most attractive and interesting genera in the Greek flora from a biogeographical point of view. According to Tzanoudakis (1977) the genus is represented in the Greek area by 4 species and two of them, viz. *P. clusii* and *P. parnassica* are Greek endemics. The other two species, viz. *P. peregrina* and *P. mascula* are more widespread in the Balkan peninsula and Europe respectively but show in Greece a distinct distribution pattern with the first occupying the central and northern continental Greece and the latter the southern part of the Greek peninsula and the Ionian and Aegean area. These distribution patterns have been modified recently by the discovery of a population of *P. mascula* on Mt. Menikion (NE Greece, province Drama). Based on morphological, cytological and ecological attributes this population is well differentiated from all the subspecies of *P. mascula* known from the Greek area and seems to belong to *P. mascula* subsp.*triternata,* a taxon new for Greece.

2. Material and methods

In the frame work of a floristic exploration of the mountains in the province of Drama undertaken by the Forestry department of the Technological Education Institute of Drama, a population of *Paeonia mascula* was collected from Mt. Menikion at an altitude of c. 1050 m.The population was very small and localized. Both herbarium specimens and living material (2 rhizomes) were collected for morphological and cytological studies. Taxonomy and nomenclature follows that of Tzanoudakis (1977) and Akeroyd (1993). Voucher specimens are deposited in Patras (UPA) as well as at the Forestry department in Drama.

The chromosomes have been studied in root tip mitosis according to Tzanoudakis (1977 & 1983).

3. Results

Morphology: As previously mentioned the *Paeonia mascula* population on Mt. Menikion is very small. Less than 10 individuals were found and most of them were sterile, i.e., consisting of a singe basal leaf. The well-developed and flowering individuals were characterized by the following characters.

Stem: 30-40 cm high, greenish and glabrous.

Leaves: 3-4, biternate, well-developed at anthesis, glabrous; the lower leaves consisting of 9 broadly elliptical to obovate leaflets, dark green above and greenish white below.

Flowers: 7-10 cm in diameter with 6(-7) narrowly obovate and almost not overlapping purplish - red petals and 2 white tomentose follicles.

Cytology: In the material cytologically investigated the diploid chromosome number of 2n=2x=10 was found. The haploid chromosome complement consisted of 2 metacentric, 2 submetacentric and one subtelocentric chromosome. In the subtelocentric chromosomes and in one of the submetacentric pairs (the smaller one) nucleolar organizers have been observed proximal to the telomere of the shorter arm (figure 1). Regarding the chromosome number similar results have been mentioned by Sopova (1971) in material of *P. mascula* (syn. *P. corallina)* from Mt. Zeden in the Former Yugoslavian Republic of Macedonia (FYROM). One of the authors (D.T.) had also found the same chromosome number in material of *P. mascula* collected by Prof. D. Phitos from the Crimea in 1978. It is well known that the Crimea is the *locus*

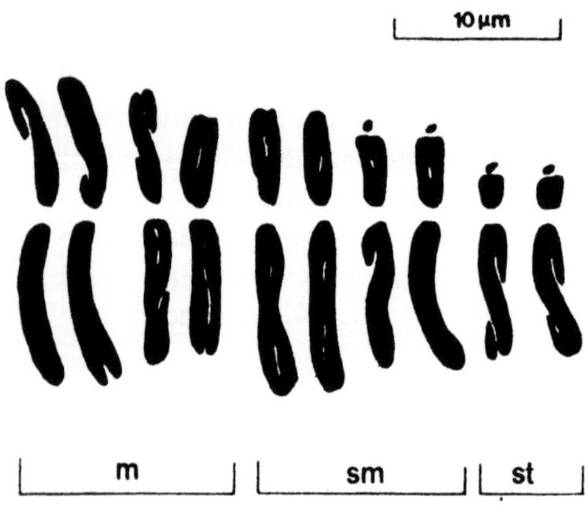

Fig. 1. Karyogram of *Paeonia mascula* subsp. *triternata* from Mt. Menikion. (North Greece Prov. Drama).

classicus of *P. daurica* Anders.(syn. *P. mascula* L. subsp. *triternata* (Pallas ex DC.) Stearn & P.H. Davis).

As far as chromosome morphology is concerned, the haploid chromosome complement in the Greek population studied seemed to be characterized by a second submetacentric chromosome (r-index values 1.7-1.9). This is the second largest chromosome in the haploid complement. According to Tzanoudakis (1977 & 1983) all diploid Greek populations of the genus have a haploid chromosome complement of 3 metacentric, 1 subtelocentric and one submetacentric chromosome.

Ecology: The population of *P. mascula* was found on the calcareous slopes of Mt. Menikion at an altitude of c. 1050 m. The area is characterized by dense *Fagus sylvatica* forest together with several other trees and shrubs such as *Cornus mas, Corylus avellana, Fraxinus ornus, Acer campestre, A. pseudoplatanus, Crataegus monogyna, Sorbus torminalis, Rubus idaeus, Ligustrum vulgare* etc.
The population flowers in the second half of May.

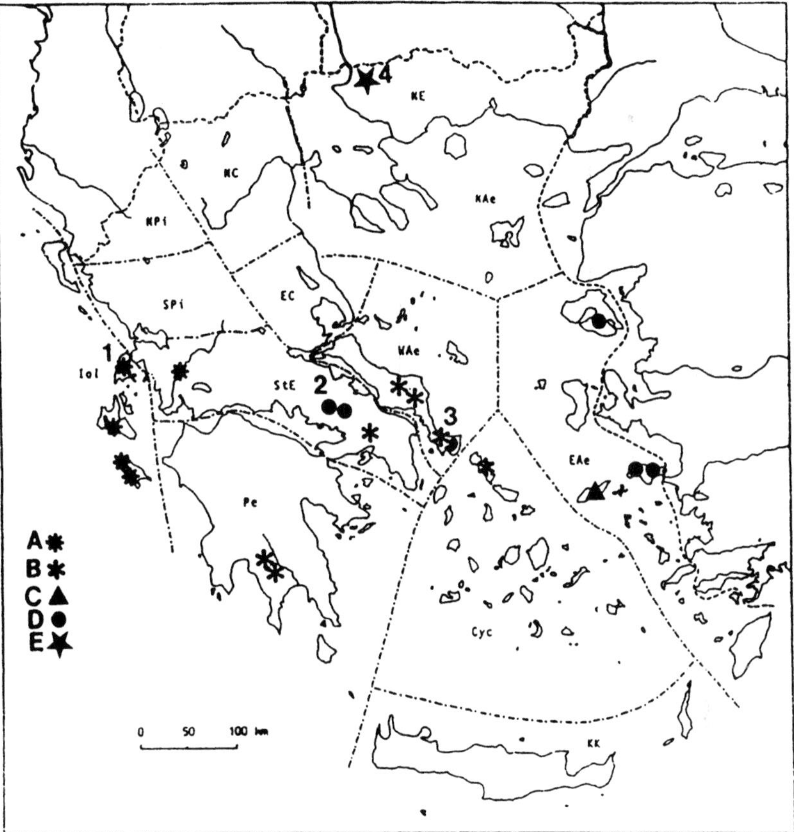

Fig. 2. Geographical distribution of *Paeonia mascula* in Greece. The number indicate localities where the distribution range of *P. mascula* meets those of *P. peregrina* (1,4) and *P. parnassica* (2) or where two different subspecies of *P. mascula* coexist (3) Phytogeographical subdivision of Greece according to Strid (1991).
A: subsp. *russoi,* **B**. subsp. *hellenica,* **C**: subsp. *icarica,* **D**: subsp. *mascula* **E**: subsp.*triternata.*

4. Taxonomic relationships

Paeonia mascula is a species which demonstrates its maximum morphological diversity in Greece. The new population found in the north of Greece not only extends the distribution range of the species in the Greek area but also the whole variation together with the problems relating to intraspecific classification. Due to its purplish - red flowers the population on Mt. Menikion is well differentiated from the white - flowered subspecies *P. mascula* subsp. *hellenica* Tzanoud. and subsp. *icarica* Tzanoud. According to the number of leaflets of the lower leaves, the population concerned is closer to subsp. *russoi* (Biv.) Cullen & Heywood but this subspecies is characterized by leaflets which are pubescent beneath, narrowly elliptical, acute, and not well developed at anthesis. Greek populations of *P. mascula* subsp. *mascula* are characterized by the lower leaves consisting of more than 9 leaflets. Another morphological character differentiating the northern Greek population of *P. mascula is* the non-overlapping almost narrowly obovate petals.

Since the northern Greek population is also cytologically well-differentiated from the tetraploid (in Greece) populations of *P. mascula* subsp. *mascula*, we decided to treat it as a different subspecies. According to the taxonomy of Stearn & Davis (1984) and Akeroyd (1993), we can classify the taxon as *P. mascula* L. subsp. *triternata* (Pallas ex DC.) Stearn & P.H. Davis. We should also mention that this classification is more or less tentative until a comparative study of the Balkan populations can be carried out. It should be noted however, that cytological results available from northern Greece, FYROM (Sopova 1971) and the Crimea favour such a taxonomical treatment.

5. References

Akeroyd, J.R. 1993. *Paeonia*: In Tutin, J., Burgess, N.A., Chater A.O., Edmondson, J.R., Heywood, V.M., Moore, D.M., Valentine, D.H., Walters, S.M. & Webb, D.A. eds: *Flora Europaea.* Second Edition, Cambridge University Press, pp.292-294.

Sopova, M. 1971. The cytological study of two *Paeonia* species from Macedonia. *Fragm. Balc. Mus. Maced. Sci. Nat, 8(16) :134-142.*

Stearn, T. W. & Davis P.H. 1984. A. taxonomic and historical survey of the genus *Paeonia* in Greece. Goulandris Natural History Museum, Athens,Greece.

Strid, A. 1991. The "Flora Hellenica" Project. *Bot. Chron. 10:81-94.*

Tzanoudakis, D. 1977. A cytotaxonomical study of the genus Paeonia in Greece PhD Thesis, University of Patras. Patras, Greece.

Tzanoudakis, D. 1983. Karyotypes of four wild *Paeonia* species from Greece. *Nord. J. Bot. 3: 307-318.*

CONTRIBUTION TO THE STUDY OF THE GREEK INSULAR FLORA: ANTIKYTHERA AND THE ISLETS AROUND KYTHERA

D. TZANOUDAKIS, G. IATROU, M. PANITSA, AND P. TRIGAS
University of Patras, Department of Biology, Division of Plant Biology, 26 500 Patras, Greece

1. Introduction

The Kythera-Antikythera island chain is one of the most interesting areas of Greece, from a floristic and biogeographical point of view. It is considered as the remnant of an old landbridge which connected Western Crete with Southern Peloponnisos. The whole area is dominated by limestone cliffs, a biotope where many Greek and Aegean floristic elements find refuge. The island chain consists of 3 main, inhabited islands, Elafonissos, Kythera and Antikythera, and a number of small and uninhabited islets (Fig. 1). Elafonissos and Kythera are considered floristically well known (see Tzanoudakis & Panitsa 1994 for references), so we focused our efforts on the floristic exploration of the islets, which remained quite unexplored, and also to Antikythera which is the most remote and isolated among the larger ones. The latter is included in "Flora Kytherea" of Greuter and Rechinger (1967), but the floristic information are based on collections made by the authors and Prof. Phitos, during the same season (first half of May) of the same year (1964).
The floristic information given in this paper should be considered as a preliminary addition to the flora of the area. It concerns mainly plant taxa new to Antikythera collected during our visits to the area in Autumn (November 1996) and early Spring (April 1997). The present page limit does not permit the publication of the complete list of species collected from the islets around Kythera. Also, the material collected during our last visit to the area (June 1997) has not been studied thoroughly, yet. As soon as it is prepared, a complete species list will be published.

2. Materials and Methods

The whole area studied is shown in Fig. 1. Material has been collected from a total of 9 islets; vouchers are deposited in Patras (UPA) and living material for cytological studies is in cultivation in the University's garden. The taxonomy and the nomenclature follow that of Flora Europaea (Tutin et al. 1964-1980, 1993) and where possible Med-Checklist (Greuter & al. 1984-1989). The same floras have been also used for chorological information regarding the taxa studied, but recent contributions to the area or to the taxa concerned have also been taken into consideration. Abbreviations used in Table 1 follow that of "Flora Hellenica" project, in addition AKT: Antikythera, KT: Kythera and EB: Euboea.

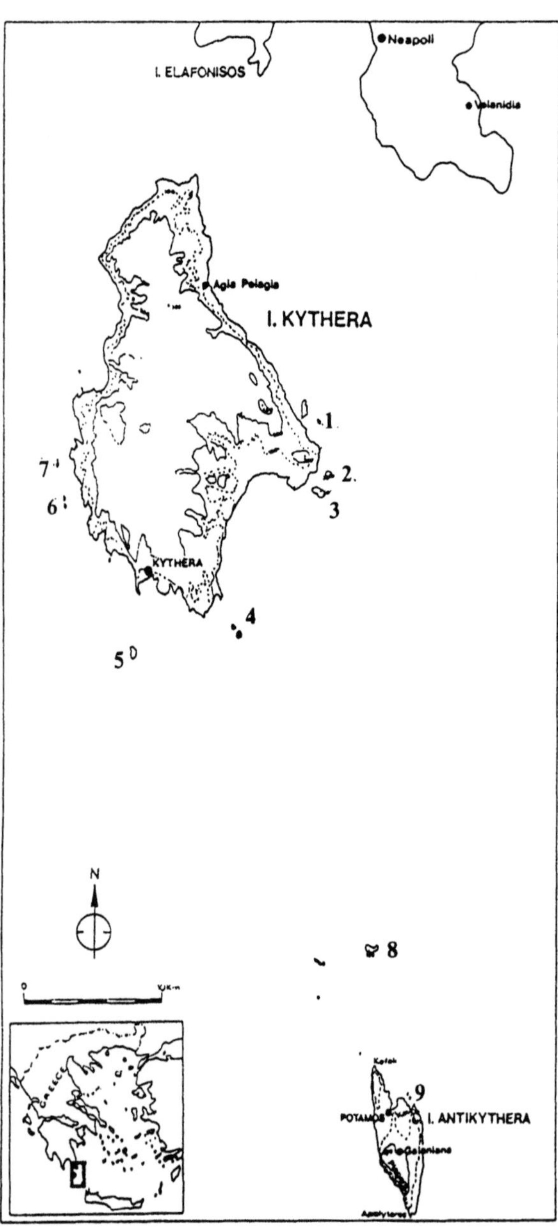

Fig 1. Antikythera and islets around Kythera.
Abbreviations: 1. Prassonisi, 2. Antidragonera, 3. Dragonera, 4. Kofinidia,. 5. Avgo (Chytra), 6. Strogylo, 7. Lydia, 8. Porion (Prassou), 9. Thimonies.

3. Results and Discussion

Islets

In a total of 9 islets studied some are close to Kythera (offshore islets 1,2,3,6,7), one is close to Antikythera (9) and the remaining (4,5,8) are much more remote and isolated from the main islands of Kythera and Antikythera (Fig.1). From the islets offshore of Kythera, a total of 84 plant taxa have been registered. Most of these islets, and especially the smaller ones, demonstrate a typical islet flora and littoral vegetation characterized by plant taxa such as: *Frankenia hirsuta, Anthemis scopulorum, Brassica cretica, Convolvulus oleifolius* subsp. *scopulorum,Allium commutatum, Suaeda vera* etc. It should be mentioned that five taxa found on these offshore islets (viz. *Anthemis rigida, A. scopulorum, Allium commutatum, Ferula communis* subsp. *glauca* and *Trigonella rechingeri*) have not been found from the main island. On the larger islets Dragonera and Antidragonera, in addition to the littoral elements, an internal vegetation zone exists, dominated by shrubby species like *Euphorbia dendroides* and *Pistacia lentiscus*. On the offshore islet "Thimonies" (NE of Antikythera) the situation is similar to that of the flora and vegetation of the small offshore islets of the Kythera area. Regarding their flora and vegetation the islets Avgo (or Chytra) and Porion (or Prassou) which are larger and much more remote and isolated than the offshore islets mentioned above, also seem to be very interesting. In both, due to their geomorphology, chasmophytic communities are well represented by a number of Greek and /or Aegean endemics. It is a pity that the first island due to its coasts which fall vertically into the sea, is almost inaccessible and its rich chasmophytic flora cannot be thoroughly explored. During our single visit to its NW side we collected or observed species like *Scorzonera cretica, Anthemis scopulorum, Centaurea cytherea, Dianthus fruticosus* s.l., *Inula candida* etc. Contrary to the islet Avgo, the islet Prassou has been explored throughout and its flora and vegetation will be published. In addition to the chasmophytic elements mentioned above, on Prassou, the presence of *Biarum tenuifolium* subsp. *abbreviatum* is also noteworthy. This taxon is new to Kythera-Antikythera area as a whole.

Antikythera

From a preliminary floristic analysis of the material collected during our three recent visits to the island, we estimate that more than 60 taxa will be added to the flora of the island (i.e. not mentioned by Greuter & Rechinger 1967), and some of them seem to be biogeographically significant (Table 1).Taxa like *Allium rubrovitatum, Valeriana asarifolia,* and *Silene integripetala* subsp. *lidenii*, were considered to be Southern Aegean (Crete-Karpathos) endemic taxa and now their distribution range is extended to Antikythera and supports the close phytogeographical relationships of this island to Crete (cf. Iatrou 1994).

Table 1. New for Antikythera and other important plant taxa of the area studied (with their total geographical distribution).

Important plant taxa new to Antikythera	
Allium rubrovitatum	AkT KK
Silene integripetela ssp. *lidenii*	AkT KK
Valeriana asarifolia	AkT KK
Allium circinatum	AkT (local form) KK Pe
Procopiana cretica	AkT KT KK Pe
Tulipa goulimyi	AkT KT KK Pe
Crocus laevigatus	AkT KT KK Pe Kik EB ST
Crocus boryi ssp *tournefortii*	AkT KK Kik EAe
Allium longanum	AkT KK (islets) Kik N.Africa
Other important plant taxa of the area studied	
Anthemis scopulorum	KT (islets) KK EAe KiK
Biarum tenuifolium ssp. *abbreviatum*	Porion (Prassou) Pe IoI
Trigonella rechingeri	KT (islets) KK (islets) KiK

The case of *Allium longanum* is also significant in that this is a very rare taxon in Greece previously known only from two islets South of Astipalea (Kik) and from the

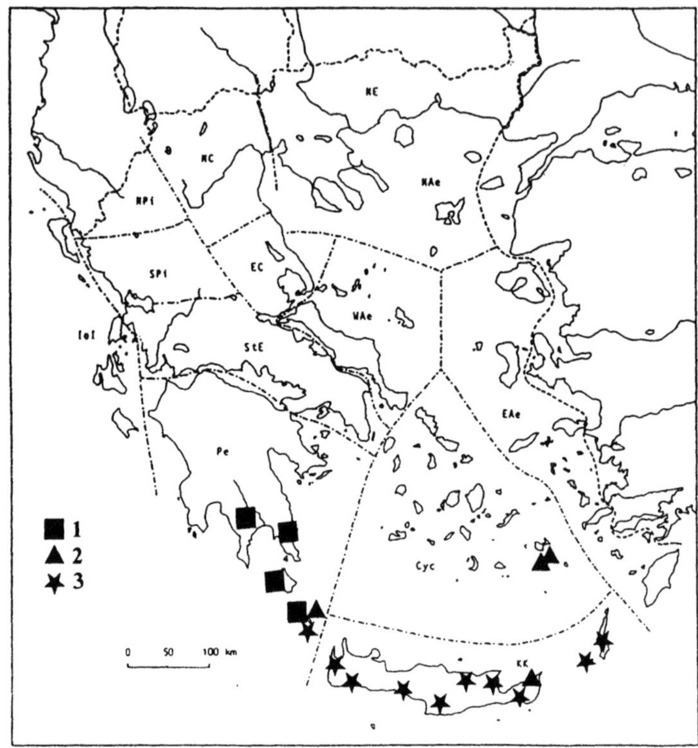

Fig. 2: Distribution ranges of *Allium gomphrenoides* (1), *A. longanum* (2) and *A. rubrovitatum* (3) in Greece.

Easternmost end of Crete, on offshore islets (Sitia). Similar is the case of *Allium circinatum,* and *Procopiana cretica,* which show a distribution centre in the Southern Aegean area, but their distribution range extends NW to Peloponnisos. Also, based on the distribution range of some species of the flora of Antikythera, it is evident that on the island species showing their distribution centre Northernmost or Southernmost of the island coexist (Fig. 2).

It is noteworthy that most of the taxa new to Antikythera are Autumn or early Spring flowering species, an observation which emphasizes once again the need for more intensive exploration of the Greek island flora during all seasons.

4. References

Greuter, W., Burdet, M.M. & Long, G. (1984-1989): *Med-Checklist*. Vols. 1, 3, 4. Geneve.

Greuter, W. & Rechinger, K.H. (1967): Flora der insel Kythera, Gleichzeitig Beginneiner nomenklatorischen Uberprufung der griechischen Gefasspflanzenarten. *Boissiera* 13.

Iatrou, G. (1994): The endemic flora of the island of Kythera. Proceedings of the 5th Sc. Conference of Hel. Bot. Soc. Delphi (Greece): 213-216.

Tzanoudakis, D. & Panitsa, M. (1994). The flora of the Greek islands. *Ecologia Mediterranea* XX (3): 1-20.

Tutin, T.G. & al. (eds.) (1964-1980): *Flora Europaea* 1-5. Cambridge University Press.

Tutin, T.G. & al. (eds.) (1993): *Flora Europaea* 1 (Second edition). Cambridge University Press.

Aknowledgements: Financial support by the Ministry of Environment and Public Works is kindly aknowledged. The authors also thank the coast guards for transport facilities.

Biochemistry, Metabolism and Bioenergenetics

CHILLING SENSITIVITY AND ANTIOXIDANT METABOLISM IN MAIZE (*Zea mays* L.)

A. H. KINGSTON-SMITH AND C. H. FOYER
Environmental Biology Department
Institute of Grassland & Environmental Research
Plas Gogerddan
Aberystwyth
Ceredigion SY23 3EB
Wales, U.K.

Abstract

Maize leaves are extremely sensitive to low temperature-induced damage. Chilling results in decreased CO_2 assimilation due to loss of key proteins, followed by photoinhibition and photo-oxidation. Destruction of proteins is linked to increased oxidation and activity of proteases. The antioxidant defence system protects the cell against damage from active oxygen species ($O_2^{\cdot -}$, H_2O_2 and $^1O_2$). Differential compartmentation of metabolism (*e.g.* photosynthesis) and fluxes of metabolites between bundle sheath and mesophyll cells are a feature of C4 metabolism. The distribution of the antioxidants in maize leaves is not uniform: these are compartmented between the bundle sheath and mesophyll according to requirement for and availability of reducing power and NADPH. Dehydroascorbate and oxidised glutathione produced in the bundle sheath have to be transported to the mesophyll to be reduced, whereas foliar H_2O_2 was predominant in the mesophyll. Chilling damage in maize leaves is severe because of an inability to acclimate to low temperatures, even when the antioxidant capacity was altered by overexpression of superoxide dismutase.

Abbreviations

AA; ascorbate, APX; ascorbate peroxidase, DHA; dehydroascorbate, DHAR; dehydroascorbate reductase, Fd; ferredoxin NADP-oxidoreductase, GR; glutathione reductase, GSH;glutathione GSSG; glutathione (oxidised form), MDHA; monodehydroascorbate, MDHAR; monodehydroascorbate reductase, PSI; photosystem I, PSII; photosystem II, SOD; superoxide dismutase.

The effect of low temperature on maize

Maize is one of the most important agricultural crops but it requires relatively high temperatures for optimal growth and is sensitive to chilling-induced damage. Various mechanisms have been proposed to account for chilling injury in plants. Persistent metabolic disturbances resulting from the imposition of sub-optimal temperatures ultimately result in imbalances of supply and demand that result in damage to the delicate cellular matrix. In maize leaves the metabolic perturbation caused by low temperatures causes damage to the chloroplast which is manifested by a light dependent inhibition of photosynthesis called "photoinhibition" (Barber and Anderson, 1992; Genty and Harbinson, 1996). This is followed by irreversible photobleaching and considerable losses in terms of quality and yield.

Many plant species are able to acclimate to low temperatures. This acclimatory process can alleviate, and even prevent chilling-induced damage and is termed "hardening" or "conditioning". Cold acclimation in spring grown cereals such as rye and wheat results in increased resistance to photoinhibition. In contrast to these species maize cannot acclimate to low temperatures. In fact the seedlings show marked decreases in viability at sub-optimal growth temperatures (Figure 1) and growth is near impossible below about 10°C. Likewise if maize plants are grown at optimal temperatures and then subjected to a brief, but severe chill of 4°C, photosynthetic carbon assimilation rates are significantly decreased although the effects on

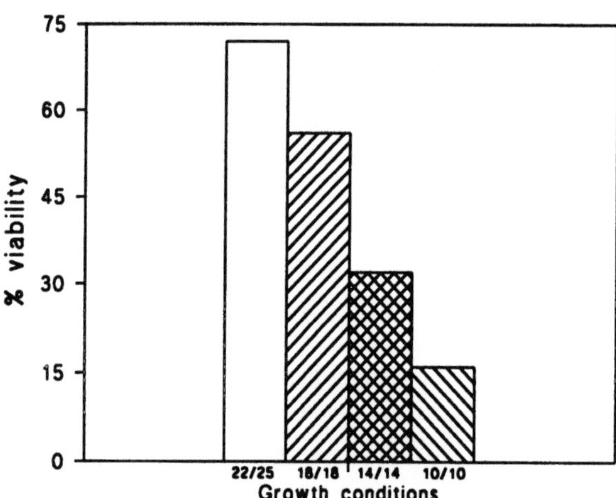

Figure 1. The effect of sub-optimal growth temperatures on the ability of maize seedlings to survive until the expansion of the second leaf.

photosynthetic electron transport are much more subtle with the electron flux through the two photosystems remaining balanced (Kingston-Smith *et al.*, 1997). In many areas such as northern Europe this chilling sensitive crop is sown and emerges in late spring when it may be exposed to low temperature stress. Improved tolerance to low temperature would therefore be of considerable benefit to the agriculture of such areas.

The antioxidant defence system

The primary effect of chilling in maize leaves is damage to the enzymes of carbon assimilation resulting in decreased rates of CO_2 fixation. Electron transport continues after chilling but the requirement for NADPH and ATP by carbon assimilation is decreased and oxygen becomes an increasingly important electron acceptor. This reaction forms the superoxide radical which while in itself is rather unreactive can initiate a cascade reaction producing other highly reactive active oxygen species (AOS); hydrogen peroxide and the hydroxyl radical (Figure 2). Enhanced hydrogen peroxide

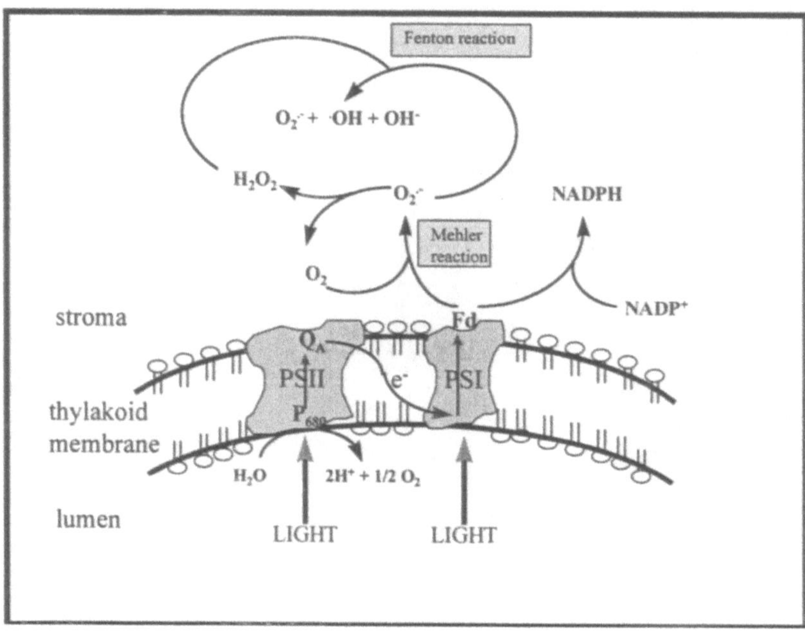

Figure 2. Generation of active oxygen species through donation to oxygen instead of NADP at PSI.

contents have been observed in higher plant leaves after exposure to low temperatures (Prasad *et al.*, 1994). It is generally accepted that AOS play a key role in photoinhibition (Richter *et al.*, 1990) and in the disruption of metabolism that occurs as a result of exposure to low temperatures (Wise, 1995).

Chloroplasts contain a complex cycle which confers the ability to decrease the foliar AOS content. This is the Mehler-peroxidase, shown in Figure 3. Through the action of superoxide dismutase (SOD) and ascorbate peroxidase (APX) superoxide is converted to water with the consumption of energy and reduced ascorbate. The oxidised ascorbate can be re-reduced enzymatically or non-enzymatically as shown in Figure 3 (Foyer and Halliwell, 1976; Miyake and Asada, 1994). The hypothesis that an inability of this cycle to match the rate of generation of superoxide radicals causes chilling damage in maize is widely accepted. In the light, maize plants grown at 15°C were found to contain over three times the foliar hydrogen peroxide content of leaves from plants grown at 20°C which corresponds well with previous information. This was accompanied by an increase in the foliar ascorbate content in low temperature grown plants, and those subjected to a short term chill.

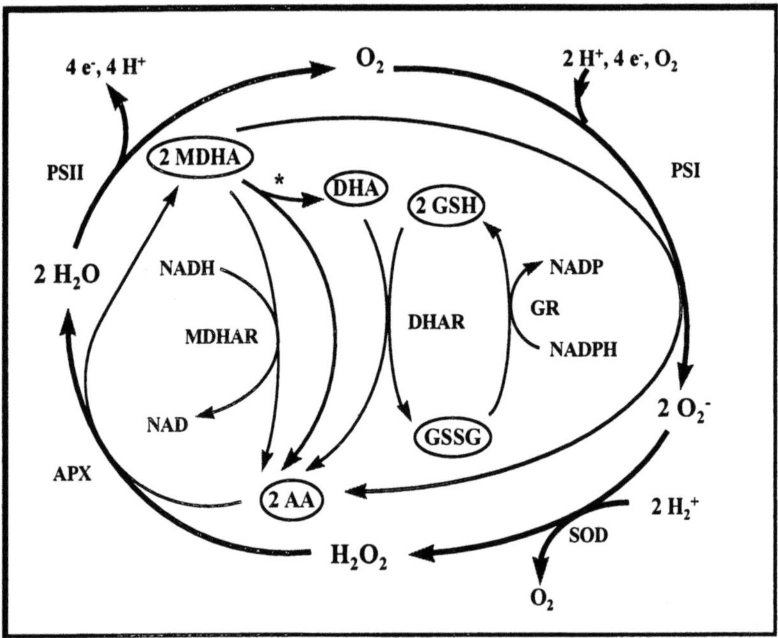

Figure 3. The Mehler peroxidase cycle (outer ring) and ascorbate-glutathione cycle (inner reactions).

Compartmentation of metabolism in maize leaves

In the C4 cycle bicarbonate is initially fixed into C4 acids in the mesophyll and these are transported to the bundle sheath where decarboxylation releases CO_2 for assimilation by the Benson-Calvin cycle (Furbank and Taylor, 1995). This spatial separation is also accompanied by a non-uniform distribution of the electron transport components. The maize mesophyll chloroplasts contain both photosystem I (PSI) and II (PSII) whereas the bundle sheath chloroplasts are deficient in PSII and ferredoxin NADP reductase (Hatch and Osmond, 1976). Therefore the bundle sheath chloroplasts are restricted in their ability to generate NADPH and are reliant on transport to the mesophyll for reduction of half the phosphoglycerate produced by CO_2 assimilation to triose phosphate (Furbank *et al.*, 1985).

We have demonstrated previously that the antioxidant defence cycle is also not equally distributed between the mesophyll and bundle sheath cells (Doulis *et al.*, 1997). Mesophyll extracts were prepared from maize leaves by an extrusion technique described by Leegood (1985). Measurements of antioxidant metabolites and enzymes revealed that most of the maize superoxide dismutase and ascorbate peroxidase activities were located in the bundle sheath whereas the majority of the glutathione reductase and dehydroascorbate reductase were localised in the mesophyll cells (Doulis *et al.*, 1997). This compartmentation is consistent with the requirement of glutathione reductase and dehydroascorbate reductase for reducing power which is plentiful in the mesophyll. This physical separation of metabolism also means that in order to reduce the oxidised forms of ascorbate and glutathione for continued operation of the cycle, these metabolites need to be returned to the mesophyll. Therefore, any disturbance in plasmodesmatal function and transport between the two cell types, such as chilling-induced closure of plasmodesmata, will cause the cycle to fail with the accumulation of oxidised ascorbate and glutathione in the bundle sheath.

Manipulation of the antioxidant defence system

The antioxidant system is central to the stress response and many studies have reported a correlation between antioxidant status and chilling tolerance. SOD overexpression has been shown to improve stress tolerance in a number of plant species (Foyer and Harbinson, 1997). Sen Gupta (1993), for example, showed that leaf discs from tobacco plants transformed to overexpress the pea chloroplastic isoform of Cu/Zn superoxide dismutase had a substantially higher resistance to photoinhibition cause by chilling temperatures and high light. Since maize mesophyll cells are deficient in superoxide dismutase and ascorbate peroxidase while bundle sheath cells lack glutathione reductase and dehydroascorbate reductase, transformed maize over-expressing the enzymes of the antioxidant system to redress this imbalance may have improved tolerance to low temperature stress (Doulis *et al.*, 1997).

Transformed maize plants were produced in which the Mn-SOD of tobacco is over-expressed constitutively (Van Breusegem, 1997). Analysis of the seed resulting from a backcross of the primary transformants and parental line yields a segregating population with a range of SOD activities (Table 1). Plants with the highest Mn-SOD activities did not have the higher total SOD activities (Table 1). When the compartmentation of the SOD activity in the transformants was studied it was clear that the superoxide dismutase activity was confined to the bundle sheath cells and this localisation was unaltered SOD overexpression (Table 2). SOD activity was not detected in the mesophyll cells of plants grown at low (18 or 14°C) temperature (data not shown).

Table 1. Total and Mn-SOD activities of maize plants expressing tobacco Mn-SOD

	Activity units mg^{-1} Chl	
Transformant	Total SOD	Mn-SOD
11	53.3	53.5
6	82.7	82.7
7	108.4	108.4
12	116.7	58.9
10	132.8	119.6
9	140.6	60.0
3	140.6	91.7
8	152.8	80.1
13	173.3	37.3
2	173.3	37.3
1	203.4	24.1
5	253.8	38.8

Table 2. Compartmentation of superoxide dismutase between mesophyll and bundle sheath of transformed maize. [a]mean ± se for 3 measurements are shown.

	SOD activity (units mg^{-1} chlorophyll)	
Transformant	**Whole leaf**	**Mesophyll extract**
7	130.1	0
3	221.3	0
5	227.8	0
25	351.9	0
15	446.1	0
24	455.6	0
13	461.5	0
4	510.6	0
control[a]	211.8 ± 5.4	0

Conclusions

The extreme sensitivity of maize leaves to low growth temperatures be related to the intercellular localisation of antioxidant defences. The overwhelming evidence for differential compartmentation of the antioxidant defence system (Doulis et al 1997) means that limitations to the transport processes may be the cause of oxidative stress, particularly in the bundle sheath cells. Transformed plants expressing a Mn-SOD cDNA in addition to the native isoforms in the chloroplast do not show an altered compartmentation profile compared with untransformed controls. Thus, transport processes are essential to the functioning of the antioxidant defence cycle in maize.

References

Barber, J. and Andersson, B. (1992) Too much of a good thing: light can be bad for photosynthesis. *TIBS* **17**, 61-66.

Doulis, A.G., Debian, N,. Kingston-Smith, A.H. & Foyer, C.H. (1997) Differential localization of antioxidants in maize leaves, *Plant Physiology* **114**, 1031-1037.

Foyer, C.H. and Halliwell, B. (1976) The presence of glutathione and glutathione reductase in chloroplasts: a proposed role in ascorbic acid metabolism, *Planta* **133**, 21-25.

Foyer, C.H. and Harbinson, J. (1997) The photosynthetic electron transport system: efficiency and control, in, C.H. Foyer and W.P. Quick (eds.), *A Molecular Approach to Primary Metabolism in Higher Plants*, Taylor & Francis, London, pp 3-39.

Furbank R.T., Stitt M., Foyer C.H. (1985) Intercellular compartmentation of sucrose synthesis, *Planta* **164**,

172-178

Furbank, R.T. and Taylor, W.C. (1995) Regulation of photosynthesis in C3 and C4 plants: a molecular approach, *Plant Cell* **7**, 797-807.

Genty, B. and Harbinson, J. (1996) Regulation of light utilization for photosynthetic electron transport, in, N.R. Baker (ed.), *Photosynthesis and the Environment, Vol 5*, Kluwer Academic Publishers, Dordrecht, the Netherlands, pp 67-99.

Hatch, M.D. and Osmond C.B. (1976) Compartmentation and transport in C4 photosynthesis, in, C.R. Stocking and U. Heber (eds.), *Encyclopedia of Plant Physiology, New Series, Volume 3,* Springer-Verlag, Berlin, pp 144-184.

Jahnke, L.S., Hull, M.R. and Long, S.P. (1991) Chilling stress and oxygen metabolising enzymes in *Zea mays* and *Zea diploperennis*, *Plant, Cell & Environment* **14,** 97-104.

Kingston-Smith, A.H., Harbinson, J., and Foyer C.H. (1997) Effect of chilling on carbon assimilation, enzyme activation and photosynthetic electron transport in the absence of photoinhibition in maize leaves, *Plant Physiology* **114,** 1039-1046.

Leegood, R. (1985) The intercellular compartmentation of metabolites in leaves of *Zea mays* L., *Planta* **164:** 163-17.

Miyake, C. and Asada, K. (1994) Ferredoxin-dependent photoreduction of the monodehydroascorbate radical in spinach thylakoids, *Plant Cell Physiology* **34,** 539-545.

Prasad, T., Anderson, M, Martin, B.A. and Steward, C.R. (1994) Evidence for chilling-induced oxidative stress in maize seedlings and a regulatory role for hydrogen peroxide, *Plant Cell,*

Richter, M., Rühle, W. and Wild. A. (1990) Studies on the mechanism of photosystem II photoinhibition II. The involvement of toxic oxygen species, *Photosynthesis Research* **24,** 237-243.

Sen Gupta, A., Webb, R.P., Holaday, A.S. and Allen, R.D. (1993) Overexpression of superoxide dismutase protects plants from oxidative stress, *Plant Physiology* **103,** 1067-1073.

Van Breusegem, F. (1997) Ph.D. Thesis, University of Gent, Belgium.

Wise, R.R. (1995) Chilling enhanced photoxidation. The production, action and study of reactive oxygen species produced during chilling in the light, *Photosynthesis Research* **45,** 79-97.

CAUSES OF DECREASED PHOTOSYNTHESIS UNDER WATER STRESS AND THE CONSEQUENCES

D.W. LAWLOR, W. TEZARA*, V.J. MITCHELL, & S.P. DRISCOLL
Biochemistry and Physiology Department, IACR-Rothamsted, Harpenden, Herts, AL5 2JQ, UK.
**Instituto de Biologia Experimental, Facultad de Ciencias, Universidad Central de Venezuela, Caracas 1041A, Venezuela.*

1. Introduction

Water stress greatly affects plant processes, including photosynthesis, decreasing dry matter production over much of the earth's surface so that harvested crop yields are below the genetic potential. Need to provide food and fibre for a large and rapidly growing human population (particularly in drought prone areas) makes the long-sought goal of selecting less sensitive crops ever more urgent. However, the mechanisms of response and ways of avoiding or over-coming stress are still not well understood. Selection breeding has not substantially improved production under stress, but would benefit - in the long run - from better understanding of the basic mechanisms. Application of molecular biological methods to improve responses to stress depends crucially upon knowledge of mechanisms: here photosynthesis is considered, to clarify the basic metabolic limitations and impacts on other stress responses.

2. Photosynthesis under water stress

Evidence of decreased net CO_2 assimilation (*A*) due to water stress abounds (see Cornic, 1994; Ort, Oxborough and Wise, 1994; Lawlor, 1995; Kramer and Boyer, 1995). With relatively rapidly stress, *A* decreases almost linearly with decreasing relative leaf water content (RWC) and water potential (Ψ). This generally correlates with decreased stomatal conductance (g_s) which limits diffusion of CO_2 to the metabolic processes in chloroplasts. However, there is strong evidence that photosynthetic metabolism is damaged by stress so that diffusion limitation is not the only or indeed primary cause of decreased *A*. Clearly, at very small Ψ and RWC there is inhibition of *A*, due to stress as CO_2 evolution occurs in the light (Lauer and Boyer, 1992; Lawlor, 1995). Photosynthetic oxygen evolution is less affected by decreasing Ψ and RWC than CO_2 assimilation (Lawlor, 1995). In order to assess the relative contributions of g_s and of metabolism to decreased *A* under stress, *A* is related to the calculated CO_2 concentration inside the leaf, C_i, (i.e. the A/C_i response). Typically the maximum rate of *A* at saturating CO_2 and light, A_{max}, and also the initial slope of the A/C_i

relationship at limiting CO_2, the carboxylation efficiency (ϕ) decrease. There is evidence that with mild stress and decreased g_s, C_i decreases as expected if A is relatively uninhibited, but with more severe stress, C_i rises, eventually leading to CO_2 efflux due to respiration when A is zero (Lawlor, 1995). These changes are regarded as caused by decreased photosynthetic metabolism: smaller A_{max} is due to decreased ribulose bisphosphate (RuBP) regeneration and decreased ϕ is due to decreased ribulose bisphosphate carboxylase-oxygenase (Rubisco) activity (see later). However, there is considerable uncertainty about the calculation of C_i.

2.1 CALCULATION OF C_i

Calculation of C_i may be erroneous due to 1) errors in measuring very small CO_2 and H_2O fluxes, temperatures etc. These are not trivial but are unlikely to grossly affect the measurements and thus alter interpretation. 2) heterogeneity of g_s ('patchy stomatal closure') across the measured leaf surface. If areas of leaf are sealed off from the atmosphere so that A is prevented then the apparent area for net CO_2 and H_2O fluxes is decreased but the calculation of C_i is not affected. Hence A appears to be reduced at a given C_i. Evidence for this is not convincing in relatively slowly water stressed plants, using a variety of methods to judge heterogeneity and metabolism (e.g. solvent infiltration, carbon fixation by $^{14}CO_2$ assimilation, dynamic measurements of chlorophyll a fluorescence). It is clearer with rapid stress and, particularly, when abscisic acid (ABA) is applied via the transpiration stream. Modelling studies suggest that the impact is small except at extreme heterogeneity (see reviews by Pospisilova and Santrucek, 1997; Weyers and Lawson, 1997). 3) Calculation of C_i assumes equivalence of CO_2 and H_2O diffusion along the flux pathway, but Boyer, Wong and Farquhar (1997) suggest that the assumption is in error with cuticular conductance to CO_2 smaller than thought and thus C_i is likely to be smaller than usually calculated. Correction brings the slope of the A/C_i relationship closer to the control values and decreases the impact of stress on ϕ. Thus, there is uncertainty about the calculated value of C_i. However, direct equilibrium measurements of C_i show it to increase and, at severe stress, to equal ambient CO_2 (Lauer and Boyer, 1992) in agreement with observations described earlier. Note that C_i is not the critical CO_2 concentration in the leaf with respect to CO_2: chloroplast stromal concentration, C_c, is. However, C_c has proved difficult to estimate. If CO_2 transport into chloroplasts decreases with stress, C_c may be underestimated by C_i. Estimation of O_2 fluxes and concentrations and the impacts of water stress on them may be similarly affected.

2.2 PHOTOSYNTHESIS UNDER ELEVATED CO_2

An approach to assess the effects of water stress on photosynthetic CO_2 and O_2 metabolism is to increase the concentration of CO_2 to overcome resistances of closed stomata and cuticle plus membranes. If the increase in the transport resistances is of the order of 10^3 then increasing CO_2 from about 300 μmol mol^{-1} to 300 mmol mol^{-1} should largely reverse the stress effects. If Fick's law applies then a progressive increase in A should result from increasing CO_2 even if the concentration reached is insufficient to fully reverse the effects.

Graan and Boyer (1990) partially reversed the decrease in A with large CO_2 concentration. However, Tezara and Lawlor (unpublished, see Tezara, 1996) measured A in an open gas systems to about 3000 $\mu mol\ mol^{-1}$ but did not reverse the decrease in A with stress Neither did measurements (using $^{14}CO_2$ as a tracer) of A at up to 20% CO_2 concentration reverse the inhibition. An important observation is that at 10-20% CO_2 A and also O_2 evolution are inhibited in sunflower. All the evidence suggests that A decreases progressively with decreased RWC etc and stops at severe stress: this is not reversed by concentrated CO_2.

2.3 FLUXES OF CO_2 AND O_2

The main flux of CO_2 is gross photosynthesis (*A*g) which is offset by the efflux of CO_2 from photorespiration (Pr) plus tricarboxylic acid cycle respiration in the light (Rd). The Pr/*A*g ratio depends on the relative concentrations of CO_2 and O_2 and the characteristics of Rubisco. With closed stomata under stress and A not inhibited, recycling of CO_2 from Pr occurs. However, when Pr plus Rd exceeds *A*g, CO_2 is evolved. Thus, at severe stress gross CO_2 assimilation ceases or is less than respiration (see Lawlor, 1995; Lauer and Boyer, 1992). The question is does stress inhibit progressively or only at a particular RWC or Ψ?

Oxygen fluxes in the leaf are production from photosynthetic water splitting in the thylakoids and consumption by the respiratory processes. As O_2 is emitted, water splitting must continue at RWC etc where A is inhibited. Thus there is a clear difference of response of CO_2 assimilation and O_2 emission under stress, suggesting that they are not stoichiometric under stress conditions and `photosynthesis' cannot be treated as a single, uniform process (as is well known). Also, sinks for electrons must change with O_2 becoming increasingly important as CO_2 assimilation decreases (Cornic, 1994).

In conclusion, mild water stress decreases A by decreasing g_s and diffusion of CO_2 and more severe stress impairs metabolism. Quantification of the relative importance of the two processes is difficult because of uncertainty in C_i but evidence that g_s becomes increasingly less important as RWC and Ψ decrease is probably largely valid. Clear CO_2 and O_2 fluxes are different in stressed leaves, with CO_2 assimilation decreases more at relatively small stresses than does O_2.

3. Impacts of water stress on photosynthetic metabolism

Evidence of the impacts of stress is considered for the following main areas of photosynthesis:

- light harvesting and electron transport
- concentrations and fluxes of substrates for the Calvin cycle viz. ATP for the phosphoglycerate kinase and phosphoribulokinase (PRK) reactions and NADPH for the triosephosphate dehydrogenase reactions
- activities and amounts of Calvin cycle enzymes
- production of RuBP, the substrate for CO_2 assimilation
- accumulation of metabolic products which affect the activity of the cycle.

A key point is that water stress causes imbalance between photosynthetic energy transduction and the functions of the Calvin cycle.

3.1 LIGHT HARVESTING AND ELECTRON TRANSPORT

For physiologically relevant water stress and other environmental conditions there is little evidence of water stress affecting the light harvesting capacity of plants, although, in the long term, senescence caused by stress may affect it. As discussed, O_2 evolution shows that even quite severe stress does not prevent water splitting. Measurements of chlorophyll a fluorescence show that electron transport continues even when CO_2 fixation stops. Thus, inhibition of primary energy transduction is not a cause of decreased CO_2 assimilation (Cornic, 1994, Lawlor. 1995).

3.2 CALVIN CYCLE SUBSTRATES

Amounts of reduced pyridine nucleotides (NADPH and NADH) per unit leaf area are maintained or increased with increasing stress (see Lawlor., 1995 for references and discussion, also Tezara, 1996). The reductant state of cell compartments under stress is not known, but the redox state of the stressed chloroplast stroma should be comparable to or more reduced than that of the unstressed. In contrast, ATP measured in the complete tissue decreases substantially and progressively with increasing stress (Lawlor, 1995; Tezara, 1996). This might be caused by ATP depletion in the chloroplast (as in the light ATP synthesis in that organelle probably dominates cellular adenine nucleotide metabolism) relative to the mitochondria or cytosol. Decrease ATP may be due to a) decreased photophosphorylation related to functions of the thylakoids and ATP synthase, b) decreased oxidative phosphorylation (respiratory ATP synthesis), c) increased consumption of ATP by processes such as ion or phloem transport or syntheses (e.g. of proteins). Considering a) Boyer (see Kramer and Boyer, 1995) suggested it as the main cause of decreased *A*. As electron transport is maintained and generation of ΔpH and membrane potential across the thylakoid membrane continues, ATP synthase (chloroplast coupling factor or CFo-CF1) may be impaired. The flux of H^+ (linked to the transthylakoid ΔpH) is coupled to rotation of an asymmetrical subunit (γ) of CF1: CF1 is exposed to the chloroplast stroma. Rotation of γ alters the configuration of the active sites in CF1 in which ATP is synthesised (Boyer, 1997). Although the mechanism is now generally understood, specific sites at which stress could impair function are not identified; such a complex molecular `rotary generator' could be sensitive to a range of conditions in the stroma. Boyer (see Kramer and Boyer, 1995) identified increased Mg^{2+} concentrations due to desiccation as a cause of inhibition. Other complex enzyme systems are affected by ionic concentration, e.g. protein synthesis requires a narrow range of K^+ concentrations as well as correct ATP concentration. b) Decreased respiratory synthesis of ATP is possible as the ATP synthases are probably the same. Respiration (O_2 uptake in darkness) continues at rates similar to those in unstressed leaves suggesting continued ATP synthesis: mitochondrial ATP synthase operates in a different environment to that of the chloroplast. c) Increased ATP consumption is a possible cause of decreased ATP, but is not easily assessed. Increased ion concentrations with stress could activate ATP-dependent ion fluxes as part of the regulatory mechanisms. However, another energy consuming process, protein synthesis, is strongly inhibited by stress. On balance, this cause of decreased ATP is unlikely.

The central role of decreased ATP concentration suggested by this analysis is not supported by measurements of CF activity, which was maintained in leaves stressed slowly in the field (Wise et al, 1990): see section 4.

3.3 CALVIN CYCLE ENZYMES

These enzymes have not been shown, unequivocally, to limit *A* under water stress. Rubisco is the most abundant cycle protein (in large concentration in the chloroplast) and might, because of its size and multi-subunit structure, be susceptible to changes in its ionic environment. However, although its activity decreases with stress and sometimes also amount (depending on the severity and duration of the stress) Rubisco is not particularly sensitive. Phosphoribulokinase (PRK) is also not affected. Impairment of a process in this autocatalytic cycle will be indicated by changes in NADPH and ATP. This is discussed in the next section.

3.4 CALVIN CYCLE PRODUCTS

With decreasing *A*, RuBP content also decreases (Gimenez et al, 1992) and neither is reversed by elevated CO_2 (Tezara, 1996). Clearly, the decrease in RuBP follows the decrease in *A* as the cycle is autocatalytic. However, this `chicken and egg' conundrum is resolved by considering what happens to NADPH and ATP at small *A* and RuBP. From the evidence above NADPH appears adequate. In unstressed transgenic tobacco with decreased activity of PRK, which causes a decrease in *A*, RuBP and ATP clearly increase (Paul et al. 1995). Such changes will occur for any lesion in the cycle unless feed-back decreases ATP synthesis or increases consumption.

Accumulation of carbohydrates from the Calvin cycle is complex under stress conditions. Relatively mild, short-term stress decreases plant growth more than it decreases *A* by decreasing the diffusion of CO_2 and so may decrease demand for assimilates, so accumulation of carbohydrates may occur and inhibit photosynthesis by feed-back regulation (Cornic, 1994). More severe stress generally decreased carbohydrates, e.g.. of sucrose which is probably consumed by continued respiration, proline synthesis, etc (Lawlor, 1995). Thus it is unlikely that feed-back regulation is important. Also, ATP concentrations would increase as a consequence of such regulation, not decrease as observed.

In conclusion: Under relatively rapid, moderate to severe stress, the simplest explanation for decreased CO_2 assimilation is that inadequate ATP limits regeneration of RuBP and impairs Calvin cycle activity. The cause of inadequate ATP is probably impaired chloroplast ATP synthase.

4. Acclimation to water stress

This model of damage to metabolism due to decreased ATP and impaired ATP synthase is not generally accepted. Cornic (1994; see his Fig. 17.7) suggests that the primary effect of stress is synthesis of ABA (mechanism unspecified) which causes stomatal closure. The resulting decreased C_i means that there must be increased thermal dissipation of electron energy, a relative increase in photorespiration and decreased activities of enzymes such as nitrate reductase and sucrose phosphate synthase. Wise et al. (1990) assert that CF was not affected by long-term water stress in the field. Decreased *A* was not thought due to physiological damage. The difficulty of comparing controlled environment and field conditions have been emphasised (Ort, Oxborough and Wise, 1994).

The views that metabolism is sensitive to stress but under field conditions is not responsible for decreased *A* and production are not unreconcilable. Because basic metabolic processes are sensitive to stress, acclimation of metabolism to avoid damage may be the key. Plants with extensive roots, grown in large volumes of soil, are able to partially replace water lost during the day and become stressed slowly. Smaller Ψ and turgor would decrease g_s and C_i, and also the growth of organs so that uptake and loss of water would be in balance and RWC maintained, especially if carbohydrates accumulate and osmotic adjustment occurred. Protective mechanisms, such as carotenoids against over-energization and energetic oxygen species, increase.

If dehydration became sufficient to affect ATP synthase then the excess of energy capture over utilization might trigger abscisic acid (ABA) synthesis (Young et al. 1997) which reinforces the hydraulic signals and triggers acclimatization responses (e.g. increased heat shock proteins). Decreased C_i and *A* may also result in energy imbalance and trigger the responses. Such changes, if more sensitive to stress than the ATP synthase would protect core metabolism. In contrast, plants grown in small volumes of soil rapidly experience persistent stress, before acclimation can occur, with core photosynthetic metabolism affected. It has long been known that field crops may suffer considerable loss of production, but more because of decreased leaf and other organ production than impaired metabolism. Barley suffering early and progressive stress produced only 50% of the biomass over the growing season of the irrigated control. However, 80% of the decrease was caused by decreased leaf area. Photosynthesis accounted for only 20% with most due to decreased g_s and little to metabolic inhibition. No accumulation of `stress metabolites' (e.g. proline) or processes characteristic of damage (e.g. changes in chlorophyll *a* fluorescence) was observed. In contrast rapid stress, late in crop development had much more effect on photosynthetic metabolism (Legg et al, 1979). Such behaviour is logical if metabolism is indeed very sensitive to stress. Severe and prolonged stress would result in death of cells and organs and in reproductive failure. Even adapted crops suffer premature loss of leaves etc. under stress. Adaptation may operates at three levels of organisation: 1) whole plant, with altered morphology, e.g. reduced evaporative surface etc., 2) organ, e.g. decreased stomatal conductance, 3) cellular, e.g. acclimation by increasing protection against stress. All three are necessary but have different response times and perhaps sensitivity but may be triggered by linked signal and response mechanisms. The result is protection of metabolism from continued and even increasing stress. The observations of Ort et al. (1994) were made at the later stage of adjustment, whereas those of Lauer and Boyer (1992) and Gimenez et al (1992), suggesting major perturbation of cellular metabolism under water stress, were made under more extreme, rapid stress. Perhaps they also occur, to a small extent, during the initial phases of adjustment to stress under field conditions.

5. Implications of ATP deficiency for metabolism and acclimation

Several characteristics of water stressed plants, namely reactive forms of oxygen, ABA and proline accumulation, and synthesis of heat shock proteins may be related to maintenance of electron transport and O_2 evolution, to decreased CO_2 assimilation and, particularly, to decreased ATP content and a large reductant/ATP ratio. Briefly, the arguments for considering these as expressions of disturbed metabolism are presented. All accumulate with severe stress, below thresholds which differ for each, when A is small or stopped and are extreme expressions of normal metabolism and may protect against further stress. Consider reactive forms of oxygen (e.g. superoxide radical) and related compounds (e.g. hydrogen peroxide). They are produced as a consequence of electron transfer from the electron transport chain and its acceptors to oxygen and so increase as CO_2 assimilation decreases Cornic (1994) has considered the processes. Synthesis of ABA may also be a consequence of the over-energized state of the electron transport mechanism and the processes of dissipation by the carotenoids (Young et al, 1997). ABA is synthesized from all-*trans*-violaxanthin via 9'-*cis*-neoxanthin and its enzymatic conversion to precursors of ABA (xanthoxin) and finally ABA (Parry, 1993). This may occur under excessive energy loads in the chloroplast when non-photochemical quenching by the xanthophyll cycle occurs. This speculative mechanism would link chloroplast energy state to ABA synthesis and to hormonal control of growth, as required for the concepts of acclimation outlined above.

Proline accumulates under more severe stress than ABA. Proline is derived from glutamate in reactions which require carbohydrates and organic acids and much reductant and enzyme synthesis is involved (Samaras et al, 1995). However, proline synthesis requires ATP in the first step of the pathway. As the process is mitochondrial possibly the localized ATP supplies the requirement. The trigger required for this 'stress metabolite' may be the highly reduced state of the system. Another indicator of the putative dependence of metabolism under water stress to altered ATP is provided by heat shock proteins (HSPs). These proteins accumulate in all organisms under a variety of stresses (e.g. heavy metals, respiratory inhibitors, anoxia) which induce ATP depletion. HSPs act as molecular chaperones to fold nascent proteins, to transport proteins across membranes and, importantly in the context of stress, to remove damaged or partial proteins from the cell. HSPs bind to damage proteins and ATP is needed for the subsequent refolding and release of the protein. HSPs also bind to a Heat Shock Factor (HSF) which is responsible for activating the HSP genes. Free HSP binds HSF and so inhibits HSP synthesis. However, if the amount of damaged protein is large, the HSP binds to it and releases the HSF, so stimulating transcription of HSP genes and translation of the mRNA. Once excess HSP is produced, HSF is again bound and an equilibrium is reached (Kabakov and Gabai, 1997). Under stress, with depleted ATP, it is likely that much damaged/incomplete protein will be produced so stimulating HSP production. However, inadequate ATP prevents the recycling of protein and HSP so large amounts of HSP accumulate. There is still a question about the ATP available for HSP synthesis under conditions where synthesis of normal proteins is strongly inhibited. Perhaps there is localization of HSP synthesis close to the site of ATP production.

These examples serve to show the possible interactions between energy and carbon metabolism with ATP deficiency determining CO_2 assimilation and the imbalance between reductant and ATP acting as a switch, in part regulating the wider behaviour of stressed cells and contributing to acclimation of the plant to water stress.

6. References

Boyer, J.S., Chin Wong, S. and Farquhar, G.D. (1997) CO_2 and water vapour exchange across leaf cuticle (epidermis) at various water potentials. *Plant Physiol.* **114**, 185-191.

Boyer, P.D. (1997) The ATP synthase - a splendid molecular machine. *Annu. Rev. Biochem.* **66**, 717-49.

Cornic, G. (1994) Drought stress and high light effects on leaf photosynthesis, in N.R. Baker and J.R. Bowyer (eds), *Photoinhibition of Photosynthesis*, Bios Scientific Publishers, Oxford, pp. 297-313.

Gimenez, C., Mitchell, V.J. and Lawlor, D.W. (1992) Regulation of photosynthetic rate of two sunflower hybrids under water stress. *Plant Phsiol.* **98**, 516-52.

Graan, T. and Boyer, J.S. (1990) Very high CO_2 partially restores photosynthesis in sunflower at low water potential. *Planta* **181**, 378-384.

Kabakov, A.E. and Gabai, V.L. (1997) *Heat shock proteins and cytoprotection: ATP-deprived mammalian cells*, R.G. Landes Company, Georgetown, Texas.

Kramer, P.J. and Boyer, J.S. (1995) Water relations of plants and soils. Academic Press Ltd, London.

Lauer, M.J. and Boyer, J.S. (1992) Internal CO_2 measured directly in leaves. *Plant Physiology* **98**, 1310-1316.

Lawlor, D.W. (1995) The effects of water deficit on photosynthesis, in N. Smirnoff (ed), *Environment and Plant Metabolism*, Bios Scientific Publishers, Oxford, pp. 129-160.

Legg, B.J., Day, W., Lawlor, D.W., Parkinson, K.J. (1979) The effect of drought on barley growth: models and measurements showing the relative importance of leaf area and photosynthetic rate. *J. Agric. Sci, Cambs.* **92**, 703-716.

Ort, D.R., Oxborough, K. and Wise, R.R. (1994) Depressions of photosynthesis in crops with water deficits, in N.R Baker and J.R. Bowyer (eds.), *Photoinhibition of Photosynthesis*, Bios Scientific Publishers, Oxford, pp. 315-329.

Parry, A.D.)1993) Abscisic acid metabolism. In P.J. Lea (ed), *Methods in Plant Biochemistry, Volume 9, Enzymes of Secondary Metabolism*. Academic Press, London, pp. 381-402.

Paul, M.J., Knight, J.S., Habash, D., Parry, M.A.J., Lawlor, D.W., Barnes, S.A., Loynes, A. and Gray, J.C. (1995) Reduction in phosphoribulokinase activity by antisense RNA in transgenic tobacco: Effect on CO_2 assimilation and growth in low irradiance. *Plant Journal* **7**, 535-542.

Pospisilova, J. and Santrucek, J. (1997) Stomatal patchiness: effects on photosynthesis, in M. Pessarakli (ed.), *Handbook of Photosynthesis*, Marcel Dekker Inc., New York, pp. 427-441.

Samaras, Y., Bressan, R.A., Csonka, L.N., Garcia-Rios, M.G., Paino D'Urzo and Rhodes, D. (1995) Proline acccumulation during drought and salimity, in N. Smirnoff (ed.), *Environment and Plant Metabolism*, Bios Scientific Publishers, Oxford, pp.161-188.

Tezara, W. (1996) Efectos del deficit hidrico y su interaccion con el suministro de CO_2 sobre la bioquimica y fisiologia de la fotosynthesis en girasol (Helianthus annuus L), PhD Thesis, University of Caracas, Venezuela.

Weyers, D.B. and Lawson, T. (1997) Heterogenicity in stomatal characteristics. *Advances in Botanical Research* **26**, 317-52.

Wise, R.R., Frederick, J.R., Alm, D.M., Kramer, D.M., Hesketh, J.D., Crofts, A.R. and Ort, D.R. (1990) Investigation of the limitations to photosynthesis induced by leaf water deficit in field-grown sunflower (Helianthus annuus L), *Plant, Cell and Environment* **13**, 923-931.

Young, A.J., Phillip, D. and Saxill, J. (1997) Carotenoids in higher plant photosynthesis, in M.Pessarakli (ed.), *Handbook of Photosynthesis*, Marcel Dekker Inc., New York, pp. 575-596.

EFFECT OF Cu(II) ON SHOOT REGENERATION IN *TRITICUM AESTIVUM* TISSUE CULTURES PROBED BY THE FAST FLUORESCENCE (JIP-TEST)

M. MOUSTAKAS[1], M. TSIMILLI-MICHAEL[2,3], L. PURNHAUSER[4], G. OUZOUNIDOU[1], L. BONA[4] AND R. J. STRASSER[2]

[1]*Department of Botany, Aristotle University of Thessaloniki, Thessaloniki 54006, Greece,* [2]*Laboratory of Bioenergetics, University of Geneva, CH-1254, Jussy-Geneva, Switzerland,* [3]*Cyprus Ministry of Education and Culture,* [4]*Cereal Research Institute, Szeged 6701, Hungary,*

1. Abstract

Copper concentrations (0.1, 1, and 100 μM) significantly enhance shoot regeneration from *Triticum aestivum L.* calli and stimulate root formation. The results of the JIP-test show that though the phenomenological activities for trapping and electron transport decrease by increasing [Cu], the corresponding specific activities increase. Thus, though a fraction of PSII reaction centers (RCs) become inactivated, the rest of the RCs become more active when [Cu] in the culture medium increases. We recognize an optimization strategy that appears to regulate at low [Cu] a high photosynthetic activity per tissue, and at high [Cu] a high specific photosynthetic activity, in parallel to the stimulation of shoot formation.

2. Introduction

Copper is an essential micronutrient, being a component of enzymes involved in several important metabolic and physiological processes (Ciscato *et al.* 1997). Its function as a plant nutrient is based mostly on the participation of enzymatically bound copper in redox reactions (Marschner 1995). Besides other functions, Cu is also involved in the photosynthetic electron transport, being a component of plastocyanine. It also probably plays an important role in the regulation of PS II-mediated electron transport (Barón *et al.* 1995, Ouzounidou *et al.* 1995, 1997).

In the present study we utilised the polyphasic Chl *a* fluorescence rise (OJIP) as a tool to investigate the behaviour of PSII in wheat (*Triticum aestivum L.*) leaves growing *in vitro* culture under different Cu concentrations (0.1, 1, and 100 μM). The use of *in vitro* cultures enables the selection of uniform individuals that can be used to study the effect of Cu on shoot regeneration.

The fluorescence transients analyzed by the JIP-test (Strasser and Strasser 1995), which is based on the theory of energy (Strasser 1981), lead to the calculation of different functional and structural parameters that offer a quantification of the PSII bahaviour. The functional parameters calculated by this test are the specific (per reaction center, RC) and phenomenological (per cross section of the sample, CS) energy fluxes for absorption (ABS), trapping (TR) and electron transport (ET). The flux ratios or the yields (the maximum quantum yield of primary photochemistry, (Φ_{Po}, the efficiency with which a trapped exciton can move an electron into the electron transport chain, Ψ_{o}, or the probability that an absorbed photon will move an electron into the electron transport chain, (Φ_{Eo}), as well as an expression for the concentration of the reaction centers (RC/CS), are the structural parameters provided by this test (Strasser and Strasser 1995, Krüger *et al.* 1997).

3. The JIP-test

All oxygenic photosynthetic material investigated so far show a polyphasic fluorescence rise during the first second of illumination. These phases have been labelled as 0, J, I and P. The OJIP transient changes its shape according to many environmental conditions, such as light intensity, temperature, drought or chemical influences. From the data stored during the first second, the following values are retained as original data and used for the calculation of several phenomenological and biophysical expressions leading to the dynamic description of a photosynthetic sample at a given physiological state: the maximal measured fluorescence intensity, F_P, which can here be denoted as F_M since the excitation intensity is high enough to permit the closure of all RCs; the fluorescence intensity at 50 μs, considered to be the intensity when all RCs are open (F_O); the fluorescence intensity at 150 μs; the fluorescence intensity at 2 ms, denoted as F_J; the time t_{Fmax} to reach the maximal fluorescence intensity; the area between the fluorescence transient and the level of the maximal fluorescence intensity. From these original data, the initial slope M_0 (in ms^{-1}, normalised over the maximal variable fluorescence $F_V = F_M - F_0$), and the relative variable fluorescence at 2 ms, V_J, are directly calculated to be used in the Equations:

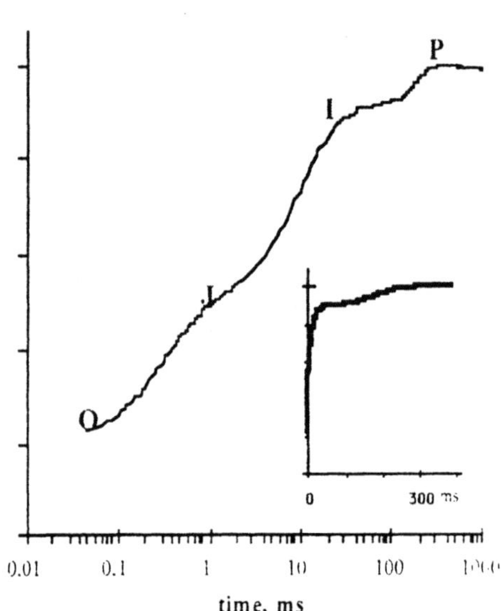

A typical ***O-J-I-P*** *fluorescence transient, here induced by a saturating illumination, plotted on a logarithmic time scale (and on a linear one, insert).*

$M_0 = 10.(F_{150}-F_0)/(F_M-F_0)$

$V_J = (F_J-F_0)/(F_M-F_0)$

Based on the theory of energy fluxes in biomembranes (Strasser 1981), formulae for the energy fluxes-both per RC and per CS-and for the flux ratios or yields have been derived using the experimental values provided from the JIP-test. The constellation of their values at any instant can be considered as expressing the function of the system.

The equations are shown below.

ABS refers to the photons absorbed by the antenna pigments Chl*. Part of this excitation energy is dissipated, mainly as heat and less as fluorescence emission, F, and another part is channelled as trapping flux, TR, to the reaction center, RC. There the excitation energy is converted to redox energy by reducing an electron acceptor Q_A to Q_A^- which is then reoxidised to Q_A thus creating an electron transport, ET, that maintains the metabolic reactions.

4. Definitions and Equations

4.1. THE YIELDS (OR THE FLUX RATIOS)

Φ_{Po}: the max. quantum yield of primary photochemistry $\equiv TR_0 / ABS = 1-(F_0/F_M)$ (Φ_{Po} for dark adapted, Φ'_{Po} after light adaptation)

Φ_{EO}: the probability that an absorbed photon will move an electron into the electron transport chain $\equiv ET_0/ABS = \{1-(F_0/F_M)\}\cdot(1-V_J)$

4.2. THE SPECIFIC FLUXES (PER RC-REACTION CENTRE)

TR_0/RC: trapping at time zero, per RC $= M_0/V_J$

ET_0/RC: electron transport at time zero, per RC $= (M_0/V_J)-M_0 = (M_0/V_J).(1-V_J)$

ABS/RC: absorption per RC $= (TR_0/RC)/(TR_0/ABS) = (M_0/V_J)/\{1-(F_0/F_M)\}$

DIS_0/RC: dissipation per RC $= (ABS/RC)-(TR_0/ABS)$

4.3. THE PHENOMENOLOGICAL FLUXES (PER CS-CROSS SECTION OF THE TISSUE CULTURE)

$ABS/CS \sim F_M$

$TR_0/CS = (TR_0/ABS).(ABS/CS) = \Phi_{Po}.(ABS/CS)$

$ET_0/CS = (ET_0/ABS).(ABS/CS) = \Phi_{EO}.(ABS/CS)$

$DIS_0/CS = (ABS/CS)-(TR_0/CS)$

4.4 THE CONCENTRATION OF THE REACTION CENTRES : RC/CS

$RC/CS = (ABS/CS)/(ABS/RC) = F_M.\{1-(F_0/F_M)\}/(M_0/V_J)$

5. Materials and Methods

5.1. PLANT MATERIAL

Triticum aestivum L. calli were initiated from immature embryos (ca. 1-1.5 mm long) on medium consisting of Murashige and Skoog's (1962) inorganic nutrients (MS),

1 mgl^{-1} thiamine-HCl, 100 mgl^{-1} meso-inositol, 2% sucrose, 0.7% Oxoid-agar, and 1 mgl^{-1} 2,4-D. Highly purified water was used for preparation of the medium. Four weeks after initiation, calli were transferred to the same medium with 0.5 mgl^{-1} 2,4-D, and 4 weeks later calli were subcultured again onto the same medium. Experiments on regeneration were then performed, using callus pieces containing compact and yellowish-green parts as described previously (Purnhauser *et al.* 1987). Calli were transferred onto MS media free of 2,4-D and supplemented with different concentrations of $CuSO_4$ (0.1, 1, and 100μM). For each subculture, calli to be transferred were cut into pieces about 5 mm in diameter (Purnhauser and Gyulai 1993).

5.2. FLUORESCENCE MEASUREMENTS

Chlorophyll a fluorescence transients were measured at room temperature and at 10 °C by a Plant Efficiency Analyser (PEA, built by Hansatech Ltd, King's Lynn, Norfolk, UK) and recorded up to 1 s, with a data acquisition rate of 10 μs for the first 2 ms and of 1 ms thereafter, and 12-bit resolution (Strasser *et al.* 1995). The transients were induced by a red (peak at 650 nm) light of 600 W m^{-2} intensity (excitation intensity) provided by an array of six light-emitting diodes. The samples were subjected to a 30-s dark interval before the measurement. The dark interval is required for the re-opening of the RCs so that the subsequent OJIP-transient will reflect their light-induced closure. On the other hand, darkness drives the sample towards the dark-adapted state. The duration of 30 s for the dark interval was experimentally found to be the optimal compromise, as it ensures the re-opening of all RCs without causing a significant disturbance of the light-adapted state (Krüger *et al.* 1997).

6. Results and Discussion

6.1.THE EFFECT OF [Cu] ON WHEAT SOMATIC CALLUS CULTURES

In wheat somatic calli, Cu dramatically increased shoot regeneration over a wide range of concentrations (0.1, 1, and 100μM). On MS medium lacking Cu the regeneration frequency was only 13%, whereas on MS medium containing 0.1 μM $CuSO_4$ 35% of the calli produced shoots. A more than two-fold further increase occurred at 100μM of $CuSO_4$. Copper also stimulated root formation from wheat callus. The root length, number of roots per callus, and the total dry root mass increased with increasing concentration of Cu. As compared with the 0.1 μM $CuSO_4$, the 100 μM treatment doubled dry root weight.

It has become evident that several heavy metals as micro elements play important roles in the regeneration of plant tissue cultures.

Copper is known to be a component or activator of many important enzymes involved in electron transport, protein and carbohydrate biosynthesis, polyphenol metabolism and so on; and thus might play an important role in plant regeneration. To optimize the nutrient requirement of tissue cultures in different species, it seems that the mineral composition, especially of microelements, should be examined to achieve maximal regeneration (Purnhauser and Gyulai 1993).

6.2. ANALYSIS OF THE FAST FLUORESCENCE RISE BY THE JIP-TEST

The results of the JIP-test show that the specific activities (per RC) trapping fluxes and electron transport increase by increasing Cu concentration. These fluxes, however,

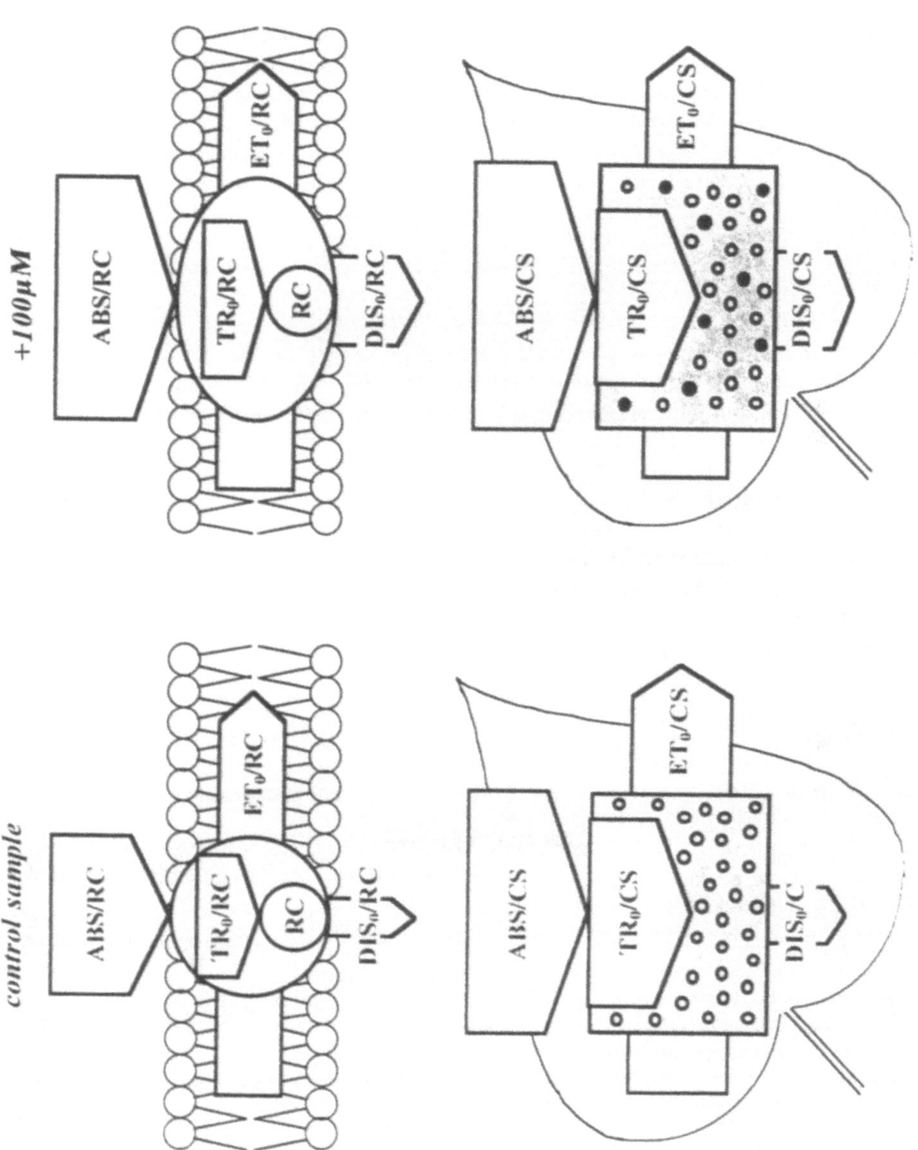
+100µM
ABS/RC
TR_0/RC
RC
ET_0/RC
DIS_0/RC
ABS/CS
TR_0/CS
ET_0/CS
DIS_0/CS
control sample
ABS/RC
TR_0/RC
RC
ET_0/RC
DIS_0/RC
ABS/CS
TR_0/CS
ET_0/CS
DIS_0/C

decreased when they are expressed phenomenologically as e.g., trapping flux or electron transported per cross section of the tissue culture. Therefore, all quantum yields like trapping per absorption Φ_{Po} or electron transport per absorption Φ_{EO} decrease by increasing [Cu]. This indicates that an increasing fraction of Rcs becomes inactivated and that the rest of the RCs become more active, however by an equal or slightly increasing chlorophyll concentration per tissue area and without compensating the overall yields. It appears that the stimulation of shoot formation by [Cu] is associated with the increase of the specific activities.

Certain metals as Cu, are constituents of a great number of proteins and enzymes and are consequently essential for maintaining optimum plant metabolism (Sandmann and Böger 1983). Copper deficiency has a direct impact on energy metabolism because it affect the synthesis of the Cu-containing electron carriers plastocyanin and cytochrome oxidase, whose depletion results in decreased photosynthesis and respiration (Bar6n *et al.* 1995, Ouzounidou *et al.* 1997).

We recognize an optimisation strategy (Tsimilli-Michael *et al.* 1995, 1996) that appears to regulate at low [Cu] a high photosynthetic activity per tissue, and at high [Cu] a high specific photosynthetic activity. Whether this strategy is controlled by mechanisms regulating the shoot formation, is worth to be further investigated.

7. Acknowledgement

This research was supported by a grant from the Greek General Secretariat for Research and Technology to M. Moustakas, by the Swiss National Foundation (SNF 31-46.860.96 and SNF 31-52.541.97) to R. J. Strasser and by the Cyprus Ministry of Education and Culture (sabbatical leave) to M. Tsimilli-Michael.

8. References

Barón, M., Arellano, J.B., and López Gorgé, J. (1995) Copper and photosystem II: A controversial relationship, *Physiologia Plantarum* **94**, 174-180.

Ciscato, M., Valcke, R., van Loven, K., Clijsters, H., and Navari-Izo, F. (1997) Effects of in vivo copper treatment on the phosyntheitic apparatus of two *Triticum durum* cultivars with different stress sensitivity, *Physiologia Plantarum* **100**, 901-908.

Krüger, G.H.J., Tsimilli-Michael, M., and Strasser, R.J. (1997) Light stress provokes plastic and elastic modifications in structure and function of photosystem II in camellia leaves, *Physiologia Plantarum* **101**, 265-277.

Marschner, H. (1995) *Mineral Nutrition of Higher Plants*, pp. 333-347, Academic Press. London.

Murashige, T., and Skoog, F. (1962) A revised medium for rapid growth and bioassays with tobacco tissue cultures, *Physiologia Plantarum* **15**, 473-497.

Ouzounidou, G., Moustakas, M., and Lannoye, R. (1995) Chlorophyll fluorescence and photoacoustic characteristics in relationship to changes in chlorophyll and Ca^{2+} content of a Cu-tolerant *Silene compacta* ecotype under Cu treatment, *Physiologia Plantarum* **93**, 551-557.

Ouzounidou, G., Moustakas, M., and Strasser, R. J. (1997) Sites of action of copper in the photosynthetic apparatus of maize leaves, *Australian Journal of Plant Physiology* **24**, 81-90.

Purnhauser, L., and Gyulai, G. (1993) Effect of copper on shoot and root regeneration in wheat, triticale, rape and tobacco tissue cultures, *Plant Cell, Tissue and Organ Culture* **35**, 131-139

Purnhauser, L., Medgyesy, P., Czako, M., Dix, P.J., and Marton L. (1987) Stimulation of shoot regenaration in *Triticum aestivum* and *Nicotiana plumbaginifolia* Viv. tissue cultures using the ethylene inhabitor $AgNO_3$, *Plants Cell Reports* **8**, 182-185.

Sandmann, G., and Böger, P. (1980) Copper-mediated lipid peroxidation processes in photosynthetic membranes, *Plant Physiology* **66**, 797-800.

Strasser, R.J. (1981) The grouping model of plant photosynthesis: Heterogeneity of photosynthetic units in thylakoids, in G. Akoyunoglou (ed.), *Photosynthesis III. Structure and Molecular Organisation of the Photosynthetic Apparatus*, Balaban International Science Services, Philadelphia, PA, pp. 727-737.

Strasser, B.J., and Strasser, R.J. (1995) Measuring fast fluorescence transients to address environmental questions: The JIP-test, in P. Mathis (ed.). *Photosynthesis: From Light to Biosphere*, Kluwer Academic Publishers, Dordrecht, Vol. 5, pp. 977-980.

Strasser, R.J., Srivastava, A., and Govindjee (1995) Polyphasic chlorophyll a fluorescence transient in plants and cyanobacteria, *Photochemistry and Photobiology* **61**, 32-42.

Tsimilli-Michael, M., Krüger, G.H.J., and Strasser, R.J. (1995) Suboptimality as driving force for adaptation: A study about the correlation of excitation light intensity and the dynamics of fluorescence emission in plants, in P. Mathis (ed.). *Photosynthesis: From Light to Biosphere*, Kluwer Academic Publishers, Dordrecht, Vol. 5, pp. 981-984.

Tsimilli-Michael, M., Krüger, G.H.J., and Strasser, R.J. (1996) About the perpetual state changes in plants approaching harmony with their environment, *Archives Science Geneve* **49**,173-203.

[illegible]

BIOTIC ACTIVITY OF ESSENTIAL OIL AND THE OTHER *JUNIPERUS COMMUNIS* L. METABOLITES

LJ. ČOMIĆ, B. RANKOVIĆ, M. TOPUZOVIĆ AND S. ĆURČIĆ
Faculty of Science, Univ. of Kragujevac, 34000 Kragujevac, Yugoslavia

1. Introduction

Numerous plant species in their tissues produce metabolites with different biotic activities, which are nowadays used in traditional medicine, phytopharmacy, biotechnology... The significance of biotically active substances is related both to plants themselves in the function of protection from phytopathogens, and for the changes which can be caused in microbiocoenosis and in ecosystem in general. Thus essential oils of numerous plant species have an important antimicrobial activity (Morris *et al.*, 1979; Janssen *et al.*, 1987; Deans and Riche, 1987) which makes them potential aseptics. *Juniperus communis* L. in its stem, needles and first of all fruits, produces a complex of volatile compounds with strong bactericidal activity (Voroncihin, 1976). The concentration of phytoncides is the highest in the fruits (above 80%) from which various products can be extracted (essential oil, alcoholic extract, water extracts, etc.). After the extraction of primary products, there remains a significant quantity of secondary metabolites which can be utilized (Matović *et al.*, 1996, Čomić *et al.*, 1996). A hectare of juniper forest during 24 hours synthesizes about 30 kg of biotically active substances. Biotic activity of metabolites depends on the biology of plant species, its physiological condition and total environmental conditions (Tokin 1951, Koleva 1980).

The aim of this research was to determine the biotic activities of various biochemical products of *Juniperus communis* L. from the region of Mt. Zlatar (Serbia), in relation to microorganisms.

2. Material and Methods

The samples of *Juniperus communis* for the extraction of metabolites were collected in the region of mountain Zlatar during 1995-96. Metabolite extraction from the collected plant material was performed at the Institute for Medicinal Plants Josif Pančić (Belgrade), by standard biochemical methods (Matović *et al.*, 1997). In these experiments, biotic activity of the following products was tested: M1 - Ground whole fruits of juniper, M2 - Ground fruit residues after alcoholic extraction; M3 - Alcoholic extract from juniper fruits from which alcohol is eliminated by evaporation; M-4 Ground juniper fruit residues after water extraction; M5 - Water extract of juniper fruits; M6 - Ground juniper residue after essential oil distillation; M7 - Essential oil of juniper; M8 - Ground juniper needles.

Biotic activity of *Juniperus communis* metabolites was tested in relation to bacterial species *Agrobacterium tumefaciens, Bacillus subtilis, Escherichia coli, Pseudomonas fluorescens, P. syringae* var. *phaseolicola* and fungi *Aspergillus niger, Doratomyces stemanitis, Mucor* sp., *Trichoderma viride, T. harzianum*. 48-hour old bacterial cultures in nutrient agar, and seven-day old fungal cultures on PDA were used in the experiment. Antimicrobial activity of *Juniperus communis* metabolites was tested by the method of diffusion in agar (Garrod, Grady, 1972). Bacteria were sawn in melted nutrient agar (Torlak, Belgrade) and fungi into potato-dextrose agar. After that, 0.3 ml of water solution of metabolite was placed into holes made in agar plates. Two concentrations of metabolites were tested: a - 1 g of plant material in 10 ml of water, and b - 1 g of plant material in 20 ml of water. Incubation of prepared media was 48 hours for bacteria and 7 days for fungi, at the temperature of 20^0C. Biotic activity was determined based on the zones of inhibited or stimulated growth around the hole with each of the metabolites, and it was given in percentage compared to bactericidal activity of penicillin (Galenika, Belgrade). i.e. fungicidal activity of nistatine. Experiments were performed in nine repetitions, and the results of measurement were statistically processed.

3. Results and disscusion

The results of microbiological research of antimicrobial activities of different juniper metabolites have been shown in Tab. 1.

Metabolic products of *J. communis* have antimicrobial properties, whose effect is conditioned by the chemical nature of the metabolite, its concentration and taxonomic characteristics of microorganisms. All the tested metabolites had a bactericidal effect to at least one tested bacterium species, whereas their fungicidal effects were not determined.

The growth of *Agrobacterium tumefaciens* was inhibited by all the tested metabolites. The intensity of bactericidal activity of metabolites on *Agrobacterium tumefaciens* was 42.53 - 62.56% compared to the effect of penicillin (Tab.1). The maximum effects were produced by alcoholic extract from juniper fruits and ground residues of juniper fruits after water extraction, which caused the zone of inhibited growth 9.66 mm wide. With reduced metabolite concentration, their inhibitory effect also decreased or disappeared completely. It was also determined that ground juniper needles in low concentrations stimulated the growth of *Agrobacterium tumefaciens* which is the result of the content and chemical structure of the specific components with antimicrobial activity on the one hand, and growth factors on the other hand.

The growth of *Bacillus subtillis* was inhibited by all the tested metabolites. Ground residues of juniper fruits after water extraction had the maximum inhibitory activity and essential oils (Tab. 1) - the minimum. The intensity of inhibitory effect of tested metabolites was proportional to their concentration.

None of the tested metabolites affected the growth of *Escherichia coli.*

Bacteria *Pseudomonas fluorescens* and *P. syringae* var. *phaseolicola* are resistant to an alcoholic extract from juniper fruits, and susceptible to the presence of

TABLE 1. Bactericidal activity of *Juniperus communis* metabolites

metab olite	*Agrobacterium tumefaciens*		*Bacillus subtilis*		*Escherichia coli*		*Pseudimonas fluorescens*		*Pseudomonas phaseolicola*	
	inh. zone (mm)	%	inh. zone (mm)	%	inh. zone. (mm)	%	inh. zone. (mm)	%	inh. zone. (mm)	%
metabolite concentration 1/10										
M_1	9,22	59,33	11,33	/	0	0	8,22	49,33	8,44	58,00
M_2	6,77	43,84	10,44	/	0	0	0	0	4,00	41,23
M_3	9,66	62,56	11,66	/	0	0	0	0	0	0
M_4	9,66	62,56	11,77	/	0	0	8,22	49,33	8,88	61,03
M_5	7,33	47,74	10,99	/	0	0	5,99	35,95	4,00	27,48
M_6	7,44	48,18	9,77	/	0	0	6,22	37,33	6,77	46,52
M_7	7,99	51,74	7,66	/	0	0	0,77	4,62	1,33	9,14
M_8	6,55	42,53	7,99	/	0	0	0	0	0	0
metabolite concentration 1/20										
M_1	7,66	49,61	9,88	/	0	0	7,44	44,65	4,77	32,78
M_2	6,11	39,57	7,10	/	0	0	2,11	12,66	0	0
M_3	0	0	0	/	0	0	7,11	42,67	0	0
M_4	8,66	56,08	10,77	/	0	0	0	0	6,77	46,52
M_5	0	0	0	/	0	0	0	0	0	0
M_6	6,55	42,42	8,99	/	0	0	7,55	45,31	5,88	40,41
M_7	3,33	21,56	5,55	/	0	0	0	0	0	0
M_8	0	0	3,88	/	0	0	0	0	0	0

M1 - Ground whole fruits of juniper; M2 - Ground fruit residues after alcoholic extraction; M3 - Alcoholic extract from juniper fruits from which alcohol is eliminated by evaporation; M4 - Ground juniper fruit residue after water extraction; M5 - Water extract of juniper fruits; M6 - Ground juniper fruit residue after essential oil distillation; M7 - Essential oil of juniper; M8 - Ground juniper needles.

other metabolites. Ground residues of juniper fruits after water extraction have the maximum inhibitory activity, 61.03% compared to a standard (Tab. 1). Ground juniper fruits after water extraction have a specific biotic effect on *Pseudomonas* spp. A zone of stimulated growth appears around the zone of inhibited growth of the bacteria. This phenomenon is conditioned by the concentration of metabolites and by different abilities of its constituents to diffuse in the medium. Antimicrobial properties of juniper result from the chemical composition of each metabolite, by the representation of biotically active components and by their interrelationship. By all means, inhibitory effect of essential oil results from a high content of α-pinene (30.736%), sabinene (19.372%) and by the presence of p-cimene (0.226%), l-limonene (4.904%), α-terpinolene (1.318%), substances whose biotic activity was previously described by Ross *et al.*, (1980), Adebajo *et al.*, (1989), Janssen *et al.*, (1987). Although many authors point to the fungitoxic properties of various plant oils and other metabolites (Diksit *et al.*, 1984, Morris *et al.*, 1979, Singh *et al.*, 1983), during this research it was not determined that *Juniperus communis* metabolites inhibited the growth of *Aspergillus niger, Doratomyces stemanitis, Mucor* sp., *Trichoderma viride, T. harzianum.* None of the tested juniper metabolites inhibited the growth of tested fungal species, but they affected the dynamics of growth and sporulation.

Comparative research of the effect of *Juniperus communis* metabolites on bacterial species *Agrobacterium tumefaciens, Bacillus subtilis, Escherichia coli, Pseudomonas fluorescens, P. syringae* var. *phaseolicola* and fungi *Aspergillus niger, Doratomyces stemanitis, Mucor* sp., *Trichoderma viride,* and *T. harzianum,* shows that ground whole fruits of juniper, ground juniper fruit residues after alcoholic extraction, ground juniper fruit residues after water extraction, water extract of juniper fruits, ground juniper fruit residue after essential oil distillation, and essential oil of juniper inhibit the growth of *Agrobacterium tumefaciens, Bacillus subtilis, Pseudomonas fluorescens,* and *P. syringae* var. *phaseolicola,* while alcoholic extract of juniper fruit and ground juniper needles inhibit *Agrobacterium tumefaciens* and *Bacillus subtilis.* The intensity of inhibitory effect of tested metabolites is proportional to their concentration.

Bearing in mind the intensity and nature of the biotic effect of juniper biochemical products, it can be concluded that their effect on growth and metabolism of microorganisms has a biological and ecological significance and therefore it justifies further research of potential application of these properties.

References

Adebajo, A.C., Otske, K. J., and Aladesanmi, A. J. (1989) Antimicrobial activities and microbial transformation of volatile oils of Eugenia uniflora. *Fitoterapia* **5**: 415-455.

Čomic, Lj., Matović, M., Ćurčić, S. (1996) Antimikrobno dejstvo biohemijskih produkata plodova kleke (*Juniperus communis*). 5. Kongres ekologa Jugoslavije. *Zbornik sazetaka*. 177.

Diksit, A., Husain,A. (1984) Antifungal action of some essential oils against animal patogens. *Fitoterapia* **3**: 171-176.

Garrod, L.P., and Grady, O.F (1972) *Antibiotic and Chemotherapy*. Churchill, Livingstone, Edinburgh, London. 3th ed.

Janssen, A. M., Cheffer, J.J.C., and Baeyheim-Svedndsen, A. (1987) Antimicrobial activity of essential oils 1976-1986. Literature rewiev. Aspect of test methods. *Planta medica.* **53**:395-398.

Koleva, P. *et al.* (1980): *Biologicni osnovi na parko-ustrojstvo*, Zemzidat. Sofija.

Matović, M., Čomić, Lj., Ćurčić, S. (1996) Primena ekologije u preradi plodova kleke (*Juniperus communis* L.). 5. Kongres ekologa Jugoslavije. *Zbornik sazetaka*. 1176.

Mimica-Djukić, N., Čomic, R., Kite, G., Jančić, R., Gašić, O., Fellows, L., Stajner, D. (1995) Comparative study of volatile constituents and antimicrobial activity on Mentha species, in Pavletich, E. Putievski (eds.), *Proceedings of International Symposium on Medical and Aromatic Plant*, Israel. 110-115.

Morris, J.A., Khetry, A., and Seitz E.W. (1970) Antimicrobial activity of aroma chemicals and essential oils. *J. Am. Oil Chem.Soc.* 560-598.

Ross, S.A., El-Kektawl, N.E., Megula S.E. (1980) Antimicrobial activity of some Egyptialaromatic plants. *Fitoterapia.* 201-205.

Singh, A.K., Dikshit A., Dixit, S.N. (1983) Fungitoxic properties of essential oil of Mentha arvensis var piperascens. *Perf. Flavour.* **8**: 55-58.

Tokin, B.P.(1951) *Fitoncidy.* Akademii medicinskih nauk SSSR, Moskva.

Voroncihin, N.Z. (1976) Prigorodnye lese I ih rolj v ozdrovlenii sredy obitanija celoveka. *V knj. Biogeografija kraevrdenie, Perm, Vyp.***4,** Moskva.

LIMITED PROTEOLYSIS OF BARLEY RUBISCO AND RECOGNITION OF THE FRAGMENTS BY MONOCLONAL ANTIBODIES

K.N. DEMIREVSKA-KEPOVA*, L.P. SIMOVA-STOILOVA*, B.V. JUPERLIEVA-MATEEVA* and ST.D. KYURKCHIEV**
*Institute of Plant Physiology, Bulgarian Academy of Sciences, 1113 Sofia, Bulgaria**
*Institute of Biology and Immunology of Reproduction, Bulgarian Academy of Sciences, 1113 Sofia, Bulgaria***

1. Introduction

Little is known about the fragmentation of Rubisco by endogenous and exogenous proteases. More sensitive to proteolysis is Rubisco large subunit (LS) and the characteristic pattern of *in vivo* degrading fragments from Rubisco is different depending on experimental conditions (5, 7, 9, 10). Under limited proteolysis of native spinach Rubisco with trypsin two short peptides were released from the N-terminus and one from the C-terminus of the LS (13). Denatured Rubisco from spinach is digested by different proteases (chymotrypsin, papain and *St. aureus* V8 protease) and the LS of the enzyme generates about 15-25 bands (1). Barley Rubisco is less sensitive to trypsinolysis compared to wheat and spinach Rubisco (6). The characterization of Mabs against Rubisco applying limited proteolysis of the enzyme has not been made so far. Monoclonal antibodies (Mabs) against Rubisco from barley leaves have been produced and partially characterized (11). All the Mabs recognise only LS in Western blotting. In the present paper an attempt was made to more thoroughly characterize the epitopes on Rubisco molecule recognized by the Mabs, using limited proteolysis.

2. Materials and Methods

Seeds of *Hordeum vulgare* L. were germinated and grown 10 days under a 12 h day with 200 mE m^{-2} s^{-1} at 27°C during day and 25°C at night. Rubisco was isolated and purified by a modified method described elsewhere (3). Purified Rubisco was stored under liquid nitrogen.

The native Rubisco was activated by 20 mM $MgCl_2$ and 10 mM $NaHCO_3$ for 30 min at 30°C. Rubisco was semi-denatured by 0.5 % w/v SDS in 0.1 M Bicine-NaOH (pH 8.2) containing 20 mM b-mercaptoethanol, and 10 % v/v glycerol. Native or semi-denatured

Rubisco was subsequently proteolyzed with trypsin or with protease from *St. aureus* strain V8. Proteolytic degradation was stopped by boiling the reaction mixture for 5 min after addition 1:1 v/v of Sample buffer for SDS electrophoresis. Soluble proteins were measured by the method of Bradford (2).

SDS-polyacrylamide gel electrophoresis was performed according to Laemmli (8). Immunoblotting was made as described by Mladjova *et al.* (11).

3. Results

After incubation of native Rubisco with trypsin (Fig.1.1A,2A,3A-left) bands corresponding to three fragments of the LS were obtained and after incubation with protease from *St. aureus* strain V8 (Fig.1.1A,2A,3A-right) only one fragment was revealed.

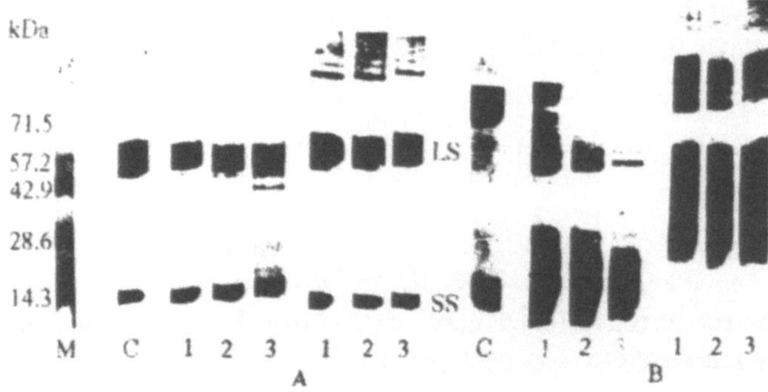

Figure 1. SDS-PAGE analysis of native (A) and semi-denatured Rubisco (B) after proteolysis with trypsin (500:1 w/w) for 15-60 min at 30°C (1,2,3-left) and with *St. aureus* V8 protease (100:1 w/w) for 30-90 min at 30°C (1,2,3-right). The positions of the LS and SS of Rubisco are indicated. Lanes: C - Control without protease; A1, A2, A3 (left) - trypsin digestion for 15, 30 and 60 min respectively; A1, A2, A3 (right) - *St. aureus* V8 protease digestion for 30, 60 and 90 min. M - molecular weight markers 14.3-71.5 (Fluka).

The electrophoretic profiles of both control and variants contained additional polypeptide bands, being probably cross linked aggregates of LS and small subunits (SS).

After incubation of semi-denatured Rubisco with trypsin (Fig.1.B1,B2,B3-left) two relatively stable fragments with molecular masses close to LS, two fragments with lower molecular masses, and many overlapping fragments below 28.6 kDa were observed. Using protease from *St. aureus* strain V8 (Fig.1.1B,2B,3B-right) relatively stable fragments were also obtained but with different electrophoretic mobility. The obtained fragments were transferred onto nitrocellulose sheet and were exposed to the four monoclonal antibodies against Rubisco (Fig. 2).

When native Rubisco was used in the presence of trypsin, all the Mabs interacted with one wide band related to LS and with Rubisco LS aggregates with larger molecular masses (Fig.2.1A,2A,3A,4A-left). We suggest that the wide polypeptide band includes non-degraded LS and LS without two short peptides released from the N- and C-termini. Hence, the determinant groups could not be situated in those termini. Mabs 2F4 and 2D11 recognized 7 fragments from semi-denatured Rubisco with the same electrophoretic mobility for both Mabs (Fig.2.2A,4A-right). Probably these were stable determinant groups during proteolysis. Mab 2G2 recognized the same fragments as above except for one (Fig.2.1A-right). Mab 2H5 recognized two fragments close to LS and only one with lower molecular mass (Fig.2.3A-right).

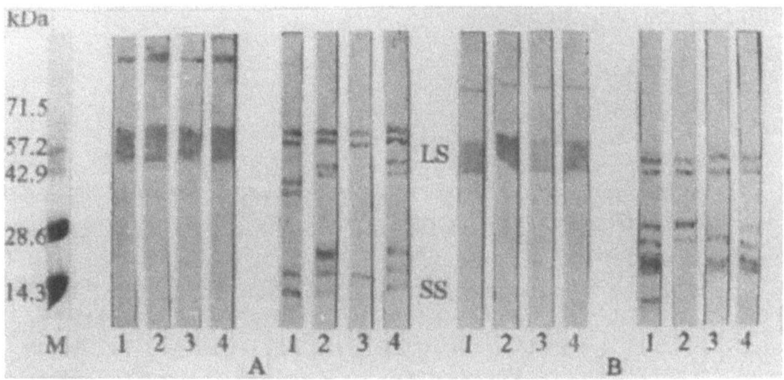

Figure 2. Immunoblotting analysis of native (A-left, B-left) and semi-denatured (A-right, B-right) Rubisco fragments obtained after proteolysis with trypsin (500:1 w/w) for 20 min at 30°C (A) and after proteolysis with *St. aureus* V8 protease (100:1 w/w) for 40 min at 30°C (B) revealed by Mabs against barley Rubisco. The positions of the molecular weight markers 14.3-71.5, (Fluka) (M) are indicated at the left side. Lanes: 1 - Mab 2G2; 2 - 2F4; 3 - 2H5; 4 - 2D11.

All Mabs recognized only LS when native Rubisco was used in the presence of protease from *St. aureus* strain V8 (Fig.2.1B,2B,3B,4B-left). The results were similar to those obtained with native Rubisco digested by trypsin. But the semi-denatured Rubisco was recognized differently by the four Mabs (Fig.2.1B,2B,3B,4B-right). Differences among Mabs appeared in the fragments with lower molecular masses as was the case of trypsin-digested semi-denatured Rubisco.

4. Discussion

Two proteases with different specificity were applied expecting different polypeptide profiles after proteolysis. Susceptibility of the Rubisco LS to both protease actions was established. Evidently, SS were discriminated from proteolytic degradation when assembled into holoenzyme. Such a differential susceptibility to proteolysis was also observed by other authors (14). Rubisco as a multimeric protein complex cannot be simply subjected to proteolytic degradation. Modification or denaturation of Rubisco is the possible mechanism triggering protein degradation (5,12). SDS seems to change the susceptibility of Rubisco to proteases by changing its conformation (9). The relative stability of fragments obtained by exogenous proteases is similar to that of fragments produced by endogenous proteases (7). When Rubisco was in native state it was more stable than semi-denatured one towards the proteases used.

Our earlier studies have shown that Mabs produced against Rubisco holoenzyme from barley recognize only the LS and that the epitopes are partially overlapping as established by the ELISA additivity test (4, 11). In the present study all the Mabs recognized many fragments obtained from the LS of Rubisco by treatment with both proteases. Probably, these determinant groups were relatively stable during proteolysis. Differences among Mabs about the recognized epitopes appeared only in the fragments with lower molecular masses, which further supports the finding that these epitopes overlap.

Acknowledgements: The work was supported by grant No. K-443 from the National Fund "Scientific Research", Bulgaria.

5. References

1. Bottomley, W. (1982) Characterization of chloroplast proteins by partial proteolytic degestion, in Edelman, M., Hallick, R.B., Chua, N.-H. (eds.), *Methods in Chloroplast Molecular Biology*, Elsevier Biomedical Press, pp. 1055-1061.
2. Bradford, M.M. (1976) A rapid and sensitive method for the quantitation of the microgram quantities 0f protein utilising the principle of protein dye binding, *Anal. Biochem.* **72**,248-254.
3. Demirevska-Kepova K. and Simova, L.(1989) [Isolation and purification of ribulose 1,5-bisphosphate carboxylase/oxygenase from barley leaves.] *Plant Physiol. (Bulg.)* **15**, 3-10. [In Bulg.]
4. Demirevska-Kepova, K., Simova-Stoilova, L., Kyurkchiev, St. and Juperlieva-Mateeva, B. (1996) Epitope mapping of Rubisco from barley leaves, *Comt. Rend. Acad. Sci. Bulg.* **49**, 93-96.
5. Desimone, M., Henke, A. and Wagner, E. (1996) Oxidative stress induces partial degradation of the large subunit of ribulose-1,5-bisphosphate carboxylase /oxygenase in isolated chloroplasts of barley, *Plant Physiol.* **111**, 789-796.
6. Gutteridge, St., Millard B.N. and Parry M.A.J. (1986) Inactivation of Ribulose-bisphosphate carboxylase by limited proteolysis, *FEBS Letters.* **196**, 263-268.
7. Hildbrand, M., Fischer, A. and Feller, U. (1994) Protein catabolism in bean leaf discs: accumulation of a soluble fragment of ribulose-1,5-bisphosphate carboxylase/oxygenase under oxygen deficiency, *J. Exp. Bot.* **45**, 1197-1204.
8. Laemmli , U.K. (1970) Cleavage of structural proteins during assembly of the bacteriophage T4, *Nature* **227**, 680-685.
9. Mae, T., Kamei, Ch., Funaki, K., Miyadai, K., Makino, A., Ohira, K. and Ojima, K. (1989) Degradation of ribulose-1,5-bisphosphate carboxylase/ oxygenase in the lysates of the chloroplasts isolated mechanically from wheat (Triticum aestivum L.) leaves, *Plant Cell Physiol.* **30**, 193-200.
10. Mitsuhashi W., Craft-Brandner St.J. and Feller, U. (1992) Ribulose-1,5-bisphosphate carboxylase/ oxygenase degradation in isolated pea chloroplasts incubated in the light or in the dark, *J. Plant Physiol.* **139**, 653-658.
11. Mladjova, N., Demirevska-Kepova, K., Kehayov, I. and Kyurkchiev St. (1992) Production and chracterization of monoclonal antibodies to barley ribulose 1,5-bisphosphate carboxylase/oxygenase, *Plant Physiol. Biochem.* **30**, 523-529.
12. Moreno, J., Peñarrubia and L., Garcia-Ferris, C. (1995) The mechanism of redox regulation of ribulose 1,5-bisphosphate carboxylase/ oxygenase turnover. A hypothesis, *Plant Physiol. Biochem.* **33**, 121-127.
13. Mulligan, R.M., Houtz, R.L. and Tolbert, N.E. (1988) Reaction intermediate analogue binding by ribulose 1,5-bisphosphate carboxylase/oxygenase causes specific changes in proteolytic sensitivity: the N-terminal residue of the large subunit is acetylated proline, *Proc. Natl. Acad. Sci. USA* **85**, 1513-1517.
14. Otto, S. and Feierabend, J. (1994) Assay and comparative characterization of the proteolytic degradation of isolated small subunit and holoenzyme of ribulose 1,5-bisphosphate carboxylase/oxygenase in chloroplasts from rye leaves, *J. Plant Physiol.* **144**, 26-33.

MEMBRANE TRANSPORT IN *Chara*: STUDY OF MACROSCOPIC CURRENTS AND EFFECT OF Gd-DTPA

M.R.DJURIŠIĆ [1], A.A.CHERKASHIN[2], P.R.ANDJUS[3], D.VUČELIĆ[4]
[1]IGPC, POB 551, 11001 Belgrade, YU, [2]Institute of Cell Biophysics, Puschino, Moscow Region, Russia, [3]School of Biology, Belgrade University, POB 52, 11001 Belgrade, YU, [4]School of Physical Chemistry, Belgrade University, POB 550, 11001 Belgrade, YU

1. Introduction

Protoplasmic droplet membrane of the green alga *Chara* contains two types of K^+ channels (Laver and Walker, 1987, Pottosin and Andjus, 1994) and a Cl^- channel (Tyerman and Findlay, 1989).

Here we measured potassium currents elicited at depolarization (Tyerman and Findlay, 1989), presumed to be mediated by the slow delayed rectifier K^+ channel, SDRC (Pottosin and Andjus, 1994). The toxicity of the complex salt Gd-DTPA i.e. traces of Gd^{3+} reported before on K^+ currents (Elinder and Arhem, 1994), which could become toxic at high concentrations, was tested on both macroscopic potassium curents and single-SDRC level.

2. Materials and methods

The cytoplasmic droplets were prepared by eluting the cytoplasm of internodal cells of *Chara* with the solution containing Ca^{2+} which is necessary for membrane formation.

The well known electrophysiological technique, patch-clamp (Hamill *et al.*, 1981), was used for the recording of currents across the whole-droplet membrane (whole-droplet configuration) or single channel currents (outside-out configuration).

3. Results

3.1.Cs^+ BLOCK

In order to test the nature of the main mediator of the macroscopic currents of the *Chara* cytoplasmic drops, the measurements were made in the presence of a known K^+ channel blocker, Cs^+ (150 mM), in the bath. As seen from figs. 1.a and 1.b, Cs^+ completely blocked inward currents, and shifted the I/V dependence of the outward currents (fig.2).

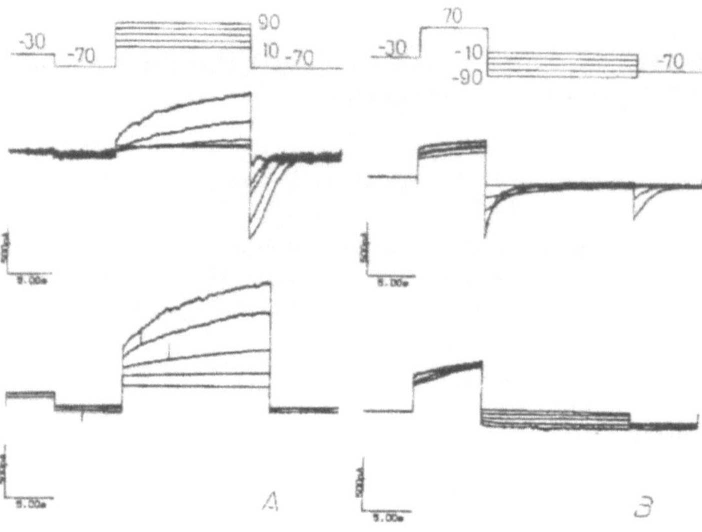

Fig 1. Recordings of the whole-droplet currents: a) control (150 mM KCl, Ca^{2+} chelated to 0.05 mM, mannitol added as osmoticum, pH 7.2) (upper) and 150 mM CsCl (lower) solution; voltage protocol for obtaining I/V curve shown above; b) control (upper) and 150 mM CsCl (lower) solution; voltage protocol for measuring voltage dependence of deactivation kinetics shown above.

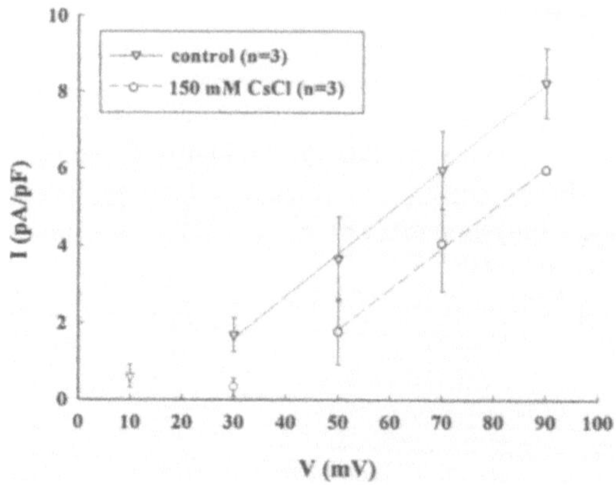

Fig.2. Cs^{+}- block of the outward macroscopic current; I/V curve was shifted by 16 mV in the positive direction in the presence of CsCl. The membrane conductance was: 105±5 and 110±3 pA/pF (conductance ± S.D. of the fit) in the presence and absence of 150 mM CsCl, respectively.

3.2. Gd-DTPA EXPERIMENTS

The voltage dependence of the macroscopic currents (fig.3) and the time constant of deactivation (τ) (fig.4) did not change in the presence of 20 and 100 mM Gd-DTPA, as compared to the control.

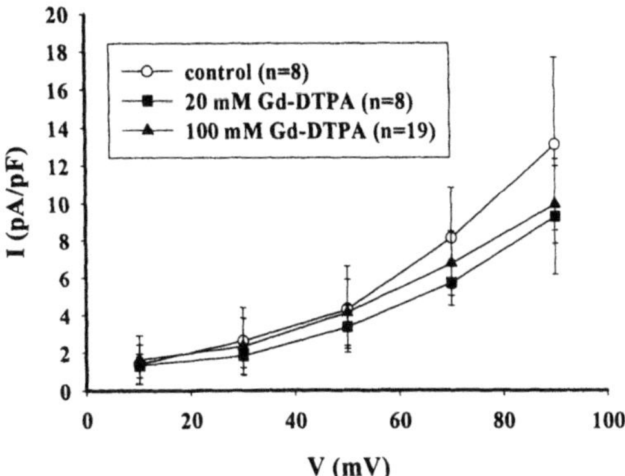

Fig.3. I/V curves of whole-droplet current obtained in the presence of 20 and 100 mM Gd-DTPA, and control solution.

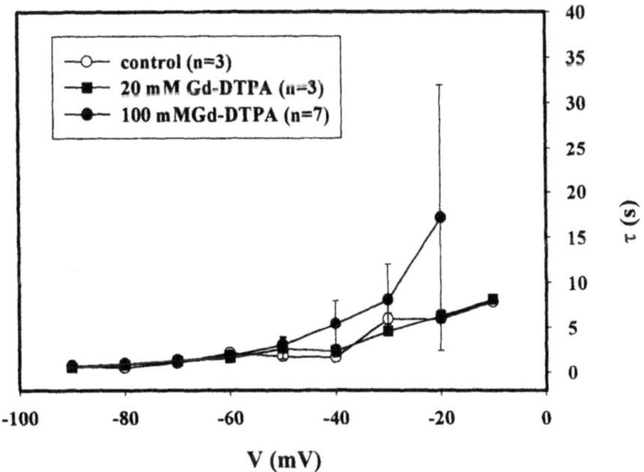

Fig.4. Voltage dependence of the deactivation time constant, τ, in the presence of 20 and 100 mM Gd-DTPA, and control solution.

Also, on the single-channel level, parameters of the SDRC channel activity did not change in the presence of 100 mM Gd-DTPA, as compared to the control (Table I).

TABLE I: Single SDR channel parameters; unitary current, *I* (pA), unitary conductance, *G (pS)*, probability of the channel opening, *p*, and channel kinetics (mean open, *mot*, and mean shut, *mst*, times)

	V (mV)	I (pA)	G(pS)	p	mot (ms)	mst (ms)
Control	50	3.46	78±6	0.56	10.8	16.6
100 mM Gd-DTPA	50	3.54	73±4	0.55	12.2	19.7

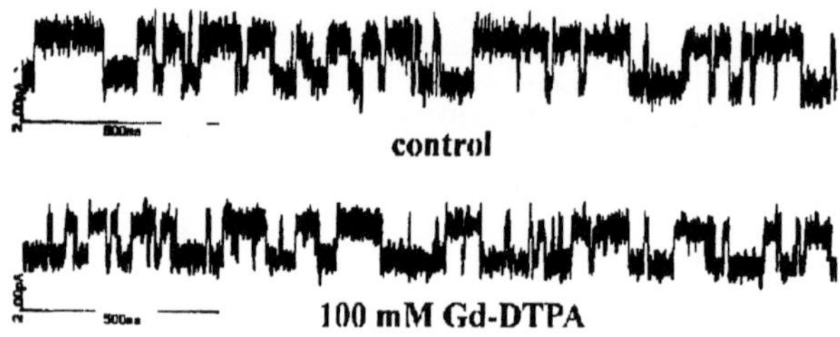

Fig. 5. Single SDRC recording in the control solution and in the presence of 100 mM Gd-DTPA

4. Disscusion

Cs^+ ion is a probe that discriminates K^+ channels from other cation channels that conduct K^+ ions. The complete block of the inward deactivating current by Cs^+ is an evidence that the positive membrane current was carried by K^+ through characteristic channels.

Of the two K^+ channels in the droplet membrane, only SDRC is activated by depolarization, (Tyerman and Findlay, 1989., Laver and Walker, 1987, Pottosin and Andjus, 1994) and can explain the depolarization activated macroscopic currents.

The effect on K^+ currents was not observed neither with 20 mM nor with 100 mM Gd-DTPA, in whole-droplet configuration as well as in outside-out patches which indicated that even at higher concentrations of the complex salt the contamination of Gd^{3+} is negligible.

5. References

Elinder, F. and Arhem, P. (1991) Properties of the Leakage Current Pathway: Effects of the Gadolinium Ion on Myelinated Axons, *Neuroreport* **2**, 685-687.

Hamill, O.P., Marty, A., Neher, E., Sakmann, B., and Sigworth, F.J. (1981) Improved Patch-Clamp Techniques for High-Resolution Current Recording from Cells and Cell-Free Membrane Patches, *Pflugers Archives* **391**, 85-100.

Laver, D.R., and Walker, N.A. (1987) Steady-State Voltage-Dependent Gating and Conduction Kinetics of Single K^+ Channels in the Membrane of Cytoplasmic Drops of *Chara australis*, *J. Memb. Biol.* **100**, 31-42.

Pottosin, I.I. and Andjus, P.R. (1994) Depolarization -Activated K^+ Channel in *Chara* Droplets, *Plant Physiol.* **106**, 313-319.

Tyerman, S.D. and Findlay, G.P. (1989) Current-Voltage Curves of Single Cl^- Channels which Coexist with Two Types of K^+ Channel in the Tonoplast of *Chara corallina*, *J. Exp. Bot.* **40**, 105-117.

INFLUENCE OF 5-AMINOLEVULINIC ACID ON CHLOROPHYLL SYNTHESIS IN BLACK PINE COTYLEDONS

G. DRAZIC AND M. BOGDANOVIC
Institute for the Application of nuclear Energy, 11080 Zemun, Banatska 31b, Yugoslavia

1. Introduction

Although it has been established long ago that some gymnosperms synthesize chlorophyll in the dark, there is only a few data on specific regulation pattern of this process: Bogdanovic (1973) for *Pinus nigra,* Michael-Wolvertz (1977) for *Pinus jeffreyi,* Krol (1978) and Kasemir *et al.* (1980) for *Pinus sylvestris,* Mariani *et al.* (1990) *for Picea abies and Larix decidua,* Ou and Adamson (1995) for *Pinus Pinea.* Our previous studies showed that the greening of black pine (*Pinus nigra* Arn.) cotyledons in the dark necessary required a contact with megagametophyte or exogenous cytokinin (Jeli6 and Bogdanovic 1988).

Chlorophyll formation is thought to be regulated at different sites along the complex biosynthetic pathway, the most important of which are formation and utilization of 5-ALA (Castelfranco and Beale 1983, Kumar *et al.* 1996).

The effect of exogenous 5-ALA and LA (competitive inhibitor of 5-ALA dehydroganase) on chlorophyll accumulation in black pine cotyledons was monitored during germination in the dark in order to understand the regulation of this process.

2. Material and methods

Pine seeds were imbibed in water during 24h in the dark, and cultivated in Petri dishes in the light (220µM $m^{-2}s^{-1}$, powertube 72 cool white 1F72T12-CW-UHO, Sylvania and incandescent bulb 25W) or in the dark on 24°C at: A - distilled water (control plants); B - 10^{-5} M 5-ALA (ALA-fed plants).

Ten days old seedlings were 24h incubated in LA or 5-ALA.

All manipulations of seedlings were performed in the dark or under a green safe light (a photolamp with a 25W bulb covered with a dark-green filter Fotokemika TZ 35).

Chlorophyll was extracted in dimethylformamide and estimated spectrophotometrically according to Moran (1982). Results represents mean value of five replications ± SD. Statistical evaluation of all measurements was made using Student's t-test.

3. Results

The chlorophyll content in 5-ALA-fed black pine seedlings grown in the dark was concentration-dependent. Low concentration of 5-ALA stimulate Chl accumulation while higher concentrations inhibit it. The highest-effect was on 10^{-5} M 5-ALA concentration (Tab. 1.).

Table 1. The effect of different 5-ALA concentrations on chlorophyll accumulation in cotyledons of black pine grown 7 days in the dark

5-ALA conc.	0	10^{-6} M	10^{-5}	10^{-4} M	10^{-3} M	10^{-2} M
Chl a+b (µg/embryo)	0.80±0.04	1.20±0.08[aa]	1.34±0.10[aa]	0.65±0.05[b]	0.60±0.04[bb]	0.46±0.02[bb]

[a] stimulation ($p < 0.05$); [aa] stimulation ($p < 0.01$);
[b] inhibition ($p < 0.05$); [bb] inhibition ($p < 0.01$)

When black pine seedlings were grown for 10 days in the dark and after that incubated in different 5-ALA concentrations, it may be noticed that only 10^{-5} M 5-ALA induced an increase of Chl accumulation in the control seedlings, while 10^{-3} M 5-ALA induced a decrease of Chl content in the control plants. In ALA-fed plants, 5-ALA induced a decrease of Chl accumulation (Tab. 2). The same changes tend to appear when the seedlings are incubated in the presence of LA, although at lower concentrations.

Table 2. Effect of 24h incubation in ALA and LA on Chi accumulation in 10 days old pine seedlings grown in the dark.

	control plants	ALA-fed plants		control plants	ALA-fed plants
5-ALA conc.	Chl µg/embryo	Chl µg/embryo	LA conc.	Chl µg/embryo	Chl µg/embryo
0	1.34±0.08	1.84±0.10	0	1.34±0.08	1.84 ±0.10
10^{-5} M	1.92±0.06[aa]	1.40±0.09[b]	10^{-6} M	1.81±0.07[aa]	1.12 ±0.02[bb]
10^{-4} M	1.30±0.11	1.16±0.07[bb]	10^{-5} M	1.14±0.05	0.82 ±0.06[bb]
10^{-3} M	1.12±0.05[b]	0.98±0.04[bb]	10^{-4} M	0.72±0.09[bb]	0.65 ±0.03[bb]

[a] stimulation ($p < 0.05$); [aa] stimulation ($p < 0.01$);
[b] inhibition ($p < 0.05$); [bb] inhibition ($p < 0.01$)

4. Discussion

The formation of 5-ALA is the rate-limiting step in chlorophyll biosynthesis in higher plants (Castelfranco and Beale, 1983). In all plant tissues investigated up till now it was shown that 5-ALA was formed by the C5 pathway. All the genes that encode the relevant enzymes of this pathway are present in at least two families, one of which is activated by light, and the other is expressed independently of light influence in all tissues (Kumar, 1996). As the chlorophyll in gymnosperm cotyledons is synthesized in the dark, although much slower than in the light, it may be supposed that in this case also a light-independent C5 biosynthetic pathway is utilized.

In this study we have tested a hypothesis that in *P. nigra* seedlings grown in the dark there accumulates a suboptimal 5-ALA concentration. Low concentrations of exogenous 5-ALA compensate for the deficiency of endogenous 5-ALA and stimulate Chl accumulation. Levulinic acid inhibits, although incompletely (Stobart 1984), ALA dehydrogenase, inducing thus an increase of 5-ALA content which exerts the same effect on chlorophyll accumulation.

At high concentrations of 5-ALA, there occurs a feed-back inhibition of POR (NADPH-protochlorophyllide oxidoreductase) which interrupts any further Chl accumulation, which was already found in other plant species (Reinbothe 1996). In *P. jeffrey* the exogenous ALA leads to the accumulation of untransformable protochlorophyll (MichaelWolvertz 1977), decreasing thus the accumulation of Chl.

Chlorophyll accumulation inhibition at high LA concentrations may be explained by a complete suppression of ALA dehydrogenase by which the biosynthetic pathway is interrupted and/or by a high ALA accumulation by which POR is being blocked.

Although the results presented here support the set hypothesis, for its absolute confirmation, determination of endogenous ALA content in seedlings *of P.nigra* would be necessary.

5.References

1. Bogdanovic, M. (1973) Chlorophyll formation in the dark. 1. Chlorophyll in pine seedlings, *Physiol Plant.* **29**, 17-18
2. Castelfranco, P. A. and Beale S.I. (1983) Chlorophyll biosynthesis. Recent advances and areas of current interest, *Annu. Rev. Plant Physiol.* **34**, 241-278
3. Jelic. G. and Bogdanovic, M. (1988) Antagonism between abscisic acid and cytokinin in chlorophyll synthesis in pine seedlings, *Plant Sci.* **61**, 197-202.
4. Kasemir, H., Schroder, S. and Steinhilber, A. (1980) Light-mediated control of chlorophyll formation in seedlings of *Pinus sylvestris* in J.de Greef, (ed), *Photoreceptors and plant development*, Antwerpen University Press, Antwerpen, pp 185-197.
5. Krol. M. (1978) The relationship between photosynthetic activities and the polypeptide pattern of pine seedling chloroplasts, Z. *Pflanzenphysiol.* **86**, 379-387.
6. Kumar, A.M., Schaub, U., Soil, D. and Ujwal, M. (1996) Glutamyl-transfer RNA: at the crossroad between chlorophyll and protein biosynthesis, *Trends in Plant Science* **1**, 371-376.
7. Mariani, P., et al. (1 990) Synthesis of chlorophyll and photosynthetic competence in etiolated and greening seedlings *oflarix decidua* as compared with *Picea abies*, J. Plant Physiol. **137**, 5-14
8. Michel-Wolvertz. M. R. (1977) Chlorophyll formation in cotyledons *of Pinusjeffteyı* during germination in the dark - occasional accumulation of protochlorophyllide forms, *Plant Sci. Lett.* **8**, 125-134
9. Moran , R. (1982) Formulae for determination of chlrophyllous pigments extracted with N,N-dimethylformamide, *Plant Physiol.* **69**, 1376-13 8 1.

10. On, K. and Adamson, H. (1995) Chlorophyll accumulation in cotyledons, hypocotils and primary needles *ofpinuspinea* seedlings in light and dark, *Phisiol. Plant* **93**, 719-724
11. Reinbothe, S. et al.(l 996) PORA and PORB, two light-dependent protochlorophyllide reducing enzymes of angiospertn chlorophyll biosynthesis, *Plant Cell* **8**, 763-769.
12. Stobart, A. K. and Hendry (1984) The turnover of the chlorophyll in greening wheat leaves, *Phytochemistry* **23**, 27-30.

CYTOCHEMICAL INVESTIGATION OF TOMATO AND CUCUMBER POLLEN AFTER CUCUMBER MOSAIC VIRUS INFECTION

I. D. GEORGIEVA AND E. STOIMENOVA
Institute of Genetics
Sofia 1113, Bulgaria

1. Abstract

It was established that CMV infection of tomato and cucumber plants is associated with disturbances in respiratory pathways in pollen - decrease in activities of enzymes taking art in aerobic respiration and increase in activities of enzymes involved in anaerobic respiration. A rise of peroxidase activity as a response to the biotic stress was pointed out. In cucumber pollen enhancement of G6PDH activity was found which is supposed to indicate "de novo" biosynthesis of virus RNA. A slight intensification of hydrolytic processes and decline of PPO activity in pollen of both plant species was observed. These metabolic disturbances were connected with drastic drop in pollen viability.

2. Introduction

Biochemical investigations of the host responses after virus infection showed changes in photosynthesis, respiration, and secondary metabolite accumulation (Zaitlin, 1987). The alterations in the oxidative pentose-phosphate pathway (PPP) connected with virus multiplication (Burketova, 1995; Sindelar and Sindelarova, 1987) as well as of peroxidase (POD) activity which is involved in the plant stress responses were reported as well (Montalbini et al, 1995; Repka and Slovakova, 1994). Little information is available about the metabolic shifts in floral organs and disturbances of reproductive processes (Repka and Jung, 1995).

The purpose of the present research was to investigate the changes in the metabolism of pollen grains of cucumber mosaic virus (CMV) infected cucumber and tomato plants by the use of cytochemical methods.

3. Material and Methods

Tomato (*Lycopersicon cheesmanii typicum*) and cucumber (*Cucumis sativus*) plants were infected with one of the two CMV strains - CMV-MB and CMV-CB respectively. The first strain was isolated from tomato and the second one from cucumber (Stoimenova, 1996). The plants were inoculated mechanically in the phase of cotyledons - first leaf. The investigation was conducted on mature pollen collected in the day of blossoming. The cytochemical investigation of some enzymes involved in different metabolic pathways was carried out as follows: the dehydrogenases (isocitrate

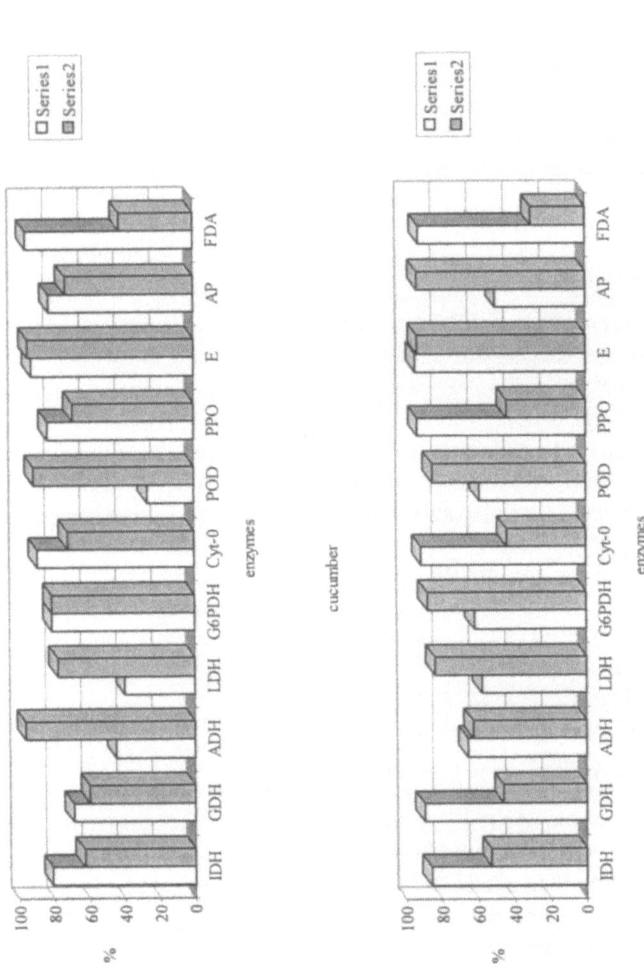

series: 1- non-infected control, 2 - CMV infected pollen.

Fig. 1. Changes in enzyme activities and viability of CMV infected tomato and cucumber pollen expressed as percent pollen grains with positive cytochemical reaction.

dehydrogenase - IDH, glutamate dehydrogenase - GDH, alcohol dehydrogenase - ADH, lactate dehydrogenase - LDH and glucose-6-phosphate dehydrogenase - G6PDH) were demonstrated by the method of tetrazolium reductases. Cytochrome oxidase (Cyt-oxidase) was proved by NADI reaction, peroxidase (POD) - by the method of Graham and Karnovski and polyphenoloxidase (PPO) - with DOPA. Hydrolases - esterase (E) and acid phosphatase (AP), were investigated by the method of simultaneous azocopulation (Lojda et al, 1979). The viability of pollen was monitored using fluorochromatic (FCR) test (Heslop-Harrison et al, 1970). The effect of CMV infection was estimated by determination of both staining intensity and percentage of pollen grains with positive reaction. Pollen samples were assessed by observing more than 300 pollen grains.

4. Results

The cytochemical investigation of IDH in CMV infected tomato and cucumber pollen anthers showed that IDH activity in pollen grains dropped and their staining became pale. After applying the cytochemical reaction for GDH demonstration it was found that the pollen grains of both infected plants stained more strongly than the pollen of controls but during the infection process the percentage of grains with negative reaction increased. Virus infection increased the percentage of cucumber pollen grains with G6PDH activity and their staining intensity as well, but did not change significantly G6PDH activity in tomato pollen. The decrease of cyt-oxidase activity in infected tomato pollen was expressed as weak staining of pollen grains and decrease of percentage of pollen with positive reaction. In the cucumber pollen the quantity of stained pollen grains dropped too. The percentage of pollen with positive LDH reaction in both plants under study rose. After infection the quantity of tomato pollen grains with ADH positive reaction increased significantly and the staining intensity became stronger whereas CMV infection nearly did not change the LDH activity in cucumber pollen.

The virus infection was connected with rise of POD activity and fall of PPO activity in both tomato and cucumber pollen (Fig.1).

The CMV infection increased slightly esterase activity in tomato pollen and nearly did not change the percentage of stained pollen grains of cucumber. It intensified the acid phosphatase reaction in tomato pollen and increased the percentage of stained cucumber pollen grains (Fig.1).

5. Discussion

The present investigation revealed that CMV infection of tomato and cucumber plants was associated with significant disturbances in respiratory pathways in pollen - decrease of pollen grains with activities of enzymes taking part in aerobic respiration (LDH, GDH and cyt-oxidase). In the same time activation of enzymes involved in anaerobic respiration (ADH and LDH in tomato pollen and only LDH in cucumber pollen) was observed. It is interesting to point out that not with staining the decrease of pollen grains with positive reaction for GDH (which accomplishes the connection between metabolism of carbohydrates and proteins) their activity was higher than the control. The negative GDH reactions of pollen grains was characteristic for the sterile empty pollen grains whose number increased during the process of infection Biochemical study of cucumber cotyledons infected with CMV have shown considerable alteration of carbohydrate metabolism too (Tesci et al., 1994).

In pollen of CMV infected cucumber plants the rise of G6PDH activity (the rate limiting enzyme of PPP) was found. It is known that PPP is the pathway producing recursors of nucleotides involved in "de novo" biosynthesis of virus PNA. It was reported that G6PDH activity increased in potato and tobacco leaves infected with PVY (Sindelar and Sindelarova, 1987) as well as in marrow cotyledons infected with CMV (Zaitlin, 1987). The increase of peroxidase activity (a stress related enzyme) in pollen grains of CMV infected tomato and cucumber pollen is probably the response of the viral pathogens as a natural stress factors. Similar results were obtained in host vegetative organs in other plant-virus systems (Montalbini et al., 1995; Repka and Slovakova, 1994; Repka and Jung, 1995).

The CMV invasion caused slight enhancement of hydrolytic processes (esterase and acid phosphatase activities) and decline of polyphenoloxidase activity in pollen of the plants studied.

All abnormalities in pollen grain metabolism following CMV infection of tomato and cucumber plants were connected with drastic drop of pollen viability and consequently with disturbances of reproductive processes.

6. References

Burketova, L. (1995) Changes in glucose-6-phosphate dehydrogenase and ribonucleases activities during PVY-RNA biosynthesis in infected potato plants, *Biologia Plantarum* **37**, 423-428.

Heslop-Harrison, J., Heslop-Harrison, Y., and Schivanna, K.R. (1970) Evaluation of pollen viability by enzymatically induced fluorescence, intracellular hydrolysis of fluorescein diacetate, *Stain Technol.* **45**, 115-120.

Lojda, Z., Gossrau, R., and Schiebler, T.H. (1979) *Enzyme Histochemistry*, Springer Verlag, Berlin.

Montalbini, P., Buonaurio, R., and Umesh Kumar, N.N. (1995) Peroxidase activity and isoperoxidase pattern in tobacco leaves infected with tobacco necrosis virus and other viruses inducing necrotic and non-necrotic alterations, *J. Phytopathology* **143**, 295-301.

Repka, V. and Jung, M. (1995) Organ-specific expression of the stress-related anionic peroxidases in cucumber flowers, *Biologia Plantarum* **37**, 523-531.

Repka, V. and Slovakova, L. (1994) Purification, characterization and accumulation of three virus-induced cucumber peroxidases, *Biologia Plantarum* **36**, 121-132.

Sindelar, L. Changes in the activity of glucose-6-phosphate dehydrogenase and some problems relating to its regulation in tobacco plant infected with potato virus Y, *Biologia Plantarum* **28**, 440-448.

Sindelar, L and Sindelarova, M. (1987) Changes in the activity of phosphoguconate dehydrogenase and its regulation in tobacco infected with PVY, *Biologia Plantarum* **29**, 150-154.

Stoimenova, E. (1996) Investigations of the strain variability of tobamoviruses and cucumovirus isolated from Bulgaria, *J. Culture Collections* **1**, 46-52.

Tecsi, L.I., Maule, A.J., Smith, A.A., Leegood, R.C. (1994) Metabolic alterations in cotyledons of *Cucurbita pepo* infected by cucumber mosaic virus, *J Expr. Bot.* **45**, 1541-1551.

Zaitlin, M. and Hull, R. (1987) Plant virus-host interactions, *Ann. Rev. Plant Physiol.* **38**, 291-315.

EXTERNAL FLAVONOID AGLYCONES IN GENUS *TANACETUM* (*ASTERACEAE*)

S. IVANCHEVA, J. CHERNEVA., B. STANCHEVA
Institute of Botany, Bulg. Academy of Sciences, 1113 Sofia, Bulgaria

1. Introduction

The tribe *Anthemideae* of the *Asteraceae* comprises a large number of species that have been still are used as medicinal plants, particularly in folk medicine. It has recently shown [1] flavonoid aglycones are accumulated in several medicinal plants, especially on their leaf and stem surfaces.

The genus *Tanacetum (Asteraceae)* comprises of about 145 species, complied mainly from the Floras of Europe, Turkey and Russia. The chemotaxonomy of *Tanacetum* has suffered great neglect in comparison with that of its "sister" genera like *Artemisia*. At stated in previous papers, chemical investigation of species of the genus *Tanacetum*-. *T. serotinum* [2], *T. parthenium* L. [2, 3], *T. macrophyllum* [4] , *T. vulgare* [5] , *T. corymbosum* L.[6] , *T. cilicium* [7] on the contents of sesquiterpene lactones and flavonoids has reveled. These compounds are capable of reacting inhibitory effects of *T. parthenium* L. (feverfew) on platelet and leukocyte secretion and be responsible for their anti-inflammatory and cytotoxic properties [3]. However such compounds are likely to be present since several flavonoid methyl ethers have been characterized from *Tanacetum* species. Thus, 6-hydroxyluteolin 6,3'-dimethyl ether (jaceosidin) and quercetagetin 3,6,3'- trimethyl ether (jaceidin) have been reported from the flowers of *T. vulgare*, while 6-hydroxyluteolin 6,7,3'- trimethyl ether (cirsilineol), 6-hydroxyluteolin 6,7,3',4'-tetramethyl ether, quercetagetin 3,6,7,4'- tetramethyl ether (casticin) have been found in aerial parts of *T. santolinoides*.[8]. More recently a unique flavonoid structure with a carboxylic acid substituent in the 7-position, 3,5,3'- trihydroxy 4'-methoxy- 7-carbomethoxyflavone has been characterized from *T. macrophyllum* together with quercetagetin 3,6,4'-trimethyl ether [9]. The present paper reports the lipophilic flavonoids identified from the leaves of *Tanacetum* species growing in Bulgaria.

2. Materials and Methods

Dried leaves and steams from wild populations and cultivated plants were used for analysis.

Voucher specimens are deposited at the Herbaria of the Institute of Botany, Sofia.

List of plant samples .

Section *Tanacetum*

T. vulgare L.: (1). Central Stara Planina Mt., Trojan ; (2). Sofia region, Tzarigradsko schosse ; (3). Sredna gora Mt., Pirdop ; (4). Rila Mt., Borovetz ; *T. balsamita* : (5).Experimental field of Institute of Botany, BAS, Sofia

Section *Pyrethrum*

T. parthenium : (6). West Stara Planina Mt., Barsia ; *T. corymbosum* L.: (7). Sofia region, Kokaljane monastery ; (8). Rodopes Mt., Chudni mostove ; (9). Sofia region, Lozenska planina ; *T. macrophyllum* L.: (10). Sofia region, Kokaljane monastery

Aerial parts (leaves and stems) were briefly rinsed with acetone to dissolve the exudate materials which was concentrated by evaporation of the solvent. The exudate was chromatographed over Sephadex LH 20 to separate the flavonoids from the dominating terpenoid consistuents. Concentrated flavonoid fractions were subjected to comparative TLC on Silica and Polyamide using several solvent systems [10]. Individual compounds were identified by co-chromatography with markers. In some cases the structures were confirmed by UV- Spectroscopy.

3.Results and Discussion

The *Tanacetum* species analyzed were belong to different systematic group. Their arrangement in Table 1. follow principally the taxonomic concept of Kuzmanov (1984). The main distinction is made between *T. vulgare* L. and *T. balsamita*, sect. *Tanacetum* (yellow flowering) and *T. parthenium* (L.) Schultz.Bip., *T. corymbosum* (L.) Schultz.-Bip., *T. macrophyllum* (Waldst. et Kit.) Schultz.- Bip. in sect. *Pyrethrum* (white flowering).

The major accumulation trend is represented by the presence of 6-hydroxyflavones, 6-hydroxyflavonols and their derivatives in the leaf exudates (Table 1.). The flavones scutellarein and 6-hydroxyluteolin are represented by a series of methyl derivatives, while the derivatives of the flavonols 6-hydroxykaempferol and quercetagetin are observed with a higher degree of methylation. Thus, within the both sections the most common aglycones were luteolin, 6-hydroxyluteolin 6-methyl ether (nepetin), 6-hydroxyluteolin 6,3'-dimethyl ether (jaceosidin) and quercetagetin 3,6- dimethoxy ether (axillarin). Further derivatives of flavones and flavonols co-occurred in varied amounts. Within sect. *Pyrethrum*, luteolin, nepetin and jaceosidin were encountered in one taxon each. Flavonoids lacking 6-substitution rarely accumulated in the leaf exudates, but luteolin and quercetin 3-methyl ether are found in both sections. Quercetagetin 3,6,4'- trimethyl ether (centaureidin) and quercetagetin 3,6,3',4'- tetramethoxyl ether (casticin) were detected only in *T. corymbosum*. In case of *T. balsamita* (cultivated) was detected only one derivative of quercetagetin, another flavonols lack. 7- substitution derivatives lack in investigated species.

TABLE 1. Flavonoid aglycones distributed in genus *Tanacetum*

Compounds	Species									
Flavones	**1**	**2**	**3**	**4**	**5**	**6**	**7**	**8**	**9**	**10**
1. Scut- 6- Me (hispidulin)	+	+	+		+		++			
2. Scut- 6, 4'- diMe (pectolinarigenin)	+	++			++					
3. Luteolin	(+)			+	+	(+)	(+)	(+)	(+)	(+)
4. Lut- 3'- Me (chrysoeriol)				+						
5. 6-OH Lut- 6- Me (nepetin)	+	+	+	(+)	+	(+)	++	(+)	(+)	+
6. 6-OH Lut- 6,3'- diMe (jaceosidin)	++	++	++	++	+		(+)			
Flavonols										
7. 6-OH Kae 3,6,4'- triMe (methylbetuletol)						+		+		
8. Qu- 3- Me	(+)			(+)			(+)	(+)		(+)
9. Queg- 3,6- diMe (axillarin)	+	++	+	+	+	(+)	++	++	(+)	+
10. Queg 3,6,3'- triMe (jaceidin)	++	+	+	++		(+)	(+)	++		
11. Queg 3,6,4'- triMe (centaureidin)									++	
12. Queg 3,6,3',4'-tetraMe (casticin)									++	

<u>Legend</u>: *T. vulgare*: (1). Central Stara Planina Mt., Trojan ;(2). Sofia region, Tzarigradsko schosse; (3). Sredna gora Mt., Pirdop ;(4). Rila Mt., Borovetz; *T. balsamita* :(5). Experimental field of Institute of Botany, BAS, Sofia; *T. parthenium* (6). West Stara Planina Mt., Barzia; *T. corymbosum* L.: (7). Sofia region, Kokaljane monastery; (8). Rodopes Mt., Chudni mostove; (9). Sofia region, Lozenska planina; *T. macrophyllum* L. (10). Sofia region, Kokaljane monastery

However, it should be noted that aglycone profiles observed within genus *Tanacetum* proved to be of more chemotaxonomic significance. The diversity in terms of flavonoid substitution trends is high and formation of exudate flavonoid aglycones is related to the habitat since most of the taxa analyzed are either meso or prefer xeric habitats [1].

4. Acknowledgments

This work was supported by National Research Foundation - project B-525/95 and Prof. E. Wollenweber, Botanical Institute, Darmstadt, Germany for authentical samples.

5. References

1. Wollenweber, E.: On the distribution of exudate flavonoids among Angiosperms, *Rev Latinoamer. Quim.* **21** (1990), 115-121.
2. Stefanovic, M., Mladenovic, S., Dermanovic, M., and Rastic, N.: Sesquiterpene lactones from domestic plant species Tanacetum vulgare L. (Compositae), *J. Serb Chem. Soc.*, **50** (1985), 236- 276.
3. Williams, C., Hoult, J., Harborne, J., Greenham, J., and Eagles, J.: A biologically active lipophilic flavonol from Tanacetum parthenium, *Phytochemistry*, **38** (1995), 267- 270.
4. Stefanovic, M., Mladenovic, S., Dermanovic, M., and Rastic, N : Chemical investigation of domestic plant species Tanacetum macrophyllum, *Bulletin de la société Chemique Beograd*, **47** (1982), 319-320.
5. Stefanovic, M., Mladenovic, S., Dermanovic, M., and Rastic, N.: Sesquiterpene lactones from domestic plant species Tanacetum parthenium L. (Compositae), *J. Serb. Chem. Soc.*, **50** (1985), 435- 441.
6. Thomas, O.: Phytochemistry of the leaf and flower oils of Tanacetum corymbosum, *Fitoterapia*, **LX** (1988), 225- 228.
7. Thomas, O.: Phytochemistry of the leaf and flower oils of Tanacetum cilicium, *Fitoterapia*, **LX** (1989), 131- 137.
8. Wollenweber, E., and Mann, K.: External flavonoid aglycones in Artemisia and some further Anthemideae (Asteraceae), *Fitoterapia*, **LX** (1988), 460- 463.
9. Ivancheva, S., and Stancheva, B.: External flavonoid aglycones from Tanacetum macrophyllum, *Comptes rendus de l'Academie bulgare des Sciences*, **50** (1997), 12-15.
10. Wollenweber, E., Valant- Vetschera, K., Ivancheva, S., and Kuzmanov, B.: Flavonoid aglycones from the leaf surfaces of some Achillea sp., *Phytochemistry*, **26**, (1987), 181-182.

CHLOROPLAST ATP SYNTHASE OF WHEAT AND ITS ALLOPLASMIC LINES

T.LEE and *S.V.PONOMARENKO
IPPGB MS-AS RK St. Timiryasev, 45, Almaty-70, 480070
**IGGC MS-AS RK, Av. Al-Faraby, 75, Almaty-60, 480060 Kazakhstan*

1. Abstract

Alloplasmic lines (AL) of cereals are convenient model for exploration of role of interaction nuclear and plastome genes for testing their morphological, physiological and biochemical characteristics. ATP synthase of wheat AL is hybrid: subunits γ, δ, J are coded by nuclear genome of Triticum aestivum (Tae), and subunits α, β, ε, F, H, I are coded by chloroplast genome from Aegilopsis cylindrica, ventricosa and also from wild wheat - T.macha, kicharae. AL with cytoplasm Ae.cylindrica have higher productive stems per plant, height of plant and drought resistance than Tae. Catalytic activity CF1 from Tae is higher than from Aegilops about in 20%, and from T.macha and T.kicharae in 40%. Difference of AL CF1 activity depends from SUγ but it is not higher than the one of their parental forms.

2. Introduction

Chloroplast DNA composes more than 1/6 part of the whole cell DNA in differential cells of mesophyll leaf gexaplody wheat [1]. The problem of interaction between plastome and nuclear genes and influence of such interaction on chloroplast and mitochondria protein properties is topical for modern physiological and biochemical genetics of cereals. Alloplasmic lines (AL) of cereals are advantageously model for exploration of role of nuclear/plastome genomes interaction for testing their morphological, physiological and biochemical characteristics. Some influences of the non nuclear genomes on morphological also reproductive characteristics is studied specially intensive on common cereals [2]. Yet the influence of the interaction between nuclear and plastome genes of distant species on photosynthesis are poorly known. In this report results of investigation of some catalytic characteristics for the chloroplast ATPase from distant cereals species having different genomes origin and their alloplasmic lines are presented.

3. Material and methods

10-12 day-old seedlings of cereals: Aegilops cylindrica (Acy), Aegilops ventricosa (Ave); Triticum aestivum (Tae) sorts Leningradka (TaeL), Saratovskaja 29 (TaeS) or Iljichevka (TaeI); Triticum macha (Tma), Triticum kicharae (Tki) and five wheat AL (Acy X TaeL) (B8), (Ave X TaeL) (B8), (Tki X TaeS) (B6), (TaeS X Tki)(B6), (Tma X TaeI) (B6). In our experiments the soluble chloroplast ATPase (CF1) was isolated from thylakoid membranes by chloroform extraction [3] The ATPase activity of was calculated from amount of inorganic phosphate (Pi) accumulated during 5-10 min at 37° C. Incubation medium contained 50 mM Tris/HCl pH 8, 2 mM ATP, 40 mM NaCl, 2 mM $MgCl_2$. Pi was detected as a phospho-molibdate complex by spectrophotometrically, protein – by Lowry.

4. Results

Variety of ATPase activity of CF1 from Aegilops species, wheat and their AL was investigated. These plants [5] and their hybrids had different genome types and ploidy (Table). Our wheat AL displayed their father's phenotype. AL (Acy X TaeL), with cytoplasm from Acy had higher productive stems per plant and drought resistance than Tae. The ATPase activity of CF1 from TaeL is higher in factor 1,3 than from Aegilopsis Acy and Ave. The parental pair TaeS and Tki had more difference in velocity of ATP hydrolysis about 1.5 (Table). Comparative research of CF1 activity from cereals is demonstrated that the variety of biochemical characteristics between cereals is lower than between wheat and dicots: pea or spinach [9].

5. Discussion of results

The chloroplast ATP synthase (CFoCF1) catalyzes the light-dependent synthesis ATP from ADP and Pi and locates in thylakoid membranes [6-7]. The enzyme consists of two subcomplexes: hydrophobic - CFo and catalytic - CF1. Subcomplex CF1 has 9 polypeptides five types in ratio: $3\alpha:3\beta:\varepsilon:\gamma:\delta$ and CFo - four subunits (SU's) I:J:F:H. Six SU's types (α, β, ε, I, F, H) of the chloroplast ATP synthase are coded by plastome genome, and three γ, δ, J - by nuclear genome[7-8]. Isolated CF1 catalyzes the ATP hydrolysis [7].

The ATP synthase of cereal AL is hybrid, because cereal chloroplasts always receive the maternal genes and their hybrids should get only the maternal plastid genome [2]. ATP synthase SU's γ, δ, J for chloroplast of (Acy X TaeL), (Ave X TaeL), (Tki X TaeS), (Tma X TaeI) are coded by nuclear genome from common wheat T.aestivum of three sorts (TaeL), (TaeS) or (TaeI), and for (TaeS X Tki) from - Tki. Other SU's α, β, ε, I, F, H of

CFoCF1 were from Aegilops cylindrica (Acy) and Aegilops ventricosa (Ave) or from wild wheat T.macha (Tma), T.kicharae (Tki) (Table). The ATPase activity of wheat AL (Acy X TaeL), (Ave X TaeL), (Tki X TaeS), (TaeS X Tki), (Tma X Tael) was not over their parental forms (Table).

TABLE. Types of nuclear and chloroplast genomes Triticum, wheat alloplasmic lines and Aegilops. Type of chloroplast ATP synthase subunits and their issue. (Abbreviation in the text)

Species	Type of nuclear genome	Type of plastid genome	Type of subunits and their issue	ATPase activity (mkmol Pi/h.mg protein)
TaeL	ABD	ABD	$\alpha_L\ \beta_L\ \gamma_L\ \delta_L\ \varepsilon_L\ F_L\ H_L\ I_L\ J_L$	22.8
Acy	CD	CD	$\alpha_C\ \beta_C\ \gamma_C\ \delta_C\ \varepsilon_C\ F_C\ H_C\ I_C\ J_C$	17.9
Acy X TaeL	ABD	CD	$\alpha_C\ \beta_C\ \gamma_L\ \delta_L\ \varepsilon_C\ F_C\ H_C\ I_C\ J_L$	18.1
Ave	DM	DM	$\alpha_V\ \beta_V\ \gamma_V\ \delta_V\ \varepsilon_V\ F_V\ H_V\ I_V\ J_V$	19.2
Ave X TaeL	ABD	DM	$\alpha_V\ \beta_V\ \gamma_L\ \delta_L\ \varepsilon_V\ F_V\ H_V\ I_V\ J_L$	18.2
TaeS	ABD	ABD	$\alpha_S\ \beta_S\ \gamma_S\ \delta_S\ \varepsilon_S\ F_S\ H_S\ I_S\ J_S$	23.0
Tki	ABD	ABD	$\alpha_k\ \beta_k\ \gamma_k\ \delta_k\ \varepsilon_k\ F_k\ H_k\ I_k\ J_k$	16.0
Tki X TaeS	ABD	ABD	$\alpha_k\ \beta_k\ \gamma_S\ \delta_S\ \varepsilon_k\ F_k\ H_k\ I_k\ J_S$	23.0
TaeS X Tki	ABD	ABD	$\alpha_S\ \beta_S\ \gamma_k\ \delta_k\ \varepsilon_S\ F_S\ H_S\ I_S\ J_k$	17.2
TaeI	ABD	ABD	$\alpha_I\ \beta_I\ \gamma_I\ \delta_I\ \varepsilon_I\ F_I\ H_I\ I_I\ J_I$	22.0
Tma	ABD	ABD	$\alpha_m\ \beta_m\ \gamma_m\ \delta_m\ \varepsilon_m\ F_m\ H_m\ I_m\ J_m$	24.8
Tma X TaeI	ABD	ABD	$\alpha_m\ \beta_m\ \gamma_I\ \delta_I\ \varepsilon_m\ F_m\ H_m\ I_m\ J_I$	22.0

In our experiments chloroplast ATPase was isolated by chloroform extraction. It was limited by SU δ (CFo-δ) [3] and consisted from eight polypeptides because SDS-EP in 15% PAAG shown four types of CF1 SU's in ratio 3α: 3β: γ: ε. So SU's α, β, ε were from Triticum and γ - is from other cereals (Table). The chloroplast DNA of high plants [1] and genes *atpA*, *atpB* for SU's α, β are very conservative [8]. SU's α, β are catalytic, have nucleotide-binding sites (NBS) and also bind phosphate [5]. These SU's and their NBS change very slowly during the species evolution [8-9]. SU γ regulates enzyme activity [10]. This polypeptide is less conservative than catalytic SU's [8]. It was shown by method of frequency restriction fragments that the chloroplast genes of Tae, Acy, Ave, Tma had some difference in operons *atpA* and *atpB* [6]. In our experiments ATPase AL (Acy X TaeL), (Ave X TaeL), (Tki X TaeS), (Tma X Tael) contained the same SU γ coded by the same nuclear Triticum aestivum genome, other SU's α, β, ε, I, F, H originated from the plastid genome different cereals (Table). Two lines Tki X TaeS and TaeS X Tki have opposite origin of ATP synthase SU's. Velocity of ATP hydrolysis by TaeS X Tki is low than Tki XTae and correlates with an origin of SUγ.

6. Conclusions

Fertile wheat AL have hybrid chloroplast ATP synthase. Its ATPase activity was compared with parental. It is possible that the influence of SUγ on the CFo-δ enzyme activity is tangible for hybrid ATPase from all hybrid plants. We suppose the stability of catalytic properties for hybrid chloroplast ATPase is resulted by interaction between nuclear and plastid genomes of plant cell. Molecular mechanism which limits change of structural and functional signs ATP synthase and also their heredity of hybrids should be studied.

7. References

1. Odinzova M.S. 1987. *Chloroplast genome: organization and expression*, VINITI, Moscow
2. Palilova A.N. 1986. *Plant genetic systems and their interaction*, Nauka i technika, Minsk.
3. Younis H.M., Winget G.D., Racker E. 1977. Requirement of the δ - subunit of chloroplast coupling factor 1 for phosphorylation *J.Biol.Chem.* 252: 1814-1818.
4. Ivashchenko A.T., Ponomarenko S.V. 1987. Effect of anions and alcohol on photophosphorylation in subchloroplast particles *Plant Physiology (Russian)* 33: 34-38.
5. Ogihara Y., Tsunewaki K. 1988. Diversity and evolution of chloroplast DNA in Triticum and Aegilops as revealed by restriction fragment analysis *Theor.Appl.Genetics* 76: 321-332.
6. Strotmann H., Bickel-Sandkotter S. 1984. Structure, function and regulation of chloroplast ATPase *Ann.Rev.Plant Physiol.* 35: 97-120.
7. Graber P., Bottcher B., Boekema E.J. 1990. The structure of the ATP-synthase from Chloroplasts *Bioelectrochemistry III.*: 247-276.
8. Walker J.E., Cozens A.L., Dyer M.R., Fearnley I.M., Powell S.J., Runswick M.J. 1987. Studies of the genes for ATP synthases in eubacteria, chloroplasts, and mitochondria: implications for structure and function of the enzyme *Chemica Scripta, 27B:* 97-105
9. Ponomarenko S.V. 1993. Evolution of nucleotide-binding sites of chloroplast ATP synthase *Izvestia AS RK. Ser. Biol.* 2: 9-14.
10. Moroney J.E., Fullmer C.S., McCarty R.E. 1981. Characterization of the cysteinyl-containing peptides of the γ–subunit of coupling factor *J.Biol.Chem.* 259: 7281-7285.

ISOLATION AND SPECTROSCOPIC CHARACTERIZATION OF CYTOCHROME c6 FROM *CLADOPHORA GLOMERATA*

N.S. SAFAROV[1], R.I. AGALAROV[1], M.N. ISAEV[1],
A. DIKIY[2], R.A. GASANOV[1]
[1]*Institute of Molecular Biology and Biotechnology, Azerbaijan Academy of Sciences, Metbuat pr.2. Baku.370073. Azerbaijan Republic*
[2]*Department of Chemistry, University of Florence, Via G.Capponi 7, I-50121 Florence, Italy*

1. Introduction

Cytochrome c6 (cyt c6) is a water soluble low-spin hemoprotein involved in electron transfer chain of oxigenic photosynthesis. It functions in photosynthetic membranes of some algae and cyanobacteria shuttling electrons between the two membrane-bond complexes, cytochrome b6/f and photosystem I [1]. In higher plants and advanced algal species cyt c6 is replaced by the blue-copper protein, plastocyanin [2]. The structures of the green algae *Chlamydomonas reinhardtii* and *Monoraphidium braunii* cyt c6 have been determined by X-ray crystallography [3, 4] and nuclear magnetic resonance (NMR) [5] techniques. Although a number of structural data are now available also for plastocyanin (for review see [2]), the peculiarities of these two evolutionarily unrelated proteins that allow them to perform the same function are not fully clarified. Detailed characterization of cyt c6 as well as plastocyanins obtained from various sources could help in understanding the evolution of these proteins. This paper reports the purification and spectroscopic properties of cyt c6 from the green algae *Cladophora glomerata*.

2. Materials and Methods

The purity and concentration of protein were monitored and optical spectra recorded on a CARY-3 UV/VIS spectrophotometer using 1-cm optical path length cuvette at room temperature.

The pH values of protein samples were measured on Orion Model MI-720 pH-meter and a MI-410 microcombination pH probe. X-band EPR spectrum was recorded on a Bruker ER-200-D SRC spectrometer at 4.7 K. The spectrum was calibrated assigning g = 2.0037 to the signal of solid DPPH.

Samples of oxidized cyt c6 for ^{1}H NMR spectroscopy at 300 K were prepared in 50 mM sodium phosphate buffer in 85% D_2O+15% H_2O or in 100% D_2O. ^{1}H 1NMR spectra were recorded on a Bruker ASL-200 spectrometer operating at 200.14 MHz.

3. Results and Discussion

3.1. PURIFICATION OF CYTOCHROME C6

Soluble proteins were extracted from fresh thalli of the green alga *Cladophora glomerata* in distilled water at 15°C for about 48 hours with stirring. All subsequent procedures were performed at 4°C. The extract was clarified by centrifugation and precipitated by ammonium sulfate (95% of saturation). The precipitate was dissolved in 10 mM potassium phosphate buffer, pH 7.0 and dialyzed. The solution of reduced cyt c6 was chromatographed on a DE-52 column equilibrated with 50 mM potassium phosphate buffer, pH 7.0. Cyt C6 was eluted with 200 mM potassium phosphate buffer, pH 7.0. Then the protein solution was fractionated by hydrophobic chromatography on a Toyopearl HW-65 using the decreasing gradient of ammonium sulfate (60% - 20% of saturation) in 5 mM potassium phosphate buffer, pH 7.0. The gradient hydrophobic chromatography was then repeated at the same conditions. The final preparation of cyt c6 obtained after dialysis and concentration by ultrafiltration was stored at-20 °C. The absorbance ratio of purified reduced form of cyt C6, A_{273}/A_{553}, was estimated to be 1.2.

3.2. OPTICAL PROPERTIES

The optical spectra of cyt c6 at pH 7.0 are shown in Fig.1. The spectra are similar to those of other algal high potential c-type cytochromes except that the α band is symmetrical in shape resembling thus *M.braunii* cyt c6 [6]. The ferrocytochrome has absorption maxima at 552.8 (α band), 522.2 (β band), 416 (γ band, Soret), 317.5 (δ band) and 276 (protein) nm with shoulders at 390, 354, 289 and 282 nm. Absorbance peaks of the ferricytochrome are at 694 (not shown) 527.2 and 409.5 nm with a shoulder at 357 nm. Main maxima and minima in difference (reduced minus oxidized) spectrum of cyt c6 are shown in Fig.1B.

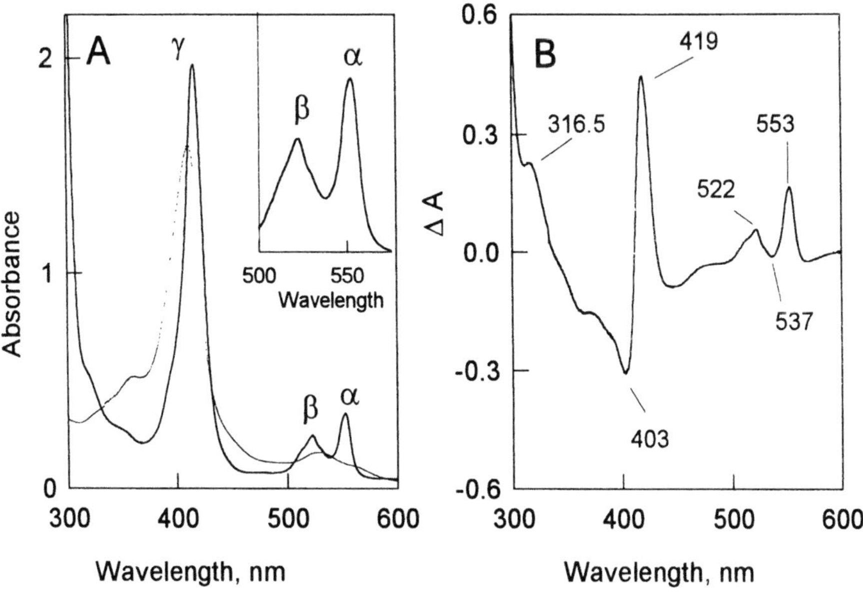

Figure 1. UV/VIS spectra of cyt c6 from *Cladophora glomerata*. A: ascorbate reduced (----) and ammonium persulfate oxidized (....) cyt c6. The inset shows the symmetry of the α band. B- difference (reduced minus oxidized) spectrum of cyt c6.

The presence of a small peak at 694 nm in the spectrum of oxidized cyt c6 suggests methionine-histidine axial coordination of the heme [6]. Oxidation of cyt

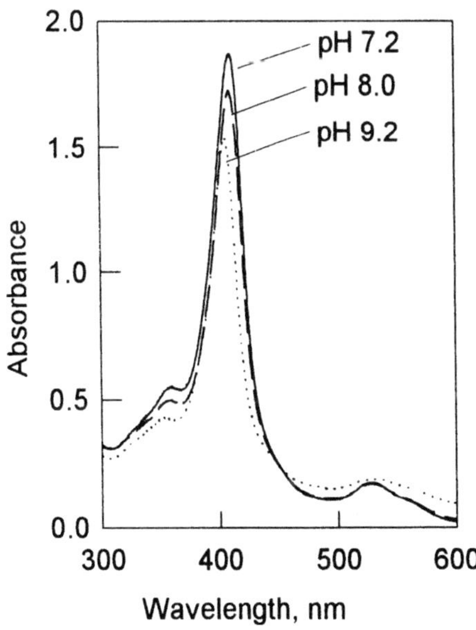

Figure 2. UV/VIS spectra of oxidized cyt c6 at three different pH.

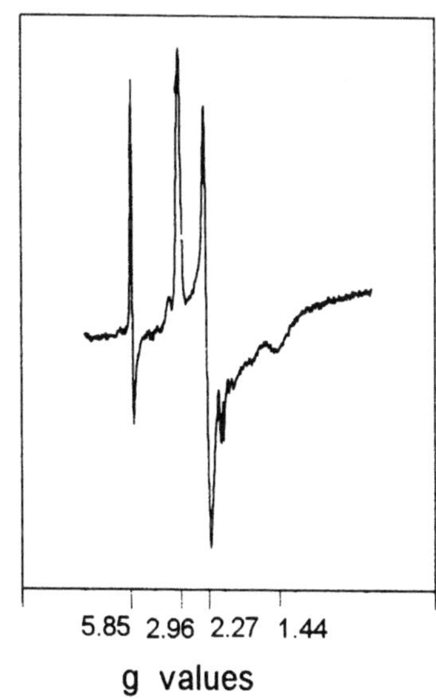

Figure 3. EPR spectrum of oxidized cyt c6 at pH 5.1. M.f.-9.6 GHz; m.p.-20.9 mW; mod. ampl. -0.9 mT.

c6 at neutral pH values leads to the shift of the β band to 409.5 nm. This band shifts further up to 406 nm at the alkaline pH values (*Fig*.2). pH-titration of oxidized cyt c6 at 410 nm (not shown) reveals a group with apparent pK_a 8.7.

3.3. NMR AND EPR SPECTROSCOPIES

The EPR spectrum of ferricytochrome C6 at pH 5.1 (*Fig.* 3) shows low-spin heme resonances. Previously Campos *et al.* [6] reported a similar spectrum for cyt c6 from *M.braunii* and suggested the presence of two interconvertible forms of the protein. Analogously, we can divide the spectrum into two components: form I with g_x=3,27; g_y=1,98; g_z not resolved and form II with g_x=2.96; g_y=2.27; g_z=1.44.The ^{1}H NMR spectrum of low spin ferricytochrome C6 shows the overall similarity with the spectra and shift patterns of other cyt C6 (data not shown). The chemical shifts as a function of pH values for downfield-shifted regions provide several pK_a and the signals are now in the process of assignment. Preliminary results of pH titration of cyt c6 analysed using ^{1}H NMR as well as optical and EPR studies suggest axial methionine and histidine.

4. Acknowledgements

This work has been supported by NATO Linkage Grant HTECH.LG 950035.
We are grateful to Prof. I. Bertini and Department of Chemistry of the University of Florence for the allowance to use the 200 Mhz spectrometer. The authors thank Prof. G. Martini for helpful discussions.

5. References

1. Wood,P.M.: Interchangeable copper and iron proteins in algal photosynthesis, *Eur. J . Biochem.* **87**, (1978), 9-19.
2. Redinbo, M.R., Cascio, D., Choukair, M.K., Rice, D., Merchant, S., and Yeates, T.O.: Plastocyanin: structural and functional analysis, *Biochemistry* **32**, (1993), 10560-10567.
3. Kerfeld, C.A., Anvar, H.P., Interrante, R., Merchant, S., and Yeates, T.O.: The structure of chloroplast cytochrome c6 at 1.9 A resolution: evidence for functional oligomerization, *J. Mol. Biol.* **250**, (1995), 627-647.
4. Frazao, C., Soares, C.M., Carrondo, M.A., Pohl, E., Dauter, Z., Wilson, K.S., Hervas, M., Navarro, J.A., De la Rosa, M.A., and Sheldrick, G.M.: *Ab initio* determination of the crystal structure of cytochrome c6 and comparison with plastocyanin, *Structure* **3**, (1995), 1159-1169
5. Banci, L., Bertini, I., Quacquarini, G., Walter, O., Diaz, A., Hervas, M., and De la Rosa,M.A.:The solution structure of cytochrome c6 from the green alga *Monoraphidium braunii*, *J. Biorg. Inorg. Chem.* **1**, (1996), 330-340.
6. Campos, A.P., Agular, A.P., Hervas, M., Navarro, J.A., Ortega, J.M.,Xavier, A.V., De La Rosa, M.A., Teixeira, M.: Cytochrome c6 from *Monoraphidium braunii*. A cytochrome with an unusual heme axial coordination, *Eur J. Biochem.* 216, (1993), 329 -341.

THE PRIMARY BIOENERGETICS PROCESSES IN OPTIMAL PHOTOSYNTHETIC TYPE OF WHEAT PLANT

O.I. SIDORENKO
Dr., Head of Laboratory for Photosynthesis
Institute of Plant Physiology, Genetics and Bioengineering, Ministry - Academy of Sciences, Timiryasev St., 45, Almaty, 480090, Kazakstan

1. Abstract

The mechanism of relation between primary bioenergetics processes of photo-synthesis and high crop production in wheat and it importance for creation of the Optimal Photosynthetic Plant Type (OPPT) Conception as the biotechnological model is not known yet. Spectrophotometric, polarographic and chlorophyll fluorescence methods were used for the determination of primary bioenergetic processes in photosystem II and chloroplasts on natural gene -controlled models of wheat. We have shown the intensification of noncyclic electron transport, total photophosphorylation, light coupling of energetic processes in chloroplast; the increasing of the chlorophyll a contents and the reaction center of photosystem II activity for OPPT with maximum productivity. Consequently, OPPT had genes controlled photosynthetic «strong» photosystem II and chloroplast. Namely the intensification of photosystem II and chloroplast function is an important factor in regulating the rate of photosynthesis CO2 - gas - changes and productivity. The mechanism of endogenous regulation of photosynthetic bioenergetic processes and the role of integrated approach in OPPT are observed.

2. Introduction

The mechanism of relation between primary bioenergetics processes of photosynthesis and high crop production in wheat and it importance for creation of the Optimal Photosynthetic Plant Type (OPPT) Conception as the biotechnological model is not known yet. But namely such conception may be used as the base for physiology - genetic increasing of photosynthetic activity on different levels of photosynthetic apparatus organization: from photosystem - to plant; determination of mechanism of relation between primary energetic processes of photosynthesis in chloroplast, high CO2- fixation in leaf and high crop production; understanding of possible mechanisms of photosynthetic apparatus adaptation to different environments and it importance for creation of maximum productivity.

3. Methods

Spectrophotometric, polarographic and chlorophyll fluorescence methods were used for the determination of primary Photosynthetic processes in photosystems II and chloroplast on vies, species, lines and heteroseous hybrids of winter wheat as objects with high yield potential. Gas - changes measurement were conducted for determination of Photosynthetic CO_2-assimilation in leaves. The measurements of leaf area and biomass distribution and redistribution by organs for study of source - sink regulation in whole plant as well as the determination of productivity elements were conducted . Photosynthetic parameters were calculated by dispersion, polygene, combinative ability and correlative analysis methods.

4. Result

We have shown the intensification of the reaction center of photosystem II activity, Hill reaction, non cyclic and separate electron transport activity, total photophosphorylation, light coupling of energetic processes in chloroplast; the increasing of the chlorophyll a contents for OPPT with maximum productivity in comparison with photosynthetic apparatus of wheat standard or parents forms in concrete environments (figure).

This increasing was not due to enlargement of the sizes of light-harvesting chlorophyll a/b protein complexes of photosystem II or number of chloroplasts in cell and suspension. These structural parameters can be for some genotypes of OPPT in wheat. The important factor in regulation of OPPT forming was optimal functional - structural interactions between activity of primary bioenergetic reaction and structural organization of photosynthetic apparatus of photosystem and chloroplast. Namely photosystem II reaction center activity, rate of electron flow, intensification of photophosphorilation and tight coupling of primary bioenergetic processes in chloroplast are the reasons for net photosynthesis rate increasing in leaf of OPPT.

Indexes of primary bioenergetic processes are controlled by genes-systems with additive and non additive effects. Effects of sub-dominance are typical. They stipulate for positive and negative heterosis in wheat hybrids which can be inherit in next generations. High photochemical activity predominates over low ones. Low activity of Hill Reaction, non cyclic and separate electron transport are inherit in by recessive type that facilitates selection by these indexes from F2 in OPPT forms. On the other hand OPPT was characterized by low number of chloroplasts in cell, but high photosynthetic activity of one chloroplast. Leaf area was characterized by polygene type of inheritance with genetics polymorphism. For nuclear genes, which controlled the development of assimilation apparatus, were inhered to sub-domination effects. OPPT showed intermediate inheritance of leaf area. Genetic variation of net photosynthesis was operated by polygene nuclear system with non - additive and less additive effects. High rate of net photosynthesis was controlled by recessive genes as rule, independently on gene-type. But OPPT always had high rate of net photosynthesis. The best donor properties of OPPT parents for obtaining of high photochemical activity posterity were demonstrated by estimation of common combine ability of different genotypes ($gi > 90$). Combine ability coefficient for net

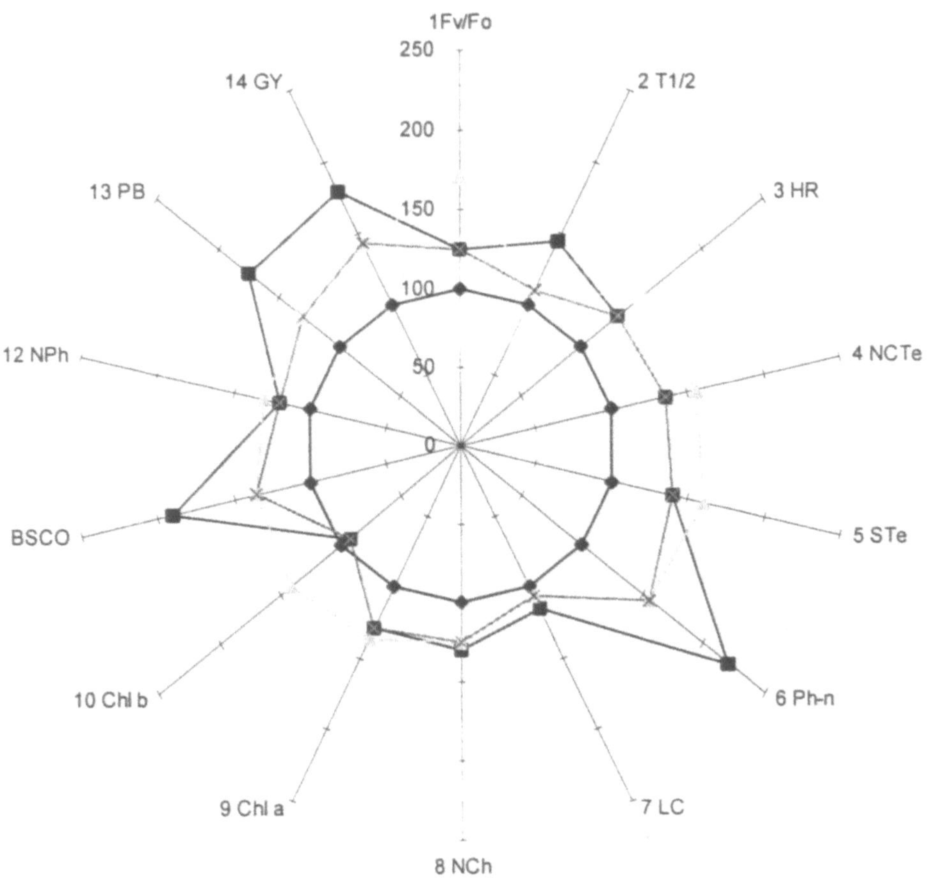

Figure. The primary bioenergetic processes in optimal photosynthetic type of wheat plant. I-III - photosynthetic indexes in different wheat genotypes (%) in comparison with the best initial form. 1 Fv/Fo - variable fluorescence of photosystem II, 2 T 1/2 - time of decay of fluorescence, 3 HR - Hill Reaction, 4 Nce - non cyclic electron transport, 5 STe - separate electron transport, 6 ph-n - photophosphorilation, 7 LC - light coupling degree of electron transport and photophosphorilation, 8 Nch - number of chloroplasts in cell, 9 Chl a - chlorophyll a content, 10 Chl b - chlorophyll b content, 11 RBSCO - RuBisCO-activity, 12 NPh - Net photosynthesis, 13 PB - Plant biomass, 14 GY - grain yield.

photosynthesis was > 3 and for leaf area > 2 in OPPT. The high positive correlation coefficients (>0,50) were calculated between activity of reaction center of photosystem II, non cyclic electron transport and net photosynthesis as well as the same ones between net photosynthesis and plant and grain biomass. The correlation coefficients between number cell, leaf area and net photosynthesis were strong negative (>>-1) and low (>-0,50) or high (>-1) negative between these structural parameters and plant and grain biomass in OPPT.

5. Conclusions

Consequently, OPPT had genes controlled photosynthetic «strong» photosystem II, chloroplast and optimal non large photosynthetic apparatus structure. Namely the intensification of photosystem II and chloroplast function is an important factor in regulating the rate of photosynthesis CO2 - gas - changes of the leaf and productivity of the whole plant. The mechanism of endogenous regulation and the role of integrated approach in OPPT can be formulated in Genetic determination study The main thesis of conception of Optimal Photosynthetic Plant Type in wheat are: have observed united chain of compatible photosynthetic processes : from photosystem to whole plant and crop production ; is determined the reproductive type of production process and the role of organs of high part of shoot («flag» leaf, steam after ear, ear's elements) in accumulation and redistribution of assimilates; is ascertained the optimal combination of extensive (structure) and intensive (function) indexes of photosynthetic activity; is confirms the relatively independent (discrete) character of revealing and inheritance photosynthetic indexes that determined their different interactions and relations; in total high photosynthetic productivity is determined both by high autonomy of photosynthetic structures of low levels of photosynthetic apparatus organization (photosystem, chloroplast, leaf) and by complicated system of integration of photosynthetic processes in the whole plant. Genetic determination study confirms the significance of photosynthetic activity integration on all of the levels of photosynthetic apparatus organization for maximum productivity and adaptability creation. The mobile system of physiology - genetic correlation between photosynthetic indexes and productivity elements in limiting factors of environmental conditions are suggested. Results of these researches can be used for development of new crop biotechnology as the model of high photosynthetic production and adaptation to different environments.

6. References

Nichiporovich A.A.(1982) Physiology of photosynthesis and productivity, in A.A. Nichiporovich (eds.), Physiology of Photosynthesis, «Science» Publishers, M., pp. 7-33.

Mokronosov A.T. (1983) Photosynthetic Function and Integration of Plant Organism, 42nd Timiryasev Reading, «Science» Publishers, M.

Krendeleva T.E. (1985) Structural-Functional Heterogeneity of Primary Photosynthetic Processes in Plant, Thesis of Doctoral dissertation, Moscow University Publishers, M.

Sidorenko O. (1996) Primary photochemical and photophysical processes in optimal photosynthetic type of wheat plant, Plant Physiology and Biochemistry, **Special Issue,** 10th FESPP Congress, Florence, Italy, Sept. 9-13, p. 87.

Sidorenko O.I., Gaziyantz S.M. (1994) Genetics aspects of photosynthesis in hybrids' wheat, Proceedings of National Academy Sciences of Kazakstan, Series Biology, **3,** 9-18.

Sidorenko O. (1997) Optimal Photosynthetic Wheat Plant Type and Environment, in XYIII Congress of SPPS, The Scandinavian Society for Plant Physiology Publishers, Uppsala, Sweden, pp. 170-171.

Sidorenko O., Shpak E. (1997) Photosynthetic Indexes and its Genetic Determination as a New Way of Biotechnology for Adaptive and Productive Crop Breeding in Wheat. XYIII Congress of SPPS, The Scandinavian Society for Plant Physiology Publishers, Uppsala, Sweden, p. 171.

Sidorenko O. (1997) The primary bioenergetics processes in optimal photosynthetic type of wheat plant. I Balkan Botanical Congress, p. 75.

Sidorenko O.I., Shpak E. (1997) Genetic determination and application of photosynthetic indexes as a real way in crop production improvement. I Balkan Botanical Congress, p. 173.

A THERMODYNAMIC INSIGHT INTO THE RELATIONSHIP: DARK TRANSPIRATION, RESPIRATION, AND GROWTH IN PLANTS

J.ST. STOYANOVA
Forest Research Institute, Bulgarian Academy of Sciences
Kliment Ohridski 132, Sofia 1756, Bulgaria

1. Introduction

The importance of respiration in biomass accumulation and productivity has increasingly been clarified in respiration-based growth models, but the precise nature of this relationship is not clear. Existing models may closely simulate aspects of plant growth and are therefore useful, but they are limited in their ability 'to test hypotheses about respiration and its link to growth and productivity' (Amthor, 1989). They cannot provide further insight until they can be placed into the context of an accurate and complete description of the character and mechanism of this relationship. The relation between the main physiological processes during the dark: transpiration and respiration is scare studied, but it is of a special interest. The existence of such a relation is experimentally established by studies in laboratory- (Prokof'ev and Katz, 1963; Jeanrenaud, 1978) and nature- (Polstcr, 1959) environment. As far as the energetics of the relationship between respiration and transpiration is concerned it is a most complicated problem. The commonly accepted vision is that the respiration is the source of energy for transpiration (Jeanrenaud, 1978). The contrary thesis although in an implicit form is covered in studies of Polster (1959) and Zholkevitch (1968). The first data that document a higher respiration rate, respectively a higher energy of respiration at a higher dark transpiration rate are reported by Stoyanov and Florov (1969). Later, in the Surface-boundary approach in biothermodynamics (Florov 1978; 1988) it is motivated the concept that the free energy of dark transpiration provides with energy the dark respiration. This report is aimed to summarise some of our experimental results concerning the responses of dark transpiration, respiration, growth and productivity of plants as provoked by air relative humidity (RH) during the dark period. This kind of study is also aimed to be an energetic insight into the mechanism of respiration-growth interdependence.

2. Experimentation

One-year old plants of annual, bush and tree species were grown from seeds or cuttings in vegetative experiments under controlled environment in phytoclimatic chamber or in greenhouse conditions. During the light period plants were grown at uniform air temperature- (about 24 °C) and humidity- (about 70-75 %) environment for photosynthesis. During the dark period two regimes of RH were maintained: low (about 50 %) and high (about 90 %) at uniform air temperatures of about 24 °C. The courses of main physiological processes during the dark, the energetic aspects of their relations as well as the final growth response as provoked by RH treatments were evaluated. The scheme of measurements is described in details in (Stoyanova, 1996).

3. Results and Discussion

3.1. ASSESSMENT OF DARK TRANSPIRATION AND RESPIRATION

Our experimental work has been provoked for gaining an increased understanding of respiration-transpiration energetics. The data for tendencies in physiological reactions and their energy equivalents at different RH during the dark are presented on Table 1. Two manners of dark respiration assessment are used. The first one is based on the CO_2 exchange rate $X(CO_2)$, expressed in energy units considering the energy equivalent of glucose. The second one is from a thermodynamic point of view. Using the linear irreversible thermodynamics applied to the vivid systems according to the Surface-boundary approach (Florov, 1988), respiration is characterized by the Gibbs' free energy exchange rate, $\Delta G/\Delta t$. It is quantified as a sum of the free energy of the main processes going on the surface boundary layer between leaf system and environment, *i.e.* of transpiration and heat-exchange. From the analyses of these data the following conclusions might be drown:
1. A higher dark transpiration rate is derived at the lower RH (50 %) in the night-time in comparison to this at the high RH; 2. A higher dark respiration rate, assessed by the CO_2 evolution, or by the free energy of Gibbs exchange rate, at the lower RH of 50 %, is derived and vice versa; 3. The energy required for evaporation LE varies significantly between species and RH-environment but it always exceeds many times (approximately 60 -70) the energy released during the dark respiration, determined either from a thermodynamic point of view $\Delta G/\Delta t$ or by the released CO_2; 4. Regardless the plant species or RH, there is a quantitative correspondence between the respiration rate, accounted on the basis of CO_2 exchange, and this accounted 'thermodynamically' ($\Delta G/\Delta t$) according to the Surface-boundary approach.

These dependencies (1 ÷ 4) are valid regardless the manner of transpiration rate regulation: by shoot- or root- environment control. They are also confirmed in experiments where the transpiration rate is provoked by root area temperature regulation (Stoyanova-Georgieva, 1993).

3.2. ASSESSMENT OF METABOLIC ACTIVITY

It is well known the primary importance of dark respiration as a source of free energy and initial substances for biomass accumulation. Metabolic aspects of physiological activity are related to the capacity of substances biosynthesis. Nucleic acids are one of the main metabolic factors of growth and morphogenesis as responsible for binding of amino acids and constructing of proteins. Included in the respiratory chain during the electron transition from the oxidizable substrate to the oxygen, the ascorbic acid is also directly related to the physiological activity. The contents of these two biochemical indexes are used to designate the metabolic activity after RH treatments (Table 2). It is derived a higher metabolic activity in term of studied aspects in plants grown at RH of about 50 % during the dark period in comparison to this at the high RH. The more effective metabolism is accompanied by a higher transpiration rate and a higher respiration rate. This problem is reported more comprehensive in Georgieva (1989).

TABLE 1. Influence of air relative humidity (RH) during the dark period on: dark transpiration rate E [x 10^{-2} g(H_2O) m^{-1} s^{-1}]; energy used for evaporation LE [x 10^{-1} J m^{-2} s^{-1}] (L = 2500 J g^{-1} - latent heat of evaporation); dark respiration rate assessed by the CO_2 evolution, expressed in energy units X(CO_2), and assessed thermodynamically (Florov, 1988) $\Delta G/\Delta t$ [x 10^{-1} m^{-2} s^{-1}]

Species	RH ~ 50 / 90 (%)			
	E	LE	X(CO_2)	$\Delta G/\Delta t$
Fraxinus oxicarpa Wild.*	0.44/ 0.17	110.0/ 42.5	2.05/ 0.82	1.97/ 0.78
Rubus idaeus L.*	0.47/ 0.22	117.5/ 55.0	2.21/ 1.21	2.10/ 1.01
Phaseolus vulgaris L.***	2.47/ 0.92	617.5/ 230.0	12.21/ 5.01	11.24/ 4.22

Differences between two RH: * - $P<0.05$, *** - $P<0.001$.

TABLE 2. Contents of ascorbic acid (AA) and nucleic acids (NA) in plant leaves as affected by air relative humidity (RH) during the dark period

Species	RH ~ 50 / 90 (%)	
	AA (mg %)	NA (%)
Fagus silvatica L.*	74.58/ 50.94	0.105/ 0.088
Rubus idaeus L.*	166.60/ 88.40	0.145/ 0.105
Phaseolus vulgaris L.*	38.20/ 18.27	0.305/ 0.223

Differences between two RH: * - $P<0.05$.

3.3. ASSESSMENT OF GROWTH AND PRODUCTIVITY

Growth reactions of plants are the final reveal of their metabolic activity. Highly significant differences in growth response to RH environment during the dark period are derived (Table 3). A clear tendency of increased seedlings' height and dry mass yield is assessed at 50 % in comparison to these parameters at 90 % RH. Productivity assessed by the number and dry weight (DW) of beans is also favoured by the low RH. Generative phase is accelerated in both annuals as the floral initiation in *Phaseolus* plants is 4 - 5 days earlier.

4. Conclusion

Summarising, the observed patterns of growth promotion or depression are accompanied by certain differences in respiration rate and metabolic efficiency. At uniform photosynthate pool these effects are obviously provoked by the differences in RH during the dark and the dry RH regime of 50 % seems to be the favourable one. From the one hand, these data extend our knowledge about the environmental factors that qualitatively and quantitatively influence the dark respiration processes. It is too possible different respiratory pathways to be used from the plant organism at different

RH environment. From the other hand, it is shown that an open question still remains about the reason of uniform responses in dark transpiration rate, respiration rate, metabolic activity in term of studied basic growth factors (nucleic acids, ascorbic acid), growth and productivity. On this basis it is reasonable a reassessment of interdependence between transpiration and respiration from an energetic point of view, namely: which one of the physiological processes - transpiration or dark respiration is the cause, and which one is the consequence. And finally, the data give the reason to put a basic question: how or by what means an environmental stimulus, such as a change in air relative humidity for example, is sensed and converts to metabolic processes which in turn modify the quantity and quality of growth.

TABLE 3. Effect of air relative humidity (RH) during the dark period on growth, dry matter yield (DWY), and productivity (bean yield)

Species	RH ~ 50 / 90 (%) Growth parameters					
	Height (cm)	DWY (g plant^{-1})			Bean yield (plant^{-1})	
		Shoot	Root	Total	Number	DW
Fraxinus oxicarpa Wild.*	10.3/ 7.2	0.32/ 0.16	0.004/ 0.001**	0.324/ 0.161**	—	—
Phaseolus vulgaris L.*	25.2/ 19.8	0.55/ 0.34	0.260/ 0.170	0.810/ 0.510	earlyer floral initiation by 4 - 5 d at 50 % RH	
Glycine max L. Merr.**	not measured	1.17/ 0.48	0.100/ 0.100^{ns}	1.270/ 0.580	22/10***	0.62/0.23

Differences between two RH: *- P<0.05, **- P<0.01, ***- P<0.001. ns - not significant differences.

5. References

Amthor, J.S. (1989) *Respiration and Crop Productivity*, Springer-Verlag, New York.

Florov, R.I. (1978) *Investigations about the application of irreversible thermodynamics in plant physiology and ecology*, DrSc. Thesis, Izd. Akad. Nauk SSSR, Moskva. [In Russ]

Florov, R.I. (1988) *Thermodynamics of Biosystems. Surface-boundary Approach in Biothermodynamics*, Izd. Bulgarian Akad. Nauk, Sofia [In Bulg.]

Georgieva, J.St. (1989) Free energy balance of leaf system in dark and metabolism of some broad-leaved species, *Fiziologiya Rastenii* **36**, 446-450.

Jeanrenaud, E. (1978) La dynamique de l'intensité de la respiration avec l'intensité de la transpiration et avec le deficit hyrique chez quelques espèces de plantes, *Rev. Roum. Biol.* **23**, 145-158.

Polster, H. (1959) *Die physiologische Grundlagen der Stofferzeugung im Walde. Untersuchungen über Assimilation, Respiration und Transpiration unserer Hauptholzarten*, Bayerischer Landwirtschafts Verlag, München.

Prokof'ev, A.A. and Katz, Y.(1963) Influence of meteorologic factors and age on the fruit and the inflorescence transpiration, *Fiziologiya Rastenii* **10**, 204-211.

Stoyanov, Zh.V. and Florov, R.I. (1969) Respiration and entropy production of leaf system in dark under different air humidity, *Dokl. Akad. Nauk SSSR* **188**, 1405-1406.

Zholkevitch, V.N. (1968) *Respiration Energetics of Higher Plants at Environment of Water Deficit*, Izd. Nauka, Moskva. [In Russ.]

Stoyanova, J.St. (1996) Relationship between transpiration and respiration in plants during the dark period, *Biologia Plantarum* **38**, 77-83.

Stoyanova-Georgieva, J.St. (1993) Influence of root temperature on the free energy balance of leaf system in dark, in L.O. Nilsson, R.F. Hüttl, U.T. Johansson and P.Mathy (eds.), *Nutrient Uptake and Cycling in Forest Ecosystems*, European Commission Ecosystem research report No 21, Brussel, pp. 225-234.

ALUMINIUM STRESS EFFECTS TO NUCLEOLYTIC ACTIVITIES OF *TRITICUM AESTIVUM*

T. YUPSANIS[1], M. MOUSTAKAS[2] AND C. COSMA[1]
[1]*Laboratory of Biochemistry*, [2]*Department of Botany, Aristotle University of Thessaloniki, Thessaloniki 54006, Greece*

1. Abstract

The effects of Al-stress on the levels of DNase and RNase specific activities were studied in the bread wheat cultivar *Triticum aestivum* cv Nestos growing in nutrient solution supplemented with three Al levels (0, 74 and 148 μM) at pH 4.3 for 48 h. The specific activities of the nucleolytic enzymes showed interesting changes in wheat roots and leaves under Al-stress. The activities of the neutral RNases in the roots as well as in the leaves increased (~ 80%) at 74 μM Al, while at higher Al levels (148 μM) decreased (~ 50%). As regard to the root-DNase activities, they decreased at both Al concentrations. Electrophoresis of the enzyme extraction on active gels polymerized in the presence of ssDNA revealed that under Al-stress one form of the root-DNases disappeared. This enzyme may be essential to the root growth, since the primary Al toxicity symptom is a reduction of root growth.

2. Introduction

Plants are responsive to various external stimuli and endogenous developmental signals that must be recognized and translated into cellular responses. Little is known about nucleases activities under abiotic stress conditions in plants. However these enzymes are important since they participate in the metabolism of nucleic acids (**1**, **2**). One of them, ribonuclease (RNase) is considered to be a sensitive marker enzyme for stress conditions. Its activity increases during senescence and after exposure to various environmental stresses such as drough, freezing etc. (**3**, **4**).

The inhibition of root elongation has been recognized as a primary symptom of Al toxicity. However, the primary lesion that leads to inhibition of root elongation has not been identified (**5**).

In this work the effects of Al-stress on the appearance of DNases and RNases as well as their activities were studied in the bread wheat cultivar *Triticum aestivum* cv Nestos.

3. Materials and Methods

3.1. PLANT MATERIAL AND GROWTH CONDITIONS

Experiments were performed on leaves and roots of bread wheat (*Triticum aestivum* L. cv Nestos) growing in nutrient solution supplemented with three Al-levels (0, 74 and

148 μM) at pH 4.3 for 48 h (**6**).

3.2. EXTRACTION OF NUCLEASES

The frozen material was ground in a pestle and mortar with 0.1 M Tris acetate pH 7.0.

3.3. ENZYME ASSAYS

Nucleolytic activity was assayed at 37° C in a total volume of 0.5ml in the presence of 0.6 mg ml^{-1} nucleic acid (DNA or RNA), 33mM ammonium acetate (pH 5.5) and 1mM $ZnCl_2$ (acid enzyme) or Tris acetate (pH 7.5) and 1mM $CaCl_2$ (neutral enzyme). Reactions were terminated by the addition of an equal volume of chilled 12.5% $HClO_4$ and the absorbance of the supernatant solution was measured at 260 nm. One unit of nucleolytic activity is the amount of enzyme that gives an increase in absorbance of 0.01 at 260 nm after 1 min of incubation.

3.4. ACTIVITY DETECTION OF ENZYMES IN PAGE

Electrophoresis of DNases/RNases on active gel polymerized in the presence of ssDNA (80 μg/ml) was performed according to Kefalas and Yupsanis (**7**). After incubation the gels were stained with ethidium bromide and visualized under UV light at 254 nm.

4. Results and Discussion

The protein content of leaves increased under Al stress while that or roots decreased. The specific activities of the nucleolytic enzymes showed interesting changes in wheat roots and leaves under Al-stress. The activities of the neutral RNases in the roots as well as in the leaves increased (~ 80%) at 74 μM Al, while at higher Al levels (148 μM) decreased (~ 50%).

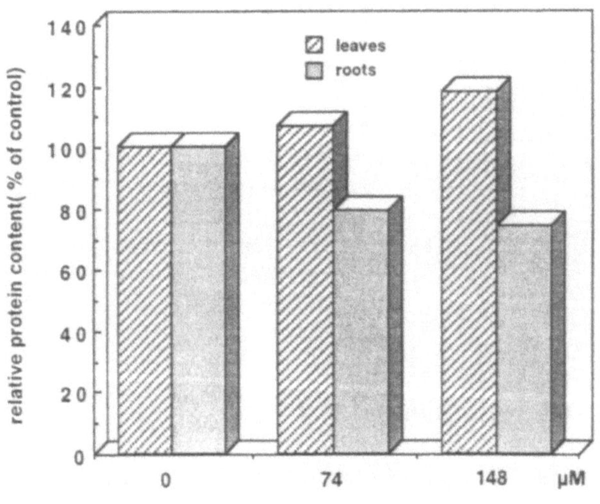

Effect of Al concentration (μM) on relative protein content (% of control) in roots and leaves of *Triticum aestivum* L.cv Nestos.

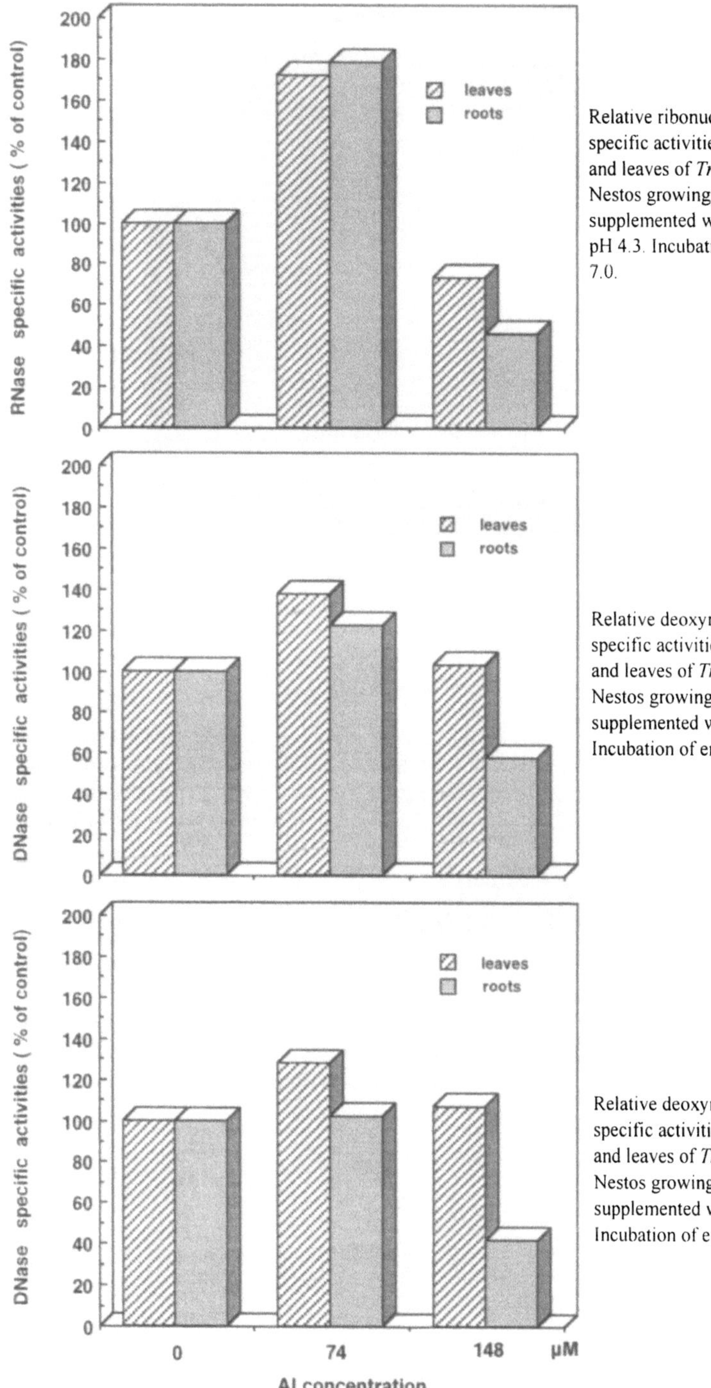

Relative ribonucleolytic (RNase) specific activities (% of control) in roots and leaves of *Triticum aestivum* L. cv Nestos growing in nutrient solution supplemented with 74 or 148μM Al at pH 4.3. Incubation of enzymes at pH 7.0.

Relative deoxyribonucleolytic (DNase) specific activities (% of control) in roots and leaves of *Triticum aestivum* L. cv Nestos growing in nutrient solution supplemented with 74 or 148 μM Al. Incubation of enzymes at pH 7.0.

Relative deoxyribonucleolytic (DNase) specific activities (% of control) in roots and leaves of *Triticum aestivum* L. cv Nestos growing in nutrient solution supplemented with 74 or 148 μM Al. Incubation of enzymes at pH 5.5.

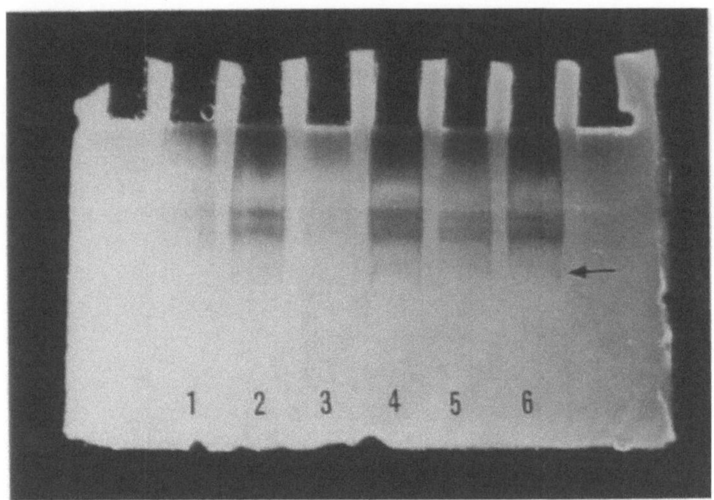

Analysis by PAGE (9.5%) gel-polymerized in the presence of ssDNA of root and leaf DNases of bread wheat (*Triticum aestivum* L. cv Nestos). Lines 1, 3 and 5 root DNases of plants growing in nutrient solution supplemented with 74, 148 or 0 μM Al respectively. Lines 2, 4, 6 leaf Dnases of plants growing in nutrient solution supplemented with 74, 148 or 0 μM Al respectively. The arrow indicates the faster migrating enzyme. The incubation time is 2 h. Two enzyme units were loaded in each case.

The increase of RNase activity under Al stress may be due to a decrease in inhibitor content or increase in activator content as has been shown for water stress and senescence (**3**). It is well known that at a stress-done overloading the vitality of the plant is progressively lost. The decrease of RNase activities under high Al level (148 μM) support the above suggestion.

The activities of both neutral and acid Dnases in leaves and roots increased at 74 μM Al, while at 148 μM Al increased slightly in leaves but decreased significantly in roots. Data presented here suggest that Al (148 μM) inhibits root nucleases activities, thereby disturbing cell division. These results are conistent with observations by others on the inhibitory effect of Al on DNA synthesis and cell division (**8**).

The DNase-electrophoretic patterns revealed that under Al-stress one form of the root-DNases disappeared. This enzyme may be essential to the root growth, since the primary Al toxicity symptom is a reduction of root growth.

5. References

1. Yupsanis, T., and Pantazaki, A.: *Journal of Experimental Botany* **207** (1988), 1469-1473
2. Yupsanis, T., Eleftheriou P., and Kelepiri, Z.: *Journal of Plant Physiology* **149** (1996), 641-649
3. Chokwang, Y., and Todd, G.W.: *Physiologia Plantarum* **46** (1979), 13-18
4. Rouxel, M.F., Singh, J.P., Beopoulos, N., Billard, J.P., and Esnault, R.: *Journal of Plant Physiology* **133** (1989), 738-742
5. Kochian, L.: *Annual Review of Plant Physiology* **46** (1995), 237-260
6. Moustakas. M., Yupsanis, T., Symeonidis, S., and Karataglis, S.: *Journal of Plant Nutrition* **15** (1992), 627-638
7. Kefalas, P.S., and Yupsanis, T.: *Journal of Plant Physiology* **146** (1995), 1-9
8. Matsumoto, H., Hirasawa, E., Torikai, H., and Takahashi, E.:*Plant Cell Physiology* **17** (1976), 127-137

ELECTROPHORETIC PATTERNS OF DIPHOSPHONUCLEOSIDE KINASES, DNases AND RNases IN *TRITICUM AESTIVUM, THINOPYRUM PONTICUM* AND THEIR HYBRIDS

T. YUPSANIS[1], L. SYMEONIDIS[2] and C. COSMA[1]
[1] *Laboratory of Biochemistry, School of Chemistry, Aristotle University of Thessaloniki, Greece*
[2] *Department of Botany, School of Biology, Aristotle University of Thessaloniki, Greece*

1. Introduction

Thinopyrum ponticum (*Agropyron elongatum* 10x=70) with its unique properties of salt tolerance and disease resistance provides a potential means for the transfer of these important and complex traits in cultivated wheat through intergenetic hybridization [1]. In this work, it has been hybridized with *Triticum aestivum* and seeds from the resulting hybrids and their parents were evaluated for diphosphonucleoside kinases (E.C.2.7.4.6), DNases and RNases.

Diphosphonucleoside kinases (NDP-Kinases) catalyse the transfer of the γ-phosphate of the nucleoside triphosphate to an NDP, showing no specifity towards the base of either the donor or the acceptor nucleoside, nor discriminating between ribo- and deoxyribonucleotides.

Plant nucleases, DNases and RNases are mainly nucleic-acids degrading enzymes [2]

2. Materials and Methods

Plant material: Seeds of hybrids, derived from the cross of *Triticum aestivum* (4FAWWON) x *Thinopyrum ponticum* and their parents were used.

Cytological observation: Somatic cells were prepared from root tips of young seedlings. The usual Feulgen technique of staining and squashing was applied for cytological observations [3].

Protein phosphorylation: Protein phosphorylation patterns were analyzed by SDS-polyacrylamide gel electrophoresis followed by autoradiography [4].

Detection of Diphosphonucleoside kinases: The TLC (Thin-Layer-Chromatography) plates were developed with 0,1 M phosphate buffer pH 6.8/ammonium sulphate/propan-1-ol (100:60:2) and autoradiographed.

Detection of DNases and RNases: Native gel electrophoresis of nucleases was performed on active polyacrylamide slab gels according to Kefalas and Yupsanis [5].

3. Results and Discussion

Triticum aestivum has been hybridized with *Thinopyrum ponticum*. Seeds from the resulting hybrids and their parents were evaluated for diphosphonucleoside kinases (NDP-kinases). DNases and RNases.

Figure 1 revealed that the endogenous phosphorylated peptides (indicated by the arrows A and B) from the hybrid seeds almost disappeared. Autoradiography of a TLC plate (Fig.2) showing the reaction products of gel exctracted peptides A and B (Fig.1) with nucleoside diphosphate proved that the above peptides are catalytic subunits of NDP-kinases.

As regard to the nucleolytic activities (DNases and RNases) of the hybrid seeds they were high because of the appearance of a novel nuclease (indicated by the arrow in Fig. 3).

Our data showed that the above hybridization had drastic results on the activity level and the appearance of the examined enzymes. As it is known these enzymes are indispensible in the metabolism of nucleic acids.

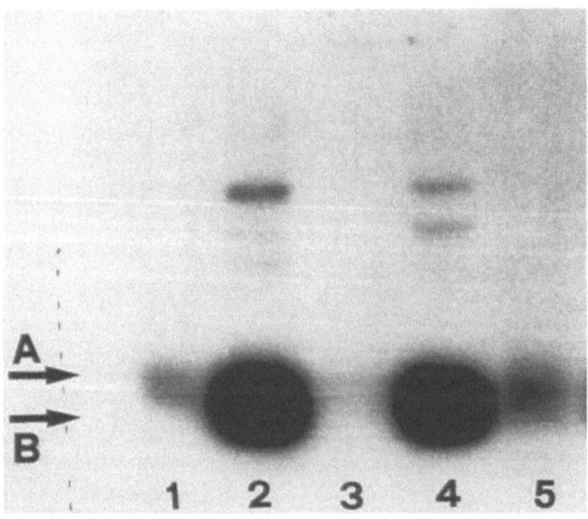

Figure 1. Analysis by SDS-PAGE and autoradiography of the endogenous phosphorylated proteins of the extractions prepared from *Thinopyrum ponticum* (track 4) wheat (track 2) and hybrid (tracks 1,3,5) seeds. Exposure time 12 h.

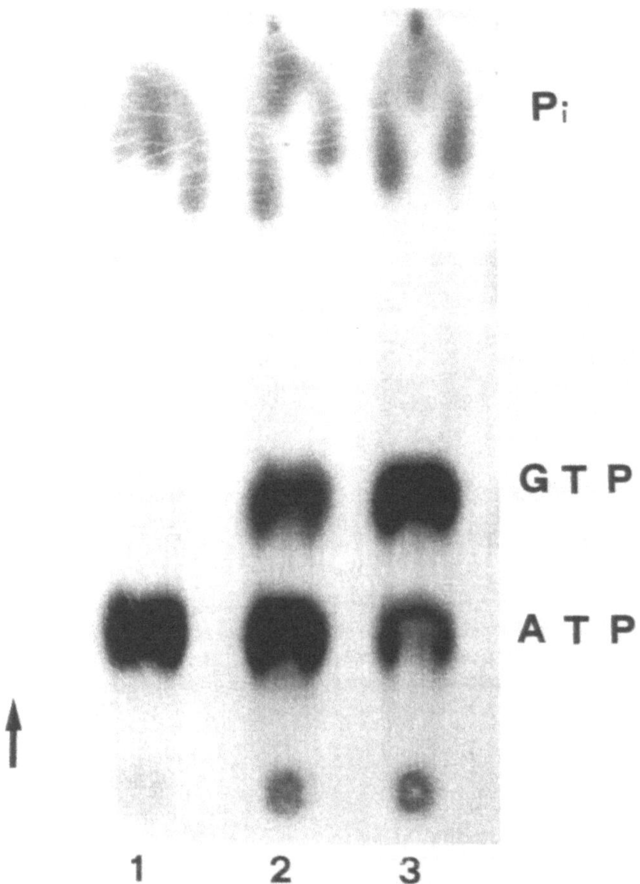

Figure 2. Autoradiography of a TLC plate showing the reaction products of gel-extracted catalytic subunits of NDP-kinases with nucleoside diphosphate (GDP). Extracted catalytic subunits of NDP-kinases were incubated at pH 7.5 in the presence of 1mM $MgCl_2$ with (γ-^{32}p)ATP and GDP. Track 1, control (GDP but no enzyme); Track 2, as 1 but with catalytic subunits of NDP-kinase (A); track 3, as 2 but with catalytic subunits of NDP-kinase (B) (see Fig. 1). Exposure time 24 h.

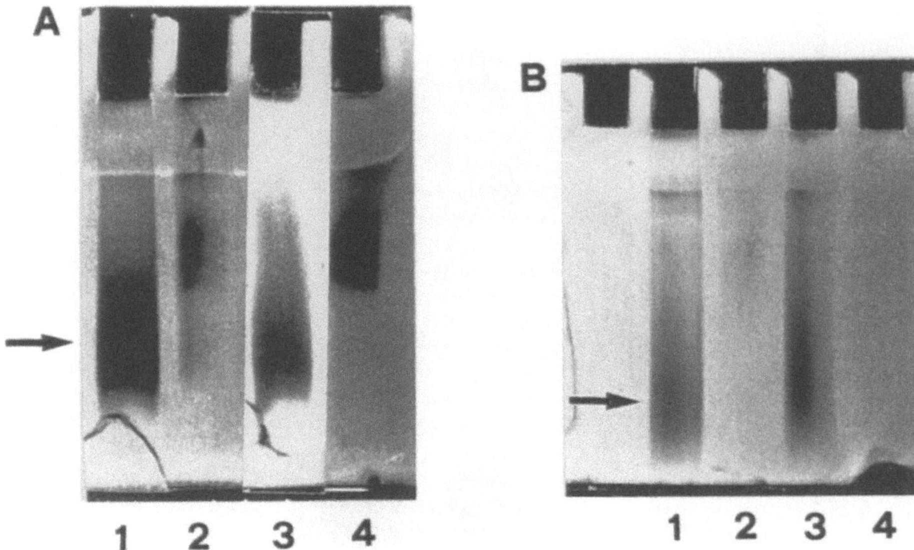

Figure 3. Analysis by PAGE on 9.5% gel-polymerized in the presence of ssDNA (A) and RNA (B) of the DNases and RNases extracted from *Th. ponticum* (track 4), wheat (track 2) and hybrid seeds (tracks 1,3). The gels were incubated at 37° C for detection of nuclease activities (dark bands). The bands were appeared after staining the gels with ethidium bromide. A) 1h incubation B) 3h incubation.

4. References

1. Zhong, G.Y., McGuize, P.E., Qualset, C.O., and Dvorak, J. (1994) Cytological and molecular characterization of a *Triticum aestivum* x *Lophopyrum Ponticum* backcross derivative resistant to barley yellow dwarf, *Genome* **37**, 876-881.
2. Yupsanis, T., Eleftheriou, P., and Kelepiri, Z. (1996) Separation and purification of both acid and neutral nucleases from germinated alfalfa seeds, *J. Plant Physiol.* **149**, 641-649.
3. Coucoli, H. and Symeonidis, L. (1980) Karyotype analysis on some Greek wild species of *Hordeum* (*marinum* group) and *Taeniatherum*, *Scientific Annals of the Faculty of Physics and Mathematics*, University of Thessaloniki.
4. Yupsanis, T., Hadjianghelu, A., and Georgatsos, J.G. (1989) A heat-stable protein kinase of germinated barley seeds active in the presence of EDTA, *Phytochemistry* **28**, 385-388.
5. Kefalas, P. and Yupsanis, T. (1995) Properties and specificity of a calcium dependent endonuclease from germinated lentil (*Lens culinaris*), *J. Plant Physiol.* **146**, 1-9.

Ecology and Ecophysiology

A DENDROCLIMATOLOGICAL STUDY ON *ABIES CILICICA* CARR. IN WEST MEDITERRANEAN REGION OF TURKEY

Ü. AKKEMİK
University of Istanbul, Faculty of Forestry, Department of Forest Botany
80895 Bahçeköy-Istanbul/Turkey

1. Introduction

The widest natural distribution of *Abies cilicica* Carr. (Cilician fir), one of the most important forest trees of Turkey, is in Taurus Mountains of Turkey with the total area of about 350,000 hectares. It constitutes both pure and mixed forests with *Pinus nigra* Arn., *Cedrus libani* A.Rich., *Juniperus foetidissima* Willd., *J.excelsa* Bieb., *Arceuthos drupacea* Ant et Kotschy.and *Qurcus cerris* L. The limits of its natural distribution are 30° 35' - 37° 18' N and 36° 12' - 38° 35' E (Bozkuş,1988).

The aim of the research is to find out the relationships between tree rings and climatic records. This species has been indicated as number 1 in crossdating index(CDI) , that means, it has a minor importance in dendrochronology and there is not enough number of studies on it(Grissino-Meyer,1993). For this reason, dendrochronological studies should also be focused on this species.

A few dendrochronological studies have been carried out in Turkey. The most important ones were done by Kuniholm(1991,1996). Especially dendroclimatological studies were done by Gassner and Cristiansen-Weniger(1948), Akkemik(1996). Since Turkey has a lot of different local climatic and environmental areas, dendroclimatological studies have a great importance.

2. Material and Method

2.1.DENDROCHRONOLOGICAL METHOD

The materials were taken from West Mediterranean Region of Turkey. Three sites were selected in the region. Their characteristics were explained in Table 1. 20 increment cores from 10 trees, two for each, in each site were extracted using a Swedish Increment Borer. The samples were taken from 1.3 m height of trees. Before measuring, the transversal surfaces of total 60 increment cores were smoothed for seeing the tree rings clearly.

After measuring the widths of the tree rings, individual chronologies in each site were obtained, and false and missing rings were eliminated. After that, the standardizing method described by Fritts(1976, formula 6.8 in page 266) was used.

TABLE 1. The characteristics of the sampled sites

	Bucak- Katran Mountain	Akseki- İmrasan Mountain	Gazipaşa- Karatepe Mountain
Site code	BKAT	AKSE	GAZI
Altitude(m)	1350	1400	1350
Latitude	37° 27'	37° 06'	36° 16'
Longitude	30° 45'	32° 00'	32° 37'
Exposure	N	N	SW
Trees sampled	10	10	10
Tree cover	90%	90%	80%
Climate type	Humid	Very humid	Very humid
Parent material type	Lime stone	Lime stone	Quarsit
Soil type	Terrarosa	Terrarosa	Sandy
Mixed species	*Abies cilicica* Carr. *Cedrus libani* A.Rich. *Pinus nigra* Arn.	*Abies cilicica* Carr. *Cedrus libani* A.Rich *Pinus nigra* Arn	*Abies cilicica* Carr. *Quercus cerris* L.

2.2.DENDROCLIMATOLOGICAL METHOD

A simple correlation function(Blasing et al. 1981, Schweingruber, 1988) was calculated to determine the response to climate, using mean monthly temperature and total monthly precipitation from October of previous year to September of current year. The climatic records were taken the nearest meteorological stations to sampled sites. The places of meteorological stations are in the provinces Bucak, Akseki and Gazipaşa.

3. Results and Discussion

After measuring and matching of individual chronologies, a master chronology of *Abies cilicica* Carr. was constituted for the West Mediterranean Region of Turkey. The master chronology is given in Figure 1.

The simple correlation coefficients between tree rings with mean monthly temperature and total monthly precipitation are shown in Figure 2. As seen in Figure 2, the relationships between temperature and tree rings are rather explicit but, they are complex for precipitation. Both temperature and precipitation have a significant influence on the width of the tree rings in some months.

Temperature in winter and summer has a negative influence and especially in period from June to September it is significantly negative. In this region there is a considerable amount of water deficit in summer. Because of this, low precipitation and high temperature in summer cause to form narrower rings. Temperature in spring has positive influence. In the beginning of the growing period trees need high temperature, therefore, temperature is positive but, this relationship is not significant.

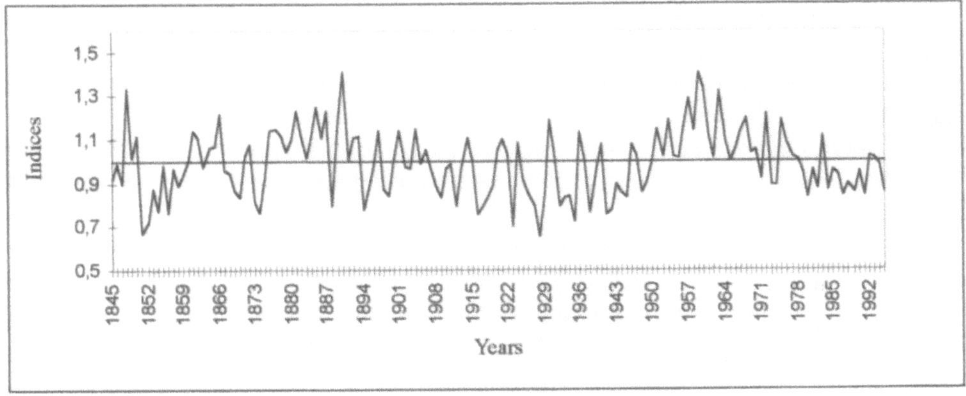

Figure 1. The master chronology of *Abies cilicica* Carr.

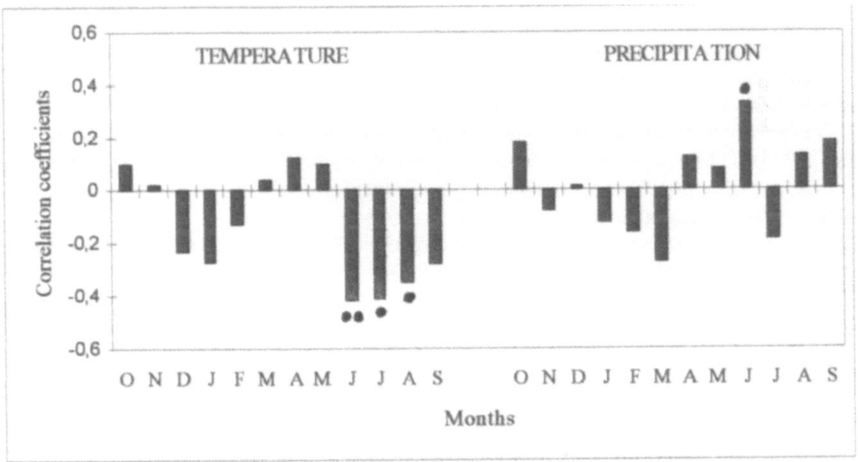

Figure 2. The correlation coefficients between tree rings and climatic records (• = $P<0.05$, •• = $P<0.01$)

Precipitation has a positive influence from April to September, except July. Annual precipitation is unsystematic in this region. It is too high in winter and too low in summer. The total precipitation from June to September, as % of annual precipitation, is 0.07 %.

Basing on this reason, it is expected that the summer precipitation has an important positive influence on the growing of the tree rings. Since July precipitation is not positive, it is rather difficult to comment this result. The sampled sites are under a foggy belt. The reason of the complexity of the relationships between tree rings and precipitation can be this foggy belt. The precipitation in winter is negative for growing, because its amount is too high, but is not significant.

The soil has a great importance for growing of the tree rings. The soil type is terrarosa and terrafusca, and parent material is lime stone in this region. A soil like this has a lot of deep crocks in parent materials, that means, the physiological depth is rather high. Because the water economy of these soils is high due to especially the deep crocks, these soils can be very efficient for growing of roots and other parts of trees.

Abies cilicica Carr. is not a long-living tree. It can reach only 200-250 years of age. The species is not useful as much as long-living trees like pine, oak species in dendrochronology. It is very difficult to constitute a long master chronology using only this species. However, the master chronology of this species was overlapped with the master chronology of *Pinus nigra* Arn in same region, and it was found an important agreement (Akkemik, 1997). Basing on this result, the reconstruction of past climatic conditions will be possible using both these two master chronology.

4. References

Akkemik,Ü.(1996): Dendroclimatology of umbrella pine(*Pinus pinea* L.) in Istanbul (Turkey), *European Dendrochronology Workshop 1996, Moudon-Swiss*

Akkemik,Ü.(1997): Dendrochronological studies on *Pinus nigra* Arn. and *Abies cilicica* Carr in West Mediterranean Region of Turkey, PhD thesis, unpublished

Blasing,T.J., Duvick,D.N. and West,D.C.(1981): Dendroclimatic calibration and verification using regionally averaged and single station precipitation data, *Tree Ring Bulletin,* **41**:37-43

Bozkuş,F.(1988): The natural distribution and silvicultural characteristics of *Abies cilicica* Carr. in Turkey, Doctorate thesis, published by The Ministry of Forestry

Fritts,H.C.(1976): Tree Rings and Climate, Academic Press London

Gassner,G. and Cristiansen-Weniger,F.(1948): Anadolu Çamlarında Yıl Halkaları Gelişmesi Üzerinde Dendroklimatolojik Araştırmalar (Translator into Turkish: Kerim Ömer Çağlar),Ticaret Dünyası Basımevi-İstanbul-Turkey

Grissino-Meyer,H.D.(1993): An update list of species used in tree ring researches, *Tree Ring Bulletin,* **53**, 17-43

Kuniholm,P.I.(1991): A 1503 year chronology for the bronze and iron ages: 1990-1991 Progress Report of the Aegean Dendrochronology Project, *VII Arkeometri Sonuçları toplantısı*, 27-31 May 1991, pp:121-130, Çanakkale-Turkey

Kuniholm,P.I.(1996): Long tree-ring chronologies for the eastern Mediterranean, *Archaeometry 1994*, The proceedings 29th International Symposium on Archaeometry pp:401-409

Schweingruber, F.H.(1988): Tree Rings, Basics and Applications of Dendrochronology, Kluwer Academic Publishers, Dordrecht

SPATIAL ORGANIZATION OF A GRASSLAND RELATED TO METALLIFEROUS SOIL VARIABILITY

D. BABALONAS[1], A.P. MAMOLOS[2] AND M. KONSTANTINOU[1]
[1]Department of Botany, School of Biology, Aristotle University of Thessaloniki, Thessaloniki, Greece; [2]Laboratory of Ecology & Environmental Protection, Faculty of Agriculture, Aristotle University of Thessaloniki, Thessaloniki, Greece.

1. Introduction

In recent decades researches, related to the effects of heavy metals and other trace elements on vegetation development, have increased, revealing that the vegetation in many heavy-metal soils is often patchy. Such patches may differ in floristic composition, even if they are close to each other. As a result, contrasting aspects on heavy-metal soils are created (Ernst 1974).

For the present paper, the vegetation of a grassland area, already known for surface copper mineralization (Babalonas & Reeves 1988), in the village Gerakario (41^0 06′ 50′′ N; 22^0 55′ 28′′ E, Northern Greece) was studied. The study area (ca. 25-30 ha), is covered with fairly dense grassy vegetation forming patches and strips of different size and floristic combinations.

Our interest is focused on the way the plant taxa are associated and on determining the soil factors responsible for the development of the patchy vegetation.

2. Material & methods

Fourty-three quadrats, 1 m^2 each, were laid out along a transect. The transect passed through physiognomically distinct vegetational patches. In each quadrat, exposition, inclination and vegetation cover were recorded. Cover-abundance and frequency estimates were made for all species occurring in the quadrats. For cover-abundance, a seven-part scale (Braun-Blanquet 1964) was used. For the frequency, each quadrat was subdivided into 25 smaller quadrats of equal size using a frame.

The vegetation data were divided into groups by the use of "Normal Association Analysis" (NAA, according to Williams & Lambert 1959).

Four sub-samples from the top soil (1-10 cm) were collected in each quadrat and combined for analysis. A total of 43 soil samples were analysed for various factors (see Table 2). Methods of soil chemical analyses are given in Babalonas et al. (1997).

Finally, a discriminant analysis was performed on the studied soil variables (Harris 1975).

3. Results & discussion

3.1. FLORISTIC DATA OF VEGETATION

Fourty-one plant taxa were recorded; most of them found in less than 20% of the quadrats (Table 1). The cover-abundance was low in most cases (r-2).
NAA revealed 5 quadrats groups (a, b, c, d, e) with *Trifolium scabrum, Linaria pelisseriana, Anthoxantum ovatum* and *Gypsophila muralis* as differentiating species (Fig. 1). The fifth group of the remaining quadrats had no differentiating species, but can still be characterized by *Minuartia hirsuta* ssp. *falcata.* The alllocation of the quadrats to vegetation groups is shown in Table 2.
The comparison of the floristic features of the vegetation in the 5 groups revealed differences in mean species number (Table 1), mean frequency value of the species (MFV), as well as mean vegetation cover value (MCV). From group a to group e a decrease in the mean species number is observed, which is possibly related to a progressively increased severity of the soil environment. Few taxa are frequent (MFV > 10) and play an important role in the physiognomy of the vegetation. The remaining species with lower frequency also contribute to the physiognomical differentiation of the vegetation along the transect, some of them being present only in one of the 5 groups, e.g. *Moenchia graeca* (group b).

3.2. SOIL - VEGETATION RELATIONSHIPS

The soil texture in relation to low pH values, low organic C content and lack of $CaCO_3$ indicate that the soils of the 5 vegetation groups are acid, xeromorphic, poor in nutrients with low productivity (Table 2). This unfavourable edaphic environment becomes even more hostile for plant growth, due to the presence of toxic metals, mainly Cu and Pb, in concentrations higher than normal. Furthermore, the Ca content of the soils is very low and the Ca/Mg ratio is less than one. This is of special importance since presence of Ca generally moderates the toxic action of heavy metals (Proctor & Woodell 1975).
As the soil analysis showed, there are clear differences in the soil Cu and Pb content among the five quadrat groups. The action of these two metals seems to be additive and should be the main reason for the contrasting physiognomical aspects of the vegetation in the area. At least for the taxa *Thymus sibthorpii* and *Minuartia hirsuta* ssp. *falcata*, their presence seems to be affected by the soil concentrations of these two metals (see Tables 1 & 2).
It should be also noticed that in the soils of vegetation groups c and e, the metals Zn,

Table 1. Presence of 41 plant taxa in 43 quadrats of the transect and their mean frequency values (MFV) in five quadrat groups (a, b, c, d, e). Numbers in bold show the highest MFV in each quadrat group.

	Presence	MFV				
	—	a	b	c	d	e
Thymus sibthorpii	37	**13.8**	12.3	**11.6**	0.5	5.4
Minuartia hirsuta ssp. *falcata*	33	1.5	0.3	9.2	12.5	**16.6**
Psilurus incurvus	26	0.9	3.5	3.5	**13.8**	0.4
Jasione heldreichii	25	5.5	4.0	4.4	8.5	0.7
Tuberaria guttata	25	9.5	10.3	8.9	0.8	2.0
Anthoxanthum ovatum	24	13.5	11.1	5.3	.	.
Sanguisorba minor	21	1.8	1.8	6.0	.	1.4
Centaurea grisebachii	21	1.8	3.0	1.1	0.3	0.8
Rumex acetosella	17	9.8	**22.3**	0.4	3.5	0.3
Poa bulbosa	14	2.6	0.3	0.2	1.8	0.1
Gypsophila muralis	9	2.8	.	.	4.3	.
Linaria pelisseriana	8	2.5	3.8	.	.	.
Vulpia myuros	8	1.5	1.8	0.1	.	.
Ornithopus compressus	8	1.0	1.1	.	.	.
Trifolium scabrum	8	9.4	.	.	.	.
Trifolium campestre	7	2.6	.	0.3	.	.
Trifolium arvense	7	1.5	.	0.2	.	.
Lotus angustissimus	6	0.9	1.3	0.5	.	.
Avellinia michelii	5	3.3	3.1	.	.	.
Cynodon dactylon	4	3.8	.	.	.	.
Anthemis arvensis	4	.	2.0	.	.	.
Moenchia graeca	4	.	5.5	.	.	.
Aira elegantissima	4	0.3	0.5	.	.	.
Chrysopogon gryllus	3	2.5	.	.	.	.
Plantago bellardii	3	3.1	.	.	.	.
Hypochoeris radicata	3	0.3	.	.	.	.
Convolvulus cantabrica	3	0.5	.	.	.	.
Crepis neglecta	3	0.3	.	.	0.5	.
Teesdalia coronopifolia	2	.	.	.	1.0	.
Koeleria macrantha	2	.	.	0.7	.	.
Taraxacum spp.	2	.	.	.	.	0.2
Trifolium angustifolium	2	1.1	.	.	.	.
Allium spec.	2	.	.	0.2	.	0.1
Cichorium intybus	2	.	0.3	0.1	.	.
Trifolium stellatum	1	0.1	.	.	.	.
Hypericum olympicum	1	0.1	.	.	.	.
Gnaphalium luteo-album	1	0.1	.	.	.	.
Agrostis spec.	1		.	.	0.5	.
Petrorhagia velutina	1	0.1	.	.	.	.
Dianthus pinifolius	1	.	0.1	.	.	.
Verbascum spec.	1	.	0.3	.	.	.
Mean species number MSV		14.8	12.3	8.4	7.3	3.6
Mean vegetation cover MCV (%)		75.6	90.0	52.7	8.8	27.4

Table 2. Allocation of the 43 quadrats to the five groups distinguished in the transect, and soil data (mean values) for the groups. a = *Trifolium scabrum*; b = *Linaria pelisseriana*; c = *Anthoxanthum ovatum*; d = *Gypsophila muralis*; e = *Minuartia hirsuta* ssp. *falcata*.

1	2	3	4	5	6	7	8	9	10	11	12	13	14	15	16	17	18	19	20	21	22	23	24	25
a	a	b	d	d	d	e	e	d	a	a	a	a	a	e	e	e	c	c	c	c	c	c	c	c
26	27	28	29	30	31	32	33	34	35	36	37	38	39	40	41	42	43							
c	e	e	e	e	e	e	e	c	c	c	b	b	b	a	c	e	e							

Group	a	b	c	d	e
Sand (%)	65.01	61.54	62.13	68.48	67.60
Silt (%)	27.16	30.94	28.65	21.47	23.08
Clay (%)	7.88	7.52	9.23	10.05	9.32
pH	4.8	4.6	4.8	4.69	4.95
$CaCO_3$ (%)	0.0	0.0	0.0	0.0	0.0
Org. C (%)	2.08	2.97	2.66	0.75	2.04
Zn (µg/g)	89	97	54	89	54
Pb (µg/g)	808	943	504	1018	626
Cu (µg/g)	484	545	1045	520	1068
Fe (%)	2.23	2.26	1.77	2.29	2.00
Mn (%)	0.09	0.05	0.02	0.08	0.02
Ca (%)	0.13	0.21	0.18	0.09	0.15
Mg (%)	0.17	0.28	0.35	0.12	0.31
K (%)	0.21	0.25	0.27	0.22	0.27
Na (%)	0.01	0.01	0.01	0.01	0.01

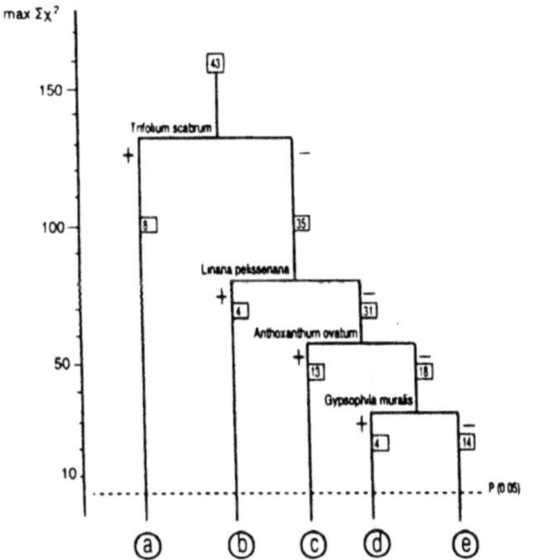

Fig. 1. Normal Association Analysis of the 43 quadrats of the study area. □ = Number of quadrats; ○ = groups.

Pb, Fe and Mn are found with the lowest concentrations, while in groups a, b and d with the highest ones. A different picture is presented by the soils for Cu, Ca, Mg, K and Na. The lowest concentrations occur in soils of vegetation groups a and d and the highest in b, c and e.

In order to assess the variation in soil properties among the quadrats of the transect, a discriminant analysis was carried out on all 43 quadrats using the 14 soil variables. The soil classification resulted in five soil groups, which are significantly well related to the five vegetation groups. Thus, existing discontinuities in soil conditions play a major role in differentiating the vegetation.

The first discriminant function (Wilks lambda = 0.01766, sig. level = 0.00000), which explains 62.24% of the variation in soil data, discriminates better between soil groups of e-b, e-a and e-d. On the first discriminant function, soil variables Zn, Cu, Mg, Na, and silt content appear to be the most important. The second discriminant function (Wilks lambda = 0.11926, sign. level = 0.00209) which explains 22.28% of the variation, discriminates better between the groups b and d, as well as c and d. On the second discriminant function, the most important variables are Cu, Mg, Na, and sand content.

Finally, the results presented here show that soil conditions play a key role in the development of the vegetation patchiness, despite the small area of the site, where the survey was carried out. Furthermore, the importance of analysing the floristic composition of the vegetation and use this characteristic for the prediction of soil conditions is emphasized.

4. References

Babalonas, D. & Reeves, R.D. (1988) Okologische Untersuchungen auf einer auf metallhaltigem Boden entwickelten Wiese, *J. Agr. & Crop Sci.* **160**, 155-162.

Babalonas, D., Mamolos, A. & Konstantinou, M. (1997) Spatial variation in a grassland on soil rich in heavy metals, *J. Veg. Sci.* **8**, 1-4.

Braun-Blanquet, J. (1964) *Pflanzensoziologie. Grundzuge der Vegetationskunde*, Springer-Verlag, Wien.

Ernst, W. (1974) *Schwermetallvegetation der Erde*, Gustav Fischer Verlag, Stuttgart.

Harris, R.J. (1975) *A primer of multivariate statistics*, Academic Press, New York.

Proctor, J. & Woodell, S.R. (1975) The ecology of serpentine soils, *Adv. Ecol. Res.* **9**, 255-384.

Williams, W.T. & Lambert, J.M. (1959) Multivariate methods in plant ecology. I. Association Analysis in plant communities, *J. Ecol.* **47**, 83-101.

CHARACTERIZATION OF THE PHOTOINHIBITION OF PSII IN WHEAT SEEDLINGS BY QUENCHING OF CHLOROPHYLL FLUORESCENCE

F. BABANI[1], H.K.LICHTENTHALER[2]

1.Biochem.Biophys.Dept. Biological Research Institute, Tirana, Albania.

2.Botanical Institute II, Karlsruhe University, Karlsruhe, Germany.

1. Introduction

The study of the light-induced chlorophyll fluorescence of green plant tissue provides the basic information on the functioning of the photosynthetic apparatus [1, 2, 4, 5, 9, 10]. Chlorophyll fluorescence induction kinetics and various parameters and ratios derived from these (F690/F735 as a stress indicator and Rfd-values as plant vitality indices) reveal valuable information on the state of the photosynthetic apparatus in plants [2]. Various variable chlorophyll fluorescence parameters (Fo, Fv), variable fluorescence ratios (Fv/Fm, Fv/Fo, ΔF/Fm') as well as the quenching coefficients (qN and qP) are applied to define the photosynthetic apparatus and its functionality [2]. These parameters allow a deeper insight into the molecular mechanisms involved. Various stress conditions may reduce the rate of photosynthesis changing the chlorophyll fluorescence parameters and variable fluorescence ratios [3, 6, 9]. Exposure of the photosynthetic apparatus to excessive radiation induces photoinhibition which is related to a quenching of variable chlorophyll fluorescence usually expressed as a decrease in the quantum yield of PSII (Fv/Fm ratio). Further changes in fluorescence parameters, including Fo, Fv, the ratio Fv/Fo as well as Rfd690 and Rfd735, may serve to characterize the mechanism of photoinhibition [4, 6, 7, 9]. The objective of the presented work was to detect and evaluate the effect of high irradiance on photosynthesis of some varieties of wheat by using chlorophyll fluorescence parameters.

2. Materials and Methods

Plant. Five winter wheats (*Triticum aestivum* L.) were grown for 20 days in a peat soil with full complements of nutrition growth (TKS2) in a 14/10 h day/night rhythm and irradiance of 200 μmol m^{-2} s^{-1}. Leaves of wheat plants were exposed to 1500 μmol photons m^{-2} s^{-1} up to 3 hours in room temperature. Chl fluorescence measurements as well as the pigment determination were performed in the middle part of third leaves of wheat.

Pigment determination. The chlorophylls and carotenoids were extracted with 100% acetone and determined spectrophotometrically (Shimadzu UV 2000) using the redetermined coefficients and equations of Lichtenthaler [5]. The values represent the mean of 6 separate extracts from small leaf discs of 10 leaves.
Fluorescence induction kinetics. The chlorophyll fluorescence induction kinetics (Kautsky effect) of predarkened leaves (15 min) were measured simultaneously at 690 nm and 735 nm by using the two wavelength fluorometer LITWaF as described in [6]. The ratio of fluorescence decrease fd=fm-fs to the steady-state fluorescence fs (Rfd=fd/fs) determined in 690 nm and 735 nm (Rfd690 and Rfd735) was evaluate according to [6, 7].
Variable chlorophyll fluorescence. The parameters of variable fluorescence: ground fluorescence Fo, variable fluorescence Fv=Fm-Fo, maximum fluorescence Fm (in the dark), ΔF=Fm'-F and Fm' (in the light) were measured using the pulse amplitude modulation chlorophyll fluorometer PAM (Walz, Effeltrich, Germany) as described in [7, 8, 9, 10]. The chlorophyll fluorescence ratios Fv/Fm (the PSII optimal quantum yield), Fv/Fo and ΔF/Fm′ (effective quantum yield) as well as the quenching coefficients qP=(Fm'-F)/(Fm'-Fo') and qN=1-(Fm'-Fo')/(Fm-Fo) and the electron transport ETP were calculated from PAM kinetisc according to [9, 10].

3. Results and discussion

Pigment content and pigment ratios. The content of chlorophylls (chl a, chl b, chl a+b) seemed to be in the same amount in normal (200 μmol m^{-2} s^{-1}) and high irradiance (1500 μmol photons m^{-2} s^{-1}) conditions (chl a+b : 1.9 - 2.4 μg/g). The content of total carotenoids exhibited a tendency to increase after treatment at high irradiance. The increase is more evident in two of the varieties Ni1488 (from 1.9 μg/g to 2.31 μg/g) and Ni684 (from 1.3 μg/g to 1.98 μg/g). The pigment ratios chl a/b and chlorophylls (a+b) to carotenoids (x+c) demonstrated in some of varieties a tendency to decrease in stress conditions. These differences are not significant.
Chlorophyll fluorescence ratios. The values of the chlorophyll fluorescence ratios calculated from the induction kinetics F690/F735 at maximum (fm) and steady-state (fs) fluorescence in the five analyzed wheat varieties remained in the same level in normal and high irradiance conditions. This can be explain due to the same amount of chlorophyll content in leaves and the established relation between these ratios and chlorophyll content in leaves [6,10]. The decrease of the Rfd values (Fig.1 A., B.) at 690 nm and 735 nm (Rfd690 and Rfd735) in stress conditions, which was more evident up to 1h, demonstrated a damage and decline of the photosynthetic function concerning photosynthetic light conversion [1,7]. The Rfd values of two wheat varieties (Ni1488 and Ni684) seemed to be more effected by high irradiance than the others (The significance values are: $p<0.01$).
Variable chlorophyll fluorescence. Maximum fluorescence Fm and variable fluorescence Fv were reduced while initial fluorescence was increased during the exposure of wheat plants to excessive radiation. High irradiance caused a decrease of

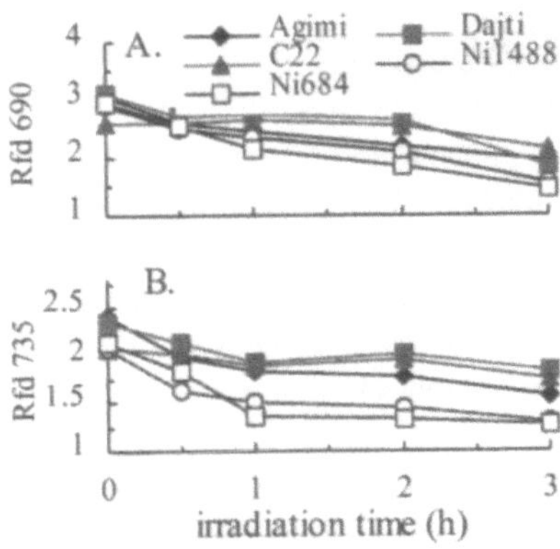

Figure 1. Effect of high irradiance on Rfd values (A) at 690 nm (Rfd690) and (B) 735 nm (Rfd735) in five wheat varieties. Mean of six determinations. std.dev.< 7%.

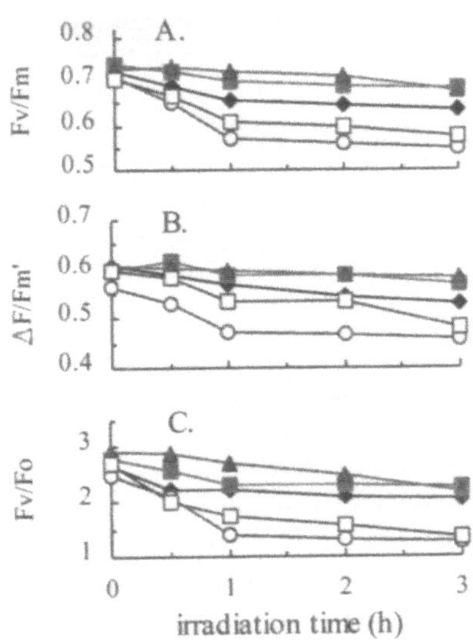

Figure 2. Effect of high irradiance (A) on optimal quantum yield Fv/Fm, (B) effective quantum yield (Fm'-F)/Fm' and (C) Fv/Fo in five wheat varieties. Mean of six determina tions, std.dev.< 5%. (Symbols as in figure 1)

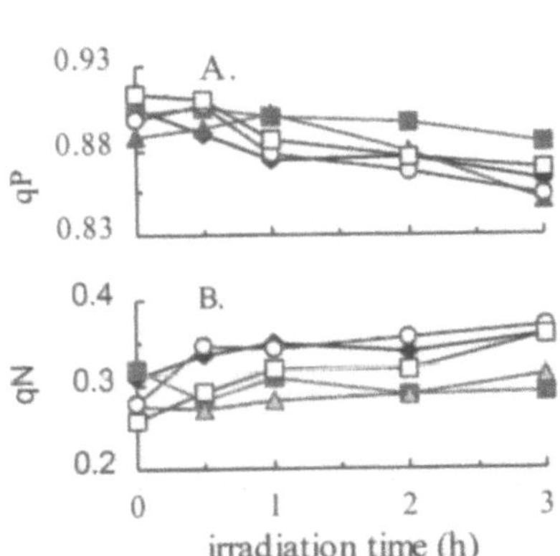

Figure 3. Effect of high irradiance (A) on photochemical quenching coefficients qP and (B) nonphotochemical quenching coefficients qN in five wheat varieties. Mea of six determinations, std.dev.< 5%. (Symbols as in figure 1)

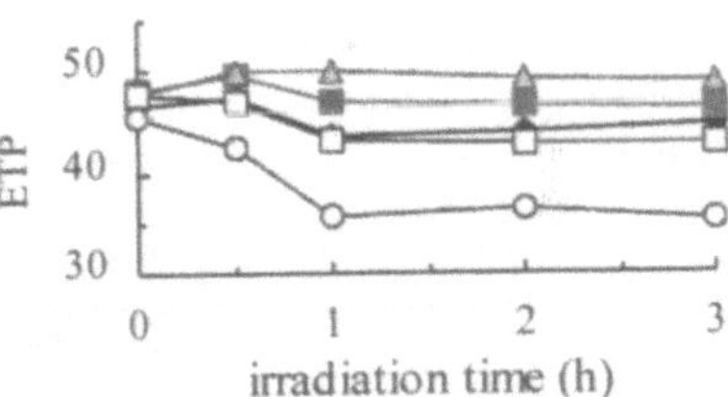

Figure 4. Effect of high irradiance on ETP (ETP = F/Fm'*PFD*0.42) in analyzed wheat varieties. Mean of six determinations, std.dev. < 5%. (Symbols as in figure 1)

PSII optimal quantum yield (Fv/Fm) (Fig.2 A.) and effective quantum yield (ΔF/Fm') (Fig.2 B.). Also decreased was the variable fluorescence ratio (Fv/Fo) (Fig.2 C.). The results indicate that the photoreduction of the primary stable electron acceptor of PSII was inhibited. The decline of the efficiency of exciton capture of open PSII centers (related to optimal quantum yield) and of the quantum yield of non-cyclic electron transport *in vivo* (related to effective quantum yield) [3,8,9] was higher up to 1h of high-light exposure of the plants. The analysed parameters exhibited a tendency for a decrease in three varities, but in two of them (Ni1488 and Ni684) the decrese was surely significant ($p<0.01$). The values of photochemical quenching coefficient qP were lower in control than in high irradiance (Fig 3 A.), this demonstrates that the relative rate of charge separation at PSII centers is affected by high irradiance[9]. The values of non-photochemical quenching coefficients qN (Fig 3 B.) were higher in stressed than control plants ($p<0.05$). The exhibited differences in fluorescence quenching at steady-state fluorescence with closed PSII centers varied among analyzed varieties. The capacity for PSII-driven electron flow was lower at high light exposure up to 1 h (Fig. 4) being in the same level thereafter. In one of varieties (Ni1488) ETP is significanty reduced after 3h ($p<0.05$).

Loss of photosynthetic activity at high-light exposure was assayed by changes of the chlorophyll fluorescence induction kinetics and the variable Chl fluorescence parameters. The decline of Rfd-values as a measure of the potential photosynthetic activity as well as the decline of the optimal quantum yield of PSII -Fv/Fm, effective quantum yield of PSII -ΔF/Fm' and Fv/Fo within the high irradiation period indicate that the plants experienced photoinhibition. The increase of nonphotochemical quenching coefficients qN, apparently by transformation of active reaction centres into photochemical inactive centres, also points to the photoinhibitory effect of high irradiance. The PSII inactivation expressed by the decrease of PSII quantum yield and of the capacity for PSII-driven electron flow depends on total photon absorbed. Changes on the Chl induction kinetics and the variable Chl fluorescence parameters allow to estimate the loss of the photosynthetic activity in high irradiance conditions. The results indicate that chlorophyll fluorescence parameters can be used to evaluate the differential high light sensitivity of the wheat varieties.

References

1. Baker, N.R. and Bradbury, M.(1981) Possible application of chlorophyll fluorescence techniques for studying photosynthesis in vivo, in H. Smith (ed.), *Plants and Daylight Spectrum*, Academic Press, London, pp 355-373.
2. Babani, F.,Lichtenthaler, H.K (1996) Light induced development of chloroplasts, *J. Plant Physiol.* 148, 555-566
3. Genty, B., Briantais, J-M.,Baker, N.R. (1989) The relationship between the quantum yield of photosynthetic electron transport and quenchers of chloropyll fluorescence, *Biochem.Biophys.Acta,* 990, 87-92.
4. Krause, G.H.,Weis, E. (1991) Chlorophyll fluorescence and photosynthesis: the basics, *Annn.Rev.Plant Physiol.Plant Mol.Biol.* 42, 313-349.
5. Lichtenthaler, H.K. (1987) Chlorophylls and carotenoids: pigments of photosynthetic biomembranes. *Methods in Enzymol.* 148, 350-382.
6. Lichtenthaler,H.K., Rinderle,U. (1988) The role of chlorophyll fluorescence in the detection of stress conditions in plants, *CRC Critical Review on Analytical Chemistry* 19 (Suppl.1), 29-85.
7. Lichtenthaler, H.K. et al. (1992) Changes in photosynthetic pigments and in vivo chlorophyll fluorescence parameters under photoinhibitory growth conditions, *Photosynthetica* 27 (3), 343-353.
8. Schreiber, U., Schliva, U., Bilger, W. (1986) Continuous recording of photochemical and non-photochemical quenching with a new type of modulation fluorometer, *Photosyn. Res*, 10, 51-62.
9. Schreiber, U.,Bilger, W. (1993) Progress in chlorophyll fluorescence research, *Progress in Botany*, vol 54, 150-175.
10. van Kooten, O., Snel, J.F.H. (1990) The use of chlorophyll fluorescence nomenclature in plant stress physiology, *Photosyn. Res.* 25, 147-150.

METABOLIC RESPONSE OF *NERIUM OLEANDER* TO AIR POLLUTION

C. BARDIS, G.P. BANILAS, M.S. MELETIOU-CHRISTOU
Section of Botany, Department of Biology, University of Athens, 15784 Athens, Greece

1. Abstract

A preliminary study was performed in order to investigate the metabolic response of *Nerium oleander* to air pollution. For this purpose, leaves from shrubs grown at six different sites of differing levels of air pollution, in the city and the suburbs of Athens, were collected monthly, during the period of one year. The concentrations of carbohydrates, total lipids and chlorophyll were measured. The fluctuations of carbohydrates were not affected by air pollution. The concentration of chlorophyll was higher in more polluted sites, while total lipids's fluctuation showed a seasonality in less polluted sites that was not found in polluted environments.

2. Introduction

Although numerous laboratory studies exist on the effects of individual pollutants on plants, in nature pollutants rarely exist alone (10). Plant sensitivity to air pollution depends not only on combinations of pollutants, but also on the species, the metabolic activity (developmental stage of the organ studied), the time of the year and on combinations with abiotic factors such as climatic conditions (10, 25). The most informative physiological responses occur when there is less visible injury (10). Air pollutants inhibit chlorophyll synthesis and photosynthesis (4, 11, 24) and alter stomatal aperture (16, 17), permeability of cell membranes (1, 7), amounts of stored carbohydrates, proteins and lipids (4), leaf morphology (6) and ultrastructure (3, 19, 20). Chlorophyll fluorescence of intact plants can be used for screening urban pollution in the field (9, 22).

Metropolitan Athens covers an arid mountain basin which is divided north-south by a line of hills. The mountains surrounding Athens reduce air circulation, causing pollution to accumulate over the city (18) and forcing vegetation to grow in a burdened environment. Monitoring air pollution has revealed that greater values of SO_2, NO_2 and CO are measured in the city center of Athens while for O_3 greater values are observed in the suburbs (14, 15).

Nerium oleander is a Mediterranean evergreen sclerophyll shrub that was selected for investigation because it could be found along an air pollution gradient in the city and the suburbs of Athens and did not show any visible symptoms of air pollution injury. The objectives of the present study were to investigate the influence of air pollution on the concentrations of carbohydrates, lipids and chlorophylls in the leaves of *N. oleander* and to test whether climatic conditions and air pollution would interact to change the concentration of the above substances.

3. Materials and methods

Leaves from shrubs of *Nerium oleander* L. (Apocynaceae) were sampled at monthly intervals between November 1995 and October 1996 from six different sites of central Athens and the suburbs, along a gradient of air pollutant concentrations (14, 15). Plant material was collected early in the morning (6 to 7 a.m.) to avoid diurnal variations (8). Due to invariant values of plastochron ratio (21) a section from the 5th up to the 10th leaf was selected for sampling. Samples were brought to the laboratory and elaborated as described elsewhere (13).

The determination of soluble sugars was done according to Dubois et al. (5). Starch was measured using the method of McCready et al. (12). Total lipids were extracted and determined according to Bligh and Dyer (2). Chlorophyll content was determined by a method described by Steubing (23) on a fresh weight basis. Determinations were repeated twice (n=3). Values reported in the present study are given as proportions of dry weight ± S.E.M.

4. Results and Discussion

The fluctuations of soluble sugars, during the year, in the leaves of *N. oleander* from the six different sites that were selected for investigation are shown in figure 1. In contrast to our hypothesis, variations, in general, did not show considerable differences among sites, except the sites of Patission (heavy polluted) and N. Smyrni (less polluted) (14, 15) where the variations were different during spring and early summer when compared to the other sites. At these other sites high values of soluble sugars were recorded in March. Nevertheless those differences in soluble sugar fluctuations among sites do not reflect the gradient of pollution levels. Starch concentrations are shown in figure 2. It seems that the degree of air pollution in Athens did not affect starch content of the leaves of *N. oleander* since seasonal variations of starch concentrations were similar for all sites. The maximum values were recorded during spring in agreement with previous results on Mediterranean species (13). Figure 3 presents the fluctuations of total lipids. A higher variation of the values was distinguished during the winter in the suburbs (N. Smyrni, Helliniko, Spata), which are less polluted (14, 15), than in the sites near the city center

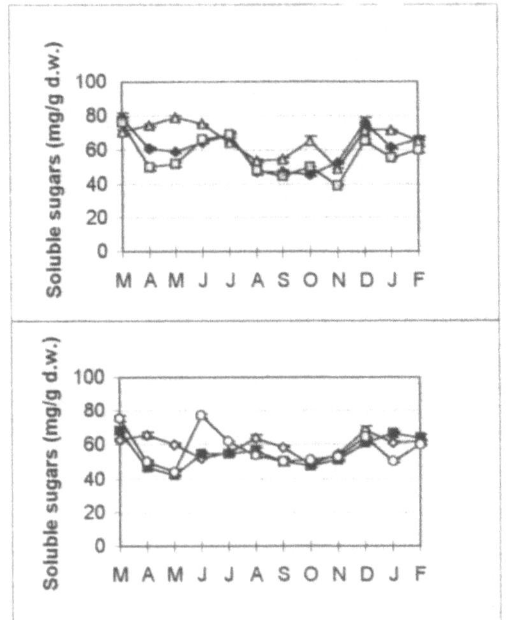

Figure 1. The total soluble sugar content in the leaves of *Nerium oleander* in different sites of the city and the suburbs of Athens during the course of a year. S.E.M. values ranged from 0.3 to 3.2 mg/g d.w. for n=3. Closed rhombi = Renti, shaded squares = Agricultural University, triangles = Patission, open rhombi = N. Smyrni, closed squares = Helliniko and cycles = Spata.

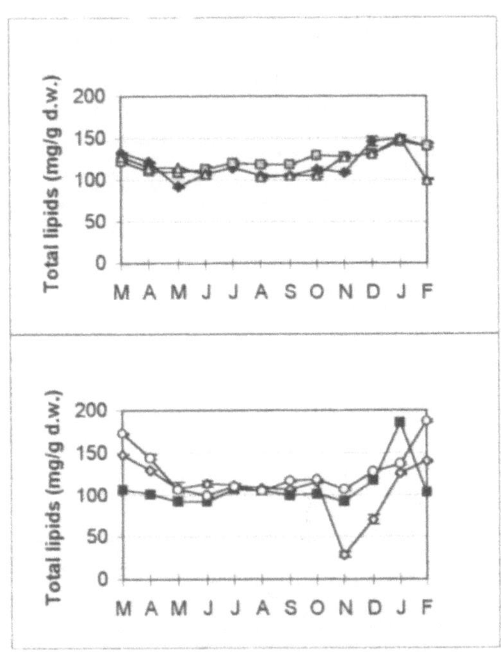

Figure 3. Total lipids in the leaves of *Nerium oleander* in different sites of the city and the suburbs of Athens during the course of a year. S.E.M. values ranged from 0.3 to 6.4 mg/g d.w. for n=3. Symbols as in Fig. 1.

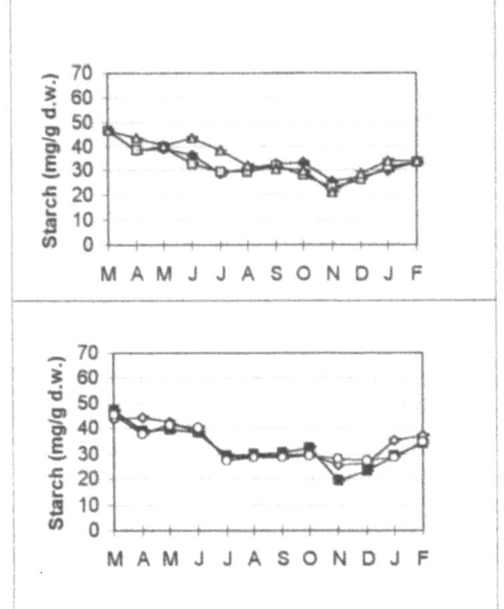

Figure 2. Starch content in the leaves of *Nerium oleander* in different sites of the city and the suburbs of Athens during the course of a year. S.E.M. values ranged from 0.1 to 0.9 mg/g d.w. for n=3. Symbols as in Fig. 1.

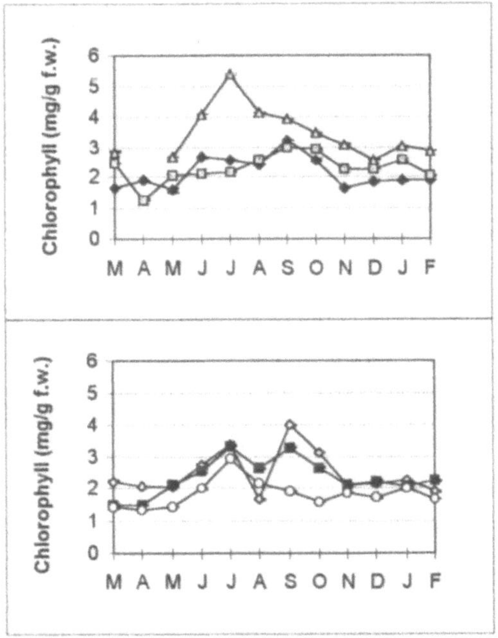

Figure 4. Total chlorophyll content in the leaves of *Nerium oleander* in different sites of the city and the suburbs of Athens during the course of a year. S.E.M. values ranged from 0.00 to 0.06 mg/g d.w. for n=3. Symbols as in Fig. 1.

(Patission, Renti, Agricultural University). According to our hypothesis in the case of total lipids air pollution interacted with climatic conditions to alter the fluctuations of those compounds. The concentrations of total chlorophyll are shown in figure 4. It can be seen that total chlorophyll reached significantly higher values at the site of Patission (city center), with a maximum value in July, than anywhere else. Taking into consideration that at the site of Patission higher concentrations of oxides of nitrogen were recorded in comparison to the other sites (14, 15) and that leaves of plants exposed to nitrogen dioxide appear greener than those of controls (26), it seems rather that the conditions of the city center favour the synthesis of chlorophyll. In the suburbs higher values of ozone concentrations were measured than in the city center sites (14, 15), which may affect chlorophyll concentration, since it have been observed that leaves exposed to ozone usually appear chlorotic (26).

5. References

1. Becherson D.W. and G. Hofstra. Can. J. Bot, 58, 451-457, 1980.
2. Bligh E.G. and W.J. Dyer. Can J. Biochem. Physiol., 37, 911-917, 1959.
3. Christodoulakis N.S. and C. Fasseas. Bull. Environ. Contam. Toxicol. 44, 276-281, 1990.
4. Constantinidou H.A. and T.T. Kozlowski. Can. J. Bot., 57, 176-184, 1979
5. Dubois M., KA. Gilles, J.K. Hamilton, P.A. Rebers and F. Smith. Analyt. Chem. 28., 350-356, 1956.
6. Eleftheriou E.P. Env. Exp. Bot., 27, 105-117, 1987
7. Heath R.L. Ann. Rev. Plant Physiol. 31, 395-431, 1980
8. Kimura M. Bot Mag. (Tokyo) 82, 6-19, 1969
9. Lanaras T., S.P. Sgardelis and J. D. Pantis, Sci. Total Environ. 149, 61-68, 1994
10. Lendzian K.J. and M.H. Unsworth In: O.L. Lange, P.S. Nobel, C.B. Osmond and H. Ziegler (Eds), Encyclopedia of Plant Physiology, vol. 12 D. Springer-Verlag, Berlin, 1983.
11. Lorenc-Plucinska G. Photosynthetica 16, 140-144, 1982.
12. McCready R.M., J. Guggolz, V. Silviera and M.S. Owens. Analyt. Chem. 22, 1156-1158, 1950
13. Meletiou-Christou M.S., S. Rhizopoulou and S. Diamantoglou. Env. Exp. Bot. 34, 129- 140, 1994
14. Ministry of Environment, Physical Planning and Public Works. Atmospheric pollution in Athens, 1995
15. Ministry of Environment, Physical Planning and Public Works. Atmospheric pollution in Athens, 1996
16. Norby R.J. and T.T. Kozlowski. Oecologia, 51, 33-36, 1981
17. Norby R.J. and T.T. Kozlowski. Oecologia, 53, 34-39, 1982
18. Profous G.V., R.A. Rowntree and R.E. Loeb. Arboric. J., 12, 83-107, 1988.
19. Psaras G.K. and N.S. Christodoulakis. Bull. Environ. Contam. Toxicol. 38, 610-617, 1987.
20. Psaras G.K. and N.S. Christodoulakis. Bull. Environ. Contam. Toxicol. 39, 474-480, 1987.
21. Rhizopoulou S., M.S. Meletiou-Christou and S. Diamantoglou. J. Exp. Bot., 42, 627-635, 1991.
22. Sgardelis S., C.M. Cook, J.D. Pantis and T. Lanaras. Sci. Total Environ., 158, 157-164, 1994.
23. Steubing L. Pflanzenokologisches Praktikum Verlag Paul Parcy, Berlin, 1965
24. Suwannapinunt W. and T.T. Kozlowski. Can. J. For. Res., 10, 78-81, 1980
25. Treshow M. and F.K. Anderson. Plant stress from air pollution. John Wiley and Sons. Chichester, New York, 1991.
26. Whitmore M.E., W.J. Davies and T.A. Mansfield. In: N.R. Baker, W.J. Davies and C.K. Ong (Eds); Control of leaf growth. Cambridge University Press, Cambridge, 1985.

CADMIUM TOLERANCE AND CADMIUM-BINDING POLYPEPTIDES IN AILANTHUS ALTISSIMA

G.BAYÇU
Univ. of İstanbul, Fac. of Science, Dept. of Biology, Div.of Botany , 34460 , İstanbul/Türkiye

1. Abstract

Cadmium-binding polypeptides (Cd-BPs) are produced by higher plants and some fungi in response to cadmium (Cd) exposure. They are presumably involved in detoxification and homeostasis. In this investigation, we analyzed Cd-BP production in the 200μM $CdSO_4$ exposed *Ailanthus altissima* (Miller) Swingle roots and we detected fast migrating low-molecular weight (LMW) Cd-BPs, most likely phytochelatins, in addition several slow migrating high-molecular weight (HMW) Cd-BPs which play role in Cd-tolerance.

2. Introduction

Heavy metal contamination is increasing in the environment due to human industrial activities. Cd is one of the major environmental contaminant and potentially toxic to plants and animals [1]. An important biological response to Cd stress in higher plants and in some fungi is the synthesis of Cd-BPs, which are also called as phytochelatins (PC), cadystins, or γ(EC)nG [2]. These small polypeptides are secondary metabolites derived enzymatically from glutathione [3]. Cd-BPs seem to play an important role in detoxification of Cd ions [4]. In this investigation we wished to obtain some information about Cd-tolerance and Cd-BP production in *Ailanthus altissima* (Miller) Swingle which appeared to have considerable tolerance to Cd and other environmental pollutants [5].

3. Material and Methods

Ailanthus altissima seedlings were first grown in sand and perlite mixture for 2 weeks. When the cultures had become established we set up hydroponic cultures of seedlings in a modified Ingestad nutrient solution bubbled with air. Plants were grown in the phytotrone chamber for 4 months (light.intensity: 20.000 lux, temperature: 24°C, light & dark period: 16h-8h, relative humidity.: 60%) and exposed to sublethal level of Cd (200 μM $CdSO_4.8/3\ H_2O$, Sigma) for a week. Extracts were prepared by grinding root samples with mortar and pestle in liquid nitrogen and icecold buffer as shown in the procedures of Rauser, W.E.[6]. The crude homogenates were filtered and cleared by high- speed centrifugation at 4°C. Supernatants were applied to QAE-Sephadex A-25 (Pharmacia) column equilibrated with a start buffer and Cd-BPs were eluted with a gradient of KCl. The Cd content of 50 eluted fractions was determined using flame AAS (SpectrAA-10 AAS, Varian). The material which contains highest Cd was then applied to a Sephadex G-75 column in order to obtain further separation of Cd-BPs.

We next tried to analyze isolated *Ailanthus* material using polyacrylamide gel electrophoresis (PAGE). The principal of this method is to incubate an extract containing Cd-BPs with $^{109}Cd^{2+}$ (^{109} $CdCl_2$, 100μCi, Carrier Free, Amersham) allowing a period of exchange and equilibration followed by non-denaturing PAGE in 40% (monomer) gels as mentioned in the methods of Abrahamson, S.L.et al.[7]. The radioactive $^{109}Cd^{2+}$BPs separated during the run (Pharmacia GE-2/4 LS Gel Electrophoresis Apparatus) are then visualized by direct autoradiography (Hyperfilm MP and Hypercassette, Amersham). Molecular weights of purified Cd-BPs were analyzed using SDS-PAGE in two different gradient gels as 5% and 16%. Gels were fixed and stained with 0.1% Coomassie Brillant Blue R-250, Fluka. Rf values of standard proteins (MW-SDS/17 S, Sigma) and Cd-BPs were detected and molecular weights were calculated on a semi-log paper.

4. Results and Discussion

Cd started emerging from QAE Sephadex A-25 column at fraction no.28 with several shoulders. The highest concentration was found in fraction no.33, which had a pinkish-grey color. Cd was apparently bound to both smaller polypeptides and to larger proteins in an unspecific manner (Fig.1). Satisfactory separations were obtained in $^{109}Cd^{2+}$ binding assay. Most samples from the *Ailanthus* root extracts contained three slow migrating HMW proteins binding $^{109}Cd^{2+}$. In addition, five fast migrating LMW compounds, most likely phytochelatins, were detected (Fig.2).

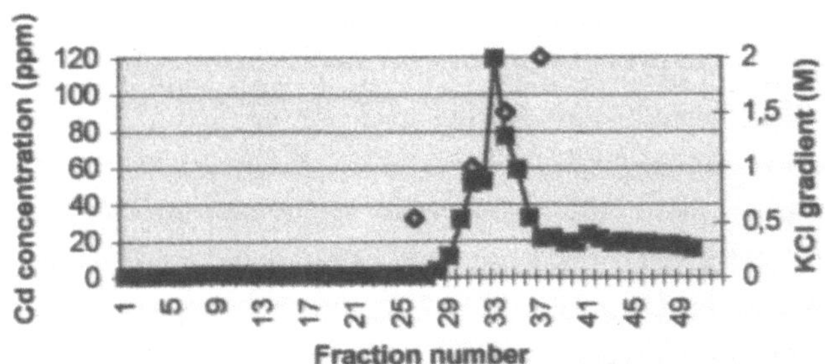

Figure 1. QAE Sephadex A-25 anion exchange chromatogram of crude extract from roots of *Ailanthus altissima*.

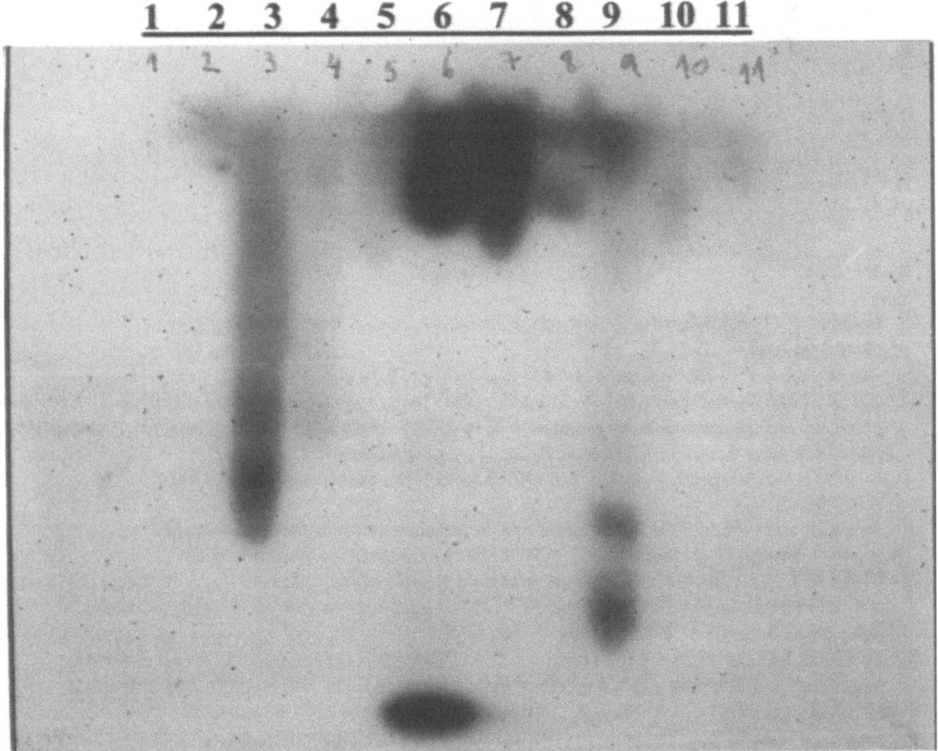

Figure 2. 40% PAGE assay of Cd-BPs. Lane 1, sample buffer($^{109}Cd^{2+}$only); Sephadex G-75 purifications: lanes 2,3,4,5, slower migrating LMW (23.5-26 kDa) Cd-BPs and a poorly resolved area of lower mobility which is seen as a smear; QAE Sephadex A-25 purifications: lane 6, HMW (65 kDa) and very fast migrating LMW (10.5 kDa) Cd-BPs; lane 7, 8, HMW (50-75 kDa) Cd-BPs; lane 9, faster migrating LMW (18-22 kDa) and HMW (75 kDa) Cd-BPs; lanes 10,11, weak signals of the same HMW Cd-BPs found in lane 9.

Molecular weights of fast migrating LMW phytochelatins and HMW Cd-BPs were found as 10.5, 18, 22, 23.5, 26, 50, 65 and 75 kDa, respectively. The heterogeneity of Cd-BPs is probably due to the difference in the number of Cd-BP chains, chain lengths, molecular weights and number of Cd^{2+} ion bounds. Tentatively, they were assumed to represent PCs of the type $(\gamma\text{-Glu-Cys})_n\text{-Gly}$ where n=2-6. However, their exact molecular structure will require further analysis. This assay performed on crude and pure samples, thus, through the use of metal isotopes the visualization of Cd-BPs via their bound Cd is allowed. $^{109}Cd^{2+}$ containing buffer without Cd-BPs showed no signal (Fig 2,lane 1). The experiments described above are consistent with previous studies (7,8,9). We can conclude that Cd-BPs which are produced by *A.altissima* may play an important role in Cd-tolerance mechanism.

5. Acknowledgments

This research was supported by University of Oslo, Department of Biology, Division of Botany, and NAVF/SEP. I would like to thank Prof. Dr. Sven Erik Rognes for his scientific supervisions and for helpful discussions during my stay in Oslo, Norway.

6. References

1. Steffens,J.C. (1990) The heavy metal-binding peptides of plants. *Annu. Rev. Plant Physiol. Plant. Mol. Biol.* **41**: 553-575.
2. Rauser, W.E.(1990) Phytochelatins. *Annu. Rev. Biochem.* **59**: 61-86.
3. Grill,E., Löfler,S., Winnacker,E.-L & Zenk,M.H. (1989) Phytochelatins, the heavy metal-binding peptides of plants, are synthesized from glutathione by a specific γ-glutamylcysteine dipeptidyl transpeptidase (phytochelatin synthase). *Proc. Natl. Acad. Sci.* USA, **86**: 6838-6842.
4. Ernst,W.H.O.,Verkleij,J.A.C.,andSchat,H.(1992) Metal tolerancein plants. *Acta Bot.Neerl..* **41**: 229-248.
5. Bayçu,G., and Önal,M., (1992) The effects and the accumulation of cadmium in Ailanthus altissima. *Physio. Plant.*, **85 (3)**, Part 2: A74. 8th FESPP Congress, Antwerpen,Belgium.
6. Rauser,W.E.(1991) Cadmium-binding peptides from plants. *Methods in Enzymology*, **205**, Part B:319-314.
7. Abrahamson,S.L, Speiser,D.M. and Ow,D.W. (1992) A gel electrophoresis assay for phytochelatins. *Analytical Biochemistry*, **200**:239-243.
8. De Knecht, J., Koevoets, P.L.M., Verkleij, J.A.C. and Ernst, W.H.O.(1992) Evidence against a role for phytochelatins in naturally selected increased cadmium tolerance in *Silene vulgaris*. *New Phytol.*,**122**: 681-688.
9. Leita, L., Contin,M. and Maggioni,A. (1991) Distribution of cadmium and induced Cd- binding proteins in roots, stems and leaves of *Phaseolus vulgaris*. *Plant Sci.* **77**:139-147.

ROOT AND SHOOT GROWTH OF COMMON AND DURUM WHEAT SEEDLINGS INFLUENCED BY LOW pH AND ALUMINUM STRESS

L. BONA[1], F. ZSOLDOS[2], Á. VASHEGYI[2], M. MOUSTAKAS[3] AND L. PURNHAUSER[1]

[1]Cereal Res. Institute, POB 391, Szeged 6701, Hungary; [2]Dept. of Plant Physiology, József A. Univ. POB 654, Szeged, Hungary; [3]Dept. of Botany, Univ. Thessaloniki, POB 109, Thessaloniki 54006, Greece

1. Introduction

Crop yield is reduced by soil acidity on ca. 30%. Much of the damage to plant production is due to aluminum (Al), the most common metal in soils. In common wheat (*Triticum aestivum* L.), genetic improvement for Al tolerance helped management practices; however, there is lack of information on durum wheat (*Triticum durum* Desf.).

In acidic soils, Al^{+3} and H^{+} ion toxicities, and deficiencies of essential elements lead to retardation of root growth and development (Marschner, 1995). In cereals, the most obvious influence of Al and low pH on root morphology is the decrease of root length and root weight when root tips turn to brown with lack of branching and smaller primary root hairs (Foy, 1992). The morphological response of common (hexaploid) wheat to Al and low pH is well documented (Baligar *et al.*, 1991; Foy, 1992; Carver and Ownby, 1995); however, limited information is available on the response of durum (tetraploid) wheat to the above stress factors (Moustakas *et al.*, 1992).

The objectives of this study were to examine changes of root and shoot growth of wheat seedlings subjected to low pH and increasing Al concentrations and to compare the responses of common and durum wheats to short term effects of Al in solution culture.

2. Materials and Methods

Common winter wheat cultivar Jubilejnaja 50 (JUB-50) and durum wheat cultivar GK Betadur (BETADUR) were used as experimental material in this study. Both cultivars are widely used in the Hungarian agriculture. Previous tests in acid soil (Bona *et al.*, 1992) showed that BETADUR is moderately sensitive to Al toxicity while JUB-50 can be considered as an intermediate one. Seeds were surface-sterilized for 15 minutes rinsed in deionized water and pregerminated at 20 °C in petri dishes. After two days, uniform healthy seedlings were selected and placed into stainless screens over glass beakers, transferred to a controlled growth chamber. Each baker contained eight seedlings and 300 ml of 0.5 mM Ca SO_4 with pH 4.0 or 6.0 and supplemented with 0, 0.001, 0.025, 0.05, 0.075, and 0.1 mM Al. All culture solutions were renewed daily and pH was adjusted by 0.1 N HCl. After six days, the plants were harvested and separated into shoots and roots. Fresh weight (FW) was measured on individual plant basis. In the second experiment, short terms (0, 3, 6, 12, and 24 h) of Al stress exposure was applied as described previously with Al concentrations 0, 0.5. 1.0, and 5.0 mM.

3. Results and Discussion

Figure 1 shows the effect of increasing Al concentration on the growth of wheat seedlings. At pH 4.0, the inhibition of root elongation appeared first at 0.001 mM Al in BETADUR, and at a higher concentration (0.01 mM Al) in JUB-50 (Figure 1).

Figure 1. Effect of increasing (0, 0.001, and 0.01 mM) levels of Al on the growth of durum (BETADUR) and common (JUB-50) wheat seedlings at pH 4.0

Root growth of BETADUR was strongly inhibited by 0.01 mM Al in the solution. We did not find any retarding effect in the shoot growth. Yet, very low concentration (0.001 mM) of Al in the solution seemed to be beneficial for shoot growth in both varieties. This confirms Foy (1992) that low concentration of Al might be beneficial for growth of plants. The poor rooting of BETADUR at pH 4.0 is in agreement with earlier observation of Bona *et al.* (1995) and may be the explanation of poor yield of durum in Al toxic, acidic soils.

Fresh weight (FW) of roots also indicates that the effect of 0.01 mM Al was very strong in BETADUR durum wheat (80% loss) while in JUB-50 caused only a 10 %

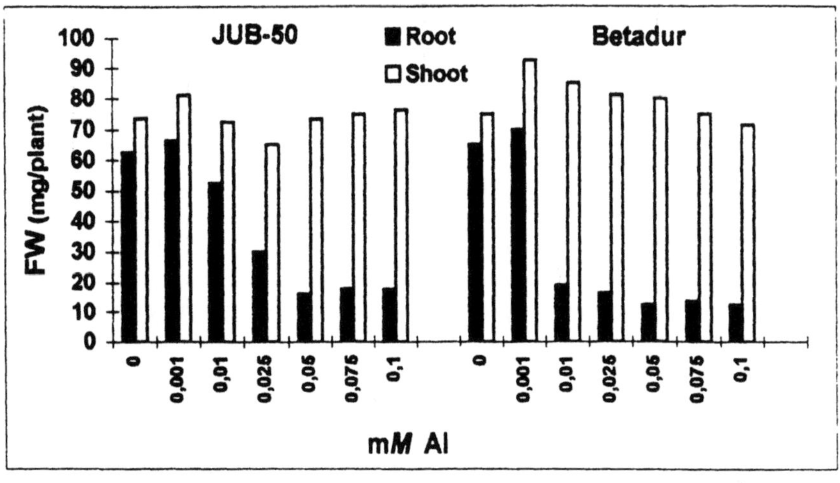

Figure 2. The effect of increasing Al concentration on root and shoot fresh weights of wheat at pH 4.0

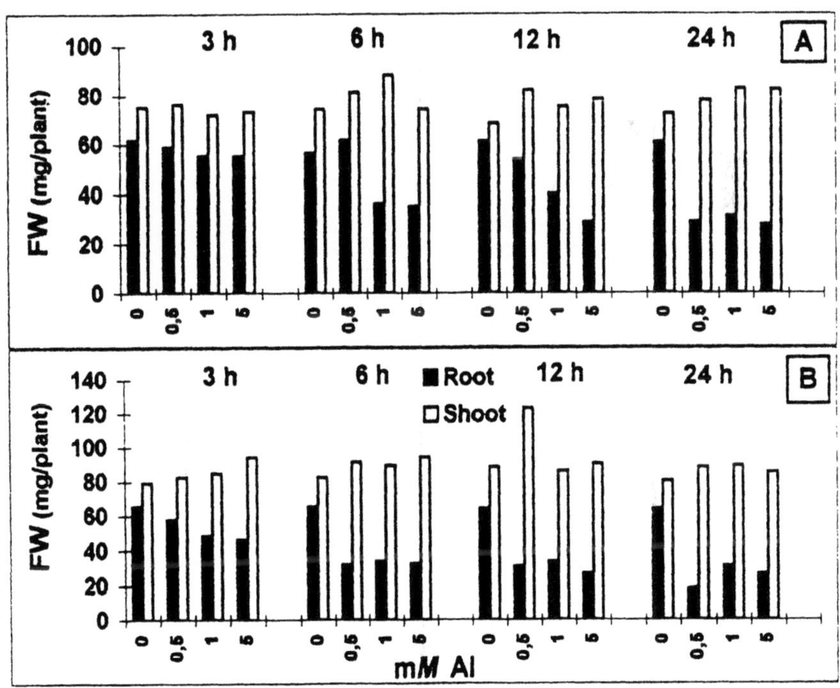

Figure 3. Short exposure times of Al on root and shoot growth of common wheat cv. JUB-50 (A) and durum wheat cv. Betadur (B)

loss, but 0.025 mM and higher Al concentrations proved to be much more toxic. In any cases, no reduction in the shoot weight was observed (Figure 2). However, at pH 6.0 (data not shown), no effect on the root or shoot growth and weight was observed at any Al concentration.

In the short time exposure experiment, the inhibitory effect of Al on root development was measurable after 3 h in durum wheat, particularly, at 1.0 and 5.0 mM Al level with a 15% reduction (Figure 3A). In common wheat (Figure 3B), significant inhibition started later, after 6 h. At the concentration level of 0.5 mM which may regularly occur naturally in acidic soils, root growth of durum and common wheat reacted differently. In durum wheat, significant decrease appeared after 6 h but in common wheat, appeared only after 24 h. Short time exposures even at higher Al concentrations did not have any negative influence on shoot growth either in durum or in common wheat. Even in some cases (BETADUR at 0.5 mM and 12 h; JUB-50 at 1.0 mM and 6 h), parallel with the root inhibition, a definite shoot stimulation was visible. Further research is needed for a better understanding of this phenomenon.

4. Conclusion

Tetraploid durum wheat exhibited higher susceptibility both to short-term and longer term exposure to Al than hexaploid common wheat. The negative effects of toxic Al was obvious in root growth but inhibition of shoot elongation was not detected. However, extremely low Al concentration might be beneficial for wheat growth.

5. Acknowledgements

This research was supported by the Hungarian Sci. Res. Fund (OTKA) No. T019922.

6. References

Baligar, V.C., Wright, R.J., Ritchey, K.D., Ahlrichs, J.L. and Woolum, B.K. (1991) Soil and soil solution property effects on root growth of aluminium tolerant and intolerant wheat cultivars, in R.J. Wright, V.C. Baligar and R.P. Murrmann (eds.) *Plant-Soil Interaction at Low pH,* Kluwer Academic Publishers, Dordrecht, pp. 245-252.

Bona, L., Wright, R.J. and Baligar, V.C. (1992) Acid soil tolerance of *Triticum aestivum* L. and *Triticum durum* Desf. genotypes, *Cereal Research Communications* **20,** 95-101.

Bona, L., Baligar, V.C. and Wright, R.J. (1995) Soil acidity effects on agribotanical traits of durum and common wheat, in R.A. Date, N.J. Gurdon, G.E. Rayment and M.E. Probert (eds.), *Plant-Soil Interaction at Low pH: Principles and management*, Kluwer Academic Publishers, Dordrecht, pp. 425-428.

Carver, B.F. and Ownby, J.D. (1995) Acid soil tolerance in wheat, *Advances in Agronomy* **54,** 117-173.

Foy, C.D. (1992) Soil chemical factors limiting plant root growth, *Advances in Soil Science* **19,** 97-149.

Marschner, H. (1995) *Mineral nutrition of higher plants,* Academic Press, New York.

Moustakas, M., Yupsanis, T., Symeonidis, L., and Karataglis, S. (1992) Aluminum toxicity effects on durum wheat cultivars, *Journal of Plant Nutrition* **15**, 627-638.

ECOPHYSIOLOGICAL CHARACTERISTICS OF *STAPHYLEA PINNATA* L. AND *SAMBUCUS NIGRA* L. FROM THE BELGRADE BOTANICAL GARDEN

D. DIMOVIĆ[1], S. MARJANOVIĆ[2], J. ŠINŽAR[2], B. STEVANOVIĆ[2]
[1]Institute for protection of nature of Serbia
III bulevar 106, 11070 Belgrade, Yugoslavia
[2]*Institute of botany and Botanical garden, Faculty of biology*
Takovska 43, 11000 Belgrade, Yugoslavia Introduction

1. Introduction

The ecophysiological investigations were concerned with the species *Staphylea pinnata* L. and *Sambucus nigra* L., of similar general appearance, both having the same life form - *fo dec MiP Scap* - Broad-leaved Deciduous Scapose Microphanerophyte. These plants differ in evolutionary age and ecological preferences for different mesophyllous forests.

Staphylea pinnata L belongs to the family *Staphyleaceae.* This species is a tertiary relict distributed in South Europe and Asia Minor (Meusel et al, 1965). Its habitats are restricted to protected sites within mesophillous forest (mostly of *Quercus petraea* and *Carpinus betulus*), up to 700 masl, chiefly on the limestone.

Sambucus nigra L. belongs to the family *Sambucaceae.* This is a common, wide-spread species inhabiting various mesophyllous forests along altitudinal gradient from lowland, to montane zone, on varying soils. It belongs to Middle-European - Mediterranean - Submediterranean - Pontic-South-Siberian floristic element (Jovanović,1994)).

The examined plants grow under outdoor conditions in the Belgrade Botanical garden. The aim of the study was to establish the ecophysiological differences or similarities between *Staphylea pinnata* and *Sambucus nigra* as two phylogenetically and taxonomically distinct species with convergent life form. In this respect the dynamics of water and osmotic potential as well as morpho-anatomical features of leaves have been measured.

2. Material and methods

Water and osmotic potentials were measured every month during the vegetational season - from April to October. Water potential was measured by pressure chamber method (Scholander et al. 1965) and osmotic potential according to Walter (1931) on the semimicroosmometer KNAUER.

Specimens for anatomical analysis were collected in spring, summer and autumn. Thickness of the leaf, mesophyll, palisade and spongy tissue, as well as the upper and lower epidermal cell size were measured on cross-sections prepared by standard method for light microscopy (Chamberlain 1921, Jensen 1962).

Quantitative data obtained were computed using Quatrro Pro for DOS 4.0.

3. Results

Staphyllea pinnata and *Sambucus nigra* have the same habit: they are shrubs with pinnately compound leaves showing both morphological and anatomical similarities, i.e. mesomorphic structural pattern. They were glabrous and dorsiventral. The ratio of palisade to spongy tissue was 1:2 in *S. nigra*, and 1:1.3 in *S. pinnata*. The mesophyll tissue cells, both of palisade and spongy parenchyma were sparse. Intercellular spaces were markedly large among the cells of spongy parenchyma. The spongy tissue in *S. nigra* leaves was thicker than in *S. pinnta*. Leaves of both species were hypostomatic, and stomata were leveled with epidermal cells.

Fig.1. Morpho-anatomical features of examined plants

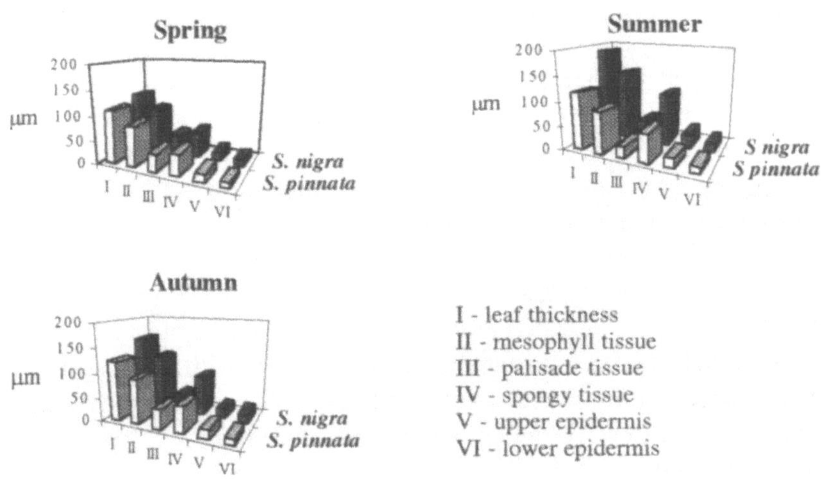

In both species the water balance was the most favourable in spring (*S. pinnata* - April: water potential -1.05 MPa, osmotic potential -1.3 MPa; *S. nigra* - April: water potential -1.67 MPa, osmotic potential -0.7 MPa). The most unfavourable water balance was observed in autumn (*S. pinnata* - October: water potential -1.75 MPa, osmotic potential -1.7 MPa; *S. nigra* - October: water potential -1.9 MPa, osmotic potential - 1.4 MPa) (Figs.2 & 3)

Fig.3. Seasonal dynamics of the water potential

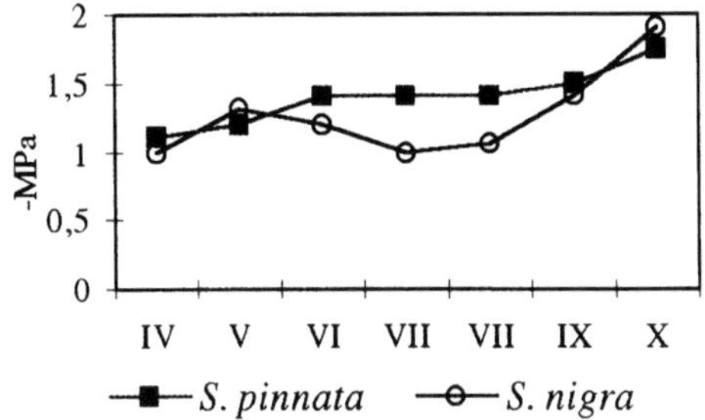

Fig.3. Seasonal dynamics of the water potential

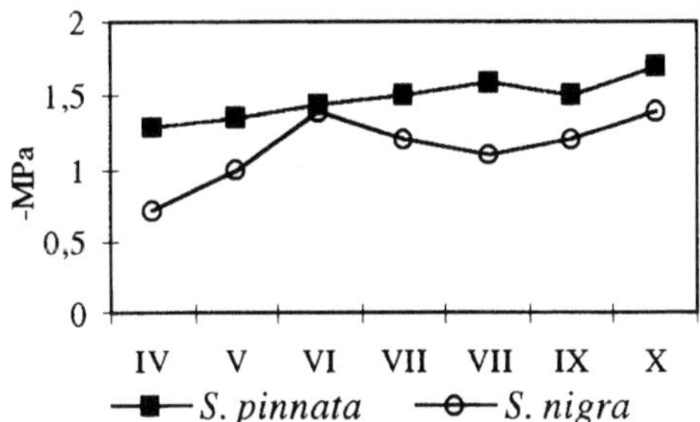

For the same species the similar results were reported under field conditions on their natural habitats (Popović et al. 1979, 1983).

4. Conclusion

In addition to having the convergent life form, the species *Staphylea pinnata* and *Sambucus nigra* have the same ecological affinity as well, which is evidenced by:

1. the similar mesomorphic pattern of leaf structure. The leaves of both species have quite identical mesophyll cell size, and ratio of palisade to spongy tissue, as well as the same volume of intercellular spaces.
2. the changes of water balance parameters, in both species, correspond to those of mesophytes:

- tertiary relict *Staphyllea pinnata* was distinguished by isohydric type of water balance, which is, otherwise, the common characteristic of the great number of relict plants in the Balkan Peninsula (Stevanović, 1996).
- *Sambucus nigra* showed greater fluctuations of water balance parameters under the same ecological conditions. The tolerance of larger losses water and greater increases in the cell sap concentration enables the species *S. nigra* to survive in wide variety of present day plant communities, from natural mesophyllous forests to ruderal associations in the habitats disturbed by human impacts.

5. References

Chamberlain, C.J. (1921) *Mikrotehnika i botanicki praktikum*, Kemozan, Zagreb

Jensen, W. (1962) *Botanical histochemistry*, W.H. Freeman and Company.

Jovanovic, S. (1994) *Ecological study of ruderal flora and vegetation in the city of Belgrade*, Faculty of Biology, University if Belgrade.

Meusel, H., Jäger, E. and Weinert, E. (1965) *Vergleichende Chorologie der Zentraleuropäiscen flora*, 1, VEB Gustav Ficher Verlag, Jena.

Popovic, R. Jankovic, M. and Dimitrijevic, J. (1979) Basic characteristics of water regime in some plant species of the community *Querco-Carpinetum serbicum* Rud. on the mountain Fruska gora, *Arch. biol. sci.* **31**, 13-30.

Scholander, P.F., Hammel, H.T., Bradstreet, E.A. and Hemingsen, E.A. (1965) Sap Pressure in Vascular Plants, *Science* **148**, 339-346.

Stevanovic, B. (1990) Značaj fiziološke ekologije biljaka za poznavanje životnih uslova endemičnih i reliktnih biljaka. - Ekologija i geografija u rešavanju problema životne sredine, *Srpsko geografsko društvo, Posebna izdanja*, **69**, 66-72.

Walter, H. (1931) Die Hydratur der Pflanze und ihre physiologisch-ökologische Bedeutung, Gustav Fischer Verlag, Jena.

Walter, H. and Kreeb, K. (1970): Die Hydratation und Hydratur der Protoplasmas der Pflantze und ihre öko-physiologische Bedeutung, *Protoplasmatologia* IIC 6, Springer Verlag, Wien, New York.

ECOPHYSIOLOGICAL PROPERTIES OF SOME FOREST TREES

M. DJUKIC, D. OBRATOV
Faculty of Forestry,
Kneza Viseslava 1, Belgrade, Yugoslavia

1.Abstract

To identify simple, cost effective methods of selection of vigorous individuals of several forest trees, some morphophysiological properties were used. The investigation included adult, superior trees of beech, spruce, birch and Balkan maple from various localities in Serbia. All the investigated morpho-anatomical and physiological properties show variability between various species as well as between individuals of the same species. on the various localities.

2. Introduction

Picea abies Karst., the main species of coniferous forests in Serbia, and *Fagus moesiaca Domin.Maly.Czeczott.* are the most widely distributed species, which are recommended for afforestation of the largest areas in Sebia. *Betula pendula Roth.*, a fast growing, pioneer species with very wide ecological adaptability and *Acer heldreichii Orph.*, an endemic and relic of Balkan Peninsula , with best quality timber, deserves wider utilization. in forestry.

Drought, extreme temperatures and other ecological factors, lead to significant modification in ecosystems and affect mainly the decrease of biomass productivity and the reduction of the number of species that can be grown.

3. Material and Methods

The research was carried out on 7 phenotypically superior spruce trees on the mountain Kopaonik (altitude 1,500 m) in the same ecological forest type, 13 beech trees on Mt. Kopaonik, Jastrebac and Miroc (alt. cca. 600 m), 11 birch trees on Mt. Jastrebac (alt. 600 m), Mt. Kopaonik (alt. 1100-1600 m) and in urban ecosystem (Belgrade), and 8 individuals of Balkan maple on Mt. Jastrebac (alt. 1,400 m).

The samples of leaves where taken in April, July and September 1997, from illuminated parts, lower third of the crown and from the middle part of the branches.

Fresh mass of leaves and water content (free and bound) were measured before and after drying at 60^0 and 105^0C, intensity of transpiration was after Stocker, water potential by pressure chamber, N, P, K elements in the leaves by Kjldahl, colorimetric and flamephotometric method. Number and dimension of stomata were determined microscopically, by making preparations by separating the lower epidermis.

4. Results and Discussion

The results of the research of leaf characters (length, width, fresh mass, number and size of stomata), parameters of water regime (transpiration intensity, water content, water potential), and the content of biogeneous elements (N,P,K), have been presented in Graphs (page 3). These are the average values of the seasonal fluctuation.

Variation of analyzed properties is evident in all trees of the same species, between different localities and between different species.

The low ratio between dimension of leaves and fresh mass of leaves, higher ratio between water content and dimension of leaves, higher transpiration intensity, higher number of stomata per unit area and smaller dimensions, lower water deficit, can point to their xeromophic character and possib!y to a better adaptation to the condition of poor water supply (Kojic, 1987).

According to these criteria, the most xeromorphic leaves are those of beech from Mt. Kopaonik and the leaves of Balkan maple are less xeromorphic.

The analysis of contents of biogeneous elements (N,P,K) in leaves show good supplies and the presence of variability.

The strategies of adaptability therefore can be slow (genetic) and quick, i.e. primarily physiological. A comparative approach to the above mechanisms reveals the fundamental principles of the genetic-physiological construction of trees and shrubs, as adaptations are, by all means, one of the significant elements of speciations and an important agent of evolution divergence. (Tucovic, Djukic, 1991.)

The aim of this research is to help the determination of simple markers for the selection of physiologically superior individuals.

The role of the early test in the selection of spruce was pointed out by Ekberg et al. (1994).

The application of the researched morpho-physiological criteria can accelerate and develop the selection procedures of seed (parent) trees of forest species for the needs of plant production in forestry.

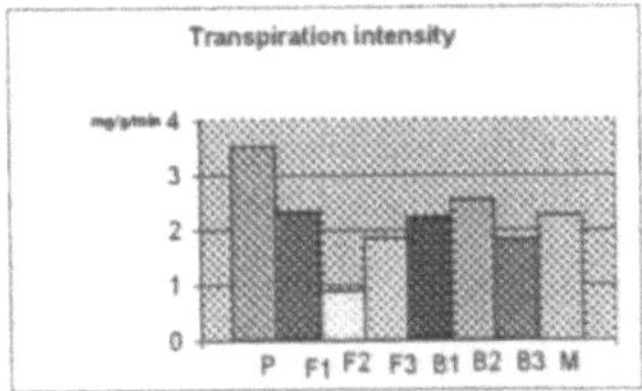

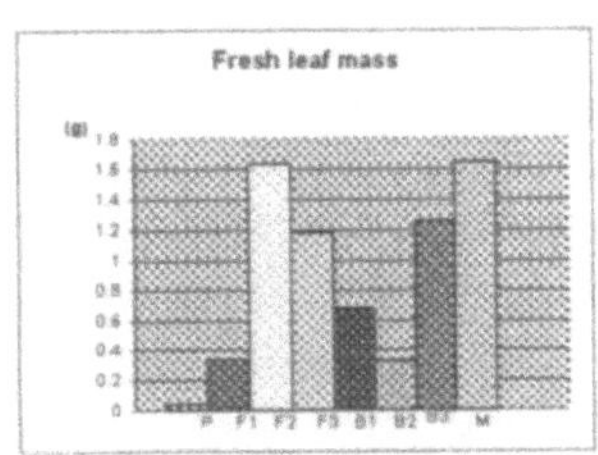

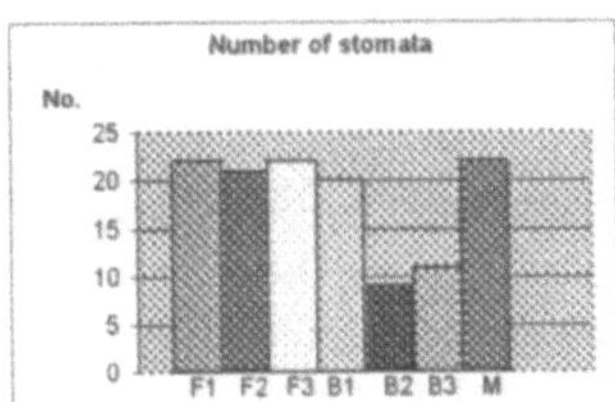

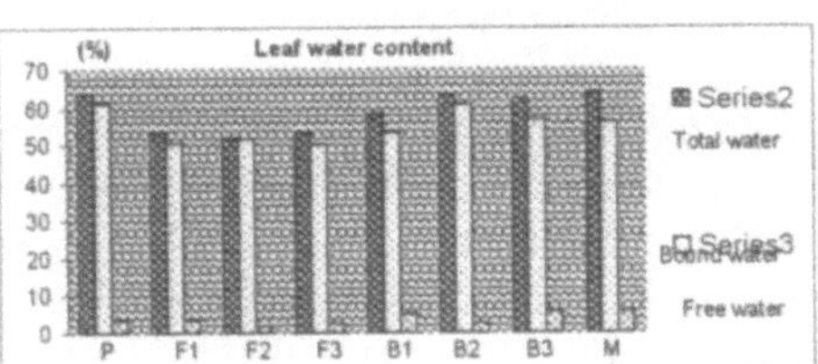

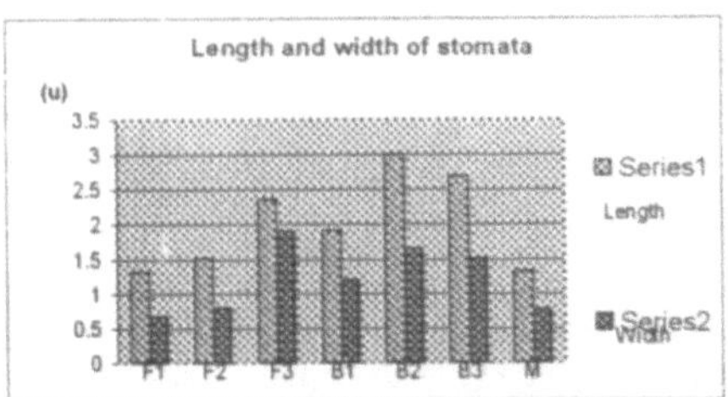

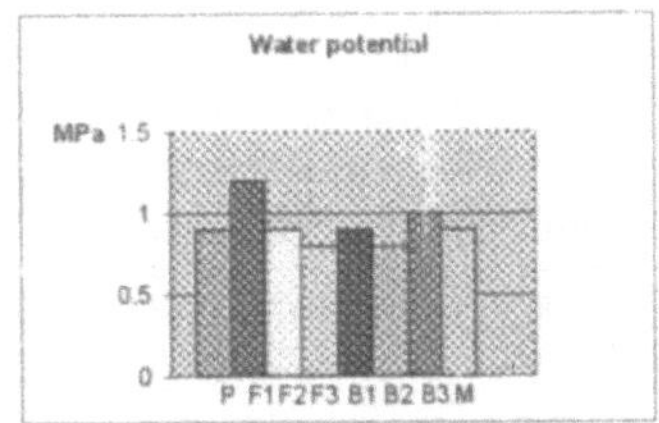

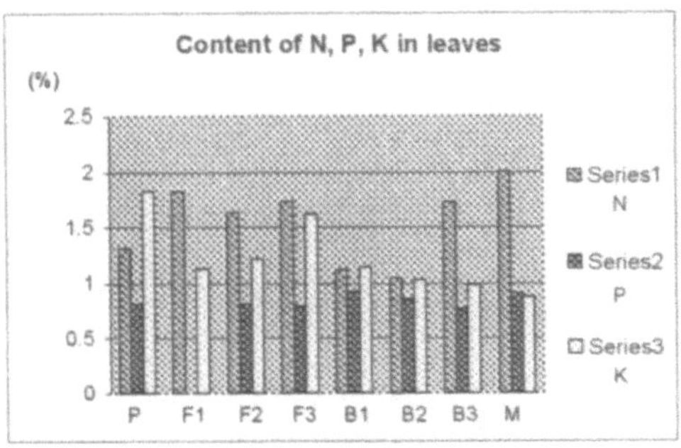

P - Picea abies (Kopaonik)
F - Fagus moesiaca
(1.Jastrebac, 2.Kopaonik, 3.Miroc)
B - Betula pendula
(1.Jastrebac,2.Kopaonik, 3.Belgrade)
M - Acer heldreichii (Jastrebac)

5. Conclusions

All the investigated morpho-physiological characters (leaf dimensions, fresh mass, number and size of stomata on the lower epidermis), water relation parameters (transpiration intensity, water content, water potential), and content of elements (N,P,K) in leaves, show evident variability among various individuals on the same and on thedifferent localities and between investigated forest species - spruce, beech, birch and Balkan maple.

It is probably possible, on the basis of the ratio between these properties, to help the selection of the individuals with more or less xeromorphic characters of the leaves and better adaptation to water deficit in the soil, or other extreme ecological factors, better biomass production, and so on.

If the selection of optimal forest trees for afforestation is to be done based on the studied parameters, it would be necessary to carry out further research, especially the raising the progeny and the checking of characters, such as capacity of biomass production, resistance to drought and other unfavorable factors of environment,,

6. References

Blake T.J., Tschaplinski, Easthm A. (1984): Stomatal control of water use efficiency in poplar clones and hybrids. *Can J. Bot.* 61.1344-1351.

Djukic,M.,Obratov, D (1986): Effect of some physiological properties of parent trees on spruce seedling growth. X FESPP Congress, Firenza, Italy. p.223.

Ekberg,I.,Eriksson, G.(1994): Genetic correlations for growth rhythm and growth capacity at ages 3-8 years in provenance hybrids of Picea abies. *Scandinavian Journal of Forest Research*, 9 (1), 25-33.

Kojic,M, (1987): *Fiziolska ekologija kulturnih biljaka.* Naucna knjiga, Beograd.

Kozlowsky,T.T. (1982): Water supply and tree growth. *Forestry abstracts, Review Article,* Vol. 4, no.2.

Tucovic A., Djukic,M. (1991): Adaptability of trees and shrubs to extreme temperatures. *Bulletin of the Faculty of Forestry*, Belgrade. No.73, p. 29-139.

ALLELOPATHY OF *ERICA CARNEA*, A DOMINANT SPECIES IN THE MIXED PINE FORESTS AT MALJEN MOUNTAIN (SERBIA)

L. DJURDJEVIC, A. DINIC, M. MITROVIC, P. PAVLOVIC
Institute for Biological Research "Sinisa Stankovic",
29. Novembra 142, 11060 Belgrade, Yugoslavia

1. Introduction

The spring heath (*Erica carnea L.)* is widespread within the community of Austrian and Scots pine (*Pinetum nigrae-sylvestris* Z. Pavlovic 1951) at the Maljen Mt (Serbia). In the most of stands this species is forming dense coenotic microcomplexes. Being the perennial dominant species, *E. carnea* influences at the composition and structure of pine forests. In the places in forest where *E.carnea* is dominating, the numerosity and degree of coverage of other herbaceous plants are considerably decreased. Let us suppose that besides the competition for water and nutrients in soil, and the allelopathic factor is influencing too, particulary phenolic compounds. Phenolics are widely distributed in soil, litter and fresh plant material (Harborne 1980; Kuiters and Denneman 1987; Djurdjevic 1991; Djurdjevic and Oberan 1995). These compounds can participate in allelopathic relations among plant species (Lodhi and Rice 1971; Chou and Muller 1972; Lodhi 1975; Lodhi 1978).

2. Materials and methods

For analyses of phenolic compounds, leaves of *E. carnea,* partially decayed litter and soil (0-20cm) were collected during the spring (April), summer (July) and autumn (October). Free forms of phenols and phenolic acids (PAs) were extracted from dried material (4 g for litter and leaves of *E. carnea*) in boiling 80% ethanol solution and transfered into ethylacetate. The bound phenols were extracted in ethylacetate after the pretreatment procedure involving the one hour boiling of dried material in 2N HCl.

Free forms of phenols were extracted from 30 g per sample of dried soil in boiling ethylacetate, during the 24 hours period. Residual soil was treated with 2N NaOH, and after boiling in 24 hours the bound phenols were determined (Hennequin and Juste 1967; Katase 1981a,b).

The quantitative and qalitative analysis of PAs was performed using the ascending two-dimensional paper chromatography in two mixtures: 1. iso-propyl alcohol, ethylacetate, NH_4OH and water (30:50:1:19) and 2. 2% acetic acid. Dry chromatograms were sprayed with p-nitroaniline and 20% sodium carbonate. Content of individual PAs was estimated by a spectrophotometric method at the corresponding wave length. Content of total phenolic compounds was determined at 600 nm using Folin-Ciocalteu reagent (Feldman and Hanks 1968).

3. Results and discussion

In the leaves of *E. carnea*, the dominant species in the herb layer, during the growing season prevail free forms of total phenolics (31,208-39,775 μg/g) in relation to the bound ones (6,557-10,108 μg/g). The amount of free forms of total phenolics is 3-5.5 times higher than bound forms. In the leaves of *E. carnea*, from PAs, were identified p-coumaric and ferulic acids in free (9.71-78.13 μg/g) and bound form (803.22-1,764 μg/g). The amount of PAs and total phenolics is the higest one at the end of the growing season.

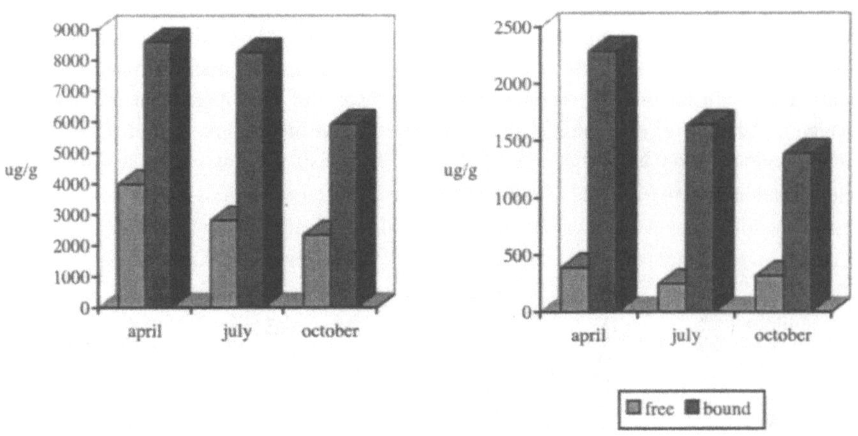

Figure 1. Seasonal variations of total phenolics (free and bound forms) in litter (A) and soil (B) under *E.carnea* L.

In the litter under *E. carnea* was measured 2,345-3,992 μg/g free and 5,924-8,591 μg/g bound phenolics and their amount is decreasing by the end of growin season (Fig.1 A). The litter within microcomplexes with the domination of *E. carnea* contains p-coumaric, ferulic, p-hydroxybenxoic and vanillic acids in free and bound forms. The bound forms of PAs prevail, but at the most vanillic acid (by 1,027.06 μg/g)(Fig. 2 A,B).

In the soil under *E. carnea* was measured 259.27-389.47 μg/g of free and 1,388-2,289 μg/g of bound phenolics. The amount of bound phenolics is decreasing by the end of the growing season (Fig. 1 B). The mentioned 4 phenoloc acids were detected in the soil in free as well as in the bound form., but in a considerably less amount in relation to the litter (Fig. 2 C, D).

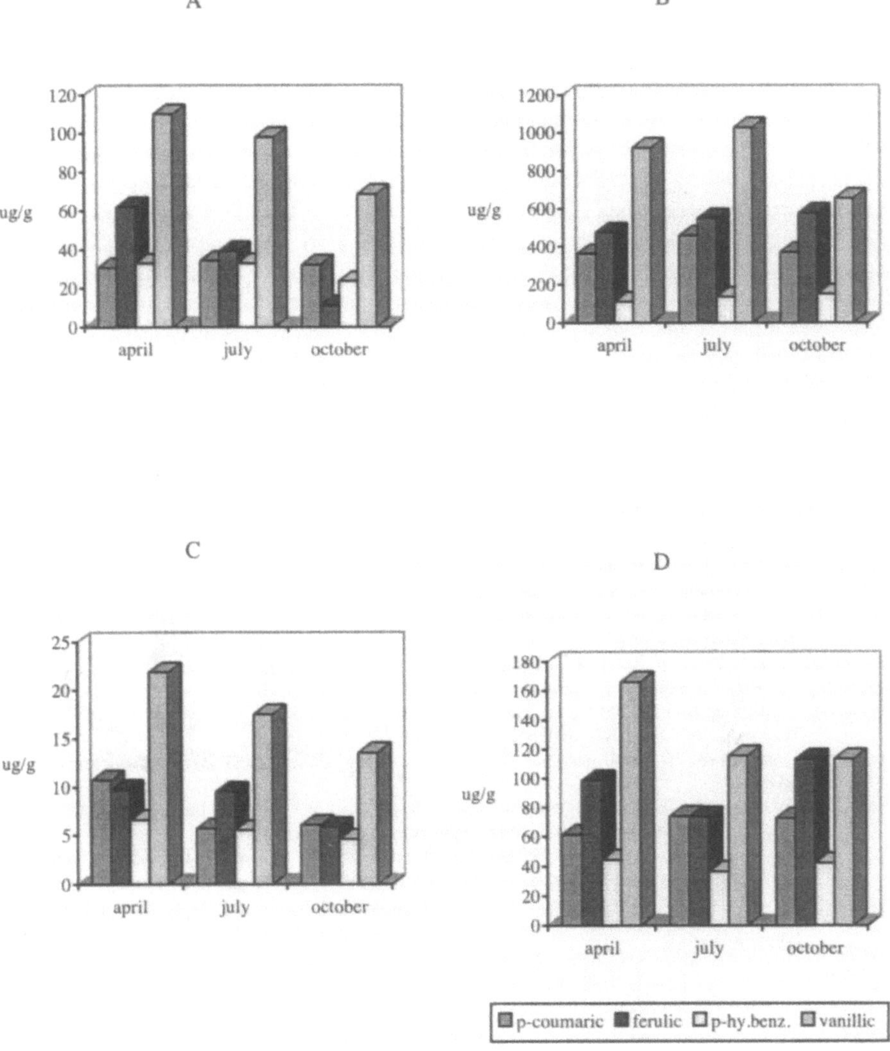

Figure 2. Fluctuations in the amount of free (A) and bound (B) PAs in litter; free (C) and bound (D) PAs in soil under *Erica carnea* L.

The allelopathic influence of dominant tree species in the phytocoenosis upon the herb species is revealing through phytotoxins of phenolic nature, which are at the most in leaves (Naib and Rice, 1971; Lodhi, 1978; Feeny and Bostock, 1968). The decreasing of free phenolics amount in litter in relation to the autumn leaves of *E. carnea* occurs due to their elution by rainfalls during the first phases of the litter decomposition (Yaavitt and Fahey 1986). PAs in litter occur, besides, due to the microbial degradation of lignin, the most frequent plant polymer in forest litter (Alexander 1977; Crawford and Crawford 1980; Kögel and Bochter 1985; Kögel 1986). In partially decomposed beech forest litter, five PAs were identified. In the spectrum of phenolics, ferulic, p-hydroxybenzoic and vanillic acids as free form were dominant (Djurdjevic and Oberan 1995).

The phenolics reach the soil through the elution by rainfalls from plants and litter accumulating within during the autumn and winter as bound forms (Kuiters and Denneman 1987). As the consequence of their accumulation in soil under dominant plant species in phytocoenosis, occurs the inhibition of seed germination and seedlings growth, and the consequence is the decreasing of number of individuals or their total elimination. In this way the phenolics, as allelopathic active substances in a great extent determine the composition and structure of plant community (Chou and Muller 1972; Lodhi 1978; Djurdjevic 1991).

In our case *Erica carnea* as dominant perennial forming dense coenotic microcomplexes, has an adverse influence upon structure and composition of pine forests, decreasing the numerosity and degree of coverage of herbaceous plants.

We suppose that *Erica carnea* through allelopathic way influences upon herbaceous plants thanks to the accumulation of phenolic phytotoxins in litter and soil.

4. References

Alexander, M. (1977) *Lignin decomposition. In:Introduction to soil crobiology* II ed. New York, Santa Barbara, London, Toronto. pp.472.

Chou, C.H. and Muller, C.H. (1972) Allelopathic mechanisms of Arctostaphylos glandulosa var. zacaensis, *Amer. Midl. Natur.* 88, 324-347.

Crawford, D.L. and Crawford, R.L. 1980. Microbial degradation of lignin. Enzyme Microb. Technol. 2:11-22.

Djurdjevic, L. (1991) Allelopathic effects of blackberry (Rubus hirtus W.K.), *Arch. des sci. biolog.* 34, 31-38.

Djurdjevic, L. and Oberan Lj. (1995) Free and bound phenolic compounds in the litter of a beech forest at Cer Mountain. *Arch. Biol. Sci.*, Belgrade, 47, 19P-20P.

Feeny, P.P. and Bostock, H. (1968) Seasonal changes in the tannin content of oak leaves, *Phytochemistry* 7, 871-88O.

Feldman, A.W. and Hanks R.W. (1968) Phenolic content in the roots and leaves of tolerant and susceptibile citrus cultivars attaced by Rodopholus similis, *Phytochemistry* 7, 5-12.

Harborne, J.B. (1980) *Plant phenolics.* In: *Secondary plant products.* Eds. Bell, E.A. and Charlwod, B.V. Springer-Verlag, Berlin, Heidelberg, New York, Encyclopedia of plant physiology, New series, 8, 329-402.

Hennequin, J. R. and Juste, C. (1967) Présence d'acides phenols libre dans le sol: Etude de leur influence sur la germination et la croissance des vegétaux, *Annual agronomy* 18, 545-569.

Katase, T. (1981a) The different forms in which p-coumaric acid exists in a peat soil, *Soil Sci.* 131, 271-275.

Katase, T. (1981b) The different forms in which p-hydroxybenzoic, vanillic, and ferulic acids exist in a peat soil, *Soil Sci.* 132, 436-443.

Kögel, I. (1986) Estimation and decomposition pattern of the lignin component in forest humus layers. *Soil Biol. Biochem.* 18, 589-594.

Kuiters, A.T. and Denneman, C.A.J. (1987) Water-soluble phenolic substances in soils under several coniferous and deciduous tree species. *Soil Biol. Biochem.* 19, 765-769.

Lodhi, M.A.K. and Rice, E.l. (1971) Allelopathic effects of Celtis laevigata. *Bulletin of Torrey Botanical Club.* 98, 83-89.

Lodhi, M.A.K. (1975) Soil-plant phytotoxicity and its possible significance in patterning of herbaceous vegetation in abottomland forest. *Amer. J. Bot.* 62, 618-622.

Lodhi, M.A.K. (1978) Allelopathic effects of decaying litter of dominant tress amd their associated soil in a lowland forest community. *Amer. J. Bot.* 65, 34O-344.

Yavitt, J.B. and Fahey, T.J. (1986) Litter decay and leaching from the forest floor in Pinus contorta (lodgepole pine) ecosystems. *J. Ecol.* 74, 525-545.

THE POSSIBILITIES OF USING SOME EXOTIC SPECIES IN URBAN CONDITIONS OF BELGRADE

D. DRAŽIĆ
Institute of Forestry,
11030 Belgrade, Kneza Višeslava 3, Yugoslavia

Abstract

In the capital of Yugoslavia, there are a great number of autochthonous and exotic species and their horticultural cultivars (ornamental forms) in the open green spaces of different types. Many of them show very successful development, health condition and degree of general vitality and decorativeness, for example: *Cedrus atlantica* Man., *Libocedrus decurrens* Torr., *Tilia argentea* Desf., *Quercus rubra* Du Roi non L., *Betula pendula* Roth., etc. In this paper, we would like to point out the possibilities of introduction of some, till now very little in use, but very interesting ornamental species. Three analyzed species (*Davidia involucrata* Baill., *Liriodendron tulipifera* L. and *Zelkova carpinifolia* K. Koch.) showed very successful results on the appropriate sites in Belgrade and special resistance under the conditions of polluted environment, which recommend them for similar sites in urban regions.

1. Introduction

In the considerably modified ecological conditions of the environment, where the constant increasing trend of air, soil, and water pollution must be taken into account as an eco-factor, in the conditions when many autochthonous species lose vitality or even disappear, the introduction of new species resistant to actual environmental conditions is a significant contribution.

By a long-term research of allochthonous flora in Belgrade and its surroundings, more than 300 species and horticultural cultivars have been recorded, analyzed and subsequently widened.

In addition to numerous species described by Jovanović (1950, 1971, 1971), Jovanović et Tucović (1962), Tucović (1954, 1957), Tucović et Žujović (1986), Žujović (1970, 1971), Vukičević (1971, 1974), Purić (1966, 1967, 1969, 1970),

Šijak (1983), Dražić *et al.* (1985, 1987, 1989), etc., this paper presents some characteristics of davidia (*Davidia involucrata* Baill.), liriodendron (*Liriodendron tulipifera* L.) and zelkova (*Zelkova carpinifolia* K. Koch.), recorded on some green spaces of Belgrade.

2. Environmental conditions

All study species in Belgrade grow in temperate-continental climate (after Kerner), in subhumid moister (C_2) to subhumid drier (C_1) climate - after Thornthwaite, with mean annual air temperature 11.5° C and average annual rainfall 669.9 mm. The warmest month is July (22.1°C), and the coldest is January (–0.3°C). Mean air temperature in the vegetation period is 18.3^{0}C. Negative air temperatures are from October to April, which can affect the tender plant species. Although annual rainfall is not high, it is significant that rainfall is evenly distributed per seasons: winter 150.5 mm, spring 169.4 mm, summer 105.1 mm and autumn 144.9 mm. The share of rainfall during the vegetation period is 380.7 mm, i.e. 56.8% of the total annual rainfall. Mean value of relative air humidity is 69.3%. The characteristic wind is from southeast - košava, rather strong, supporting dry spells and affecting evaporation from soil surface, which is a negative moment for vegetation. Northwest and southwest winds are also prevalent.

The analyzed species occur on deposols – natural soil disturbed by man to a greater or a lesser degree.

3. Results of research

3.1. *DAVIDIA INVOLUCRATA* BAILL., Fam. *Nyssaceae*, Genus *Davidia* Baill.

Davidia, pocket-handkerchief tree, the only representative of the genus of the same name is native to SW China. It was discovered by the French missionary David in 1869, introduced to Europe in 1904 (Wilson). It can thrive in our conditions successfully.

The flowering and fruit-bearing specimens were recorded already in 1943 in Paris. In the former Yugoslavia, it was described only in the region of Zagreb, where it was registered in the botanical garden (Jurković, 1980, 1984).

In Belgrade in 1984/85, one specimen was observed in the representative park of Stari Dvor and White Dvor at Dedinje (Dražić *et al.*).

Aged 20, very vital and extremely ornamental, without any insect pests or disease agents, it has attained the height of 6.5 m and dbh 12.5 cm.

The tree has a regular conical habit, reddish-brown bark. Buds are large, 1.2-1.5 cm long. Leaves are ovate-cordate, sharply toothed, up to 15 cm long, bright green on the adaxial side and silvery white on the abaxial surface. This specimen flowers abundantly in mid May. Flowers, globular clusters about 1.5 cm, are composed of one female flower surrounded by numerous male flowers, color of a ripe morello. They have no petals, but are surrounded by two pendent bracts of unequal size, the largest being about 18 cm. Fruits, first green with a pattern of reddish-brown, sized 3.2-3.5 cm, ripen in Oct/Nov.

Davidia is a very decorative tree of medium height, attractive throughout the year. Resistance to low temperatures, polluted air and unfavorable urban environment make it suitable for green spaces of different categories.

3.2. *LIRIODENDRON TULIPIFERA* L.

Order *Magnoliales*, Fam. *Magnoliaceae*, Genus *Liriodendron* L.

Tulip tree is much more represented in different categories of green spaces in Belgrade. Specimens of different age and height are vigorous and extremely ornamental.

By its regular tree with light gray bark, rich crown, unusual leaves formed like a lire and by large yellow tulip-shaped flowers, tulip tree attracts attention and the resistance to insects and plant diseases recommends it for wide use. This species was applied in the recultivation of opencast coal mines in the vicinity of Belgrade. Its development in the specific, relatively unfavorable environment confirmed the positive experience in N America, where it is often used in recultivation (Medvick, Ch., Hill, R.D. etc.).

3.3. *ZELKOVA CARPINIFOLIA* K. Koch.

Order *Urticales*, Fam. *Ulmaceae*, Genus Zelkova Spach.

Native to E Mediterranean and Caucasus, this species was recorded in Belgrade in a courtyard of an elementary school. There are two specimens 32 years old.

Tree 1 has mean diameter 38.5 cm, stem height 2.10 m, crown 11.50 m, total height 13.60 m, crown width 12.20 m. Tree 2 has a short stem, characteristic of this species, forking already at 0.60 m in three branches of the first order, diameters 22, 27 and 11 cm. Crown is 12.80 m high, 12.50 m wide, and total height is 13.40 m.

Bark is one of the main ornamental details of this species. It is ash gray, smooth in youth and later on, as in the study trees, it flakes forming the characteristic yellow-orange-brown mosaic. The same phenomenon appears on larger branches, which are numerous and form a decorative crown.

Leaves are elliptic elongate, with a crenate leaf margin. Their size, depending on the position, varies 2-12 cm, which is larger than described in literature. Number of veins varies between 6-12. Leaves are alternatewith short stalks, mildly rough on the upper surface, and on the abaxial surface finely downy along the nerves.

Flowers appear in late May/early June. They are formed on younfer branchlets, in leaf axils, light green and mildly scented. Fruit is a one-seed dry nut varying in length 3.8-5.7 mm, averagely 4.7 mm, width 2.9-4.4 mm, averagely 3.8 mm.

These trees do not show any signs of either of biotic or abiotic damage

The characteristic silhouette of a bizarre form, decorative stem and branches with flaking bark creating gray-yellow nuances, leaves with interestingly teethed margins, make zelkova attractive throughout the year, and its good adaptation to environment recommend its wider application.

4. Conclusion

All analyzed species in urban conditions of Belgrade and its near surroundings, showed very good development results, vigor, and especially decorativeness, which recommends them for wider application in similar climatic and other conditions in urban environments. In this way, the floristic fund of park and other green spaces of different categories and intentions, would be significantly enriched, which would contribute to the enrichment of the environment.

5. References

Dražić, D., Ilić, D., Marković, D. et Vitas, I. (1985) *Davidia involucrata* Baill. - a new decorative exotic species in flora of Belgrade. *Zbornik radova Instituta za šumarstvo i drvnu industriju*, **XXIV-XXV**, Beograd, 99-105

Dražić, D. et Ilić, D. (1987) Dendroflora of the park surrounding old and white court on Dedinje. *Zbornik radova Instituta za šumarstvo i drvnu industriju*, **XXVIII-XXIX**, Beograd, 231-251

Dražić, D. et Vitas, I. (1989) *Zelkova carpinifolia* K. Koch - another contribution to the study of exotic trees and shrubs of Belgrade. *Zbornik radova Instituta za šumarstvo i drvnu industriju*, **XXXII-XXXIII**, Beograd, 161-169

Hilliers manual of trees and shrubs (1981) David and Charles Newton Abbot. London.

Jovanović, B. (1950) Nesamonikla dendroflora Beograda i okoline. *Glasnik Šumarskog fakulteta*, **1**. Beograd.

Jovanović, B. (1971) Dalji prilog proučavanju egzotične dendroflore u Beogradu. *Hortikultura*. Split

Jovanović, B. et Tucović, A. (1962) Dalji prilog poznavanju alohtone dendroflore Beograda i okoline. *Glasnik Šumarskog fakulteta*. Beograd.

Jurković, M.(1980) *Davidia involucrata* Baill. *Hortikultura*, **4**, Split.

Krussman, G. (1977) *Handbuch der Laubholzkunde* I-II. Berlin.

Phillips, R. (1978) *Trees in Britain, Europae and North America*. London

Tucović, A. et Žujović, K. (1986) Dendroflora Beograda i okoline i dejstvo ekstremnih temperatura tokom zima 1984/85. i 1985/86. godine. *Glasnik Šumarskog fakulteta*, **66**, serija C. Beograd.

IDENTIFICATION OF Fe AND Mn DEFICIENCIES BY CHLOROPHYLL FLUORESCENCE TECHNIQUE

M. EYLETTERS[1], P. TOURLIERE[2], C. BITAUD[2], C. BROU[1], A. BENOMAR,[1] M. MELLAS[1] AND R. LANNOYE[1]

[1] *Université Libre de Bruxelles, Laboratoire de Physiologie et d'Agrotechnologies Végétales, Av F. Roosevelt 50 CP169, 1050 Bruxelles, Belgium*

[2] *S.C.P.A. Centre de recherche, Aspach-Le-Bas, 68700 Cernay, France*

1. Abstract

Minor elements deficiencies such as Fe and Mn lead to disturbances of some physiological and biochemical mechanisms.

Indeed, these ions are directly invol ded in the primary photochemical process of photosynthesis. An unavoidable consequence is a perturbation of the PSII activity and a slow down of electron transport.These destructive effects on PSII were quantified by chl a fluorescence measured on samples of different plants known for their own sensibility (grape, maize, potato, colza). We have observed that each deficiency can be characterized by a specific variation of a fluorescence parameter.

2. Introduction

Iron and manganese are micronutrients involved directly in photosynthetic activities of plants. Iron influences the ultrastructure of chloroplasts (Stocking *et al.*, 1975) and, protein and lipid composition of thylacoid membranes (Abadia *et al.*, 1989). The Fe deficiency leads to a decreased activity of electron transport whereas iron is a constituent of electron transport chains in both mitochondrion and chloroplast and diminishes non cyclic ATP formation (Terry, 1980).

Manganese is also a micronutrient transition metals and activates a number of enzymes in vitro, particularly decarboxylases and dehydrogenases (Marschner, 1986). The most well known function of Mn in green plants is its involvement in photosynthetic O_2 evolution (Amesz, 1983). It is believed that photosystem II (PSII) contains a manganoprotein which catalyses the early stages of O_2 evolution.The Mn deficiency impairs the first step of the electron transport chain of the light reaction and has negative effects on subsequent reactions such as photophosphorylation and the reduction of CO_2.

The purpose of the present research were to investigate the characterisation and the quantification of chl a fluorescence fast kinetic for these different micronutrients deficiencies and the possibility to diagnose the deficiencies in different plants.

3. Material and methods

3.1. PLANT MATERIAL

Greenhouse experiments were carried out on different plants: colza (*Brassica napus* L.), maize (*Zea mays* L.), potato (*Solanum tuberosum* L.), vine (*Vitis vinifera* L.) grown hydroponically in a nutrient solution (Terry, 1980) without or with precited micronutrient.These young, rapidly expanding leaves were used for all measurements.

3.2. CHLOROPHYLL A FLUORESCENCE MEASUREMENT - FAST FLUORESCENCE KINETICS

The leaf attached discs were dark-adapted for 30 min. The chl a fluorescence induction curve was monitored by a Plant Efficiency Analyser (PEA, manufactured by Hansatech Ltd King's Lynn, Norfolk, England) with 600 W.m-2 of ligh intensity. Illumination was provided by an array of 6 light emitting diodes (peak 650nm) focused on the sample surface to provided a homogenous illuminated light-spot of about 4 mm in diameter as described in detail by Strasser et al (1995). The fluorescence curve of a 1 second illumination was digitized into 1198 datapoints. From each fluorescence transient of the type OJIP, different values were selected: t_{Fm} = time to reach the maximal fluorescence intensity; Area = summation of all values (F_M - F_t) for t=50µs to $t=t_{Fm}$; Fo = Initial fluorescence intensity, calculated by the instrument; F_M = maximal measured fluorescence intensity ; $F_V = F_M$ - Fo = variable fluorescence.
The experimental values of Fo and F_M are used for the statistical calculations.

4. Results

Chlorophyll a fluorescence transients provides informations on the filling up of the plastoquinone (PQ) pool and the associated heterogeneity, but they are affected by both the electron donor and the acceptor sides of PSII of oxygen evolving organisms (Krause and Weis, 1991). In figure 1, the Fe-deficient plants show a very rapid rise in Fo compared to control plants, whereas F_M exhibited no significative difference. The initial chl a fluorescence yield at O level reflects the minimal fluorescence yield when all Qa are in the oxidized state. The Fo increase indirectly reflects differences on the electron acceptor side of PSII related to (1) the rate of electron transport and (2) the PSII heterogeneity. A direct consequence of this Fo increase is a decrease of $\varphi_{po} = 1-$ Fo/F_M (the maximal quantum yield of PSII photochemistry) which indicates a decrease of the maximal efficiency that an absorbed photon is trapped by a reaction center of

PSII. The decrease in the area shows a different and complex kinetics of the fluorescence rise for control and deficient plant. Vine and maize plants show the same behavoir when they are submitted to Fe deficiency.

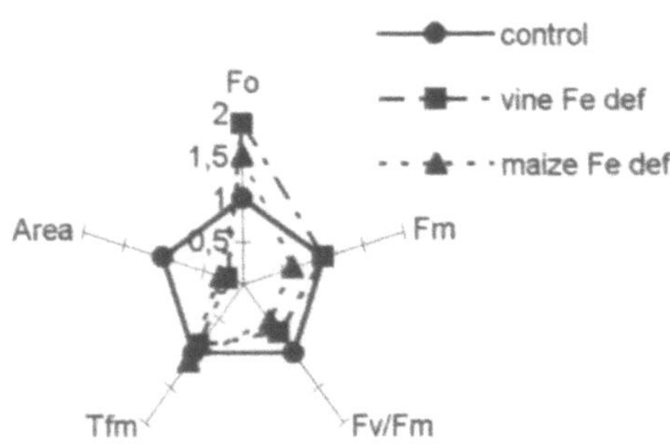

Figure 1. Effects of Fe deficiency on the maize and vine plants. The fluorescence parameters are expressed in % of variation to the control plant (=1): t_{Fm} = time to reach the maximal fluorescence intensity; Area = summation of all values (F_M - F_t) for t=50μs to t=t_{Fm}; Fo = Initial fluorescence intensity, F_M = maximal fluorescence intensity.

For both plants (colza and potato), the Mn deficiency (figure 2) has reducted the quantum efficiency for photochemistry which suggested that Mn was somehow involved in the early processes of photosynthesis. We can observe a specific response to Mn deficiency compared to Fe deficiency: an increase in time to reach the maximal fluorescence (t_{FM}) and in area.

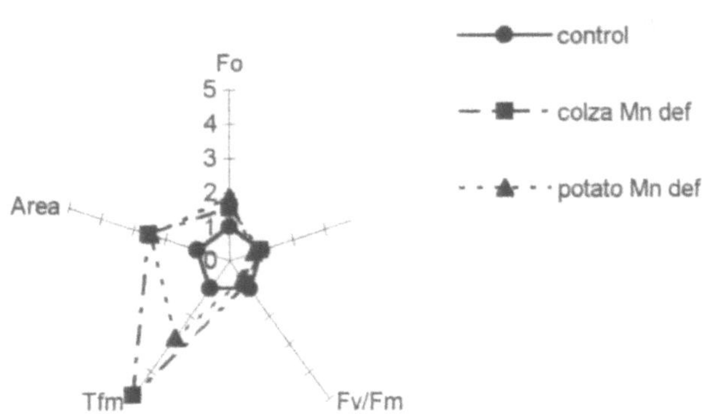

Figure 2. Effects of Mn deficiency on the colza and potato plants. The fluorescence parameters are expressed in % of variation to the control plant (=1): t_{Fm} = time to reach the maximal fluorescence intensity; Area = summation of all values (F_M - F_t) for t=50μs to t=t_{Fm}; Fo = Initial fluorescence intensity; F_M = maximal fluorescence intensity.

5. Discussion and conclusion

The fast fluorescence rise, within the measuring time of only one second, is a powerful tool to screen in a short time many samples under *in vivo in situ* conditions. According to the proposed data handling routines (JIP-test) some estimations about structural and functional changes in the photosynthetic apparatus can be calculated. Fluorescence data show differences between control and deficient plants. We can observe a specific behavior for the Mn and Fe deficiency with typical changes of the fluorescence signals. Fe deficiency is characterized by a drastic increase in the initial fluorescence intensity (Fo) inducing a decrease in the maximal quantum yield of the PSII photochemistry.
Mn deficiency has shown a different kinetic of the fast fluorescence due to different mechanistic causes as revealed by specific activities.
The studied plants show a similar behaviour according the Fe and Mn deficiency.
When making use of normalized fluorescence ratios and indices, it is possible to differentiate between Mn and Fe deficiency at the level of PSII activity and to obtain a diagnostic of these differencies.

6. References

ABADIA, A., LEMOINE, Y., TREMOLIERES, A., AMBARD-BRETTEVILLE, F., REMY, R., 1989. Iron deficiency in pea: effects on pigment, lipid and pigment-protein complex composition of thylacoids. Plant Physiol. Biochem. 27(5). 679-687.

AMESZ, J. (1983). The role of manganese in photosynthetic oxygen evolution. Biochimica et Biophysica Acta, 726.1-12.

KRAUSE, G.H., WEIS, E. 1991. Chlorophyll fluorescence and photosynthesis: The basics. Ann. Rev. Plant. Physiol. 42. 313-349.

MARSCHNER, H. (1986)*Mineral nutrition of higher plants*, Academic Press.

STOCKING, C.R., 1975. Iron deficiency and the structure and physiology of maize chloroplasts. Plant Physiol. 55. 626-631.

STRASSER, R.J., SRIVASTAVA, A., GOVINDJEE, 1995. Polyphasic chlorophyll a fluorescence transient in plants and cyanobacteria. Photochem. Photobiol.61.32-42.

TERRY, N., 1980. Limiting factors in photosynthesis. Use of iron stress to control photochemical capacity in vivo. Plant Physiol. 65. 114-120.

ECOLOGICAL EVALUATION AND MANAGEMENT OF THE ECOSYSTEMS OF ACHERON RIVER'S ESTUARIES AND GORGE (W. GREECE).

TH. GEORGIADIS, G. DIMITRELLOS, M. PANITSA, M. THEOCHAROPOULOS AND S. CHOCHLIOUROS
Botanical Institute-Plant Ecology, Division of Plant Biology, Department of Biology, University of Patras, GR 265 00, Patras-Greece.

1. Indroduction

The site we have to manage is located in the south-western part of Epeiros (Western Greece) and consists of two separate areas which are linked by the River Acheron: a) The coastal wetland of the Delta area which consists of alluvial deposits and is surrounded by low hills with hard limestone and flysch deposits with medium to steep gradients.b) The area of the gorge (hereafter referred to by its Greek name of Stena) is characterized by large limestone rocks with steep gradients and lies about 17 km from the delta, close to the source of the river.
The climate of the site is Mediterranean and belongs, from a bioclimatical point of view, to the humid bioclimatic zone with mild winters.

2. Biological and Cultural Environment

Flora-Fauna: The total number of plant taxa in both the areas of Delta and Stena reaches 509, of which 415 taxa are recorded from the Delta site and 204 from the Stena site, with 110 common species (Georgiadis & al. 1997). The bird fauna of Acheron Delta consists of 155 bird species. Mammals, reptiles, amphibians and the most important invertebrate taxa have also been recorded. The plant and animal species (except birds) which could be used as indicators of the ecological condition of the habitats have been determined.
Habitat types: a) The Delta plain is characterized by a high number of habitat types and intense human impacts. Within the plain, there are 19 natural habitat types and 3 semi-natural and artificial habitat types. b) The Stena area is characterized by a smaller number of habitat types, which are in better condition, a smaller variety of soil types, and important species especially, plants, birds and other animals (Georgiadis & al. -E.K.B.Y. 1996).
Land use - Human impacts: According to the Corine Land Cover classification, and local study of the Acheron Delta and Stena, we have registered 14 and 4 categories of

land use, respectively, while a total of 14 different categories of land use are referred for the area. Land reclamation work for the allocation of larger areas of the Delta to agriculture, and changes in land use which followed during the last three decades, have resulted in the reduction of the area of natural habitats, their division into fragmented units and the increase of artificial habitats. All the positive and negative human activities and their impacts to the area have been recorded.

3. Ecological evaluation

For the ecological evaluation, among the criteria suggested by Alexander (1993), Ratcliffe (1986), Sutherland & Hill (1995) and Beeby (1996) et al. for sites of special scientific interest, we have chosen the following criteria that we believe to be useful in this case: size, diversity, naturalness, rarity, fragility, typicalness, potential value and restoration capacity, as suggested by the Country Council of Wales (1996). The ecological evaluation is examined for the area as a whole and includes the evaluation of plant and animal species and habitat types occurring in the area.
a) Plant and Animal Species: The majority of the 19 plant species which have been evaluated from the delta area (89%) have a satisfactory and 11% of them have a sufficient global assessment. From the 11 important plant species which have been evaluated from Stena, 36% have excellent value and 54% have been evaluated as satisfactory and 10% as sufficient. As concerns the evaluated animal species, the majority are protected according International Conventions.
b) Habitat types: According to our data, it can be concluded that within the delta area there are no habitat types of excellent value. 58% of the habitat types of the delta valley have sufficient value, while the remaining 42% have satisfactory value. At Stena, three categories are present: the largest percentage belongs to the category with satisfactory value (2), while categories with excellent and sufficient value are represented by equal percentages.
The total evaluation of the species and habitat types shows that the Delta area has satisfactory ecological value which has been seriously disturbed by intense human influences. It also indicates the great ecological value of the Stena area and the possibility of conservation of its characteristics with the correct management measures.

4. Management

The management objectives must describe the favourable condition of the future and can be identified as follows: a) Conservation and protection of the natural resources of the area.(riparian forests, water resources etc), b) Reduction of the agricultural and stock - breeding pressures on the natural ecosystems, c) Amelioration of degraded habitats, d) Development through targeted ecotourist activities, e) Localization of changes (in space, structure and function) of the main features of the site by monitoring parameters and environmental indicators, f) Environmental education and increased awareness. For the realization of the objectives, a status of special management is required, based on the definition of “Management Zones”. The following zones have been determined for the Acheron Delta and Stena.

ZONE I.NATURE PROTECTION: These habitat types are characterized as having high ecological value. Hence their protection, conservation and amelioration is necessary for the continuation of the function of the wetland.

ZONE II.PROTECTIVE NATURAL LANDSCAPES - ECODEVELOPMENT: This zone includes the habitat types which are characterized by lower ecological value because of degradation, and which can be ameliorated and developed within the frames of ecotourist development.

ZONE III.RATIONAL MANAGEMENT: These habitat types are characterized as human ecosystems, which influence the functions of this wetland formation more than any other factor, and the objective is to arrange their functions in order to reduce the pressures on the natural ecosystems.

5. Proposed measures

The proposed measures concerning the whole studied area are outlined as follows: Protection measures concerning the whole studied area. Elaboration of an education and increased awareness program. Monitoring and evaluation of the wetland’s functions. Permanent guarding and monitoring of the area. Determination of the management service. Development of collaboration possibilities (Management service with national services and research educational institutions. The prohibition of hunting throughout the studied area. Management measures have been proposed for each individual zone:

Delta Acheronta

ZONE I.Nature Protection: 1.Extension of the riparian vegetation. Compensation for the farmers. 2. Preservation of the existing drainage network without extension or creation of another for the drainage of wetland areas. 3. Small agricultural areas surrounded by habitat types should be given up and enlisted to the natural vegetation. 4. Preservation of the existing roads without creating new ones. Dirt roads through the habitat types of Zone I. should not be used by vehicles and should be included in the network of small paths for ecotourist tours in the wider area. 5. The demolition of arbitrary buildings. 6. A marina at the gulf of Odysseas and the neighbouring gulfs should not be constructed. The prohibition of free camping in the coastal habitat types.

ZONE II. Protective natural landscapes ecodevelopment: 1. Immediate removal of the stable infrastructure for stockbreeding outside the studied area. 2. Suppression and movement of the areas of refuse disposal and restoration of the environment by tree planting. 3. Reconstitution of the stands of *Quercus ithaburensis* subsp. *macrolepis* by establishing seed-sown stands and enclosing them. 4. Reforestation of the surrounding hills. 5. Restoration of the landscape around redundant quarries. 6.

Creation of observation towers and posts (Axoutis, Adilalos, Glossa and Alonaki). 7. Planning and tracing of tour paths and connection of these with paths in the wider area. 8. Reduction of grazing land and amelioration of pastures. 9. Prohibition grazing after fire.

ZONE III.Rational Management:1. Rational management of the edaphic sources. 2. Application of the subsidized practice of "letting land lie". 3. Limitation of the use of pesticides and fertilizers (2080/92 EEC) and to inform the local population about the provision of financial support in return for the harmonization of agricultural practices with those of environmental protection. 4. Reconstitution of natural hedges, about 2-5m wide, at the borders of cultivated fields. 5. The application of plans for alternative agriculture (aromatic plants, etc). 6. Choice of the suitable place for the establishment and function of an area for the sanitary burial of rubbish outside the studied area, as referred in the Law 69269/90. 7. Creation of a unit for biological waste treatment within the settlement of Ammoudia, according the Law 69269/90. 8. Development of the historical, archaeological and cultural sites. 9. Construction of buildings (lodgings, hospices, centres of environmental education, etc.) in the inhabited borders.

Stena

ZONE I.Nature protection: 1. Dissuasion of any kind of interference in the frames of the ecotourist evaluation of the area. 2. Construction of a tour path (research, recreation, environmental education). 3.Examination of the possibility of enriching waters with progeny of trout (*Salmo trutta fario*) from the mountain waters.

ZONE II.Protective landscapes-development: 1.Management reports should not be falsified and should be adapted to the new data of land use and management of the area. 2.Prevention of grazing. 3.Building of view and observation towers. 4. Construction of tour paths (view places, areas with historical and cultural interest).

6. References

Alexander, M. (ed) (1993) *Management planning for nature conservation*, 265 p.

Beeby, A. (1993) *Applying Ecology*, Chapman & Hall, 441p.

Countryside Council for Wales (1996): *A guide to the production of management plans for nature reserves and protected areas.*

Georgiadis, Th. & coll. (1996) Special Management Plan of the area "Delta and Stena Acheronta", M.G.P.H. - E.K.B.Y. & University of Patras, 228 p.

Georgiadis, Th., Dimopoulos, P. & Dimitrellos, G. (1997) The flora and vegetation of the Acheron delta (W. Greece) aiming at nature conservation, *Phyton,* **37** (1), 31-60.

Ratcliffe, D.A. (1986) Selection of important areas for wildlife conservation in Great Britain, in M.B. Usher (ed.), *Nature Conservancy Council' s approach in Wildlife Conservation Evaluation*, Chapman & Hall, pp. 135-159.

Sutherland, W.J. & Hill, D.A. (1995) *Managing Habitats for Conservation*, Cambridge University Press, Cambridge, 399 p.

PHOTOSYNTHETIC RESPONSES OF GERBERA TO DIFFERENT SUBSTRATES AND HEATING SYSTEMS

M. ISSA[1], G. OUZOUNIDOU[2], H. MALOUPA[3] and H.A. CONSTANTINIDOU[4]

[1]*Mediterranean Agronomic Institute of Chania, Crete, Greece*
[2]*Dept. of Botany, Aristotle University of Thessaloniki, Greece*
[3]*National Agricultural Research Foundation, Thermi, Greece*
[4]*Dept. Agriculture, Aristotle University of Thessaloniki, Greece*

1. Introduction

Interest has recently been heightened for the production of good cut-flower cultivars, thus gerbera is presently a major greenhouse crop in the European countries. Zeolite and perlite are widely used for soilless culture because of their specific characteristcs [1]. We examined the effect of (a) Zeolite, perlite and their 1:1 mixture and (b) simultaneous application of either an upper (i.e. heating pipes running at culture bag level) or a lower (i.e. pipes running on the floor at a 50cm distance below the culture bags) heating system on several physiological parameters of two gerbera varieties (*Cyprus* and *Heart Breaker*) throughout the day and the year.

2. Materials and Methods

Two varieties (*Cyprus* and *Heart Breaker*) of *Gerbera jamesonii* were studied. Two locally available substrates were used for soilless cultivation: zeolite, perlite and their 1:1 mixture (Z+P). Sun protection was offered by covering the bags with coextruded black and white polyethylene film (40% transmittance). Substrate heating was supplied through a heated pipe network inside the premises in which warm water was running either at the upper (i.e. heating pipes running at culture bag level) or at the lower (i.e. pipes running at a 50cm distance below the bags) substrate level. Heating was applied from 1 December 1994 till the end of April 1995. Shade offered by installing a black net (May 14- September 14). Gas exchange measurements throughout the day and year (November 1994-September 1995) were performed with a portable gas exchange system (Li-6200, LiCor, Inc. Lincoln, NE, USA) using a 1/4 L leaf cuvette.

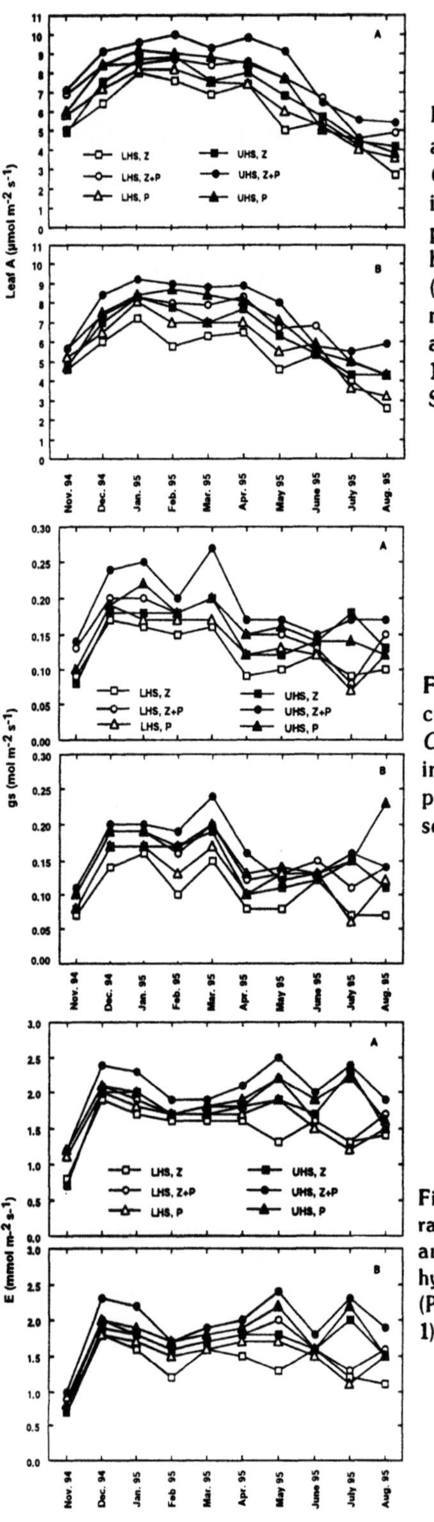

Fig. 1: Seasonal courses of CO_2 assimilation rates of *Gerbera jamesonii* cv. *Cyprus* (A) and *Heart Breaker* (B) cultivated in hydroponic substrates of zeolite (Z), perlite (P) and mixture (Z+P). Two types of heating systems, upper (UHS) and lower (LHS), were applied. Each data point is a mean of 6 measurements. Substrate heating application: 1 December 1994-end April 1995. Black net shade installation: May 14 September 14 1995.

Fig. 2: Seasonal courses of stomatal conductance of *Gerbera jamesonii* cv. *Cyprus* (A) and *Heart Breaker* (B) cultivated in hydroponic substrates of zeolite (Z), perlite (P) and mixture (Z+P). (For details see Fig. 1).

Fig. 3: Seasonal courses of transpiration rates of *Gerbera jamesonii* cv. *Cyprus* (A) and *Heart Breaker* (B) cultivated in hydroponic substrates of zeolite (Z), perlite (P) and mixture (Z+P). (For details see Fig. 1).

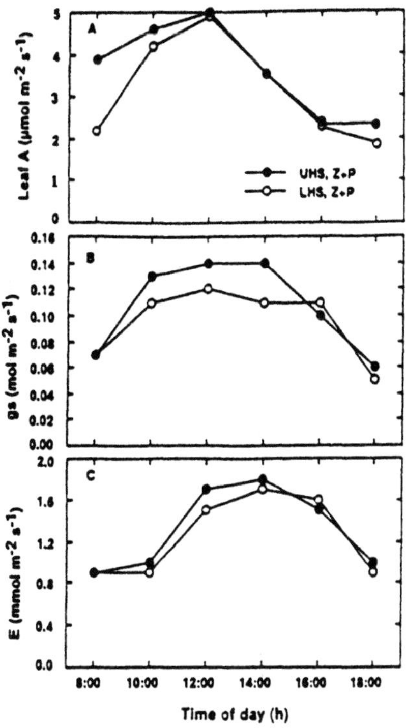

Fig. 4: Diurnal courses of photosynthetic rates (A), stomatal conductance (B) and transpiration rates (C) of *Gerbera jamesonii* cv. *Cyprus* followed on 12 September 1995. Plants were grown hydroponically on Z+P mixture medium under the influence of either a lower (LHS) or an upper (UHS) heating system. Each data point is a mean of 8 measurements. Black net shade installation: May-14 September 14 1995.

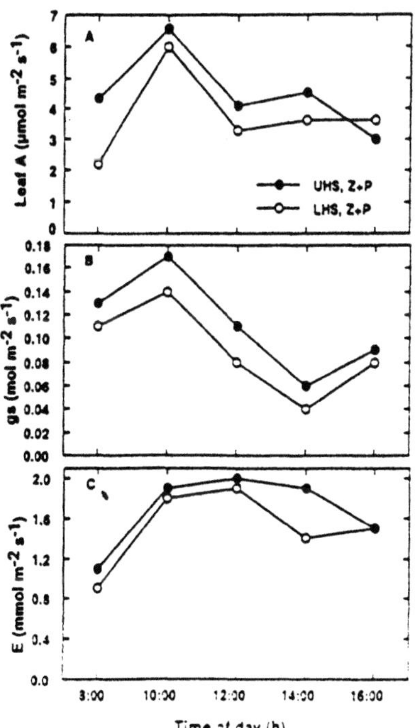

Fig. 5: Diurnal courses of photosynthetic rates (A), stomatal conductance (B) and transpiration rates (C) of *Gerbera jamesonii* cv. *Cyprus* followed on 18 September 1995. (For details see Fig. 4).

3. Results-Discussion

a. The type of substrate along with the type of heating system used in our study, imposed a dominant influence on the net CO_2 assimilation rates of the examined gerbera varieties (Fig 1). This might be attributed to photosynthetic responses to the differential root system development and/or root capacity for water supply, as affected by the thermal conditions at the root zone, the physical properties of substrates in providing water, or both [2]. Lower heating system-induced reduction in photosynthesis can be partly associated with stomatal closure [3].

b. *Cyprus* revealed higher CO_2 assimilation and stomatal conductance than *Heart Breaker,* while transpiration was almost the same (Figs 1, 2, 3). Seasonal course of transpiration was closely related to evaporative demand. Stomatal opening and closing rates are altered directly by air temperature, vapour pressure difference and PAR [4, 5].

c. Based on both, stomatal conductance and transpiration, the zeolite-perlite mixture substrate exerted the most beneficial effect on the soil-plant water system.

d. Overall, measurements of A, gs and E on 12 September 1995 (under black net influence) and on 18 September 1995 (black net removed) indicated that under shade, lower but more stable values of the above parameters were obtained compared to the rather abrupt and high values under ample sun (Figs 4, 5).

e. Diurnal assimilation rate, followed for *Cyprus*, was reduced significanlty between morning and late afternoon possibly as a result of stomatal conductance reduction between these two time periods. The practical outcome with respect to different substrates, heating systems and varieties was that *Cyprus* grown on the mixture medium and subjected to an upper heating system (i.e. with heating pipes running at the culture bag level) exhibited the best physiological responses.

4. References

1. Maloupa, E., Mitsios, I., Martinez, P.F., and Bladenopoulou, S. (1992). Study of substrates used in gerbera soilless culture grown in plastic greenhouse. *Acta Horticulturae* **323**, 139-144.
2. Romero-Aranda, R., and Martinez, P.F. (1993). Diurnal course of gaseous exchange and leaf water potential in relation to substrate heating of *Gerbera*. *Photosynthetica* **29**, 95-101.
3. Moustakas, M., Ouzounidou, G., Eleftheriou, E., and Lannoye, R. (1996). Indirect effects of aluminium stress on the function of photosynthetic apparatus. *Plant Physiology and Biochemistry* **34**, 553-560.
4. Barradas, V.L., Jones, H.G., and Clark, J.A. (1994). Stomatal responses to changing irradiance in *Phaseolus vulgaris* L. *Journal of Experimental Botany* **45**, 931-936.
5. Moustakas, M., Ouzounidou, G., Symeonidis , L., and Karataglis, S. (1997). Field study of the effects of excess copper on wheat photosynthesis and productivity. *Soil Science and Plant Nutrition* **43**, 531-539.

MORPHOLOGICAL CHARACTERISTICS OF *CHRYSANTHEMUM INDICUM*, CULTIVATED UPON DIFFERENT RATES OF NITROGEN FERTILIZATION

V. IVANOVA
Higher Institute of Agriculture, Plovdiv, Bulgaria

1. Introduction

From the studies carried out so far about nutritional requirements of chrysanthemum was observed the best growth and yield when it was supplied at adequate level (1). Maintaining high N level during the early growth period was essential (2). The optimal term for application and optimal N rates for well growth and cut flower production with good quality have been determined (3, 4, 5). Little is known about influence of different rates of fertilization on morphological characteristics of poly-stemmed formatted plants. The aim of present investigation is to study the effect of increasing rates of N nutrition on morphological and decorative characteristics of three-stemmed chrysanthemum plants.

2. Materials and Methods

The investigations were conducted during the period 1993-1995 in unheated plastic greenhouse with two cultivars: small-flowered cv. Refur and large-flowered cv. Dark Westland (D. W.), on experimental field of Higher Institute of Agriculture, Plovdiv, Bulgaria. The plants were planted in the end of July, 49 plants per sc.m. The pinching was done two weeks later for three stem formatting. The six nitrogen rates /0,60,80,100,120 and 140 kg/ha/ were tested. Untreated plants were used as a control. The treatment was done as soil dressing, with NH_4NO_3, every week during the period of active vegetative growth - total - four times. Indices determining the morphological characteristics and ornamental qualities of plant were estimated. The variants analysis was made by the Fishers method. For the similarity in the results averaged values are given.

3. Results and Discussion

Observations indicated that in three-formatting plants with increasing of N rates stem length increased until determined levels /tabl.1/. In cv. D.W. with highest stem length in the first branch differed these, treated with 100 or 120 kg N/ha - respectively 71,0 and 71,6 cm. In cv. Refur the stem length of first, second and third branch was highest in plants, treated with 100 and 120 kg N/ha - respectively 74,4 cm, 66,3 cm and 50,4 cm in var. 4 and 72,9 cm, 71,7 cm and 61,4 cm in var. 5. In cv. D.W. very positive effect on basic stem diameter gives dressing with 100 kg N/ha - 0,57 cm in the first branch; 0,55 cm in the second and 0,54 cm in the third branch. In cv. Refur with biggest basic diameter are plants, dressed with 120 kg N/ha, var.5 - 0,72 cm, 0,64 cm and 0,62 cm respectively in the first , second and third branch. The number of one plant branches in cv. Refur increased with increasing N rates and was highest in variant dressed with 120 kg N/ha. The number of leaves in all variants in cv. D.W. modify in very wide limits - from 9,0 nr of leaves in third branch, treated with 140 kg N/ha, to 16. 0 nr in first branch, treated with 60 kg N/ha (tabl. 2). The plants from cv. Refur kept tendency for increasing of leaves number with increasing of N rates until 120 kg N/ha. In cv. D.W. the date about leaf size shown that in the first branch the differences between the control and the tested variants was unimportant - 20,2 cm^2 in the best var. 2 against 18,8 in the control. In cv. Refur the tendency for increasing the leave size with increasing the N rates have an exception in all three stem - var. 3, treated with 80 kg N /ha. In cv. D.W. with largest flowers were plants in var. 5 - dressing with 120 kg N/ha - 13,29 cm; 13,05 cm and 13,0 cm - respectively for first second and third branch (tabl. 3). In cv. Refur the flower size increased with increasing of N rates, but until determined level. The variant, dressed with highest N rate - var. 6, dressed with 140 kg N/ha - had smaller flower diameter than previous variants - var. 5, dressed with 120 kg N/ha - respectively - 11,7 cm; 10,7 cm and 9,3 cm. With biggest number of flowers in cv. Refur were plants, dressed with higher N rates - var. 5 dressed with 120 kg N/ha - respectively 31,9; 30,4; 29,9 number; var. 6 - dressed with 140 kg N/ha - respectively 31,7; 30,7 and 24,3.. Tendency for decreasing of roots volume with increasing of dressing rates was observed in both cultivars (fig.1). This fact probably due to high salt concentration in roots area. Number of ligules was affected strongly by N rates in cv. D.W. - it was increased with increasing of N nutrition (fig. 2). The difference between the control - untreated plants, and the best variant - var. 6 - dressed with 140 kg N/ha was 291 nb about 90 % more. Analogycal was tendency in case with cv. Refur. Fig. 2 was apparent that N rates over 80 - 100 kg N/ha did not affect on diameter of inside disk in cv. D.W.. In cv. Refur the diameter of inside disk in all treated variants was about the same.

4 Conclusions

The flower stem length, basic diameter of stem in both cvs. - Dark Westland and Refur and number of branches of cv. Refur have the highest value in the variants, dressed with 100 - 120 kg N/ha. In cv. Refur at this level of N nutrition is observed some balance in this indices between three branches of plants. In cv. Dark Westland the total number of leaves per plant are higest in variants, treated with 60-80 kg N/ha. For cv. Refur the value of this character is the most in variants, treated with 100 -120 kg N/ha. The leaves surface in both cvs. have the highest value in the variants, dressed with 120 - 140 kg N/ha. Dressing with different increasing rates of N gives the best results

TABLE 1 **MORPHOLOGICAL INDICES IN CHRYSANTHEMUMS STEM CVS. DARK WESTLAND AND REFUR**

Variant	Stem														
	lenght - cm						diameter - cm						number of branches		
	Dark Westland			Refur			Dark Westland			Refur			Refur		
	I	II	III	I	II	III	I	II	III	I	II	III	I	II	III
1.	65,7	62,0	60,4	21,1	18,8	14,7	0,50	0,44	0,33	0,29	0,28	0,21	12,7	15,4	9,6
2.	66,7	65,6	63,7	33,0	18,5	13,9	0,52	0,49	0,32	0,50	0,39	0,33	7,8	9,0	9,2
3.	63,9	63,3	61,4	42,3	26,1	16,5	0,51	0,48	0,32	0,48	0,43	0,39	14,2	10,0	21,0
4.	71,0	69,9	68,5	74,4	66,3	50,4	0,57	0,55	0,54	0,68	0,54	0,53	22,5	20,0	20,0
5.	71,6	70,8	69,8	72,9	71,7	61,4	0,51	0,47	0,46	0,72	0,64	0,62	26,4	22,4	18,4
6.	66,5	65,6	65,3	43,4	41,6	33,7	0,54	0,47	0,30	0,65	0,36	0,34	17,9	20,5	11,0
GD 5%	4. 33	4. 08	3. 18	4,36	4,66	3,41	0. 02	0. 3	0. 01	0,05	0,3	0,05	4,36	2,64	3,19
GD 1%	4. 9	4. 81	4. 3	5,8	6,25	4,72	0. 03	0. 5	0. 07	0,07	0,5	0,06	5,8	3,53	4,45
GD 0,5%	6. 51	6. 11	6. 51	7,55	8,25	6,51	0. 2	0. 8	0. 1	0.1	0,7	0,09	7,55	4,66	6,09

TABLE 2 **MORPOLOGICAL INDICES IN CHRYSANNTHEMUM' S LEAVES, CVS. DARK WESTLAND AND REFUR**

Variant	Leaves											
	Number						Surface					
	Dark Westland			Refur			Dark Westland			Refur		
	I	II	III	I	II	III	I	II	III	I	II	III
1.	12,7	11,2	10,2	11,5	11,1	9,7	18,8	14,4	10,2	11,5	11,7	9,2
2.	16,0	13,5	10,8	14,9	15,1	11,1	20,2	16,4	12,56	22,7	20,1	20,4
3.	15,3	12,3	10,8	18,9	20,0	16,5	16,4	10,2	14,3	10,8	12,1	13,3
4.	13,9	12,7	12,0	28,7	29,7	22,7	19,5	17,3	16,4	26,1	22,9	20,1
5.	12,7	12,5	12,1	28,8	29,9	21,4	18,8	20,3	19,2	28,0	25,3	22,1
6.	14,2	12,9	9,0	23,5	21,4	17,1	18,1	16,1	11,4	30,5	24,6	19,8
GD 5%	2. 32	3. 11	3. 4	2,83	3,11	3,2	2. 41	2. 18	4. 31	2,83	4,36	5,62
GD 1%	3. 8	4. 2	4. 3	3,76	4,18	4,43	3. 81	3. 42	6. 41	3,76	5,85	7,79
GD 0,5 %	4. 7	5. 1	6. 13	4,9	5,51	6,12	3. 81	6. 43	9. 11	4,90	7,72	10,75

Table 3 **DIAMETER OF CHRYSANTHEMUM'S FLOWERS, CVS. DARK WESTLAND AND REFUR**

Variant	Flowers								
	Diameter - cm						Number		
	Dark Westland			Refur			Refur		
	I	II	III	I	II	III	I	II	III
1.	10,2	9,2	8,9	5,4	4,1	3,9	14,3	19,1	10,9
2.	13,7	10,2	9,5	9,0	8,7	7,2	8,9	12,1	10,0
3.	11,3	11,5	11,8	6,9	8,0	7,9	16,1	10,8	21,7
4.	12,6	12,2	12,1	10,9	9,5	7,8	30,7	29,5	27,8
5.	13,3	13,05	13,0	11,7	10,7	9,3	31,9	30,4	29,9
6.	11,9	10,5	6,3	10,0	9,8	8,6	31,7	30,7	24,3

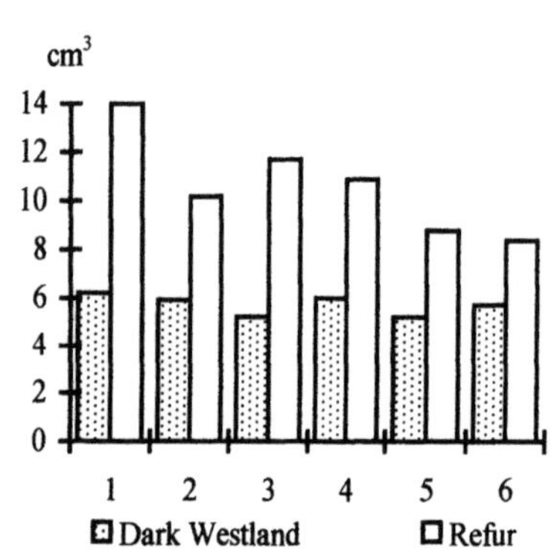

Fig.1 Volume of roots

Fig, 2 Number of ligules and diameter of inside disc

in flower diameter at 120 kg N/ha level in both cvs., as in Dark Westland there is some balance in this character at levels 80 -100-120 kg N/ha. Number of flowers in cv. Refur is the most at 120-140 kg N/ha rates. Number of ligules in cv. Dark Westland is the biggest in the highest rate 140 kg N/ha, while the same indice in cv. Refur the best results are obtained in 120 kg N/ha level. N rates over 80-100 kg N/ha did not affect on diameter of inside disk in cv. Dark Westland. In cv. Refur the affect on diameter of inside disk is the same from all different increasing N rates. The volume of roots in both cultivars is the smallest in variant, treated with highest N rates.

5 References

1.Wolz, S. S. 1979 St. on the nutr. requir. of chr. Proc. Fla. St. Hort. Soc. 69; 352-356.
2. Lunt, O. R., 1958. N and K nutr. of chr. Proc. Am. Soc. Hort. Sci. 72; 487-497.
3. Waters, W. 1962, Eff.of fert. rates and freq. on pot chr. Proc. Fla. St. Hort. S.75;460.
4. Magie, R. O. 1956. G. botrytis control. Poc. Fla. State Hort. Soc. 69; 337-343.
6. Mastalerz, J.1972. N level, grown t and keeping quality of fl. NY St. Flw. Gr. 88; 2.

MORPHOLOGICAL CHARACTERISTICS OF CHRYSANTHEMUM PLANTS, ONE-; TWO-OR THREE STEM FORMATED

V. IVANOVA
Higher Institute of Agriculture, Plovdiv, Bulgaria

1. Introduction

Morphological characteristics of chrysanthemum are at the roof of quality of production. Lunt and Kofranek(1) found influence of nutritional regime of several chrysanthemum varieties on growth and yield. Waters, W.E. (2) and Hobbs, E.L. and Waters, W.E. (3) have determined the optimal light regime for well growth and cut flower production with good quality Little is known concerning mode of plant formatting on chrysanthemum behaviours. The aim of present investigation is to study the effect of one-; two-or three stem plant formatting on vegetative and decorative characteristics of two chrysanthemum cultivars.

2. Materials and methods

The investigations were conducted during the period 1994-1996 in unheated plastic greenhouse with two cultivars - large flowered cv . Dark Westland and small-flowered cv. Refur, on experimental field of Higher Institute of Agriculture, Plovdiv, Bulgaria. The rooted cuttings were set about July 20, pinched two weeks later, and pruned to one-; two-ore three stems per plant four week after setting. One stem formatted plants were used as a control. Induces, determining the ornamental qualities and morphological characteristics of plant were estimated. The variants analysis was made by the Fisher's method. For the similarity in the results averaged values are given.

3. Results and discussion

In the table 1 is shown that with the highest stems in cv. Dark Westland was first branch of three- stem formatted plants - var. 3 - 65,7 cm. With the lowest value of this indice was the second branch of two- stemmed plants - var. 2 - 58, 1 cm. In cv. Refur it make impression the low values of this character in comparison with the control, while in the third branch of three-stemmed plants - var. 3 - reduction was more than 100 %. In basic stem diameter with the lowest value was third branch of three-stemmed plants var. 3 - 0,33 cm in cv. Dark Westland and 0,21 cm in cv. Refur. The date about number of branches per plant shown that first branch of two-stemmed plants (var. 2) had very little number - 4,8; and first and second branches of three-stemmed plants had very high numbers -where the difference in comparison with control were respectively more than 100 % and 150 %. In cv. Dark Westland was observed tendency for increasing the total leaves number per plant with increasing number of branches - in var. 2 - two-stemmed plant- 26,5; in var. 3 - three-stemmed plants - 34. In the same time it makes impression bigger leaves number of separate branches in two-stemmed plants, where in first branch the differences in comparison with the control (var.1) and first branch of three-stemmed plants (var.3) were respectively 1,6 and 0,6. Analoogycal tendency was observed in cv. Refur. In cv. Dark Westland the leaf surface vary from 30,9 cm^2 in the control (var. 1) to 10,2 cm^2 in third branch of three-stemmed plants (var. 3). In cv. Refur- the differences were considerable bigger - 11, 5 cm^2 in first branch of three-stem formatted plants (var. 3); 18,6 cm^2 in the same branch of two-stemmed plants (var. 2) against the control 30,4 cm^2. With the largest flower diameter in cv. Dark Westland was var. 2- first branch of two-stemmed plants - 12,0 cm against the control 11,2 cm (table 3). In cv. Refur influence of mode of formatting on flower diameter almost was not observed. The number of flowers was influenced considerable of mode of plant formatting - it was observed the tendency for increasing of flower number with increasing of branch number. With the biggest number of flowers was second branch of three-stemmed plants (var. 3) - 19,7, as increase against the control was 177,5 %. The date from fig.1

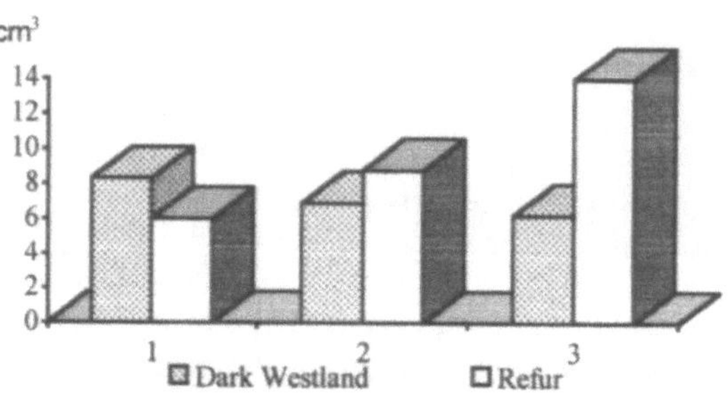

Fig1. Morphological indices in chrysanthemum's roots

was shown underlined tendency for decreasing the roots volume with increasing the branch number in cv. Dark Westland. The differences against the control of two-or three-stemmed plants were respectively 1,4 cm^3 and 2,1 cm^3. The opposed tendency was observed in cv. Refur increase of root volume with increasing of branch number. While in control the root volume was 6,7 cm^3, in two and three-stemmed plant it was respectively 8,8 cm^3 and 14,0 cm^3, i.e. increase was with 31,3 % and 109 %. In cv. Dark Westland it is visible underlined tendency for decreasing of ligules number with increasing of branch number - 505; 376 and 323 respectively for one- ; two- or three stemmed plants (fig 2). In cv. Refur influence of mode of plant formatting on number of ligules did not observed. The same tendency was estimated in diameter of inside disk - in both cultivars this character did not influenced by branch number (fig. 3).

Table 1 **Morphological indices in chrysanthemum 's stem, cus. Dark Westland and Refur**

Variants	Stem														
	Lenght - cm						Diameter - cm						Number of branches		
	Dark Westland			Refur			Dark Westland			Refur			Refur		
	I	II	III	I	II	III	I	II	III	I	II	III	I	II	III
1.	60,6			30,0			0,54			0,40			6,1		
2.	60,2	58,1		22,4	19,2		0,46	0,45		0,40	0,24		4,8	9,9	
3.	65,7	62,0	60,4	21,1	18,8	14,7	0,50	0,44	0,33	0,29	0,28	0,21	12,7	15,4	9,6

Table 2 **Morphological indices in chrysanthemum 's leaves cus. Dark Westland and Refur**

Variants	Leaves											
	Number						Surface					
	Dark Westland			Refur			Dark Westland			Refur		
	I	II	III	I	II	III	I	II	III	I	II	III
1.	11,7			10,7			30,9			30,4		
2.	13,3	13,2		13,8	13,9		23,6	16,9		18,6	18,5	
3.	12,7	11,2	10,2	11,5	11,1	9,7	18,8	14,4	10,2	11,5	11,7	9,2

Table 3 **Morphological indices in chrysanthemum 's flower, cus. Dark Westland and Refur**

Variants	Flower								
	Diameter						Number		
	Dark Westland			Refur			Refur		
	I	II	III	I	II	III	I	II	III
1.	11,2			5,1			7,1		
2.	12,0	9,1		4,1	3,9		5,2	10,2	
3.	10,2	9,2	8,9	5,7	5,0	4,3	14,3	19,7	10,1

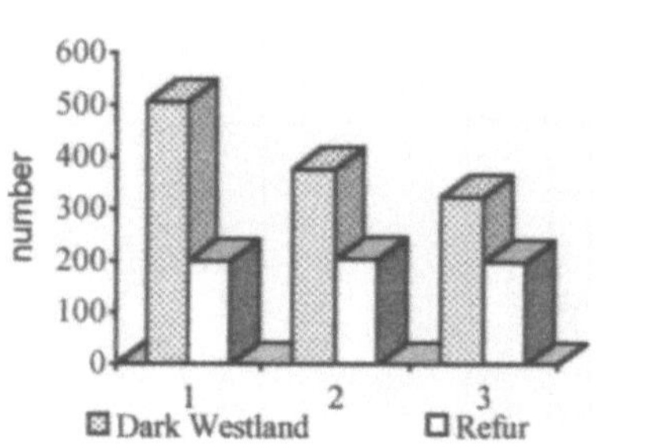

Fig2. Number of ligules

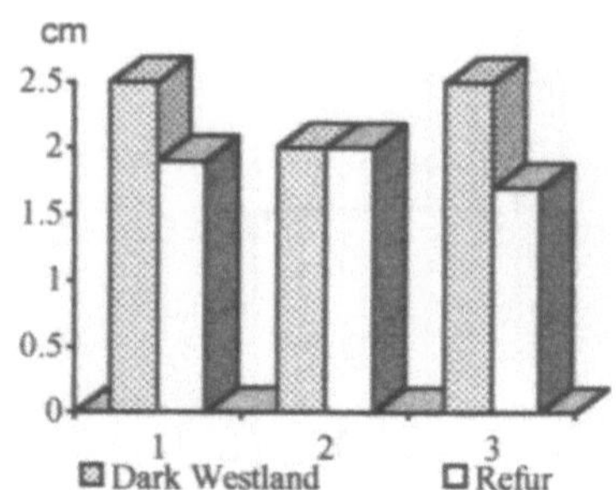

Fig. 3. Diameter of inside disk

4. Conclusion

1. In cv. Dark Wesland with highest stem length were three stem formatted plant. Increasing of brunch number cause negative effect on stem length in cv. Refur. In both cultivars one-stemmed plant had the biggest basic stem diameter. Considerably positive effect had increasing of brunch number on number of brunch per plant in cv. Refur.

2. The total leaves number per plant was highest in three-stemmed plants in both cvs. - Dark Westland and Refur. The leaf size underlined decreased with increasing of number of branch.

3. The flower diameter in both cultivars almost did not influenced by mode of plant formatting. This fact permit obtaining bigger quantity of production with equal quality. In cv. Refur positive effect of increasing of branch number on number of flowers was considerably. Influence of mode of plant formatting on number of ligules in cv. Refur, and on diameter of inside disc in both cultivars were unimportant. Negative effect of increasing of brunch number on number of ligules in cv. Dark Westland was observed.

4. With the highest value in root volume were one-stemmed plants in cv. Dark Westland. while in cv. Refur three-stemmed plant had the biggest root volume.

5. References

Lunt, O. R. and A.M. Kofranek, 1958. Nitrogen and potassium nutrition of chrysanthemums. Proc. Americ. Soc. Hort. Sci. 72; 487-497.

Waters, W. E. 1962, Effects of fertilization rates and frequencies on potted chrysanthemums production. Proc. Fla. State Hort. Soc. 75; 460-463.

Hobbs, E. L. and W. E. Waters, 1964. Influence of N and P on susceptibility of chr. morifolium to B. cinerea Phytopath. 54; 674-676.

PHOTOSYNTHETIC RESPONSE OF SPECIES THAT ARE INCLUDED IN A GROUP OF VERNAL EPHEMERALS

Z. JOVANOVIC, V. SASO-JOVANOVIC, R. POPOVIC, L. DJURDJEVIC, A. MIJOVIC, G. NABIL AND B. KARADZIC
Institute for Biological Research "Sinisa Stankovic", 29 Novembra 142, Belgrade, Yugoslavia

1. Introduction

The term *ephemerals* denotes an ecological group of species that complete their life cycle during a short period, when environmental conditions are suitable for existence and reproduction. The short life cycle succeeding a period of dormant stage represents a powerful adaptation to periodically disturbed environments. The periodic disturbances involve seasons with unfavorable environmental factor such as drought, extreme temperatures, reduced light intensity *etc*.

This group of species may be divided in two subgroups: 1) The true ephemerals are *therophytes* (annual plants) that survive unfavorable season in the form of seeds. 2) The vernal ephemerals (or "*ephemeroids*") are perennial plants, surviving the unfavorable season in the form of modified stems (rhizomes, tubers, bulbs). Their short vegetative phase is well adapted to the specific light climate of temperate, broad-leaved deciduous forests.

The competition intensity among two or more species is proportional to the overlapping of their ecological niches (Pianka, 1983; MacArthur, 1981). The vernal ephemerals may be considered as a guild of species with similar ecological requirements, and therefore they are in strong competition for environmental sources. Difference in pheneology (or temporal stratification) may reduce competition of ecologically similar species. Phenological stages of vernal ephemerals are mutually well separated (Ellenberg, 1986), so the temporal stratification enables their coexistence. The aim of this paper was to detect whether the temporal stratification is linked with physiological adaptations of vernal ephemerals or not.

2. Material and methods

The investigations were performed in a beech-dominated forest on Avala mountain, near Belgrade. The forest is located at 450 m a. s. l., on north-exposed slope. Five species (*Dentaria bulbifera, Arum maculatum, Ficaria verna, Corydalis cava and Scilla bifolia*) have been included in ecophysiological analyzes. For each species we measured both, the net photosynthesis rate, and respiration rate in the dark using the LiCOR 6200 infrared gas analyzer. Measurements were performed during 3 months, in approximately two-weeks intervals. Simultaneously, we sampled plants from five .25 x .25 m^2 quadrates, in order to assess the specific leaf area (leaf area per leaf dry weight). Leaves of sampled species were scanned, and their area was detected using the "AREAMETER" software (Popović *et al.*, 1996). There are numerous models that assume photosynthetic response of plants to irradiance (Drake & Read, 1981; Leverenz, 1995; Kubiske & Pregitzer, 1994). In this article we used a rectangular hyperbola as a function connecting the net photosynthetic rate and irradiance:

$$P = \frac{a(I - c)}{1 + b(I - c)}$$

where P is the net photosynthetic rate, I is irradiance, c is the photon compensation point, whereas a and b represent constants.

3. Results and discussion

Net photosynthesis of investigated plants was analyzed with respect to light environment. In order to avoid effects of transpiratory reduction and down regulation of gas exchange during midday period, the measurements of net photosynthesis were performed in morning hours (between 9 and 11 h). The photosynthetic response of plants to different light intensity is represented in Fig. 1. The shape of fitted curves

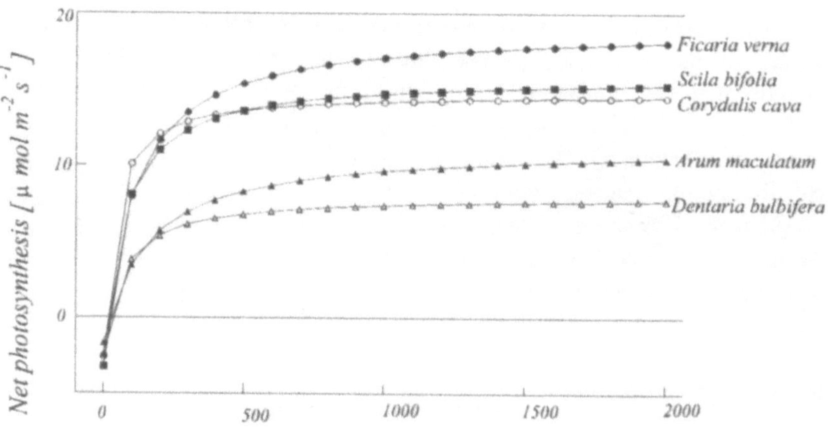

Fig1. Photosynthetic response of analyzed species to irradiance

connecting net photosynthesis and light intensity clearly indicates differentiation of analyzed species. The photosynthetic saturation level of *Dentaria bulbifera and Arum maculatum* is significantly lower than that of *Corydalis cava, Scila bifolia and Ficaria verna.* In order to assess differences among species with respect to the net photosynthesis, we performed ANOVA, including only measurements of net photosynthesis at irradiance above 400 μ mol quanta m^{-2} s^{-1}. Above this light level all species reach the light-saturated photosynthetic maxima. The results of ANOVA indicate that investigated species differ significantly with respect to net photosynthesis ($F_{(4,168)}$ = 10.22; $p<0.01$). The light-saturated photosynthetic rates of *Dentaria bulbifera* (7.2 μ mol m^{-2} s^{-1}), and *Arum maculatum* (9.04 μ mol m^{-2} s^{-1}) were significantly lower than that of other species. *Ficaria verna* had the greatest value of the light-saturated photosynthetic rate (16.74 μ mol m^{-2} s^{-1}).

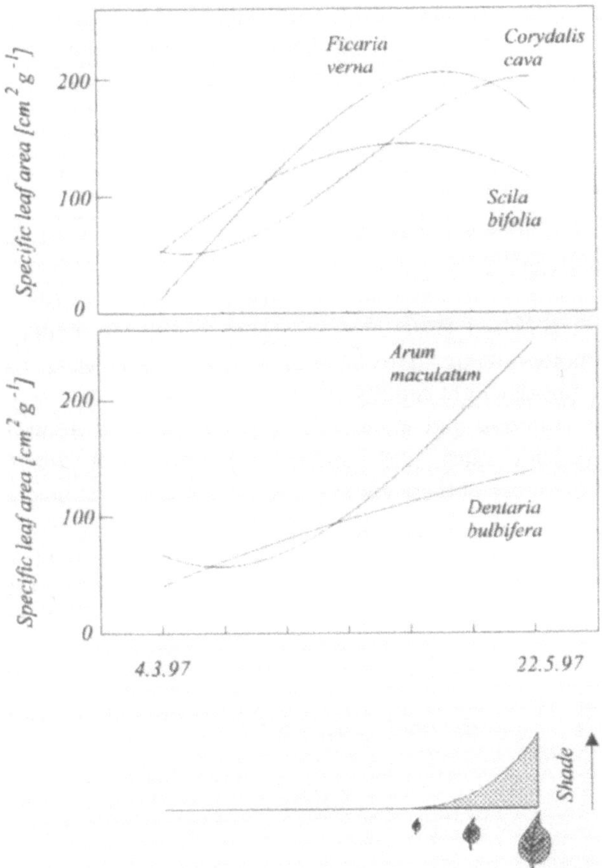

Fig. 2 Seasonal change of Specific Leaf Area (SLA) of investigated species. Leaves flushing of dominant trees and corresponding change of light climate is indicated at lower diagram.

Observed variation of net photosynthetic rates is well correlated to life cycles of analyzed species. *Scila bifolia, Corydalis cava and Ficaria verna* complete their life cycles much faster than *Dentaria bulbifera and Arum maculatum.*

Life cycles of plants is strongly correlated with dynamics of the specific leaf area (SLA). This parameter represents a measure of density of assimilating organs in plants. The primary production of plants depends on leaf density and therefore the SLA may be used as an indicator of functional state of plants.

The change of SLA during investigated period was assessed using the 3-rd degree polynomial regression (Fig. 2). The species with high net photosynthetic rates (*Corydalis cava, Scilla bifolia, Ficaria verna*) reach the peak before the flushing of leaves of dominant trees. After significant increase of shade that is caused by flushing of *Fagus sylvatica* and other trees, SLA of these species decreased. Decrease of SLA was correlated with disappearance of these species.

On the other hand, SLA of *Arum maculatum and Dentaria bulbifera* continuously increased, regardless on increased shade in the forest.

3. Conclusions

It may be concluded that the analyzed species are both, ecologically and physiologically well separated in two groups. The first group (*Corydalis cava, Scilla bifolia, Ficaria verna*) are sun-adapted species which finish their life-cycle immediately after increased shade that is caused by the development of leaves of dominant trees. Photosynthetic characteristics of these plants clearly indicate that they are adapted to the full-day light conditions.

Dentaria bulbifera and *Arum maculatum* have much lower levels of light-saturated photosynthetic rates, and smaller respiration rates. Such characteristics enable prolonged existence of these species, even in conditions of increased shade.

4. References

Drake, B. G. & Read, M. 1981. Carbon dioxide assimilation, photosynthetic efficiency and respiration of a Chesapeake Bay salt marsh. *J. Ecol.* 69, 381-391.

Ellenberg, H. 1986. *Vegetation mitteleuropas mit den Alpen.* Ulmer, Stuttgart.

Popović, R., Jovanović, Z., Šašo-Jovanović, V., Djurdjević, L., Karadžić, B. Fotosintetička efikasnost i primarna produkcija kod nekih prolećnica. *5. Kongres ekologa Jugoslavije, zbornik sažetaka* 79-80. Belgrade.

Kubiske, M. E. & Pregitzer, K. S. 1996. Effects of elevated CO_2 and light availability on the photosynthetic light response of trees of contrasting shade tolerance. *Tree Physiology* 16, 351-358.

Leverentz, J. W. 1995. Shade shoot structure of conifers and the photosynthetic response to light at two CO_2 partial pressures. *Functional ecology* 9, 413-421.

MacArthur, R. H. & Levins, R. 1967. The limiting similarity, convergence and divergence of coexisting species. *Amer. Natut.* 101, 377-385.

Pianka, E. R. 1983. *Evolutionary ecology.* Harper & Row, New York.

DEGRADATION OF PLANT COVER AND SOILS OF KYZYLKYM DESERT

L. KAPUSTINA
Institute of Botany, Laboratory of Geobotany, F. Khodzaeva 32, Tashkent, Uzbekistan, 700143

1. Abstract

The degradation of the desert vegetation and the resulting acceleration of soil degradation and erosion processes contitute major elements of desertification in the Kyzylkym. Mapping and assessment were used to study the relationship among ecological conditions, land use, and degree of antropogenic changes of the natural pastures.

2. Introduction

Land degradation processes- which imply a reduction of the potential productivity of the land (e.g. soil degradation and accelerated erosion, reduction of quantity and diversity of natural vegetation) are widely spread in the Mediterranen region (Hill,Sommer,Mehl and Megier, 1995). The region of our investigation belongs to the East part of Mediterranen. Natural vegetation of the Uzbekistan is 27 million hectares dominated by year round natural pastures. 96% of pastures are in the desert zone. Population is 21,9 million. Livestock of sheep are 38,6 million. 138 sheep are for each 100 hectares of agricultural land. Intensive human activity and utilization of natural resources often bring about the disturbance of the ecological equilibrium which triggers desertification processes. These are the reasons of the degradation of the vegetation and soils of Uzbekistan.

3. Methods

Studing of the modern state vegetation and its mapping will be carried out by generally accepted of geobotanic investigation (The Field Geobotany, 1960, 1972). The map of the current status of desertification was composed accoding to the monograph " Assessment and mapping of Desertification Processes" (Babaev and etc.,1993).
Desertification processes on the 1476600 hectares of South Kyzylkum desert were studied by methods of geobotanic investigation, lage-scale mapping and analysis of "Vegetation maps" which were drawn in the

preceding years. The method used permits to find out conventionally klimaksis association and their antropogenic modification and to differentiate between five desertification types: desertification by degrading vegetation, desertification caused by wind erosion, desertification caused by water erosion, by soil salinisation and by technics.

4. Results

4.1 DESERTIFICATION BY DEGRADING VEGETATION

It is showed by four variants: (Table 1)

On loamy soils the degraded vegetation is 179207 hectares.

On shallow loamy soils of eroded mountains and hilly premountains the most important factors of degraded vegetation are overgrazing and clearing of *Salsola arbuscula and Artemisia turanica.* Under slowly ongoing desertification of hemipetrophytic (dominated by annual halophytes like *Climacoptera lanata, Gamanthus gamocarpus, Girgensohnia oppositiflora, G. diptera* on tops and slopes of chalky deposits, and by *Salsola arbusculiformis, Artemisia turanica* and *A. diffusa* in depressions) and petrophytic variants of plant associations (*Salsola arbusculiformis, Artemisia turanica*) ruderal species like *Peganum harmala* and *Ceratocarpus utricolosus* intrude.

On loamy soils of premountainous plains the main factors of degraded vegetation are overgrazing and clearing *Artemisia diffusa* and *Salsola orientalis*. The succession changes according to the pelitophytic row wich probably starts with a society of *Artemisia diffusa, S. orientalis, Carex physodes and Poa bulbosa,* into wich initiatially *Peganum harmala* intrudes (weak desertification). Later a strong desertification is characterized by societies dominated by *Peganum harmala* and *Artemisia diffusa* or *Peganum harmala* alone.

On the rather sandy grey - brown soils the vegetation is degraded on 337323 hectares caused by overgrazing and clearing of *Artemisia diffusa, Convolvulus hamadae, Astragalus villosissimus, Salsola arbuscula* and *Haloxylon aphyllum.* Trees and shrubs are used for fuels. 7000 half-shrubs are cut for waving silkworm (*Bombyx mori*) cocoons per year. Here the desertification starts with psammophytic societes intruded by *Eremopyrum orientale, E .distans, Vexibia pachycarpa* and *Peganum harmala.* In the last stage of anthropic sucessions, societies of Arthratherum pennata, Iris songorica, Peganum harmala and Eremopyrum orientale are dominating in degraded, sandy places.

On desert sands the degraded vegetation comprises 283359 hectares. A low degree of desertification is most often characterised by the introduced weeds *Arthratherum pennata, Ammothamnus lehmanii* and *Peganum harmala* in psammophytic societies which are dominating *Haloxylon persicum, Calligonum setosum, C. ellatum, Carex physodes, Artemisia diffusa, Salsola arbuscula* and *Astragalus paucijugus*. In places with strong desertification on less hardened, furrowed, and lacunose sands as well as sand plains societies of dispersed *Ammodendron conollyi, Arthratherum pennata, A.karelinii, Ammothamnus lehmanii*, with *Agriophyllum latifolium, Peganum harmala* and *Eremopyrum orientale* are dominating.

Tabl 1. Degree and scale of the desertification (hectares)

Class of the desertification	Types of the desertification: Degradation of vegetation, grey-brown soils, loam	sandified	desert sand soil	solinisation of soil	water erosion	wind erosion	tecnogenic desertification	Forest planting
Slight	88 197	78 223	172400	52 413	43 960	332468	2 960	
Moderate	53 860	60 600	62 370	23 470	7 090	169413	6 590	14 800
Severe	37 150	198500	48 586	19 290	19 180	268236	4 099	
In all	179207	337323	283359	95 173	70 230	770017	13 649	14 800

4.2 DESERTIFICATION CAUSED BY WIND EROSION

770017 hectares are affected by wind erosion. There is close correlation between the two types of desertification caused by degrading vegetation and by wind. The more the vegetation is destroed the more places of deflation rise. The wind erosion leads to a reduced thickness of the loamy soils and uncovers the gypsum layer. There are dominating annual halophytes like *Ceratocarpus utricolosus, Girgensohnia oppositiflora* and *G. diptera* on degraded loamy soils. Deflation of the desert sands and sandy grey-brown soils is causing sand dunes. More then 50% this area belong to strongly desertificated places around settlements and water wells.

4.3 DESERTIFICATION CAUSED BY WATER EROSION

70230 hectares are undergoing by water erosion.

On hilly premountains and slopes of the desert mountains dominate sediments which can easily be washed out. They underline water erosion also if precipitation is low. Such types of desertification correlates with desertification caused by wind erosion and degradation of hemipetrophytic and petrophytic variants of the plants societies of the Kuldzhuktau and

Kokchatay mountains and their hilly premountains. Desertification caused by water erosion is visible too on the dry waterways of the old rivers which fills up with water in spring time only.

4.4 SOIL SALINISATION

It caused by artificial lake Shorkyl and by water and by wind erosions is on the 95173 hectares.

5. Conclusion

The given analysis of desertification processes on the territory of the South Kyzylkym detected 54% of degraded vegetation (23% is on the sandy grey-brown soils, 19% is on desert sands and 12% is loamy soils). Wind erosion causes aeolic accumulative processes on the 52% of territory, water erosion is 5%, soil salonisation- 6% and technogenic desertification - 1%. Man-made reservoirs are reason for salinisation of soils and degradation of vegetation on 19560 hectars. Largest changes in the plant cover took place under the influence of heavy grazing. Çutting of the half-shrubs for silkworm cocoons is the reason for degradation of vegetation and wind erosion on the 38380 hectares.

6. References

1. *The Field Geobotany* (1960, 1972), Volume 2-4, Moskow- Leningrad.
2. Babaev A.G., Kharin N.G., Orlovsky N.S. (1993) *Assessment and Mapping of Desertification processes a metodological guide.* ESCAP/ UNEP Research and Training Centre of desertification control. Ashgabad, p. 71.
3. Hill, J., Sommer, S., Mehl, W. and Megier, J. (1996). Use of earth observation satellite data for land degragation mapping and monitoring in Mediterranean ecosystems: towards a satellite-observatory. Enviromental Monitoring and Assessment. Vol. 37. N 1-3, S. 143-158

DEVELOPMENT, STRUCTURE AND PRODUCTIVITY OF ARTIFICIALLY ESTABLISHED LIME STANDS ON BEECH SITE

M. KRSTIĆ[1], P. ALEKSIĆ[2] and A. STAMENKOVIĆ[2]
[1] *Faculty of Forestry, 11030 Belgrade, Kneza Višeslava 1, Yugoslavia*
[2] *JP »Srbijašume«, 11030 Belgrade, Djure Djakovića 31, Yugoslavia*

1. Introduction

Limes (Tilia spp.) do not cover large forest areas in Serbia, so they are not classified as economically valuable forest tree species. Consequently, limes have not been paid much attention to and they have been neglected from economic aspect, but they are valuable and widely used as ornamental and park species.

Large-leaved lime (*Tilia grandifilia* Ehrh.), as an accessory species, occurs as individual trees in forests of lowland beech - *Fagetum montanum serbicum* Rud. (Černjavski and Jovanović, 1953). The sites are usually below 500 m asl., on cooler and moister exposures. Being a mesophillous species, lime was not considered suitable for bareland afforestation and it was not applied in such works, despite its rather fast growth and beneficial effects in soil reclamation (Radulović, 1953). Therefore, the aim of this paper is to emphasize the significant role of large-leaved lime in the afforestation of bare land on beech sites.

2. Site conditions and Method

The research was carried out in lime stands (*Tilia grandifolia*) established by the afforestation of clearings in a beech site (*Fagetum submontanum moesiacum s.l.*) on the mountain Jastrebac in Serbia. The afforestation was carried out on terraces and bench terraces 1 m wide, spaced 5 m, with plant spacing 0.4 x 0.4 m, for the purpose of erosion control.

Altitude is 400 m, exposure northeast and slope up to 20 degrees.

Parent rock is gneiss and the soil is shallow to deep acid brown.

Climatic conditions at this elevation characterised by: mean annual air temepratture 10.4° C, average annual rainfall 572 mm and relative humiditty 77 %. According to Thornthweite's classification, the climate is moist subhumid – C_2.

The usual method applied in permanent sample plots has been implemented.

3. Results of research

The main data of the researched stand productivity have been shown in Table 1.

Table 1. The main data of the researched stand productivity

SP	Nt	Ba	V	Iv	Dm	Hm	Dmd	Hmd	Vm
	pcs/ha	m^2/ha	m^3/ha	m^3/ha	cm	m	cm	m	m^3
I	3728	62.0	673	14.9	14.5	19.7	21.8	21.9	0.18
II	2832	45.7	420	8.3	14.3	16.2	19.6	16.9	0.15
III	2076	31.4	247	6.1	13.9	13.3	19.1	14.5	0.12

SP – Sample plots Nt – Number of trees Ba – Basal area
V - Volume Iv – Volume increment Dm – Mean diameter
Hm – Mean height Dmd – Mean diam. of domin. trees
Hmd – Mean height of domin. Trees Vm - Mean volume (V/Nt)

The main data of tree increment have been presented in Tab. 2. and Fig. 1-4.

Table 2. The main data of tree increment

SP	Dominant trees aged 42							
	Dmd		Id max		Dmh		Ih max	
	Cm	%	mm	%	M	%	M	%
I	17.3	100	6.9	100	21.3	100	1.08	100
II	15.1	87.3	6.5	94.2	17.3	81.7	0.70	64.8
III	13.9	80.3	6.7	97.1	15.1	71.0	0.57	52.8

Id – Diameter increment Ih – Height increment

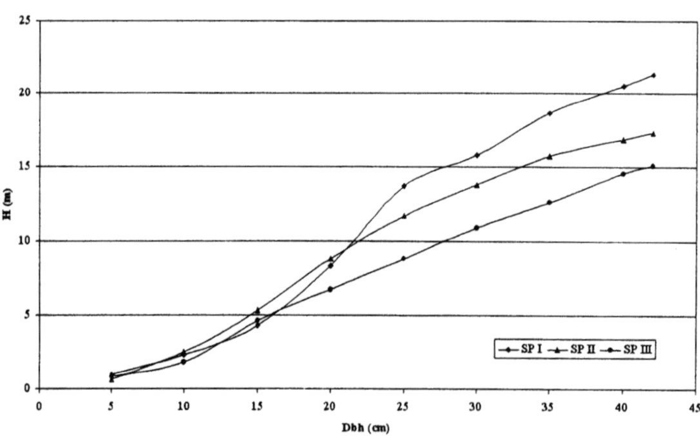

Figure 1. Development of dominant tree heights

The development of dominant trees (20 % largest dimaeters in the stand) aged 42, shows that they attain the height of 15.1-21.3 m, and dbh 18.0-21.8 cm. The trees on the best sites attain up to 20 % larger diameter and 30 % larger height than trees on the inferior conditions. Current height increment culminates about the age of 20, with the maximum value 1.08 m (in the best conditions), current diameter increment culminates between the age of 15 and 20, maximum value 6.9 mm.

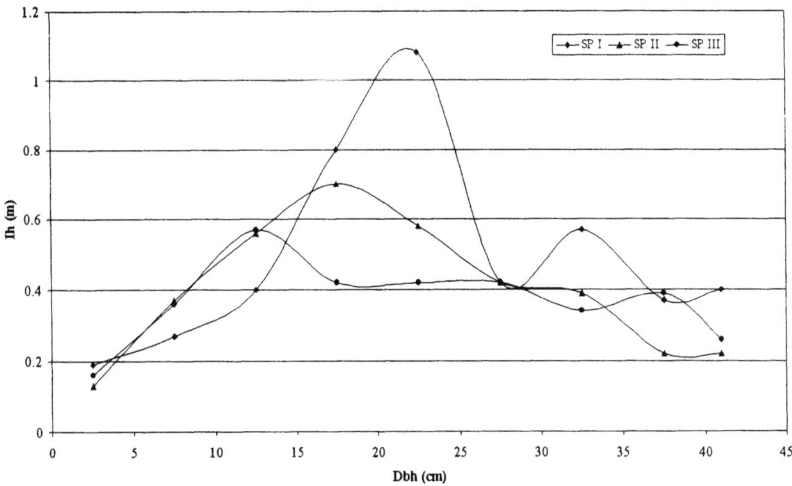

Figure 2. Current height increnment

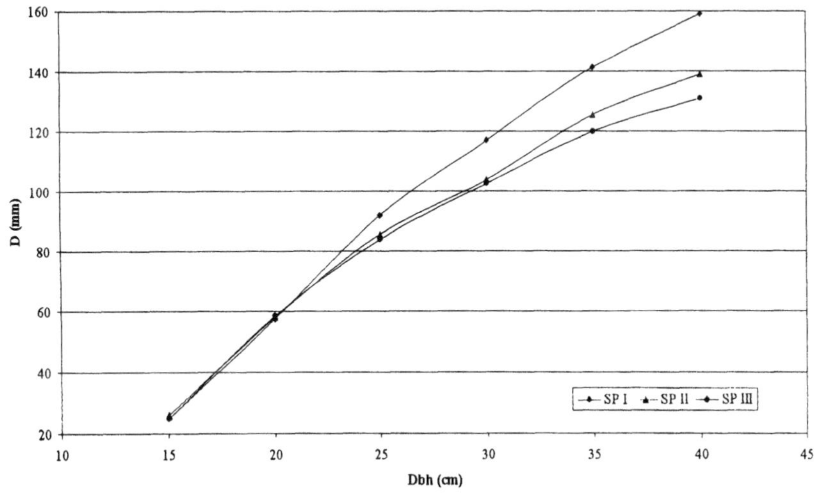

Figure 3. Development of dominant tree diameters

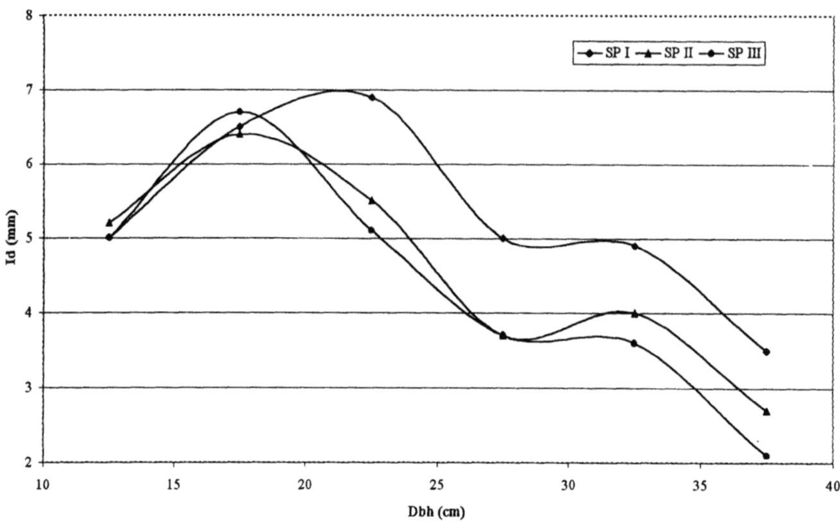

Figure 4. Current diameter increment

4. Conclusion

This research proves that the aim of afforestation has been multiply achieved: soil erosion has been controlled, timber production is considerable. The conclusions of this research are significant for lime stand management and utilization of lime in bareland afforestation.

5. References

Bobinac, M. (1996) Proučavanje uzgojnih potreba u sastojinama lipe na Fruškoj gori. *Šumarstvo* **1-2**, Beograd, 36-38.

Černjavski, P. and Jovanović, B. (1950) Šumska staništa i odgovarajuća dendroflora u Srbiji. *Posebno izdanje SAN-u,* knj. **CLIX**, Beograd, 1-48.

Jovanović, B. (1985) *Dendrologija.* Univerzitet u Beogradu, Beograd.

Jovanović, M. (1959) Potrebno je naučno istražiti lipu. *Šumarstvo* **6**, Beograd.

Radulović, S. (1953) Lipe i njihov značaj za obnovu šuma kod nas. *Šumarstvo* **2**, Beograd, 108-115.

PRELIMINARY STUDY OF METAL CONCENTRATION IN *ULVA RIGIDA* C. AGARDH FROM THERMAIKOS GULF, GREECE

P. MALEA AND S. HARITONIDIS
University of Thessaloniki, School of Biology, Department of Botany, P.O. Box 109, GR 54006 Thessaloniki, Greece

1. Abstract

The relative abundance of metals in *Ulva rigida* of Thermaikos Gulf decreased in the order: Fe>Zn>Pb>Cu>Cd. The seasonal variation was significant for Cd and Zn The mean values increased from spring to fall becoming maximum in October or June (only for Cu). These variations were discussed in relation to growth dynamics, tissue age etc. The concentrations of all metals in the algal were distributed uniformly among the 5 stations.

2. Introduction

Algae have been considered as a valuable indicator for assessment of heavy metal in coastal areas, because of their accumulation capacity (e.g. Försberg A *et al.*, 1988).

There is little information on metal content in *Ulva rigida, e.g.* Güven K.C. *et al.*, 1992). In Thermaikos Gulf the accumulation of Co, Ni and Cr in *U. rigida* was determined (Haritonidis S. & Malea P., 1995). The cosmopolitan green algal *U. rigida,* which is widespread along the Greek shores and is common along the Mediterranean Sea, is usual in coasts polluted by domestic sewage and high nutrient levels (e.g. Ho Y.B., 1987). Metals such as Fe, Cu, Zn, Pb and Cd are associated with sewage and industrial influents (Marcantonatos G., 1990). *Ulva lactuca* has widely been used as heavy metal indicator (such as Fe, Mn, Pb, Cu, Zn) (e.g. Ho Y.B., 1990).

This paper reports on local ans seasonal variation of Fe, Pb, Zn,Cu and Cd in *U. rigida,* from Thermaikos Gulf (Greece).

3. Material and Method

The study was carried out in Thermaikos Gulf (Greece) (Fig. 1). Untreated sewage from the city of Thessaloniki is discharged into the Gulf (Symbols t_1, t_2, t_3 of Fig. 1). From the industrial zone treated or partially treated effluents are also released (symbol I of Fig. 1). Seasonal samplings of *Ulva rigida* were carried out at stations Micro Embolo, Nea Krini, Biamyl, Perea and Agia Triada (Fig. 1). Three subsamples were treated through method descriped by Malea P. & Haritonidis S. (1995). The metal content was determined by AAS using a HGA 72-Perkin Elmer Atomizer. Non parametric statistical analysis was used.

4. Results

Mean metal concentrations (±SE) in *U. rigida* decreased in the order: Fe>Zn>Pb>Cu>Cd. There was significant positive correlation (Spearman coefficient) between Pb-Cu, Fe-Cu ($p<0.05$) and Cd-Pb ($p<0.01$) in *U. rigida*.

Kruscal-Wallis one-way analysis of variance demonstrated that the seasonal variations were significant for Cd and Zn ($p<0.01$). The seasonal variation of all metals followed the same pattern. Mean values (±SE) increased from spring to fall becoming maximum in October or June (only for Cu); Pb, Zn and Cu displayed their minimum values in spring, whereas Cd and Fe a little bit earlier (in January) (Fig. 2). Kruscal-Wallis one-way analysis demostrated that concentrations (±SE) of each metal in *U. rigida* showed no significant difference within 5 locations ($p>0.05$) (Fig. 3). Lead, Cu, Zn and Fe concentrations were higher in Biamyl station or coming second in the order whereas Cd content was minimum in Micro Emvolo and maximum in Perea and Nea Krini (Fig. 3).

5. Discussion

The high levels of some metals in the plant (Fe, Zn, Pb) reflect the high bioavialability of the metals in the environment (unpublished data) and the capacity of the alga to take them up (Karez C.S. *et al.*, 1994).

The seasonal variation of Cd, Pd, Cu, Zn and Fe content in *U. rigida* was associated with growth dynamics. The mean concentrations were higher in fall when biomass was at its minimum (Haritonidis S., 1978) and lower in spring (except for Cd and Fe, where their minima presented earlier), when biomass exhibited its maximun values (Haritonidis S., 1978). Increase in the biomass of *U.rigida* resulted in the decrease of metals per mass unit due to their dilution (e.g. Haritonidis S., & Malea P., 1995). Concentrations of Cd, Pb, Cu, and Fe were low in winter, after the appearence of the young individual of *U. rigida* (in fall) (Haritonidis S., 1978) and generally, increased with plant age. With an increase in tissue age, irreversible adsorption of metals occur and more binding sites are formed the older parts have higher dry weight and also in the older parts some contamination with fine particles occur (Bryan G.M., & Hummerstone L.G., 1973). Increases in Fe, Cd, Zn with age have been observed in several macrophytes (e.g. Lyngby J. E. & Brix H., 1989; Malea P., 1994). Similar seasonal variation with us showed Cd and Zn concentrations in *U. lactuca* (Catsiki V.A. & Papathanassiou E., 1993), Cd, Pb and Fe concentrations in other macroalgae (Shiber J.G., 1980) and Ni concentrations in *U. rigida* in the study area (Haritonidis S. & Malea P., 1995).

The uniform distribution of all metals at the 5 stations in *U. rigida* can be attributed to the uniform distribution of these metal in the environment (unpublished data). This means that metals are transported into the system from the principal metal sources and remain in available form in seawater (Karez C.S. *et al.*, 1994). The relatively high levels of Pb, Cu, Zn and Fe in *U. rigida* in Biamyl station seems to follow the high levels of these metals in the sediment (unpublished data). This is probably due to the transport of industrial sewage towards this station by sea currents (Ganoulis J. & Krestenitis J., 1982) or the sediment of Biamyl station presents by nature high levels of these metals.

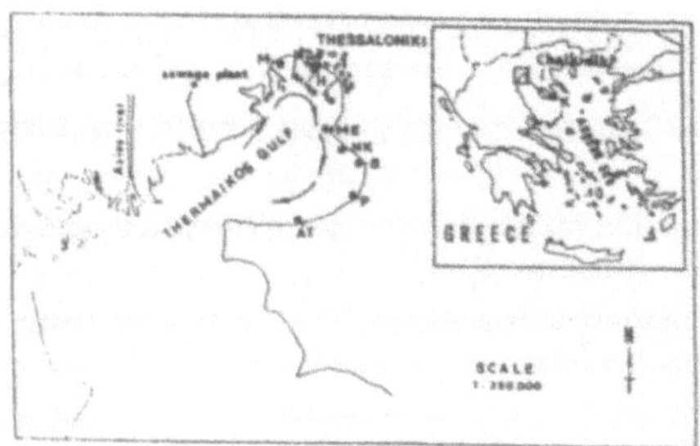

Fig. 1. Sampling stations in Thermaikos Gulf: M=Mamidakis; H=Harbour; ME=Micro Emvolo; NK=Nea Krini; B=Biamyl; P=Perea; AT=Agia Triada.

Fig. 2. Seasonal variation of metal concentrations in *Ulva rigida* (μg g^{-1} dry weight). Each dot represents the mean of the values recorded in a season at all sampling stations; bars represent the standard error.

Fig. 3. Local distribution of metal concentrations in *Ulva rigida* (μg g^{-1} dry weight) Each column represents the mean of the values recorded at a station during all seasonal samplings; bars represent the standart error.

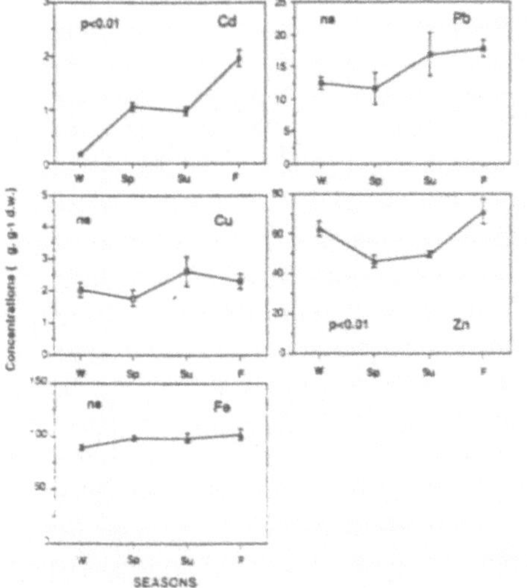

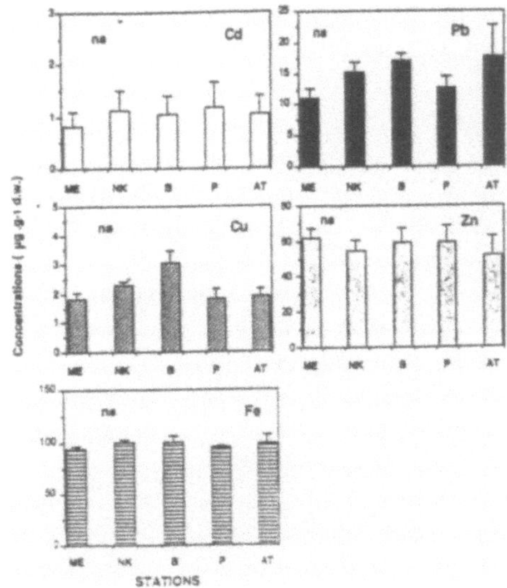

6. References

Bryan, G.W. & Hummerstone, L.G. **(1973).** Brown seaweed as an indicator of heavy metals in estuaries in south-west England. *J. Mar. Biol. Ass.* U.K. **53,** 705-720.

Catsiki, V.A. & Papathanassiou, E. **(1993).** The use of the chrorophyte *Ulva lactuca* (1) as indicator organism of metal pollution. *Proc. of the Cost-48 Symposium of Sub. Group III. Macroalgae, Eutrophication and Trace Metal Cycling in Estuaries and Lagoons,* Thessaloniki, 24-26 Sep., pp 93-105.

Försberg, A., Söderlung, S., Frand, A., Petersson, L.R & Pedresén, M. **(1988).** Studies on metal content in the brown seaweed *Fucus vesiculosus,* from the Archipelago of Stockholm. *Environ. Pollut.* **49,** 245-63.

Ganoulis, J. & Krestenitis, J. **(1982).** Choix du site optimum d' evacuàtion des eaux usées dans une région côtiére. *Societé Hydrotechnique de France,* 17 Journeés de J' Hydraulique.

Güven, K.C., Topcuogvlu, S., Kut, D., Esen, N., Erenturk, N., Saggi, N., Cevher, E., Guvener, B. & Ozturk, B. **(1992).** Metal uptake by Black Sea algae. *Bot. Mar.,* **35,** 337-40.

Haritonidis, S. **(1978).** Contribution to the research of marine machrophyceae (Chloro-, Phaeo- and Rhodophyceae) of Thermaikos Gulf. Ph D Thesis, Thessaloniki, Greece, 175 pp (in Greek).

Haritonidis, S. & Malea, P. **(1995).** Seasonal and local variation of Cr, Ni, and Co concentrations in *Ulva rigida* C. Agardh and *Enteromorpha linza* (Linnaeus) from Thermaikos Gulf. *Environ. Pollut.,* **89(3),** 319-27.

Ho, Y. B. **(1987).** (Chlorophyta, Ulvales) in Hong Kong intertidal waters - its nitrogen and phosporus contents and its use as a bioindicator of eutrophication Asian. *Mar. Biol.* **4,** 97-102.

Ho, Y.B. **(1990).** *Ulva lactuca* as bioindicator of metal contamination in intertidal waters in Hong Kong. *Hydrobiol.,* **203,** 73-81.

Karez, C.S., Magalhaes, V.F., Pfeiffer, W.C., Amado Filho G.M. **(1994).** Trace metal accumulation by algae in Sepetiba Bay, Brazil. *Environ. Pollut.* utog **83,** 351.

Lyngby, J.E. & Brix, H. **(1989).** Heavy metals in eelgrass (*Zostera marina L.*) during growth and decomposition. *Hydrobiol.,* **176/177,** 189-96.

Marcantonatos, G. **(1990).** Treatment and disposal of aquatic wastes. Sewage, Industrial Effuents, Animal Residue, Athens, 738 pp.

Malea, P. **(1994).** Seasonal variation and local distribution of metals in the seagrass *Halophila stipulacea* (Forsk.) Aschers in the Antikyra Gulf, Greece. *Environ. Pollut.* **85,** 77-85.

Shiber, J.G. **(1980).** Trace metals with seasonal considerations in coastal algae and mollusis form Beirut, Lebanon. *Hydrobiol.,* **69(1-2),** 147-62.

MORPHO-ANATOMICAL ALTERATIONS IN LEAVES OF THE GENUS *ACER* L. SPECIES IN URBAN HABITATS

M. MITROVIC[1], P. PAVLOVIC[1], R. POPOVIC[1], Z. DAJIC[2], R. MATIC[1]
[1]Institute for Biological Research "Sinisa Stankovic",
29 Novembra 142, 11060 Belgrade, Yugoslavia
[2]Faculty of Agriculture
Nemanjina 6, Zemun, 11080 Belgrade, Yugoslavia

1. Introduction

In big trade and industry urban centres, like Belgrade, a whole complex of unfavourable ecological conditions influences the vegetation. The overall negative effect of the altered urban environment conditions, including pollution, causes the destructive changes on plants and shortening of the vegetation period. Atmospheric dust and precipitation containing heavy metals lead to a significant accumulation of toxic materials on the leaves' surfaces (Smith 1975) which, going along with the gas pollutants, contribute to a general retardation of physiological processes in plants. Injuries of peripheral protection structures enable penetration of pollutants and intoxication of the internal tissues (Turunen *et al.* 1997). The leaves of woody plants are able, owing to their relatively large surface and the active exchange of gases, to improve the quality of air in cities by absorbing and bonding pollutants (Freer-Smith *et al.* 1996). They also can be useful as the atmospheric pollution indicators (Markert 1993). Environmental conditions, age of the trees, and the specific features of the species themselves, are some of the elements that determine the character of pollutant uptake as well as the symptoms of injuries.

2. Materials and methods

The examined species belong to the group of deciduous trees widely spread in Europe. They are found mostly in mixed forests, and are seen quite often in city parks and avenues. These species are mesomorphic, adapted to the moderate environmental conditions. The examinations of the morphological and anatomic characteristics and visible symptoms of injuries of leaves of the maple *Acer pseudoplatanus* L., and the common maple *Acer campestre* L. lasted from 1989 to 1995. The examinations included the observation of crowns, branches and leaves, as well as the leaf cross-section analysis in several city parks and control habitats on Fruška Gora and Maljen. The permanent preparations of leaf cross-sections, of thickness up to 20μm, were obtained through a standard procedure (Blazencic 1988), and analyzed under a light microscope.

3. Results

As early as in July, there can be observed a necrosis of marginal leaf parts on the *A. pseudoplatanus* trees in city parks. Later on, as the vegetation season progresses, the alterations become more noticeable taking the appearance of intravenous chlorosis and dry, necrotic patches. Young maple leaves, which appear later during the summer, are chlorotic, and on certain places there appear some black spots, probably because of action of the pathogens. On leaves of the species *A. campestre* L. during the season there appear chlorotic patches along the nerves evolving into necrotic injuries which spread gradually all over the surface of the leaves. The degree of the morphological alterations in both species is higher in the exposed sun leaves. There have not been noticed any morpho-anatomic alterations of maple leaves from the control habitats.

The leaves of the examined species of genus *Acer* L. are dorsiventral. The mesophyll generally includes one layer of palisade parenchyma cells and spongy tissue permeated by vast intercellular spaces. Epidermis of the leaves is partly or entirely composed of mucilaginous cells. Cuticle is thin and, most often, wrinkled. Lower epidermis of the leaves consists mostly of the papillous cells (in this particular case, in the species *A. pseudoplatanus* L.) (Fahn 1990). Stomata are on the lower epidermis of the leaves, tiny and of ranunculoid type. In the mesophyll there are crystals easily visible, small or very big, single as incrustations. The hairs are monocellular, uniseriate, with a tendency of branching (Metcalfe 1950). The described structure characterizes the leaves of both species from the control habitats.

Urban leaves of species *A. pseudoplatanus* are covered by atmospheric dust. The structure of the upper epidermis peripheral protection is damaged leaving a free passage for the gas pollutants and pathogens causing the mesophyll parenchyma necrotization. The most serious injuries are recognized in the palisade parenchyma (Figure 1.).

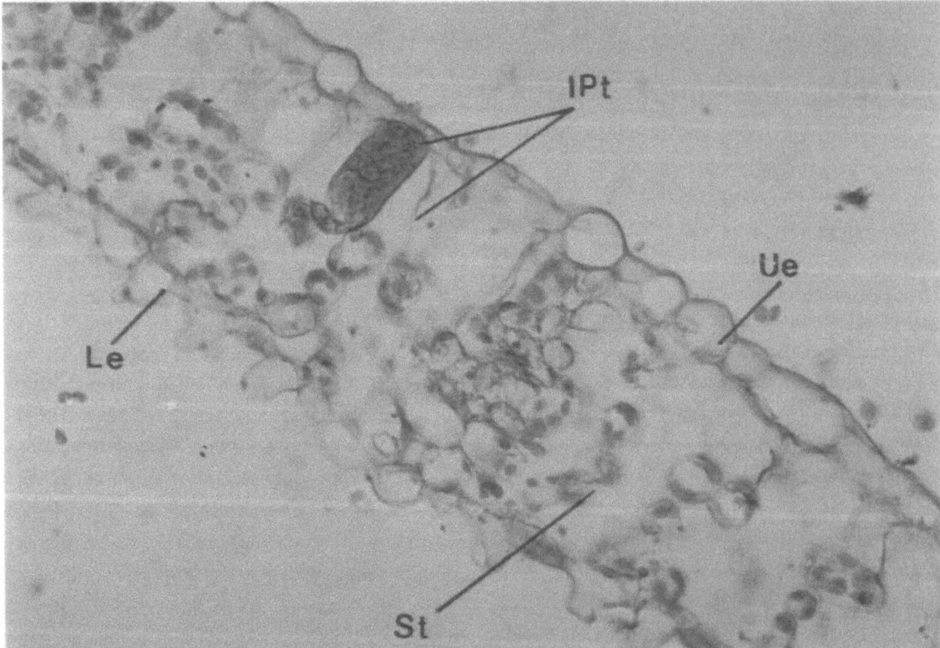

Figure 1. Cross-section of urban leaf of *A. pseudoplatanus* L. species (Ue-upper epidermis, Le-lower epidermis, IPt-injuried palisade tissue, St-spongy tissue)

At the species *A. campestre*, the urban leaves are considerably thicker in comparison with the rural ones, and covered with a layer of dust as well. Unlike the other maple species, in this case there occur the lower epidermis injuries. Cuticle is eroded, and thus the dust particles are enabled to penetrate through it, especially into the stomata guard cells. These spongy tissue injuries, as results of the described changes, indicate that this maple species absorb the pollutants through its stomata (Figure 2.).

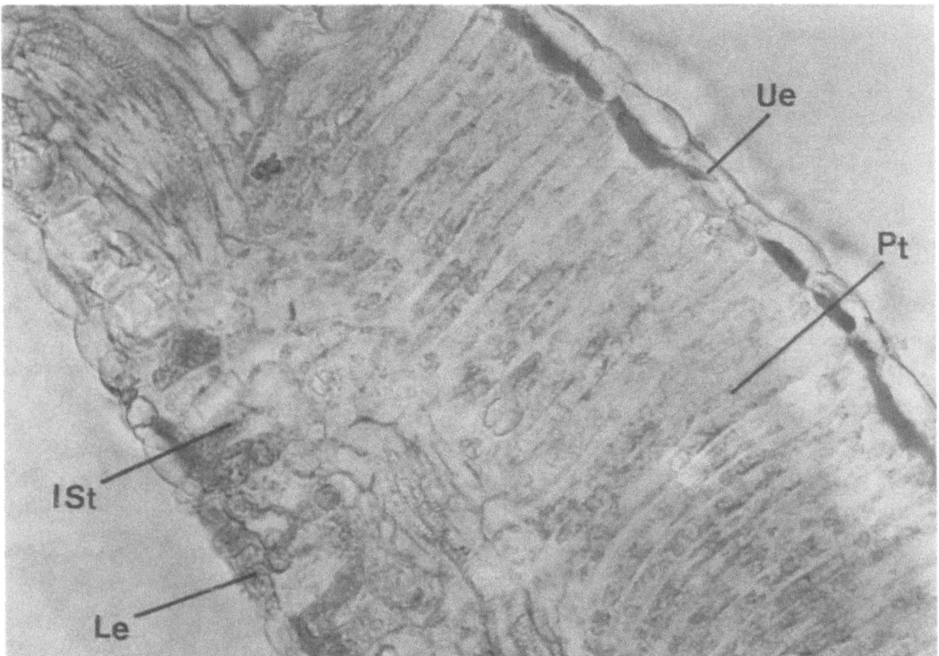

Figure 2. Cross-section of urban leaf of *A. campestre* L. species (Ue-upper epidermis, Le-lower epidermis, Pt-palisade tissue, ISt-injuried spongy tissue)

4. Discussion

Morpho-anatomical structure of the urban leaves of the examined species of the genus *Acer* is characterized by xeric adaptations. The urban leaves are heavier, smaller, thicker and of a brighter colour than the rural ones. Thickening especially occurs on the palisade tissue level in both species (Table 1.), and then on the peripheral protection structure level (thicker epidermis and cuticle). In addition to it, the tissue gets more compressed, that is the intercellular space gets smaller which, according to Elephtheriou (1987), means the adaptation to urban environment stress. Structural adaptations are more prominent in the species *A. campestre*.

Maple trees are exposed to injuries in city avenues and parks. The decline of a maple is easier to recognize by symptoms then by particular causes (Burns and Manion 1984). Decline is provoked by a complex action of a series of detrimental agents: land and air pollution, deficiency of nutrients, drought, radiation, pathogens. The symptoms of the injuries to tree branches and leaves are reflections of the long-lasting stress effects on the above habitats.

The authors (Smith and Dorchinger 1976) think that the examined tree species belong to the group of the species that are tolerant to action of the most common pollutants. On the basis of the former examinations, we would classify them as the intermediate species. Tolerance of plants to industrial pollutants is generally associated to the species that are eurivalent for the extreme environmental conditions. However, the sensitivity is linked with the specific complex of morpho-anatomic and physiological injuries.

5. References

Blazencic, J. (1988) Praktikum iz Anatomije biljaka. Naucna knjiga. Beograd. pp. 200.

Eleftheriou, E. (1987) A comparative study of the leaf anatomy of olive trees growing in the city and the country. *Environmental and Experimental Botany* 27, 105-117.

Freer-Smith, P.H., Holloway, S., Goodman, A. (1997) The uptake of particles by an urban woodland: site description and particulate composition. *Environmental Pollution* 95, 27-35.

Markert, B. (1993) Plants as Biomonitors. Ed. Markert, B. VCH. Weinheim, New York, Basel, Cambridge. pp. 644.

Metcalfe, C.R., Chalk, L. (1965) Anatomy of Dycotiledones. Vol.I. Clarendon Press, Oxford.

Smith, W.H. and Dorchinger, L.S. (1976) Capability of metropolitan trees to reduce athmospheric contaminats. In: H. Gerhold, F. Santamor, and S. Little (Eds.), Proc. *Better Trees for Metropolitan Landscapes*, USDA. Forest Service, Gen. Tech. Report No. NE-22, Upper Darby, Pennsylvania, pp. 49-59.

Smith, W.H. (1981) Air Pollution and Forests. Springer-Verlag, New York, Heidelberg, Berlin, pp. 379.

Turunen, M., Huttunen, S., Percy, K.E., McLaughlin, C.K. and Lamppu, J. (1997) Epicuticular wax of subarctic Scots pine needles: response to sulphur and heavy metal deposition. *New Phytologist* 135, 501-515.

EVALUATION OF ALUMINIUM TOLERANCE IN WHEAT

M. MOUSTAKAS[1], L. BONA[2], F. ZSOLDOS[3], G. OUZOUNIDOU[1] AND L. PURNHAUSER[2]
[1]Department of Botany, Aristotle University of Thessaloniki, POB 109, Thessaloniki 54006, Greece, [2]Cereal Research Institute, POB 391, Szeged 6701, Hungary, [3]Department of Plant Physiology, Jozsef A. University POB 654, Szeged, Hungary

1. Abstract

A total of 60 wheat cultivars (44 of common wheat, 16 of durum wheat) was tested for Aluminium (Al) tolerance using hematoxylin staining method and chlorophyll fluorescence analysis. Pre-germinated seedlings were grown in nutrient solution containing a range of Al concentrations at pH 4.0. Results show that the higher Al concentration, the decreasing number of tolerant genotypes observed. Fifty percent of the common wheat cultivars were no or partially stained by hematoxylin at 180μM Al. Thus, 50% of the common winter wheat varieties tested possess an appropriate level of tolerance to Al. These genotypes may be suitable in less toxic acid soils. The chlorophyll fluorescence kinetics results confirmed the hematoxylin staining results that there were five genotypes with higher level of tolerance to Al, suggesting that these cultivars may be useful under highly acidic conditions. In almost all cases, the percentage fluctuations of the chlorophyll fluorescence parameters were shown to correlate very closely with the assessment of Al-tolerance evaluated by the hematoxylin staining method.

2. Introduction

Common wheat (*Triticum aestivum* L.) and durum wheat (*Triticum durum* Desf.) are among the most important crop species in agriculture. New, intensive type wheat varieties may improve the grain yield in rich soils, but the performance of intensive type genotypes may be very poor on infertile, acidic or dry soils. Soil acidity is a major growth limiting factor for plants in many parts of the world. Currently, many of these areas are either undeveloped for agriculture or of very low productivity (Taylor 1988).

Aluminium toxicity, the major phytotoxic factor in these soils, is not always economically correctable with conventional liming or other soil management practices. Plants differ in their reaction to Al toxicity, and variability is found between plant species as well as between cultivars (Bona *et al.* 1992, 1993, Moustakas *et al.* 1992, 1995, Bona 1996).

The reaction of hematoxylin with Al-stressed roots was used by Polle *et al.* (1978) and Carver *et al.* (1988) as the basis of a rapid and simple assay for Al tolerance in wheat cultivars, with Al-sensitive cultivars exhibiting more intense staining than tolerant cultivars along the vertical axis of the root.

Chlorophyll-a fluorescence emitted from the chloroplast thylakoid membranes reflects the primary processes of photosynthesis (Lichtenthaler and Rinderle 1988). The yield of chlorophyll fluorescence is influenced in many ways by the state of the photosynthetic apparatus and indirectly senses the secondary changes induced at the membrane level in the course of photosynthetic reactions. Thus, changes in fluorescence of chlorophyll-a have the potential for serving as a very early, non-destructive sign of stress injury (Lichtenthaler and Rinderle 1988, Krause and Weis 1991, Moustakas *et al.* 1995, 1996).

In this study we used the above two inexpensive diagnostic methods in the evaluation of Al stress tolerance of wheat cultivars.

3. Materials and Methods

3.1 PLANT MATERIAL

A total of 60 wheat cultivars (44 of common wheat, *Triticum aestivum* L. and 16 of durum wheat, *Triticum durum* Desf.) was tested for Al tolerance using hematoxylin staining method and chlorophyll fluorescence analysis. Control cultivars with known tolerance level were also included as checks in the experiments. Seeds were obtained from Greece (Cereal Institute of Thessaloniki), Hungary (breeding programs of Szeged and Martonvasar) and Slovakia (Solary).

3.2 HEMATOXYLIN STAINING METHOD

Pre-germinated seedlings were grown in nutrient solution and adjusted to pH 4.00 with 0.1 M HCl. After 24 h, the nutrient solutions in the containers were replaced with nutrient solution supplemented with 0 180 360 and 720 μM Al. Then roots were stained in hematoxylin solution (2 g hematoxylin and 0.2 g $NaI0_3$ dissolved in one litre of distilled water). The roots were rinsed again for 1 h and visually evaluated for the degree of staining as suggested by Carver *et al.* (1988). At each level of Al, approximately 5 mm of the primary-root tip of an entry was scored as completely (C), partially (P), or not (N) stained and considered as susceptible (S), intermediate (I) or tolerant (T), respectively. According to this classification, a genotype was considered (T) at a particular Al level if no stained (N) and intermediate (I) if partially stained (P).

3.3 CHLOROPHYLL FLUORESCENCE METHOD

Pre-germinated seedlings were grown in nutrient solution and adjusted to pH 4.00 with 0.1 M HCl. After 24 h, the nutrient solutions in the containers were replaced with nutrient solution supplemented with 0 180 360 and 720 μM Al. *In vivo* chlorophyll fluorescence was measured, after 7 days of Al treatment, on the upper surface of the second leaf with a fluorometer (Plant Productivity Fluorometer model SF-10, Richard Brancker Research Ltd, Ottawa, Canada). Prior to measurement of fluorescence, plants were dark adapted for at least 30 min at room temperature before the initial (F_o), intermediate (F_I) and peak (F_P) fluorescence parameters were measured at room temperature (Moustakas *et al.* 1995).

4. Results and Discusion

4.1 EVALUATION OF AL TOLERANCE USING HEMATOXYLIN STAINING METHOD

Results show that the higher consentration of toxic Al, the decreasing number of tolerant genotypes observed in the study. Fifty percent of the common wheat cultivars were no or partially stained by hematoxylin at 180μM Al. Thus, 50% of the common winter wheat varieties tested possess an appropriate level of tolerance to Al. These genotypes may be suitable in less toxic acid soils. There were found five wheat (*Triticum aestivum* L.) genotypes with high level of tolerance (Yecora E, Mv Vilma, Mv 9, Jbj 50, GK Kende) suggesting that these varieties may be useful for farmers under highly acidic conditions. Also, these tolerant varieties might serve as mating parteners in wheat breeding programs.

4.2. EVALUATION OF Al TOLERANCE USING *IN VIVO* CHLOROPHYLL FLUORESCENCE METHOD

The chlorophyll fluorescence kinetics results confirmed the hematoxylin staining results that there were five genotypes with higher level of tolerance to Al. It was concluded that genetic variation to Al stress can be easily seen by *in vivo* chlorophyll fluorescence measurements. Although the photosystems may not be the direct location of injury, fluorescence appears to be quite sensitive to cellular disruptions, including those caused by Al. The strong relationship between Al-tolerance determined via the hematoxylin staining method and fluorescence technique supports the inference that rapid reductions in fluorescence reflect actual tissue injury.

Cultivar improvement by selecting for increased Al stress-tolerance constitutes another practical application of chlorophyll fluorescence technique. Screening for Al-tolerance by chlorophyll fluorescence has obvious advantages as a pre-selection technique prior to long-term field trials. The Kautsky effect can serve to follow defects in photosynthetic activity and for plant breeders *in vivo* chlorophyll fluorescence assay provides an easy method for rapid detection of tolerant genotypes.

The value of chlorophyll fluorescence analysis is that it allows the breeder to screen plants at all stages of growth without necessarily use expensive growth chambers or phytotrons. Since it is non-destructive and very rapid, large populations can be quickly screened and those plants that are identified as having exceptional aluminum tolerance can be retained and bred from during that growing season.

5. Acknowledgements

This research was supported by the Hungarian Sci. Res. Fund (OTKA) No. T019922 and the Greek General Secretariat for Research and Technology.

6. References

Bona L. (1996) Tolerance of Hungarian winter wheat genotypes to aluminum toxicity, Abstracts IV International Symposium Plant-Soil Interactions at Low pH. Belo Horizonte, Brazil, p. 25.

Bona, L., Wright, R.J., and Baligar, V.C. (1992) Acid soil tolerance of *Triticum aestivum* L. and *Triticum durum* Desf. genotypes, *Cereal Research Communications* **20**, 95-101.

Bona, L., Wright, R.J. Baligar, V.C., and Matuz, J. (1993) Screening wheat and other small grains for acid soil tolerance, *Landscape and Urban Plannning* **27**, 175-178.

Carver, B.F., Inskeep, W.P., Wilson, N.P., and Westerman, R.L. (1988) Seedling tolerance to aluminium toxicity in hard red winter wheat germplasm, *Crop Science* **28**, 463-466.

Krause, G. H., Weis, E. (1991) Chlorophyll fluorescence and photosynthesis: The basics, *Annual Review of Plant Physiology and Plant Molecular Biology* **42**, 313-349.

Lichtenthaler, H.K., Rinderle, U. (1988) The role of chlorophyll fluorescence in the detection of stress conditions in plants, *CRC Critical Reviews in Analytical Chemistry* **19** (Suppl. 1): s29-s85.

Moustakas, M., Yupsanis, T., Symeonidis, L., and Karataglis, S. (1992) Aluminum toxicity effects on durum wheat cultivars, *Journal of Plant Nutrition* **15**, 627-638.

Moustakas, M., Ouzounidou, G., and Lannoye, R. (1995) Aluminum effects on photosynthesis and elemental uptake in an aluminum-tolerant and non-tolerant wheat cultivar, *Journal of Plant Nutrition* **18**, 669-683.

Moustakas, M., Ouzounidou, G., Eleftheriou, E.P., and Lannoye, R. (1996) Indirect effects of aluminium stress on the function of the photosynthetic apparatus, *Plant Physiology and Biochemistry* **34**, 553-560.

Polle, E., Konzak, C.F., and Kittrick, J.A. (1978) Visual detection of aluminium tolerance levels in wheat by hematoxylin staining of seedling roots, *Crop Science* **18**, 823-827.

Taylor, G J. (1988) The physiology of aluminum phytotoxicity, in H. Sigel and A. Sigel, (eds.), *Metal ions in biological systems,* vol. 24. Marcel Dekker Inc., New York, pp. 123-163.

GROWTH AND PHYSIOLOGICAL RESPONSES OF TOBACCO AND COTTON EXPOSED TO $AgNO_3$ AND AgI

G. OUZOUNIDOU[1] and H.A. CONSTANTINIDOU[2]
[1]*Dept. Botany, Aristotle University of Thessaloniki,, Greece*
[2]*Dept. Agriculture, Aristotle University of Thessaloniki, Greece*

1. Introduction

The normal concentration of Ag in non-contaminated soils is lower than 0.1μg g^{-1} dry wt [1], while the typical concentration of Ag in plants is reported at 0.5μg g^{-1} dry wt [2]. Silver concentrations differ greatly among plant species and stage of growth. Locally, the Hellenic Agricultural Insurance Organization has for several years been involved in a weather modification program aiming at reducing hail damage through cloud nucleation with AgI. Although silver per se is among the most toxic heavy metals, the low solubility of AgI in the water as well as the immobilization in biotic and abiotic complexes minimizes its toxicity [3]. Thus, there is an urgent need for additional research to establish the effects of various levels and forms of silver on plant growth, physiology and metabolism. With these considerations in mind, an evaluation of the response of tobacco (*Nicotiana tabacum*) and cotton (*Gossypium hirsutum*) to the supply of different concentrations and chemical forms of Ag, was attempted.

2. Materials and Methods

1μg L^{-1} Ag (as AgI) or 1μg L^{-1}, 10^3μg L^{-1}, 5.10^3μg L^{-1} (as $AgNO_3$) were added for 20 days in 30 day-old tobacco and cotton plants grown in Hoagland solution. At the end of Ag exposure, the plants of the highest Ag concentration were transferred for 4 days to Hoagland solution. Growth, chlorophyll fluorescence and photosynthesis were measured [4].

3. Results and Discussion

The effect of Ag on root length growth differed strongly between the two plant species (Tab. 1). On exposure to 1μg L^{-1} Ag as AgI physiology and growth were unaffected, while they were reduced under 1μg L^{-1} $AgNO_3$. A

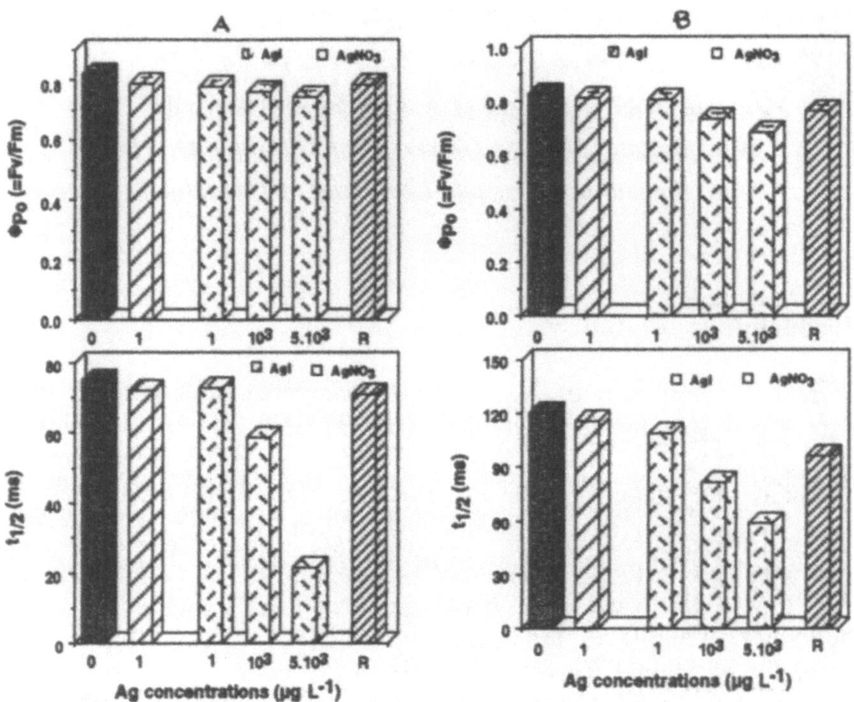

Fig. 1: Changes of the maximum quantum yield of PSII (Φp_0) and the half time rise ($t1_{/2}$) in the younger second leaf of tobacco (A) and cotton (B) during either a 20 days exposure to Ag or recovery (R) in Hoagland solution for 4 days (n=5, ±se).

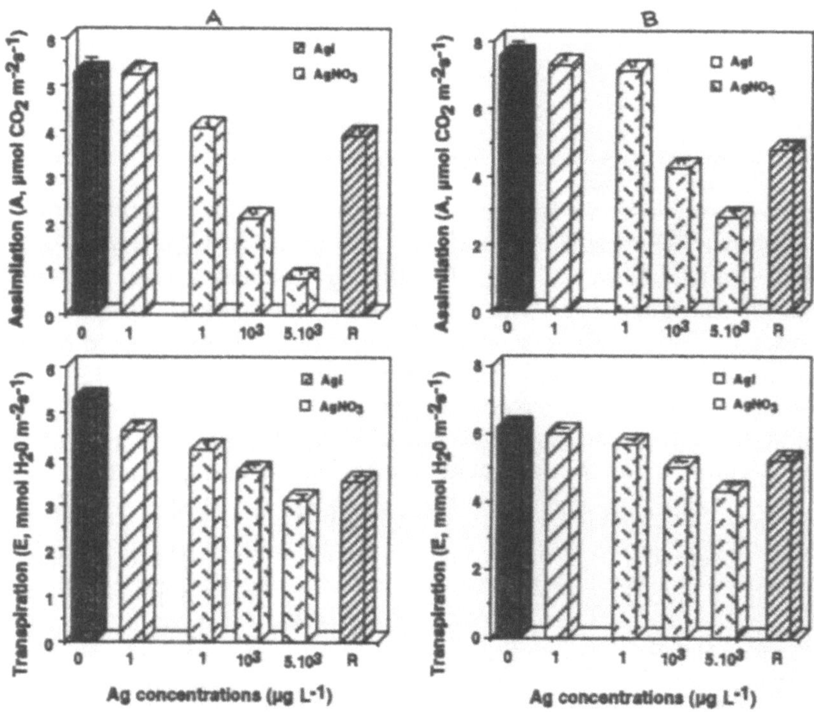

Fig. 2: Changes of the net photosynthesis (A) and the transpiration rate (E) in the younger second leaf of tobacco (A) and cotton (B). For details see Fig. 1.

TABLE 1: Root elongation (n=21, ±SE) and Ag concentration in tissues (n=3, RSD=10%) in 30 day-old *Nicotiana tabacum* cv. Virginia 944 and *Gossypium hisrutum* cv. Acala SG2 during either a 20 days exposure to Ag or recovery in Hoagland solution for 4 days.

Ag-Treatments	**Root elongation (cm)**		**Ag concentration (μg Kg^{-1} DW)**			
			root		**leaf**	
(μg L^{-1})	**Tobacco**	**Cotton**	**Tobacco**	**Cotton**	**Tobacco**	**Cotton**
0 (control)	4.80±0.7	4.30±0.9	89	198	19	73
AgI 1	3.50±0.4	3.80±0.3	100 (+12%)	324 (+1.6 fold)	30 (+58%)	87 (+20%)
$AgNO_3$ 1	2.30±0.2	3.85±0.4	158 (+78%)	408 (+2 fold)	44 (+2 fold)	136 (+2 fold)
$AgNO_3$ 10^3	0.90±0.5	2.15±0.3	640 (+7 fold)	1588 (+8 fold)	107 (+5.5 fold)	774 (+10 fold)
$AgNO_3$ 5.10^3	0.52±0.2	1.20±0.4	2169 (+25 fold)	5343 (+27 fold)	197 (+10 fold)	825 (+11 fold)
Recovery	0.86±0.4	1.50±0.3				

marked defference between the two species was observed with regard to the uptake and translocation of Ag. In all Ag concentrations, a higher percentage of the total silver was taken up by the roots of cotton rather, than of tobacco (Tab. 1). Root growth reduction might have occured by Ag-inhibition in cell division and elongation (mutagenic action) [2]. The increase of Ag^+ accumulation in roots might be attributed to changes of the membrane permeability leading to nutrient loss [5]. A slight reduction of the maximum quantum yield for photochemistry, Φ_{Po} (=Fv/Fm) and the half time rise, ($t_{1/2}$) values under 1 μg L^{-1} Ag (supplied as AgI) in cotton and tobacco was observed (Fig 1). Φ_{Po} and $t_{1/2}$ showed the highest quenching under the higher Ag concentration (5.10^3 μg L^{-1} as $AgNO_3$) for cotton and tobacco ($P<0.005$, Fig 1). The inhibition of the maximum quantum yield for photochemistry and the dysfunction of thylakoids and antennae molecules of PSII resulted in reducing photochemistry conversion. [6, 7]. Silver supplied as $AgNO_3$ induced large drop in the assimilation rates of cotton and tobacco, while the transpiration rate was apparently more tolerant to Ag stress revealing a lower inhibition (Fig 2). Net CO_2 fixation displayed a significant recovery for cotton and tobacco, while transpiration rate showed a slower recovery rate for either species (Fig 2). The reduction of photosynthetic and transpiration rates was due to stomatal closure and not to lower PSII activity. Finally, a moderate recovery rate for all parameters measured was detected (reversible effects).

4. References

1. Adriano, D.C. (1986) Trace element in the terrestrial environment. Springer-Verlag, New York.
2. Smith, I.C. and Carson, B.L. (1977) Trace Metals in the environment. Ann Arbor Science pp. 362.
3. Underwood, E.J. (1977) Trace Elements in Human and Animal Nutrition. Academic Press, pp 444.
4. Ouzounidou, G., Moustakas, M. and Eleftheriou, E.P. (1997) Arch. Environ. Contam. Toxicol. **32**, 1-7.
5. Ouzounidou, G. (1994) Environ. Exp. Bot. **34**: 165-172.
6. Ouzounidou, G. (1996) Plant Sci. **113**, 229-237.
7. Moustakas, M., Ouzounidou, G., Eleftheriou, E., and Lannoye, R. (1996) Plant Physiol. Biochem. **34**, 553-560.

Acknowledgments: This work was financially supported by the Hellenic Agricultural Insurance Organization

SEED QUALITY AND SOME PHYSIOLOGICAL BEHAVIOURS IN PRESOWING TREATMENT OF CARROT SEEDS

N. PANAYOTOV AND N. STOEVA
Higher Institute of Agriculture 4 000 Plovdiv, Bulgaria

1. Introduction

Akimov (1988) recommended 5 - time irradiation with laser energy of seeds of carrots for increase of yield. The stimulation effect on germination of carrot seeds, Ludilov (1995) observed in treatment with electromagnetic field with hyperhigh frequency in range-1250- 5000w/sec. Cantliffe *et al.* (1995) pointed out, that the effect of temperature treatment of carrots seeds was greater than PEG - independently by or with the addition of GA_3, but in comparasion with nontreated seeds strong positive influence was achieved. Sanders (1990) pointed out increase with 50 % of yield and weight of the carrot roots after the use of some biostimulators. Most of the carried out studies for the presowing treatment of carrots seeds are on self -nfluence of different agents, but not on comparative evolution between them.The main purpose of this investigation was to carry out study of different means and agents for presowing processing on some sowing qualities and physiological behaviours and upon the development of the plants and the yield.

2. Materials and Methods

The experiments were carried out with seeds of carrot - cultivar Nanatski 3. The treatments were done: trough soaking in 1.0 %, 2.0 %, 3.0 % KCL and 10.0 mg/l, 20.0 mg/l, 40.0 mg/l 3 -IAA and GA_3 for 20 hours with periodical stir; through 5-time, 7-time, 9-time irradiation with heli - neon laser with length of wave 632.8 nm and power of the exit 20 mv/s and period before sowing from 19 days and through ultrasound for 1, 5 and 10 minutes. The germinative energy and the germinability were by Bulgarian State Standart (BSS-601-85); the rate of germination by Piper (1952) ; the simultaneity of germination by Strona (1966) ; the fresh matter of sprouts of one seed, were established. The watersabsorbs ability was established by the quantity of absorbing water from 3.0 g seeds during 1 hour intervals. The intensity of respiration, activity of enzyme peroxidase (Bojarkin, 1951), activity of enzyme catalase (Ermakov, 1972), content of total protein (Lowry, 1951), content of row fat by Sucsle (Stancev *et al.*, 1968), absolutely dry matter (weightly) of the seeds were determinated. The statistic analysis was made after the method of Fisher (1958). Because of the similarly in the results obtained during the three years period of investigation, average values are given.

3. Results and Discussion

The germinative energy (Table 1.) after soaking in KCL was highest in concentration 1.0 % - 75.6 %. The used phytohormons increased this characteristic in comparative only with dry control (65.3 %) and the highest values were in 10 mg/l 3 - IAA - 67.7 % and in 20.0 mg/l GA_3 - 68.0 % and for control-water - 73.0 %. The energy of germination changed between 66.0 % to 70.0 %, for 9- and 7-time irradiation with laser, respectively. Among all investigation agents for processing of carrot seeds the highest germinative energy was obtained in 5 minutes ultrasound - 79.6 %. The greatest increase of germinability under influence of KCL was observed in concentration 1.0 % - 80.3 % too. All concentrations of 3 - IAA and GA_3 induced inhibition, with the exeption of 10.0 mg/l 3 -IAA - 78.6 % with 2.3 % and 3.0 % above both controls. The laser energy increased the sowing quality, the best results were obtained in 5-time irradiation - 80.0 %. The germinability in all variants was highest also in 5 minutes ultrasound - 89.3 % and in 10 minute - 83.0 %. What is impressive is, that through treatment with laser and above all with ultrasound i.e. by means of physical character, the sowing qualities were improved to the greatest extent. The similar influences in favour of processing through physical agents were established and in the fresh matter of sprouts of one seed. The sprouts of the seeds processed by soaking developed lower matter in comparison with the control, with the exception of 1.0 % KCL. Strong stimulation was observed in treatment with ultrasound and laser and the highest matter was of the sprouts in 5 minutes ultrasound - 8.26 mg and in 9-time irradiation with laser - 7.46 mg. From the estimated improvement of the vitality processes of the seeds it could be pointed out, that the carrot seeds responsed positively to a great extent when direct energy is brought in them i.e. by placing them in surroundings of electromagnetic and sound waves. The seeds after all means of treatment were with higher rate of germination and the biggest it was in 2.0 % KCL - 4.92 day (twenty - four hours). The average concentrations of the used chemical agents improved to the greatest extent the rate and simultaneity of germination. In irradiation with laser and ultrasound the highest rate of germination was observed in 9-time - 5.25 days and in 1 minute - 5.05 days. The greatest differences in favour of the physical agents were observed in the simultaneity of germination. The highest rate for all variants was from 7-time laser influence - 12.50 %, followed from 5-time and from 1 and 10 minutes ultrasound - 12.20 %.The waterabsorbing ability (Fig.1.) in the earlier hours was highest in 3.0 % KCL - on the 2 hours was absorbed 103.3 ml water. The absorbtion of water was considerable too in 10.0 mg/l GA_3 - 43.3 ml on the 3-rd hour, 7-time laser - 66.6 ml and 5 minutes ultrasound - 63.3 ml on the 4-th hour. Constant weight was established earliest - on the 3-rd hour in 20.0 mg/l 3 - IAA and in the 40.0 mg/l 3 - IAA - on the 4-th hour - i. e. this means of treatment precipitated the passing of the first phase of the germination - absorbtion of water. In all variants watersabsorbtion finished on the 5-th hour with the exception of KCL for the two controls. The content of absolutely dry matter (Table 2.) was not influenced considerably by the presowining processing of the seeds. The treated seeds had lower intensity of respiration. This process was most intensive in 2.0 % KCL, and in physical agents from 1 minute ultrasound. In such variants, the activity of enzyme peroxidase was lowest. Inhibition effect on peroxides activity induced 40.0 mg/l 3 - IAA and GA_3. The enzyme catalase was more active in all variants in comparison with the controls. This was very sharp in the seeds treated

Table 1. Sowing quality of seeds of carrot in different means of presowing processing.

	Variants	Germinative energy-%	Germinability-%	Fresh matter of sprouts-mg	Rate of germ.- day	Simultaneity of germination -%
1	Contol	65.3	76.3	6.76	5.79	12.17
2	Control-water	73.0	75.6	6.56	5.65	10.43
3	KCL - 1.0 %	75.6	80.3	6.96	5.59	10.37
4	KCL - 2.0 %	70.0	75.6	6.40	4.92	11.67
5	KCL - 3.0 %	67.6	74.3	6.26	5.28	11.67
6	3-IAA -10.0 mg/l	67.6	78.6	5.73	5.34	11.10
7	3-IAA -20.0 mg/l	60.6	72.6	4.66	5.00	11.80
8	3-IAA -40.0 mg/l	66.6	76.0	6.50	5.06	10.37
9	GA_3 - 10.0 mg/l	66.0	72.6	6.25	5.41	10.37
10	GA_3 - 20.0 mg/l	68.0	72.0	6.20	5.18	11.57
11	GA_3 - 40.0 mg/l	65.3	69.3	4.06	5.20	10.37
12	Laser 5 - time	69.0	80.0	7.13	5.39	12.20
13	Laser 7 - time	70.0	78.3	6.93	5.89	12.50
14	Laser 9 - time	66.0	77.0	7.46	5.25	11.80
15	Ultrasound-1 min	66.6	75.3	6.80	5.05	12.20
16	Ultrasound-5 min	79.6	89.3	8.26	5.30	10.37
17	Ultrasound-10min	68.6	83.0	7.36	5.28	12.20
	5.0 %	10.2	10.3	1.74	0.57	1.31
	GD - 1.0 %	13.7	13.8	2.33	0.76	2.10
	0.1 %	18.1	18.2	3.08	1.01	3.10

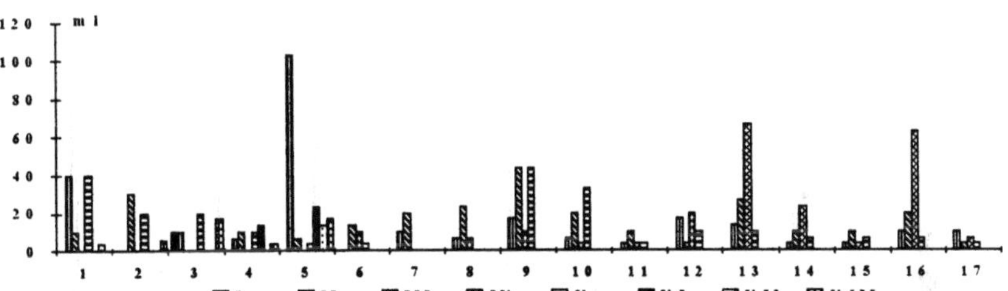

Figure 1. Watersabsorbs ability of seeds of carrot in treatment with different agents - ml absorbing water.
I - VIII - hours for establisment; 1 17 - variants as in Table 1.

with physical agents. The presowing treatment of the carrot seeds decreased the content of total protein. It was higher than the control only after treatment with 40.0 mg/l 3 - IAA and in 9-time laser. The lowest values were obtained in 5-time ultrasound. In this variant the sowing quality of the seeds and fresh matter of the sprouts were higher. It can be supposed that the decrease was a result of the degradation of the proteins and their use as a building and energy materials during the germination and the growing of the sprouts.The content of row fat was increased in the treated seeds. This tendency was observed in the chemical and in the physical means of processing. In the physical agents it was about 60 % above the control.

4.Conclusion

The germinative energy, the germinability, the simultaneity of germination and the fresh matter of sprouts of the carrot seeds were highest in presowing treatment with ultrasound (5 and 10 minutes) and with laser - in agents with physical nature. The influence of the investigated means of processing with chemical agents (KCL, 3 - IAA and GA_3) was lower. Watersabsorbtion ability was improved by the investigated means for presowing processing of the seeds of carrot, and slight preponderance was observed in soaking in chemical agents. The presowing treatment of the carrot seeds did not induce significant influence on absolutely dry matter, intensity of respiration and on peroxidase activity, but increased catalase activity most in laser and ultrasound treatment. As a result of the processing, the content of total protein decreased and was the lowest in the variant with the highest germinability and the highest fresh matter of sprouts, and the content of raw fat increased. To sum up, in the investigated treatments of carrot seeds, the effect was more significant than the use of agents with physical character (laser and ultrasound) in comparison with the chemical means (KCL,3-IAA and GA_3).

Table 2. Physiological characteristics of seeds of carrot after presowing treatement.

N	Variants	Absolutely dry matter-%	Intensity of respiration- $mgCO_2/g$	Peroxidase activity -E/g f.m.	Catalase activity- cm^3O_2	Total protein -γ ml	Row fat - %
1	Contol	92.90	0.216	168.4	3.52	6.45	15.7
2	Control-water	92.69	0.193	170.3	3.62	6.68	15.4
3	KCL - 1.0 %	92.88	0.174	180.6	4.34	5.25	20.1
4	KCL - 2.0 %	93.21	0.183	152.0	6.28	4.56	22.4
5	KCL - 3.0 %	92.89	0.162	164.0	5.80	5.88	21.7
6	3-IAA -10.0 mg/l	92.71	0.169	186.3	5.40	6.25	24.9
7	3-IAA -20.0 mg/l	91.95	0.113	198.0	4.86	5.16	22.2
8	3-IAA -40.0 mg/l	93.12	0.140	162.0	5.50	6.88	21.4
9	GA_3 - 10.0 mg/l	93.40	0.123	175.6	4.98	6.46	25.3
10	GA_3 - 20.0 mg/l	93.15	0.138	178.0	4.60	5.90	25.3
11	GA_3 - 40.0 mg/l	93.21	0.126	162.0	5.45	5.35	21.7
12	Laser 5 - time	92.87	0.125	160.0	5.78	4.18	24.8
13	Laser 7 - time	93.15	0.118	152.8	7.06	4.92	23.7
14	Laser 9 - time	92.98	0.128	146.3	4.35	6.85	24.3
15	Ultrasound-1 min	93.27	0.192	158.0	6.35	6.05	25.4
16	Ultrasound-5 min	93.41	0.168	166.0	7.18	3.58	26.1
17	Ultrasound-10min	92.88	0.162	181.3	7.04	4.44	25.2

5. References

Akimov V.I. (1988) Vlian. lazer. obluc. sem.na rost prod. morkovi ,Puti intens.v lesostepi, Konfer. Kuibisev, 2-5.
Boiarkin A.N. (1951) Biochimia, Tehnika, Moskva.
Contliffe D.J. and Balla M. El (1995) Improve germination of carrots at sreessful high tempreture by seeds priming. Proceeding of the Florida State Horticulture Society, IFAS. Univ. of Florida. USA. 121 - 128.
Ermakov A. I. (1972) Metodii biohimiceckogo izsledvania rastenii, Kolos,Moskva.
Fisher R.E. (1958) Statisticeskie metodi dlia izsledovatelej, Nauka, Moskva.
Lowry R.N. (1951) Protein measurement with Foliu-phenol reagent, J. of Biolog. chem., 2 , 193, 256 - 257.
Ludilov V. A. (1995) Obrab. semian v EMP SVC ulucsaetsia posevnie kacestva, J. Kartofel i ovosti.,N 2, 8- 9.
Piper H. (1952) Das Saatgut, Zweite Auflage., Berlin.
Sanders D. C. (1990) Improv. of carrot stonds with plant biostimul.and fluid drilling., HortScience., 25, 81-83.
Stancev L., Gurov G., Masev N. (1968) Rakov. za himicni analizi na rastenia, pocvi i torove, H. Danov, Plovdiv.
Strona I. G. (1966) Obstee semenovodstvo polevnih kultur, Kolos, Moskva.

EFFECTS OF Cu EXCESS ON OREGANO [*Origanum vulgare* subsp. *hirtum* (Link) Ietswaart]. II PHYSIOLOGICAL EFFECTS

E. PANOU-FILOTHEOU, A. M. BOSABALIDIS and S. KARATAGLIS
Department of Botany, School of Biology, Aristotle University, 54006 Thessaloniki, Greece.

1. Abstract

The physiological effects of increased copper concentrations (up to 25.5 µM / g of soil) on oregano plants noted after copper stress were the following: Water potential increased by 58%, transpiration remained about the same (11 µM / m^2.s), but the water potential / transpiration ratio decreased by 62%. Stomatal conductance decreased by 39% and so was the difference CO_2atm-CO_2int. (by 11%). Photosynthetic rate and chlorophyll content (a+b) underwent a reduction by 82% and 66%, respectively.

2. Introduction

Copper is an essential microelement for the plant. It plays an active role in enzymes performing key functions in respiration and photosynthesis. Cu-proteins have been implicated in lignification, anabolic metabolism, cellular defense mechanisms and hormone metabolism (Bussler 1981). However, excessive concentrations of copper are highly toxic and cause physiological constraints that can decrease the vigor of the plants and even totally inhibit plant growth.

The object of this work was to study several physiological parameters associated with the growth of oregano plants in Cu rich soil.

3. Materials and Methods

Oregano plants [*Origanum vulgare* subsp. *hirtum* (Link) Ietswaart] were grown in a green house in soils containing various levels of copper [control (0.3), 13.0, 17.0, 19.0, 22.0, 24.5 and 25.5 µM Cu / g of soil].

Soil from rhizosphere of plants and plant samples were oven-dried at 70° C to constant weight. Determination of copper content in the samples was measured with an atomic absorption devise Perkin-Elmer 2380 (Karataglis et al. 1982). An index of tolerance (IT) was estimated as described by Karataglis (1980). Water potential was estimated by using a pressure chamber (Turner 1981). To determine the total chlorophyll (a+b) content in leaves the method described by Ouzounidou at al. (1992) was used.

The photosynthetic rate was measured by a closed system (chamber 0,25L) with an infrared gas analyzer (IRGA, LI-COR Model 6200) which contained also sensors for temperature, humidity, photosynthetically active radiation and console system for registration and elaboration of the elements.

4. Results and discussion

The first obvious effect of Cu excess on oregano plants was a decline in the number and size of the stems. Leaves also appeared to underwent a progressive reduction in size while they simultaneously became chlorotic. Measurements of Cu accumulation in oregano organs showed that Cu content was the highest in the roots, followed by the inflorescences, the leaves and stems (Table 1).

Table 1: Copper concentrations in different parts of the plant (μM / g d.w.)

Part of the plant	n	Cu soil concentrations (μM / g of soil) 0.3 (control)	13.0	17.0	19.0	22.0	24.5	25.5
Root	4	0.22±0.14	9.92±1.11	13.71±1.73	14.58±1.96	15.54±2.01	16.38±2.33	18.85±2.41
Lower part of stem	4	0.13±0.09	3.09±0.09	3.31±0.94	3.65±1.12	3.87±0.98	3.97±0.74	4.25±0.95
Upper part of stem	4	0.10±0.07	2.55±0.41	2.93±0.55	3.43±0.69	3.50±1.05	3.62±0.95	3.91±1.12
Mature leaves	4	0.16±0.09	3.53±0.89	3.87±0.75	4.32±1.12	4.63±1.05	5.10±1.09	5.58±0.95
Young leaves	4	0.10±0.01	2.74±0.23	3.12±0.35	3.31±0.54	3.69±0.61	4.03±0.84	4.38±0.79
Inflo-rescences	4	0.11±0.02	2.87±0.31	3.05±0.28	3.50±0.33	3.72±0.65	4.01±0.88	5.23±0.42

±: standard deviation
n : number of replications
All the averages of the treatments are statistically significant from the control at the level 0.001.

Copper is binded particularly to the cell walls of roots (Turner 1969), suggesting that the walls may function as a metal-immobilizing compartment. Copper, also may induce the production of phytochelatins and metallothionein that could restrict the transport of the metal from the root to the shoot (Wainwright and Woolhouse 1977). The greatest length of the roots decreased dramatically (Fig. 1) with increasing soil Cu concentration and thus the indices of tolerance underwent a reduction from 72% to 10%. This phenomenon can be explained by the toxic effect of Cu on the root system as described by Karataglis et al. (1982). Transpiration (Fig 2) from leaves and canopies

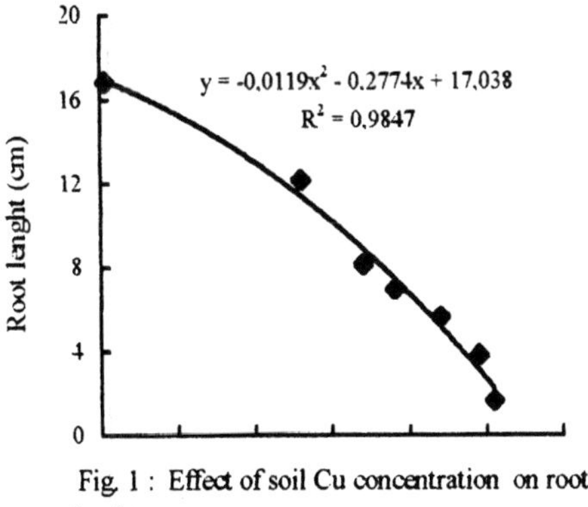

Fig. 1 : Effect of soil Cu concentration on root lenght.

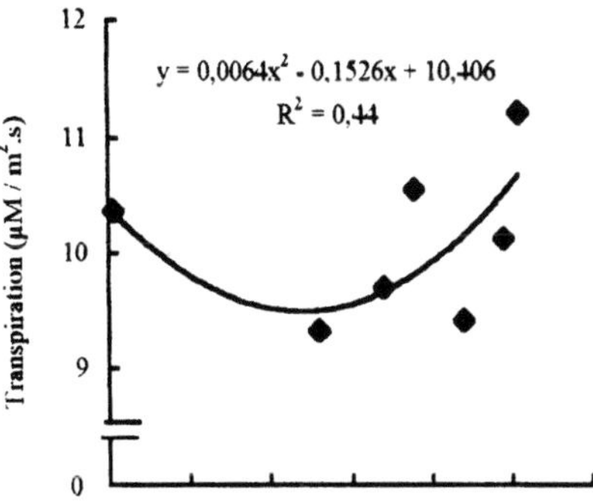

Fig. 2: Effect of soil Cu concentration on transpiration.

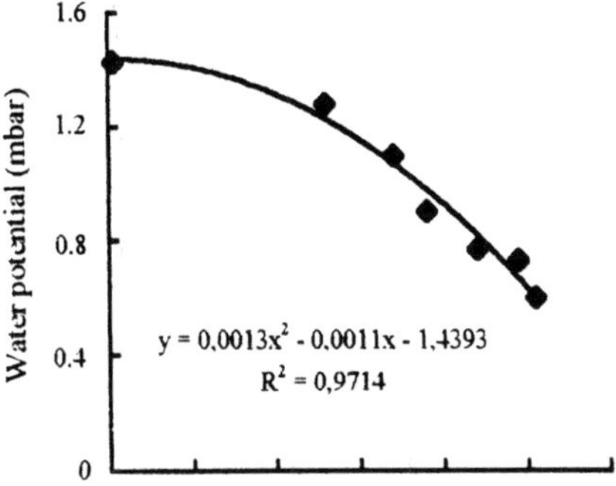

Fig. 3: Effect of soil Cu concentration on water potential.

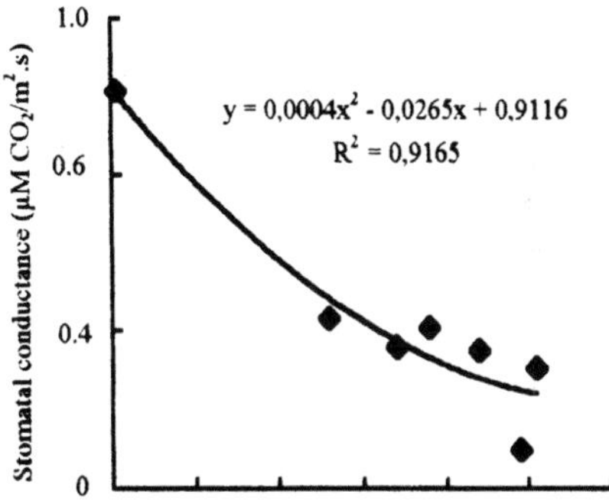

Fig. 4: Effect of soil Cu concentration on stomatal conductance.

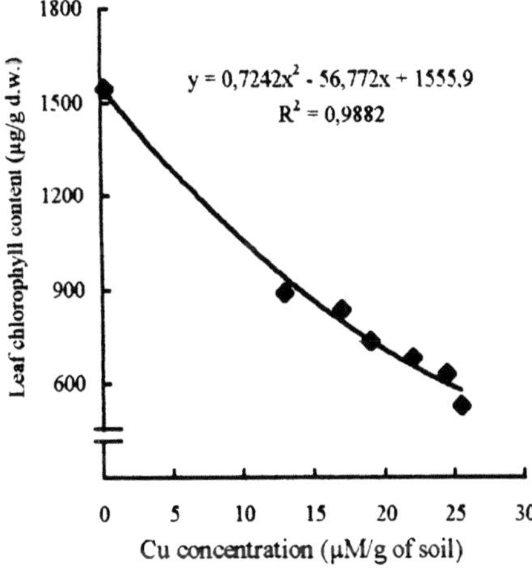

Fig 5: Effect of soil Cu concentration on leaf total chlorophyll content.

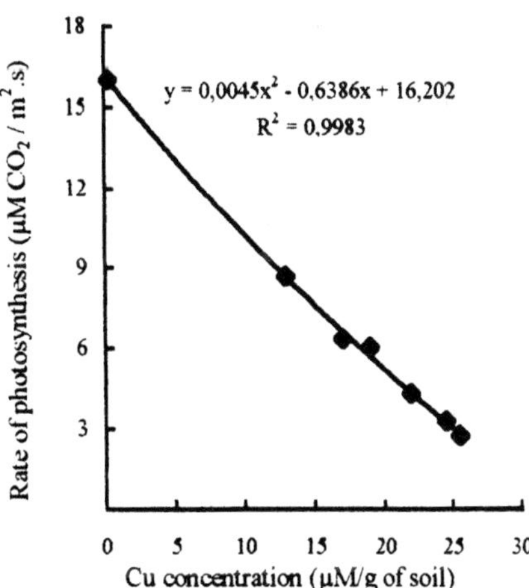

Fig. 6: Effect of soil Cu concentration on rate of photosynthesis.

remained about the same (11 μM /m^2.s), in spite of the decreased stomatal conductance and lamina surface. That can be explained by the decrease in intercellular space volume, the increase in xylem elements and also the increase in glandular and non-glandular trichomes per leaf unit area. The water potential increased by 58% (Fig 3). This may be due to the Cu effect on absorption, transpiration and water retaining capacity. As mentioned, the root system altered, the transpiration remained about the same, but the relative volume of the xylem increased (consequently the conservation of the water became greater). This is evident by the ratio water potential / transpiration which increased by 62%. Although the number of stomata per leaf unit area increased with increasing Cu concentration, the stomatal conductance became decreased by 39% (Fig. 4) and so the difference CO_2atm.- CO_2int by 11%. This may be explained by the shorter guard cells and the increased K concentration. The chlorophyll content (a+b) decreased by 66% (Fig 5). Copper restricts the absorption of the Fe-ion that participates in chlorophyll biosynthesis (Baker 1990). It has been also proposed that a Cu-mediated lipid peroxidation in chloroplasts may cause the distruction of chloroplast membranes (Sandmann and Boger 1980). Physiological observations demonstrated that copper caused reduction of the photosynthetic rate by 82% (Fig. 6). This reduction is probably indebted to the following reasons: the decreased lamina surface, the diminished chlorophyll content, the altered structure of the chloroplasts, the reduction of the interior CO_2 (by 11%) and the Cu inhibitory effects on the electron transfer systems in two different regions around photosystems I and II .

5. References

Baker D.E. 1990. *Heavy Metals in soils.* Alloway B. J., John Wiley & Sons, New York.

Bussler W. 1981. *Copper in soils and Plants.* Academic Press.

Karataglis S., Babalonas D. and Kabasakalis V., 1982. I. The ecology of plant populations growing on serpentine soils. II. Ca/Mg ratio and the Cr, Fe, Ni, Co concentrations as development factors of *Buxus sempervirens* L. *Phyton* 22, 317-327.

Ouzounidou G., Eleftheriou E. P. and Karataglis S., 1992. Ecophysiological and ultrastructural effects of copper in *Thlaspi ochroleucum* (Cruciferae). *Can. J. Bot.* 70, 947-957.

Sandmann, G and Boger P., 1980. Cu-mediated lipid peroxidation process in photosynthetic membranes. *Plant Physiol.* 66, 797-800.

Turner, R.G., 1969. Heavy metal tolerance in plant, In Robinson I.H. (Ed.), *Ecological aspect of the mineral nutrition of plants.* Oxford, Edingburgh: Blackwell.

Turner, N.C., 1981. Techniques and experimental approaches for the measurements of plant water status. *Plant Soil,* 58, 339-366.

Wainwright, S.J. and Woolhouse H.W. 1977. *J. Exp. Bot.* 28 (105), 1029 Woolhouse, H. W. 1983. Toxicity and tolerance in the responses of plants to metals. In. Pirson, A. and M. H. Zimmerman (eds): *Encyclopedia of Plant Physiology*, Vol. 12C, 245-300, Springer-Verlag, Berlin.

GAs LEVELS AND THEIR RELATION TO GROWTH OF OREGANO PLANTS [*Origanum vulgare* subsp. *hirtum* (Link) Ietswaart] GROWN IN COPPER RICH SOILS

E. PANOU-FILOTHEOU[1], M. KOUKOURIKOU-PETRIDOU[2], A. M. BOSABALIDIS[1] and S. KARATAGLIS[1]
[1] *Department of Botany, School of Biology*, [2]*Lab. of Plant Biology School of Agronomy, Aristotle University, 54006 Thessaloniki, Greece.*

1. Abstract

Oregano plants were grown in soils containing various levels of copper [0.3 (control), 13.0, 17.0, 19.0, 22.0 and 24.5 μM Cu / g of soil]. Treated plants showed a reduction in height which was concentration dependent. In spite of this effect, the levels of GAs in the stem apices of the treated plants were higher than those of controls. Application of 30 μMol GA_3 (drops put on stem apical meristem) resulted in a significant increase of the height of the control plants. GA_3 treated plants grown in copper rich soil exhibited an increase in height not statistically significant. The negative effect of copper on the growth of oregano plants is considered not to be due to the level of GAs but probably to other parameters (toxicity).

2. Introduction

The agricultural soil is continuously contaminated with copper, residues of cultivation (fungicides, pesticides, fertilizers, sewage sludge, pig slurries, municipal composts etc). Copper is specifically absorbed or fixed in soils, making it one of the trace metals which moves the least. Copper is an essential microelement for plants in the order Fe>Mn>B>Zn>Cu >Mo>Cl (Baker 1990) but in higher concentrations it is toxic and causes inhibition of growth (especially of height growth) (Eleftheriou and Karataglis 1989, Karataglis and Babalonas 1985). Gibberellins are responsible for height growth in almost all plants (Metraux 1988). For this reason the effect of increasing concentration of copper in soil on gibberellin content of stem apices and of mature leaves of oregano plants was studied.

3. Materials and Methods

Oregano plants [*Origanum vulgare* subsp. *hirtum* (Link) Ietswaart] were grown in soils containing various levels of copper [0.3 (control), 13.0, 17.0, 19.0, 22.0 and 24.5 μM Cu / g of soil]. After thirty five days the height of the plants was measured. At the same time samples of stem apices and leaves (of the fifth node) were collected and lyophilized. The dried samples were analyzed for their gibberellin content according to the following diagram:

Dried sample 500 mg dry weight
↓
Extraction at 0° C in MeOH 30 min X 3
Filtration
↓
Filtrate
Addition of 5 ml H_2O
Evaporation till water phase in a rotary-evaporator
Addition of 0.5 M K_2HPO_4 (pH 8.0)
↓
Separation with petroleum ether X 1
↓ organic phase discarded
Water phase
Addition of 0.28 M H_3PO_4 (pH 2.5)
Separation with diethyl ether X 3
↓
Organic phase
Evaporation till dryness
↓
Addition of 1 ml K_2HPO_4
↓
PVC
pH 8.0 ↓elution with 160 ml K_2HPO_4 solution
Reduction in 10 ml
Addition of H_3PO_4 (pH 2.5)
Separation with diethyl ether X 3
↓
Organic phase
Evaporation till dryness
↓
Paper chromatography
↓
Lettuce hypocotyl bioassay

For the quantitative determination of gibberellins the method of «lettuce hypocotyl» was used. Lettuce seeds var. «Grand rapids» were sterilized and placed on wet filter paper in petri dishes. They were germinated and grown in dark at 23° C for 40 h. Each of ten pieces of chromatograms were placed in small beakers with 0.5 ml distilled water and ten lettuce seedlings. The beakers were transferred in a growth chamber with continuous light and 23° C. The measurement of hypocotyl length was done 3 days later.

In a second experiment one drop of GA_3 solution was applied to the stem apices of oregano plants grown in soils with various copper concentrations for two months. Two months after gibberellin application, the height of GA_3-treated plants as well as of the control was measured.

4. Results and discussion

Experiment 1. The gibberellin content (ng/g d.w.) in stem apices and leaves of oregano plants is shown in Fig. 1, a and b. Although the morphological features of the treated plants have not demonstrated any excess of gibberellin, as described by Bozabalidis and Exarchou (1995), there were much more gibberellins in plants (that were smaller than the control), (Fig. 2) grown in soils enriched with various quantities of copper and it was copper concentration depended. The small height of these plants probably was not due to low gibberellin content but to the toxicity of the copper which caused an inhibition in the physiological parameters (respiration, photosynthesis etc) of the plant (Clisters and Van Assche 1985).

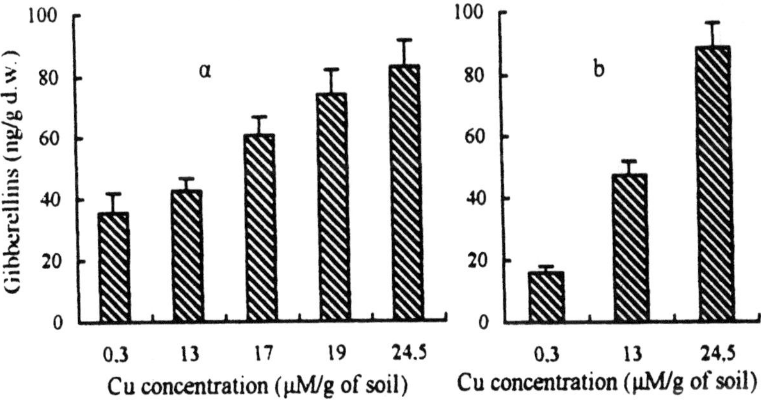

Fig. 1. Effect of soil Cu concentration on gibberellin content (ng/g d.w.) a: apices. b: mature leaves.

Experiment 2. An increase in the height of plants treated with GA_3 was observed that was greater than control (Fig. 3). This is another evidence that the reduction of height was not related to gibberellin content. In dwarf pea varieties, the gibberellin content was high as in the regular ones (James and Reid 1988). Treatment of dwarf pea plants with GA_3 caused an increase in stem height. In this case it was believed that there was no GA_1 which was responsible for stem growth.

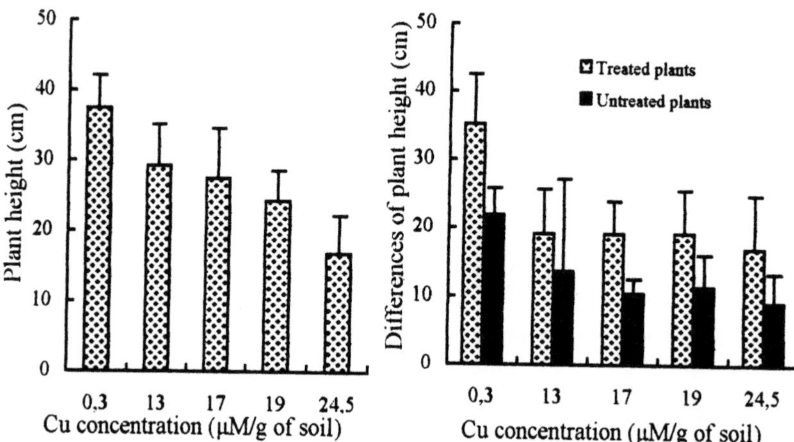

Fig 2 : Effect of soil Cu concentration on plant height.

Fig 3 : Effect of applied gibberellic acid (GA_3) on plant height two months after treatment.

3. References

Baker D.E. 1990. *Heavy Metals in soils*. Alloway B. J., John Wiley & Sons, New York.

Bosabalidis A.M. and Exarchou F. 1995. Effect of NAA and GA3 on leaves and glandular trichomes of *Origanum X intercedens* Rech.: Morphological and anatomical features, *Int. J. Plant. Sci.* 156(4): 488-495.

Clisters H. and F. Van Assche, 1985. Inhibition of photosynthesis by heavy metals, *Photosynth. Res.* 7, 31-40.

Eleftheriou, E.P. and Karataglis S. 1989. Ultrastructural and morphological characteristics of cultivated wheat growing on Cu-polluted fields. *Bot. Acta* 102: 134-140.

James B. and Reid 1988. The genetic control of growth via hormones. In : Davies, P.J. (Ed.), *Plant Hormones and their role in plant growth and development*, Dordrecht. Netherlands.

Karataglis S. and Babalonas D., 1985. The toxic effect of copper on the growth of *Solanum lycopersicum* L. collected from Zn- and Pb- soil. *Angew. Botanik* 59, 45-52.

Metraux J.-P. 1988. Gibberellins and plant cell elongation. In : Davies, P.J. (Ed.), *Plant Hormones and their role in plant growth and development*, Dordrecht, Netherlands.

EFFECTS OF INTERMITTENT COOLING ON GROWTH OF *LAVANDULA STOECHAS*.

I. PAPAGEORGIOU[1] AND E. MALOUPA[2]
[1]*Mediterranean Agronomic Institute of Chania, Chania 73100, Greece,*
[2]*Agricultural Research Centre of Macedonia & Thrace, Thermi 57001, Greece*

1. Abstract

The effect of intermittent low temperatures (3 °C, 7 °C or 10 °C) for different time periods (1, 2 or 3 h) on growth of *Lavandula stoechas* was tested. Short time exposure to 10 °C accelerated shoot elongation by 22.51 % (3.4 cm) on average, whereas exposure to 3 or 7 °C accelerated shoot elongation by 4.6 % or 9.9 % (0.7 or 1.5 cm), respectively. It appears that *L. stoechas* plants can tolerate low night temperatures for a short period indicating that cultivation in a greenhouse (without constant heating) during winter may be possible.

2. Introduction

Commercial production of native Mediterranean species offers, to Mediterranean horticulture, the possibility to confront low price mass plant production of North European countries and importation (Caballero and Cid, 1993). Recent market studies showed that European consumers are ready to accept native Mediterranean pot plants because they are exotic, water stress tolerant and demand little care (LaMalfa and Leonardi, 1993). Due to these characteristics, native Mediterranean plants are also ideal for landscaping in arid Mediterranean zones.

Lavandula stoechas is a perennial shrub of about 50 cm in height, native to many Mediterranean regions, flowering from May to June. It grows very well in sunny habitats with poor and dry soils. Its ecological and morphological characteristics make it an interesting native plant in arid zones below 1500 m (Scroubis, 1985; Herrera, 1991).

The introduction of *L. stoechas* in the garden shrub market is currently under investigation; its natural growth habit corresponds well to the commercial standards of this market. The main commercial period for garden shrubs lasts from March to May, suggesting that the plants must be grown either outside, from spring to autumn one year before commercialisation, or in greenhouses during winter. The economic feasibility of

greenhouse production depends mainly on the temperature requirements of the plants, greenhouse heating being the major expenditure during winter (Grafiadellis, 1987).
Growth response, of various plants, to low temperature has been widely studied (Baker et al., 1988; Graham and Patterson, 1982; Kneinendorst and Brouwer, 1972; Lyons, 1973; Miedema, 1982; Sowinski and Maleszewski, 1989). In this study the effect, of temperatures of 3 °C, 7 °C or 10 °C for 1, 2 or 3 hours, on the elongation and node production rate of lateral shoots of *L. stoechas* was investigated experimentally.

3. Materials and methods

Rooted cuttings of *L. stoechas* plants, were transplanted (May 5th) in 5.5 L pots, containing a mixture of soil (50%), peat moss (30%), gravel (15%) and zeolite (5%). The pots were placed in a plastic greenhouse, equipped with dynamic ventilation, pad cooling system and a shading net (which was placed 17 days later) The plants were pinched to 6 nodes, 13 days after transplanting, and were cold-treated 19 days after pinching.
The experiment was set to test the effect, of three low temperatures (3, 7 or 10 ^{0}C) at different time durations (1, 2 or 3 hrs), on the organogenesis and elongation of the lateral shoots of *L. stoechas* and consisted of 10 treatments (3 time spans x 3 temperatures + 1 as control) replicated 7 times. The plants were arranged on benches in a greenhouse in 7 groups (blocks) of 10 plants each (randomised complete block design). In each block, the plants of every treatment were picked up using random numbers.
Cold treatments started 19 days after pinching in growth chambers (SHEER, Rheen with height of 170 cm, length of 130 cm and width of 75 cm) equipped with heating and cooling units for temperature regulation, and light sources for photosynthetic (12 parallel ramps, 130 cm long, of 115 W Sylvania Cool White, F48 T12 CW VHO lamps) and photoperiodic (4 Philips ML 160 W lamps) control.
After the cold treatments the plants were placed back to their previous position in the greenhouse. The number of nodes and the length of the first three randomly selected lateral shoots (one per node of the main shoot) of each plant were measured at pinching, 20-25, 39 and 54 days after pinching.

4. Results and discussion

Climatic conditions recorded in the plastic greenhouse during the growing period (May to July) are shown in table 1. Mean day light integrals and maximum incident were reduced during June and July because of the greenhouse shading net. Mean length (DLS_s) and node number (DNN_s) net increase, elongation rate and node production rate of the lateral shoots of *L. stoechas*, recorded 35 days after treatment, are shown in table 2. At the beginning of the experiment the lateral shoots were 0.6±0.7 cm long and had 7±2 nodes. Control plants, 35 days after treatment, had lateral shoots 15.1 cm longer and

TABLE 1. Temperature and light conditions in the greenhouse during the growing period: absolute minimum and maximum temperatures (T_{mn}, T_{mx}), mean day light integrals $\Sigma(PPFD)$ and maximum incident $PPFD_{mx}$.

Month	T_{mn} °C	T_{mx} °C	$\Sigma(PPFD)$ mol m^{-2} d^{-1}	$PPFD_{mx}$ µmol m^{-2} s^{-1}
May	14.3	29.2	26.5	1490.6
June	17.8	28.2	16.0	793.9
July	18.8	31.5	15.9	800.6

TABLE 2. Net increase of length (DLS_s) and node number (DNN_s) and elongation and node production rate of lateral shoots of *L. stoechas* plants, grown in pots in a plastic greenhouse for 54 days, subjected to 0 (control), 3, 7 or 10 °C for 1, 2 or 3 hr.

Temperature (°C)	Duration (h)	DLS_s (cm)	DNN_s	elongation rate (cm/day)	node production rate (nodes/day)
Control	0	15.1 [ab]	9 [ab]	0.43 ±0.03	0.25 ±0.01
3	1	13.5 [a]	8 [a]	0.39 ±0.03	0.23 ±0.01
	2	15.7 [ab]	10 [ab]	0.44 ±0.03	0.26 ±0.02
	3	15.9 [abc]	9 [ab]	0.46 ±0.03	0.27 ±0.02
7	1	15.3 [ab]	9 [ab]	0.43 ±0.04	0.24 ±0.03
	2	17.7 [bc]	10 [ab]	0.51 ±0.04	0.28 ±0.02
	3	16.8 [abc]	9 [ab]	0.48 ±0.04	0.27 ±0.02
10	1	17.8 [bc]	10 [ab]	0.51 ±0.04	0.28 ±0.02
	2	20.3 [c]	11 [b]	0.55 ±0.04	0.28 ±0.03
	3	17.4 [abc]	10 [ab]	0.50 ±0.03	0.29 ±0.02

* Mean values were compared with the Duncan's multiple range test at the 5% significance level. Means with the same letter are not significantly different.

* Means are of 3 shoots per plant for 7 plants.

produced 9 nodes more. Lateral shoots of plants treated at 10 °C for 2 h were significantly longer (5.2±1.3 cm) compared to control, having 11±1 nodes (2 nodes more than the control on average), whereas lateral shoots of plants treated at 3 °C were not significantly longer compared to control (table 2).

The intermittent cooling of *L. stoechas* plants at 10 °C for a short time showed to increase shoot elongation (table 2). It is possible that a temperature of 10 °C for 2 h accelerates enzymatic processes, involved in internal hormone production, leading to an increase of shoot length, but further experiments should be performed in order to investigate such an effect. From a practical point of view, it would be possible during the growing period, to maintain the plants at 3 °C for at least 3 h without any consequence on their growth. Grafiadellis (1987) reported that, fuel saving under these conditions, in commercial greenhouses compared to the maintenance of the commonly used night setpoint of 14 °C, ranges from 19 g m^{-2} (glass or double layer) to 35 g m^{-2} (single polyethylene). In addition, it appears that, short term storage of plants (during transportation) at relatively low temperatures, may in fact have a positive outcome on their growth.

5. References

1. Baker, N.R., Long, S.P. and Ort, D.R. (1988) Photosynthesis and temperature, with particular reference to affects on quantum yield, in S.P. Long and F.I. Woodward (eds.), *Plants and temperature*, The company of Biologists Limited, Cambridge, pp. 347-375.
2. Caballero, M. and Cid, M.C. (1993) Development of new ornamentals from Canary islands native plants as alternative crops for Mediterranean areas, in P.f. Martinez (eds.), *Environmental constrains in protected cultivation for new growing techniques and crops*, Report EUR.15123, pp. 163-168.
3. Grafiadellis, M. (1987) *Modern Greenhouses*, Gartaganis publications, Thessaloniki Greece, pp. 328.
4. Graham, D. and Patterson, B.D. (1982) Responses of plants to low, unfreezing temperatures: Proteins, metabolism and acclimation, *Ann. Rev. Plant Physiol.* **33**, 347-372.
5. Herrera, J. (1991) Allocation of reproductive resources within and among inflorescences of *Lavandula stoechas* (*Lamiacea*), *Amer. J. Bot.* **78(6)**, 789-794.
6. Kneinendorst, A. and Brouwer, R. (1972) The effect of local cooling on growth and water content of plants, *Neth. J. Agric. Sci.* **20,** 203-217.
7. LaMalfa, G. and Leonardi, C. (1991) New crops for plastic GH in Medit. Area: Why, which and how?, *Agriculture*, pp. 145-161
8. Lyons, J.M. (1973) Chilling injury in plants, *Ann. Rev. Plant Physiol.* **24,** 445-466.
9. Miedema, P. (1982) The effects of low temperature on *Zea mays*, *Advances in Agronomy* **35,** 93-128.
10. Skroubis, B.G. (1985) *Aromatic Plants and Essential Oils*, pp. 116-123.
11. Sowinski, P. and Maleszewski, S. (1989) I. Growth and functioning of shoots and roots, *Acta Physiologiae Plantarum* **11,** 165-171.

DECOMPOSITION OF *QUERCUS FARNETTO* TEN. AND *QUERCUS CERRIS* L. LEAF LITTER UNDER NATURAL FOREST CONDITIONS

P. PAVLOVIC, M. MITROVIC, R. POPOVIC
Institute for Biological Research "Sinisa Stankovic", 29 Novembra 142, 11060 Belgrade, Yugoslavia

1. Introduction

Decomposition, with the level of significance attributed to it, can be compared to primary production as a fundamental process of ecosystems. Theoretically speaking, with the aid of this process organic matter is reduced to inorganic components used further by green plants in autotrophic production of organic matter. In this way the same limited quantity of organic matter circles around in ecosystems, and this is depicted as the cycle of circling matter.
Decomposition is a complex phenomenon and various processes contained in it contribute to the degradation of complex organic compounds (Swift *et al.* 1979). Release of nutrients in forest litter is affected by climatic conditions (Meentemeyer and Berg, 1986), soil properties (Raulund-Rasmussen and Vejre, 1995) and litter quality (Vesterdal *et al.* 1995).
The idea of this research is to study the decomposition of two species of oak, that are dominant plants of the *Quercetum farnetto-cerris* Rud. (1940) 1949., community, thus obtaining a clearer picture of functioning and stability of this ecosystem, as well as the effects of examined species litter on the process of decomposition in these forests.

2. Materials and methods

The analysed community of Italian oak and Turkey oak is located on the Maljen mountain (Western Serbia). The experimented area is exposed to the north-west, with an altitude of about 500 m, terrain inclination of 25°. Its geological base consists of hydro-thermally altered vulcanite, and distric cambisole makes the pedological base. Its stand is around 80 years old.
The experiment was conducted under natural conditions where the organic fall of *Quercus farnetto, Quercus cerris* and a combination of the two, had the 1:1 ratio. Organic fall was placed in bags woven out of plastic threads with 1x1 mm openings. There were five repetitions. Three stages were considered: initial position (in autumn), after a six-month period of decomposition and after 12-months decomposition. The concetration of carbon and nitrogen during the process of organic matter decomposition was determined on the same sample using the Anstat method modified by Ponomareva and Nikolaeva (1975). The C:N ratio was determined by means of calculation.
Decomposition rate of examined species organic matter and its combinations together with the prognosis of the decomposition process of examined ecosystems have been described using the Olson decomposition model and Olson loss constant:

$$M/M_0=e^{-kt} \quad (1)$$

The decomposition rate prognosis of organic fall of dominant plants of examined ecosystems was drawn based on the above-mentioned parameters, namely: the 95% decomposition was calculated using 3/k, 99% decomposition using 5/k (Olson, 1963).

3. Results and discussion

Decomposition analysis has proved that the fall of Italian oak decomposes more readily at all stages of analysis than that of the Turkey oak (*Fig. 1.*). After six months of decomposition organic matter of Italien oak decomposed 45.644 % (±10.034), whereas Turkey oak decomposed 21.228 % (±3.562). After 12 months of decomposition the gradient remained unchanged (Italian oak 51.256 %, ±5.354; Turkey oak 28.522 %, ±8.591). The mixture of organic fall of the two species lost 37.250 % (±5.839) after 6 months of decomposition, and 48.330% (±7.455) after 12 months. Olson's loss constants also confirm the same decomposition gradient (Tab. 1).

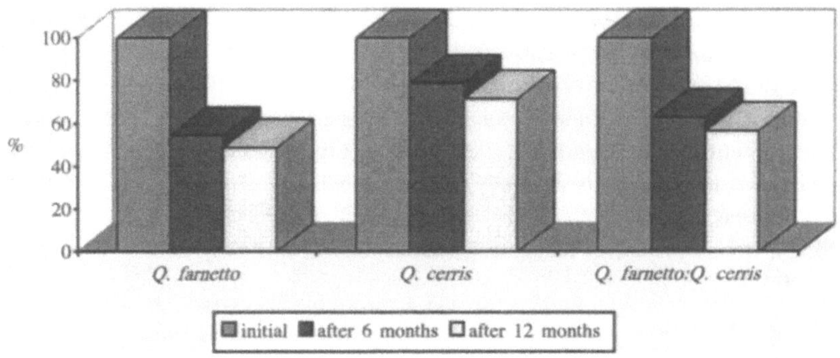

Figure 1. Decomposition dynamics of examined species organic matter during the process of decomposition

Analysing results obtained, one notices at first sight that there occurred a more intensive decomposition during the first 6-month period as opposed to the second, at all stages of analysis. The reason for this definitely lies in the fact that readily dissolved organic compounds are the first to decompose in the initial stages, thus providing micro-organisms easier access to organic carbon and nitrogen, needed for tissue synthesis. Resistant organic compounds lag behind in the second stage of analysis thus making it difficult for micro-organisms to access their nutrition supply. The whole process during the research has an exponential layout. Every next step of decomposition has a slower development than the former, due to the increased lagging of resistant matter. Some 50% of initial organic matter (in the case of Italian oak and the blend of Italian and Turkey oak), becomes mineralised already during the first year, and 28% in the case of Turkey oak. During further processes of fermentation and humification unmineralised parts of organic matter become transformed soil organic matter of a specific nature, known as "humus".

Bearing in mind that the decomposition processes of Italian oak and Turkey oak organic fall develop in this specific case in natural conditions of oak communities and that indentical abiotic conditions regulate this process in both cases, one cannot but conclude that the differences in decomposition rate of these species are determined by the differences in their chemical composition.

TABLE 1. Statistical analysis of decomposition rate (k, 3/k, 5/k) of dominant plant organic fall, and its mixtures in the *Quercetum farnetto-cerris* forest

Species	k M (SD)	F	*Q.farnetto*(F)	*Q. cerris* (F)	*Q. farnetto:Q. cerris* (F)
Q. farnetto	0.718 (0.152)	12.629**		13.884**	0.422
Q. cerris	0.336 (0.122)		13.884**		26.714***
Q. farnetto:Q. cerris	0.660 (0.109)		0.422	26.714***	
	3/k M (SD)	F	*Q.farnetto*(F)	*Q. cerris* (F)	*Q. farnetto:Q. cerris* (F)
Q. farnetto	4.176 (0.962)	10.395**		9.841*	0.674
Q. cerris	8.920 (3.478)		9.841*		12.031**
Q. farnetto:Q. cerris	4.546 (0.665)		0.674	12.031**	
	5/k M (SD)	F	*Q.farnetto*(F)	*Q. cerris* (F)	*Q. farnetto:Q. cerris* (F)
Q. farnetto	6.966 (1.603)	10.481**		9.839*	0.884
Q. cerris	14.886 (5.798)		9.839*		12.275**
Q. farnetto:Q .cerris	7.578 (1.260)		0.884	12.275**	

* P < 0.05, ** P < 0.01, *** P < 0.001

Statistical analysis of decomposition rate of dominant plants and their mixtures in the examined ecosystem prove that there are considerable differences in the decomposition rates of the examined species and their mixtures after a period of 12 months ($P < 0.01$), Tab. 1. If the second resolution level is analysed, ie if these types of organic fall are mutually compared, it is noted that there are virtually no significant differences between Italian oak and the mixture of Italian oak and Turkey oak at all stages of analysis (Tab. 1). This leads to the conclusion that Italian oak organic fall has crucial influence on the litter decomposition of this ecosystem during the first year when 50 % of organic matter of Italian oak and the mixture of Italian oak and Turkey oak get decomposed. Indirectly this leads us to the assumption that during the later stages of decomposition with the presence of a considerable remainder of resistant matter, the litter decomposition process of this ecosystem is controlled by Turkey oak organic matter.

TABLE 2. The dynamics of C (%) and N (%), and the C:N ratio in organic matter during decomposition

Species	Onset of the experiment		
	C (%)	N (%)	C:N
Q. farnetto	48.00	1.44	31.04
Q. cerris	39.02	0.93	41.95
Q. farnetto:Q. cerris	44.70	1.29	34.65
	After 6 months of decomposition		
Q. farnetto	44.70	1.87	25.66
Q. cerris	36.77	1.08	34.04
Q. farnetto:Q. cerris	44.10	1.73	25.49
	After 12 months of decomposition		
Q. farnetto	40.50	2.30	17.60
Q. cerris	34.83	1.45	24.02
Q. farnetto:Q. cerris	43.30	1.87	23.22

The C:N ratio is an indicator which is usually taken into account during the first level of discussion and analysis of the process of decomposition. It is believed that the process of decomposition develops smoothly if the C:N ratio is lower than 25 in the initial organic source,whereas if it is greater than 25 decomposition takes on a slower pace(Swift *et al.*1979). Analysis of carbon and nitrogen concentrations and their dynamics during the process of decomposition in the forests of Italian oak and Turkey oak proved that in the very beginning the C:N ratio was the most favourable in the organic fall of Italian oak, that of the Italian oak and Turkey oak mixture, and the least favourable ratio was noted in the organic matter of

Turkey oak. The above-mentioned gradient of this ratio indicated at the very start the decomposition rate gradient of organic matter of these species. The C:N ratio narrows down during decomposition at all stages concerned. This is the result of oxidation and mineralisation firstly of carbon on the one hand, and the increase of nitrogen concentration on the other. The nitrogen increase in litter during decomposition is associated with the immobilisation of nitrogen content in bodies of micro-organisms during leaf humification (Swift *et al.* 1979). A further analysis of this relation also indicated that the decomposition process had a quicker pace during the first six months (when a sudden narrowing of the C:N ratio occurred) as opposed to the second six months of decomposition, in line with the same decomposition gradient already noted (Italian oak, mixture of Italian oak and Turkey oak, Turkey oak).
The theoretical scenario of future decomposition development would prove that organic matter of Itelian oak would be the first to decompose (95% in 4.17 years), organic matter of the Italian oak and Turkey oak mixture in 4.54 years and the longest period would be required by Turkey oak to decompose (8.92 years). Statistical analysis has shown that there are no statistical differences only between the Italian oak litter and that of the Itelian oak and Turkey oak litter mixture, which stresses the significant role of the Italian oak litter as part of the *Quercetum farnetto-cerris* ecosystem litter, which contributes to its decomposition rate, ie a quicker release of nutrients and their quicker inclusion in the processes of circling. The same gradient was noted in the case of complete mineralisation (99 %, 5k), which also stressed the importance of Italian oak organic matter in the improvement of the quality of chemical composition of the nutrition supply of Italian oak and Turkey oak litter (Tab. 1).

4. Conclusion

Based on the above analyses the conclusion is that organic matter of Italian oak is quicker to decompose than that of Turkey oak. This gradient of decomposition rate is only confirmed by the C:N ratio in organic matter of the examined species. The decomposition rate time prognosis has proved that Italien oak litter is the first to mineralise, then the litter of the Italien oak and Turkey oak mixture, and the longest period is required for Turkey oak litter mineralisation. The fact that abiotic conditions remained unchanged at all levels of analysis leads to the conclusion that the above-mentioned gradient of decomposition first depended upon the differences in chemical composition of the examined species.

5. References

Meentemeyer, V., Berg, B. (1986): Regional variation in rate of mass loss *Pinus sylvestris* needle litter in Swedish pine forests as influenced by climate and litter quality. *Scand. J. For. Res.*, 1, 167-180.

Olson, J. S. (1963): Energy storage and the balance of producers and decomposers in ecological systems. *Ecology*, 44, 322-331.

Ponomareva, V. V., Plotnikova, T. A. (1975): Odnovremenoe opredelenie obscego sodrzanija ugleroda i azota v torfjanih pocvah metodom Anstata v modifikacijii Ponomarevoj i Nikolaevoj. In: *Metodiceskie ukazanija po opredeleniju sodrzaja i sostava gumusa v pocvah (mineralnih i torfjanih)*, 79-83. Vsejuzna Akad. Selkohoz. nauk V. I. Lenina i Centr. Muzej Pocvoved. V. V. Dokucaeva. Leningrad.

Raulund-Rasmussen, K., Vejre, H. (1995): Effect of tree species and soil properties on nutrient immobilization in the forest floor. *Plant and Soil*, 168-169, 345-352.

Swift, M. J., Heal, O. W., Anderson, J. M. (1979): *Decomposition in Terrestrial Ecosystems.* Studies in Ecology, Vol. 5. 372 pp. Blackwell Scientific Publications, Oxford, London.

Vesterdal, L., Dalsgaard, M., Felby, C., Raulund-Rasmussen, K., Jørgensen, B. B. (1995): Effects of thinning and soil properties on accumulation of carbon, nitrogen and phosphorus in the forest floor of Norway spruce stands. *Forest Ecology and Management*, 77, 1-10.

PHOTOSYNTHESIS OF OAK SPECIES UNDER NATURAL DROUGHT

K. RADOGLOU
Forest Research Institute,
Vassilika, Thessaloniki 57006, Greece

1. Introduction

The genus *Quercus* includes several species distributed all over Europe and the Mediterranean area with scientific interest for their productivity, wood quality, and ability to overcome the environmental constrains. Oak forests in Greece cover about 22 % of the total forested area. Oak trees, occurring naturally in Greece, form a group of eleven (11) different species, adapted to a wide range of habitats differing widely in moisture availability. Drought affects plant physiological processes particularly in Mediterranean ecosystems, hence, it is important to understand the physiological mechanisms involved in the response of oak trees to drought and to explore differences between species. Differences in drought tolerance among species may be partly due to the differential sensitivity of the photosynthetic process in leaves. However, the physiological response of oak species to drought, has not been investigated with comparative studies on trees growing in natural conditions.

The aim of the present study was to improve our knowledge on ecophysiological mechanisms related to drought adaptation and to identify differences in drought adaptation among oak species. An additional aim of the study was to assess the respective role of stomatal and non stomatal processes in the decline of CO_2 assimilation during drought.

2. Materials and methods

The experimental area was located close to the Forest Research Institute near Thessaloniki (Latitude 40 35', Longitude 22 58' and Altitude 10 m). It consisted of 30 years old separate stands of *Q. macrolepis Kotschy, Q. pubescens Willd* and *Q. trojana Webb.* The mean annual precipitation was 409 mm (range 211-

704 mm). The dry period coincides with summer, beginning in June and lasting until October. The area belongs to the *Ostryo-carpinion* vegetation zone (Dafis, 1973), where *Q.pubescens* grows naturally.

Leaf water potential was measured with a pressure chamber on three leaves per sampled tree, harvested between 11.30 to 14.00 hours. Gas exchange measurements were made using a portable photosynthesis system (Li 6400, Li-Cor, Inc. Lincoln, NE USA), on two mid canopy and south facing branches, on each of the nine (9) selected trees per species. Observations were made during midday on seven different dates throughout the growing period.

3. Results - Discussion

Precipitation was high until mid -June, followed by a dry season until the end of October. During that period soil moisture declined progressively. The seasonal progress of midday leaf water potential (Ψ) of the three oak species showed no significant differences in Ψ among species during the study period. Values of Ø at the beginning and the end of the study period were high (0.5-1.0 MPa), coinciding with relatively high soil moisture, whereas Ψ declined rapidly from early June and reached values of less than -4 MPa in mid-September.

The minimum values of Ψ for the oak species are comparable to those of other *Quercus* spp. growing under water stress; values of -3.5, -2.8 and -3.4 MPa, were recorded for other oak species and for other trees of the Mediterranean ecosystem (Tenhunen *et al*, 1987; Nunes *et al*, 1992).

The seasonal pattern of net assimilation (A) was similar in all study species. During May, under high soil moisture conditions, the assimilation rates at midday were high in all species and particularly in *Q. macrolepis*. As the soil dried, the assimilation rate decreased in all species. In middle June, when Ψ was around -2 MPa, assimilation had declined in all species with *Q. macrolepis* showing the highest assimilation rate and *Q. trojana* the lowest. By the mid-August the lowest A values were recorded, a month before the lowest Ψ values were observed, with *Q. macrolepis* showing the highest assimilation rate and *Q. trojana* the lowest. Low values of assimilation were observed until October. Assimilation rates recovered in all species by early November after drought had ended when *Q. pubescens* had the highest values *and Q. trojana* the lowest. In general, the highest values of assimilation were observed during early November. The maintenance of a high photosynthetic rate for most of an oak leaf's life is important. Consequently the ability of oak species to keep a high photosynthetic rate in November is advantageous for tree growth. The recorded values were found to be in the range reported for oaks species (Ceulemans and Saugier, 1991; Goulden., 1996).

No significant differences in stomatal conductance were observed among species during the growing season, except in November when the *Q. trojana* had higher values. Throughout the season low values of stomatal conductance were recorded, with the exception of early May and November, indicating that stomata were fully open for a limited period only. Recorded values of stomatal conductance were similar to those reported for leaves from other *Quercus* species (Ni and Pallardy, 1991) and *Ceratonia siliqua* in the Mediterranean area (Nunes *et al*, 1992).

Drought induces stomatal closure and has usually been considered a primary cause of decreased assimilation rate as stomatal closure reduces diffusional CO_2 supply to the chloroplasts and water diffusion out of the leaf (Roupsard *et al*., 1996; Dickson and Tomlinson, 1996). Physiological response to moisture stress associated with carbon fixation can be conveniently divided into stomatal and nonstomatal response. To remove stomatal effects on assimilation the results are also expressed in terms of intercellular CO_2 concentration (Ci) .

Our results showed that drought leads to a decreasing net assimilation rate and an increase or stabilisation in intercellular CO_2 concentration (Ci), in all study species. Various nonstomatal responses such as light energy reaction, mesophyll resistance to CO_2 diffusion, rubisco, carbon fixation and other enzyme reactions may be affected by water stress and decreased assimilation rates. Our results confirmed the view that the limitation of assimilation rate could not be ascribed to the stomatal closure alone as values of intercellular CO_2 concentration (Ci) remained high during stress.

4. Conclusion

This study demonstrates photosynthetic characteristics along with leaf water potential measurements in 30 years old stands of *Q. macrolepis*, *Q. pubescens* and *Q. trojana* during one growing period.

There were no significant differences in leaf water potential among study species during the measured period.

The seasonal patterns of assimilation rate were similar in all study species, although *Q. macrolepis* showed the highest values. Increased values of assimilation rate were recorded for all study species in May and in November after rain events. The ability of the oak species to sustain a high photosynthetic rate in November is advantageous for tree growth.

No significant differences in stomatal conductance were observed among the species during the growing season. Throughout the growing season low values of stomatal conductance were recorded, except during early May and

November

Decrease in assimilation rate caused by drought is accompanied by stability and even increase in intercellular CO_2 concentration (Ci). The limitation of assimilation rate in *Q. macrolepis Q. pubescens* and *Q. trojana* could not ascribed to the stomatal closure alone.

5. References

Ceulemans, R. & Saugier, B. (1991) Photosynthesis, in A. S., Raghavendra (ed) *Physiology of trees* , John Wiley & Sons, New York, pp. 21-49.

Dafis, Sp. (1973) Classification of forest vegetation in Greece, *Sci. Ann. School Agric. For.* **15**, 57-91. (in Greek).

Dickson, R. E. & Tomlinson , P.T. (1996) Oak growth, development and carbon metabolism in response to water stress, *Ann. Sci. For.* **53**, 181-196.

Goulden, M..L. (1996) Carbon assimilation and water-use efficiency by neighboring Mediterranean - climate oaks that differ in water access, *Tree Physiology* **16**, 417-424.

Ni. B.R. & Pallardy, S.G. (1991) Response of gas exchange to water stress in seedlings of woody angiosperms. *Tree Physiology* **8**, 1-9.

Nunes, M.A., Ramalho, J. & Pijo, P. (1992) Seasonal changes in some photosynthetic properties of *Ceratonia siliqua* under natural conditions. *Plant Physiology* **86**, 381-387.

Radoglou, K. (1996) Environmental control of CO_2 assimilation rates and stomatal conductance in five oak species growing in field conditions in Greece. *Ann. Sci. For.* **53**, 269-278.

Roupsard, O. Gross, P. & Dreyer, E. (1996) Limitation of photosynthesis activity by CO_2 availability in the chloroplasts of oak leaves from different species and during drought, *Ann. Sci. For.* **53**, 243-254.

Tenhunen, J. D Beyschal, W. Lange, O.L. & Harley, P.C. (1987) Changes during summer drought in leaf CO_2 uptake rates in macchia shrubs growing in Portugal: limitation due to photosynthetic capacity carboxylation efficiency and stomatal conductance, in F. M., Tenhunen *et al. (eds), Plant response to Stress*, Nato ASI Series G, Vol. Heidelberg, pp. 305-328.

EFFECT OF HILLING ON OVERWINTERINGOF SEED SUGAR BEET

M. RAJIC [1] AND B. MARINKOVIC[2]
[1]*Institute of field and vegetable crops, 21000 Novi Sad, Maksima Gorkog 30, Yugoslavia*
[2]*Faculty of Agriculture, 21000 Novi Sad, Yugoslavia*

1. Abstract

An investigation was conducted in the period 1986 m1996 in two locations, Novi Sad and Kikinda, to study overwintering of seed sugar beet with and without hilling. The planting was done at the same time using the same seed rate. Percentage of winterkilling was determined in early spring, by counting the number of survived plants. In the course of winter, soil temperature was monitored at two depths, 5 and 10 cm. The percentage of winterkill varied from year to year. The highest percentages (61-100%) were recorded in the period 1989 - 1993. In these years, negative mean monthly temperatures occurred in two consecutive winter months, in both locations and at both depths. Snow blanket was either missing or a thin blanket (0-5 cm) lasted for only a few days. In the other years, when the winterkilling percentage ranged between 3 and 35%, the remaining number of seed sugar beet plants was satisfactory. The ten-year investigation made it clear that climate is the only factor important for success of overwintering. It was noticed that extensive winterkilling occurs in the Vojvodina Province in one out of three years. No significant differences could be found between the hilled and non-hilled plots regarding the number of overwintered plants.

2. Introduction

When large amounts of water turn into ice, they release a large amount of latent heat energy. This is why waterlogged soils freeze to a smaller depth than less moist sandy soils (Ciric, 1984). Snow blanket acts as a heat insulator. Snow-covered soils are warmer 3-4ACthan those without snow cover. The depth of soil freezing depends on the thickness of snow cover (Antic *et al.*,1982).

3. Material and method

Overwintering performance of hilled and non-hilled seed sugar beet was studied in the period 1986 m1996 in the locations, Kikinda, on the black sandy soil of humic-gley origin, calcareous or limeless, and Srbobran, on south-Pannonian calcareous chernozem soil. The planting of all lines used in the experiment was done at the same time using the same seed rate.

4. Results

During the ten-year investigation period, in both regions, four years were distinguished for high percentages of winterkill (61 m100%): 1989/90, 90/91, 91/92 and 92/93 (Table 2). In these years the seed plots were plowed under. In the other years, the percentages ranged from 4 to 35%.

The highest and the lowest percentage of winterkill per location and year did not correspond exactly to the highest and lowest number of plants per plot because these numbers were determined by the number of plants per plot that entered the winter period.

In the months with negative mean monthly temperatures, there either was no snow blanket at all or a thin blanket of 1-2 cm lasted for short spells of 1-3 days. Also, the mean monthly temperatures for the soil layer 0-5 cm were lower than those for the layer 5-10 cm, in both locations.

The number of plants before the winter period was similar in all years save 1987/88, 1993/94 and 1994/95, when it was higher or lower. However, the differences in planting rate bore no effect on overwintering performance of the hilled and non-hilled seed sugar beet.

5. Discussion

In our investigation conducted in two locations and on two soil types there was no effect of hilling on the overwintering performance of seed sugar beet.

The 10-year monitoring singled out only one factor that definitely affected the overwintering performance of seed sugar beet m the climate. The Vojvodina Province has a moderately warm climate and moderately warm winters (Savic, 1979). Such climate is suitable for sugar beet seed production by the one-year procedure

Tab.1- Number of plants of seed sugar beet

Year	Hybrid	Srbobran			Hybrid	Kikinda		
		Before winter	After winter			Before winter	After winter	
			Non-hilled	Hilled			Non-hilled	Hilled
			000/ha				000/ha	
1986/87	NS-Hy-11	153	139	141	KW Perla	141	116	113
87/88	NS-Hy-11	120	108	107	PN mono 4	107	91	94
88/89	NS-Hy-11	155	135	140	KW Duka	140	109	105
89/90	KW Perla	158	10	10	KW Maja	110	10	10
90/91	KW Perla	153	47	46	KW Maja	146	57	37
91/92	PN mono 4	154	31	31	KW Maja	131	0	0
92/93	PN mono 4	159	30	29	NS-Hy-11	130	43	42
93/94	PN mono 4	210	176	178	Dana	170	116	116
94/95	NS-Hy-11	126	120	121	Dana	188	127	120
95/96	Dana	130	92	99	Nera	140	99	109

of Brewbaker (1960). Extensive winterkilling occurred in four out of the ten years of study. It means that, when the one-year procedure is applied in the Vojvodina Province, one in each three years is risky for sugar beet seed production. In those unsuitable years, there occurred two consecutive months with negative mean monthly temperatures in the soil layers 0-5 and 5-10 cm. The periods of a negative soil temperature lasted up to of the soil 7-9 days: during these periods, the mean daily temperature ranged from -7 to -9AC.

High seed yields were obtained in the first three years, 1.70-2.45 t/ha, low seed yields in the last three years, 0.78-0.91 t/ha. There were no significant differences in yield level between the variants with and without hilling.

In our investigation, the percentage of winterkill was similar in six years to the results obtained by Jasem *et al.* (1981) and Slavov (1983), i.e., the seed plots retained their status since the percentage of winterkill was below 30%.

6. Conclusions

The ten-year investigation of the effect of hilling on overwintering of seed sugar beet showed that, in the Vojvodina Province, the risk of large-scale freezing occurs in one out of each three years. Hilling had no effect on the overwintering performance of seed sugar beet; the dominant effect was exhibited by climatic factors which, however, change from one year to another. The number of plants before the beginning of winter had no effect on the overwintering performance of seed sugar beet regardless of hilling. The practice of hilling had no effect on yield of seed sugar beet.

7. References

1. Jassem M., Joachemjak A. (1981): Uproscennye metody vyrascivanija semjan obnorostkovych sortov saharnoj svekly, *JHAR-Bygoszsz*, pp. 11-12.
2. Slavov K. (1983): Prauchvane vahrkny njakoi ot uslvijata za prezimuvane na zakharnoto sveklo za semeproizvodstvo po bezrazsadnija, *Rast Nauki*, pp.71-77, Sofija.
3. Antic M., Jovic N., Avdalovic V. (1982): *Pedologija*, Toplotne osobine i toplotni rezim zemljista, "Naucna knjiga" pp.92-96, Belgrade.
4. Brewbaker H.E. (1960): *Final report*, Belgrade, pp. 14-16.
5. Ciric M. (1984): *Pedologija*, Toplotne osobine zemljiLta, "Svjetlost", 135-139, Sarajevo.
6. Savic S. (1979): *Klimatska klasifikacija Jugoslavije* po Koppenu. Prilozi poznavanja vremena i klime SFRJ, Savezni hidrometeoroloLki zavod, Beograd, Sv. 7, pp. 1-36.

EFFECT OF Cd AND Pb ON PHOTOSYNTHESIS IN CUCUMBER

L. SYMEONIDIS[1], I. KIRALY[2], E. SARVARI[2],
E. CSEH [2], Z. SZIGETI [2] and F. LANG [2]
[1] *Department of Botany, School of Biology, Aristotle University of Thessaloniki, Greece* [2]*Department of Plant Physiology, Eötvös University, Budapest, Hungary*

Heavy metal pollution strongly affects plant metabolism including photosynthesis. The effects of Cd and Pb (10 μM) on photosynthetic apparatus were investigated at different leaf storeys of hydroponically cultured (4 μM Fe chelated with either EDTA or citrate) cucumber plants. Cd given from the fourth-leaf stage of plants retarded their growth, particularly that of roots and reduced the amount of all pigment containing complexes in thylakoids, the PSII core being a little more resistant. The organization of treated thylakoids seemed to be very similar to that of iron deficient plants. Pb caused similar but moderate changes. EDTA exerted a protective effect. Fluorescence induction parameter Fv/Fm showed a slight change. However, severe inhibition of CO_2 fixation was observed at all leaf storeys.

1. Introduction

Pb and Cd are major pollutants accumulating in the environment and can enter the food chain by absorption through plant roots and leaves (Foy et al., 1978; Greger et al., 1993). The symptoms of these metals in plants are the inhibitions of growth and photosynthetic activity (Krupa and Baszynski, 1995; Titov et al., 1995). The aim of the present work was to investigate the in vivo effects of Pb and Cd in the presence of EDTA or citrate as iron (III) chelatin-agents on the growth and photosynthetic activity of more or less developed cucumber leaves.

2. Materials and Methods

Cucumis sativus L. plants were grown hydroponically in ¼ strength Hoagland solution with 4 μM Fe supplied as Fe-EDTA or Fe-citrate. Heavy metals Pb $(NO_3)_2$, $Cd(NO_3)_2$ were added in 10 μM concentration starting in four-leaf stage of plants. 6-week-old plants bearing 7 leaves were harvested. The chlorophyll content was determined as in Porra et al. (1989). Chloroplasts, thylakoids and chlorophyll-containing bands were isolated by Deriphat PAGE (Sarvari and Nyitrai, 1994). Light-induced CO_2 fixation was determined according to Lang et al. (1985). Fluorescence induction parameters were measured with a PAM fluorometer.

3. Results and Discussion

The influence of Fe-chelator on plant growth was slight, but in some cases apparent, with Fe-EDTA playing a protective role (Fodor et al., 1996). The retardation of root development was stronger than that of shoot, especially in the case of Cd, increasing the ratio of shoots to roots (on a fresh weight basis). Changes in the chlorophyll content (expressed in $\mu g\ cm^{-2}$) of treated leaves were in parallel with the developmental stage of the leaf at the beginning of the treatment, decreasing chlorophyll content only in those leaves which developed under heavy metal stress (Fig. 1). Pb treatment had only slight effect.

Cd and Pb caused decrease in chlorophyll containing complexes, in order of PSI>LHCII>PSII core complex (Fig. 2), except of the Pb-treated plants if iron was supplied as Fe-EDTA. In more developed leaves these heavy metals did not induce growth retardation or destruction of chlorophyll (Fig. 2). The organization of thylakoids in these Cd and Pb treated plants was in many respects similar to those of iron deficient plants (Fodor et al., 1995).

Fluorescence induction parameter Fv/Fm of treated plants showed a slight change even in the young leaves treated with Cd in the presence of Fe-citrate (Fig. 3).

However, photosynthetic activity measured as CO_2 fixation was relatively strongly influenced in the leaves which developed under heavy metal stress (Fig. 4). Cd caused higher inhibition in less developed leaves, while Pb caused moderate inhibition that was somewhat higher in more developed leaves.

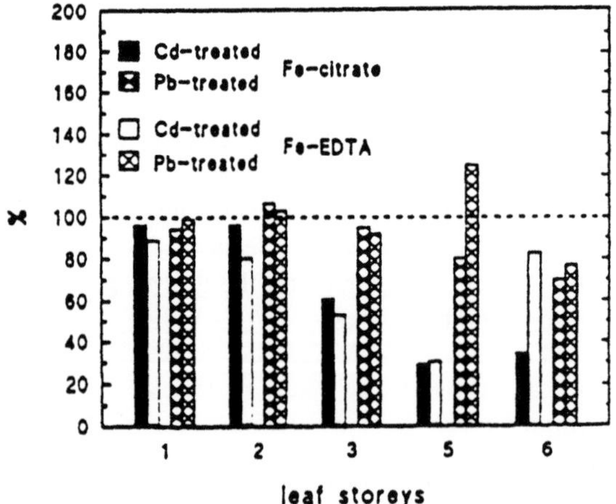

Figure 1. Effect of Pb and Cd on the chlorophyll content of differently treated cucumber leaves. Control values (100%) of all leaf storeys were around 35-40 $\mu g\ cm^{-2}$.

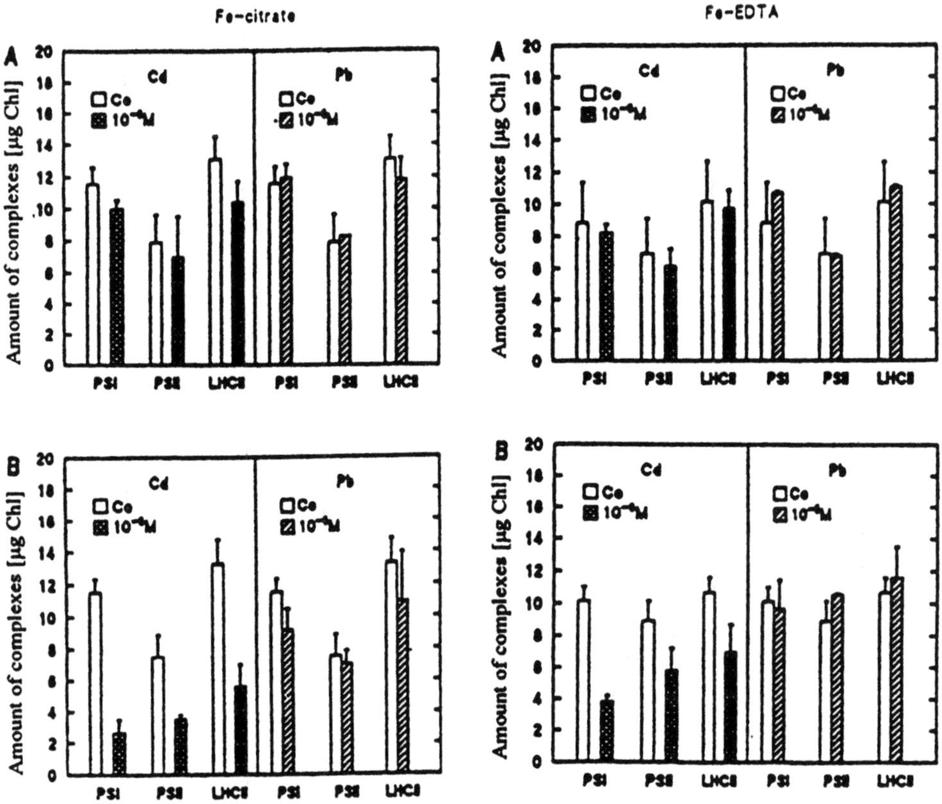

Figure 2. The absolute amounts of chlorophyll-containing complexes in differently treated cucumber leaves: second and third leaves (A), and fifth and sixth leaves of plants (B). The amount of a given complex shown here was calculated by summing up the denistogram bands corresponding to this complex. Then the chlorophyll content of leaves was shared among the complexes according to their relative proportions.

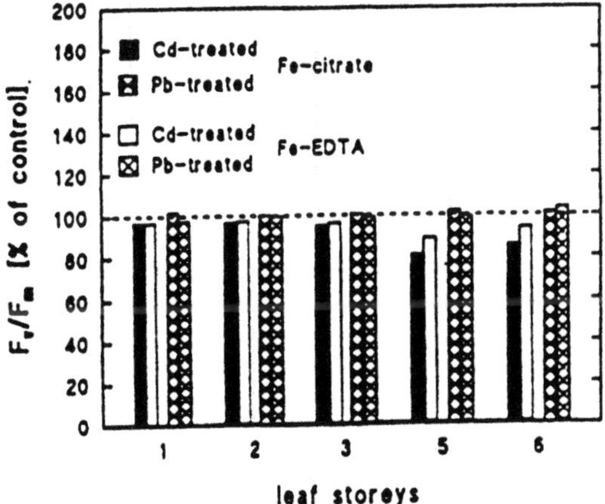

Figure 3. Effect of Pb and Cd treatment on the variable fluorescence (Fv/Fm) of different leaf storeys of differently treated cucumber plants. Control value was 0.8.

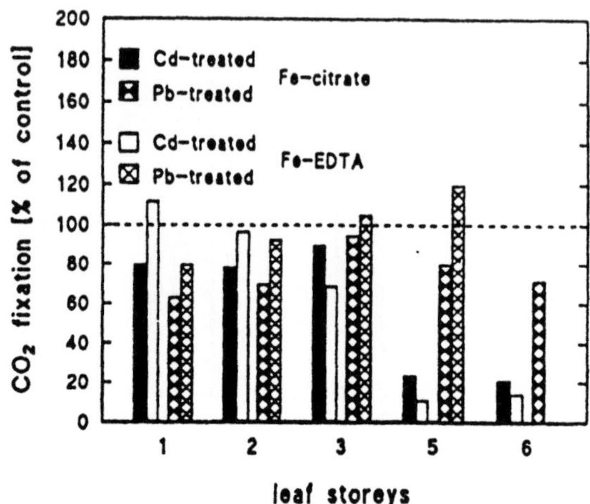

Figure 4. **Effect of Pb and Cd on the $^{14}CO_2$ fixation of differently treated cucumber plants.**

According to some authors the inhibiting effect of heavy metals on CO_2 assimilation can be related to the slower rate of electron transport of PSI and PSII (Krupa et al., 1993; Siedlecka and Baszynski, 1993).

4. References

Fodor, F., Boddi, B., Sarvari, E., Zaray, Gy., Cseh, E., and Lang, F. (1995) Correlation of iron content, spectral forms of chlorophyll and chlorophyll-proteins in iron deficient cucumber (*Cucumis sativus*), *Physiol. Plant.* **93**, 750-756.

Fodor, F., Sarvari, E., Lang, F., Szigeti, Z., and Cseh, E. (1996) Effects of Pb and Cd on cucumber depending on the Fe-complex in the culture solution, *J.Plant Physiol.* **148**, 434-439.

Foy, C.D., Chaney, R.L., and White, M.C. (1978) The physiology of metal toxicity in plants. *Annu. Rev. Plant Physiol.* **29**, 511-566.

Greger, M., Johansson, M., Stihl, A., and Hamza, K. (1993) Foliar uptake of Cd by pea (*Pisum sativum*) and sugar beet (*Beta vulgaris*), *Physiol. Plant.* **88**, 563-570.

Krupa, Z. and Baszynski, T. (1995) Some aspects of heavy metals toxicity towards photosynthetic apparatus – direct and indirect effects on light and dark reactions. *Acta Physiol. Plant.* **17**, 177-190.

Krupa, Z., Oquist, G., and Huner, N.P.A. (1993) The effects of cadmium on photosynthesis of *Phaseolus vulgaris* – a fluorescence analysis, *Physiol. Plant.* **88**, 626-630.

Lang, F., Sarvari, E., and Szigeti, Z. (1985) Apparatus and method for rapid determination of photosynthetic CO_2 fixation of leaves, *Biochem. Physiol. Pflanzen* **180**, 333-336.

Porra, R.J., Thompson,W.A., and Kriedmann, P.E. (1989) Determination of accurate extinction coefficients and simultaneous equations for assaying chlorophyll a and b extracted with four different solvents: verification of the concentration of chlorophyll standards by atomic absorption spectroscopy, *Biochim. Biophys. Acta* **975**, 384-394.

Sarvari, E. and Nyitrai, P. (1994) Separation of chlorophyll-protein complexes by Deriphat polyacrylamide gradient gel electrophoresis, *Electrophoresis* **15**, 1068-1071.

Siedlecka, A. and Baszynski, T. (1993) Inhibition of electron flow around photosystem I in chloroplasts of Cd-treated maize plants is due to Cd-induced iron deficiency, *Physiol. Plant.* **87**, 199-202.

Titov, A.F., Talanova, V.V., and Boeva, N.P. (1995) Growth responses of barley and wheat seedlings to lead and cadmium, *Biol. Plant.* **38**, 431-436.

SOME PHYSIOLOGICAL AND BIOCHEMICAL CHANGES IN *Solanum melongena* L. GENOTYPES GROWN UNDER SALT CONDITIONS

R. TIPIRDAMAZ[1] AND Ş. ELLİALTIOĞLU[2]
[1] Hacettepe University, Faculty of Science, Department of Biology Beytepe Ankara - TURKEY, [2]Ankara University, Faculty of Agriculture Department of Horticulture, Ankara- TURKEY

1. Abstract

Six eggplant (*Solanum melongena* L.) genotypes from different origins were grown under controlled conditions in 50% Hoagland culture solution for a period of 3 weeks and later they were subjected to 150 mM NaCl for one month. Proline content, plant height stress index, dry matter stress index, chlorophyll stability index, membrane damage index, inhibitions in chlorophyll and carotenoid shynthesis and the changes in osmotic potential were determined. The interrelationsships between these parameters were discussed for the screening of eggplant genotypes for salinity tolerance.

2. Introduction

Salt stress is one of the major stresses affecting agricultural productivity especially in arid lands. Salinity is known to affect almost all aspects of plant growth, development and metabolism and it induces changes in their morphology and anatomy [1]. However, the type of physiological response and metabolic changes vary among plant species and as well as in between different organs of the plant,and also genetic variations for this biological process is known to exist among different genera, species or varieties of the same species[2]. These changes may make possible the use of some parameters in salt tolerance screening. The physiological parameters that are mostly used in stress tolerance screening are; accumulation of different inorganic ions, organic osmotica that play an important role in osmoregulation in relation to salinity stress[3]. Changes in leaf photosynthetic parameters and in the photosynthetic pigment composition of plants in response to salinity [2]; membrane injury index [4]; dry matter stress index (DMSI) and plant height stress index (PHSI) [5].
The research has been undertaken to examine the effect of salinity stress on the some parameters based on the physiological responses of eggplants to salinity that could be useful for screening.

3. Material and Methods

Eggplant seeds were germinated and grown in vermiculite which was moistened with half-strenth Hoagland solution in growth chamber with 12000 lux light at a temperature of 27±2 °C, 70 % RH, and a photoperiod of 16/8 h light/dark. The seeds of six different eggplant vaieties (Sinampiro, Antou Nasu, Bhangur, K 510, Yamashina, Nepali Local) were obtained from The Germplasm Storage Center of National Institute of Agrobiological Resource (NIAR) Ibaraki, Japan. The three-week old seedlings were transferred to aerated 50% Hoagland solution containing 150 mM NaCl concentrations and they were grown under salt condition for 4 weeks. After four weeks, plant height and plant weight were measured. Stress indices were plant height stress index (PHSI) and dry matter stress index (DMSI), and were calculated according to the method of Bouslama and Schapanagh [5].Proline, total chlorophyll, carotenoid and osmotic potential (OP) of leaf samples (2nd and 3rd from apex) and injury index of root samples were estimated from stressed and non-stressed plants. Proline was estimated according to the method of Bates et al. [6]. Chlorophyll was extracted in cold 80% acetone and estimated according to Arnon [7]'s method. Chlorophyll stability index (CSI) was calculated as the ratio of chlorophyll in plants under stress to control plants and it was expressed in percentage[8]. The changes in carotenoid content were calculated by using the formula of Kirk and Allen [9]. Ion leakage of root tissue was measured by following the method of Marsh [10]. NaCl injury to the membranes (membrane injury index) were expressed as % of control values. The sap was extracted from frozen leaf tissue in liquid nitrogen by following the method of Ashraf and Waheed [3] and osmotic potential was measured with Cryoscopic Osmometer.

4. Results and Discussion

Salinity decreased the plant's height, its dry matter and chlorophyll content, but increased the proline content. Membranes were also adversely affected from salinity. The six genotypes responded in widely different ways to salinity. Differences among the genotypes were significant. Antou Nasu, K 510 and Nepali Local significantly exhibited higher DMSI, PHSI and CSI and lower percent injury indicating higher membrane stability (Table 1, Fig.1). The reports indicate that environmental stresses reduced total chlorophyll [11].

TABLE 1. DMSI, PHSI, CSI, Membrane injury (%) in six eggplant varieties

Varieties	DMSI(%)	PHSI(%)	CSI(%)	Mebrane injury (%)
Sinampiro	63.04 c	72.05 cd	73 a	56.91 c
Antou Nasu	81.41 a	84.57 b	95 c	46.96 a
Bhangur	47.62 d	77.57 bc	74 b	51.89 c
K 510	81.99 a	91.23 a	84 cd	44.92 bc
Yamashina	27.03 e	68.18 d	69 ab	55.09 bc
Nepali Local	74.54 b	92.82 a	87 d	42.00 ab

Means with the same letters in each column do not differ significantly at the 1% level

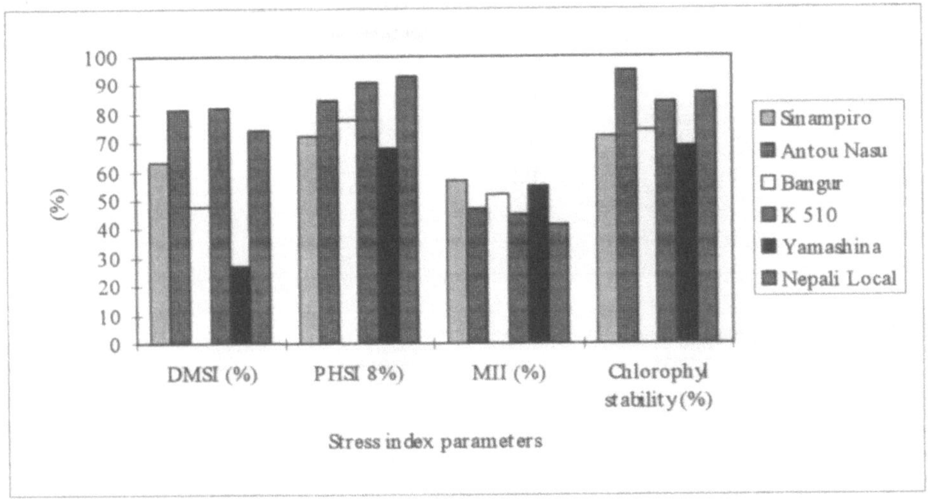

Figure 1. The effects of salinity on DMSI, PHSI, CSI, membrane injury (%) in six eggplant varieties

According to Blum [4], injury index signifies the degree of membrane damage due to stress. The maintenance of integrity of cell membrane is an important factor in developing a greater tolerance to environmental stresses [12]. Salinity decreased total chlorophyll and carotenoid content when compared to the controls in all of the genotypes. As shown in Table 2, the inhibitions in chlorophyll and carotenoid contents of Antou Nasu, K 510 and Nepali Local were lower than those in other genotypes. On the other hand, salt stress increased proline content in leaf tissue in all genotypes (Table 3). There were significant differences among the genotypes. While Sinampiro, Bhangur and Yamashina varieties accumulated 2.24, 3.43 and 3.62 times more proline than the control plants, Antou Nasu, K 510 and Nepali Local varieties accumulated 16.42, 27.60 and 14.36 time more proline than the controls. Perez et al. [13] reported a high relationship between proline level and stress tolerance.

TABLE 2. The effects of 150 mM NaCl on the chlorophyll and carotenoid synthesis of eggplant varieties

Varieties	*Chlorophyll*			*Carotenoid*		
	δ^{A}663 Control	*δ^{A}663 Stress*	*inhibition (%)*	*δ^{A}480 Control*	*δ^{A}480 Stress*	*inhibition (%)*
Sinampiro	0.977	0.690	29.37	0.430	0.284	33.76
Antou Nasu	0.968	0.888	8.13	0.341	0.309	9.20
Bhangur	1.280	0.965	24.60	0.347	0.264	23.85
K 510	1.256	1.041	17.11	0.346	0.297	14.05
Yamashina	1.019	0.718	29.53	0.376	0.284	24.62
Nepali Local	0.957	0.809	16.00	0.335	0.285	15.00

Leaf osmotic potential (Table 3) of all genotypes decreased significantly under salt conditions. Antou Nasu, K 510 and Nepali Local had significantly lower leaf osmotic potential than the other genotypes. Correlations among the stress parameters were significant, except from the relationship between CSI and PHSI, CSI and OP, and between CSI and proline increasing to be within 5% error limits (Table 4).

TABLE 3. Mean leaf osmotic potential (-MPa) and proline content (μMol / g fresh weight) in six eggplant varieties.

Varieties	*Proline content*		*Osmotic potential*	
	Control	*Stress*	*Control*	*Stress*
Sinampiro	29.775	66.839	1.07	1.54 bc
Antou Nasu	2.424	39.826	0.86	2.02 a
Bhangur	15.237	52.294	0.93	1.85 ab
K 510	2.597	71.688	0.71	2.03 a
Yamashina	11.082	40173	0.84	1.47 c
Nepali Local	3.809	32.552	0.85	2.08 a

Table 4. Correlations among six stress indices in eggplant.

	DMSI	PHSI	Membran injjury (%)	CSI	OP
PHSI	0.782**				
MI (%)	-0.706*	-0.955**			
CSI	0.580*	0.497	-0.635*		
OP	0.769**	0.924**	-0.914**	0.546	
Proline increasing	0.722**	0.673*	-0.652*	0.534	0.639*

*,** Significant at 0.05 and 0.01 level of propability, respectively.

5. References

1. Sharma, S.K., and Gupta, I.C. 1986. *Saline environment and plant growth.* Agro-Bot. Publ. (India):Bikaner.
2. Belkhodja, R., Morales, F., Abadia, A., Gomez-Aparisi, J., and Abadia, J. 1994. Chlorophyll flourescence as a possible tool for salinity tolerance screening in barley (*Hordeum vulgare* L.). *Plant Physiol.* **104**: 667-673.
3. Ashraf, M, and Waheed, A. 1993. Organic solute status and water relations of some salt tolerant and salt sensitive accessions of lentil (*Lens culinaris*). *Acta Bot.* **42**(1): 63-72.
4. Blum, A. 1985. Breeding crop varieties for stress environments. *CRC Critical Rev. in Plant Sci.* **2**(**3**): 199-238.
5. Bouslama, M., and Schapanagh, W.T. 1984. Stress tolerance in soy beans. I. Evaluation of three screening techniques for heat and drough tolerance. *Crop Sci.* **24**: 933-937.
6. Bates, L.S., Waldren, R.P., and Teare, I.D. 1973. Rapid determination of free proline for water stress studies. *Plant and Soil* **39**: 205-207.
7. Arnon, D.I. 1949. Copper enzymes in isolated chloroplasts. Polyphenoloxydase in *Beta vulgaris. Plant Physiol.* **24**: 1-15.
8. Deshmukh, P.S., Sairam, R.K., and Shukla, D.S. 1991. Measurement of ion leakage as a screening technique for drough resistance in wheat genotypes. *Ind. J. Plant Physiol.* **34** (1): 89-91.
9. Kirk, J.T.O., and Allen, R.L. 1965. Dependence of chloroplast pigment synthesis on protein synthesis: Effect of actidione. *Biochem. and Biophys. Res. Com.* **21** (6): 523-530.
10. Marsh, L.E., Davis, D.W., and Li, P.H. 1985. Selection and inheritance of heat tolerance In the common bean by use of conductivity. *J. Amer. Soc. Hort. Sci.* **110**: 680-683.
11. Motsan, Z., Lerner, H.R., and Reinhold, L. 1988. Dependence of chlorophyll synthesis on NaCl or osmotic stress in NaCl adapted tobacco cells. *Plant Physiol. Biochem.* **26**: 29-34.
12. Chaudhuri, K., and Choudhuri, M.A. 1993. Effects of short term NaCl salinity stress on free radical mediated membrane damage in two jute species. *Indian J. of Experimental Biology* **31**: 327-331.
13. Perez, A.F., Estan, M.P., Santa Cruz, A., and Bolarin, M.C. 1993. Effects of salinity on nitrate, total nitrogen, soluble protein and free aminoacid levels in tomato plants. *J. Hort. Sci.* **68** (6): 1021-1027.

PRELIMINARY DATA ON FOOD PREFERENCES OF THE SEA URCHIN PARACENTROTUS LIVIDUS LAMARCK (ECHINOIDEA REGULARIA)

E. TSIAGA[1], S. HARITONIDIS[1] and E. LAZARIDOU[2]
[1]*Institute of Botany, School of Biology, University of Thessaloniki, 540 06 Thessaloniki, Greece* [2]*Greek Biotope/Wetland Centre, 14th km Thessaloniki - Mihaniona, 57001 Thermi, Greece*

1. Abstract

Food preferences of sea urchin *Paracentrotus lividus* Lamarck have been studied under laboratory conditions. Four different experiments were carried out using tanks filled with fresh seawater. Six individuals were placed in each tank and they were offered pairs of algal species to select and feed upon. Gut content analysis showed that it fed selectively on these algae. The range of feeding preference from "preferred" to "non preferred" was: *Ulva rigida* > *Dictyota dichotoma* *Enteromorpha linza* *Cymodocea nodosa* *Codium fragile* *Cladophora prolifera* *Padina pavonica* *Halopteris scoparia* *Cystoseira spinosa* *Posidonia oceanica* *Halimeda tuna* *Digenea simplex*.

2. Introduction

In infralittoral communities, *Paracentrotus lividus* Lamarck among herbivores is a determining factor which controls the dynamic of seaweeds and seagrass. On rocky substrates, when it is at high densities, it eliminates the erect stratum of photophilic algae and induces the formation of a bare patch dominated by encrusting algae (Verlaque & Nedelc, 1983). In laboratory conditions few experiments have been done in order to determine the food preferences of *P. lividus* (Knoepffler - Pefuy, 1987). In this investigation we tried to determine the food preferences *of P. lividus* in purpose of the acquaculture and the commercial harvesting of this echinoid.

3. Material and method

Sea urchins, benthic macrophytes and sea water used for the experiments were collected in shallow water of Potidaia (Kassandra, Chalkidiki peninsula).

Four different experiments were carried out from November 1996 to May 1997. At the laboratory, sea urchins were placed in 13 especially designed tanks filled with seawater. Each tank was divided into two equal compartments using a plastic grid. Temperature and photoperiod simulated natural levels. Aeration of each tank was achieved by the aid of an air pump. Six sea urchins were placed per tank. However, each compartment was stocked with different size classes. The sea urchins after being for 2.5 days on starvation were given pairs of algal species to select and feed upon. Ten grams of each macrophyte were offered to the individuals of each compartment (Knoepffer - Peguy et al. 1987, Frantzis 1992). After three days of feeding, the echinoids were removed from the tanks and subsequently preserved in formalin seawater (10%). 12 macrophytes were used: *Digena simplex, Cytoseira spinosa, Dictyota dichotoma, Halopteris scoparia, Padina pavonica, Cladophora prolifera, Codium fragile, Enteromorpha linza, Halimeda tuna, Ulva rigida, Cymodocea nodosa, Posidonia oceanica.*

Cut contents were analyzed using the "method of contacts" (Jones 1968). 100 contacts were identified in each gut content (Knoepffer - Peguy et al. 1987). In order to determine the mean food preference for the macrophytes examined, the results were analyzed by STATISTICA 5.0. Moreover, to test differences in mean food preferences between size classes, data of the four experiments was analyzed by using one-way analysis of variance (ANOVA) (Zar 1984).

4. Results and discussion

In total 186 individuals were studied. Mean food prefrence for the macrophytes examined is presented in figure 1. Data was analyzed by using one-way ANOVA in order to classify the echinoids in size classes and to test statistical differences between size classes and mean food preferences (Figs 2-3).

Gut content analysis showed that *Paracentrotus lividus* fed selectively on the macrophytes examined. Food preference decreased as follows: *Ulva rigida* *Dictyota dichotoma* *Enteromorpha linza* *Cymodocea nodosa* *Codium fragile* *Cladophora prolifera* *Padina pavonica* *Halopteris scoparia* *Cystoseira spinosa* *Posidonia oceanica* *Halimeda tuna* *Digenea simplex.*

The results showed that *Paracentrotus lividus* is a herbivore feeding mainly upon erect algae and seagrasses. These results are in agreement with similar studies (Fernandez 1989, Frantzis 1992, Knoepffler - Peguy et al. 1987, San Martin 1987). According to Grantzis (1992) P. lividus mostly preferred feeding upon erect algae illustrating an "apostatic generalist" behaviour. Knoepffler - Peguy et al. (1987) observed that P. lividus mostly preferred *Cystoseira mediterranea* and *Ulva rigida*; San Martin (1987) observed that *P. lividus* mostly preferred *Undaria pinatifida, Solieria chordalis* and *Ulva rigida.*

It seems that food preference may significantly affected by plant nutrient status and plant defensive characteristics. Frantzis (1992) analyzed food quality by measuring water content, organic matter, energy content, proteins, carbohydrates and phenolics of macrophytes examined. It seems that *P. lividus* is efficient in digesting proteins and soluble carbohydrates but incapable of digesting proteins and soluble carbohydrates but incapable of digesting structural polysaccharides. As a result, the relative proportion of a given macrophyte should affect its digestibility. In addition, phenolics and terpenoids contribute to chemical defenses in Rhodophyta and in Chlorophyta and Phaeophyta, respectively (Hay et al. 1988).

Furthermore, data analyzed showed that food preference of *Paracentrotus lividus* varied between size classes. The individuals of the first size class mostly preferred the algae with soft thalli, such as *Ulva rigida, Enteromorpha linza, Dictyota dichotoma* whereas the bigger echinoids preferred the robust filamentous and abrustive algae, like *Cystoseira spinosa* and *Cladophora prolifera.* According to Fernandez (1989)

Fig. 1. Total mean values, st. error and st. deviation of food preferences for urchins (CSCi) fed upon different macrophyte

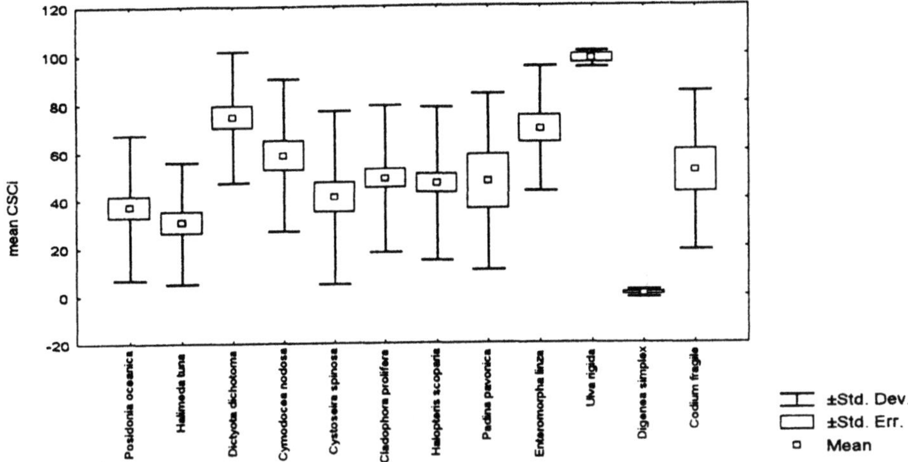

Fig. 2. Mean values of food preferences (CSCi) per different size class of urchins fed upon different macrophytes.

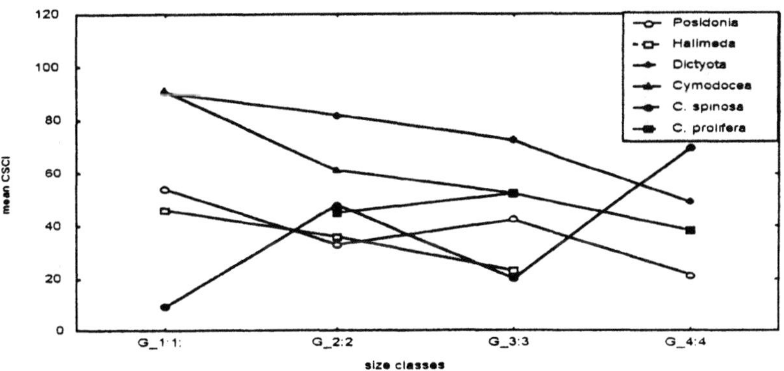

Fig. 3. Mean values of food preferences (CSCi) per different size class of urchins fed upon different macrophytes.

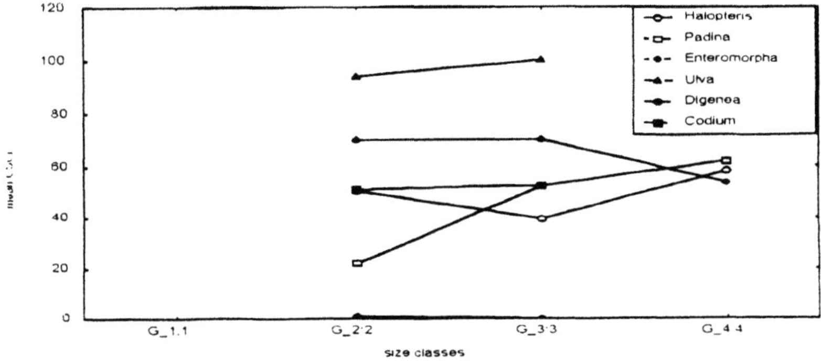

differences in food preference between size classes could be explained in relation with i) plant structure and capacity of younger individuals to browse it, ii) plant defensive compounds and the equivalent tolerance levels of younger and bigger echinoids.

In summary, because of its particular feeding behaviour, *P. lividus* is a key species controlling the photophilic sublittoral phytobenthic communities (San Martin, 1987; Verlaque, 1987; Verlaque & Nedelec 1983; Benedetti & Cinelli, 1995; Sala & Zabala, 1996).

5. References

Benedetti, L. & Cinelli, F. (1995) Habitat heterogenity, sea urchin grazing and the distribution of algae in littoral rock pools on the west coast of Italy (Western Mediterranean). Mar. Ecol. Prog. Ser., 126, 2-3-212.

Fernandez, C. (1989) Contribution a l'elaboration des bases scientifiques de l'aquaculture de l'oursin violete *Paracentrotus lividus* (Lamarck.). Mem. M.S.T. Val. Ress. Nat., Univ. Corse, Fr:: 1-40.

Frantzis, A. (1992). Etude experimentale des niveaux de consommation et d'utilization des macrophytes et des detritus derives par deux inverterbes benthique: *Paracentrotus lividus* (Echinoide, herbivore) et *Abra ovata* (Bivale, depositivore). These, Univ. Aix-Marseille II, 320p.

Jones, R.S. (1968) A suggested method for quantifying gut contents in herbivorous fishes. Micronesia, Guam. 4, 369-371.

Knoepffler-Peguy, M., Maggiore, F., Boudouresque, C. & Dance, C. (1987) Compte rendu d'une experience sur les preferenda alimentaires de *Paracentrotus lividus* (Echinoidea) a Banyuls - sur - Mer. Colloque international sur *Paracentrotus lividus* et lesoursins comestibles, C.F. Boudouresque ed., GIS Posidonie publ., Marseille, Fr., 1987, 59-64.

Sala, E. & Zabala, M. (1996) Fish predation and the structure of the sea urchin *Paracentrotus lividus* populations in the NW Mediterranean. mar. Ecol. prog. 140, 71-81.

San Martin, G.A. (1987) Comportement alimentaire de *Paracentrotus lividus* (Lmk) (Echinodermata: Echinidae) dans l'Etang de Thau (Herault, France). Coll. int. sur Paracentrotus lividus et les oursins comestible, Boudouresque C.F., ed., GIS Posidonie publ., Marseille, Fr: 37-57.

Verlaque, M. (1987) Relations entre *Paracentrotus lividus* Lamarck et le phytobenthos de medierranee occidentale. Coll. in. sur *Paracentrotus lividus* et les oursins comestible, Boudouresque C.F., ed., GIS Posidonie publ., Marseille, Fr: 5-36.

Verlaque, M. & Nedelec, H. (1983) Biologie de *Paracentrotus lividus* Lamarck sur in substrat rocheux en Corse (Mediterranee, Grance): Alimentation des adultes. Vie et Milieu 33, 191-201.

Zar, J.H. (1984) Biostatistical analysis, 2nd ed n Prentice- Hall, Englewood Cliffs, NJ.

PRELIMINARY RESULTS OF FIELD AND CULTURE STUDIES OF *PORPHYRA* ATT. AD. *LEUCOSTICTA* THURET IN LE JOLIS FROM THE GULF OF THESSALONIKI.

TSIRIKA,A.[1],ORFANIDIS,S.[2],HARITONIDIS,S.[1]
[1]*Department of Botany, Aristotle University of Thessaloniki,54006 Thessaloniki, Hellas.*
[2]*Fisheries Research Institute, 64007 Nea Peramos Kavala, Hellas.*

1. Abstract

Porphyra att. ad. *leucosticta* has been studied in the field, biotope of Epanomi, Gulf of Thessaloniki and under laboratory conditions. This species has been occured in the field from December 1996 to April 1997. The reproductive mode of the blade phase has been studied by using light microscope. Blade phase produced three types of spores: zygotospores, spermatia and monospores. The number of zygotospores and spermatia produced by each mother cell was 8 (a/2, b/2, c/2) and 32 (a/4, b/2, c/4), respectively. In exceptional cases the number of spermatia in each spermatangium was 64 (a/4, b/4, c/4). The zygotospores developed in culture by a unipolar growth mode into conchocelis plants and by a bipolar growth mode into blade plants. The monospores developed in culture by a bipolar growth mode into blade plants.

2. Introduction

The red macroalga *Porphyra leucosticta* is a species of the eulittoral and upper sublittoral zone with a wide spread to the temperate coasts of the North Atlantic Ocean and the Mediterranean Sea (Orfanidis 1990). This species occurs in the Gulf of Thessaloniki as an epiphytic species on macroalgae with big blades (Haritonidis 1978).

The typical life history of *Porphyra* species includes a conspicuous blade phase which alternates with a microscopic filamentous conchocelis phase (Cole and Conway 1980). Reproductive cells of both phases include monospores, spermatia and zygotospores (blade phase), monospores and conchospores (conchocelis phase) (Guiry 1990). The life history of a strain of *P.* att. ad. *leucosticta* from the Gulf of Thessaloniki has been studied by Orfanidis *et al.* (1996).

The aim of this study was to investigate the reproductive mode of *Porphyra* att. ad. *leucosticta* collected monthly from the field and to find out what the spores develope into in culture.

3. Materials and Methods

Mature blades of *Porphyra* att. ad. *leucosticta* were monthly collected from the field and brought to the laboratory for further examination. The blades were examined microscopically for the presence/absence of reproductive cells. Blade areas with monospores and others with spermatia and zygotospores were put in petri dishes filled with sterilized enriched with Von Stosch seawater (Guiry and Cunningam, 1984). The petri-dishes were put under optimal light (cool white fluorescent tubes, 15-25 μmol m^{-2} s^{-1}) and temperature (10-15°C) conditions (see Orfanidis *et al.* 1996).

Other blade areas were cut out and fixed in a solution of glutaraldehyde, sodium cacodylade buffer and sucrose. Washes were done using sodium cacodylade buffer plus sucrose and then pure buffer. The samples were postfixed with 1 % osmium tetroxide, dehydrated in ethanol (30%, 50%, 7O%, 90% and 100% for 15 minutes and 100% for 1 hour) and gradually embedded with resin. The samples were kept in pure resin for two days and then embedded in resin polymerized at 60°C for 24 hours. For light microscopy semi- thin sections were cut on glass knives using a microtome and then stained with toluidine blue in distilled water.

4. Results and Discussion

The blade phase of *Porphyra leucosticta* behaves as a pronounced summer alga in the cold temperate geographical regions (Kornmann and Sahling 1977) and as a pronounced winter/spring alga in the warm temperate regions (Gargiulo *et al.* 1994). *Porhyra* att. ad. *leucosticta* has been occured in the biotope of Epanomi, Gulf of Thessaloniki from December 1996 to April 1997. At the same period of time has this species also been collected in the biotope of Epanomi by Haritonidis (1978) and Orfanidis *et al.* (1993) .

In the laboratory, the zygotospores collected in December'96 developed by bipolar growth mode into blade plants. Form January to March'97 the zygotospores developed by a unipolar growth mode into conchocelis plants and by a bipolar growth mode into blade plants. In April'97 the zygotospores developed by a unipolar growth mode into conchocelis plants (Table 1). Development of zygotospores into blade and conchocelis phase has also been observed by Kornmann (1961), Cargiulo *et al.* (1994) and Orfanidis *et al.* (1996). However, the influence of time of collection on the development or zygotospores in culture, as far as we know, has not been mentioned. The monospores developed in culture by a bipolar growth mode into blade plants (Table 1).

The number of zygotospores and spermatia produced by each mother cell was 8 (a/2,b/2,c/2) (Fig. 1) and 32 (a/4, b/2, c/4) respectively (Fig. 2). In exceptional cases the number of spermatia in each spermatangium was 64 (a/4, b/4, c/4). The number of zygotospores and spermatia produced by each mother cell was 16 and 32 for strains from eastern costs of United States (Kapraun *et al.* 1991), and 16-32 and 128 for Uruguay strains (Coll and de Oliveira Filho 1977), respectively. Strains from Northern Europe and Mediterranean formed 4-8 or 8 zygotospores and 64 spermatia by each mother cell

Table 1. Type of spores found in the blade phase and their development after cultivation.

	Spores found in the blade phase	Development of spores after cultivation	
		Blade phase	Conchocelis phase
December'96	zygotospores	+++	---
	spermatia	+++	---
	monospores	+++	---
January'97	zygotospores	+++	+++
	spermatia	+++	+++
	monospores	+++	---
February'97	zygotospores	+++	+++
	spermatia	+++	+++
	monospores	+++	---
March'97	zygotospores	+++	+++
	spermatia	+++	+++
	no monospores found		
April'97	zygotospores	---	+++
	spermatia	---	+++
	no monospores found		

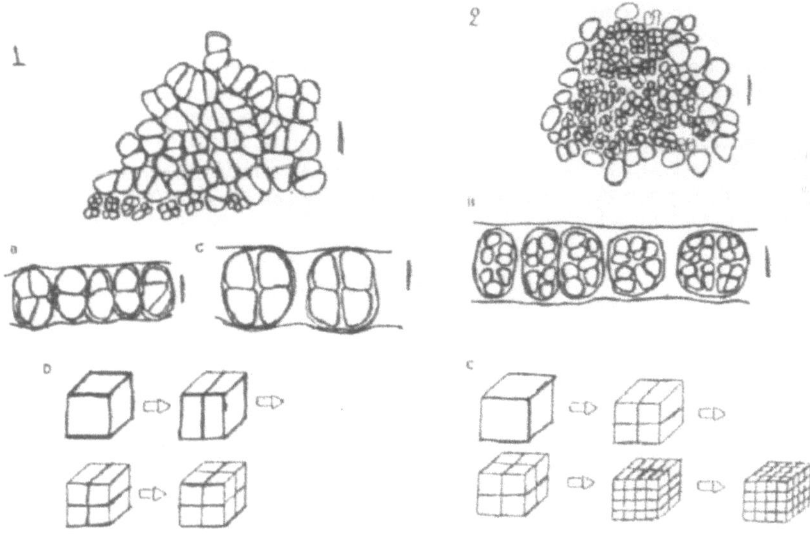

Fig.1. A: Surface view of thallus with zygotosporangia. Scale bar=30 μm. B: Cross section of thallus with zygotosporangia of different maturity. Scale bar= 14 μm. C: Cross section with mature zygotospoprangia. Scale bar=10 μm. D:Division sequence in the formation of a mature zygotosporangium.

Fig.2. A: Surface view of thallus with spermatangia. Scale bar=30 μm. B: Cross section of thallus with spermatangia of different maturity. Scale bar=14 μm. C: Division sequence in the formation of a mature spermatangium.

(Gargiulo *et al.* 1994). It seems that *Porphyra leucosticta* consists of regional populations that deserve recognition as separate species (Kapraun *et al.* 1991, Orfanidis *et al.* 1996).

5. References

Cole, K., Conway, K. (1980). Studies in the Bangiaceae. Reproductive modes. Botanica mar. 23: 545 - 553.

Coll, J., Oliveira Filho, E. C. De (1977). The nuclear State of reproductive cells of *Porphyra leucostica* Thuret in Le jolis (Rhodophyta, Bangiales). Phycologia 16: 227 -229

Gargiulo, G. M. , De Masi, F., Tripodi, G. (1994). Karyology and effects of temperature and protoperiod on the life-cycle of *Porphyra leucosticta* Thuret in Le Jolis (Bangiales, Rhodophyta) from the Mediterranean Sea. Jap J. Phycol. 42: 271-280.

Guiry, M. D., Cunningham, E. M. (1984). Photoperiodic and temperature responses in the reproduction of north - eastern Atlantic *Gigartina acicularis* (Rhodophyta: Gigartinales). Phycologia 23: 357-367.

Guiry, M. (1990), Sporangia and spores. In Biology of the red algae, eds. K. Cole and R. Sheath, pp. 347-376. Cambridge University press.

Haritonidis, S. (1978). A survey of the marine algae of Thermaikos Gulf, Thessaloniki, Greece. I. Distribution and seasonal periodicity. Botanica mar. 21: 527-535.

Kapraun, D. F. , Hinson, T. K. and Lemus, A. J. 1991. Karyology and cytophotometric estimation of inter- and intraspecific nuclear DNA variation in four species of *Porphyra* (Rhodophyta). Phycologia 30: 458-466.

Kornmann, P. (1961). Die entwicklung von *Porphyra leucosticta* im Kulturversuch. Helgolander Meeresunters. 8: 167-175.

Kornmann, P., Sahling, P. H. (1977). Meeresalgen von Helgoland. Benthische Grun-, Braun-, und Rotalgen. Helgolander wiss. Meeresunters. 29: 1-289.

Orfanidis, (1990). Temperature and photoperiodic responses of several Mediterranean macroalgae in relation to their distribution in the North Atlantic Ocean. PhD, Thessaloniki.

Orfanidis, S., Tsekos, J., Haritonidis, S. (1993). Temperature and photoperiodic responses of the red alga *Porphyra leucosticta* Thuret in Le Jolis (Bangiales, Rhodophyta) in Relation to its seasonality in the Gulf of Thessaloniki. Proc. 4th Nat. Symp. Ocean. & Fish., pp. 238-241.

Orfanidis, S., Tsekos, J., Haritonidis, S. (1996). Karyology and culture studies of *Porphyra* att. ad. *leucosticta* from the Gulf of Thessaloniki, Greece. 1st European Phycological Congress, Cologne. August 11-18, 1996. P: 32.

Structure and its Dynamics

TOWARDS A MOLECULAR MAP OF THE ALGAL BASAL APPARATUS

M. MELKONIAN, S. GEIMER, J. STEINKÖTTER,
K.-F. LECHTRECK
*Botanisches Institut, Universität zu Köln, Gyrhofstr. 15,
D-50931 Köln, FRG*

1. Introduction

The flagellar apparatus of flagellate algal cells is a complex cytoskeletal system consisting of several hundred polypeptides which together form the flagella (one to sixteen depending on the cell type) and the structures anchoring the flagella inside the cell, termed the basal apparatus. The basal apparatus comprises the basal bodies and fibrous material associated with the basal bodies. It is thought to be the functional equivalent of the centrosome of other eukaryotic cells since it is the main microtubule organizing center (MTOC; Pickett-Heaps, 1969) of the flagellate cell: the basal bodies function as a template for the assembly of the axoneme (the principal cytoskeletal component of a flagellum); additionally from the basal bodies originate the flagellar root microtubules which extend throughout the cell and mediate interactions with different cell organelles (Melkonian, 1984). Often these root microtubules also organize additional cytoskeletal microtubules which are collectively termed 'secondary cytoskeletal microtubules' (Pickett-Heaps, 1975). The function of the basal apparatus of algal cells as an organizer of the mitotic spindle (the genuine function of centrosomes; Kalnins, 1992) is less certain, during mitosis the basal apparatuses are often near or at the spindle poles but in some algal taxa they are located far from the spindle poles (Mattox and Stewart, 1984). Despite their common function, the structure of centrosomes varies considerably within eukaryotes, and as examples we refer to the trilamellar spindle pole bodies of yeast and related fungi, the acentriolar 'flexible centrosomes' of flowering plants or the typical centrosomes of mammalian cells consisting of two centrioles embedded in pericentriolar material (review by Kalt and Schliwa, 1993). It is therefore very likely that in addition to having many proteins in common different types of centrosomes will also display unique sets of proteins. Although considerable progress has been made in recent years to unravel the biochemical complexity of centrosomes (reviews: Kalt and Schliwa, 1993; Lange and Gull, 1996), in no system has this goal been achieved so far. In *Chlamydomonas reinhardtii* several mutants with defects in the structure of the basal apparatus have been characterized (Dutcher and Lux, 1989; Taillon et al., 1992; Ehler et al., 1995; Vashishtha et al., 1996) and γ-tubulin (a typical constituent of MTOCs; Joshi, 1994) was localized to the basal apparatus in interphase cells of *C. reinhardtii* (Dibbayawan et al., 1995). A different approach towards the identification of basal apparatus proteins depends on the isolation and biochemical characterization of the basal apparatus and its structural components. In this short review we describe the progress in the biochemical identification and (immuno)localization of algal basal apparatus proteins starting with the discovery of the first algal basal apparatus protein, centrin, to the most recent methodological advances involving highly purified basal apparatuses from the green alga *Spermatozopsis similis* (Geimer et al., 1997a). These studies will ultimately lead to the establishment of a molecular map of the algal basal apparatus.

2. Centrin and the Isolation of Flagellar Roots

Besides γ-tubulin, centrin is associated with many MTOCs in phylogenetically diverse organisms (reviews: Melkonian et al., 1992; Salisbury, 1995; Schiebel and Bornens, 1995). Comparative analyses of centrin coding sequences revealed that centrin belongs to a closely related protein family within the larger superfamily of Calcium-modulated EF-hand proteins (Kawasaki and Kretsinger, 1995). Considerable progress has recently been made to understand the molecular function of this unusual cytoskeletal protein (Wiech et al., 1996; Geier et al., 1996; Middendorp et al., 1997). Centrin was first characterized as a component of the striated flagellar roots (system II fibers) in the basal apparatus of the flagellate green alga *Tetraselmis striata* (Salisbury et al., 1984) and it was thus the first basal apparatus protein identified in an alga. System II fibers (review: Lechtreck and Melkonian, 1991) are often very prominent components of algal basal apparatuses and especially in the genus *Tetraselmis* they are massive, cross-banded fibers of several μm length extending from the basal bodies to the plasma membrane in the mid-part of the cell. The strategy that led to the discovery of centrin has been summarized previously (Melkonian et al., 1992) and it is therefore sufficient here to state that the system II fibers were isolated and partially purified from *T. striata* in quantities which allowed identification of their major protein (i.e. centrin) by SDS-PAGE. A polyclonal antibody raised against this protein was localized to the system II fibers *in situ*, verifying that the protein is a principal component of these fibers (Salisbury et al., 1984). This and other antibodies against centrin have subsequently been used to localize centrin in other structural components of the flagellar apparatus of green algae, namely in the distal connecting fiber (McFadden et al., 1987) and the flagellar transitional region (Schulze et al., 1987). Although the function of centrin in specific systems is still incompletely known, in general centrin is thought to play a fundamental role in centrosome duplication and contraction of centrin-based fibers (Schiebel and Bornens, 1995). Current centrin research centers around the identification of proteins which functionally interact with centrin in the basal apparatus.

3. SF-Assemblin and *In Vitro* Reassembly of Basal Apparatus Components

The second protein discovered in the green algal basal apparatus was SF-assemblin (Lechtreck and Melkonian, 1991). The methodological approach chosen was somewhat different from the one which led to the discovery of centrin: whereas with centrin, because of the conspicuous system II fibers, it was not necessary to isolate the basal apparatus intact, the cells being simply disintegrated by ultrasonication, with SF-assemblin, isolation of intact cytoskeletons (of *Spermatozopsis similis*) preceded their mechanical disintegration and the isolation of basal apparatus components. Whereas flagella can be easliy isolated from algal cells by amputation, the isolation of intact basal apparatuses is often complicated by the presence of a cell wall. Additionally, the preparations tend to be contaminated with cell debris. Basal apparatuses have been successfully isolated from several flagellate green algae naturally lacking cell walls or from cell wall-deficient mutants or after enzymatic removal of the cell wall (Gould, 1975; Snell et al., 1974; Stearns and Brown, 1979; Marano et al., 1985; Lechtreck et al., 1989). Enriched fractions of basal apparatuses from *C. reinhardtii* consist of more than 100 polypeptides, most of them distinct from polypeptides found in flagella as revealed by 2D-PAGE analysis (Dutcher, 1995). *Spermatozopsis similis* has been the subject of a series of studies describing its cellular architecture, the flagellar apparatus, and the isolated cytoskeleton (Melkonian and Preisig, 1984; Preisig and Melkonian, 1984; McFadden et al., 1987; Lechtreck et al., 1989; Geimer et al., 1997b). Because of the complete lack of a cell wall, this organism is particularly well suited for isolation of functionally intact cytoskeletons and for purification of structurally intact basal apparatuses in large quantities (McFadden et al., 1987; Lechtreck and Melkonian, 1991; Geimer et al., 1997a). The isolated basal apparatuses of *S. similis* have been used as a source for centrin and SF-assemblin (Höhfeld et al., 1994; Lechtreck and Melkonian, 1991). SF-assemblin is the major component of the striated microtubule-associated fibers (SMAFs, also known as system I fibers) in flagellate green algae and was first isolated from *S. similis* (Lechtreck and Melkonian, 1991; Lechtreck et al., 1996). SMAFs consist of highly ordered bundles of 2nm filaments. SF-assemblin (31-33 kD) has a pronounced coiled-coil forming ability since it contains an extended α-helical rod domain (Weber et al., 1993). It was identified as the principal protein component of the SMAFs because it was possible to enrich this protein severalfold through cycles of *in vitro* disassembly and reassembly of paracrystals followed by SDS-PAGE analysis (Lechtreck and Melkonian, 1991).The paracrystals obtained by *in vitro* reassembly resemble the SMAFs in overall polarity and striation pattern (Patel et al., 1992). SF-assemblin and related proteins (i.e. the giardins of the diplomonad protist *Giardia lamblia* and the kinetodesmal proteins of ciliates; Holberton et al., 1988; Peattie et al., 1989; Sperling et al., 1991) appear to be widespread among unicellular organisms, the lack of cross-reactivity of the antibodies currently available, however, has until now eluded their detection in mammalian cells. SF-assemblin and related proteins are representatives of a new class of α-helical proteins characterized by their ability to form special segemented coiled-coils and to assemble both *in vivo* and *in vitro* into striated fibers of 2nm protofilaments. Their function may be related to dissipation of mechanical stress generated by flagellar beating. A model of the SMAFs based on staggered protofilaments consisting of overlapping 36 nm subunits is presently being developed using mutated, recombinant SF-assemblin (Lechtreck, unpubl. observations).

4. Detection of Minor Basal Apparatus Proteins Using Highly Purified Basal Apparatus Preparations

Centrin and SF-assemblin were essentially identified as basal apparatus proteins because of their ability to form structurally unique macromolecular assemblies, i.e. cross-striated fibers of conspicuous size. These striated fibers could be recognized by electron microscopy through successive purification steps thus facilitating enrichment of their major protein constitutents. Most basal apparatus components, however, are much less prominent than these fibers. They cannot be readily isolated and their protein components are only minor constituents of the basal apparatus. To identify minor basal apparatus proteins highly purified basal apparatuses have been prepared from *S. similis* by sucrose density centrifugation of preparations enriched in basal apparatuses (Geimer et al. 1997a). SDS-PAGE revealed the absence of histones indicating the removal of nuclear contamination. Tubulin, centrin, SF-assemblin and many predominantly high-molecular mass proteins were enriched in the purified basal apparatuses. Purified basal apparatuses were used to produce monoclonal antibodies (mAbs). Several of the mAbs obtained stained in immunofluorescence one or two dots in the basal apparatus of *S. similis*. Three of the mAbs were further characterized (Geimer et al. 1997a). A 210 kD component (mAb BAS 1.4) was localized by immuno-electron microscopy to the flagellar transitional region. Using a different approach several protein bands were excised from SDS-PAGE gels of purified basal apparatuses and polyclonal antibodies were generated. In some cases no antibodies could be generated. A polyclonal antibody raised against a 95 kD protein labelled the sf-fiber (i.e. the fiber linking the basal bodies with the s-root microtubules; Beech and Melkonian, 1993; Geimer et al. 1997b) by immuno-electron microscopy. Screening of a cDNA library with this antibody resulted in a full length clone coding for a protein which was predicted to form a series of coiled-coils interrupted by short spacer elements (Geimer et al. submitted). Striated fibers with a periodicity resembling that of the native sf-fiber could be reassembled *in vitro* from the bacterially expressed protein. A slightly smaller protein was subsequently identified as a constitutent of the corresponding df-fiber of *S. similis* (unpubl. results). Based on the sequence of the corresponding cDNA clone this protein also has coiled-coil forming capabilities, but is structurally unrelated to the 95 kD protein of the sf-fibers.

5. Towards a Molecular Map of the Algal Basal Apparatus - The Future

The basal apparatus of flagellate green algae contains in excess of 100 different polypeptides (basal apparatus proteins: BAPs). Presently only 7 of them (α- , β-, and γ-tubulin, centrin, SF-assemblin, and two BAPs of around 90-95 kD) are known to any degree. It is anticipated that identification of the next seven BAPs will require much less time than the 15 years it took to discover the first BAPs. Reasons for optimism are: the much improved procedures to isolate purified basal apparatuses (from *S. similis*) in large quantities, high resolution 2D-protein gels of BAPs, generation of a large number of poly- and monoclonal antibodies against BAPs, and successful use of these antibodies to isolate cDNA clones encoding BAPs from well-designed cDNA libraries. In addition methods have been introduced to study protein-protein interactions by sandwich techniques using blot overlays of cDNA clones expressing BAPs (Steinkötter, unpubl. results). The latter methods will not only accelerate the identification of cDNA clones coding for new BAPs, but provide also indirect information on the location of new BAPs within the basal apparatus and thus complement the direct localization of BAPs using immunogold techniques.

6. Acknowledgements

The experimental work in our laboratory discussed in this review was supported by the Deutsche Forschungsgemeinschaft.

7. References

Dibbayawan, T.P., Harper, J.D.I., Elliott, J.E., Gunning, B.E.S., and Mare, J. (1995) A γ-tubulin that associates specifically with centrioles in HeLa cells and the basal body complex in *Chlamydomonas*, *Cell Biol. Int.* **19**, 559-567.

Dutcher, S.K. (1995) Flagellar assembly in two hundred and fifty easy-to-follow steps, *Trends Genet.* **11**, 398-404.

Dutcher, S.K. and Lux, F.G., III (1989) Genetic interactions of mutations affecting flagella and basal bodies in *Chlamydomonas*, *Cell Motil. Cytoskeleton* **14**, 104-117.

Ehler, L.L., Holmes, J.A., and Dutcher, S.K. (1995) Loss of spatial control of the mitotic spindle apparatus in a *Chlamydomonas reinhardtii* mutant strain lacking basal bodies, *Genetics* **141**, 945-960.

Geier, B.M., Wiech, H., and Schiebel, E. (1996) Binding of centrins and yeast calmodulin to synthetic peptides corresponding to binding sites in the spindle pole body components Kar1p and Spc110p, *J. Biol. Chem.* **271**, 28366-28374.

Geimer, S., Teltenkötter, A., Plessmann, U., Weber, K., and Lechtreck, K.F. (1997a) Purification and characterization of basal apparatuses from a flagellate green alga, *Cell Motil. Cytoskeleton* **37**, 72-85.

Geimer, S., Lechtreck, K.-F., and Melkonian, M. (1997b) The cytoskeleton of the naked green flagellate *Spermatozopsis similis* (Chlorophyta): analysis of isolated basal apparatuses in the parallel configuration, *J. Phycol.* **33**, 241-253.

Gould, R.R. (1975) The basal bodies of *Chlamydomonas reinhardtii*: formation from probasal bodies, isolation, and partial characterization, *J. Cell Biol.* **65**, 65-74.

Holberton, D., Baker, D.A., and Marshall, J. (1988) Segmented α-helical coiled-coil structure of the protein giardin from the *Giardia* cytoskeleton, *J. Mol. Biol.* **204**, 789-795.

Höhfeld, I., Beech, P.L., and Melkonian, M. (1994) Immunolocalization of centrin in *Oxyrrhis marina* (Dinophyceae), *J. Phycol.* **30**, 474-489.

Joshi, H.C. (1994) Microtubule organizing centers and γ-tubulin. *Curr. Opin. Cell Biol.* **6**, 55-62.

Kalnins, V.I. (1992) *The Centrosome*, Academic Press, San Diego.

Kalt, A. and Schliwa, M. (1993) Molecular components of the centrosome, *Trends Cell Biol.* **3**, 118-128.

Kawasaki, H. and Kretsinger, R.H. (1995) Calcium binding proteins 1: EF-hands, *Protein Profile* **2**, 305-490.

Lange, B.M.H. and Gull, K. (1996) Structure and function of the centriole in animal cells: Progress and questions, *Trends Cell Biol.* **6**, 348-352.

Lechtreck, K.-F., McFadden, G.I., and Melkonian, M. (1989) The cytoskeleton of the naked green flagellate *Spermatozopsis similis*: Isolation, whole mount electron microscopy, and preliminary biochemical and immunological characterization, *Cell Motil. Cytoskeleton* **14**, 552-561.

Lechtreck, K.-F. and Melkonian, M. (1991) Striated microtubule-associated fibers: Identification of assemblin, a novel 34-kD protein that forms paracrystals of 2-nm filaments in vitro, *J. Cell Biol.* **115**, 705-716.

Lechtreck, K.F., Frins, S., Bilski, J., Teltenkötter, A., Weber, K., and Melkonian, M. (1996) The cruciated microtubule-associated fibers of the green alga *Dunaliella bioculata* consist of a 31 kDa SF-assemblin, *J. Cell Sci.* **109**, 827-835.

Marano, F., Santa-Maria, A., and Krishnawamy, S. (1985) The flagellar apparatus of *Dunaliella*: Isolation of basal body-flagellar root complex, *Protoplasma* **127**, 82-92.

Mattox, K.R. and Stewart, K.D. (1984) Classification of the green algae: a concept based on comparative cytology, in D.E.G. Irvine and D.M. John (eds.), *Systematics of the Green Algae*, Academic Press, London, pp. 29-72.

McFadden, G.I., Schulze, D., Surek, B., Salisbury, J.L., and Melkonian, M. (1987) Basal body reorientation mediated by a Ca^{2+} -modulated contractile protein, *J. Cell Biol.* **105**, 903-912.

Melkonian, M. (1984) Flagellar root-mediated interactions between the flagellar apparatus and cell organelles in green algae, in W. Wiessner, D. Robinson and R.C. Starr (eds.), *Compartments in algal cells and their interactions*, Springer, Berlin-Heidelberg, pp. 96-108.

Melkonian, M., Beech, P.L., Katsaros, C., and Schulze, D. (1992) Centrin-mediated cell motility in algae, in M. Melkonian (Ed.), *Algal Cell Motility*, Chapman & Hall, New York, pp. 179-221.

Melkonian, M. and Preisig, H.R. (1984) Ultrastructure of the flagellar apparatus in the green flagellate *Spermatozopsis similis*, *Plant Syst. Evol.* **146**, 145-162.

Middendorp, S., Paoletti, A., Schiebel, E., and Bornens, M. (1997) Identification of a new mammalian centrin gene, more closely related to *Saccharomyces cerevisiae* CDC31 gene, *Proc. Natl. Acad. Sci. USA* **94**, 9141-9146.

Patel, H., Lechtreck, K.-F., Melkonian, M., and Mandelkow, E. (1992) Structure of striated microtubule-associated fibers of flagellar roots. Comparison of native and reconstituted states, *J. Mol. Biol.* **227**, 698-710.

Peattie, D.A., Alonso, R.A., Hein, A., and Caulfield, J.P. (1989) Ultrastructural localization of giardins to the edges of disk microribbons of *Giarida lamblia* and the nucleotide and deduced protein sequence of α-giardin, *J. Cell Biol.* **109**, 2323-2335.

Pickett-Heaps, J.D. (1969) The evolution of the mitotic apparatus: an attempt at comparative ultrastructural cytology in dividing plant cells. *Cytobios* **3**, 257-280.

Pickett-Heaps, J.D. (1975) Structural and phylogenetic aspects of microtubular systems in gametes and zoospores of certain green algae, in J.G. Duckett and P.A. Racey (eds.), *The biology of the male gamete*, Academic Press, London, pp. 37-44.

Preisig, H.R. and Melkonian, M. (1984) A light and electron microscopical study of the green flagellate *Spermatozopsis similis* sp. nova, *Plant Syst. Evol.* **146**, 57-74.

Salisbury, J.L. (1995) Centrin, centrosomes, and mitotic spindle poles, *Curr. Opin. Cell Biol.* **7**, 39-45.

Salisbury, J.L., Baron, A., Surek, B., and Melkonian, M. (1984) Striated flagellar roots: isolation and partial characterization of a calcium-modulated contractile organelle, *J. Cell Biol.* **99**, 962-970.

Schiebel, E. and Bornens, M. (1995) In search of a function for centrins, *Trends Cell Biol.* **5**, 197-201.

Schulze, D., Robenek, H., McFadden, G.I., and Melkonian, M. (1987) Immunolocalization of a Ca^{2+} -modulated contractile protein in the flagellar apparatus of green algae: the nucleus-basal body connector, *Eur. J. Cell Biol.* **45**, 51-61.

Snell, W.J., Dentler, W.L., Haimo, L.T., Binder, L.I., and Rosenbaum, J.L. (1974) Assembly of chick brain tubulin onto isolated basal bodies of *Chlamydomonas reinhardi*, *Science* **185**, 357-360.

Sperling, L., Keryer, G., Ruiz, F., and Beisson, J. (1991) Cortical morphogenesis in *Paramecium*: A transcellular wave of protein phosphorylation involved in ciliary rootlet disassembly, *Dev. Biol.* **148**, 205-218.

Stearns, M.E. and Brown, D.L. (1979) Purification of cytoplasmic tubulin and microtubule organizing center proteins functioning in microtubule initiation from the alga *Polytomella*, *Proc. Natl. Acad. Sci. USA* **76**, 5745-5749.

Taillon, B.E., Adler, S.A., Suhan, J.P., and Jarvik, J.W. (1992) Mutational analysis of centrin: An EF-hand protein associated with three distinct contractile fibers in the basal body apparatus of *Chlamydomonas*, *J. Cell Biol.* **119**, 1613-1624.

Vashishtha, M., Walther, Z., and Hall, J.L. (1996) The kinesin-homologous protein encoded by the *Chlamydomonas FLA10* gene is associated with basal bodies and centrioles, *J. Cell Sci.* **109**, 541-549.

Weber, K., Geisler, N., Plessmann, U., Bremerich, A., Lechtreck, K.-F., and Melkonian, M. (1993) SF-assemblin, the structural protein of the 2-nm filaments from striated microtubule associated fibers of algal flagellar roots, forms a segmented coiled coil, *J. Cell Biol.* **121**, 837-845.

Wiech, H., Geier, B.M., Paschke, T., Spang, A., Grein, K., Steinkötter, J., Melkonian, M., and Schiebel, E. (1996) Characterization of green alga, yeast, and human centrins - specific subdomain features determine functional diversity, *J. Biol. Chem.* **271**, 22453-22461.

NEW DEVELOPMENTS IN PLANT COATED VESICLE RESEARCH

D.G. ROBINSON[1], T. BRAULKE[2], J. DENECKE[3], N. HAPPEL[1], A. MOVAFEGHI[1], P. PIMPL[1]
[1]*Albrecht-von-Haller-Institut für Pflanzenwissenschaften, Universität Göttingen, Untere Karspüle 2, D-37073 Göttingen, Germany*
[2]*Biochemie II, Zentrum Biochemie und Molekulare Zellbiologie, Universität Göttingen, Goßlerstr.12d, D-37073 Göttingen, Germany*
[3]*The Plant Laboratory, Dept. Biology, University of York, Heslington, York YO1 5DD, UK*

1. Introduction

Many steps in the intracellular transport of proteins are facilitated by microvesicles (i.e vesicles with a diameter of less than 100 nm) which have a protein coat on their cytosolic surface. Morphologically speaking coated vesicles belong to one of two major classes: clathrin coated vesicles (CCV), which in transverse section have regular horn-like projections, and in tangential section or in surface view reveal a highly characteristic polygonal, honeycomb-like structure, and the COP-coated vesicles which have an unstructured, compact nap-like coat (Schmid and Damke, 1995). In mammalian cells these two vesicle types are also found at different locations: CCV are present at the plasma membrane (PM), where they act as the internalization vehicles for receptor-mediated endocytosis, and at the *trans*-Golgi network(TGN) and endosomes reflecting their participation as specialized agents for the transport of acid hydrolases and their receptors (Braulke, 1996). By contrast, COP-coated vesicles are formed at the endoplasmic reticulum (ER) and the *cis*- and medial cisternae of the Golgi apparatus, and are responsible for protein transport between these individual compartments (Schekman and Orci, 1996).

Endocytosis occurs via CCV in plant cells (Low and Chandra, 1994), even if proof that it is a receptor-mediated event remains unconvincing (Hawes et al., 1995). CCV also form at the TGN in plants, and although not involved in storage protein transport (Hohl et al., 1996), they are enriched in the vacuolar protein receptor BP-80 (Hinz et al., 1998) indicating that they may, like their counterparts in animal cells, carry hydrolases to lytic rather than storage vacuoles.

Mainly because of cytoplasmic streaming ER and Golgi are not always found in proximity to one another and this also does not allow the visualization of ER-derived vesicles waiting to fuse with Golgi membranes as frequently documented for animal cells. Vesiculation profiles at the ER, even in cells engaged in massive intracellular protein transport e.g. storage parenchyma cells of legume cotyledons, have rarely been depicted in higher plant cells. Convincing micrographs of COP-coated vesicles associated with plant Golgi membranes are also rare e.g Robinson et al. (1997). Nevertheless, as pointed out by Staehelin and Moore (1995), plants possess many proteins homologous to the COP proteins of mammals and yeast, so that it is very likely that COP-coated vesicles should be present in plants, as indeed our own investigations (reported below) confirm.

2. CCV coat polypeptides and their interaction with membrane proteins

The literature up to June 1996 on the components of the coat of plant CCV was reviewed by one of us (Robinson, 1996) so that we will only consider here data which has been reported since then.

The adaptins play a pivotal role in CCV function in that they interact, on the one hand with the cytosolic domains of trans-membrane receptors, and on the other with the clathrin cage surrounding the vesicle (Robinson, 1997). In mammalian cells the adaptins fall into three size groups: the $\alpha,\beta_1,\beta_2,\gamma$ adaptins with molecular mass around 100 kDa, the μ_1 and μ_2 adaptins of around 50 kDa, and the σ_1 and σ_2 adaptins at 20 kDa. They form two different tetrameric complexes: the AP-1 at the TGN and the AP-2 at the PM, each with a single 20, 50 and two 100 kDa adaptins. Evidence is accumulating which indicates that it is the μ adaptins which interact with membrane receptors by recognizing tyrosine-containing motifs of the YXXØ type in the cytoplasmic domain (Marks et al., 1997).

As far as plant adaptins are concerned a ß-type adaptin polypeptide has been identified and localized at the PM, but with properties more akin to a Golgi adaptin (for references see Robinson, 1996). Full-length cDNA clones for an α-adaptin (Schledzewski et al., 1997), a μ-type adaptin (Happel et al., 1998), σ_1- and σ_2-adaptins (Maldonado-Mendoza and Nessler, 1996; Roca et al., 1998), have been recently isolated, all with around a 50% identity to their animal counterparts. A cDNA clone for a γ-adaptin is not present in the currently available EST databases.

Experiments to examine the possibility of an interaction between adaptor and membrane receptors using plant extracts have also been initiated. We have attempted to show that the cytoplasmic (c) domain of the mammalian 300 kDa mannosyl-6 phosphate receptor (MPR), which has a tyrosine containing internalization motif (see Braulke, 1996), can associate with coat proteins from plant CCV. To do this clathrin cages were reassembled *in vitro* in the presence of the MPR c-domain; the cages were centrifuged out and the presence of the MPR c-domain determined by immunoblotting (Fig. 1a). Binding of the c-domain was significantly reduced in the presence of an excess of a 15-mer peptide (P1), corresponding to the amino acids 17-31 of the c-domain of human MPR-300 (RRSSNVSYK<u>YSKV</u>NK). The P1 peptide contains the internalization motif (underlined), and, because of its size, runs out of the gels used to detect the cage-associated 23 kDa MPR-300 c-domain. Parallel experiments with sporamin, a soluble vacuolar storage protein, suggest that the MPR c-domain binding was specific (Fig. 1b,c). Unfortunately, in these experiments the possible unspecific binding of the MPR c-domain to components of the clathrin coat other than the adaptor complexes cannot be ruled out. Similar binding experiments have been performed by Butler et al. (1997). However, they have used the c-domain of BP-80, which also possesses a tyrosine motif (Paris et al., 1997), anchored to an Affigel column. Adaptor-containing fractions from bovine brain and pea cotyledons were shown to bind specifically to the column. Such heterologous experiments are very encouraging, but clearly a specific binding between defined components from plant CCV must be demonstrated before a similar molecular mechanism for coat-membrane interactions to that described above for mammalian CCV can be generalized throughout the eukaryotes.

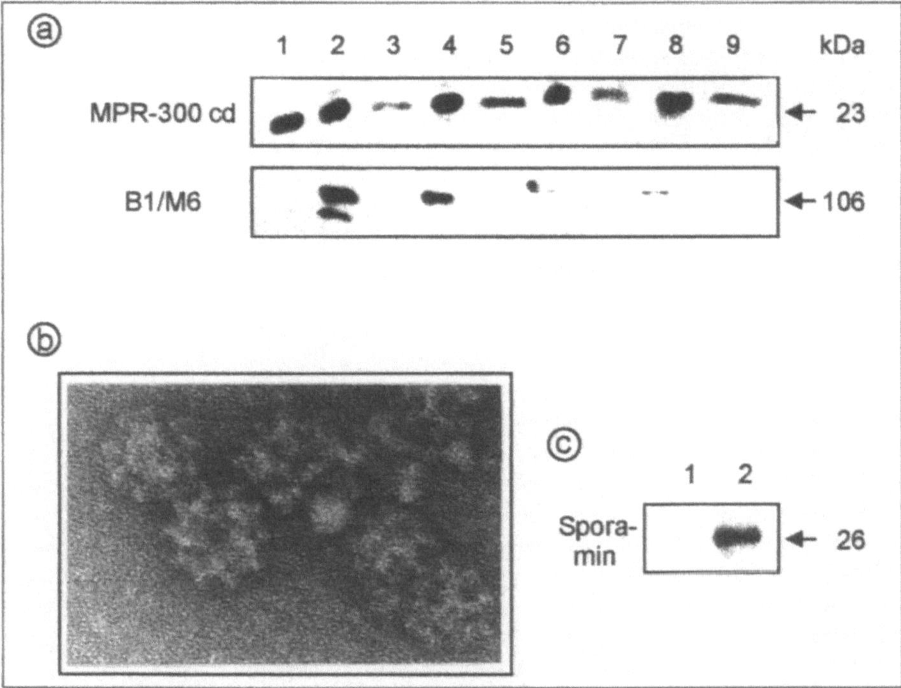

Fig.1. Analysis of the interaction between coat polypeptides isolated from CCV prepared from bovine brain or zucchini hypocotyls and the 164 amino acid long c-domain of the MPR-300 (23 kDa) through an *in vitro* reassembly assay. 50 µg coat proteins were dialysed together with 4 µg MPR-300 cd in a reassembly buffer containing 3mM $CaCl_2$ at pH 6.5 for 48 h at 4 °C. Reassembled „cages" were pelleted out of the dialysis bag by centrifugation at 110,000g for 1 h taken up in sample buffer and subjected to SDS-PAGE then immunoblotting with either antibodies against the MPR-300 cd or the ß-adaptin (B1/M6). Competition experiments were performed with a 500 molar excess of the „P1" peptide; control experiments were carried out with 10 µg sporamin and zucchini coat proteins. a) lane 1: MPR-300 cd, lane 2: brain cages plus MPR-300 minus P1, lane 3: supernatant from lane 2, lane 4: brain cages plus MPR-300 plus P1, lane 5: supernatant from lane 4, lanes 6-9 as for lanes 2-5 but with zucchini cages. b) negatively stained zucchini cages.c) sporamin does not reassemble into zucchini cages (lane 1) but stays in the supernatant (lane 2).

3. COP-coated vesicles

COP-coated vesicles were discovered as a result of experiments designed to characterize the properties of an *in vitro* protein transfer between Golgi membranes from wild type and mutant animal cells infected with vesicular stomatitis virus (reviewed by Rothman, 1994). Essentially, in addition to donor and acceptor membranes, the reaction mixture requires the presence of a cytosolic extract (the source of the COP-proteins), an ATP-regenerating system and the non-hydrolyzable GTP analogue, GTPγ S (which prevents the dissociation of the COP-coat). The events that are considered to take place are given below, in Fig. 2. The docked COP-coated vesicles, which accumulate at the surface of the acceptor membrane, are then removed by 250 mM KCl and separated from the donor/acceptor membranes by ultracentrifugation. This is the strategy which we have followed with plant membrane and cytosolic fractions.

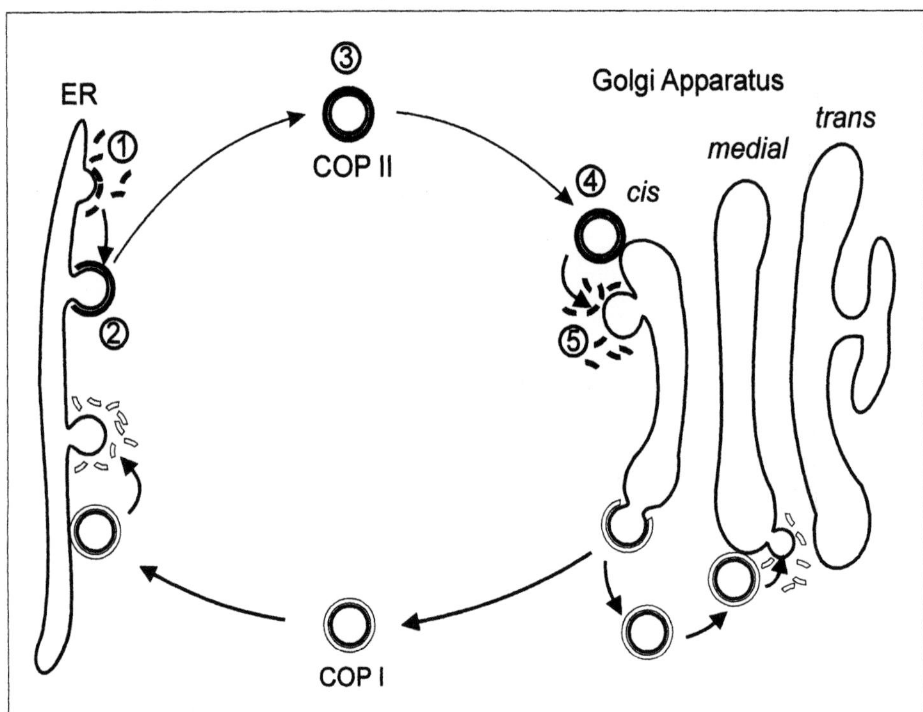

Fig. 2 Intracellular transport mediated by COP-coated vesicles. 1. Association of coatomers at donor membrane. 2. Budding of COP-vesicles at donor membrane. 3. *in vitro* transport. 4. Recognition and attachment at acceptor membrane. 5. Dissociation of coatomers and vesicle fusion.

In order to be able to induce COP-coated vesicles from plants four prerequisites have to be fulfilled: a. ER/Golgi fractions must be readily available, b. an easy-to-assay cargo molecule is necessary, c. a source of plant COP-proteins is required, and, d. antibodies against one or more of these proteins need to have been raised. By using leaves from tobacco plants expressing barley amylase we have met requirements a. and

b. The cytosol from cauliflower inflorescence appears to be a good source for COP proteins, not only because it is available in large quantities, but also because it has low endogenous protease levels and taxonomically is very close to *Arabidopsis*. An EST dbase search for cDNA sequences in *Arabidopsis* which are homologous to those corresponding to sec21p (a COP I coat protein) and sec23p (a COP II protein), their expression in *E.coli* as GST-fusion proteins and the production of antibodies against these proteins has enabled us to probe post-microsomal fractions for COP-coated vesicles. Our isolation protocol for a mixed ER/Golgi fraction is presented in Fig. 3A, and the procedure for inducing and separating COP-coated vesicles is given in Fig. 3B. Latent α-amylase activity was measured spectrophotometrically using a blocked nitrophenyl-maltoheptaoside as substrate in the presence and absence of detergent.

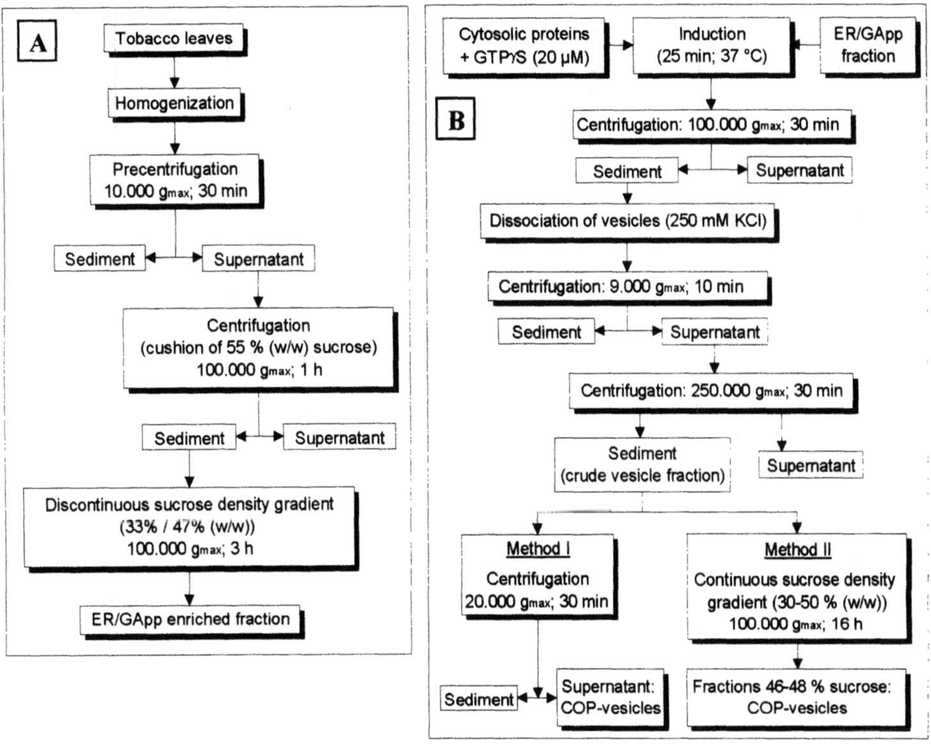

Fig. 3. Flow charts for the *in vitro* induction of COP-coated vesicles. A. Preparation of a mixed ER/Golgi fraction. B. Isolation of COP-vesicles.

Sedimentation of the mixed tobacco ER-Golgi membranes after incubation in the presence of cauliflower cytosol and GTPγ S leaves a post-microsomal supernatant in which latent α-amylase activity is measurable. In the absence of GTPγ S, or when heat-denatured cytosol is used this activity is considerably reduced. When the post-microsomal supernatant was layered onto a linear sucrose gradient and centrifuged to equilibrium conditions, latent α-amylase activity peaked at around 48% (w:w) sucrose.Only these fractions in the gradient cross-reacted with the *Arabidopsis* COP-antisera, and, when fixed and sectioned, 50 nm vesicles were detected in them (Fig. 4).

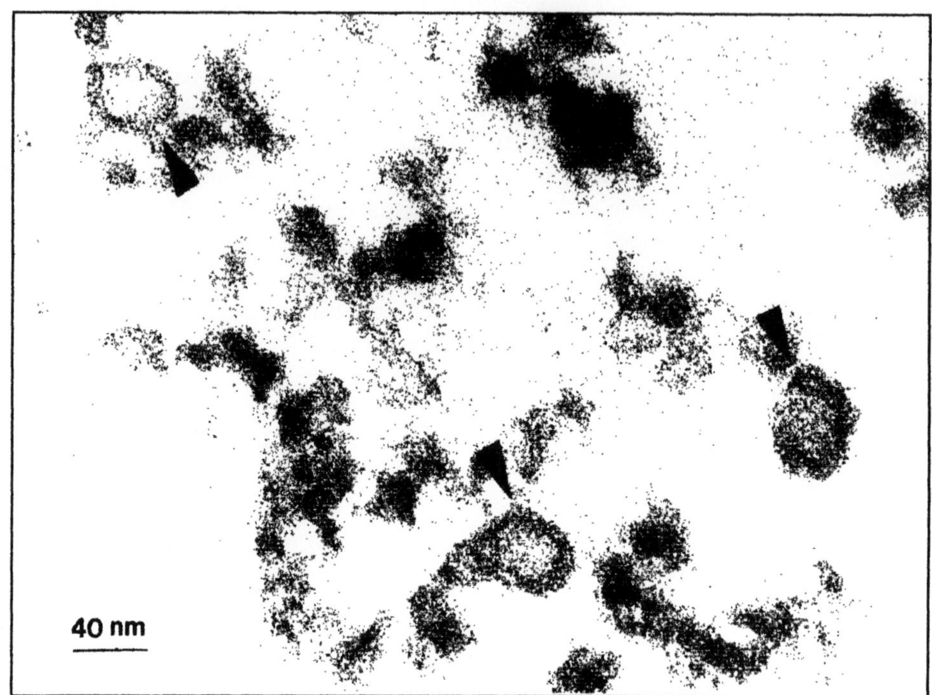

Fig. 4. Thin section through isolated COP-vesicles induced *in vitro*.

Using another strategy we have probed fractions from isopycnic sucrose density gradients prepared from cauliflower homogenates under high and low Mg^{2+} conditions. Again fractions equilibrating around 48% (w:w) sucrose cross-reacted strongly with the Atsec21p and Atsec23p antisera. ER-containing fractions also reacted with Atsec23p antibodies, whereas Golgi membranes were recognized by Atsec21p antibodies, observations which are in keeping with the production of COP II vesicles at the ER and COP II vesicles at the Golgi apparatus. Thus, we have good reason to conclude that COP-coated membranes are also a feature of plant cells, and that their types and properties may well be analogous to those in mammalian and yeast cells.

Acknowledgements. Financial support of the Deutsche Forschungsgemeinschaft (SFB 523) and the European Community (Contract# CHRX-CT94-0590) is greatly acknowledged. AM is the recipient of a scholarship of the Iranian Government.

References

Braulke, T. (1996) Origin of lysosomal proteins, in J.B .Lloyd and R.W. Mason (eds.), *Biology of the Lysosome, Subcellular Biochemistry* **27**, 15-49. Plenum Press, New York.

Butler, J.M., Kirsch, T., Watson, B., Paris, N., Rogers, J.C., Beevers, L. (1997) Interaction of the vacuolar targeting receptor BP-80 with clathrin adaptors,

Plant Physiol. **114**, Suppl.Abstract # 1210.

Happel, N., Robinson, D.G., Holstein, S.E.H .(1997) An *Arabidopsis thaliana* cDNA clone (Acccession No. AF 009631) is homologous to the microadaptins of CCV adaptor complexes, Plant Physiol. (in press).

Hawes, C., Crooks,K.,Coleman,J.,Satiat-Jeunemaitre,B. (1995) Endocytosis in plants: fact or artefact? *Plant Cell Environ.* **18**, 1245-1252.

Hinz, G., Bäumer, M., Hohl, I. (1998) Vacuolar storage proteins and the vacuolar sorting receptor BP-80 are sorted into different transport vesicles in the Golgi apparatus of developing pea cotyledons, *The Plant Cell* (submitted).

Hohl, I., Robinson, D.G., Chrispeels, M.J., Hinz, G. (1996) Transport of storage proteins to the vacuole is mediated by vesicles without a clathrin coat, *J. Cell Sci.* **109**, 2539-2550.

Low, P.S. and Chandra, S. (1994) Endocytosis in plants, *Ann.Rev.Plant Physiol. Plant Mol.Biol.* **45**, 609-631.

Maldonado-Mendoza, I.E. and Nessler, C.L. (1996) Cloning and expression of a plant homolog of the Golgi-associated clathrin assembly protein AP19 from *Camptotheca acuminata, Plant Mol. Biol.*

Marks, M.S., Ohno, H., Kirchhausen, T., Bonifacino, J.S. (1997) Protein sorting by tyrosine-based signals: adapting to the Ys and wheretofores, *Trends in Cell Biol.* **2**, 124-128.

Paris, N., Rogers, S.W., Jiang, L., Kirsch, T., Beevers, L., Phillips, T.E., Rogers, J.C. (1997) Molecular cloning and further characterization of a probable plant vacuolar sorting receptor. *Plant Physiol.* **113**, 29-39.

Robinson, D.G. (1996) Clathrin-mediated trafficking, *Trends in Plant Sci.* **1**, 349-355.

Robinson, M.E. (1997) Coats and vesicle budding, *Trends in Cell Biol.* **7**, 99-102.

Robinson, D.G., Bäumer, M., Hinz, G., Hohl, I. (1997) Ultrastructure of the pea cotyledon Golgi apparatus: origin of dense vesicles and the action of brefeldin A, *Protoplasma* (in press).

Roca, R., Stiefel, V., Puigdomenech, P. (1998) Characterization of the coding sequence for the clathrin coat assembly protein AP17 associated to the plasma membrane from *Zea mays* and constitutive expression of its gene, *Gene* (in press).

Rothman, J.E. (1994) Mechanisms of intracellular protein transport, *Nature* **372**, 55-63.

Schekman, R. and Orci, L. (1996) Coat proteins and vesicle budding, *Science* **271**, 1526-1533.

Schledzewski, K., Brinkmann, H., LaBrie, S.T., Crawford, N.M., Mendel, R.R. (1997) Gamma-adaptin from *Arabidopsis thaliana*:: molecular cloning and characterization of the gene, in T. Baskin and J.C. Rogers (eds.), *Current Topics in Plant Biochemistry, Physiology and Molecular Biology* , **16**th Annual Missouri Symposium. pp. 59-60.

Schmid, S. and Damke, H. (1995) Coated vesicles: a diversity of form and function, *FASEB Journ.* **9**, 1445-1453.

Staehelin, L.A. and Moore, I. (1995) The plant Golgi apparatus: structure, functional organization and trafficking mechanisms, *Ann. Rev. Plant Physiol. Plant Mol. Biol.* **46**, 261-288.

Plant Physiol. 114, Suppl. Abstract # 1510.

[illegible] adaptor complexes, Plant Physiol., in press.

Hawes, C., Crooks, K., Coleman, J., Satiat-Jeunemaitre, B. (1995) Endocytosis in plants: fact or artefact? Plant Cell Environ. 18, 1245-1252.

Hinz, G., Hillmer, S., Bäumer, M., Hohl, I. (1999) Vacuolar storage proteins and the vacuolar sorting receptor BP-80 are sorted into different transport vesicles in the Golgi apparatus of developing pea cotyledons. The Plant Cell (submitted).

Hohl, I., Robinson, D.G., Chrispeels, M.J., Hinz, G. (1996) Transport of storage proteins [illegible] J. Cell Sci. 109, 2539-2550.

Low, P.S. and Chandra, S. (1994) Endocytosis in plants. Ann. Rev. Plant Physiol. Plant Mol. Biol. 45, 609-631.

[illegible]

[illegible]

THE BIOSYNTHESIS OF CELLULOSE IN RED ALGAE

I. TSEKOS
Institute of Botany, School of Biology,
Aristotle University of Thessaloniki, GR-540 06
Thessaloniki, Macedonia, Greece

1. Introduction

The understanding of the function of biological membranes requires detailed knowledge of their structural organization. Freeze-fracture investigations of most biological membranes has revealed intramembranous particles which are unevenly distributed between the protoplasmic (PF) and the exoplasmic (EF) fracture faces. Some intramembranous particles are aggregated to form specific complexes (for review, see Herth, 1985; Brown, 1996). In addition there is strong indirect evidence that the synthesis of cellulose microfibrils involves "terminal" enzyme complexes (such as rosettes of six particles, linear arrays of particles) in the plasma membrane of growing plant cells (Giddings et al., 1980; Herth, 1985; Quader, 1991; Brown, 1996). In a few cases rosettes are also found on the PF of Golgi cisternae, Golgi vesicles (Haigler and Brown, 1986) and ER elements (Rudolph, 1987). Detailed reviews describe the biosynthesis (Delmer and Amor, 1995; Brown, 1996, Tsekos, 1996), structure (Brown, 1982), fuction (Krassig, 1993), and secretion (Tsekos, 1996) of cellulose. The question remains whether similar complexes, which could be involved in the synthesis of the cell wall microfibrils occur in the membranes of red algae (Tsekos and Reiss, 1992, 1994).

The red algal cell consists of a microfibrillar (crystalline) phase embedded in a more amorphous phase (the matrix) (Tsekos, 1996). The microfibrillar phase is composed mostly of cellulose or xylan (Mukai et al., 1981; Gretz and Vollmer, 1989; Okuda et al., 1994), while the amorphous mucilaginous matrix is usually a sulphated galactan polymer, such as agar, carrageenan, funoran, furcellaran or porphyran (Craigie, 1990). Additional components of the matrix of red algae are glycoproteins (Tsekos, 1985).

2. Plasma membrane particle complexes (linear terminal complexes, TCs) associated with the synthesis of cellulose microfibrils

Randomly distributed linear protein particle aggregates have been observed in the plasma membrane in *Erythrocladia* (Fig. 1), *Erythrotrichia* and *Porphyra* (Tsekos and Reiss, 1992, 1994; Tsekos et al., 1993). These linear TCs were often associated with the ends of microfibril impressions. TC length varied widely, while the width was almost constant (Fig. 1). It has been clearly shown that some microfibrils start at the ends of the TCs (Tsekos and Reiss, 1992, 1994; Okuda et al., 1994; Tsekos et al., 1996). TCs appear to be formed by the repetition of transverse rows of four (*Erythrocladia*; Fig. 2, *Erythrotrichia*) (Tsekos and Reiss, 1992; Okuda et al., 1994), three or two (*Porphyra yezoensis* and *Porphyra leucosticta*) (Tsekos and Reiss, 1994) TC subunits. The number of transverse rows varies between 8-35 (*Erythrocladia*) and 3-11 (*P. yezoensis* and *P. leucosticta*), corresponding with variation in TC length (Fig. 1) and microfibril width (Figs. 3, 5).

In *Hypoglossum woodwardii* (Fig. 4) (family Delesseriaceae) and *Laurencia obtusa* (family Rhodomelaceae), particle complexes consisting of six subunits (two rows of three closely arranged particles) seem to be associated with the ends of microfibril imprints (Tsekos et al., 1993), which could indicate that they are linear microfibril-terminal synthesizing complexes (TCs).

In the plasma membrane of the red algae *Ceramium rubrum, Spermothamnion johannis, Laurencia pinnadifita* and *Alsidium corallinum* particle strings (Tsekos et al., 1993) are found, similar to those in the brown algae *Pelvetia* (Peng and Jaffe, 1976), *Ectocarpus, Hincksia, Tilopteris, Sphacelaria* and *Asteronema* (Katsaros et al., 1996; Reiss et al., 1996). It was postulated that these particle complexes were involved in cellulose synthesis (Katsaros et al., 1996; Reiss et al., 1996).

3. Microfibril structure and shape

The structure and morphology of marine red algal microfibrils (Figs. 3, 4) (Tsekos et al., 1993) can be distinguished as two types: the cylindrical and the flat, ribbon-like type.

The marine red alga *Erythrocladia* synthesized cellulose microfibrils (cellulose I), shown by cellobiohydrolase I (CBH I)-gold labelling, X-ray, and electron diffraction analyses (Okuda et al., 1994; Tsekos, 1996). The cellulose microfibrils in *Erythrocladia* and *Porphyra* cells are quite thin, ribbon-shaped structures, 1-1.5 nm thick (constant) and 5-70 nm wide (variable) (Figs. 3, 5). The microfibrils seem to be organized by the lateral association of fibrillar (minicrystal) components (Fig. 8).

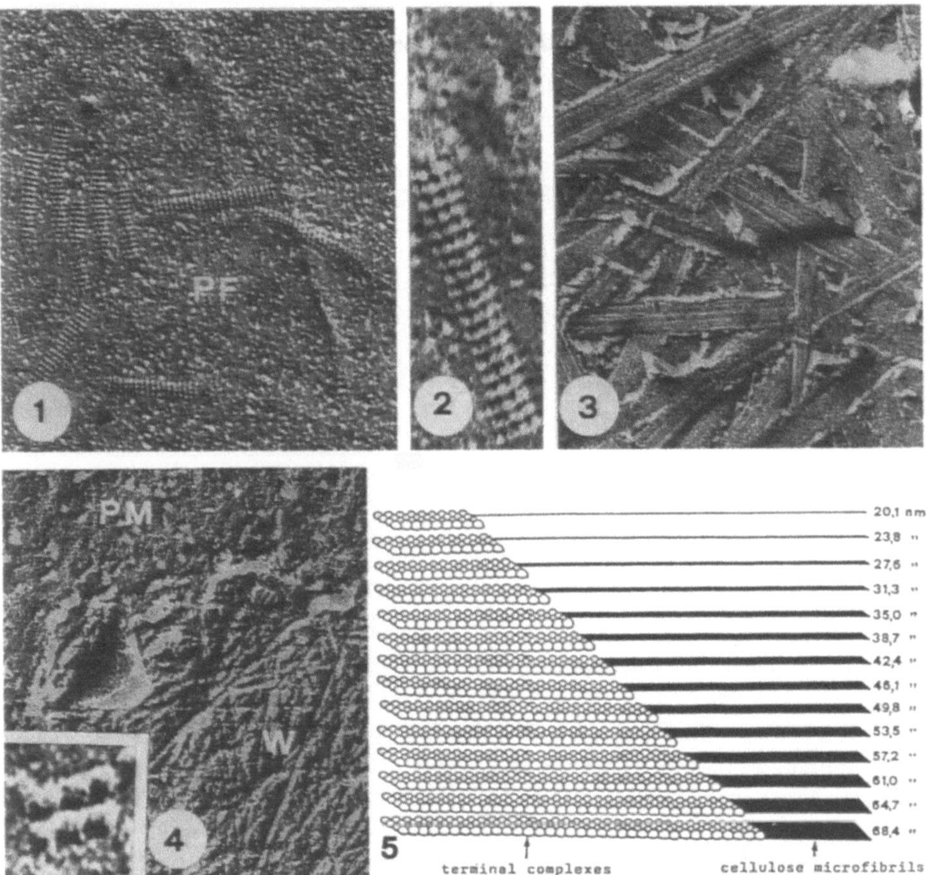

Fig. 1. *Erythrocladia subintegra*. Fracture through a freeze-etched cell at the stage of cell wall synthesis showing the PF face of the plasma membrane. Several linear TCs randomly oriented are visible. X 67.500.

Fig. 2. *Erythrocladia subintegra*. A TC at higher magnification. X 194.000

Fig. 3. *Erythrocladia subintegra*. Fracture through the cell wall showing randomly distributed, ribbon-shaped microfibrils. Some fibrils clearly have several linear subcomponents. Note that the width of microfibrils is variable. X 114.000.

Fig. 4. *Hypoglossum woodwardii*. Fracture through a freeze-etched cell showing the cell wall (W) and the EF face of the plasma membrane (PM). Note the randomly distributed cylindrical microfibrils. X 110.000. Inset: Six subunit complex on the PF face of the plasma membrane. X 325.000.

Fig. 5. Relationship between TC length and cellulose microfibril width. The number of transverse particle rows gradually increases, by 2 rows per increase. The TC with the smallest length has 9 transverse rows, while the TC with the greatest length has 35. Since each transverse particle row synthesized 12 glucan chains, which constitute a mini-crystal component, two transverse particle rows will form two minicrystal components 0.53 nm X 7 = 37.1 nm wide. Thus, width of the microfibril produced by the first TC (above) with 9 transverse particle rows will be 20.1 nm, that of the last TC (below) with 35 transverse particle rows 68.4 nm. Intermediate TCs will have analogous values shown to the right of the diagram.

Based on the above information it is evident that, in red algae, the linear terminal complexes (TCs) determine to a large extent the organization of the cellulosic component of the cell wall.

4. The origin of the linear TCs in *Erythrocladia* cells

Based on experimental data, from rotary and unindirectionally shadowing (freeze-etching technique) it is proposed a concept of intracellular flow for the origin of linear TCs in *Erythrocladia* cells (Tsekos et al. 1996). In Golgi membranes and the plasma membrane of *Erythrocladia* cells the following types of intramembranous protein particles or particle complexes have been observed: (1) single intramembranous particles (9.1 nm in diameter), (2) multisubunit globules with central hole and peripheral ridge (15-21 nm in diameter), (3) "tetrads" (complexes composed of four particles, 8 nm in diameter), and (4) linear terminal complexes (linear arrays of particles).

It is possible that the globules initially formed in the Golgi complex (Tsekos et al., 1996) constitute zymogenic particles (zymogen precursor of the TC) responsible for cellulose synthesis. The globules are transported to the plasma membrane via Golgi vesicles, which undergo exocytosis. In the plasma membrane more globules consisting of more subunits aggregate, swell and unfold, contributing to TC assembly by close insertion and arrangement of the unfolding large globules (Tsekos et al., 1996; for the formation of the *Vaucheria* TC, see Mizuta and Brown, 1992).

It should be noted that the globules are inserted into the plasma membrane with "tetrads" (Tsekos and Reiss, 1993) and intramembrane particles by the fusion of Golgi vesicles. The "membrane tetrads" seem to be involved in the synthesis of the amorphous matrix polysaccharides of *Erythrocladia* cell wall. The simultaneous presence of globules and "tetrads" in Golgi cisternae and Golgi vesicles supports earlier suggestions that synthesis of the non-crystalline and crystalline components of the cell wall is a co-ordinated cell activity (Tsekos et al., 1996).

After the synthetic function of the linear TC is finished, the structure may be recycled back into the cytoplasm via an endocytotic process (Tsekos et al., 1996) (breakdown of TCs allows protein conservation by the cell) (cf. also Northcote, 1991).

5. Relationship between synthesizing complexes (TCs) and microfibril shape and dimensions in red algal cells

The mean dimensions of the microfibrils (1,2 nm) X (27,6 $\pm$ 0,8 nm) (Okuda et al., 1994; Tsekos et al., 1996) and the interplanar d-spacings of the cell wall of

Erythrocladia [6.0 Å (101), 5.3 Å ($\bar{1}$01), 3.9 Å (002) (Okuda et al., 1994)] suggest that they contain about 159 glucan chains. Furthermore, the fact that the mean total number of subunits in each TC is about 50,4 ± 1,7 (see Fig. 6) indicates the possibility that each single TC subunit may contribute to form 3 glucan chains (Figs. 5, 8).

It seems that each transverse row of four TC subunits contributes to a fibrillar (minicrystal) component which contains 12 glucan chains, with dimensions 1.2 nm X 1.59 nm (Fig. 8). The number of fibrillar (minicrystal) components assembled by a given TC and associated laterally equals the number of transverse particle rows. The laterally associated minicrystal components contribute to the variation in microfibrillar width (Figs. 5, 7, 8) (Tsekos and Reiss, 1992; Okuda et al., 1994). It has been suggested that elementary fibrils fasciate during biosynthesis to form microfibrils in such a way that each microfibril behaves as a single crystal.

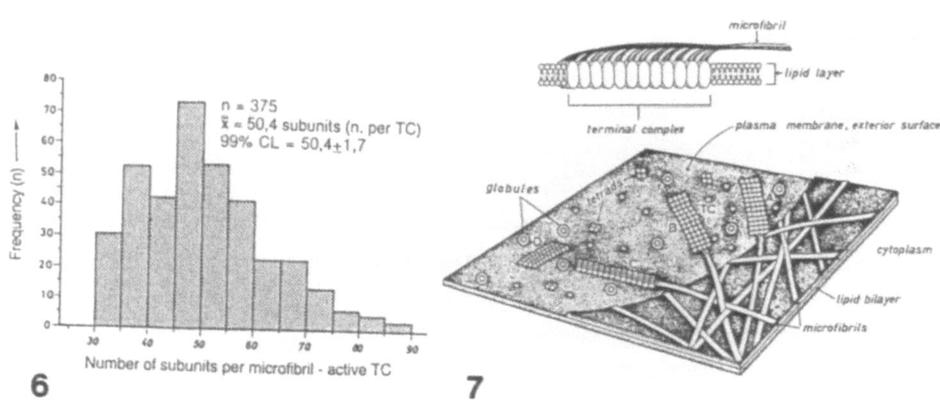

Fig. 6. Frequency histogram for the number of subunits per microfibril-active TC in *Erythrocladia* cells. Data from measurements of 375 TCs on the PF face. The confidence limits are 99%. It should be noted that TCs of 30 or less subunits are generally unassociated with microfibril impressions.

Fig. 7. Model of microfibril formation in *Erythrocladia*. Top, side view; bottom, surface view. Single TCs give rise to randomly oriented microfibrils. A-C microfibril-active TCs, D microfibril-inactive TC.

Brown (1996) for *Erythrocladia* has suggested that each single TC subunit contain only one catalytic site for glucan chain polymerization; it means that each single TC subunit may contribute to form 1 glucan chain. From the schematic representation for *Erythrocladia*, provided by Brown (1996, Fig. 8), microfibril thickness of 1,809 nm occurs, whereas measurements by Okuda et al. (1994) yield a value of 1.0-1.5 nm (see Fig. 2F by Okuda et al., 1994). It should be noted here that while Brown (1996) has used the same measurements as Okuda et al. (1994), he, however, gives a microfibril thickness of 1.5-2.0 nm.

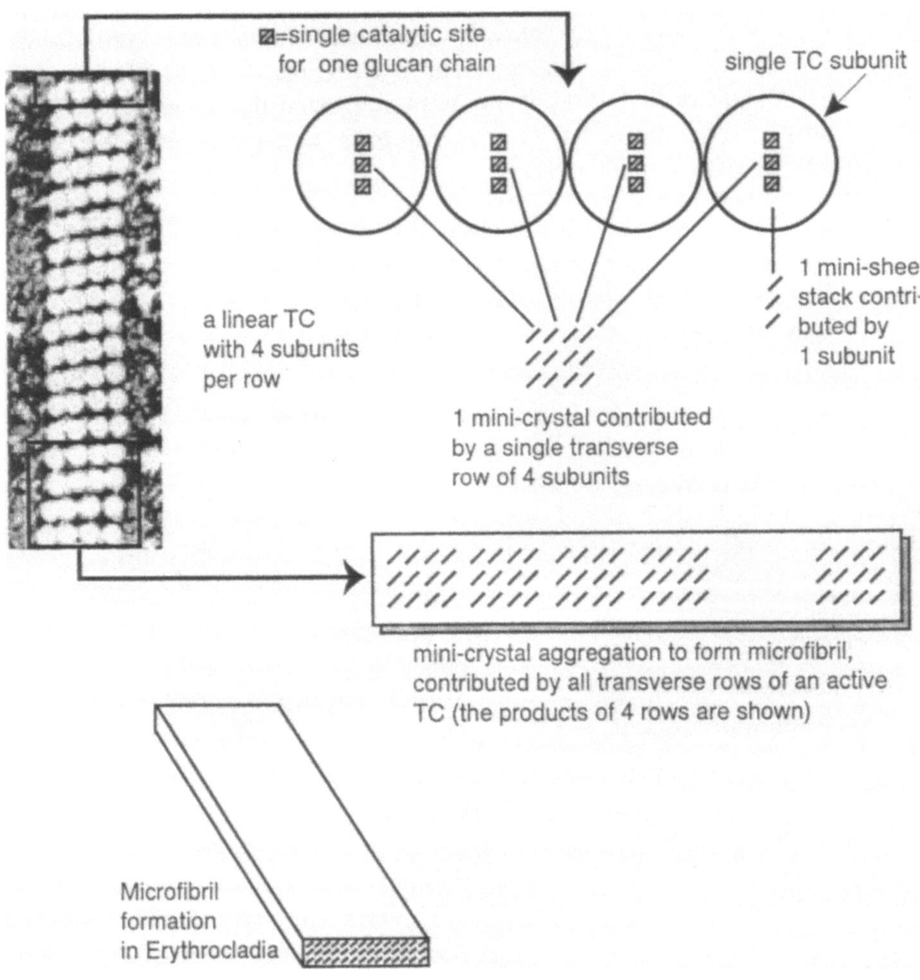

Fig. 8. Semi-schematic representation of cellulose microfibril assembly during wall formation in *Erythrocladia*. Each single TC subunit is believed to contain 3 single catalytic sites for glucan chain polymerization. In the TCs of *Erythrocladia* there is a precise arrangement of four subunits of the transverse row. One subunit contributes to a thin mini-sheet as depicted. The four subunits of the transverse row are required to generate a glucan chain minicrystal. As each row of transverse subunits adds laterally minicrystals to the microfibril, its width is increased.

Given that the mean dimension of microfibril width amounts to 27,6 ± 0,8 nm (Tsekos, 1996; Tsekos et al., 1996) in a microfibril with mean dimensions (1,809) X (27,6 ± 0,8 nm), as depicted in Brown's drawing (1996) 212 glucan chains would approximately be included. As already stated the mean total number of single subunits in each TC is about 50.4 ± 1.7 (Fig. 6, where 375 microfibril-active TCs have been measured). Thus it could be gathered (212 glucan chains: 50,4 mean subunits of a TC = 4,2) that each single subunit would contribute to the formation of 4 glucan chains and not of 1 as suggested by Brown (1996).

Taking into account current data on green, yellow-green, and brown algae, mosses, ferns and spermatophytes in relation to the results on red algae, it is concluded that the linear terminal complexes (TCs) in the marine red algae *Erythrocladia subintegra*, *Erythrotrichia* sp., *Porphyra leucosticta* and *Porphyra yezoensis*, represent morphological equivalents of plasma membrane bound multienzyme complexes involved in the biosynthesis, assembly, and orientation of the cellulose-microfibrils. The specific arrangement and consolidation of the terminal complex subunits determine to a large extent the size and shape (cylindrical or flat, ribbon-like) of the microfibril, its crystallinity, as well as its intramicrofibrillar associations.

Acknowledgements

This work was supported by the "Stiftung Volkswagenwerk", the Alexander von Humboldt-Stiftung, the Hellenic General Secretary of Research and Technology and the European Communities (COST-49).

6. References

Brown, R.M. Jr. (ed.) (1982). Cellulose and other natural polymer systems: biosynthesis, structure, and degradation, Plenum, New York.

Brown, R.M. Jr. (1996). The biosynthesis of cellulose. *J. Macromol. Sci. – Pure Appl. Chem.*, ***A33** (10)*, 1345-1373.

Craigie, J.S. (1990). Cell walls, in K.M. Cole and R.G. Sheath (eds.), *Biology of the red algae*, Cambridge University Press, Cambridge, pp 221-257.

Delmer, D.P. and Amor, Y. (1995). Cellulose biosynthesis, *The Plant Cell* **7**, 987-1000.

Giddings, T.H. Jr., Brower, D.L. and Staehelin, L.A. (1980). Visualization of particle complexes in the plasma membrane of *Micrasterias denticulata* associated with the formation of cellulose fibrils in primary and secondary cell walls, *J. Cell Biol.* **84**, 327-339.

Gretz, M.R. and Vollmer, C.M. (1989). Cellulosic cell walls of red algae, in C. Schuerch (ed.), *Cellulose and wood: chemistry and technology*, Wiley, New York, pp 623-637.

Haigler, C.H. and Brown, R.M. Jr. (1986). Transport of rosettes from the Golgi apparatus to the plasma membrane in isolated mesophyll cells of *Zinnia elegans* during differentiation to tracheary elements in suspension culture. *Protoplasma*, **134**, 111-120.

Herth, W. (1985). Plant cell formation, in A.W. Robards (ed.), *Botanical microscopy*, Oxford University Press, Oxford, pp. 285-310.

Katsaros, C., Reiss, H.-D. and Schnepf, E. (1996). Freeze-fracture studies in brown algae: putative cellulose-synthesizing complexes on the plasma membrane, *Eur. J. Phycol.* **31**, 41-48.

Krassig, H.A. (1993). *Cellulose: structure, accessibility and reactivity*, Gordon and Breach, Montreaux (Polymer monographs, vol. 11).

Mizuta, S. and Brown, R.M. Jr. (1992). High resolution analysis of the formation of cellulose synthesizing complexes in *Vaucheria hamata*, *Protoplasma* **166**, 187-199.

Mukai, L.S., Craigie, J.S. and Brown, R.G. (1981). Chemical composition and structure of the cell walls of the conchochelis and thallus phases of *Porphyra tenera* (Rhodophyceae), *J. Phycol.* **17**, 192-198.

Northcote, D.H. (1991). Site of cellulose systems, in C.H. Haigler and P.J. Weimer (eds.), *Biosynthesis and biodegradation of cellulose*, Marcel Dekker, New York, pp 165-176.

Okuda, K., Tsekos, I. and Brown, R.M. Jr. (1994). Cellulose microfibril assembly in *Erythrocladia subintegra* Rosenv.: an ideal system for understanding the relationship between synthesizing complexes (TCs) and microfibril crystallization, *Protoplasma* **180**, 49-58.

Peng, H.B. and Jaffe, L.F. (1976). Cell wall formation in *Pelvetia* embryos. A freeze-fracture study, *Planta* **154**, 184-188.

Quader, H. (1991). Role of linear terminal complexes in cellulose synthesis, in C.H. Haigler and P.J. Weimer (eds.), *Biosynthesis and biodegradation of cellulose*, Marcel Dekker, New York, pp. 51-69.

Reiss, H.-D., Katsaros, C. and Galatis, B. (1996). Freeze-fracture studies in the brown alga *Asteronema rhodochortonoides*, *Protoplasma* **193**, 46-57.

Rudolph, U. (1987). Occurrence of rosettes in the ER membrane of young *Funaria hygrometrica* protonemata, *Naturwissenschaften* **74**, 439.

Tsekos, I. (1985). The endomembrane system of differentiating carposporangia in the red alga *Chondria tenuissima*: occurrence and participation in secretion of polysaccharidic and proteinaceous substances, *Protoplasma* **129**, 127-136.

Tsekos, I. (1996). The supramolecular organization of red algal cell membranes and their participation in the biosynthesis and secretion of extracellular polysaccharides: a review, *Protoplasma* **193**, 10-32.

Tsekos, I. and Reiss, H.-D. (1992). Occurrence of the putative microfibril-synthesizing complexes (linear terminal complexes) in the plasma membrane of the epiphytic marine red alga *Erythrocladia subintegra* Rosenv., *Protoplasma* **169**, 57-67.

Tsekos, I. and Reiss, H.-D. (1993). The supramolecular organization of red algal vacuole membrane visualized by freeze-fracture, *Ann. Bot.* **72**, 213-222.

Tsekos, I. and Reiss, H.-D. (1994). Tip cell growth and the frequency and distribution of cellulose microfibril-synthesizing complexes in the plasma membrane of apical shoot cells of the red alga *Porphyra yezoensis*, *J. Phycol.* **30**, 300-310.

Tsekos, I., Reiss, H.-.D. and Schnepf, E. (1993). Cell-wall structure and supramolecular organization of the plasma membrane of marine red algae visualized by freeze-fracture, *Acta Bot. Neerl.* **42**, 119-132.

Tsekos, I., Okuda, K. and Brown, R.M. Jr. (1996). The formation and development of cellulose-synthesizing linear terminal complexes (TCs) in the plasma membrane of the marine red alga *Erythrocladia subintegra Rosenv.*, *Protoplasma* **193**, 33-45.

THE CORM'S STRUCTURE AT FERN: ATHYRIUM FILIX-FEMINA (L.) ROTH. TEND.

R. BERCU
"Ovidius" University, 124 Mamaia Ave. Constanta, Romania

1. Abstract

After a short necessary morphological caracterisation, the paper present, the anatomical structure of fern: *Athyrium filix - femina* (L). Roth. Tend., with special references to the modifications of the stele, along all it's corme: root, stem, leaf, in succesive transections. A mathematical theoretical model has been built, of the meristeles (or protostele) distribution in the central cylinder.

2. Material and Methods

For the anatomical study of the corm· at this fern I used classical method of paraffin inclusion. Samples of 1 cm lenght, were cut from 2 in 2 cm, along the root, rhisome, leaf's petiol and rachis at a Minot microtome at 10 µm [2]. The sections were coloured with Toluidin blue solution 0,05% in Mc Ilvaine buffer (pH 3). The sections were mounted in Canada balsam. The microphotos were done at a Nikon microscope.

3. Results and Discussions

3.1 MORPHOLOGICAL ASPECTS

Athyrium filix - femina, is a common fern which often can be seen in Romanian mountains, in forests, on the cliffs and near the waters. The fern has in the ground a short, thick, black rhisome, covered with many squames. The leaves are 20-25 cm in lenght, but sometimes arrive at 40 cm. They are green in colour and 2-3 pennate sectioned. On the dorsal part of the leaf, there are many round little sori, with indusium [10].

In our fack medicine, the porridge of the rhisome is used against the intestinal wormes and also the fern is used to decorate the rooms of the houses [3]; [9].

3.2 ANATOMICAL ASPECTS

The root (Figure 1), is an adventive root, which starts from the inferior part of the stem, in the ground and is composed of: epidermis cortex, stele. Epidermis is at the surface of the root, but often is broken, because of the friction, with the soil particles. Under the epidermies is cortex. The external part of it is often sclerotic, the rest is a fundamental tissue in whose center is the stele which is a protostele, surrounded by a well-differentiated endodermis. Under it, is the pericycle. The vessel tissue is represented by: xylem and phloem. Protoxylem is diarch in development and is formed of scalariform tracheids. Phloem surround the xylem. The stem (Figure 2) is an underground one; is a rhisome. At it suface, is the epidermis, covered by cuticle, which often is broken, because the friction with the soil. Under it, is the cortex. Some of the cortex cells are selerotic (3-4 layers) and the rest form a fundamental tissue, in which there are 2 vessel fascicles (2 meristeles), each of them surrounded by their own endodermis and pericycle. Between them, there are parenchymatic cells. The xylem vessel have a windding shape, which is imposed to the whole meristele. The stele at this level is a distele, resulted probably from an amfiphloic solenostele, and the leaf gapes cut it in 2 meristeles [5]; [7].

The leaf. The rhisome developes at the surface of the soil, the leaves. Along the petiol and rachis, the 2 meristeles from the rhisome and get closer and closer, surrounded by a single endodermis and pericycle, and at last they join together (Figure 3) formming a protostele, in fact an actinostele with 2 branches, in shape of a ribon (or upset down letter M) [8].

4. Conclusions

From the anatomical point of view, and in conection with the morphology of each

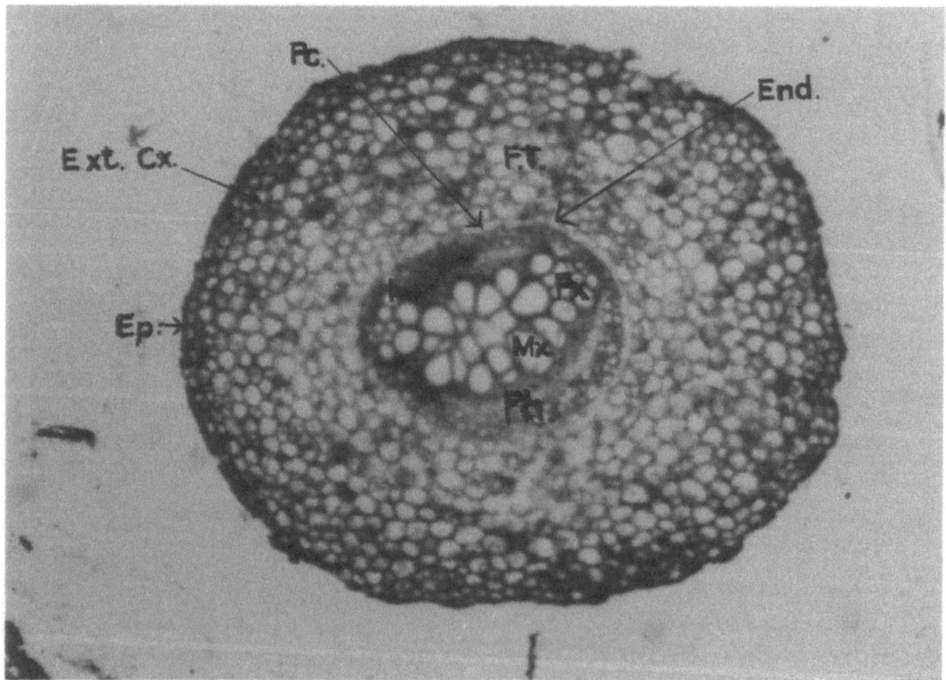

Figure 1. Transection through the root at *Athyrium filix - femina:* Ep = epidermis, Ext. ex. = external cortex, F.T. = fundamental tissue, End = endodermis, Pc = pericylcle, Mx = metaxylem, Px = protoxylem, Ph = phloem

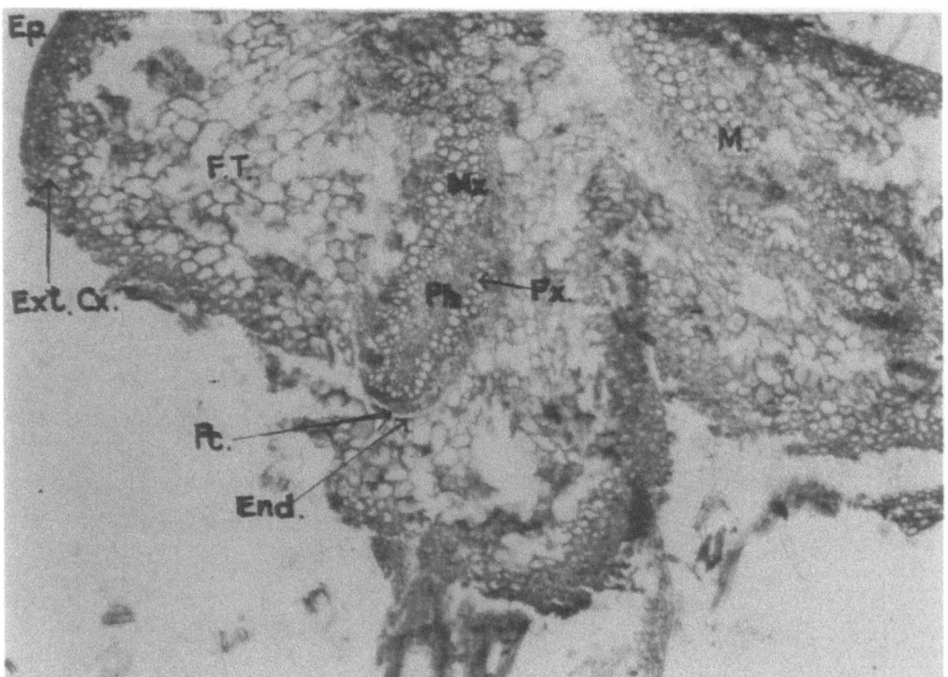

Figure 2. Transection through the rhisome at *Athyrium filix - femina:* Ep = epidermis, et = cuticle, Ext. cx. = external cortex, F.T. = fundamental tissue, M = meristele, End = endodermis, Pc = pericylcle, Mx = metaxylem, Px = protoxylem, Ph = phloem

organ of this plant, important differences appear at the level of the stele. From the root to the top of the leaf is rachis, dominant is the monostele (protostele), excepting the rhisome, where is a distele (2 meristeles).

A mathematical, theoretical function, of the meristeles, distribution, can be decluced, in the stele in even or add number, in which h, can be: 0; or <0; or >0, by the position of the vegetal organs, with respect to the soil. All this things, together with the ecological medium of the plant, can be expressed through an exponential distribution function of the meristeles, (or protostele where is the case), given by the relation:

– adventive root: $h > 0$: $N_{AR} = (2n + 1)\, K^{At.ft} \dfrac{D}{e^{h}}$

– rhisome: $h = 0$: $N_R = 2n\, K^{At.ft} D$

– leaf: $h < 0$: $N_L = (2n + 1)\, K^{At.ft} \dfrac{D}{e^{-h}}$

where:

N_{AR}, N_R, N_L = the distribution of the exposential function for each meristele (or protostele), number by each organ (AR = adventive root, R = rhisome, L = leaf);
2n = even number of meristeles in the central cylinder;
2n+1 = add number of meristeles in the central cylinder;

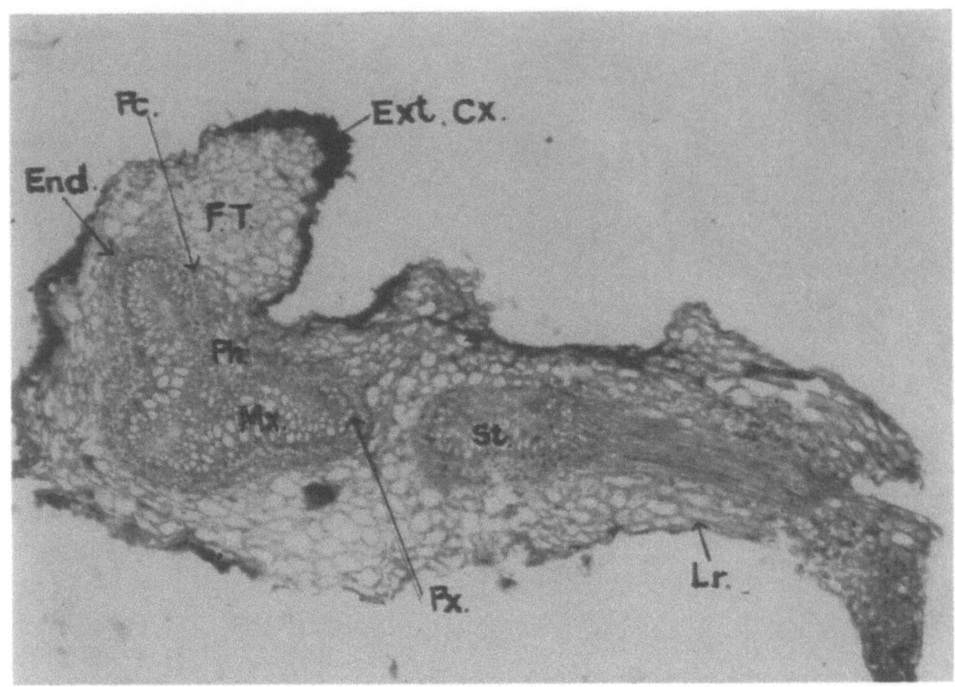

Figure 3. Transection through the leaf's rachis at *Athyrium filix - femina:* Ep = epidermis, Ct = cuticle, Ext. cx. = external cortex, F.T. = fundamental tissue, End = endodermis, Pc = pericylcle, Mx = metaxylem, Px = protoxylem, Ph = phloem

$K^{At.ft}$ = ecological constant of the plant *Athyrium filix - femina;*
D = the diameter of the vegetal organ;
e = natural logaritm's base (2,7);
h = the high of the vegetal organ to which was performed the section, with the respect to the soil.

5. References

1. Andrei, M. (1978) *Anatomia planetor*, E.D.P., Bucuresti.
2. Anghel, I. & colab. (1988) *Practicum de biologie - Biologie vegetala*, vol. I, Ed. Tipografia Univ., Bucuresti.
3. Avram, R., Andronescu, E., Fuezi, I. (1981) *Botanica farmaceutica*, E.D.P., Bucuresti.
4. Batanouny, K.H. (1992) *A Text book of Botany / Plant Anatomy,* 4th Edition, Cairo Univ. Press.
5. Bercu, R. (1997) *Structura cormului la unele Polypodiaceae*, Analele Univ. "Ovidius", seria Biologie - Ecologie, vol. I; Constanta, pp. 100-106.
6. Bierhorst, D.W. (1971) *Morphology of Vascular Plants*, The Macmillan Company, New York, Collier Macmillan Ltd. London.
7. Bold, H., Alexopoulos, C., Delevoryas, TH. (1987) *Morphology of Plants and Fungi*, 5th, Harper & Row Publichers, New York.
8. Ogura, Y. (1938) *Anatomie der Vegetationsorgane der Pteridophyten, Im Linnsbauer "Handbuch der Pflanzenanatomie"*, Gebruder Bortraegen; Berlin.
9. —— (1993) *Farmacopeea Romana*, Ed. Medicala, Bucuresti.
10. —— (1950) *Flora R.P.R.*, vol. I, Ed. Acad. Bucuresti.

GROWTH AND STRUCTURAL RESPONSES OF MAIZE PRIMARY ROOTS TO ALUMINIUM STRESS

S. BUDÍKOVÁ, M. ČIAMPOROVÁ
Institute of Botany, Slovak Academy of Sciences
Dúbravská cesta 14, 842 23 Bratislava, Slovak Republic

1. Introduction

The higher Al content in soils has become an important growth limiting factor for many plants. Identifications of Al uptake sites in the roots were studied using hematoxylin (Polle et al. 1978) or morin staining (Vitorello and Haug 1996). The addition of Al induced drastic structural changes of the peripheral root cells including swelling of the root tip, sloughing off the rhizodermis and stimulation of callose deposition (de Lima and Copeland 1994, Delhaize and Ryan 1995). Structural responses of root cells were intensified with increased exposition to Al (Eleftheriou et al. 1993, de Lima and Copeland 1994). Within 12 h the majority of the rhizodermal cells collapsed (Danilova et al. 1992, Eleftheriou et al. 1993, de Lima and Copeland 1994). The surviving cells reveal ultrastructure alterations (Danilova et al. 1992, Ikeda and Tadano 1993). Generally, the degree of root cell damage depends on their distance from the root surface and developmental stage.

In the present study we describe root elongation, morphological and ultrastructural responses in the root tip cells of maize. We localise Al by two different methods in root tissues and report that Al^{3+} can stimulate a non-constitutive callose formation.

2. Material and Methods

Seedlings of maize *Zea mays* L., cv. TO 360, with mean root length 4.5 cm (3 days after sowing) grew in 0.1 mM $CaCl_2$ (control) or 50 μM $AlCl_3$ in 0.1 mM $CaCl_2$ (pH 4.5). After 0, 24 and 48 h primary root lengths were measured. Using ink marks, seven apical segments A (0 to 1st mm) to G (6th to 7th mm) were distinguished. Increases in their lengths were measured after 24 h interval.

After 24 h treatment, root tips were fixed with glutaraldehyde and osmium tetroxide, dehydrated and embedded in Epon-Araldite. Ultrathin sections were stained with uranyl acetate and lead citrate, and examined with Tesla BS 500 electron microscope.

For callose detection free-hand sections of the root tip were stained in aniline blue. The fluorescence staining for Al was performed according to Vitorello and Haug (1996). Intact roots were stained in morin, rinsed in DMSO and cut with a razor blade. Both applied fluorochromes were examined with UV excitation. Aluminium localisation with hematoxylin was performed according to Polle et al. (1978). All reactions were done after 24 h of Al treatment.

3. Results and discussion

Inhibition of the primary root growth increased gradually during the 48 h application of 50 μM concentration of aluminium (Fig. 1). The proximal boundary of the growing part of the root is 7 mm behind the maize root tip. Within this part, cell division is confined to the apical 1 to 2 mm (Webster and MacLeod 1996). Differences in the pattern of the elongation of the first three apical segments were found between control and Al-treated roots (Fig. 2). Under control conditions, the gradient of length increase rate was in the segment order B>D>C>E>F>G>A while the order of the most rapidly elongating segments was B>C>D under Al stress. The length of the segment C increased more in Al treated than in the control roots. This root segment was constructed predominantly of the cells in their postmitotic isodiametric growth stage (Baluska et al. 1990). Their specific growth and metabolic properties (Baluska et al. 1995) might be affected by aluminium. The strongest reduction of elongation, i.e. down to 40% occurred in the more proximal segment D in which maximum elongation occurs (Baluska et al. 1990) confirming the high sensitivity of elongating cells to Al stress.

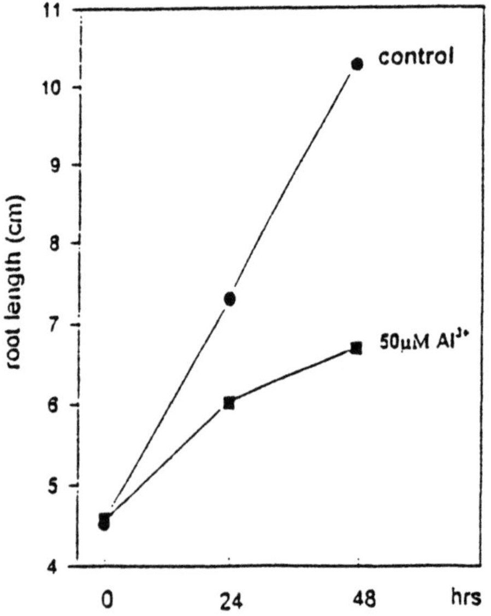

Fig. 1. Inhibition of the primary root growth of maize (*Zea mays*. L., cv. TO 360) exposed to 50 μM concentration of aluminium for 48 h.

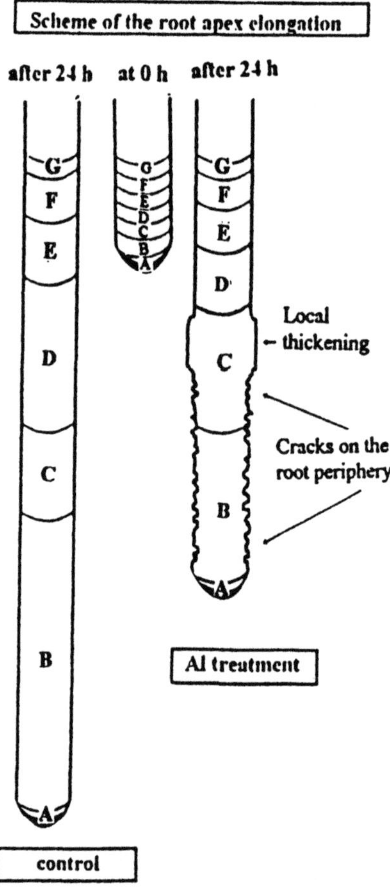

Fig. 2. Elongation pattern of the apical root segments of control and Al-treated plants.

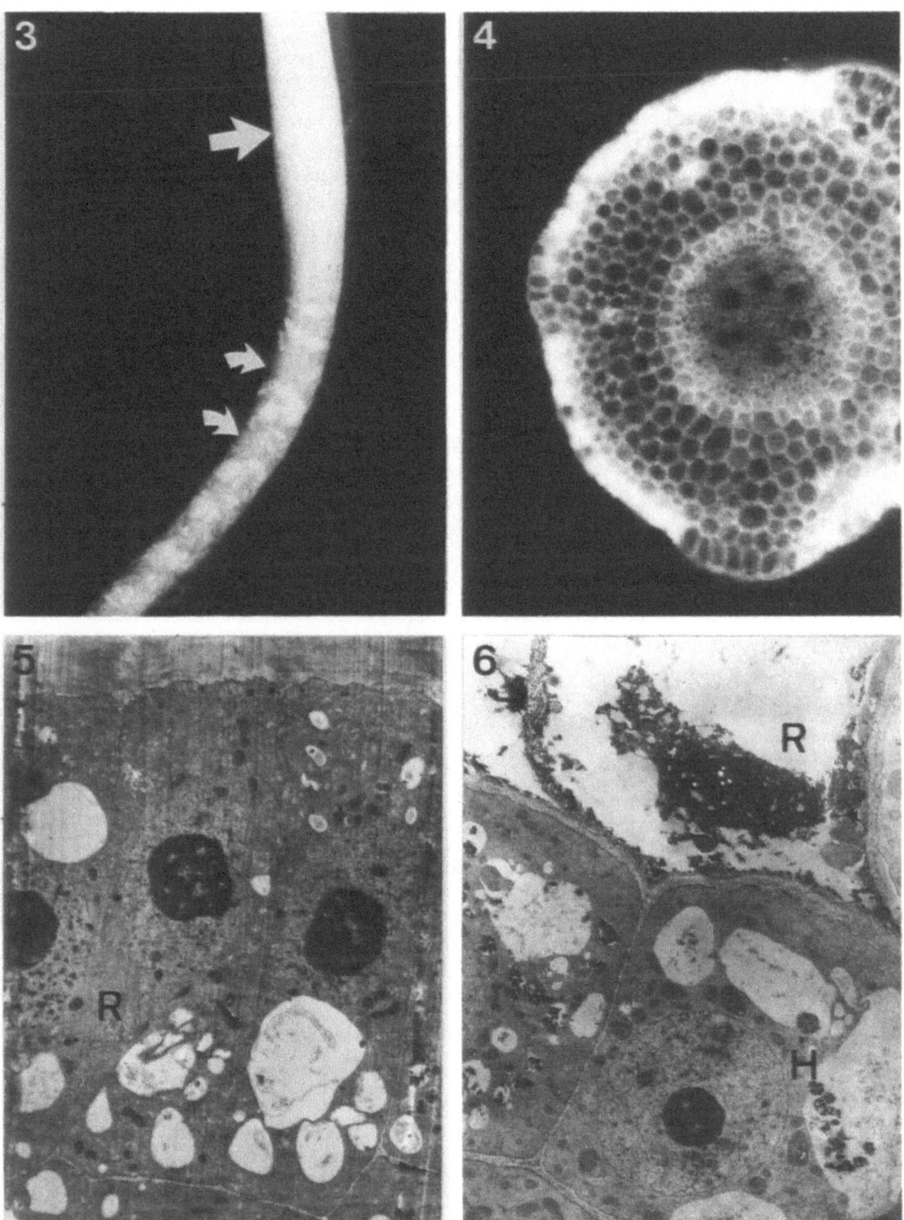

Figs. 3, 4. Histochemical detection of Al in roots. (3) Positive reaction with hematoxylin (small arrows) and local thickening in the next proximal part of the root (large arrow). (4) Positive reaction with morin in the peripheral root tissues.

Transversal cracks extending over the surface of the apical region up to about 20 mm behind the root tip appeared due to Al toxicity (Fig. 3), similarly as in wheat roots (Delhaize and Ryan 1995). This cracked part corresponded with the most rapidly elongating region of the root. Proximally to this part, a thickened zone of about 5 mm appeared which originated from the segment C of the root tip. The root surface of this zone as well as of the rest of the root remained smooth and similar to control.

Unlike control roots, in Al-treated root apex, both hematoxylin and morin gave positive reaction in outer cap cells, rhizodermis and peripheral tissues up to the fifth cortical layer of the "cracked" region (Fig. 4). The more basal root parts including the swollen region showed no positive staining for Al.

In comparison to ultrastructure of control root cells (Fig. 5), the cytoplasmic content of rhizodermis, hypodermis and the next cortical layer was disintegrated (Fig. 6). The severe damage could be related with Al detected by both histochemical tests in these tissues. The more internal cortical cells were vacuolated and contained large deposits similar to dark material supposed to contain Al ions (Eleftheriou et al. 1993) in root cells of *Thinopyrum* species. Increased vacuolation as a non-specific plant cell response to various stresses indicates accelerated differentiation or senescence. Wall ingrowths and irregular thickenings found in rhizodermal and hypodermal cell walls might be associated with formation of callose detected within the cracked region, in rhizodermal cell walls and all cortical layers including endodermis. Increased callose synthesis can serve as an adaptive mechanism providing positive charges for Al^{3+} binding.

5. References

Baluska, F., Hauskrecht, M. and Kubica, S. (1990) Postmitotic 'isodiametric' cell growth in the maize root apex. *Planta* **181**, 269-274.

Baluska, F., Barlow, B.W. and Kubica, S. (1995) Importance of the post-mitotic isodiametric growth (PIG) region for growth and development of roots. In: Baluska et al. (eds.), *Structure and function of roots.* Kluwer Academic Publishers, Dordrecht, Netherlands, pp. 41-51.

Danilova, M.F., Kravkina, I.M., Crang, R.F.E. (1992) The effects of aluminum on root tissue ultrastructure in soybean seedlings. In: L. Kutchera et al. (eds.), *Root ecology and its practical application*, 3rd ISRR Symp., Verein für Wurzelforschung, Wien, pp. 25-28.

Delhaize, E., Ryan, P.R. (1995) Aluminum toxicity and tolerance in plants. *Plant Physiol.* **107**, 315-321.

Eleftheriou, P.E., Moustakas, M. and Fragiskos, N. (1993) Aluminate-induced changes in morphology and ultrastructure of *Thinopyrum* roots. *J. Exp. Bot.*,**44**, 427-436.

Ikeda, H. and Tadano, T. (1993) Ultrastructural changes of the root tip cells in barley by a comparatively low concentration of aluminium. *Soil Sci. Plant Nutr.* **39**, 109- 117.

Lima, M. L. de and Copeland, L. (1994) Changes in the ultrastructure of the root tip of wheat following exposure to aluminium. *Aust. J. Plant Physiol.* **21**, 85-94.

Webster, P.L. and MacLeod, R. (1996) The root apical meristem and its margins. In: Y. Waisel, A. Eshel and U. Kafkafi (eds.), *Plant Roots. The hidden half.* Marcel Dekker, Inc., New York Basel Hong Kong, pp. 51-76.

MITOSIS AND CYTOKINESIS IN BROWN ALGAE SHOWING DIFFERENT THALLUS ORGANIZATION

I. DIMITRIADIS, C. KATSAROS AND B. GALATIS
University of Athens, Department of Biology, Division of Botany, Athens 157 84, Hellas

Microtubule (Mt) organization and nuclear behaviour during the cell cycle was studied using anti-tubulin immunofluorescence and Hoechst staining of the DNA [1], in vegetative apical and intermediate thallus cells of three brown algal species, showing different thallus organization and belonging to different orders, i.e. *Syringoderma phinneyi* (Syringodermatales) [2], *Tilopteris mertensii* (Tilopteridales) [3] and *Ectocarpus siliculosus* (Ectocarpales) [4].

1. Results

1.1. INTERPHASE - CENTROSOME DUPLICATION

Cytoplasmic and perinuclear Mts radiate out from the centrosome site. In intermediate thallus cells of *E. siliculosus* (Fig. 5), as well as in all cells of *S. phinneyi* and *T. mertensii*, the Mt cytoskeleton consists of Mt bundles evenly radiating from the centrosome(s) towards all the directions of the cell cortex. In apical cells of *T. mertensii*, and particularly in those of *S. phinneyi*, Mts show a polar organization, with many thin bundles directing towards the apical dome (Figs. 1-4). In interphase cells centrosome can be single or duplicated and separated in opposite sites. Single centrosomes are mainly observed in differentiated thallus cells, while in meristematic, rapidly dividing cells double centrosomes are more common. However, before or during the separation process, centrosomes appear changing position, or the nucleus-centrosome complex rotates, giving a variety of images of separating centrosomes (Figs. 1-6).

1.2. MITOSIS

In all species, prophase is characterized by the development of intensely fluorescent Mt asters in the nuclear poles. In *T. mertensii* and *E. siliculosus*, two opposite independent systems of spindle Mts, also radiating from the centrosomes sites, surround the prophase nucleus (Figs. 8-11). In *S. phinneyi*, perinuclear Mts are disassembled first, while many cytoplasmic Mts are retained during prophase (Fig. 7).

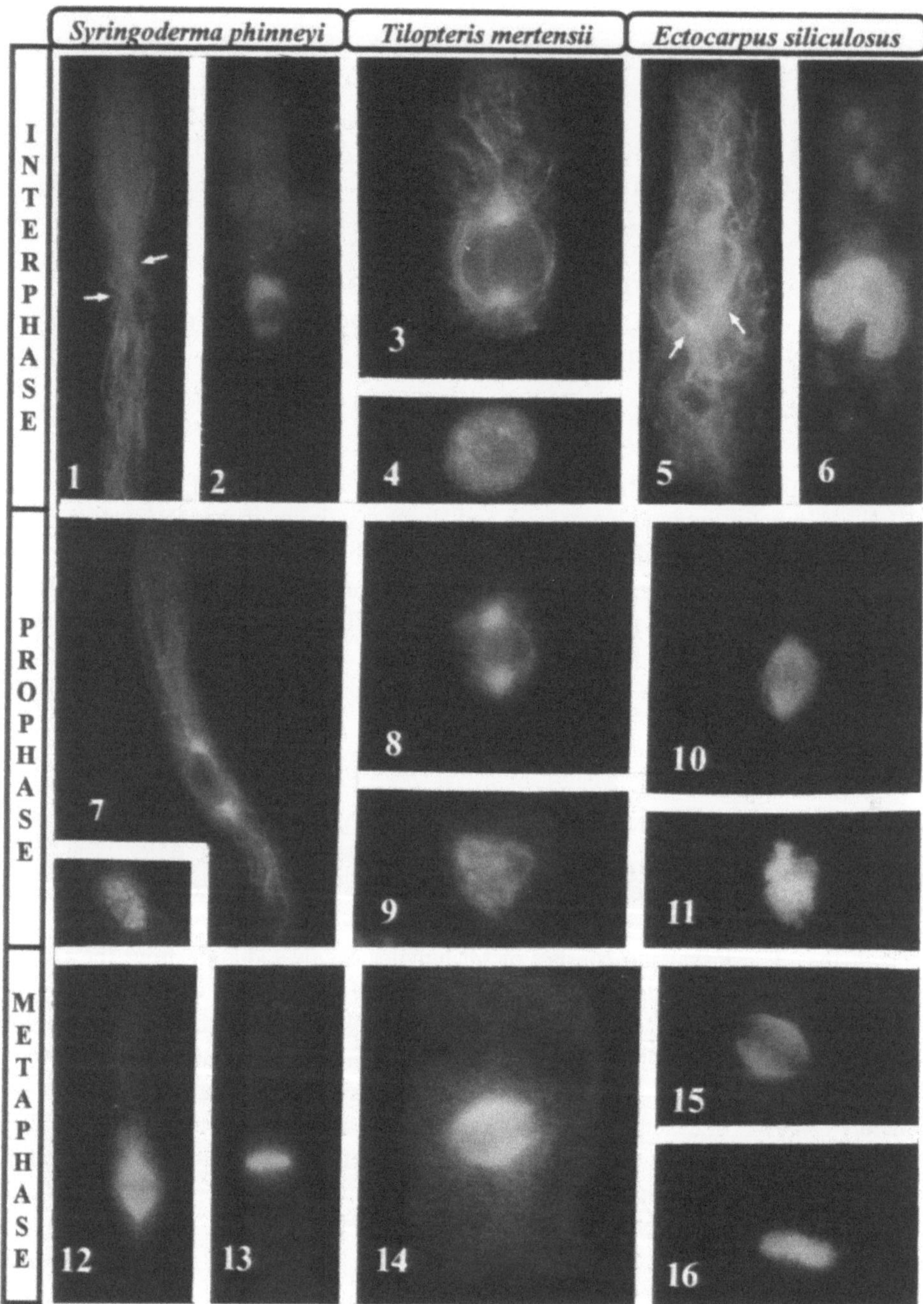

Figures 1, 3, 5, 7, 8, 10, 12, 14, 15, 17, 19, 20, 22, 24, 25, 27, 29 and 31: Anti-tubulin immunofluorescence.
Figures 2, 4, 6, 9, 11, 13, 16, 18, 21, 23, 26, 28, 30 and 32: DNA after staining with Hoechst 33258.
Figures 1-9, 12, 13, 17, 18 and 22-28: ×100.
Figures 14-16, 19, 24, 29 and 30-32: ×150.

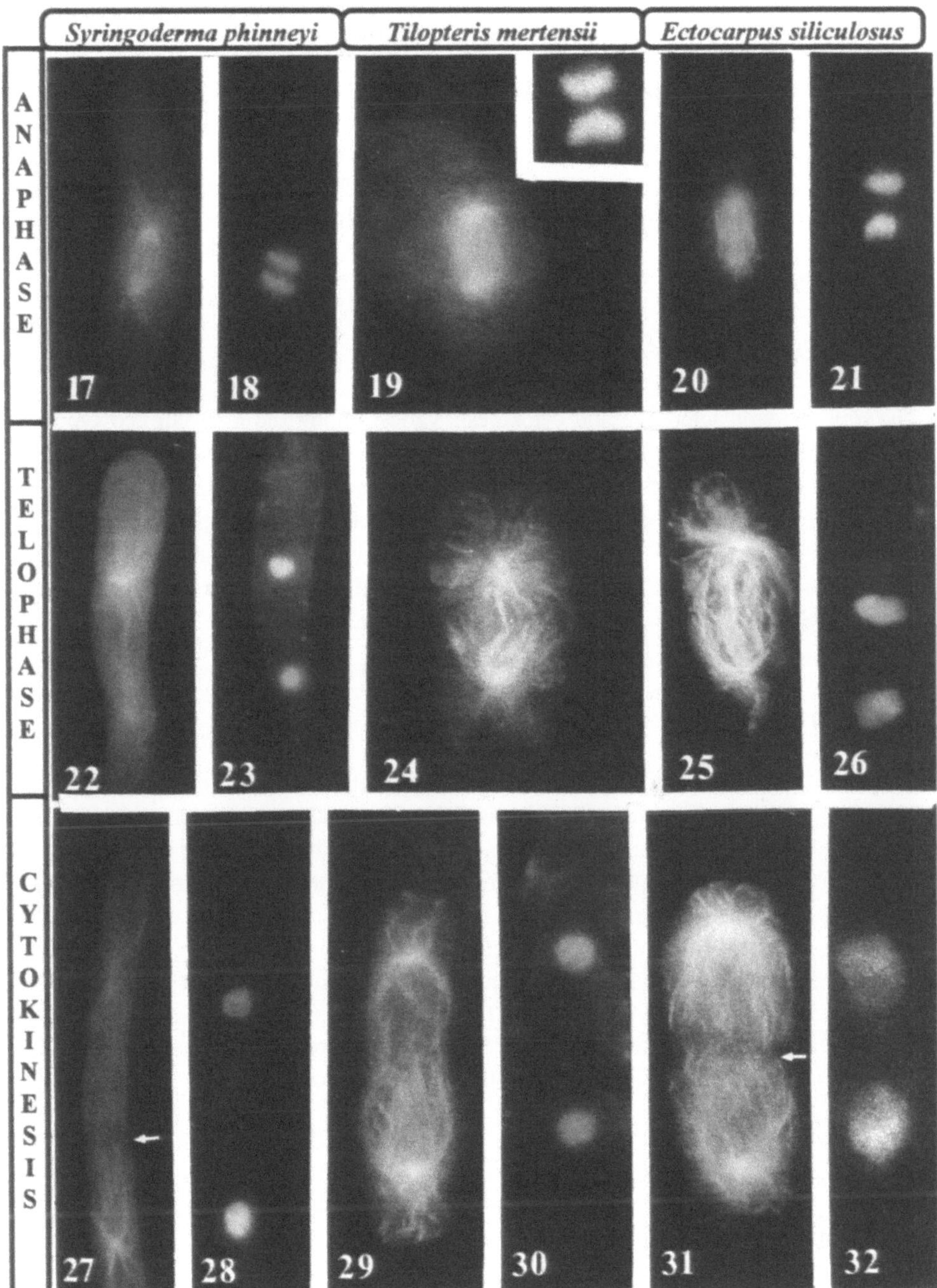

Figures 10, 11, 20 and 21: ×200.
Figures 1 and 5: Arrows point to centrosome sites.
Figure 7: Inset prophase nucleus.
Figure 19: Inset separated chromosomes.

Aster Mts disappear in metaphase cells of *T. mertensii* and *E. siliculosus*, and only spindle Mts are visible (Figs. 14 and 15). In contrast, in *S. phinneyi* distinct Mt bundles still radiate towards the cell cortex (Fig. 12). During anaphase, in cells of *T. mertensii* and *E. siliculosus* aster-like cytoplasmic Mts are formed and the increase in length of the interzonal Mts causes spindle elongation (Figs. 19 and 20). In elongated apical cells of *S. phinneyi*, the reforming cytoplasmic Mts are more in the apical part of the cell, and the interzonal Mt system is conspicuous (Fig. 17). In all species telophase is characterized by the depolymerization of the spindle and the organization of two independent Mt systems overlapping each other and enclosing the daughter nuclei (Figs. 22-26).

1.3. CYTOKINESIS

The time between telophase and cytokinesis varies among the different cell types as well as among different species. In all species the two Mt systems move apart and the separating diaphragm is formed in the space between them (arrows, figs. 27 and 31). In elongated apical cells of *S. phinneyi* the distance between the daughter nuclei is long and eventually the apical nucleus becomes larger than the basal one. In intermediate cells of *T. mertensii* or in those of *E. siliculosus*, the daughter nuclei occupy the apical and basal part of the cell close to the wall (Figs. 30 and 32).

2. Conclusions

Microtubule organization is related to the growth pattern of apical cells of all species examined, but is more intensely expressed in *S. phinneyi*, which shows a tip-like growth, like Sphacelariales. The cell cycle stage in which centrosome duplication occurs, seems to depend on the division rate of the cell. The site of centrosome duplication is also different between the above cell types. In most cases the centrosome migrates towards the basal pole of the nucleus where it duplicates. During mitosis spindle and cytoplasmic Mts show different organization in evenly growing cells compared to tip-growing ones. The mitotic and cytokinetic apparatus in *E. siliculosus* and *T. mertensii* have common features, while they both show differences from *S. phinneyi*. This suggests a stronger relationship between the two former species.

3. References

1. Katsaros, C. (1992). Immunofluorescence study of microtubule organization in some polarized cell types of selected brown algae. *Bot. Acta* **105**, 400-406.
2. Henry, C.E. and Müller, G.D. (1983). Studies on the life history of *Syringoderma phinneyi* sp. nov. (Phaeophyceae). *Phycologia* **22** (4), 387-393.
3. Kuhlenkamp, R. and Müller, G.D. (1985). Culture studies on the life history of *Haplospora globosa* and *Tilopteris mertensii* (Tilopteridales, Phaeophyceae). *Br. Phycol. J.* **20**, 301-312.
4. Müller, G.D. (1979). Genetic affinity of *Ectocarpus siliculosus* (Dillw.) Lyngb. from Mediterranean, North Atlantic and Australia. *Phycologia* **18**, 312-318.

BETULA PENDULA: RESPONSE TO OZONE IN LEAF, STEM, AND ROOT CELLS.

M.S. GÜNTHARDT-GOERG*, and C. J. MCQUATTIE**
**Author to whom correspondence should be addressed*
Swiss Federal Institute of Forest, Snow, and Landscape Research, Zürcherstrasse 111, CH-8903 Birmensdorf, Switzerland
***USDA Forest Service Laboratory, 359 Main Road, Delaware, Ohio 43015, USA*

Abstract: Cell response induced by low ozone concentrations, namely cell wall thickness increased by pectin exudation, changes in tannin structure and formation of other phenolic compounds is traced from leaf mesophyll cells to small leaf veins, stem and coarse root bark of birch cuttings.

1. Introduction

Earlier investigations reported that low continuously applied 50-90 nL L^{-1} ozone (O_3) concentrations induce structural acclimation of leaf and cell size or stomatal density in flushing leaves, whereas continued exposure to ozone leads to visual leaf symptoms which progress from light green dots to necrotic points (Günthardt-Goerg 1993, 1996), and which precede reduced assimilation and growth (Maurer et al. 1997). Thickened cell walls and wall projections formed by pectin exudation adjacent to intercellular air spaces indicate that ozone has caused a slow intercellular oxidative process to occur at the cell walls (Günthardt-Goerg et al. 1997). Cell to cell transport is interrupted by shrinking and finally collapsing cells. Less starch is produced and inhibited transport is indicated by starch accumulation in mesophyll cells along small veins (Günthardt-Goerg et al. 1993). In the present report we observe additional cell reactions which may link the visual leaf symptoms to reduced root growth.

2. Methods

Twenty cuttings of birch (*Betula pendula* Roth) were exposed, before leaf formation until autumnal leaf fall, in 20 field fumigation chambers to filtered air (control) or filtered air plus O_3 (day/night 90/40 nL L^{-1}), with low and high (10-fold) fertilization supplied by watering the plants. Visible leaf symptoms and cellular injury were observed in the leaves throughout the season, but stems and coarse roots were harvested at the end of the season for investigations by light (LM), fluorescent light (FM) and transmission electron (TEM) microscopy (experimental and methodological details see Günthardt-Goerg et al. 1997).

Figure 1: Foliage of *Betula pendula* cuttings from June to September, influenced by: C = filtered air, O_3 = added ozone day/night = 90/40 nL L^{-1}, each under low and high (10-fold) fertilization regime (means + SE).

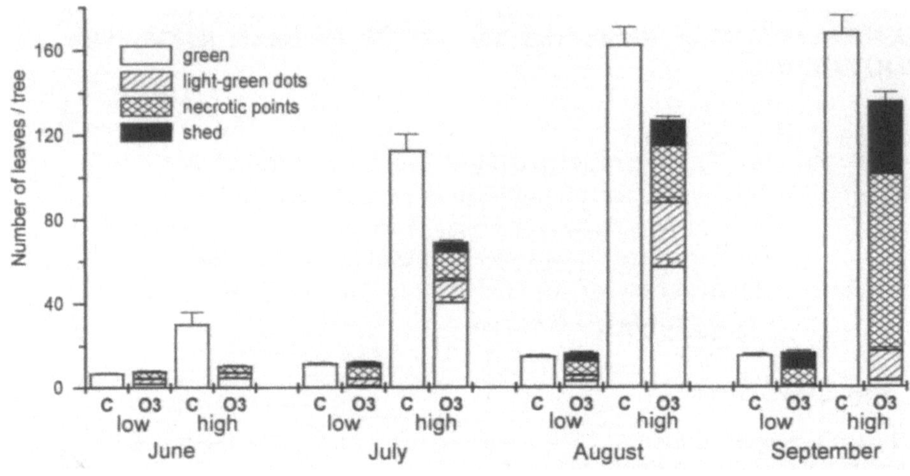

Figure 2: Left A, C, E, G, I, K=control. Right B, D, F, H, J, L=cell injury through ozone. Cb=cambium, c=cytoplasm, cc=companion cell, e=phloem element, lb=lipid body, pl=plasmodesmata, s=starch, t=tannin, v=vacuol, (y)x=(young) xylem element. Arrow head=oxidised cell with phenolic compounds, arrow=sieve plate. A-F=birch leaf vein cross sections. A,B=FM, 340-380 nm excitation. C-F=TEM. G-J=stem phloem. G,H=LM phase contrast. I,J=FM, callose luminescent by Aniline Blue. K,L=coarse root.

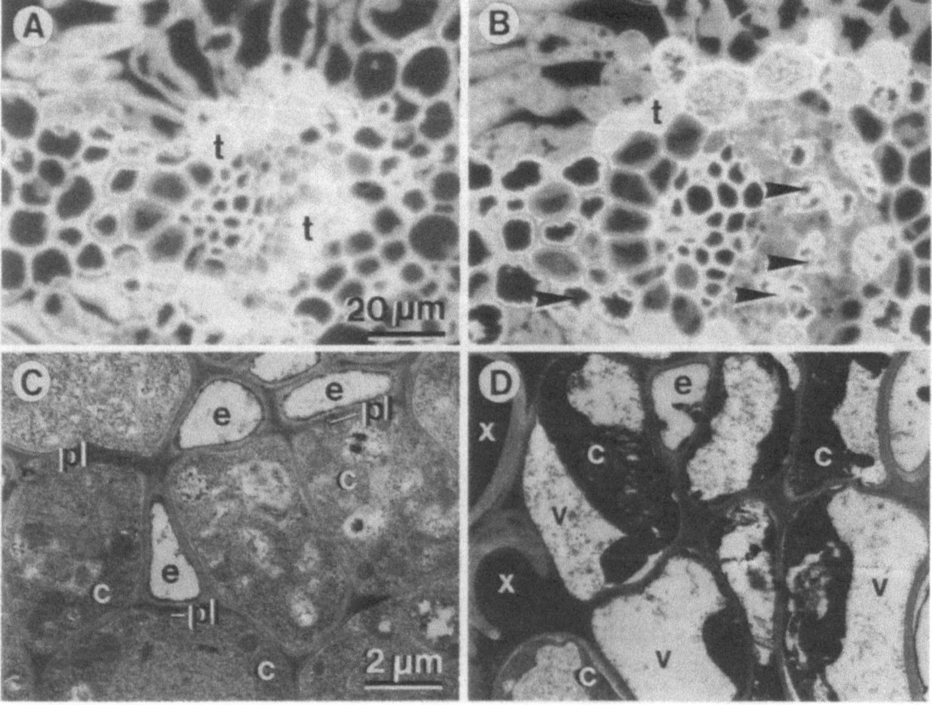

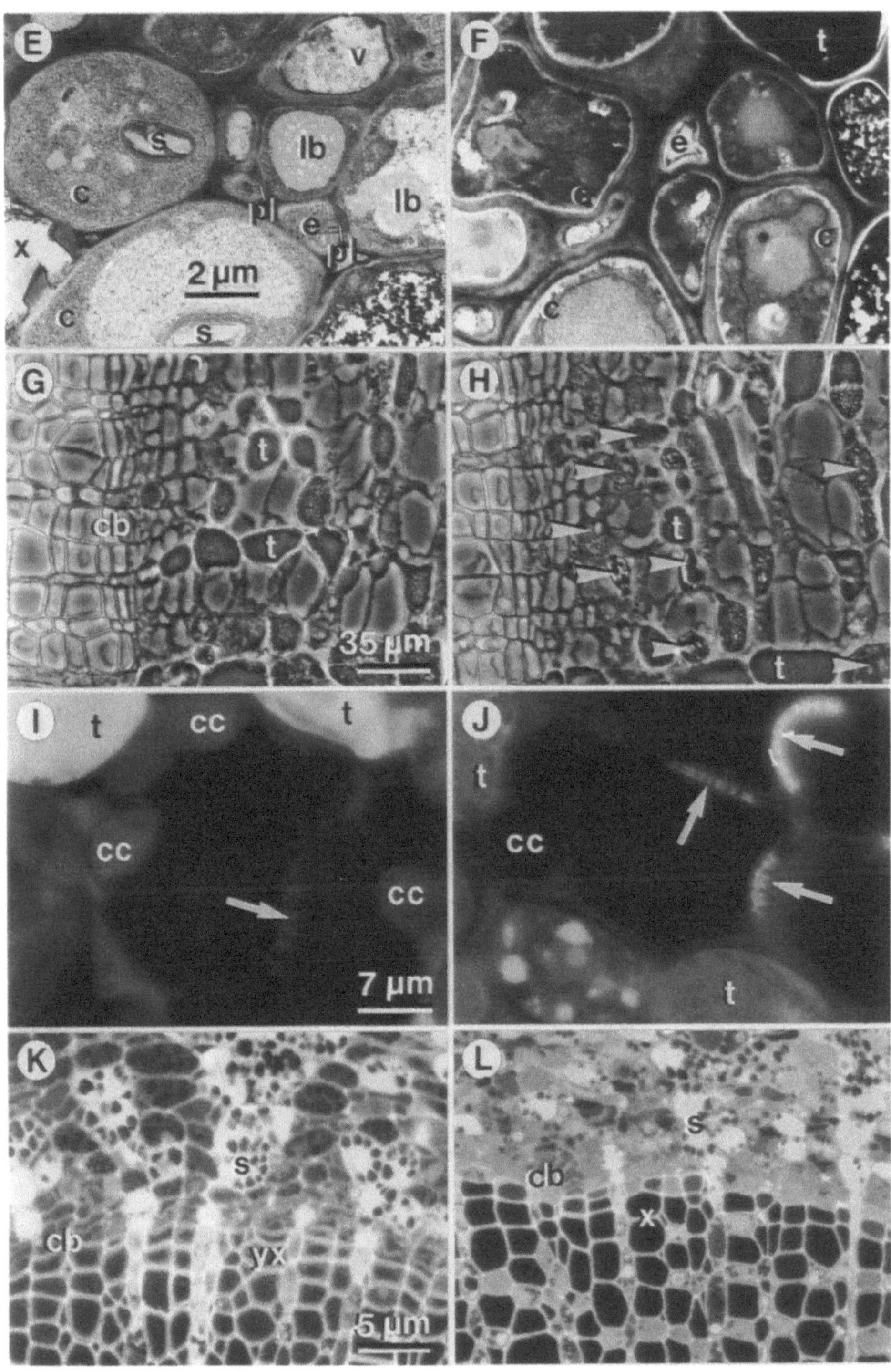
E
v
s
lb
c
pl
lb
e
x
pl
2 µm
c
s
F
t
e
c
c
c
t
G
t
cb
t
35 µm
H
t
t
I
t
cc
t
cc
cc
7 µm
J
t
cc
t
K
s
cb
yx
5 µm
L
s
cb
x

3. Results

Low fertilization, especially in the presence of added ozone, reduces tree growth, but visible leaf symptoms are induced by O_3 irrespective of the fertilization regime 25 days after flush. Therefore, tree foliage is highly symptomatic in July under low fertilization, whereas high fertilized trees form new green leaves (Fig. 1). When low ozone concentrations are used (as in our experiment with about a double ambient concentration), and when the cuttings are exposed before flush to give them the possibility to acclimate, visible ozone symptoms and cell injury develop slowly. Because ozone enters the leaf intercellular space by the stomatal openings at the lower leaf side, cell response to oxidative stress (pectin cell wall protrusions and thickening; darkened nucleus and mitochondria in comparison to the chloroplasts, which decrease in electron opacity; protoplast detaching from the cell wall) starts in the spongy mesophyll cells and later spreads to the palisade parenchyma cells. Frequent phenolic compounds indicate oxidative deterioration of cell content not only in mesophyll cells, but also in small leaf veins (arrow heads Fig. 2 B vs. A), which are similarly exposed to ozone and its reactive products. In 14 day old leaves, where cell injury due to ozone treatment is hardly visible by eye, the leaf vein cell structures (Fig. 2 D vs. C) are less clearly defined, mitochondria and cytoplasm are darkened, plasmodesmata to small phloem elements are rare, vacuoles are large or fragmented, and some xylem elements are filled with electron dense material, which appears phenolic by FM. In 35 day old leaves (Fig. 2 F vs. E), which show visible ozone injury symptoms, the cell walls in leaf veins are thickened, the cytoplasm tends to detach from the cell wall, vacuoles are filled with dark or coarse tannin precipitation in contrast to the sandy tannin appearance in the control, and lipid bodies are rare. Similar cell injury occurs in the stem phloem (arrows indicate disorganized and oxidized phloem parenchyma cells Fig. H vs. G). Before onset of autumnal leaf fall in early November, the stems from the control treatment still show open sieve pores, whereas the sieve pores in stems of the trees in the ozone treatment, which have already partially lost their foliage, are narrowed by callose and companion cells appear dead without luminescent cytoplasm (Fig. 2 J vs. I). By such earlier shut down of phloem transport trees influenced by ozone are forced to wait for the next growing season to renew growth. The above observations in leaf veins and stems can be traced down to disorganized phloem parenchyma cells, fewer starch granules, crushed cambium cells and consequently missing young xylem elements in coarse root cells (Fig. 2 L vs. K).

4. References

Günthardt-Goerg M.S., Matyssek R., Scheidegger C., Keller T. (1993): Differentiation and structural decline in the leaves and bark of birch (*Betula pendula*) under low ozone concentrations. Trees 7, 104-114.

Günthardt-Goerg M.S. (1996): Different responses to ozone of tobacco, poplar, birch, and alder. J. Plant Physiol. 148, 207-214.

Günthardt-Goerg M.S., McQuattie C.J., Scheidegger C., Matyssek R., Rhiner C. (1997): Ozone-induced cytochemical and ultrastructural changes in the leaf mesophyll cell wall. Can. J. For. Res. 27, 453-463.

Maurer S., Matyssek R., Günthardt-Goerg M.S., Landolt W., Einig W. (1997): Nutrition and the ozone sensitivity of birch (*Betula pendula*), I. Responses at the leaf level. Trees (in press).

5. Acknowledgments: We are grateful to P. Bleuler, Dr. J. Bucher, I. Kälin, Dr. W. Landolt, D. Mathies, Dr. R. Matyssek, Dr. S. Maurer, and C. Rhiner for the excellent teamwork.

EFFECTS OF Pb AND Cd ON THE ULTRASTRUCTURE OF WHEAT ROOTS

M. MOUSTAKAS[1], M. CIAMPOROVA[2], G. OUZOUNIDOU[1] AND I. TSEKOS[1]
[1]*Department of Botany, Aristotle University of Thessaloniki, Thessaloniki 54006, Hellas,* [2]*Institute of Botany, Slovak Academy of Sciences, 842 23 Bratislava, Slovak Republic*

1. Introduction

Structural responses of plant cells to heavy metals can indicate sites of action and/or sites of accumulation of toxic metal ions. Accumulation of Pb in the cell walls and vacuoles or root cells was demonstrated using autoradiography (**1**). At lower concentrations, Cd deposits occur in the vacuoles while at higher concentrations they are found in the cell walls (**2, 3, 4**). Studies of cellular responses can help to understand the mechanism of heavy metal effects on plants. The aim of our work was to reveal structural responses of wheat root cells to cadmium and lead, and to find out the degree of their toxicity.

2. Material and Methods

Seedlings of *Triticum aestivum* cv Dio, growing in the modified Hoagland solution, were exposed to 1.0 mM Pb supplied as $Pb(NO_3)_2$ and 1.0 mM Cd supplied as $Cd(NO_3)_2\ 4H_2O$ for 7 days. Root tips were fixed in glutaraldehyde and osmium tetroxide dehydrated and embedded in Spurr's resin. Ultrathin sections were taken from the cells in root meristem (at the level of the last proximal root cap cells), and for Pb from the root cap and the more differentiated cells (3 mm behind the root tip). The sections were stained with uranyl acetate and lead citrate, and investigated with Zeiss 9 S-2 electron microscope.

3. Results

3.1. CONTROL CELLS

Root epidermis (E), hypodermis (H), and the next cortical layers (C2 to C5) including endodermis (En) reveal structural characteristics of meristematic cells: large nuclei, dense cytoplasm, and small vacuoles (V). Cell divisions (asterisk) occurred in all tissues.

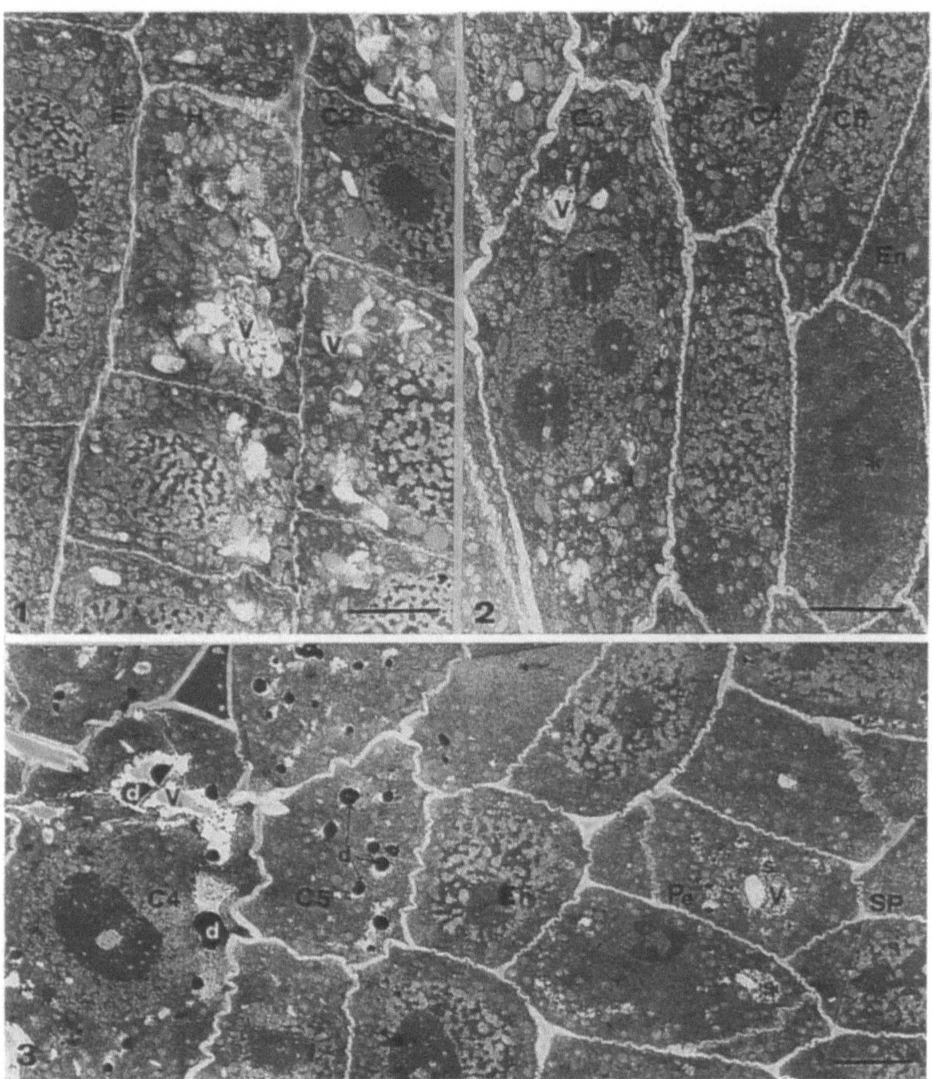

Figs. 1 and 2. Control. Cross section of the meristematic cells of root epidermis (E), hypodermis (H) and the next cortical cell layers (C2 to C5) including endodermis (En) with a mitotic figure (asterisk). **Fig. 3.** Pb treatment. Internal layers of root cortex (C4 and C5) with increased number of vacuoles, endodermis (En), and cells of central cylinder: pericycle (Pe) and stelar parenchyma (SP). Vacuoles (V) contain inclusions of various density (s,d).

Figs. 4 to 6. Pb treatment. **(4)** Plastid (P) with starch grain (arrow) and vacuoles (V) with dense inclusions (s,d). **(5)** Part of the cell wall between endodermis (En) and pericycle with a large Pb deposit (arrowhead). P = plastid, V= vacuole with a dense inclusion (d). **(6)** Pb deposit in the longitudinal cell wall (cw) of stelar parenchyma at 3 mm distance from the root tip.

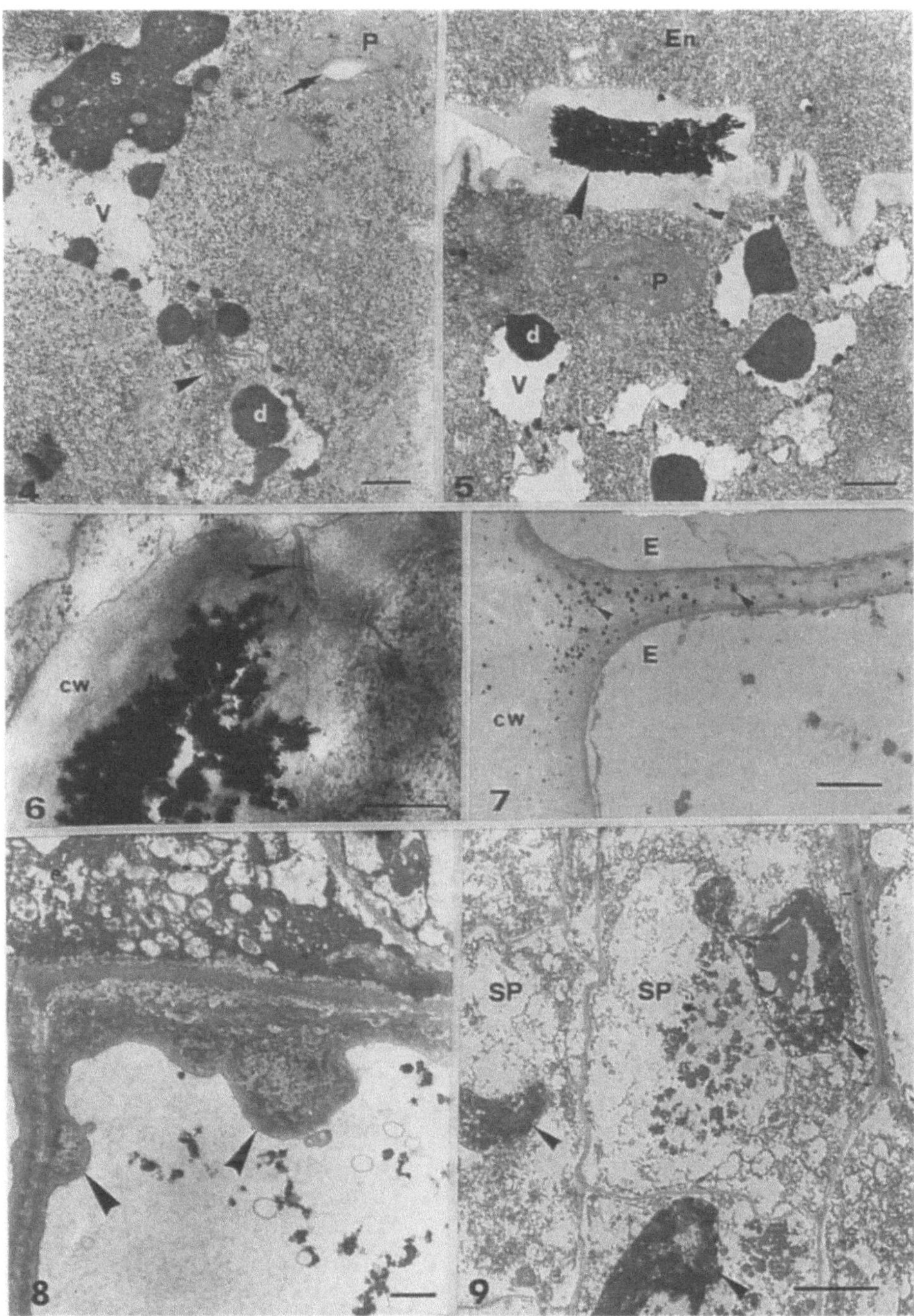

Figs. 7 to 9. Cd treatment. **(7)** Dense Cd deposits in the outer cell wall (cw) of the root epidermis (arrowheads). **(8)** Large Cd deposits (arrowheads) in the cell walls, and cytoplasm disintegration in the cortex. **(9)** Severe damage to the stelar parenchyma (SP) cells with remnants of nuclei (arrowheads).

3.2. LEAD

Mitoses were not found in the treated roots. Cell organelles were not changed with the exception of starch grain appearance in the plastids (Fig. 4, arrow). Vacuolation slightly increased (Fig. 3) and, similarly to control, it was higher in the cortex but lower in endodermis (En) and central cylinder (Pe = pericycle, SP = stelar parenchyma). Inclusions of various densities (s,d, Figs. 3 to 5), representing probably accumulation of Pb, appeared in the vacuoles. Deposits of Pb were present in cell walls of root cap, root epidermis, cortex including endodermis (Fig. 5), and of the pericycle. The Pb deposits appeared also in the longitudinal cell walls of more differentiated stelar parenchyma 3 mm behind the root tip (Fig, 6).

3.3. CADMIUM

Complete disintegration of cytoplasm was induced in root epidermis (Fig, 7) and cortex (Fig. 8). Severe injury was induced also in the cells of central cylinder (Fig. 9) where remnants of damaged nuclei (arrowheads) could be distinguished. Deposits of Cd occurred in the cell walls of epidermis (Fig. 7, arrowheads), and they were enormous in the cortex (Fig. 8, arrowheads). Cell ultrastructure in all Cd-treated tissues corresponded with criteria for dead cells (**5**).

4. Conclusions

1. The structural integrity of wheat root cells was preserved after 1.0 mM Pb while it was disintegrated after 1.0 mM Cd treatment. Thus, based on ultrastructural results, the order of toxicity was Cd>Pb.
2. Accumulation of Cd and Pb in the cell walls, and of Pb also in the vacuoles agreed with sequestration of other toxic metals in these compartments.
3. The amount of Pb and Cd deposits decreased gradually from peripheral tissues towards the central cylinder. This gradient correlated with the degree of vacuolation induced by Pb as well as with the damage to cell structure induced by Cd.
4. High Pb and Cd accumulation in root cortex might serve to prevent the tissues of central cylinder from toxic effects of the metal ions, confirming thus the importance of the root cortex for the tolerance of heavy metal stress in plants.

5. Acknowledgements

This work was supported by a grant from the Greek General Secretariat for Research and Technology.

6. References

1. Wierzbicka, M. (1987) Plant Cell and Environment **10**, 17-26.
2. Vazquez, M.D., Poschenrieder, Ch. and Barcelo, J. (1992) New Phytologist **120**, 215-226.
3. Barcelo, J., Vazquez, M.D., and Poschenrieder, Ch. (1988) New Phytologist **108**, 37-49.
4. Rascio, N., Dalla Vecchia, D., Ferretti, M., Merlo, L., and Ghisi, R. (1993) Archives of Environmental Contamination and Toxicology **25**, 244-249.
5. Davies, K.L., Davis, M.S. and Francis, D. (1991) Plant Cell and Environment **14**, 399-406.

ANATOMY OF ADVENTITIOUS ROOT FORMATION IN STEM CUTTINGS OF THE EASY-TO-ROOT (GEMLİK) AND HARD-TO-ROOT (DOMAT) OLIVE CULTIVARS

M. T.ÖZKAYA[1], M.ÇELİK[1] AND G.ALGAN[2]

[1] *Department. of Horticulture., Faculty. of Agriculture., Ankara University, 06110 Ankara Turkey,* [2] *Department. of Biology, Faculty. of Science., Ankara University, Ankara Turkey*

1. Abstract

The semi-hardwood cuttings of one of the most important black and green olive cultivars of Turkey, Gemlik and Domat, were treated with IBA or by wounding or by wounding+IBA and placed for rooting into Shaded Polyethylene Tunnel (SPT) for 90 days. At 15-day intervals, samples were collected from SPT and examined for morphological variations in the cuttings during rooting. Anatomical studies done in which slides of transverse and longitudinal sections of tissue taken from 1 cm above the site of the cutting. Samples were collected every 15 days during rooting. No morphological differences related with rooting were observed in the histological studies between sample intervals. It was also observed that the sclerenchyma does not appear to be the main source of difficulties in rooting of the Domat cultivar. Increases in metabolic activity may cause callus formation in preference to root formation. The effect of wounding alone did not improve rooting, but did increase the effects of IBA. Also no correlation was found for root initiation and wounding in cuttings. It was observed that in olive cuttings, adventitious roots were seen to emerge from the cambial zone (cambium, ray or undifferentiated secondary xylem cells).

2. Introduction

The phenomenon of adventitious root formation is still not fully understood, nor is it clear why cuttings of different cultivars have different rooting potentials. Adventitious roots in stem cuttings generally originate in the young secondary phloem (Davies and Hartmann,1988), although they may also arise from vascular rays, cambium or pith (Hartmann and Kester, 1983). Mackenzie et al (1986) observed that adventitious root formation in stem cuttings of M26 apple rootstock originate in the basal callus or wounding callus. They also found that wounding improved rooting, but only by splitting the base of cutting. Among the many processes that occur in woody plants that may make them hard-to-root are condense sclerefication, lignification in the sclerenchyma, and suberization of the cortex (Ciampi, 1964; Goodwin, 1965; Beakbane, 1969; Edwards and Thomas 1980) and Williams el al, 1984). Some researchers believe that such barriers to adventitious root formation can be overcome.

In this study, our objectives were to study how adventitious roots initiate and develop in these cultivars, and also to determine if anatomical events are associated with rooting difference between Gemlik and Domat olive cultivars.

3. Materials and Methods

Olive cuttings of Gemlik (easy-to-root) and Domat (hard-to- root) varieties, which are the most important black and green ripened table olive cultivars in Turkey, were prepared at the end of June for the years 1994 and 1995. Cutting were about 15 cm long and featured 5-6 leaves. Cuttings were treated with 4000 ppm IBA for 5 seconds, by wounding the base of the cutting with two vertical incisions about 1.5 cm length, and by a combination of wounding plus IBA. Rooting was carried out under SPT (Shaded Polyethylene Tunnel) (Çelik et al, 1993 and Özkaya and Çelik, 1993). The cuttings for each cultivar and treatment were collected from SPT, in 15-day intervals up to 90 days. The bottom 5 cm of the cuttings were fixed in solution of 5% formalin plus 5% glacial acetic acid in ethanol (50%) (FAA solution). In each cutting, transverse sections were obtained with a hand microtome, while longitudinal sections were obtained using a razor blade. Visualization of starch in tissues was done by staining with IKI. Images were recorded using a video documentation system (VDS) equipped with a black-and-white video camera mounted on a Zeiss Axiolab light microscope. When these images were examined from monitor of the computer, no significant differences were observed between the tissues of the various treatment groups and the days post-rooting when the sample was taken. Sites where adventitious root were observed to emerge were examined in an attempt to assess reasons for rooting difficulties. Lignified tissues as sclerenchyma were stained with safranin for this purpose. Color photographs were taken of microscope images with a camera mounted on the instrument.

4. Results and Discussion

It was observed that the sclerenchyma ring located in cortex of Domat cuttings was continuous. Even if the root primordia pass through this barrier as a result of the barrier becoming broken by chance or by wounding, the root could nonetheless pass through the barrier easily. But while the root may pass through, one difficulty may be seen in the inability of the root to expand. This was also apparent in visual observations of cuttings collected from rooting medium without aid of the microscope. The structure of this root (short and thick root) looks like the root of Kalamata cuttings, which is another hard-to-root olive cultivar (a Greek variety) (Özkaya, 1992). It would appear that the sclerenchymal and cortical structures would not explain the main reason for this olive cultivar being hard-to-root, although others suggest that the opposite about the effects of these histological features (Ciampi and Gellini, 1963; Beakbane (1969) and Williams et al (1984). It was observed that there appears to be a weakness in root initiation and in condense callus formation from the parenchymatous cells of secondary phloem and cortex in Domat cuttings. This suggests that Domat is capable of cell proliferation without showing an ability to de- differenentiate tissues for rooting.

No correlation was found between root initiation and wounding alone, although some researchers recommend wounding since it breaches potential histological barriers. Use chemical agents such as IBA improves root formation when used in combination with wounding. Nahlawi et al (1975), Mackenzie et al (1986) and Kantarcı and Gülşen (1987) have found that the rooting of cuttings is improved during wounding only with the use of an auxin. A problem with wounding is that is may do more to promote callus formation as the vascular cambium, filled between cortex and pith, proliferates, and this may result in a closure of the path by which vascular rays important to rooting can emanate toward the periphery.

Root primordium in stems of olive originates from the cambial zone (cambium cells or ray cells). This primordium must pass to the periphery by extending through phloem fibers and then the sclerenchymal ring, and then to emerge from the bark. As a result, there is no morphological difference in cuttings related with adventitious root formation the between easy-to-root Gemlik and hard-to-root Domat cultivars. Further research will likely focus on root initiation processes.

5. References

Beakbane,A.B. 1969. Relationships between structure and adventitious rooting. *Proc. Inter. Plant Prop. Soc.* **19**, 192-201.

Ciampi,C. and Gellini,R. 1963. Onsorgenza e sviluppo delle radici avventizie in *Olea europaea* L.: importanza della struttura anatomica alla effetti dello sviluppo delle radichette. *G.Bot.Ital.*, **70**, 62-64.

Ciampi,C. 1964. Ontogenesie e struttura della guaina sclerenchimatica nelle talee di olivo. *Atti Delle Giornate di Studio su la Propagazione delle Speci Legnose.* 94-106. 26-28 Oct. 1964. Italia.

Çelik,M., Özkaya,M.T. and Dumanoğlu,H. 1993. The research on possibilities of using Shaded Polyethylene Tunnel (SPT) on the rooting of olive (*Olea europaea* L.). *Acta Horticulturae*, **356**, 21-23.

Davies,F.T.Jr. and Hartmann,H.T. 1988. The physiology basis of adventitious root formation. *Acta Hort.* **227**, 113-120.

Edwards,R.A. and Thomas,M.B. 1980. Observations on physical barriers to root formation in cuttings. *Plant Prop.* **26**, 6-8.

Goodwin,J.P. 1965. Anatomical changes associated with juvenil to mature growth phase transition in Hedera. *Nature* **208**, 504-505. In: F.T.Jr. Davies and Hartmann,H.T. (eds.) The physiology basis of adventitious root formation. *Acta Hort.* **227**, 113-120.

Hartmann,H.T. and Kester,D.E. 1983. *Plant Propagation. Principles and Practices.* Prentice-Hall, Inc. New Jersy. p647.

Kantarcı,M. and Gülşen,Y. 1987. Değişik yaralama yöntemlerinin ve çelik tipinin tombul fındık çeşidi çeliklerinde köklenme üzerine etkileri. *Ankara Üniversitesi Ziraat Fakültesi Yıllığı* **38**, 1-2, 1-11.

Mackenzie,K.A.D., Howard,B.H. and Harrison Murray,R.S. 1986. The anatomical relationship between cambial regeneration and root initiation in wounded winter cuttings of the apple rootstock M26. *Annals of Botany.* **58**, 649-661.

Nahlawi,N., Rallo,J., Caballero,J.M. and Eguren,J. 1975. Aptitude a l'enracinement de cultivars d'olivier en bouturage herbace sous nebulisation. *Olea* **6**, 11-25.

Özkaya,M.T. 1992. Propagation of Olive (*Olea europaea* L.) by tissue culture. (M.Sc. Thesis (unpublished)). Mediterranean Agronomic Institute of Chania, Greece. 80p.

Özkaya,M.T. and Çelik,M. 1993. The effect of the rooting environment and the combination of auxin and polyamine on the rooting ability of Turkish olive cultivars Gemlik and Domat. *Acta Horticulturae*, **356**, 31-33.

Williams,R.R., Jati,A.T. and Bolton,J.A. 1984. Suberization and adventitious rooting in australian Plants. Aust.J.Bot. 32:363-366. In: F.T.Jr. Davies and Hartmann,H.T. (eds.) The physiology basis of adventitious root formation. *Acta Hort.* **227**. 113-120.

EFFECTS OF Cu EXCESS ON OREGANO [*Origanum vulgare* subsp. *hirtum* (Link) Ietswaart]. I. MORPHOLOGICAL EFFECTS

E. PANOU-FILOTHEOU, A. M. BOSABALIDIS and
S. KARATAGLIS
Department of Botany, School of Biology,
Aristotle University, 54006 Thessaloniki, Greece.

1. Abstract

Oregano plants were grown in soil containing 0.3, 13.0, 17.0, 19.0, 22.0, 24.5 and 25.5 μM Cu /g of soil. High Cu concentrations resulted in a decrease of the volume of the root system by 86% and of the height of the plants by 30%. In the stem, the relative volume of the primary cortex became reduced by 71% and that of the phloem by 12%. The relative volume of the xylem increased by 52% and of the pith by 42%. The surface of the leaf blade underwent a parallel decrease by 55%, while simultaneously the leaf exhibited sings of chlorosis. The essential oil-producing trichomes became by 137% more numerous and by 34% smaller. Non glandular trichomes increased in number by 283% and stomata by 137%, but guard cells became shorter. The relative number of chloroplasts per mesophyll cell section (leaf cross section) decreased by 75%. In the leaf, the relative volume of the upper epidermis decreased by 17% and that of the lower by 3%. The volume of the palisade parenchyma cells also decreased by 22% and so was the volume of its intercellular spaces (by 59%). The relative volume of the spongy parenchyma cells increased by 74%, whereas that of its intercellular spaces decreased by 50%.

2. Introduction.

Copper is an essential element for the plants. Normally, its concentration in the soil ranges from 0.01 to 0.6 μM/g, but different kind of pollution sources may increase this amount 10-20 fold. Major sources of soil Cu pollution are the fungicidal sprays and the sewage sludges used in agriculture. Other sources of importance are the atmospheric residues containing wastes of smelters, flying ash (coal burnt for electricity, volcano ash), as well as various industrial products, like dyes and textiles (Tiller and Merry 1981). Sources of minor significance might be the Cu mines, fertilizers, herbicides, etc (Baker 1990).

In the present study the response of oregano, a plant of high commercial value, to artificially created Cu-soil pollution, i.e. to increasing concentrations of soil copper, was investigated, from the morphological, anatomical and morphometrical points of view.

3. Materials and Methods

Oregano plants [*Origanum vulgare* subsp. *hirtum* (Link) Ietswaart] were grown in Cu containing soils [control (0.3), 13.0, 17.0, 19.0, 22.0, 24.5 and 25.5 μM Cu / g of soil]. The leaf area was measured by using an MK_2 area meter connected to a TC7000 camera. In order to study the anatomy of plant organs, small pieces of roots, stems and leaves were prefixed with a mixture of 2.5% glutaraldehyde and 2% paraformaldehyde (Karnovsky 1965) buffered with 0.05 M sodium cacodylate (pH 7.2). After washing in buffer, the specimens were postfixed for 3 h in 1% osmium tetroxide similarly buffered. Tissue was dehydrated in an alcohol series and embbedeed in Spurr's resin. Semithin sections (1 μm thick) of plastic-embedded tissues were obtained in a Reichert Om U2 ultramicrotome, stained with 1% toluidine blue O, and examined in a Zeiss III photomicroscope. Intact leaves were observed and photographed in a Zeiss M63A stereoscope. Volume fractions of leaf and stem tissues were estimated by point counting analysis, as described by Toth (1982). A transparent acetate sheet bearing intersecting lines spaced at 1 cm was used as a probe.

4. Results

Treatment of oregano plants with increasing concentrations of Cu resulted in a reduction of the height of the plants as well as of the total root volume (Fig.1).The stem became thicker and densely covered with long non-glandular hairs. In cross section, the annual stem of control plants exhibited a square profile and appeared composed of a relatively thin cortex (with collenchyma poles at the corners), a vascular system with secondary structure and a compact pith. Cross sections of Cu treated stems disclosed a disorganized cortex and an increased proportion of xylem and pith. The central pith cells became destroyed leaving an empty cavity. Increasing concentrations of Cu had further a negative effect on leaf lamina surface (Fig 2), while simultaneously leaves bent and appeared chlorotic.

The leaf lamina became additionaly thicker (Fig 2) and densely covered with long non-glandular hairs (Fig. 3). Thickening of the lamina was mainly due to the development of the spongy parenchyma. The number and size of the mesophyll chloroplasts became dramatically reduced. The glandular hairs producing the essential oil increased in number on both leaf sides (Fig 4), but their secretory head (along with the above it oil accumulating bladder) was smaller (Fig 5). Stomata on both leaf surfaces became more numerous (Fig. 6) with shorter guard cells. The effect of Cu excess on oregano roots was mainly expressed by extensive deformations. The major roots developed large local dilations. When deformed roots were sectioned at the swollen regions, the presence of new emerging roots at meristematic state could be discerned. These embryonal roots later develop to create new lateral roots. Control roots are diarch with normal sructure of the cortex, endodermis and pericycle. Some cortical cells had a dark content.

4. Discussion

Excess of Cu in the soil resulted in characteristic alterations in the normal appearance of oregano plants. Leaves and stems became highly hairy and densely covered with essential oil-producing glandular trichomes. The leaves decrease in size and became chlorotic. This is probably due to the effect of copper on the chloroplasts. Leaf cross sections showed that the number and size of the chloroplasts was greatly reduced. It has been found that about 70% of the total leaf Cu is bound to the chloroplasts particularly in plastocyanin which forms part of the electron transport chain linking the two photochemical systems of photosynthesis. Additional evidence suggests that Cu may play a role in chlorophyll stability by substituting Fe (required for chlorophyll synthesis), reflecting thus sings of chlorosis in the leaves (Mengel and Kirkby 1982). The decrease in the volume of the mesophyll intercellular spaces, the reduction of the lamina surface and the bending of the leaves are presumably associated with reduced transpiration, since deformation of roots does not allow enough water to be taken up (water balance). On the other hand the increase in the number of stomata (parallel to shortening of the guard cells) is probably related to a better control of water loss.

The translocation of Cu ions within the plants occurs in the xylem as well as in the phloem where the metal is bound to amino acids. Concentrations of 1.5-2.0 μM Cu in the xylem and 3.0-140 μM in the phloem have been reported (Loneragan 1987). In our material treatment with Cu resulted in an increase of the volume of the xylem, rather than of the phloem, a fact pobably associated with an effort of the plant to move more water and nutrients to the leaves.

Striking was also the effect of excess of soil Cu on the roots. The latter became deformed exhibiting local dilations from which small lateral roots emerged. Measurements showed that roots are the sites with the major accumulation of Cu. Half or more of the total root Cu may be bound to the cell walls (Jarvis 1978). Much of the apoplastic Cu is strongly attached to multidentate organic ligans involving N, S and O atoms (Graham 1981). Absorption of Cu by plants may be affected by pH of the soil solution. In our experiments the pH of the soil solution fluctuated from 6.8 (control) to 5.01(mine soil). A decrease in pH enhances Cu absorption because of increased solution activities of Cu related to dissolution of minerals and solid-phase absorption of Cu.

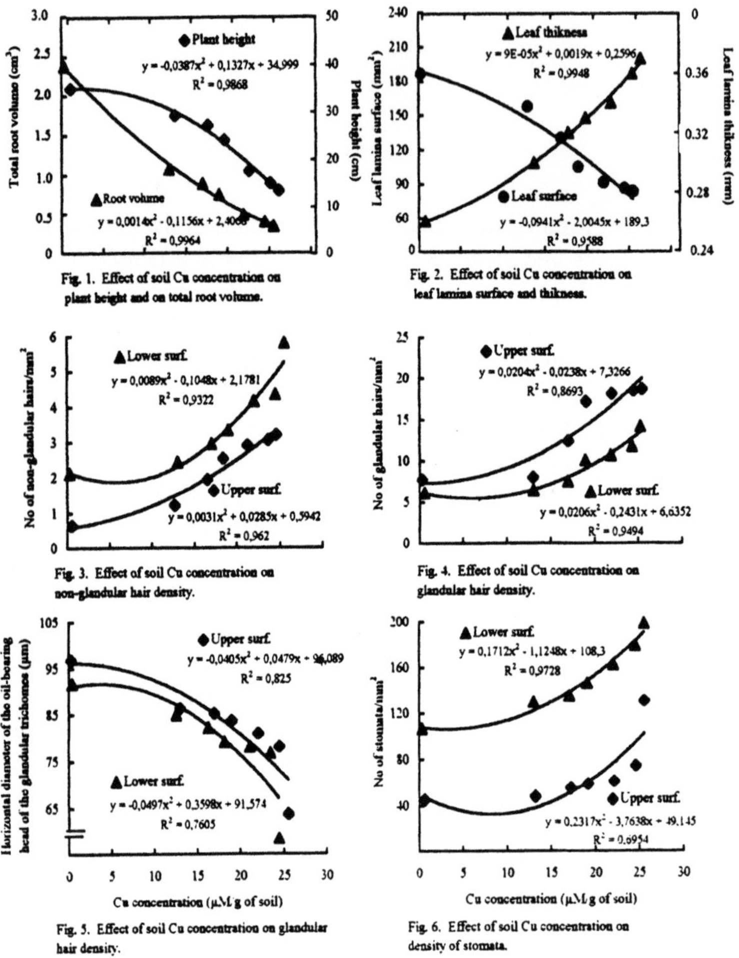

Fig. 1. Effect of soil Cu concentration on plant height and on total root volume.

Fig. 2. Effect of soil Cu concentration on leaf lamina surface and thikness.

Fig. 3. Effect of soil Cu concentration on non-glandular hair density.

Fig. 4. Effect of soil Cu concentration on glandular hair density.

Fig. 5. Effect of soil Cu concentration on glandular hair density.

Fig. 6. Effect of soil Cu concentration on density of stomata.

References

Baker . D.E. 1990. *Copper.* In: B.J. Alloway, *Heavy metals in soils.* John Wiley, New York.

Graham, R.D. 1981. *Absorption of copper by plant roots.* In J.F. Lonegaran, A.D. Robson and R.D. Graham. *Copper in soils and plants.* Acad. Press, New York, London.

Jarvis, S.J. 1978. *J. Sci. Fd. Agric.* 29, 12-18.

Karnovsky, M.J. 1965. A formaldehyde-glutaraldehyde fixative of high osmolarity for use in electron microscopy. J. *Cell Biol.* 27: 137A-138A.

Loneragan, J.F. 1987. *Copper in soils and plants.* Acad. Press, New York.

Mengel, K. and E.A. Kirkby. 1982. *Principles of plant nutrition.* Intern. Potash Instit. Bern.

Tiller, K.G. and R.H. Merry. 1981. *Copper pollution of agricultural soils.* In: Lonegaran, A.D. Robson and R.D. Graham. *Copper in soils and plants.* Acad. Press, New York, London.

Toth. R. 1982. An introduction to morphometric cytology and its application to botanical research. *Amer. J. Bot.* 69, 1694-1706.

MORPHOLOGICAL-ANATOMICAL CHARACTERISTICS OF NEEDLES OF FIR ABIES ALBA MILL. NARROW FASTIGIATE FORM

M. TOŠIĆ, D. VILOTIĆ
Faculty of Forestry, University in Belgrade
1 Kneza Višeslava, 11030 Belgrade

1. Abstract

Narrow-fastigiate fir (*Abies alba* Mill. var. *elegantissima* Tošić) on the mountain Ogorijevac (Community Sjenica) in Serbia was created by the spontaneous generative mutation. Numerous differences in morphological (habit, branching, sizes and forms of needles, cones and seeds), as well as anatomical characteristics (size of resin channels, number of epithelial cells, width of hypodermal and sclerenchymous layer of cells), compared to the typical variety of fir, have been determined by research.

2. Introduction

On mountain Ogorijevac in Serbia, at the altitude of 1100-1250 m, there are several trees of narrow fastigiate fir (Tošić, 1963).

Upon the evaluation of significant differences in morphological and physiological characteristics of the previously identified fastigiate variety (*Abies alba* var. *pyramidalis* Carr.) and the narrow fastigiate fir on Ogorijevac, the latter was described as a new variety (Tošić, 1995).

3. Material and Method

Needles were taken from 5 trees (no. 1-5) of vegetative and 7 trees (no. 6-12) of generative origin. Morphological characters were determined by morphometric analysis of 50 second-year needles taken from each tree, as well as from one tree of typical fir variety (control). Anatomical structure was analyzed on cross sections of second-year needles (the widest part) obtained by paraffin method. The sizes of resin channel, epithelial cells, widths of hypodermal and sclerenchyma layers of cells were determined.

4. Results and Discussion

Tree crown consists of numerous branches growing upwards under a very sharp angle. The needles differ both morphologically and anatomically from the typical variety. They are dark green, mainly shorter and narrower, but considerably thicker and firmer.

Trees of vegetative origin have shorter needles ($\bar{x}$ =16.83 mm) than firs of generative origin ($\bar{x}$ =18.43 mm) and firs of typical variety ($\bar{x}$ =25.48 mm). The needles of narrow fastigiate firs are spirally arranged and are not in two directions like the needles of the typical variety. The trees of vegetative origin have much more expressed two silvery rays of stomata on the lower surface of the needles. These rays, considering the arrangement of branches and needles, give the silvery appearance to the tree and enhance considerably its esthetical value.

There are differences in anatomical structure of needles between the trees of generative and vegetative origin. The needles on trees produced by grafting are wider ($\bar{x}$ =1.85 mm) and thicker ($\bar{x}$ =0.75 mm) than the needles of generative origin (width $\bar{x}$ =1.59 mm and height $\bar{x}$ =0.63 mm). The trees of vegetative origin have larger resin channel diameters of needles ($\bar{x}$ =129.28 μm) (Fig. 1) than those of generative origin ($\bar{x}$ =89.4 μm) (Fig. 2). Epithelial cells of needles of vegetative origin are larger than those of generative origin.

All the mature trees bear seed on the natural site. The phenotype of the generative progeny from free fertilization is more or less similar to parent trees in a high percentage (Tošić, 1995),which is the result of genetic mutation.

Permanent protection of natural rarities, in case of spontaneous mutants especially of conifers, should not be based only on the prohibition of felling – vegetative propagation is required for the permanent protection of the mutant genotype. In this aim, JP "Srbijašume" – Forest Estate "Golija" at Ivanjica, prepares the establishment of plantations (live archives) of this fir variety.

5. Conclusion

The analyzed fir varieties on the mountain Ogorijevac are the generative progeny of the generative mutant with specific biological and esthetic properties that make it attractive for landscape architecture.

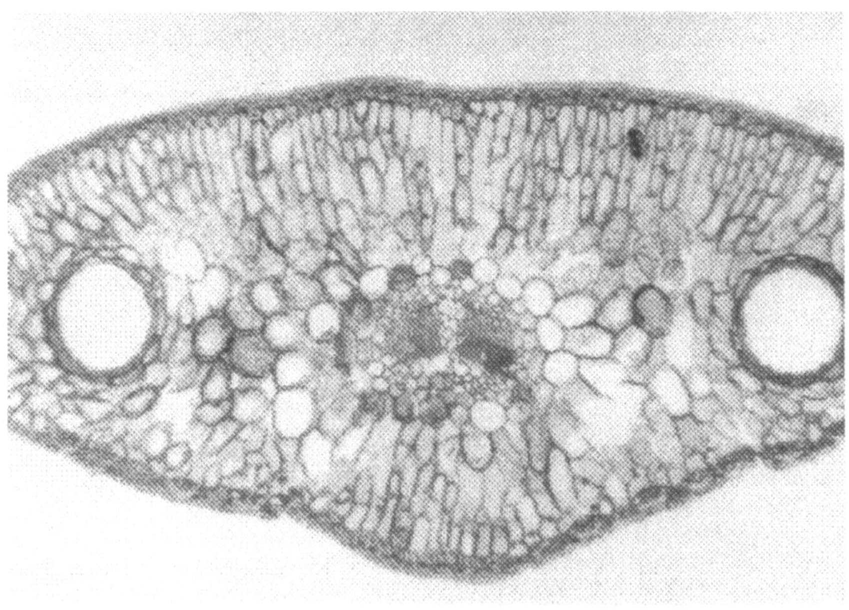

Figure 1. Anatomic section of needles on tree no. 4

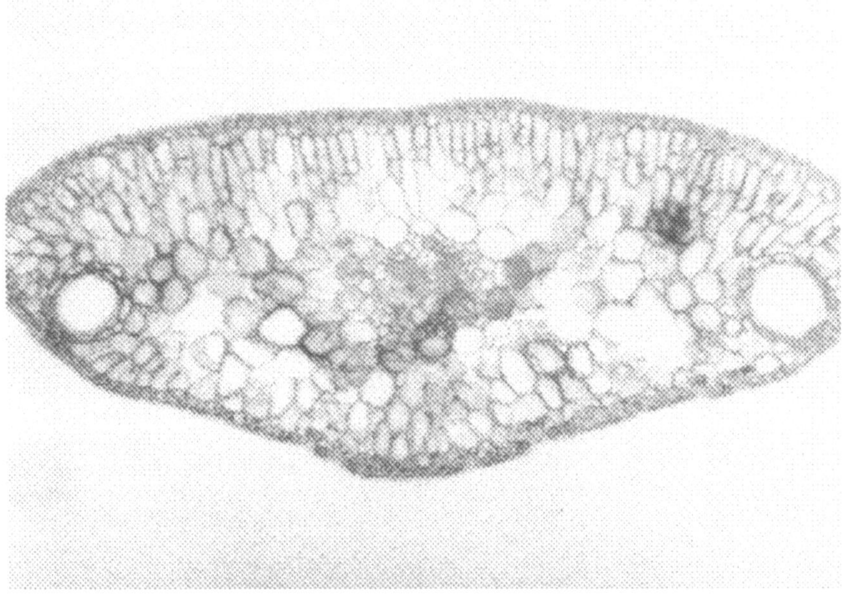

Figure 2. Anatomic section of needles on tree no. 10

6. References

1. Carriere, E. A., 1867 Traite general des Coniferes. 2. ed. Paris.
2. Tošić, M., 1963 O nalazu piramidalnog varijeteta jele (Abies alba Mill.) u okolini Sjenice. Šumarstvo, Beograd, 10 – 12: 387 – 392.
3. Tošić, M., 1995 Novi varijeteti šumskog drveća i mogućnosti njihovog korišćenja. Drugi simpozijum o flori Srbije. Vranje: 35 – 43.

MORPHOLOGICAL CHARACTERISTICS AND ANATOMIC STRUCTURE OF THREE SPONTANEOUS MUTANTS OF SPRUCE (PICEA ABIES KARST) IN SERBIA

D. VILOTIĆ, M. TOŠIĆ,
Faculty of Forestry, University in Belgrade
1 Kneza Višeslava, 11030 Belgrade

1. Abstract

Three spontaneous mutants of spruce *(Picea abies* Karst.), with specific morphological and anatomical characteristics, grow on the mountain Golia in Serbia. Being natural rarities, they have been protected by Law. Thanks to their esthetical and other significant properties, they can be used as the initial material for the production of valuable ornamental cultivars. There are essential differences in the morphological-anatomical structure of needles, both compared to common spruce on the same site and beetween the three mutants under study.

2. Introduction

Three spontaneous mutants of spruce (*Picea abies* Karst.) were recorded on the mountain Golija in Serbia. They have specific morphological and anatomical characteristics and, taking into account their extremely modified phenotype, they were described as separate varieties (*Picea abies* Karst. var. *gracilis* Tošić, *Picea abies* Karst. var. *gajicii* Tošić, and *Picea abies* L. Karst. var. *compacta* /Kirchn./Nach). Being natural rarities, they have been protected by law, and their aesthetic properties make them attractive for utilization especially in amenity horticulture.

3. Material and Method

Morphological and anatomical properties of needles of the above three spruce varieties were analyzed. Morphological characteristics were determined by morphometric analysis of second-year needles taken from the lower and middle parts of the crown, at the south exposure, of the analyzed and control trees. Anatomical structure was studied on cross sections of second-year needles (the widest part), obtained by paraffin method. The following elements were measured: needle sizes, number and size of resin channels, number of layers of hypodermal cells, layers of sclerenchyma cells under the phloem part of the vascular bundle.

4. Results and Discussion

4.1. *PICEA ABIES* (L.) KARST. VAR. *GRACILIS* TOŠIĆ

Trees with a very fastigiate, specifically formed crown. One tree is distinguished by its size and beauty of habit. It is 38 m high with the diameter at breast height 75 cm (in 1994). Its crown is very slender (*gracilis* !), composed of thin branches. Branchlets are thin, in the upper part they are short, branched and lignified; those close to the stem are very long (150 cm and more), very thin (5 – 8 mm) and not branched, because they had only terminal buds, so they hang in a vertical position. The tree bears seed, so its generative progeny was raised in the Forest Nursery at Ivanjica.

The needles are radially arranged on the branchlets, yellow-green and with rounded tips. They are 13 - 23 mm long, mean length 16.76 mm.

The width ($\bar{x}$ =1.32 mm and height $\bar{x}$ =1.18 mm) of needles in this variety is much greater compared to var. *gajicii* and var. *compacta*. Resin channels ($\bar{x}$ =41.25 μm) are significantly smaller than those of the other two varieties. The layers of sclerenchyma cells under phloem tissue are well developed (3 - 4 rows)(Fig.1).

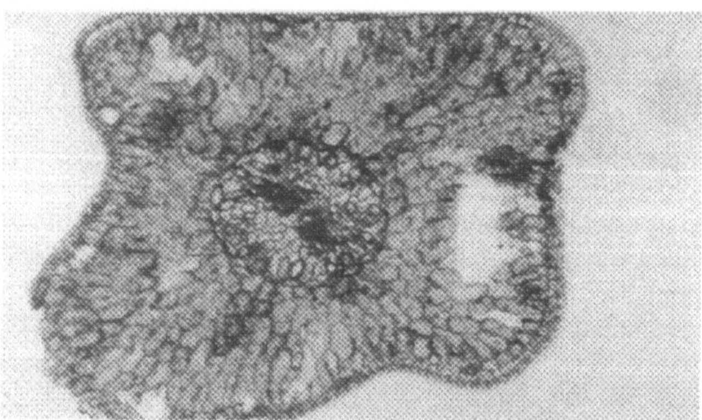

Figure 1. *Picea abies* (L.) Karst. var. *gracilis* Tošić: needle anatomic structure

4.2 *PICEA ABIES* (L.) KARST. VAR. *GAJICII* TOŠIĆ

Trees with notably fastigiate columnar crown composed of short and thin branches, slightly drooping, with abundant needles 10 to 15 mm long. Each branch consists of many thin and short branchlets very densely arranged, so that they look like a witch's broom. The most attractive tree reminds of a *Picea omorika* Pančić, so it was described as *Picea abies* f. *omorikoides* Marković. Forty years ago, two such trees died. The tree bears seed occasionally. The needles of this variety, by their cross sectional sizes: width (1.04 mm) and height (0.66 mm), differ considerably from both the typical spruce and the other two analyzed varieties. There are 1 to 2 rows of layers of sclerenchyma cells.(Fig.2).

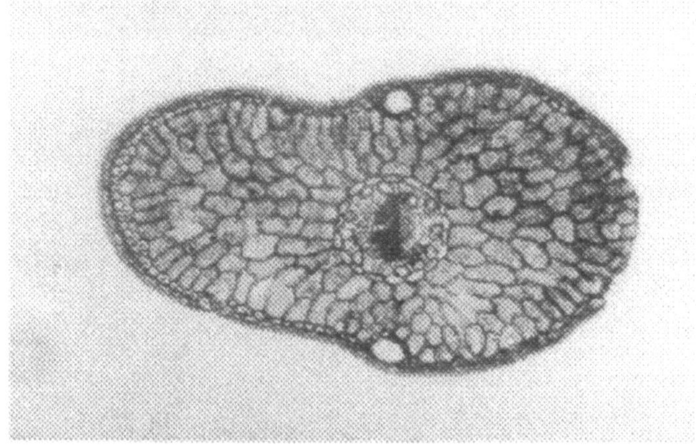

Figure 2. Picea. abies (L.) Karst. var. gajicii Tošić f. omorikoides Marković: needle anatomic structure

4..3 *PICEA ABIES* (L.) KARST. VAR. *COMPACTA* (KIRCHN.) NACH

On the mountain Golija in Serbia, there is a spruce tree with a compact crown composed of relatively large branches and numerous small branchlets with very short needles (9 - 12 mm, $\bar{x}$ =10.6 mm). The surrounding trees had an adverse effect on the development of the crown, so it is irregular.

This variety is distinguished by the special size of resin channels ($\bar{x}$ =75 μm), as well as by layers of sclerenchyma cells under the phloem tissue (1 to 2 rows)(Fig.3).

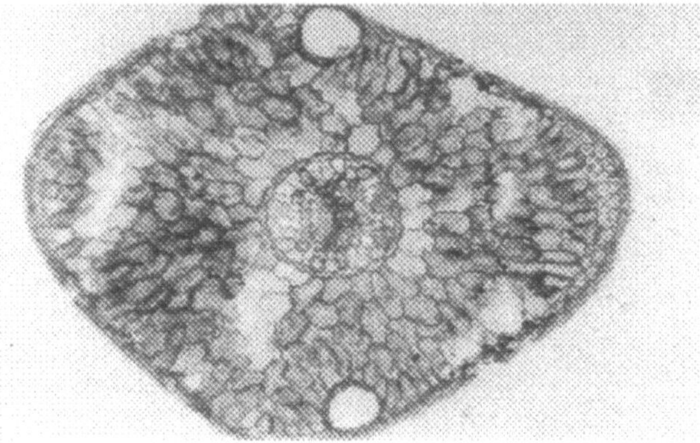

Figure 3. *P.icea abies* (L.) Karst. var. *compacta* (Kirchn.) Nach: needle anatomic structure

5. Protection of Spontaneous Mutants of Forest Trees

To protect the interesting specimens of forest trees, they are protected by law as natural rarities. However, this important measure is not sufficient in case of spontaneous mutants of conifers. An active and dynamic measure of protection is also necessary, which includes vegetative propagation in the aim of permanent conservation of mutated genotypes, so that they can be utilized in the process of further breeding for the needs of forestry and horticulture. In this aim, preparatory works are carried out by JP "Srbijašume" – Forest Estate "Golija" at Ivanjica (Serbia), to establish clonal and seedling live archives of spruce mutants under study.

6. Conclusion

Significant differences of needle anatomic structure were recorded both compared to common spruce in the same site, and compared to the three varieties under study. They are: sizes of resin channels, number of layers of hypodermal cells under epidermis and sclerenchyma cells under the phloem part of the vascular bundle.

The analyzed varieties are undoubtedly spontaneous mutants with specific biological and esthetical characteristics, which make them attractive in landscape architecture.

In the aim of permanent protection and adequate utilization, these and other spontaneous mutants with valuable characteristics should be protected as natural rarities, and their genotype should be fixed by vegetative propagation.

For amenity horticulture, the study mutants can be used as the base material for the production of ornamental cultivars.

Timely discovery, protection and research of spontaneous mutants with valuable characteristics has a scientific and practical significance in forestry and horticulture. These initial results justify the hope for a complete success.

7. References

Tošić,M.(1995) Novi varijeteti šumskog drveća i mogućnosti njihovog korišćenja. Drugi simpozijum o flori Srbije. Vranje, pp. 35-43.

Genetics, Plant Breeding and Biotechnology

ASSESSMENT OF F_1 HYBRIDS IN DIFFERENT WHEAT AND BARLEY LINES USING RAPDs.

A. ALTINKUT [1], M. TÜRET [2] and N.GÖZÜKIRMIZI [1,3]
TÜBİTAK, MRC, Research Institute for Genetic Engineering and Biotechnology Gebze-Kocaeli, TURKEY
[2] University of Bosphorous, Bebek-İstanbul, TURKEY
[3] University of İstanbul, Faculty of Science, Vezneciler-İstanbul, TURKEY

1. Abstract

Leaves of twelve wheat (*Triticum aestivum*) and barley (*Hordeum vulgare*) lines and their hybrids were obtained from Agricultural Research Institute from Eskişehir-Turkey. Both wheat and barley parental groups included drought resistant and drought sensitive lines. Total number of seven such F_1 hybrids from wheat and eigth F_1 hybrids from barley were used to isolate leaf genomic DNA for RAPD analysis. We presented here the data obtained using RAPD analysis of total thirthyeight primers. Among those primers only five distinguished polymorphic bands which were useful in the assesment of the true hybrids. Establishment of true F_1 hybrids groups is essential for future analysis of F_2 populations in identification of a RAPD marker linked to drought resistance using bulk-segregation analysis.

2. Introduction

One of the most important drawbacks of classical plant breeding is the lack of rapid and sensitive assays for the determination of true-type hybrids in F_1 generation without contamination of self-fertilized seeds. In comparison to several DNA based marker assays random amplified polymorphic DNAs (RAPDs) (Welsh and McClelland 1990; Williams *et al.* 1990) presents an ease to overcome this problem (An *et al.* 1995).

In the frame of EUREKA-1322 "Cerealstresstol " Project; our group in Turkey aim to develop rapid and sensitive assay to identify molecular markers related with drought resistance. As a first step in the project drought resistant and sensitive genotypes of wheat and barley lines were hybridized to obtain F_1 seeds. In the data presented here, we aim select true type F_1 hybrids to eliminate risk of contamination of self fertilized seeds. These true type F_1 hybrids will be used to obtain F_2 populations to identify molecular marker linked to drought resistance genes using parents and bulked F_2 individuals.

3. Material and Methods

3.1. PLANT MATERIAL

Leaves of twelve wheat (*Triticum aestivum*) and barley (*Hordeum vulgare*) lines were obtained from Agriculture Research Institute Eskişehir/Turkey. The combinations of families which consist parents of together with F_1 hybrids for each cross were given in Table 1 and Table 2. DNA isolation protocol by Walbot;1988 was followed..

3.2. AMPLIFICATION CONDITIONS

Various oligonucleotides (Table 3) that are 10, 15, 18, 19 20 bases in lenght according the literature data (R1, E3, E4, S1, S2, S3, S4 primer sequences were obtained from Weining and Langridge, (1991); H5, H8, G19 primer sequences from Tinker *et al.* (1993); A05, A09, A11, A14, A15, A17 primer sequences from He *et al.* (1992) or randomly choosen) were used as primers to produce different fingerprints. Amplification reactions were performed in volumes of 25 µl containing 25-50 ng genomic DNA, 100 mM dATP,dCTP, dGTP, dTTP mixture, 50 ng primer, 1 unit Taq DNA polymerase in 10XPCR buffer (10mM Tris HCl pH 8.3, 50 mM KCl, 0.01 % gelatin), 2mM $MgCl_2$.The first four cycles were at 94°C for 1.5 min, at 34°C for 1.5 min and at 72°C for 3 min. while the following 36 cycles were at 95°C for 1 min, 34°C for 1 min and 72° C for 2 min. Amplification products were analysed by electrophoresis in 1.5 % agarose gels and detected by staining with ethidium bromide.

TABLE 1. Combinations of families for barley lines, S: Drought sensitive, R: Drought resistant

Families	Parents			F_1 hybrids
	Female		Male	
1	Tokak (R)	X	ST 5819 (S)	F_1 51
2	Tokak (R)	X	ST 5820 (S)	F_1 53
3	Obruk (R)	X	Alaska (S)	F_1 62
4	TWW85-44 (S)	X	ST 5820 (S)	F_1 101
5	Rihane/Lignee 640//ICB107766 (S)	X	Tokak (R)	F_1 137

TABLE 2. Combinations of families for wheat lines , S: Drought sensitive, R: Drought resistant

Families	Parents			F_1 hybrids
	Female		Male	
1	RSK/NAC (S)	X	Kıraç 66 (R)	KF_1 108
2	Burvestnik//Obriy//B 16 (S)	X	Gerek 79 (R)	KF_1 126
3	Atay 85 (S)	X	Kıraç 66 (R)	KF_1 209
4	ES 14 (S)	X	Kıraç 66 (R)	KF_1 210
5	Sultan 95 (S)	X	Kıraç 66 (R)	KF_1 211

4. Results

The experiments performed were resulted in observation of RAPD bands that can be used for the identification of true type F_1 hybrids. These hybrids were obtained from crosses between drought sensitive and drought resistant parents of five wheat and barley families. RAPD patterns with all 38 primers tested, 15 primers showed clear patterns of amplification but did not exhibit any polymorphism in both wheat and barley families and were therefore monomorphic. Succesful polymorphic bands were also observed with R1 (5'TCGTGCCTGACTTACCTG3')

TABLE 3. Primers used in this study

Tested Primers	Sequence	Tested Primers	Sequence
A 15	5' TTCCGAACCC 3'	G1	5'ATGAAGACCTTACTCATCCT 3
A 17	5' GACCGCTTGT 3'	G2	5'ACATACACGTTGCACATG 3 '
A 18	5' AGGTGACCGT 3'	G19	5' GTCAGGCGAA3'
E 3	5'GAATTCCAGGTAAGT 3'	A02	5' TGCCGAGCTG 3'
E 4	5' GGAATTCCACCTGCA 3'	A05	5' AGGGGTCTTG 3'
H 5	5' AGTCGTCCCC 3'	A09	5' GGGTAACGCC3'
H 8	5' GAAACACCCC 3'	A11	5' CAATCGCCGT 3'
R 1	5"TC GTGC CTGA CTTACCT G 3"	A14	5' TCTGTGCTGG 3'
OPA 11	5' CAATCGCCGT 3'	OPC 09	5' CTCACCGTCC 3'
OPA 12	5' TCGGCGATAG 3'	OPC 15	5' GACGGATCAG 3'
OPB 12	5' CCTTGACGCA 3'	OPC17	5' TTCCCCCCAG 3'
OPC 06	5' GAACGGACTC 3'	OPC18	5' TGAGTGGGTG 3'
OPC 08	5' TGGACCGGTG 3'	OPC19	5' GTTGCCAGCG 3'
S1	5' GCACGCCGGCGGGT GGTAC 3''	OPC20	5' CATTCGCCAC 3'
S2	5' GC CGGCTTGCCGTACTTGG 3 '	TUB 340	5' TGGTCACCGA 3'
S3	5' G A GCCCAGAACGA C GCC C G G 3 '	TUB 357	5' TGCTCTGCCC 3'
S4	5' GAATTCTGAATGGAC 3 '	TUB 360	5' TGGGGGACTC 3'
OPC 16	5' CACACTCCAG 3'	TUB 363	5' TGATCCCTGG 3'
GB08	5' GTCCACACGG 3'	TUB 376	5' CAGCACCCAC 3'

(Figure1a,b) and A 14 (5'TCTGTGCTGG 3') primers (Figure 1c) .The polymorphic band patterns were interpreted by looking at the presence of a common band between male parent and F_1 hybrid of that family while such band was absent in the female parent. While R1 primer was very efficient in identification of 3 F_1 hybrids in wheat and 4 F_1 hybrids in barley families, this primer has consensus sequences for the exon-intron splice junctions of plants (Brown *et al.* 1986) and was used for detecting and mapping polymorphisms between wheat and barley (Weining and Langridge 1991), could be seen several polymorphic bands. A 14 primer was tested in wheat for detection of polymorphism and found to be monomorphic (Joshi and Nguyen 1993). In our experiments on barley, produced a selective band (app. 2.2 kb) on family 3 and family 5. 11 primers did not amplify any detectable products in both wheat and barley lines. The experiments were repeated number of times , in each case, the amplification pattern was the same for all tested primers.

Scientists have begun to use RAPD markers as a tool for characterizing germplasm (Tingey *et al.* 1992). RAPD data have serious implications for the use of this technique in less controlled applications such as trueness-to-type or genetic diversity analysis (Weeden *et al.* 1992). Welsh *et al.*(1991) have used RAPD technique to determine parentage in maize hybrids. An *et al.*(1995) also used this technique to determine true type hybrids in barley. Results indicated that each genotypes needs different molecular markers to develop. In the investigation presented here we tested reliability of the RAPD-PCR system for the determination of the true type F_1 hybrids in
wheat and barley families. In future, these true type F_1 hybrids will be used for the generation of F_2 population . Such F_2 population will be screned for the identification of drought resistance related molecular markers using bulk segregation analysis.

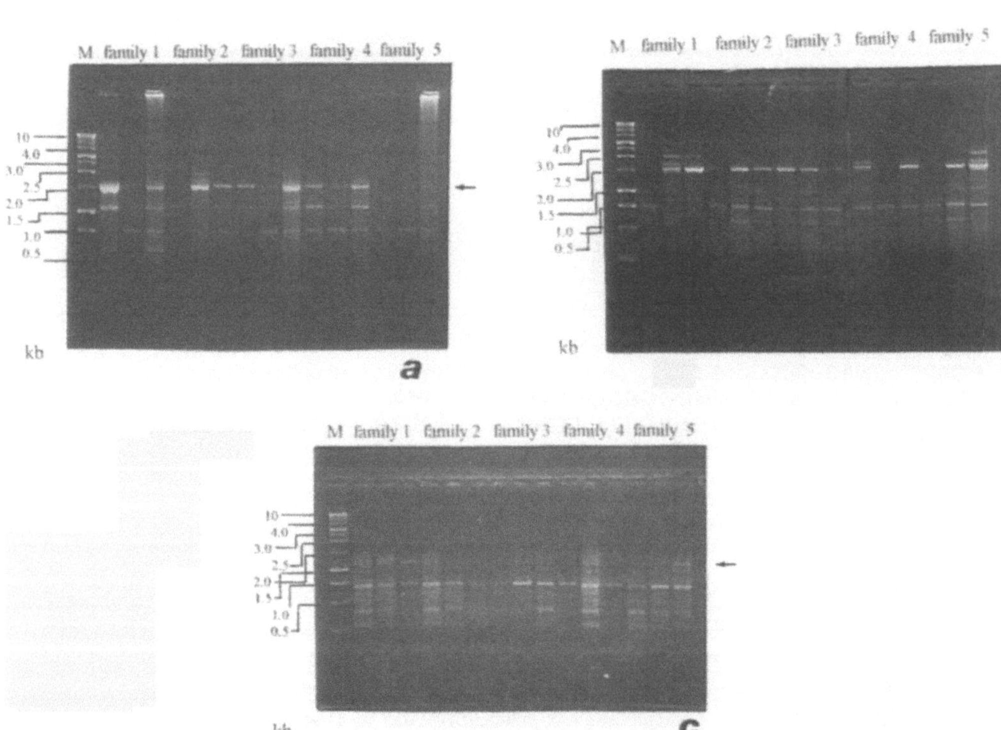

FIGURE 1 a) RAPD analysis of F_1 hybrids from 5 families of wheat using R1 primer, b) RAPD analysis of F_1 hybrids from 5 families of barley using R1 primer, c) RAPD analysis of F_1 hybrids from 5 families of barley using A14 primer.M: 1 kb DNA Ladder.

ON THE ESTIMATION OF GENOME LENGTH IN PERENNIAL WOODY ANGIOSPERMS

F. A. ARAVANOPOULOS
Lecturer
Laboratory of Forest Genetics and Tree Breeding
Department of Forestry and Natural Environment
Aristotle University of Thessaloniki, P.O. Box 238
Thessaloniki GR54006

1. Abstract

The usefulness of identifying genetic linkage is paramount in breeding in order to study QTLs and employ MAS selection. Before the construction of a complete linkage map, genome length and the number of markers necessary to cover the genome have to be estimated. Nevertheless, in perennial woody angiosperms there is absence of appropriate genetic families, therefore total genome length has to be assessed from partial linkage data only. It is proposed to estimate genome length by extending the use of the moments and the Poisson probability methods. In addition, it is proposed to employ in both approaches a common estimate of mapping distance, by using the θ value of the locus pair chosen from all linkage groups as having the maximum estimated LOD score over a specified threshold Z. The rational of this proposition is presented along with a discussion of pertinent theory.

2. Introduction

Genetic linkage maps are powerful tools in plant genetics. The usefulness of identifying instances of joint segregation is especially paramount in breeding. Knowledge of the positions of marker loci can permit the study and manipulation of important genes either single or of quantitative nature, within and among species. The development of: (1) analytical tools for the study of QTLs and (2) powerful molecular techniques which allow for the employment of a large number of genetic markers, have revolutionized the way molecular genetics interact with breeding and have given rise to the new area of molecular breeding. Linkage information assists in answering questions pertaining to the organization and establishment of genetic variation in natural populations, permits the assessment of the distribution of available loci in the genome, and provides references to locate unmapped genes on chromosomes leading to saturated gene mapping. The understanding of linkage relationships is essential for population models dealing with the maintenance of balanced polymorphisms through epistatic selection. An expected response to selection includes functions of the linkage between pairs of involved loci. Model simulations involving two or more loci showed that the behavior of multilocus systems cannot be predicted from single locus theory. In addition, linkage relationships can assist in molecular phylogeny through the study of instances of synteny above the species level.

3. Rational

Linkage maps are currently under development in many plant species. When the size of these maps is not modified by the addition of new genetic markers then the whole genome of the species under study is considered to have been reached. The total genome length is then obtained as the sum of the individual mapped intervals. Nevertheless it is important before the construction of a complete linkage map, to predict the number of markers necessary to cover the genome and to evaluate the overall relationship between physical and genetic maps as measured by the number of kilobases per centimorgan. Therefore total genome length has to be estimated even from partial linkage data.

In perennial woody angiosperms, the main obstacle for the development of linkage maps is the absence of suitable breeding material. Despite their economic importance, perennial woody angiosperms have not been subjected to detailed genetic analysis in the past. High order filial structures are usually absent and when they exist the sample sizes within full-sib families are very small; inbred lines do not practically exist and backcrosses are very few. The considerable time needed in perennial woody angiosperms for reaching sexual maturity and the paucity of information on reproductive biology, coupled with the practical difficulties in executing and handling large sets of particular mating designs, have contributed in the present scarcity of appropriate families. The data sets that are currently available are characterized by small numbers of markers and progeny sizes, as well as incomplete mating designs. However the advent of molecular markers has renewed the interest on QTL mapping and MAS selection, and in light of the potential breeding applications, new controlled crosses are being executed in many species. New experiments in forest genetics and breeding, are always time-consuming and present a long-term research commitment. Therefore, there is a need to develop estimates of genome length and marker coverage in order to establish the proper experimental parameters.

4. Theory

Two methods could be employed in genome length estimation from partial linkage data sets: the method of moments (Hulbert *et al.* 1988) and the Poisson probability method (Fjellstrom and Parfitt 1994). Both methods assume: (1) there is no interference, (2) chromosomes are of equal length. The former method can be justified as follows. Let M denote the number of informative pairs of loci and the detection of linkage be associated with a particular LOD score threshold (Z). LOD score is the decimal logarithm of the ratio of the likelihood calculated under the hypothesis that the loci are linked and the likelihood calculated under independent assortment. Let K be the number of locus pairs where LOD>Z. Then K/M denotes the probability that a locus pair chosen at random will be declared linked. This probability can also be expressed as a function of genome length G and the map distance X between two loci for which a LOD score of Z is expected. Then genome length G is shown to be:

$$G = (2MX)/K \tag{1}$$

The genetic distance X is related to the recombination fraction θ via the Haldane (1919) formula:

$$X = -[\ln(1-2\theta)]/2 \tag{2}$$

The derivation of θ can be done formally following Lander and Botstein (1986):

$$Z = n\{\theta \log 2\theta + (1-\theta)\log[2(1-\theta)]\} \tag{3}$$

where n is the number of gametes studied particular to the locus pair considered. However, since linkage data in perennial woody angiosperms are expected to be obtained from a series of progenies stemming from unbalanced mating designs, n will vary according to the specific locus pair in a particular cross. Then a separate value for X (X_i) has to be calculated for all locus pairs and then MX of (1) will be replaced by ΣX_i. Nevertheless since this approach assumes zero interference it is likely to underestimate the true genome length, because in most plant species interference has been found to be positive (Swanson *et al.* 1981). Mapping distance in higher plants is usually being estimated by the Kosambi (1944) function which assumes intermediate levels of positive interference. In addition, the nature of the filial structures analyzed in perennial woody angiosperms and of their generally small number, makes more desirable to allow for an estimate of θ and X less stringent than the above 'exact' method. I propose the use of an approximation described in Chakravarti *et al.* (1991). For a given Z there are a number of locus pairs with LOD>Z; the maximum estimated θ value associated with a locus pair is chosen and this value is used in (2).

The method of the Poisson probability is based on an additional assumption: that randomly-chosen genetic markers will have a Poisson distribution in the genome. This method allows for corrections for the reduced coverage of markers at chromosome ends. When k markers are available, the kth marker will fall outside a range of the other k-1 markers with a Poisson probability dependent on the average chromosome length (L). The equation of the Poisson probability:

$$p = e^{-[1-(1-d/l)^2(k-1)]/n} \tag{4}$$

which leads to an estimation of G is based on: p: proportion of k markers unlinked at distance d; n: chromosome haploid number, d: genetic distance according to Haldane (1919). I propose to use for the same linkage data set, the distance $d = X$, by employing the X approximation as previously described. The rationale for using this approximation was presented above. In addition the use of $d = X$, will lead to both G estimation procedures becoming directly comparable and amenable to further treatment.

The genome length G will take the form:

$$G = nd/[1-\sqrt{1+[n\ln(p)/(k-1)]}] \tag{5}$$

The estimate based on the method of moments has been considered as potentially inflating true genome length on theoretical grounds (Chakravarti *et al.* 1991), while the Poisson probability method has been proposed as more likely to generate more accurate estimations of genome length (Fjellstrom and Parfitt 1994), albeit more conservative ones. Nevertheless, it seems that the level of accuracy of both methods which present particular attributes, is not well known yet, while there is also the possibility that specific crosses may be better suited for a particular method (for example in the case of interspecific hybrids). I therefore propose to estimate genome length with both methods and use the estimate of their mean value as an initial approximation of the genome length of perennial woody angiosperms when partial linkage data are available. Then the number of markers needed to cover a particular genome can be estimated from:

$$k = [-n\ln(1-P)]/[1-(1-d/L)^2] \tag{6}$$

where P = proportion of the genome needed to be covered, d = specified distance between the markers (usually at a length of 10 -20 cM), L = average chromosome length (Fjellstrom and Parfitt 1994).

5. Conclusions

The approach presented above can provide a unified procedure in confronting genome length and marker coverage estimation in perennial woody angiosperms where only partial linkage data are available. The above methodology can aid in the design of molecular breeding research in *Betula, Cornus, Eucalyptus, Fagus, Populus, Quercus* and *Salix* where molecular marker development has already started.

6. References

Chakravarti, A., Lasher L.K. andReefer J.E. 1991. A maximum likelihood method for estimating genome length using genetic linkage data. Genetics 128: 175-182.

Fjellstrom, G. and Parfitt D.E. 1994. RFLP inheritance and linkage in walnut. Theoretical and Applied Genetics 89: 665-670.

Hulbert, S.H., Illot T.W., Legg E.J., Linkloln S.E., Lander E.S. and Michelmore R.W. 1988. Genetic analysis of the fungus *Bremia lactucae* using RFLPs. Genetics 120: 947-958.

Haldane, J.B.S., 1919. The combination of linkage values and the calculation of distance between the loci of linked factors. Journal of Genetics 8: 299-309.

Kosambi, D.D., 1944. The estimation of map distances form recombination values. Annual Eugenics London 12: 172-175.

Swanson, C.P., Merz T. and Young W.J. 1981. Cytogenetics. Prentice-Hall Inc., New Jersey.

TRANSFER OF HyHel10 scFv GENE TO PLANTS

Ş. ARI [1], S. ŞAHİN[1], F. GÜREL[1], S. D. GÜRLE[1], B. COCKBURN[2], N. GÖZÜKIRMIZI[1,3]

1. University of İstanbul, Fac. of Science, Dept. of Biology, 34459, İstanbul TURKEY

2. Plant Antibody Group, Dept. of Botany, University of Leicester, Leicester UK

3. TÜBİTAK, MRC, RIGEB, P.O.BOX 21, 41470, Gebze, Kocaeli TURKEY

1. Introduction

Transgenic plants are emerging as an important system for the expression of many recombinant proteins from heterolog sources including bacteria, insects, mammals and viruses, especially those intended for biotechnological and therapeutic purposes. Two basic reasons for the production of such heterologous proteins in plants are (I) "factory farming", in which unique properties of plants can be exploited in the production of desirable proteins, and (II) "phenotype modification", in which the heterologous protein may be used to alter the characteristics of the plant itself. One of the many possible kinds of proteins of potential biotechnological importance is antibodies. Production of antibodies and antibody derived molecules such as scFv's by plant cells offers a variety of new possibilities for basic research in plant biology as well as large scale production of antibodies for use as therapeutic, diagnostic or affinity reagents and for manipulating agronomic traits (*Whitelam et al., 1993*). Single chain fragment variable molecules (scFv's) incorporate a designed linker peptide to bridge isolated immunoglobulin variable domains (V_L and V_H) into a single polypeptide (*Whitlaw et al., 1991, Owen et al., 1992*).

The aim of this study is to transfer an animal protein gene (HyHel10) which encodes scFv protein, a synthetic antibody, against hen egg-white lysozyme to tobacco and barley. *Agrobacterium* and microprojectile bombardment mediated methods were used for the transformation of tobacco and barley, respectively.

2. Materials And Methods

2.1. CLONING OF HyHel10 scFv GENE INTO PLANT TRANSFORMATION VECTOR (pROK2):

HyHel10 scFv gene construct was supplied from University of Leicester, Dept. of Botany in *pC3* phagemid vector. PCR was performed with the aim of cloning the gene HyHel10 into plant transformation vector, pROK2. Intact pC3 :HyHel10 vector, which was isolated with the method based on Birnboim and Dolly (*Birnboim and Dolly, 1979*) and *HindIII* and *SpeI* digested pC3 :HyHel10 were used as template DNA in PCR. After enzyme digestion, HyHel10 scFv gene was purified with Prep-A-Gene DNA Purification System (BIORAD) in order to use in PCR. Primers containing complementary sequences to restriction sites of enzymes *BamHI* and *SacI*, which are at the cloning region of the plasmid pROK2, were constructed.

Primer1(Forward):5'ATACGCGGATCCGGTATCGATAAGCTTATGGATGTCCAGCTGCAG-3'
BamHI *HindIII*
Primer 2 (Reverse): 5'-CCAGCTC ACTAGTTTACTTGATCTCCAG-3'
SacI *SpeI*

After PCR, DNA was detected after running on 0.8% agarose gel. Amplified gene bands were cut under UV light and then purified with Prep-A-Gene DNA Purification System. After the preparation of the DNA's of the gene and the vector (pROK2) they were digested by *BamHI* and *SacI* respectively. After each restriction digestion, DNA samples were purified as described before. Linear vector and gene DNA's were left to ligation at 15°C O/N using T4DNA ligase. Ligation mixture was transformed by heat-shock method (*Inoue et al, 1990*) into super competent *E.coli DH5*α obtained in our laboratory. For selection of recombinant clone LB medium containing 50μg/ml *Kanamycine* (Km) was used. After the isolation of recombinant plasmid, restriction digestions and PCR were done for determining the transferred gene.

2.2. TRIPARENTAL MATING :

For the transfer of HyHel10 gene to tobacco, pROK2:HyHel10 plasmid was transferred to *A.tumefaciens* by Triparental Mating (*Shaw, 1988*). Conjugated clones containing pROK2:HyHel10 were selected on LB+50mg/l Km+200mg/l Strp (*Streptomycine*), and conjugation was also confirmed by PCR using Primer1 and Primer2.

2.3. PLANT TRANSFORMATION

2.3.1. *Agrobacterium Mediated Transformation of Tobacco:*

Plants were generated from leaf discs, after co-cultivating with *A.tumefaciens LBA4404* carrying pROK2:HyHel10 using the protocol of Horsch et al (1985). For selection of putative transgenics MSD_4 medium (MS containing 1mg/l BAP and 0.1mg/l IAA-*Murashige and Skoog*, 1962-) supplemented with 100mg/l Km and 200mg/l *Cephotaxime* (Cx) was used. MS supplemented with 0.1 mg/l IAA was used as rooting medium.

2.3.2. *Microprojectile bombardment mediated transformation of barley:*

Mature embryos of barley (*Hordeum vulgare L. cv. Zafer 160*) were cultured on MS medium supplemented 1mg/l 2,4D and bombarded after two days (*Sanford et al, 1987*). The plasmids pBI121 (Clontech Inc.) and pROK:HyHel10 were used in transformation of barley. pBI121 contains β-glucuronidase (GUS) and neomycin phosphotransferase II (npt II) genes under the control of 35S CaMV (Cauliflower mosaic virus) and nopalin synthase (nos) promoters, respectively. Biolistic PDS-1000/He (BioRad) version was used in gene transfer. A mixture including 3.0 mg gold particles (1.6μm in diameter), 10μg each plasmid DNA, 1.0 M $CaCl_2$ and 15 M spermidin was prepared. DNA and gold particles were incubated 30 minutes at room temperature. The mixture was washed with 70% ethanol and suspended in 99% ethanol. 10μl of this mixture was loaded onto center of each macrocarriers. Petri dish including embryos was positioned at 5.0 cm from the stopping screen. 1100 psi helium pressure and two shots were applied to each petri dish including 30 embryos.

2.3.2.1.*GUS Assay*: Half of embryos in each petri dishes were used in GUS histochemical assay. After 48 hours, X-Gluc substrat solution [0.3% 5-Bromo-4-chloro-3-indolyl-β-D-glucuronic acid (X-Gluc, Sigma), 0.9 % Triton-X-100, 1.5mM potassium ferrocyanid, 1.5mM potassium ferricyanid] was applied on the embryos and incubated O/N at 37°C (Jefferson, 1987).

3. Results And Conclusion

In this study a recombinant plasmid, pROK2: HyHel10 containing an animal protein gene (HyHel10) which encodes scFv protein -a synthetic antibody- against hen egg-white lysozyme was constructed for the transfer to tobacco and barley. PCR product of HyHel10 scFv gene was used for cloning of the gene into plant transformation vector. The use of gene amplified by PCR is one of the ways for gene cloning (*Whitlaw et al.,1991*) and primer construction is an important part of cloning (*Newton* and *Graham,1994*).The HyHel10 scFv gene was successfully amplified with the constructed primers containing complementary sequences to restriction sites *BamHI* (5`end) and *SacI* (3`end) which are at the cloning region of the plasmid pROK2.

More than 150 transformant *E.coli* colonies were screened with restriction digestion and PCR. Positive clone grown on selective medium (LB+50μg/mlKm) was determined by a fragment (0,8kb) after *HindIII* digestion. The presence of the gene on the recombinant plasmid (pROK2:HyHel10) was also confirmed by PCR using primer 1 and primer 2 (Figure1). pROK2:HyHel10 described as recombinant was conjugated into *Agrobacterium* using triparental mating with the aim of transferring of HyHel10 scFv gene to tobacco. Conjugated *Agrobacterium* strains were grown on selective medium. Triparental conjugation was successfully applied instead of direct DNA transformation (*Lichtenstein et al., 1985*) and the conjugation was confirmed by PCR of the plasmid isolated from *Agrobacterium* (Figure 2). Regeneration frequency of tobacco plantlets after *Agrobacterium* transformation in selection medium was about 95%. Tobacco plantlets were transferred to rooting medium.

In our experiments, mature embryos of barley were bombarded with two plasmids for co-transformation. Histochemical GUS expression was assayed in order to detect the presence of plasmids in embryo cells. We detected some cell clones expressing β-glucuronidase (GUS) gene on embryo surface (Figure 3). This may be an indication of presence of pROK2 plasmid which was co-transferred by pBI121. According to our preliminary results: regenerated plantlets seem to be transgenics. These putative transgenics are being investigated for further molecular analysis (PCR, Southern Blotting, DIG labelling, etc.).

Acknowledgement: This work was supported by NATO-TU BIOTECH II Project.

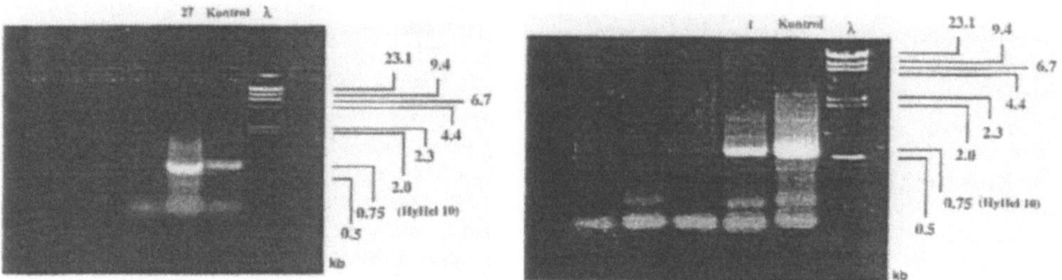

Figure 1 Figure 2

Figure 1: Determining of HyHel10 gene on recombinant clone 27 by PCR. pC3 :HyHel10 was used as control.

Figure 2: Determining of HyHel10 gene by PCR on pROK2 plasmid, transferred to *Agrobacterium* from *E.coli* by Triparental Mating 1: Positive *Agrobacterium* clone , pC3 :HyHel 10 plasmid was used as control.

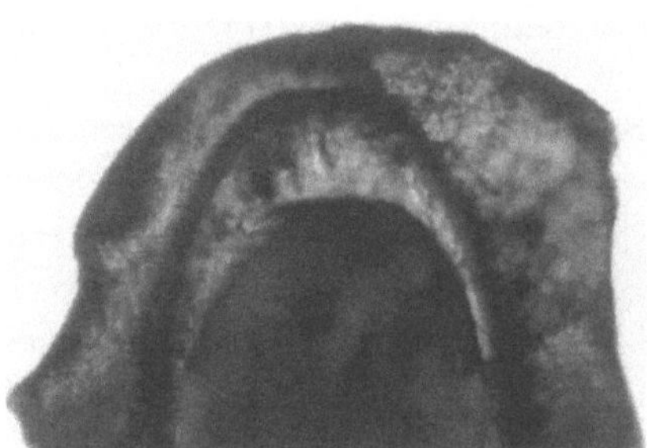

Figure 3: Cell clusters express GUS gene on the mature embryo after 48 hours.

Birnboim H.C., Dolly J. (1979), A rapid alkaline extraction procedure for screening recombinant plasmid DNA, *Nucleic Acid Research*. **7(6)**, 1513-1523

Hiatt A.(1990), Antibodies produced in plants. *Nature* **344**, 469-470

Horsch R.B., Fry J.E., Hoffmann N.L., Eicholtz D., Rogers S.G. and Fraley T.A. (1985), A simple and general method for transferring genes into plants, *Science* **227**,1229-1231

Inoue H., Nosima H. and Okayama H. (1990), High efficiency transformation of E.coli with plasmid, *Gene* **96**, 23-28

Jefferson R.A. (1987) Assaying chimeric genes in plants: The GUS Gene Fusion System, *Plant Molecular Biology*, Rep. **5**: 387-405

Lichtenstein C. and Draper J. (1985), In DNA Cloning- A Practical Approach, Glower D.M. (ed) IRL Press, Oxford, **II**: 67

Murashige T. and Skoog F. (1962), A revised medium for rapid growth and bioassay with tobacco tissue culture *Physiol.Plantarum* **15**, 473-497

Newton C.R. and Graham A. (1994), In PCR Bios. Scientific Publisher Ltd. UK, 20-25

Owen M., Gandecha A., Cockburn B. and Whitelam G. (1992),Synthesis of a functional anti-phytochrom single chain Fv protein in transgenic tobacco.*Bio/Technology* **10**: 790-794

Sanford J.C., Klein T.M., Wolf E.D. and Allen N.(1987) Delivery of substances into cells and tissues using a particle bombardment process. *J.Part. Sci.and Tech*. **5**:27-37

Shaw C.H. (1988) In Plant Molecular Biology: A Practical Approach, Ed., IRL Press, Oxford, Washington

Whitelam G.C., Cockburn B., Gandecha A.R. and Owen M.R.L. (1993), Heterologous Protein Production in Transgenic Plants. *Biotechnology and Genetic Engineering Reviews* **11**: 1-29

Whitlaw M. et al (1991), Single chain Fv Proteins and Their Fusion Proteins. *Methods Enzymology*, **2**: 97-105

REGENERATION AND TRANSFORMATION OF *Populus tremula* WITH ANTISENSE pOMT8 FROM *Stylosanthes humilis*

K.BAJROVIC[1], Z. İPEKÇİ[1], T. OĞRAŞ[1], N. GÖZÜKIRMIZI[1], M. BOYDAK[2], T. TANK[2], T. AKALP[2], Ö. ÖZDEN[2], M. ÇALIKOĞLU[2], K. TUNÇTANIR[3], M. TULUKÇU[3], H. BALKAN[4] and H. TANRIYAR[4]

[1]TÜBITAK, MRC, Research Institute for Genetic Engineering and Biotechnology Gebze-Kocaeli, Turkey, [2]University of Istanbul, Faculty of Forestry, Istanbul, Turkey, [3]Poplar Research Institute, İzmit, Turkey, [4]Turkish Paper and Pulp Mills, İzmit, Turkey

1. Abstract

Populus tremula was transformed using *Agrobacterium tumefaciens* with antisens O-metyltransferase gene from *Stylosanthes humilis* (pOMT8), which is showing 83% homology with COMT gene from poplar. In four out of ten transgenic plants OMT activity has been downregulated for 30-39%, however peroxidase activity was 100% higher compared with controls. While the lignin amount in transgenic plants was not changed, lignin composition were modified. Results indicate that it is possible to downregulate endogenous OMT activity and modified lignin composition through the introduction of heterologous consract from *Stylosanthes humilis* in *Populus tremula*.

2. Introduction

Most of the Poplar species are propagated vegetatively and poplar cells display developmental plasticity *in vitro* culture (Kajita et al. 1994). These traits coupled with susceptibility to *Agrobacterium* mediated transformation make poplar an attractive model system for studies of the molecular biology of forest trees. In addition poplar species are an important forest trees world-wide, which are also most planted tree in Northern Europe.

Lignins are complex phenolic heteropolymers associated with the polysaccharide components of the wall in specific plant cells. From a quantitative point of view, lignins represent the second most abundant organic compound of earth after cellulose, accounting for approximately 25% of plant biomass. They are present in great quantities in secondary cell walls of particular cells, fibre, xylem vessels. They provide rigidity and structural support to plant tissues and water proof the vascular elements allowing the conduction of water and solutions. In periderm they located in lower amount where they play a protective role in associated with suberin. In addition, defence lignins are formed in plants in response to pathogens and mechanical stress (Boudet and Grima-Petteneti, 1996).

In lignin biosynthesis O-methyltransferase (OMT, EC 2.1.1.68.) is one of the key enzymes, which plays an important role in plant growth and development. In the

phenylpropanoid pathways from which monolignols are derived, OMTs catalyse methylation of caffeic acid to ferulic acid and 5-hydroxyferulic acid to sinapic acid. Both gymnosperm and angiosperm dicotyledoneous OMTs can catalyse the methylation of caffeic acid whereas only angiosperm dicotyledoneous OMTs can catalyse the methylation of 5-hydroxyferulic acid efficiently (Whetten and Sederoff, 1995).
Peroxidases play an integral role in secondary cell wall biosynthesis by polymerisation of cinnamyl alcohols into lignin and by forming rigid cross-links between cellulase, pectin, hydroxypiroline-rich glycoproteins and lignin. They are also critical in wound healing in which they act in the formation of a water-tight barrier over the wound, by the deposition of polymeric aliphatic and aromatic compounds. Peroxidases may also have a role in auxin catabolism and defence of the plant against pathogen's attacks (Lagrimini, 1991).
In general the efficiency of wood pulping is directly proportional to the amount of syringyl residues in lignin. Lignin has negative impact on the utilisation of the biomass, during paper pulping for example lignin must be removed from wood by chemical treatments that are expensive and environmentally unfriendly. In this paper we described the introduction of antisense O-methyltransferase from *S. humilis*, pOMT8, in poplar with aim to downregulate the expression of OMT activity and to analyse the effect of pOMT8 on lignin content and composition, which could be useful in paper and pulp industry.

3. Materials and Methods

For this study tree years old cuttings of *Populus tremula*, Dursun bey 2 clone, were obtained in the spring from Poplar Research Institute, Izmit, Turkey. Regeneration and micropropagation of the shoots was observed on WPM containing 1 mg/l zeatine (WPMZ); root formation was observed on WPM containing 0,5 mg/l indol butyric acid (WPMB) (Han et al. 1994). For transformation of *Populus tremula*, O-methyltransferase gene isolated from *S.humilis* (pOMT8) in antisense orientation was inserted into the binary vector pGA643, behind the CaMV-35S promoter and transferred to Agl-1 strain of *A. tumefaciens*. This plasmid carrying also NPT II gene under NOS promoter. Transformation was performed by co-cultivation of stem explants (internodes) with *A. tumefaciens*. DNA used for PCR and Southern blot, was extracted from leaves of poplar plants. PCR has been performed using NPT II primers; NPT II gene labelled with DIG-pUTP used as probe for Southern blot (Sambrook *et. al.* 1989).
OMT activity was assayed in a volume of 100μl protein extracts, containing 50 mM tritiated S-adenosyl-L-methionine and 3 mM caffeic acid (Edwards and Dixon, 1991). Peroxidase activity was assayed spectrophotometricaly at 470 nm using guayacol and H_2O_2 according to Krsnik-Rasol (1991). To assay lignin amount and composition thioglycollic acid (TGA) assay and infrared spectroscopy (IR) method have been performed. For TGA stem explants were homogenized than dryed at 105°C. Dry material (50 mg) were put in 1, 5 ml tubes containing 750 μl water, 250 μl concentrated HCl and 100 μl TGA for 3 hours at 80°C and then centrifuged at 12000g for 10 min. Pellet were washed with water and resuspended in 1 ml of 1 M NaOH for 16 hours at +4°C.

4. Results

Transformation efficiency of *Populus tremula* on selective WPMZ after 15 days from the four separate experiments was 62%. After transformation from each responded explants 5-6 shoots has been obtained. These shoots were rooted on WPMB rooting media in 100%. There was not any phenotypic change obtained between transgenic and control plants. Also no morphological abnormalities were observed among the different 3 months old greenhouse grown transgenic lines.

Since pOMT8 gene shows sequence similarity to the corresponding poplar COMT gene transgenic and control plants were analysed on DNA level by PCR for the presence of NPT II gene. PCR analysis showed that DNA from transgenic plants contained NPT II gene. Those results provide evidence that the antisense pOMT8 gene was integrated in the genome of the transgenic poplar plants. Presence and integration of the T- DNA were confirmed by Southern blot analyses Fig. 1).

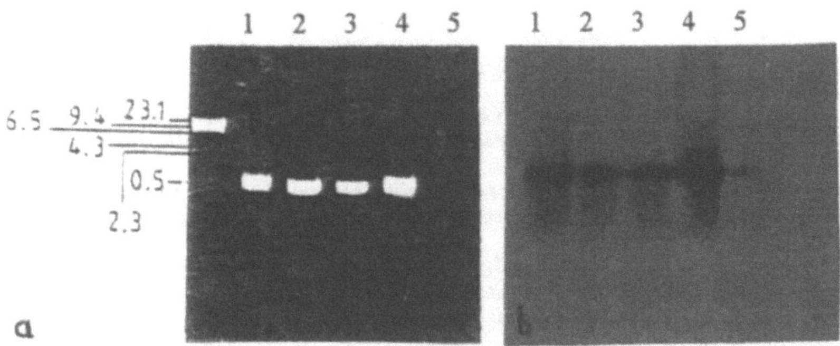

FIGURE 1. PCR amplification of the NPT II gene from total DNA (a) and Southern hybridisation of the DNA probed with DIG labeled NPT II gene (b). 1-4, transgenic lanes; 5, control.

OMT activity in four out of 10 tested transgenic plants was reduced for 30-39% in comparison to control. In addition, OMT activity in adult stage were compared with those measured of the plantlets stage, but no differences were observed. While the total lignin amount in transgenic plants was not changed, lignin composition was modified (syringyl units were decrised) without different changes in aromatic ring in transgenic plants. Similarly Van Doorsselaere et.al (1995) have decreased OMT activity for 50% in transgenic poplar using antisense OMT, but monomeric composition of lignin was only slightly modified. Dwivede et al. (1994) have transformed tobacco with OMT from aspen and reported slight reduction of both lignin content and composition in some of transformed plants.

Total peroxidase activity of four transgenic plants was 100 % higher than that of controls. Increases of peroxidase activity was confirmed using native gel electrophoresis. Similarly, increased peroxidase activity were obtained as answer to stress; Tsutsumi and Sakai (1993) have suggested that the water stress conditions stimulate wall-bound type of peroxidsases.

The increased peroxidase activity can be observed during the infection with the phytopathogenes also (Harrison et al. 1995).
This study showed the possiblity of downregulation of the endogenous OMT activity, and modification of lignin composition trought the introduction of antisense pOMT8 gene in poplar.

Acknowledgements

This study was a part of the project No. 62-02-001 TÜBİTAK-MAM-GMBAE named *Investigation on biotechnologically developed poplar clones as a raw material for pulp and paper industry* anf financially supported by General Management ot Turkish Pulp and Paper Mills.

5. References

Boudet, A.M., Grima-Pettenati, J. (1996) Lignin Genetic Engineering. *Molecular Breeding* **2**, 25-39.
Dwivedi, U.N., Capbell, W.H., Yu, J., Datla, R.S.S., Bugos, R.C., Chiang, V.L., Dodila, G.K. (1994) Modification of Lignin Biosynthesis in Transgenic *Nicotiana* Through Expression of an Antisense O-Methyltransferase Gene from *Populus*. *Plant Molecular Biology* **26**, 61-67.
Edwards, R., Dixon, R.A. (1991) Purification and Characterisation of S-Adenosyl-1-Methionine: Caffeic Acid 3-O-methyltransferase from Suspension Culture of Alfalfa (*Medicago sativa L.*). *Archives of Biochemistry and Biophysics* **287** (2), 372-379.
Han, K.H., Bradshaw, H.D., Gordon, M.D. (1994) Adventitious Root and Shoot Regeneration *In Vitro* is Under Major Gene Control in an F2 Family of Hybrid Poplar (*Populus trichocarpa x P. deltoides*). *Forest Genetics* **1** (3), 139-146.
Harrison, S.J., Curtis, M.D., McIntyre, L., Maclean, D.J., Manners, J.M. (1995) Differential Expression of Peroxidase Isogenes During The Early Stages of Infection of The Tropical Forage Legume *Stylosanthes humilis* by *Colletotrichum glooesporioides*. *Molecular Plant Microbe Interaction* **3**, 398-406.
Kajita, S., Osakabe, K., Katayama, Y., Kawai, S., Matsumoto, Y., Hata, K., Morohoshi, N. (1994) *Agrobacterium*-Mediated Transformation of Poplar Using a Disarmed Binary Vector and The Overexpression of a Specific Member of a Family of Poplar Peroxidase Genes in Transgenic Poplar Cell. *Plant Science* **103**, 231-239.
Krsnik-Rasol, M. (1991) Peroxidase as a Developmental Marker in Plant Tissue Culture. *Int. J. Dev. Biol.* **35**, 259-263.
Lagrimini, M.L. (1991) Wound-Induced Deposition of Polyphenols in Transgenic Plants Overexpressing Peroxidase. *Plant Physiol.* **96**, 577-583.
Sambrook, J., Fritsch, E.F., Maniatis, T. (1989) Appentixes Molecular Cloning, a Laboratory Manual Book 3, Second Edition, 087969-309-6, Cold Spring Habor Press. U.S.A., A.10 E.S.
Tsutsumi, Y., Sakai, K. (1993) Lignin Biosynthesis in Woody Angiosperm Tissue I. Lignification and Peroxidase activity Stimulated in Water-stressed *Populus* Callus Culture. *Mokuzai-Gakkaishi-Journal of the Japan Wood Research Society* **39**(2), 214-220.
Van Doorsselaere, J., Baucher, M., Feuillet, C., Boudet, M., Van Montagu, M., Inze. D. (1995) Isolation of Cinnamyl Alcohol Dehidrogenase cDNAs from Two Important Economic Species: Alfalfa and Poplar. Demostration of a High Homology of The Gene Within Angiosperm. *Plant Ohysiol. Biochem.* **33** (1), 105-109.
Whetten, R.W., Sederoff, R.R. (1995) Lignin Biosynthesis. *The Plant Cell* 7, 1001-1013.

REGENERATION OF *PICEA OMORIKA* AND *PINUS HELDREICHII* BY TISSUE CULTURE METHODS

S. BUDIMIR[1], D. STOJIČIĆ[2], LJ. ĆULAFIĆ[3] AND R. VUJIČIĆ[1]
[1]*Institute for Biological Research "S. Stanković", 29. novembra 142;* [2]*Institute of Forestry, Kneza Višeslava 3;* [3]*Institute of Botany, Faculty of Biology, University of Belgrade, Takovska 43; 11000 Belgrade, Yugoslavia*

1. Introduction

Serbian spruce, *Picea omorika* (Panč.) Purk. And whitebark pine, *Pinus heldreichii* (Christ.) are Tertiary relic and endemic species. *Picea omorika* is endemic to the Balkan Peninsula and is restricted to small localities in Serbia and Bosnia. As a decorative tree it is frequently planted in parks and gardens of many European countries. *Pinus heldreichii* is endemic to the Balkan Peninsula and Southern Italy. This conifer species grows on limestone cliffs and slopes of high mountain regions up to 2200 m, even under most severe conditions. Traditional methods of vegetative propagation for both conifer species have been only partially successful. Therefore *in vitro* methods could be used for preservation and propagation of these two relic and endemic species.

2. Somatic embryogenesis in *Picea omorika* culture

Somatic embryogenesis in conifers is generally divided into several phases: embryogenic tissue initiation and maintenance, maturation and germination of somatic embryos, and field establishment of plantlets. In *Picea omorika*, seedling explants were used for somatic embryogenesis initiation. Explants were grown on the modified LP (Von Arnold and Eriksson, 1981) medium supplemented with 22.5 μM benzyladenine (BA), for 2 weeks and then transferred to the medium without growth regulators (Budimir and Vujičić, 1992). In explants grown on growth regulator-free medium, white, translucent, embryogenic tissue was formed, after 4-6 weeks. Isolated embryogenic tissue was maintained on medium containing 2,4-dichlorophenoxyacetic acid (9 μM) and BA (4.5 μM), in the dark. On the maintenance medium the tissue consisted of numerous, loosely packed embryos at early stages of development. With regular 2-week subculturing somatic embryos proliferated by cleavage polyembryogenesis. When culture period was longer than 4 weeks cultures turned brown. The frequency of embryogenic tissue formation was apparently dependent on genetic factors, since considerable variation was observed among explants originating from seeds collected from different trees (Vujičić and Budimir, 1995). In the presence of abscisic acid (12 μM) proliferation of new embryos was arrested and further development of somatic embryos occurred. Differentiation of somatic embryos was evident by the appearance of several cotyledon primordia followed by hypocotyl elongation and radicle formation. For germination and plantlet development, mature somatic embryos were individually grown on half-strength medium. Regenerated plantlets of *Picea omorika* were further transferred to soil. Induction of somatic embryogenesis and plant recovery from somatic embryos is now possible in a range of spruce species: *P. abies, P. glauca, P. mariana, P. sitchensis* (Vujičić and Budimir, 1995).

3. Axillary shoot formation in *Pinus heldreichii* culture

Surface sterilized seeds of *Pinus heldreichii* were placed on four times diluted MS (Murashige and Skoog, 1962) medium to germinate. After 30 days in culture seedling explants (consisting of shoot apex, cotyledons and hypocotyl) were excised. Explants were pulse-treated with BA at 0; 4.4; 44 and 222 μM concentrations plus 0.1% dimethyl sulfoxide (DMSO) for 1 hour. Pulse-treated explants were further grown on three different media containing 0.5% activated charcoal. Axillary bud formation as a function of either culture medium or hormone concentration is shown in Table 1.

TABLE 1. Effect of culture medium and BA concentrations on axillary bud formation, 4 weeks after pulse-treatment

Culture medium	BA (μM)	% of explants with buds	Average no. of buds ($\bar{x}$ ± SE)
	0.0	0.0	0.0
GD	4.4	20.0	0.20 ± 0.09
	44	76.9	1.25 ± 0.35
	222	85.7	4.57 ± 0.60
	0.0	0.0	0.0
LP	4.4	0.0	0.0
	44	50.0	1.0
	222	50.0	2.0
	0.0	0.0	0.0
MS	0.0	0.0	0.0
	44	Nt*	-
	222	0.0	0.0

Nt*- not tested; SE - standard error

GD=Sommer *et al.* (1975); LP=Von Arnold and Eriksson (1981); MS= Murashige and Skoog (1962)

The greatest number of explants formed axillary buds on GD, Greshoff and Doy medium, modified as Sommer *et al.* (1985). Survival rate of explants grown on LP medium was low, while not a single explant formed buds on MS medium. The most effective concentration of BA was 222 μM. Similar results were obtained in *Pinus elliottii* seedling explant culture (Austin Burns *et al.* 1991). Excised axillary shoots of *Pinus heldreichii* formed roots on GD half-strength medium. Rooted adventitious shoots were transferred to soil were they continue to grow.

4. Adventitious bud formation in *Pinus heldreichii* culture

Zygotic embryos of *Pinus heldreichii* were isolated from surface sterilized seeds and placed on tree different media (GD, LP and MS). Media were supplemented with BA at 0.22, 4.44 and 22.2 μM. Cultures were grown on BA containing medium, for 4 weeks, and then transferred to the same half-strength medium supplemented with 0.5% activated charcoal. Cotyledons and hypocotyls of embryos grown on BA containing medium elongated and became swollen and nodular in appearance after 2 weeks in culture. Radicles became necrotic soon after formation and were cut off before next subculture. Adventitious bud primordia were first visible at the swollen hypocotyl region during the forth week in culture, while they appeared afterwards along the cotyledons. To promote further development of bud primordia and the formation of adventitious shoots it was necessary to transfer explants to growth regulator-free medium. The effect of media and BA concentration on adventitious bud formation is presented in Table 2.

TABLE 2. Effect of culture medium and BA concentrations on adventitious bud formation. Cultures were grown 4 weeks on BA containing medium and then transferred to growth regulator-free medium, for 4 weeks.

Culture medium	BA (μM)	% of explants with buds	Average no. of buds ($\bar{x} \pm$ SE)
GD	0.22	0.0	0.0
	4.44	100	16.3 ± 1.51
	22.2	26.2	8.40 ± 3.49
LP	0.22	0.0	0.0
	4.44	81.3	3.00 ± 0.48
	22.2	27.3	1.00 ± 0.52
MS	0.22	0.0	0.0
	4.44	40.9	2.40 ± 0.78
	22.2	12.5	1.00 ± 0.69

SE -standard error

GD=Sommer et al. (1975); LP=Von Arnold and Eriksson (1981); MS= Murashige and Skoog (1962)

In explants treated with 0.22 μM BA adventitious buds formation was not observed. The most effective concentrations of BA in respect to percentage of explants with adventitious buds and average number of buds per explant was 4.44 μM. Adventitious shoots with average lengths greater then 5 mm were isolated and transferred to the GD half-strength medim with naphtaleneacetic acid (NAA) at 0.27 and 1.08 μM to induce root formation. However, the treatment with NAA was not effective. *In vitro* rooting of *Pinus heldreichii*, as for most conifer species needs further study.

5. Conclusion

In *Picea omorika* seedling explant culture, cytokinin was sufficient to induce somatic embryogenesis, as the sole growth regulator. Somatic embryogenesis is expressed in explants transferred to the medium without growth regulators. Conditions for embryogenic tissue proliferation, somatic embryo maturation and germination are similar to other conifer species. In *Pinus heldreichii* culture adventitious and axillary bud formation was dependent on culture medium and BA concentration. These results suggest that tissue culture could be used as a method for preservation and propagation of *Picea omorika* and *Pinus heldreichi*. However, for large-scale plant production culture conditions should further be improved for both conifer species.

6. References

Austin Burns, J., Schwarz, O.J. and Schlarbaum, S.E. (1991) Multiple shoot production from seedling explants of slash pine (*Pinus elliottii*, Engelm.), *Plant Cell Rep.*, **10**, 439-443.

Budimir, S. and Vujičić, R. (1992) Benyzladenine induction of buds and somatic embryogenesis in *Picea omorika* (Panč.) Purk., *Plant Cell Tiss. Org. Cult.*, **31**, 89-94.

Sommer, H.E., Claud, L.B. and Kormanik, P.P. (1975) Differentiation of plantlets in longleaf pine (*Pinus palustris* Mill.) tissue cultured *in vitro*, *Bot. Gaz.* **136**, 196-200.

Murashige, T. and Skoog, F. (1962) A revised medium for rapid growth and bioassays with tobacco tissue cultures, *Physiol. Plant.* **15**, 473-497.

Von Arnold, S. and Eriksson, T. (1981) *In vitro* studies of adventitious shoot formation in *Pinus contorta*, *Can. J. Bot.*, **59**, 870-874.

Vujičić, R. and Budimir, S. (1995) Somatic embryogenesis and plant regeneration in *Picea omorika,* In Jain S., Gupta P. and Newton R. (eds), *Somatic embryogenesis in woody plants*, **vol 3**, Kluwer Academic Publishers, Printed in Netherlands, pp. 81-97.

CHROMOSOME DAMAGE INDUCED BY SEED AGING IN *CREPIS CAPILLARIS L.*

B. DIMITROV
Institute of Genetics, Bulgarian Academy of Sciences
Sofia 1113, Bulgaria

1. Introduction

Induction of chromosomal aberrations during seed aging was observed in various plant species (see ref. 3,8). However, the most important aspects of the mutation process related to aging seeds are not fully understood. It is not sufficiently clear what factors are involved in this process, what types of damages they induce to DNA, during aging or during germination the lesions leading to formation of chromosomal aberrations are induced. The occurrence of chromosome-type aberrations is a proof that they are induced during seed aging (G_1) and the presence of chromatid-type aberrations indicates that the damages are induced at S or G_2 of the cell cycle. In this respect the data of the different investigators are controversial. Chromosome-type aberrations were found in various plant species, whilst in other the aberrations of the chromatid-type are predominant (see ref. 3,8). It was proposed (9,4) that the chromosome-type aberrations may be due to errors arising from the scoring of the cells in second mitotic cycle along with the cells in first mitotic cycle. Therefore, a method for differentiation of the mitotic cycle must be applied for precise analysis of the type of the aberrations. In the present investigation analysis of the frequency and the type of the chromosomal aberrations and the frequency of the sister chromatid exchanges in aged *Crepis capillaris* seeds was made by differentiating of the individual mitotic cycles by means of a colchicine block. No attempt has been made up to now to investigate the sister chromatid exchanges in aging seeds.

2. Material and Methods

The experiments were conducted with *Crepis capillaris* seeds. Samples of 500 seeds were stored at laboratory conditions in glass containers at 22°C for 1, 3, 5 and 7 years. The seeds were allowed to germinate in Petri dishes on filter paper moistened with distilled water, in a thermostat at 24°C. To differentiate the separate mitotic cycles, the seeds were placed in 0,01% colchicine 6 hr after the beginning of soaking, where they were left until fixation time. Germinating meristems were fixed every 6th hr from the

4th to 40th hr after the beginning of germination. The roots were fixed in alcohol-acetic acid (3:1) for one night, and were transferred for storage in 70% alcohol. Analysis of chromosomal aberrations was made in metaphase cells using squash preparations stained with Schiff, after Feulgen, in a mixture of Schiff and acetocarmine.

A visualization of SCEs was performed by a FPG technique developed for *Crepis capillaris*. Primary roots 1-2 mm long, were incubated for 12 hr in a solution of 10^{-4}M bromodeoxyuridine, 10^{-8}M fluorodeoxyuridine, 10^{-6}M uridine. Thereafter they were treated for 12 hr with a second solution, containing 10^{-4}M thymidine and 10^{-6}M uridine. The roots remained in 0,05% colchicine for 2 hr before fixation. The methods of slide preparations and SCE analysis are described in details in our earlier paper (2).

3. Results and Discussion

Data from the analysis of chromosomal aberrations induced during seed aging and observed in first mitosis are presented in Tables 1 and 2. Data of separate fixation times were combined. Fresh seeds (2-4 months old) were used as controls. The data in Tables 1 and 2 indicate that during seed aging the frequency of the chromosomal aberrations largely increases as compared to the controls. The increase correlates with seed age.

TABLE 1. Frequency of chromatid-type aberrations observed in the first mitosis of aged *Crepis capillaris* seeds

Aging time (years)	Number of cells scored	Cells with aberrations	Chromatid breaks	Isochromatid breaks	Interchanges	Duplication-Deletions + Deletions	Aberrations per 100 cells (±SE)
0	400	2	1	1	-	-	0.5 ± 0.28
1	400	3	1	2	-	1	1.0 ± 0.27
3	400	5	2	3	1	1	1.5 ± 0.23
5	400	7	2	4	1	1	2.0 ± 0.38
7	400	8	2	4	2	1	2.3 ± 0.34

TABLE 2. Frequency of chromosome-type aberrations observed in the first mitosis of aged *Crepis capillaris* seeds

Aging time (years)	Number of cells scored	Cells with aberrations	Fragments	Dicentrics + Rings	Symmetric Translocations	Aberrations per 100 cells (±SE)
0	400	1	1	-	-	0.3 ± 0.29
1	400	3	1	2	-	0.8 ± 0.27
3	400	18	8	9	4	5.3 ± 1.19
5	400	26	14	14	4	8.0 ± 0.96
7	400	37	15	21	7	10.8 ± 1.22

The lowest increase occurred in the 1 yr. old seeds. The highest increase was observed after 7 yr. of age. These data show that the increase in the frequency of both chromatid-type and chromosome-type aberrations was pronounced to different degree. The chromatid-type aberrations increase 2-4,5 times in comparison with the control. At all times of seed aging the increase in the aberration frequency is mainly due to chromatid and isochromatid breaks (Table 1). At the same time the increase in the frequency of chromosome-type aberrations from the variant of the long time of seed aging was much more strongly expressed than that of the chromatid-type aberrations. For the different periods of aging 1,8-36 times higher frequency was observed (Table 2). The many times larger increase in the amount of the chromosome-type aberrations in first mitosis clearly shows that the primary lesions leading to the formation of chromosomal aberrations are induced predominantly during the seed aging, i.e. in G_1, because, as it is well known, the cells in the dry seeds are arrested in G_1 (1). Table 3 shows the SCE frequency observed in the aging seeds. It is evident that at all times of aging the SCE frequency does not change significantly as compared to the control. The highest frequency of SCEs was observed after 5 yr. of aging, but the difference was not significant as compare to the control and is in the limits of the variation of the spontaneous SCEs frequency in *Crepis capillaris* (2).

TABLE 3. Frequency of SCEs observed in root meristem cells aged *Crepis capillaris* seeds

Aging time	Number of cells scored	SCEs/cell mean ± S.E.
0	50	8.4 ± 0.9
1	50	8.8 ± 0.7
3	50	9.4 ± 1.4
5	50	10.2 ± 1.2
7	50	9.3 ± 1.6

Our results clearly show that in aging seeds of *Crepis capillaris* mainly chromosome-type aberrations were induced. These results are in contrast with the results of other authors obtained in investigations with various plant species. In *Triticum durum* (6), *Hordeum vulgare* (9), *Pisum sativum* (4) aged seeds predominantly chromatid-type aberrations were observed. Some authors (7,4) have suggested that when chromosome-type aberrations were observed in aged seeds they are not due to chromosome lesions induced in G_1 during aging, but rather to scoring of aberrations in cells in second mitotic cycle where chromatid aberrations of first mitosis after DNA replication appear as chromosome-type aberrations. The differentiation of the first mitotic cycle from the second one reject the probability the chromosome-type aberrations observed in our investigations to be a result of replicated chromatid aberrations.

The general question is what factors induced the lesions underlie the formation of chromosome-type aberrations in aged seeds. As is known this type aberrations arise as a result of DNA double strand breaks (DNAdsb) induced before the DNA synthesis. DNAdsb are induced by ionizing radiation and by only some chemical agents, such as bleomycin and 8-ethoxycafeine (2).

It is believed that chromosomal aberrations arising during seed aging are a result of DNA lesions induced by chemical substances with mutagenic properties, produced in the seeds during the aging and designated as automutagens. According to the generally accepted concept for molecular mechanisms of inducing of chromosomal aberrations, the chemical mutagens do not directly induce breaks to DNA but produce initial lesions, which are transformed into breaks after DNA replication (5). These lesions affect only one of the chromatids and lead to the formation of chromatid-type aberrations from misrepair. In this way all chromasomal aberrations induced by chemical mutagens are of chromatid-type. As mentioned above chromosome-type aberrations arise as a result of DNAdsb. Formation predominantly of chromosome-type aberrations during seed aging suggests that the automutagens induce mainly DNAdsb which seems not to be reasonable since ionizing radiation as well as radiomimetic agents induce both DNAdsb and DNAssb. Our results on the SCE frequency also are not consistent with this suggestion. The observed lack of induction of SCEs do not support the idea of presence of active chemical mutagens in aged seeds since it is known that chemical mutagens are efficient inducers of SCEs. Ionizing radiation despite is not so efficient as chemical also are able to induce SCEs. It is probable therefore that other factors may be involved in the production of chromosomal aberrations in aged seeds. The changes in the efficiency of repair enzymes might play a role in aging mutagenesis. Data are available that fragmentation of DNA occurs during aging (10). Normally the basic part of the DNA breaks are repaired by DNA-polymerase. The polymerase activity was greatly reduced during seed aging (11). It is reasonable to admit that the reduction of enzyme activity correlates with the time of aging. As a consequence of the latter the unrepaired breaks increase. Our results correspond to this suggestion.

4. References

1. Deltour, R.: Nuclear activation during early germination of the higher plant embryo, *J. Cell Science* **75** (1985), 43-83.
2. Dimitrov, B.: Relationship between sister-chromatid exchanges and heterochromatin or DNA replication in chromosomes of *Crepis capillaris, Mutation Research* **190** (1987), 271-276.
3. Dimitrov, B.: Types of chromosomal aberrations induced by seed aging in *Crepis capillaris, Environmental and Molecular Mutagenesis* **23** (1994), 318-322.
4. Dourado, A.M. and Roberts, E.H.: Chromosome aberrations induced during storage in barley and pea seeds, *Ann. Bot.* **54** (1984), 767-779.
5. Evans, H.J.: *Progress in Genetic Toxicology*, Elsevier, Amsterdam, 1977.
6. Floris, C., and Anguillesi, M.C.: Ageing of isolated embryos and endosperms of durum wheat: An analysis of chromosome damage, *Mutation Research* **22** (1974), 133-138.
7. Morikawa, T., and Inomata, N.: Effects of seed storage and presoaking dry treatment on chromosome aberrations in mitosis of *Alluim fistulosum L., Jap. J. Reed* **28** (1978), 320-328.
8.. Murata, M.: *Chromosome engineering in plants: Genetics, Breeding, Evolution*, Elsevier, Amsterdam, 1991
9. Murata, M., Tsuchiya, T., and Roos, E.E.: Chromosome damage induced by artificial seed aging in barley. II. Types of chromosomal aberrations at first mitosis, *Bot. Gaz* **143** (1982), 111-116
10. Osborne, D.J.: *The Physiology and Biochemistry of seed Dormancy and Germination*, Biochemical Press, Amsterdam, 1980.
11. Yamaguchi, H., Naito, T., and Tatara, A.: Decreased activity of DNA polymerase in seeds of barley during storage, *Jap. J. Genet.* **53** (1978), 133-135.

GENOTOXIC EFFECTS OF FIVE PESTICIDES IN *CREPIS CAPILLARIS* ROOT MERISTEM CELLS

B. DIMITROV AND P. GADEVA
Institute of Genetics, Bulgarian Academy of Sciences
1113 Sofia, Bulgaria

1. Introduction

Large amounts of pesticides are daily emitted in the environment due to their wide utilization. Some of the most widely spread pesticides are subjected nowadays to intensive genotoxic investigations because of their mutagenic effect reported by many authors. The existing data reveal that some pesticides induce DNA damages and /or aneuploidy, both being involved in carcinogenesis (3).

In the present study the genotoxic effects of the pesticides Reglon (dipyridiliums), Aphalon, Dicuran and Patoran (phenylcarbamates) and the insecticide Pyrinex (organophosphorous compounds) are investigated in plant cells *(Crepis capillaris*) by using two short-term tests-chromosomal aberrations and micronuclei. The plant test-system has some advantages compared to the other test-systems. These are low cost easy to handle experiments and the most important is the ability of the plant organism to activate specifically some chemicals into active mutagens (4). The latter imposes the necessity of the plant test-system involvement in genotoxic investigations. The simultaneous use of the two tests, chromosomal aberrations and micronuclei, allows to assess both clastogenic and aneugenic effects of the pesticides studied.

2. Material and Methods

All experiments were performed with *Crepis capillaris* root meristems, 1-2 mm long, treated for 2 hours. Half of the material was treated with colchicine solution (0,05%) 2 hours before fixation, which is necessary for the metaphase analysis of the chromosomal aberrations. The other half was fixed without pretreatment for scoring the frequency of the chromosomal aberrations and the lagging chromosomes in anaphase as well as that of the micronuclei. The roots were fixed in ethanol-acetic acid (3:1). The chromosomal aberrations and the micronuclei were scored in squash preparations stained after Feulgen in a mixture of Shiff's reagent and acetocarmine in a ratio 1:1. Morphometric analysis of the micronucleus diameters of 100 cells of root meristems treated with Reglon, Aphalon and Dicuran was carried out along with two controls, positive (colchicine) and negative (maleic hydrazide). The results were statistically analysed by Student's t-test.

3. Results and Discussion

The results of the metaphase chromosomal aberration frequency after treatment with Reglon, Aphalon, Dicuran, Patoran and Pyrinex are presented in Table 1.

Table 1. Frequency of chromosomal aberrations observed in metaphase in root meristem cells of *Crepis capillaris* after treatment with pesticides. 1200 cells per fixation time were scored

Pesticide	Con-cen-tration (%)	Number of cells with aberra-tions	Chromosomal aberrations: Breaks: Chromatid fragments	Breaks: Chromo-some fra-gments	Exchanges: Intra--chromo-somal	Exchanges: Interchromo-somal: Asy-metric	Exchanges: Interchromo-somal: Syme-tric	Aberrations per 100 cells % (± S.E.)
Control	0	4	4	-	-	-	-	0.33 ± 0.25
Reglon	0.01	6	3	1	2	-	-	0.50 ± 0.27
	0.05	7	3	2	1	1	-	0.58 ± 0.28
	0.1	6	2	-	4	-	-	0.50 ± 0.27
Aphalon	0.5	5	3	1	1	-	-	0.42 ± 0.26
	0.7	7	2	4	-	1	-	0.58 ± 0.28
	0.9	6	1	4	-	1	-	0.50 ± 0.27
Dicuran	0.7	6	3	2	1	-	-	0.50 ± 0.27
	1.0	6	2	3	1	-	-	0.50 ± 0.27
	1.3	7	2	3	1	1	-	0.58 ± 0.28
Patoran	0.025	5	3	2	-	-	-	0.42 ± 0.26
	0,5	6	2	2	1	1	-	0.50 ± 0.27
	0.1	5	1	3	1	-	-	0.42 ± 0.26
Pirinex	0.0375	5	2	2	1	-	-	0.42 ± 0.26
	0.075	5	3	-	2	-	-	0.42 ± 0.26
	0.15	6	2	3	1	-	-	0.50 ± 0.27

The data show that the chromosomal aberration frequency observed in all pesticides tested does not differ from that of the control. The data from the anaphase analysis of the chromosomal aberrations are similar (Table 2). In this case it is also found out that none of the pesticides investigated causes any increase in the frequency of the chromosomal aberrations. At the same time the frequency of the lagging chromosomes is considerably increased by Reglon, Aphalon and Dicuran. Patoran and Pyrinex do not induce lagging chromosomes. These data prove the lack of clastogenic effect for all the pesticides tested. However, our results of the chromosomal aberrations after Aphalon treatment are inconsistent with those of Wuu and Grant (6) who established some increase in the frequency of the metaphase and anaphase chromosomal aberrations. The discrepancy may be due to various factors such as differences in the methods of scoring or use of aged seeds. It is shown that during seed aging the frequency of chromosomal aberrations significantly increases (1), so that the use of seeds in different age may be the reason for the differences in the results reported. The results of the micronucleus frequency after Reglon, Aphalon, Dicuran, Patoran and Pyrinex treatment indicate that only the first three of them induce micronuclei. The data are presented in Table 3. It is evident from Table 3 that the micronucleus frequency correlates to the increase of pesticide concentrations. The data differ statistically significantly from those of the control (P<0,001).

TABLE 2. Frequency of chromosomal aberrations observed in ana-/telophase in root meristem cells of *Crepis capillaris* after treatment with pesticides. 1200 cells per fixation time were scored.

Pesticide	Con-centra-tion (%)	Number of cells with aberra-tions	Chromosomal aberrations: Fragments		Bridges		Aberrations per 100 cells %(±S.E.)	Lagging chromosomes	
			S	D	S+D	S+D+ Frag-ments		Number	%
Control	0	4	4	-	-	-	0.3 ± 0.25	-	-
Reglon	0.01	6	1	-	5	-	0.5 ± 0.27	25	2.1
	0.05	6	1	-	5	-	0.5 ± 0.27	36	3.0
	0.1	7	-	1	6	-	0.6 ± 0.28	42	3.3
Aphalon	0.5	6	2	-	4	-	0.5 ± 0.27	19	1.6
	0.7	7	1	-	5	1	0.6 ± 0.28	24	2.0
	0.9	7	-	1	6	-	0.6 ± 0.28	36	3.0
Dicuran	0.7	5	1	-	4	-	0.4 ± 0.26	42	3.5
	1.0	6	-	1	5	-	0.5 ± 0.27	55	4.6
	1.3	7	1	-	6	-	0.6 ± 0.28	69	5.7
Patoran	0.025	4	-	-	4	-	0.3 ± 0.25	-	-
	0.05	4	1	1	2	-	0.3 ± 0.25	-	-
	0.1	5	-	-	4	1	0.4 ± 0.26	-	-
Pirinex	0.0375	4	1	-	3	-	0.3 ± 0.25	-	-
	0.075	5	1	1	3	-	0.4 ± 0.26	-	-
	0.15	4	-	1	3	-	0.3 ± 0.25	-	

S - single; D - double

TABLE 3. Frequency of micronuclei in root meristem cells of *Crepis capillaris* after treatment with pesticides. 12 000 cells per fixation time were scored

Pesticide	Concentration (%)	Cells with micronuclei: Number	% ± S.E.	Total number of micronuclei	Micronuclei per cell
Control	0	25	0.21 ± 0.04	25	1.0
Reglon	0.01	171	1.42 ± 0.11	176	1.0
	0.05	222	1.85 ± 0.12	230	1.0
	0.1	217	1.81 ± 0.12	232	1.1
Aphalon	0.5	56	0.47 ± 0.06	57	1.0
	0.7	78	0.65 ± 0.07	79	1.0
	0.9	106	0.88 ± 0.08	111	1.0
Dicuran	0.7	200	1.67 ± 0.12	227	1.1
	1.0	314	2.62 ± 0.14	335	1.1
	1.3	511	4.26 ± 0.18	574	1.1

The highest micronucleus frequency is induced by Reglon and the lowest by Aphalon. The lack of acentric chromosome fragments and, at the same time, the appearance of lagging chromosomes in our experiments with Reglon, Aphalon and

Dicuran prove that the micronuclei originate from entire chromosomes. To elucidate the mechanism of micronuclei formation induced by the three above mentioned pesticides the micronucleus/macronucleus size ratio was determined. The data of the morphometrical analysis of the micronuclei induced by Reglon, Aphalon and Dicuran are presented in Table 4. The data reveal that the size of the micronuclei induced by these pesticides and by colchicine was similar (P>0,05). All of them are approximately 2.5 times larger than that caused by maleic hydrazide (P<0,001). These data are consistent with those obtained in animal cells (5,2). It can be stated that the micronucleus induction by Reglon, Aphalon and Dicuran is due to partial disturbances in mitotic spindle leading to appearance of lagging chromosomes from which micronuclei are formed.

TABLE 4. Size ratios of micronucleus/macronucleus after treatment with pesticides, colchicine and maleic hydrazide of *Crepis capillaris* root meristems.

Agent	Concentration (%)	Analyzed cells	Average (± S.E.) Size ratio (%)	Variation of size ratio
Reglon	0.10	100	0.23 ± 0.02	0.10 - 0.36
Aphalon	0.50	100	0.21 ± 0.02	0.11 - 0.30
Dicuran	1.00	100	0.23 ± 0.02	0.11 - 0.33
Colchicine	0.05	100	0.26 ± 0.03	0.10 - 0.37
Maleic hydrazide	0.008	100	0.09 ± 0.02	0.03 - 0.25

4. References

1. Dimitrov, B.: Types of chromosomal aberrations induced by seed aging in *Crepis capillaris, Environmental and Molecular Mutagenesis* **23** (1994), 318-322.
2. Ferguson, L.R., Mocombe, P. and Triggs, C.N.: The size of cytokinesis-blocked micronuclei in human peripheral blood lymphocytes as a measure of aneuploidy induction by Set A compounds in the EEC trial, *Mutation Res.* **287** (1993), 101-112.
3. Murphy, S.D.: *Toxicology: The Basic Science of Poisons*,. Macmillan, New York, 1986.
4. Plewa, M.J., Seo. K.Y., Ju, Y.H., Smith, S.R. and Wagner, E.D.: *Plants for Environmental Studies* . Lewis. publishers, New York, 1997.
5. Vanparys, P., Vermeiren, F., Sysmans, M. and Temmerman, R.: The micronucleus assay as a test for the detection of aneugenic activity, *Mutation Res.* **244** (1990), 95-103.
6. Wuu, K.D. and Grant, W.F. Morphological and somatic chromosomal aberrations induced by pesticides in barley (Hordeun vulgate), *Can. J. Genet. Cytol.* **2** (1966), 481-501.

TOBACCO PLANTS EXPRESSING MODIFIED SYNTHETIC BACTERIAL (*BACILLUS THURINGIENSIS*) CryIA(a) AND CryIA(c) GENES

S. DUKIANDJIEV[1], Tz. GUETCHEV[1], N. DIMITROVA[1], R. SARDANA[2], M. GIBAND[3] AND I. ALTOSAAR[2]

[1]Department of Biochemistry, University of Plovdiv, 4000 Plovdiv, Bulgaria, [2]Department of Biochemistry, University of Ottawa, K1N 6N5, Ottawa, Ontario, Canada; [3]CIRAD-CA, Laboratoire de Biologie Cellulair, INRA, 78026 Versailles, France

1. Abstract

We have developed a long term programme to exploit the potential of the soil bacteria *Bacillus thuringiencis* to produce toxin proteins with a specific insecticidal activity against a number of economically important pests. We have pursued an effective application of this phenomenon by expressing bacterial genes into plants. For synthesis of fully modified versions of two bacterial Bt genes, CryIA(b) and CryIA(C), we have established an effective protocol using a PCR procedure based on the recursive principle. Put under the control of the maize ubiquitin promoter, CryAI(c) was proved to be expressed in a maize endosperm culture system. Constructs containing these genes in combination with HPTII were used to obtain transgenic plants from three Bulgarian tobacco cultivars via *Agrobacterium* mediated transfer.

2. Introduction

During the last decade there was an increasing effort to exploit the potential of the B. thuringiensis (Bt) cells to produce strong insecticidal proteins (Estruch et al.). The main attention so far was focused on the bacterial proteins synthesised during the sporulation phase which form crystalline intracellular inclusions. The crystal proteins show specific activities and provide the opportunity for a selective control of many important insects.

Constructs containing wild genes cryIA(a), cryIA(b) or cryIA(c) isolated from the lepidopteran active Bt subsp. *kurstaki* were introduced into tobacco, tomato or potato (Adang et al. 1993; Delannay et al. 1989; Fischfoff et al. 1987; Koziel et al. 1993a). The analysis showed a poor expression of bacterial genes, resulting in limited degree of protection against the pests. The extended molecular analysis of the Bt genes and the corresponding toxin proteins revealed the possibilities for improvement of Bt genes in plants trough the modification of the bacterial coding sequences. Partially modified genes show 5- to 10-fold higher expression than the wild-type gene. Much higher quantity of toxin proteins are found in transgenic plants, obtained with fully modified genes. The higher expression of these genes is based on the elimination of putative polyadenylation signals, change of the "bacterial" codons with "plant" and increasing of the stability of the mRNA with higher G+C content (Perlak et al. 1991; Salm et al., 1994).

We have developed a programme aimed at the production of insect tolerant transgenic plants based on the insecticidal potential of Bt genes. Here we are presenting the results related to the production of transgenic tobacco plants by the transfer of CryIA(a) and CryIA(c) Bt genes in three economically important Bulgarian tobacco cultivars.

3. Materials and methods

Plant material. Three *N. tabacum* cultivars were used for the experiments: Plovdiv 7, Plovdiv 187, and Spage 28, kindly provided by the Institute of Tobacco, Plovdiv.
Bacterial strains. The CryIA(b) and CryIA(c) genes were inserted into pBin19, *A. tumefaciens C58C1Rif* ^{r} (Fig. 2), under the control of the 70S (double 35S) promoter, and the nptII gene was driven by Tnos promoter.
Synthesis of Bt genes. The modified genes were synthesised using an improved PCR procedure based on the recursive principles (Prodromou and Pearl, 1992) and applying a slow cooling period (Sardana et al.,1966).
DNA cloning manipulation and sequencing was done using standard procedures.

4. Results and discussions

4.1. CONSTRUCTION OF SYNTHETIC TOXIN GENES

For the synthesis of CryIA(b) and CryIA(c) DNA sequences coding for insecticidal proteins, we developed an improved PCR procedure similar to the previously described by Prodomrou and Pearl (1992). To build correctly the largest possible DNA sequences we have useful to apply a two-phase PCR reaction compared to the one step approach described earlier. In addition, a slow cool step was also introduced in PCR synthesis to improve the product quality.

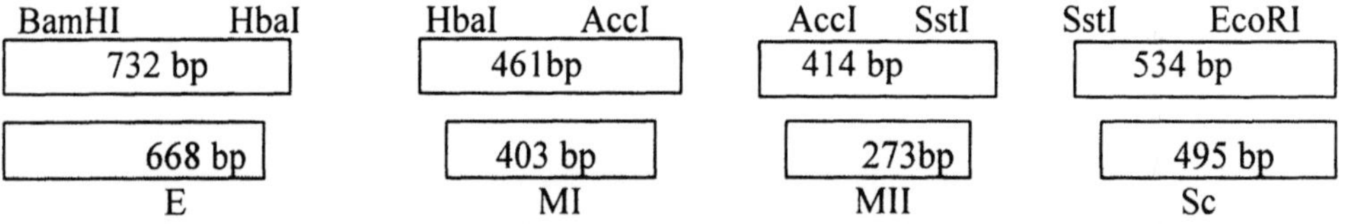

Fig.1. The four synthesized fragments (up) and their corresponding restriction fragments (down) ligated to obtain the entire coding sequence of Bt genes.

The total DNA sequences of 1845 bp for each of the two genes were created initially in form of four blocks. The three blocks E, MI and MII were common for the two genes. The blocks Sb and Sc represent the specific parts unique to the CryIA(b) and CryIA(c) genes respectively (Sardana et al. 1996). The blocks contained 30-40 bp overlapping ends to enable the restriction producing steacky ends at the convenient sites /Fig 1/. Two phase PCR reaction was performed. For the second phase, 5 μl of the first phase PCR reaction was used as template and amplification was done by adding 30 picomoles of outer oligos serving as primers. To favor high fidelity during annealing, PCR conditions were designed to allow a slow cooling period of 30 minutes from 94°C to 55°C. An improvement in the qualitative nature of expected size fragments is observed.

Finally, the desired fragments of sizes 732, 461, 414, 534 bp were excised from the gel. They were cut with desired restriction endonucleases. The resulted blocks were cloned and sequenced. On average, we found 1-2 point mutations per block (MI block was error free). The point mutations were repared by site-directed mutagenesis.

4.2. TRANSFORMATION OF TOBACCO

4.2.1. *Construction of CryIA(b) and CryIA(c) binary vectors.*

The coding sequence of the codon-optimized cryIA(b) and cryIA(c) genes were first transferred to vector pCa2NOS (unpublished), which contains the CaMV 35S promotor with a duplicated enhancer (P70) and the nopaline synthase terminator (NosTer). In this vector, the start of translation has also been optimised for expression in plant cells). The P70-synthetic cryIA(b)/(c)-NosTer cassette (Fig.2) was then transferred as a HindIII fragment into the corresponding site of the binary vector pBIN19 to yield plasmids pBINCryIA(b) and pBINCryIA(c). These vectors were then introduced into *A. tumefaciens C58C1Rif*^{r}

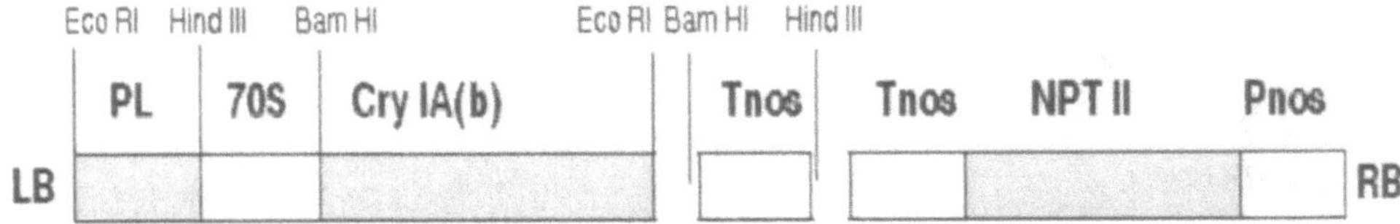

Fig. 2. The structure of P70-CryIA(b)-Pnos-NPT II of the pBINCryIA(b) plasmid construct, introduced into *A. tumefaciens C58C1Rif*^{r}

4.2.2. *Production of transgenic plants.* A number of Bulgarian tobacco cultivars were a subject of our transformation programme. Finally, based on their good responsiveness we have chosen three of them to extend the experimental work: Spage 28, Plovdiv 7, and Plovdiv 187. For each transformation experiment, leaf discs were inoculated with the bacterial suspension for 10 minutes, blotted onto sterile filter paper, and plated on MS solid medium with B5 vitamins, 2% sucrose, 0.1 mg/l NAA, and 1 mg/l BAP.

After two days the discs were transferred to the same medium supplemented with Kanamycin 100 μg/ml for selection of recombinants and Cefotaxime 500 μg/ml to kill Agrobacteria. Discs were then subcultured every two weeks. The effectiveness of the transformation is presented at the following table.

Table 1. Effectiveness of the transformation of the three tobacco cultivars

cultivar	Number of explants for transformation	Number of explants with shoots	effectiveness of the transformation %
Plovdiv 187	30	25	83.3
Plovdiv 7	70	43	61
Spage 28	60	34	57

The shoots were rooted on MSO+Km 100 μg/ml. Forty plantlets have been transferred to soil. They gave normal fertile plants. The seeds are collected for further investigation. PCR was used at this stage to prove the presence of the P70-CryIA(b)-Pnos-NPT II into plants. Using both the CryIA(c) and NPTII specific primers we have found after gel electrophoresis the bands corresponding to the expected PCR fragment products.

Earlier, we have proved the transient expression of CryIA(c) under the monocot specific Ubiquitin promoter in maize endosperm cell in culture (Sardana et al., 1996). The results that we are reporting here show that the two synthetic Bt genes, CryIA(b) and CryIA(c) can be used to obtain transgenic tobacco plants resistant to insect pests. The preliminary data of the bioassay show that the plants produce toxin proteins killing the larvae of fruit borer (*Chloridea obsuleta F*) and Cabbage moth (*Mamestra brassi*). The analysis and bioassay of the F1 and F2 plants continue.

5. References

Adang M.J., Brody M.S., Cardineau G., Eagan N,. Roush R.T., Shewmaker C.K., Jones A., Oakes J.V., Delannay X., La Vallee B.J.,Proksch R.K., Fuchs R.L., Sims S.R., et al. 1989. Field performance of transgenic tomato plants expressing the *B. thuringiensis var. Kurstaki* insect control proteins.Bio/Technology 7:1265-1269

Estruch J.J., Carozzi N.D., Duck N.B., Warren G.W. and Koziel M.G. 1997. Transgenic plants: An emerging approach to pest control. Nature Biotechnology, 15:127-141

Fischfoff D.A., Bowdish K.S., Perlak F.J., Marrone P.G., McCormick S.M. et al. 1987. Insect tolerant transgenic tomato plants. Bio/Technology 5:807-813

Koziel M.G., Beland G.L., Bowman C., Carozzi N.B., Crenshaw R., Crosslandl. Et al. 1993a. Field performance of elite transgenic maize plants expressing an insecticidal protein drived from B.Thuringiensis. Bio/Technology 11:194-200

Koziel M.G., Carozzi N.B.,CurrierT.C., Warren G.W., and Evola S.V. 1993b. The insecticidal crystal proteins of B. Thuringiensis:past, presentand future uses. Biotechnol and Genet. Engineer. Rev.11:171-228

Perlak F.J., Fuchs R.L., Dean D.A., McPherson S.L., And Fichhoff D.A. 1991. Modification of the coding sequence enhances plant expression of insect control protein genes. Proc. Natl. Acad. Sci. USA 78:2893-2897

Prodromou, C. and Pearl L.H. 1992. Protein engineering 5:827-829

Salm T., Bosch D., Honee G., Feng L., Munsterman E., Bakker P., Stiekema W., and Visser B. 1994 Insect resistance of transgenic plants that express modified Bacillus thingiensis cryIA(b) and cryIC genes: a resistance management strategy. Plant Mol. Biol. 26: 51-59

Sardana R, Dukiandjiev S, Giband M, Cheng X, Cowan K, Sauder C, and Altosaar I. (1996) Construction and rapid testing of synthetic and modified toxin gene sequences CryIA(b&c) by expression in maize endosperm culture. Plant Cell Reports, 15, 677-681

COMPARATIVE ANATOMICAL ANALYSIS OF THE VASCULAR SYSTEM IN SELECTED *HELIANTHUS ANNUUS* GENOTYPES

S. DULETIĆ-LAUŠEVIĆ[1] and M. MIHALJČEVIĆ[2]

[1] *Institute of Botany, Faculty of Biology, University of Belgrade, Takovska 43, 11000 Belgrade, Yugoslavia*

[2] *Institute for Field and Vegetable Crops, Maksima Gorkog 30, 21000 Novi Sad, Yugoslavia*

1. Introduction

The sunflower disease caused by the fungal pathogen *Diaporthe/Phomopsis helianthi* Munt.-Cvet. et al. was first found and described in Yugoslavia in 1980 (Mihaljčević *et al.*, 1980). Afterwards it spread to neighboring countries.

During the investigation of the morphological, physiological and phytopathological characteristics of the pathogen and its life cycle, the leaf-petiole-stem host invasion route of the fungus was demonstrated in a histological study (Muntañola-Cvetković *et al.*, 1989, 1991).

Considering the mode of infection and the hyphal spreading through the conductive system of the lamina and the petiole, and the supposition that some morphological and anatomical characteristics may influence the expression of resistance to the pathogen, our attention was focused on the epidermal and vascular tissues of the leaf.

A comprehensive study of the anatomy of the leaf lamina and petiole of the susceptible genotype HA-89 and the resistant genotype RHA-179 was carried out in order to try to provide evidence that the differences in the epidermal and vascular systems are responsible for the different reactions the genotypes exhibit in the presence of the pathogen *P. helianthi.*

2. Materials and Methods

Anatomical structure was studied in histological sections of the genotypes HA-89, RHA-179 and their hybrid. Small segments of the leaf (midrib, lateral veins, margin) and the petiole were sampled at the beginning of anthesis, at four levels (5th, 10th, 20th and apical leaf).

Leaves and petioles were prepared according to the standard paraffin methods. After counterstaining, the sections were analyzed, cells measured and photographed using a Reichert microscope Diastar™.

Data processing was done using the statistical package SYSTAT (modules Stats, MGLH and Sygraph).

3. Results

The lamina. Epidermis consists of a single layer of rectangular thick-walled cells. The thin cuticle which covered both leaf surfaces was usually thicker on the adaxial side. The mesophyll consisted of a single layer of palisade parenchyma cells next to the upper leaf surface and spongy parenchyma cells. The size of epidermal cells and xylem elements of the progenitors and the progeny is presented in Table 1.

The progeny of the susceptible and the resistant genotype showed the values for the size of epidermal cells between the parents, but closer to the susceptible one. Only the height of adaxial epidermal cells was slightly lower than that of the susceptible parent (Table 1).

The PC analysis of the adaxial and abaxial epidermal cells showed similar values in the resistant parent and the F_1 hybrid. Significant differences in the size of epidermal cells were not recorded.

The midrib. In transverse section, surface cells were rectangular, with thickened outer walls and a cuticle above. Collateral vascular bundles were placed among thin-walled parenchymatous cells. The size of epidermal cells and xylem elements was measured (Table 1).

Epidermis. The progeny had the height of epidermal cells between the parents, but the average cell width was smaller than those in both parents. There were no significant differences between the resistant and susceptible parent; the differences between the parents and the hybrid were similarly non-significant.

The PC analysis showed the highest variability in RHA-179. No significant differences could be found among RHA-179, HA-89 and the hybrid. The hybrid was closer to RHA-179 than HA-89.

Xylem elements. The size of xylem elements in large vascular bundles was perceptibly smaller in the progeny than in the progenitors.

There were no significant differences in variability among the genotypes studied.

The hybrid had the largest xylem elements in small vascular bundles. However, the differences were not statistically significant.

The PC analysis of xylem elements showed similar values of the parents and the hybrid.

The petiole. In histological sections, the epidermis on the surface of the petioles had rectangular cells with thickened outer walls and a cuticle above. The collateral vascular bundles were imbedded in parenchyma. The size of epidermal cells and xylem elements was measured (Table 1).

Epidermis. The height of epidermal cells in the progeny was close to that of RHA-179, but slightly lower. The width of the cells was larger in the hybrid than in either parent.

High variability was established for the parents as well as the hybrid, but the differences were non-significant. The hybrid was closer to the susceptible parent HA-89.

Xylem elements. The size of xylem elements in large vascular bundles was larger in the hybrid than in either parent, but the differences were not significant.

Table 1. Characteristics od *H. annuus* genotypes

PLANT ORGAN	PLANT TISSUE	GENOTYPE	CHARACTE-RISTICS		RANGE		PRINCIPAL COMPONENTS	
			height	width	height	width	Y1	Y2
leaf	adaxial epidermis	*H. annuus* HA-89	19,53	25,44	B	BC	0,456	0,806
		H. annuus RHA-179	20,29	29,34	B	BC	0,89	0,455
		F1	19,27	26,48	B	BC	0,887	0,459
	abaxial epidermis	*H. annuus* HA-89	16,15	22,77	C	BC	0,793	0,512
		H. annuus RHA-179	13,53	17,35	C	C	0,971	0,237
		F1	16,29	21,43	C	BC	0,952	0,305
midrib	epidermal cells	*H. annuus* HA-89	23,3	26,4	BCD	AB	0,783	0,576
		H. annuus RHA-179	19,81	22,95	CD	BC	0,985	0,071
		F1	20,19	20,67	CD	CD	0,995	0,005
	large vascular bundles	*H. annuus* HA-89	52,11	41,8	AB	A	0,986	0,15
		H. annuus RHA-179	51,63	41,34	AB	A	0,99	0,122
		F1	42,39	32,86	CD	BC	1	0,003
	small vascular bundles	*H. annuus* HA-89	34,56	25,04	AB	B	0,997	0,073
		H. annuus RHA-179	33,15	24,67	B	B	0,989	0,141
		F1	37,62	30,48	AB	AB	0,998	0,066
petiole	epidermal cells	*H. annuus* HA-89	20,79	23,62	B	BC	0,671	0,628
		H. annuus RHA-179	22,57	22,76	B	BC	0,473	0,865
		F1	22,48	27,43	B	AB	0,987	0,141
	large vascular bundles	*H. annuus* HA-89	52,14	40,55	ABC	AB	0,826	0,562
		H. annuus RHA-179	53,24	40,77	ABC	AB	0,979	0,203
		F1	59,34	47,91	A	A	0,973	0,227
	small vascular bundles	*H. annuus* HA-89	38,1	28,67	A	AB	0,953	0,262
		H. annuus RHA-179	36,01	30,67	A	A	0,995	0,099
		F1	40,01	31,88	A	A	0,996	0,087

There were no differences in variability between the susceptible HA-89 and the resistant RHA-179. The hybrid was close to the latter parent.

The hybrid had non-significantly larger xylem elements in small vascular bundles than the parents.

The PC analysis showed small differences in variability between the parents as well as between parents and the hybrid. The progeny was very close to the resistant parent.

4. Discussion

The progenitor-progeny relationship in this study was different from those in the previous studies (Duletić-Laušević and Mihaljčević, 1996, 1997) in which *H. tuberosus* and *H. argophyllus*, which had been used as donors of resistance to *P. helianthi*, were compared against susceptible *H. annuus* genotypes and their hybrids. Regarding the size of epidermis and vascular system of the lamina and the petiole, the interspecific hybrids had been much closer to the susceptible parents than to the resistant ones. In this study, the hybrid was much closer to the resistant parent in all characteristics studied, except the epidermis of the petiole.

Considering the aim of the study, it was concluded that the size of epidermal cells of the lamina and the petiole can not be used as a reliable characteristic for the differentiation of breeding materials regarding the resistance to *P. helianthi*. Furthermore, difference in the size of xylem elements can not be considered as a factor which could have a significant influence on the variability in resistance between sunflower genotypes. The ability of the pathogen to penetrate the host tissue of different genotypes seems to be more important than the size of epidermal and vascular tissues of sunflower plants.

5. References

Duletić-Laušević S. and Mihaljčević M. (1996) Anatomical comparisons among *Helianthus annuus* L., *H. tuberosus* L. and their progenies, XIV Int. Sunfl. Conf. - Symp. Disease Tolerance, Beijing, China, 12-22.06.1996., pp. 89-93.

Duletić-Laušević S. and Mihaljčević M. (1997) A comparison of the anatomical structure of susceptible *Helianthus annuus* L., resistant *H. argophyllus* and their progeny, *Helia* **20(26)**, 17-28.

Mihaljčević M., Petrov M. and Muntañola-Cvetković M. (1980) *Phomopsis* sp., a new parasite of sunflower in Yugoslavia, *Savr. poljoprivreda* **28,** 531-539.

Muntañola-Cvetković, M., Vukojević, J., Mihaljčević M. (1989) Pathohistology of sunflower stems attacked by *Diaporthe helianthi*, *Can. J. Bot.* **67**, 1119-1125.

Muntañola-Cvetković M., Vukojević J. and Mihaljčević M. (1991) The systemic nature of the sunflower disease caused by *Diaporthe helianthi*, *Can. J. Bot.* **69,** 1552-1556.

CHROMOSOME NUMBERS AND KARYOTYPE ANALYSIS OF *ERANTHIS HYEMALIS* (L.) SALISB.

A. N. GÖMÜRGEN
Hacettepe University, Faculty of Science, Botanic Department, Beytepe, Ankara, Turkey

1. Abstract

The genus *Eranthis* is presented in Turkey only by *Eranthis hyemalis* (L.) Salisb . In this present paper the chromosome numbers and karyotype analysis of this species are going to be represented. Chromosome numbers of *Eranthis hyemalis* (Winter aconite) was found to be 2n=16. Six pairs of chromosomes have centromeres at median region (m) and two pairs have centromeres at sub median region (sm) . An idiogram, a karyogram and a microphotograph of somatic chromosomes of *E. hyemalis* in metaphase are also going to be given.

2. Introduction

Eranthis Salisb. (Ranunculaceae family) is a perennial herb with tuberous rhizomes. This genus has 7 species natives of Europe and Asia (Komarov 1937,Tutin et al. 1964) but it is presented in Turkey only by *Eranthis hyemalis* (L.) Salisb. (winter aconite) (Davis 1965). This plant is one of the important commercial and exporting geofits of Turkey (DHKD,1996). The collection of the flower bulbs and rhizomes in the wild populations have grown rapidly (Ekim 1991).
Zimmer and Girmen have stated that the export of *E. hyemalis* buds and tubers from Turkey is endangering the genetic resources of the region (Zimmer and Girmen 1987a).*Eranthis stellata* was assessed as being endangered in Tokuyosan national park (KyongBin et.al. 1994).
KyongBin and his colleagues studied the *Eranthis hyemalis* seed's fatty acids because of their chemotaxonomic evidence (KyongBin et al . 1996). There are also various investigations about the effect of temperature on *Eranthis* (Hejnowicz et al. 1994, 1995, Zimmer and Girmen 1987b). Winter aconite is important while its an abundant source of nectar in Denmark (Svendsen, 1987).
In this study it is aimed to find chromosome numbers and make karyotype analysis of *E. hyemalis* which is native in Turkey

3. Material and Method

E. hyemalis were collected by Prof. Dr. Demirsoy from Afyon region and were planted in the garden of Hacettepe University, Biology Department garden. Seeds of these plants were collected at May 1995.

There were some problems for the germination of the seeds .A study was made by Tıpırdamaz and Gömürgen in order to solve the germination problem (Tıpırdamaz and Gömürgen, in press).At the end of the study seeds were germinated on wet filter papers at 4^{0}C at darkness 90 days.

Roots in other words radicula were excised when 1.5- 2 mm long. Root tips were pretreated with hydroxyquinoline (0.002 Molar aqueous solution) for 3 hours, fixed in GAA (glacial acetic acid) for half an hour and refrigerated in 70 % ethanol. Hydrolization was carried out with 1 N HCl (hydrochloric acid) for 11 minutes at 60^{0}C. Materials were squashed in Feulgen for chromosome staining. Permanent preparations were made with Canada balsam (Gömürgen 1996).Important metaphase plates were photographed.

Five cells with well spread metaphase chromosomes were used for constitution of karyotype. The eight homologous chromosome pairs were designated by the numbers 1 to 8, according to decreasing length.

The detection of the homologous chromosomes and the determination of their position in the karyogram were carried out by following the method of Levan et al. (1964).

4. Results

Chromosome numbers were obtained from somatic mitosis of root tips of the radicula. Karyomorphological details of the species are given in Table 1. The diploid chromosome numbers of *Eranthis hyemalis* (Winter aconite) was found to be 2n=16 (*Figure 1.*). The karyotype of *E. hyemalis* was characterized by six pairs of metacentric chromosomes (m) (1.-6.) and two pairs of submetacentric chromosomes (sm) (7.and 8.) (Table 1, *Figure 2*).

TABLE 1. Morphometric data on chromosomes of *Eranthis hyemalis*

Chromosome pairs	Long arm (μm) (Standart error)	Short arm (μm) (Standart error)	Total length (μm)	Relative length (%)	Arm ratio L/S	Centrometric index	Chromosome types
I	5.39 (± 0.17)	4.59 (± 0.19)	9.98	161.96	1.17	46.08	m
II	5.59 (± 0.29)	3.83 (± 0.28)	9.42	152.87	1.46	40.65	m
III	5.04 (± 0.21)	3.84 (± 0.14)	8.88	144.1	1.31	43.29	m
IV	4.79 (± 0.26)	3.95 (± 0.23)	8.74	141.84	1.21	45.25	m
V	4.16 (± 0.22)	2.84 (± 0.21)	7.00	113.6	1.46	40.65	m
VI	3.86 (± 0.29)	2.59 (± 0.19)	6.45	104.67	1.49	40.16	m
VII	3.98 (± 0.17)	1.96 (± 0.1)	5.94	96.4	2.03	33.00	sm
VIII	3.66 (± 0.33)	1.55 (± 0.2)	5.21	84.55	2.37	29.67	sm

Figure1. -Metaphase chromosomes of *Eranthis hyemalis.(Bar 10* μm.)

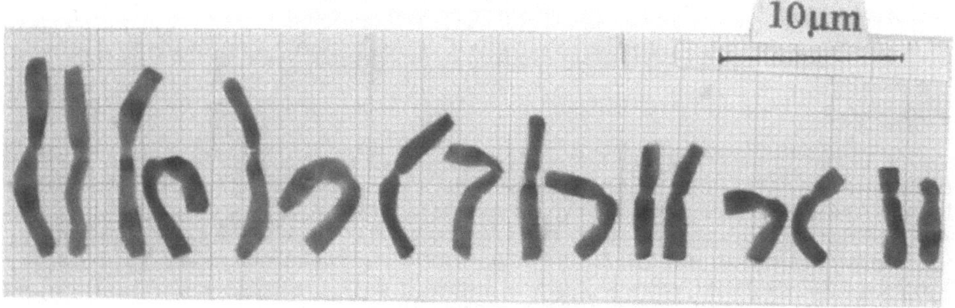

Figure2. -A karyogram of *Eranthis hyemalis.*

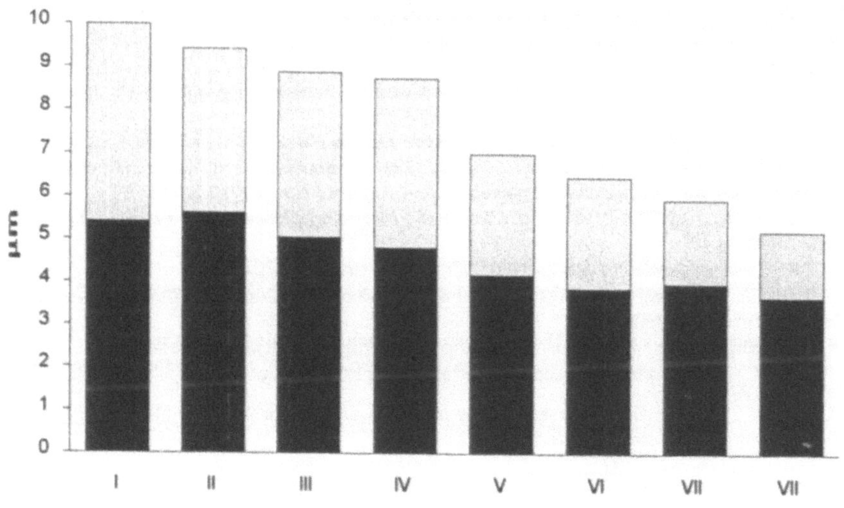

Figure3. -An idiogram of *Eranthis hyemalis.*

There was a difference in total chromosome length between the first and last pair of the chromosome complement which varied from 9.98 μm and 5.21 μm. The last chromosome pair was roughly twice as short as the first one. The chromosome pairs of 1., 2., 3. and 4.were longest chromosomes of the chromosome set and did not show an important difference in total length (9.98 μm - 8.74 μm). The last two pair of chromosomes were shortest chromosomes of the chromosome set. From the absolute chromosome length of *E. hyemalis* an idiogram *was* drawn (*Figure 3*).

5. Discussion

Our result on the root tip meristems confirms the chromosome number given in Illustrierte Flora von Mitteleuropa (Rechinger,K.H.,Damboldt,J.,1975). The karyotype of *E. hyemalis* collected from Turkey is given in this present work for the first time.

References

Davis,P.H.,(1965) *Flora of Turkey and the East Aegean Islands,* vol.1, Edinburgh University Press, Edinburgh.

DHKD, (1996), *Soğanlı bitkiler*, Doğal Hayatı Koruma Derneği ,İstanbul.

Ekim,T., Koyuncu,M., Güner,A., Erik,S., Yıldız, B.,Vural,M.,(1991) *Türkiye'nin ekonomik önem taşıyan Geofitleri üzerinde taksonomik ve Ekolojik araştırmalar.* T.C. Tarım Orman ve Köy İşleri Bakanlığı Orman Genel Müdürlüğü, Ankara.

Gömürgen,A.N.,(1996) Meiotic analysis of selected material of Sainfoin and its Progeny with Branched and unbranched peduncles. *Tr.J. of Botany,* **20** ,399-409.

Hejnowicz,Z., Trebacz,K., Sievers,A., (1994)Temperature dependent changes of membrane potentials in cells of thermonastic tepals of Eranthis hyemalis (L) Salisb., *Acta Societatis Botanicorum Poloniae.* **63**: **3-4,** 219-227.

Hejnowicz,Z., Trebacz,K., Sievers,A.,(1995) Temperature dependent changes of electric potential differences on opposite sides of the tepals of Eranthis in relation to thermonastic responses. *Plant, Cell and Environment* .**18:4**,471-474.

Levan,A., Fredga, K.and Sandberg, A.A., (1964) Nomenclature for centromeric position on chromosomes. *Hereditas* **52**: 201-220.

Komarov,V.L.,(1937) *Flora of the U.S.S.R.* vol.VII. Izdatel'stvo Akademii Nauk USSR, Moskva-Leningrad.

KyongBin,Y.,YongShik,K.,SeungHoon,C.,SunHee,K.,Whi,K.,(1994), Conservation status of rare and endangered plant species in Tokyusan National Park., *Journal of Korean Applied Ecology*. **7: 2,** 112-117.

Rechinger,K.H.,Damboldt,J., (1975) *Gustav Hegi Illustrierte Flora von Mitteleuropa, Verlag Paul Parey,* Berlin und Hamburg .

Svendsen,O., (1987) Eranthis hyemalis (winter aconite). *Tidsskrift-for Biavl.,* 121:2,69.

Tutin,T.G., Heywood,V.H., Burges,N.A., Valentine ,D.H., Walters,S.M., Webb,D.A., (1964) *Flora Europaea,* vol.1.Cambridge at the University press

Tıpırdamaz,R.T., Gömürgen,A.N.,(in press) The germination of Eranthis hiemalis. Research article.

Zimmer,K., Girmen,M.,(1987) Garden cultivation instead of overexplotation of natural resources. *Deutscher-Gartenbau.* **41: 16,** 988-989.

Zimmer,K., Girmen,M.,(1987) Temperature dependency of the development of Anemone blanda and Eranthis hiemalis.*Gartenbauwissenschauft.* **52:6,** 263-265.

INDIVIDUAL VARIABILITY OF REGULARITY, ABUNDANCE AND MORPHOMETRIC PROPERTIES OF SERBIAN SPRUCE (*Picea omorika* /Panč. / Purkyne) MICROSTROBILES

B. GRBOVIC
Institute of Forestry,
Kneza Višeslava 3, Belgrade, Yugoslavia

Abstract

In Serbian spruce seed orchard Bela Zemlja near Užice (Serbia), 25 test trees were selected as representative. The analysis of microstrobile regularity and abundance indicates the existence of individual variability, both between individuals during one year and between the same individuals in two years of research. The analyzed micro-strobile morphometric properties are: size, mass, moisture, and pollen quantity. The confirmed variability in flowering abundance and analyzed strobile properties enables the choice of trees – pollinators, and the assessment of the available pollen quantity for further breeding of this Balkan endemic and Tertiary relic.

1. Introduction

Serbian spruce (*Picea omorika*/Panč./Purkyne) as a Balkan Peninsula endemite and Tertiary relic is very important for the reconstruction of plant life evolution in these lands. Its ecological plasticity has been proved by a high degree of adaptation to often rugged and inaccessible land and conditions where other species cannot survive. It also thrives in the regions of higher air pollution and it has been introduced to the countries of Northern Europe (Nymoen, H. 1978 after Isajev, V. 1987). It was only recently that it was discovered that Serbian spruce could grow spontaneously also on flat land with deep soil layer, in increased moisture conditions – almost in marshes, such as the site of Serbian spruce on Mitrovac na Tari, Serbia. Both heterosis and the data on intra- and interspecies hybridization offer the opportunity of utilization of the genetic potential of this valuable species in the aim of breeding (Vidaković, M. 1963).

2. Material and method

The research was carried out on 25 test-trees in Serbia spruce plantation at the site Bela Zemlja, near Užice, Serbia. The plantation was established in 1959 from the seed collected in natural Serbian spruce populations on Mt. Tara, at the altitude of 680 m, on the site of *Quercetum farnetto cerris*.

During two successive years 1991-92 in the optimum of the flowering phenophase in the first decade of May, the data on regularity and abundance of micro-strobiles were taken on all flowering test trees.

Quantitative characteristics were measured in laboratory on a statistically justified sample of mature strobiles, collected in the phase immediately before pollination in both study years: sizes (length, diameter, form coefficient = diameter/length × 100, strobile weight before drying, weight of the same pollen after strobile drying and strobile moisture lost during drying.

The applied method of individual selection is, along with hybridization and induced mutations, one of the main methods of forest tree breeding.

The data of quantitative analysis of flowering and microstrobiles were processed by variation-statistical method.

3. Results

3.1 VARIABILITY OF MICROSTROBILE REGULARITY AND ABUNDANCE

Flowering of all analyzed trees was evaluated by the above criterion for the assessment of miscrostrobile abundance. Serbian spruce is a forest species distinguished by not so much expressed periodicity of flowering. This has been proved by the results in Table 1, which show that 1991 was the year of abundant flowering. Tree number 10 excelled in both study years by its high degree of flowering abundance. There is an expressed variability both between analyzed trees in one analyzed year and between the same ones in two study years.

TABLE 1. Abundance of microstrobiles of analyzed test-trees

Criterion for the assessment of flowering abundance	1991		1992	
	Tree number	%	Tree number	%
+ (<500)			15,22	12.5
1 (500-1000)	15	4.0	5,18,21,24,25	31.2
2 (1000-2000)	2,5,6,7,16,18,22,24	32.0	1,4,13,14,17,23	37.5
3 (2000-3000)	3,8,9,21,25	20.0	2,11	12.5
4 (3000-4000)	1,4,11,20,23	20.0	10	6.3
5 (>4000)	10,12,13,14,17,19	24.0		
	Mean grade: 4	100,0	Mean grade: 2	100,0

3.2 VARIABILITY OF MICROSTROBILE QUANTITATIVE CHARACTERISTICS

Mean values for length of 9.5 mm and width of 6.2 mm, obtained in this analysis, classify Serbian spruce compared to other species of the same genus as the species with finer microstrobiles of oval form (Table 2).

TABLE 2. Boundary and mean values of microstrobile quantitative characteristics

	Value parameter	Dimensions: Length (mm)	Dimensions: Diameter (mm)	Form coefficient (%)	Mass (g): Strobile before drying (g)	Mass (g): Pollen after drying (g)	Moisture (%)
1991	Minimum	8.4	5.6	53.57	0.1204	0.0040	26.55
	Tree no.	25	25	17	25	25	13
	Maximum	12.3	6.9	70.22	0.2407	0.0363	96.31
	Tree no.	17	14	8	24	20	25
	Mean value	10.3	6.3	62.21	0.1822	0.0186	74.11
1992	Minimum	6.0	4.9	61.95	0.0559	0.0073	81.62
	Tree no.	25	25	1	25	25	8
	Maximum	9.8	6.9	81.70	0.1855	0.0220	90.23
	Tree no.	1	10	25	13	8	12
	Mean value	8.6	6.0	70.42	0.1215	0.0165	86.51

The results of variation-statistical processing of microstrobile quantitative characteristics confirm the justification of differences, at a high level of significance, between study years, individual trees, and the effect of these two factors on all analyzed characteristics. The effect of environment, i.e. year factor was not statistically justified only in the analysis of dry pollen weight, which indicates its genetic determination (Table 3).

TABLE 3. "F" values with the level of significance of the analysis of variance of microstrobile quantitative characteristics

Year	Factor of variability	Analyzed factors – Dimensions: Length (mm)	Dimensions: Diameter (mm)	Form coefficient (%)	Mass (g): Strobile before drying (g)	Mass (g): Pollen after drying (g)	Moisture (%)
1991	Tree	16.804***	15.012***	16.380***	19.957***	22.214***	214.827***
1992	Tree	18.220***	17.059***	20.511***	15.057***	5.152***	5.978***
1991/	Year	380.327***	63.066***	347.850***	243.609***	1.035	1,203.981***
1992	Tree	22.211***	23.699***	21.833***	19.446***	21.787***	170.163***
	Interaction	21.878***	11.250***	20.332***	13.978***	14.576***	161.840***

* $\alpha = 0.05$ ** $\alpha = 0.01$ *** $\alpha = 0.001$

By relating the results of microstrobile abundance with the average quantity of pollen per strobile, the values of pollen quantity produced by each test tree per year were calculated. In the year of abundant flowering when the average quantity of pollen per tree was 60.7 g, even more than 200 g of dry pollen could be collected from the trees with most abundant flowering (Figure 1).

FIG. 1 Quantitative analysis of microstrobiles and pollen

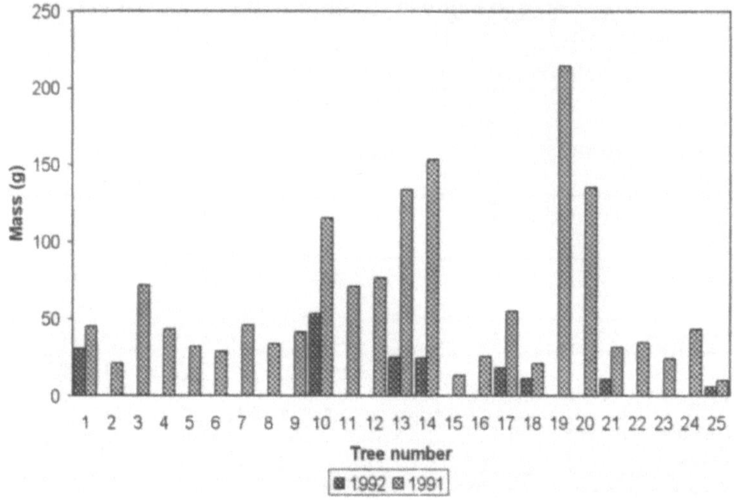

4. Conclusion

The results of individual variability of the analyzed Serbian spruce plantation can be utilized in the assessment of the quantity of available pollen during the optimum of the tree physiological maturity. The selected trees - pollinators, producers of the maximum quantity of pollen, should be taken into account in the further program of breeding of this interesting species.

5. References

Isajev, V. (1987) Oplemenjivanje omorike (*Picea omorika* /Panč./ Purkyne) na genetičko selekcionim osnovama, Doktorska disertacija, Beograd.

Vidaković, M. (1963) Međuvrsno križanje Pančićeve omorike (*Picea omorika* /Panč./ Purkyne) sa sitkanskom smrčom (*Picea sitchensis* /Bong./ Carr.), *Šumarstvo* **10-12/63**, Beograd.

Čolić, D. & Gigov, A. (1958) Asocijacija sa Pančićevom omorikom (*Picea omorika* Panč.) na močvarnom staništu, *Biološki institut NR Srbije*, Posebna izdanja, knjiga 5, Beograd.

ANTIOXIDANT ENZYMES AND CHILLING STRESS IN TOBACCO

T. GUETCHEV[1], H. WILLEKENS[2], M. VAN MONTAGU[2], S. DUKIANDJIEV[1], D. INZE[2] AND W. VAN CAMP[2]
[1]*University of Plovdiv, 24 Tsar Assen str., Plovdiv 4000, Bulgaria tel. +359 32 262716, fax +359 32 274865*
[2]*Laboratorium voor Genetica, Ledeganckstraat 35, Gent 9000, Belgium; tel. +32 9 2645010, fax. +32 9 2645349*

1. Abstract

The effect of chilling temperatures combined with elevated light (5 ^{0}C, PPFD 360 μmol m^{-2} sec^{-1}) on the activity of six antioxidant enzymes and on glutathione peroxidase (GPx) protein accumulation were studied in tobacco *Nicotiana tabacum* c.v. Petit Havana SR1. At these conditions, transient wilting but no photooxidative damage was observed. As a control, only elevated light treatment was applied (25 ^{0}C, PPFD 360 μmol m^{-2} sec^{-1}). Catalase (EC 1.11.1.6) showed the most dramatic reduction (five-fold) in activity after 4 days of chilling. Dehydroascorbate reductase (DHAR, EC 1.8.5.1) showed 30% reduction in activity while superoxide dismutase (SOD, EC 1.15.1.1) , ascorbate peroxidase (APx, EC 1.11.1.11), glutathione reductase (GR, EC 1.6.4.2), and monodehydroascorbate reductase (MDHAR, EC 1.6.5.4) were not affected by the chilling treatment. These results indicate that catalase and DHAR are affected by the chilling treatment.
Attempts to enhance catalase activity during chilling by using transgenic plants overexpressing catalase increased catalase levels 2-fold during chilling.

2. Introduction

Active oxygen species (AOS), such as superoxide ($O_2^{\cdot -}$) and hydroxyl ($^{\cdot}OH$) radicals, hydrogen peroxide(H_2O_2), and singlet oxygen ($^1O_2$), were shown to play a very important role in photooxidation and photoinhibition induced by chilling temperatures (Wise and Naylor, 1987; Prasad et al. 1994). The enzymes involved in the scavenging of these AOS are superoxide dismutase (SOD, EC 1.15.1.1), ascorbate peroxidase (APx, EC 1.11.1.11), monodehydroascorbate reductase (MDHAR, EC 1.6.5.4), dehydroascorbate reductase (DHAR, EC 1.8.5.1), glutathione reductase (GR, EC 1.6.4.2), and catalase (Cat, EC 1.11.1.6). Superoxide dismutates convert the superoxide radicals to O_2 and H_2O_2. APx, MDHAR, DHAR, and GR form so-called ascorbate-glutathione cycle (Foyer et al. 1994; Jimenez et al. 1997; Bunkelman et al. 1996). H_2O_2 produced by SOD is converted to water by APx using ascorbate as a substrate. The product of ascorbate oxidation, monodehydroascorbate radical, can either directly be reduced to ascorbate by MDHAR or can dismutate to AsA and DHA. DHA is recycled to ascorbate by DHAR, an enzyme utilising reduced glutathione as electron donor. GR reduces the oxidized glutathione using NADPH. Catalase is mainly found in the peroxisomes or glyoxysomes, where it scavenges H_2O_2 produced from photorespiration and oxydation of fatty acids, respectively. Our main aim was to investigate chilling response of antioxidant enzymes in order to find which enzyme is affected most and could be a potential target to genetoc engineering. We have previously created transgenic plants overexpressing catalase under a constitutive promoter (Chamnongpol et al. 1996). In this study we investigated the difference in activity levels at chilling temperatures between nontransgenic and transgenic plants overexpressing catalase.

3. Materials and methods

Tobacco plants *Nicotiana tabacum* cv. Petit Havana SR1 were grown 14 h light/ 10 h dark period, temperature of 25 ^{O}C, PPFD 100 μmol m^{-2} sec^{-1}(low light), and relative humidity 70%. Mature preflowering plants were taken for chilling and light treatments. Both treatments were performed in a plant growth chamber with light intensity PPFD 360 μmol m^{-2} sec^{-1}, 16 h light/ 8 h dark period, and relative humidity 70%. The temperature was 25±1 ^{O}C for the light and 5±1 ^{O}C for the chilling treatment. Catalase was determined according to Beers and Sizer (1952) $mM^{-1}.cm^{-1}$, APx according to Chen and Asada (1989), GR according to Schaedle et al. (1977), DHAR according to Asada (1984), MDHAR according to Polle et al. (1990), and SOD according to Beauchamp and Fridovich (1971). Protein concentration was determined by the method of Bradford (1976) with a kit supplied by Bio-Rad.

4. Results and discussions

Chilling treatment significantly decreased catalase activity. Catalase is known to be very photolabile enzyme (Volk et al. 1989, Feiarabend et al. 1992). Compared with the non-stressed plants, catalase activity in chilling-treated plants decreased with 44% after one day of chilling and 80% after four days of chilling (Fig. 1).

Within the first 3 days of the chilling treatment DHAR remained relatively unchanged. The first significant loss of activity was observed only after four days of chilling (Fig. 2). In addition, DHAR is not very active at low temperatures (Hakam et al. 1996). Plants with low ascorbate/dehydroascorbate ratio have decreased antioxidant capacity; in addition, DHAR is not very stable and can decompose to toxic products with possible citotoxic effects. However, the consequence of DHAR reduction at low temperatures remains unclear.

Chilling had no effect on SOD, APx, MDHAR, and GR activity levels (data not shown). This data indicates that these enzymes are not limiting for AOS scavenging at low temperatures.

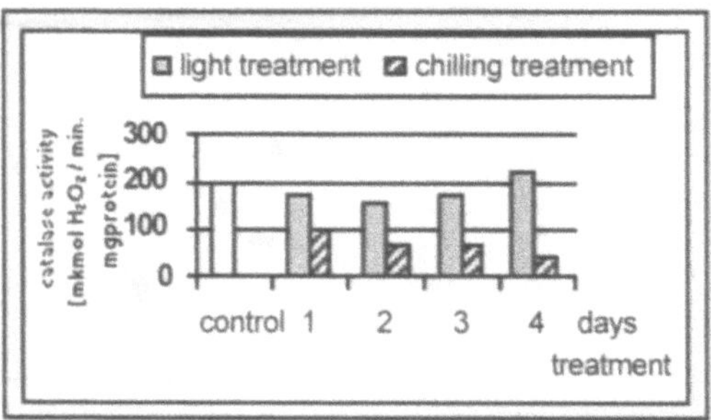

Figure 1. Catalase activity in control, non-stressed plants (opened bars) and plants given light stress (grey bars) and light+chilling stress (black bars).

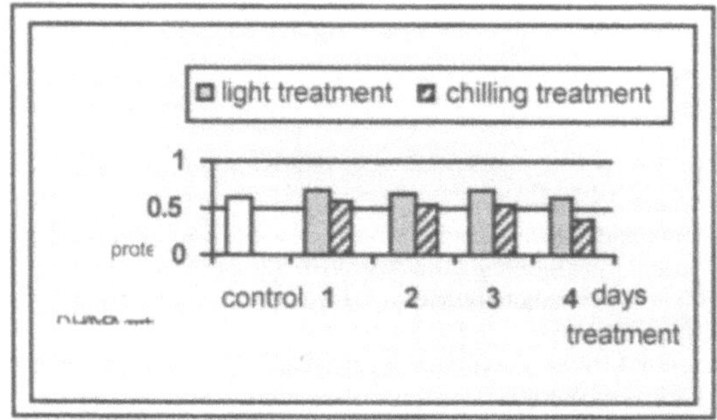

Figure 2. DHAR activity in control, non-stressed plants (opened bars) and plants given light stress (grey bars) and light+chilling stress (black bars).

A number of studies have indicated that an overexpression of particular antioxidant enzymes can eventually lead to better stress-tolerance (SenGupta et al 1993a; Sen Gupta et al. 1993b; Allen et al. 1995; Foyer et al. 1995). Catalase is very important for removal of H_2O_2 in C_3 plants (Chamnongpol et al. 1996). Studies on chilling revealed that chilling-resistant plants have elevated levels of catalase which correlated with decreased lipid peroxidation (Zhang et al. 1995). One of our goals was to minimize the reduction of catalase by overexpressing additional catalase gene under control of 35 S promoter. In addition to nontransgenic SR1 plants, we investigated catalase activity levels in catalase overproducing plants (cc25, Chamnongpol et al. 1996). During the chilling in both plants catalase activity was decreased but in catalase overproducers that decrease was less pronounced (Fig. 3). After 4 days of chilling the activity in the transgenics was about two- fold higher than in the controls. This indicates that using catalase overproducers can partially overcome catalase deficiency during chilling temperates and possibly increase chilling tolerance.

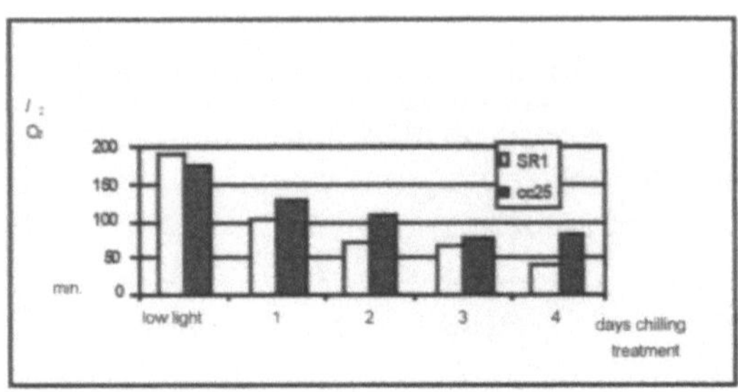

Figure 3. Catalase activity of non-transgenic (SR1, white bars) plants and catalase overproducers (cc25, dark bars) in non-stressed plants (grown at low light) and plants given light+chilling treatment.

5. References

Asada K (1984) Chloroplasts: formation of active oxygen and its scavenging. Methods Enzymol 105: 422-429

Beauchamp C, Fridovich I (1971) Superoxide dismutase: improved assays and assay applicable to acrilamide gels. Anal Biochem 44: 276-287

Bradford MM (1976) A rapid and sensitive method for the quantification of microgram quantities of protein utilizing the principle of protein-dye binding. Anal Biochem 72: 248-254

Beers RF, Sizer IW (1952) A spectrophotometric method for measuring breakdown of hydrogen peroxide by catalase. J Biol Chem 195: 133-140

Bunkelmann JR, Trelease RN (1996) Ascorbate peroxidase. A prominent membrane protein in oilseed glyoxysomes. Plant Physiol 110: 589-598

Chamnongpol S, Willekens H, Langebartels C, Van Montagu M, InzÈ D, and Van Camp W (1996) ransgenic tobacco with a reduced catalase activity develops necrotic lesions and induces pathogenesis-related expression under high light. Plant J 10: 491-503

Chen GX, Asada K (1989) Ascorbate peroxidase in tea leaves: occurance of two izozymes and the differences in their enzymatic and molecular properties. plant Cell Physiol 30:987-998

Foyer CH, Lelandais M, Kunert KJ (1994) Photooxidative stress in plants. Physiol Plant 92: 696-717

Hakam N, Simon JP (1996) Effects of low temperatures on the activity of oxygen-scavenging enzymes in two populations of the C_4 grass *Echinochloa crus-galli*. Physiol Plant 97:209-216

Jimenez A, Hernandez JA, Del Rio LA, Sevilla F (1997) Evidence for the presence of the ascorbate-glutathione cycle in mitochondria and peroxisomes of pea leaves. Plant Physiol 114: 275-284

Polle A, Chakrabarti K, SchurmannW, Renneberg H (1990) Composition and properties of hydrogen peroxide decomposing systems in extracellular and total extracts from needles of Norway spruce (*Picea abies* L., Karst). Plant Physiol 94: 312-319

Schaedle M, Bassham JA (1977) Chloroplast glutathione reductase. Plant Physiol 59:1011-1012

Prasad TK, Anderson MD, Martin BA, Steward CR (1994) Evidence for chilling- induced oxidative stress in maize seedlings and a regulatory role for hydrogen peroxide. The Plant Cell 6:65-74

Wise R, Naylor AW (1987a) Chilling-enhanced photooxidation. Plant Physiol 83: 272-277

Zhang J, Gui S, Li J, Wei J, Kirkham MB (1995) Protoplasmic factors, antioxidant responses, and chilling resistance in maize. Plant Physiol Biochem 33:567-575

MICROPROPAGATION OF SOME VEGETATIVE ROOTSTOCKS OF SWEET AND SOUR CHERRY BY MERISTEM CULTURE

H. GÜLEN AND A. KÜDEN
Çukurova University, Agricultural Faculty, Department of Horticulture, 01330, Adana - TURKEY

1. Abstract

This study was carried out to determine best culture media for the in vitro propagation of sweet and sour cherry rootstocks. Two different rootstocks which are called Edabriz and Damil were cultured in this study. Meristems showed better development when they are cultured in MS medium supplemented with GA_3. 2 mg/l BAP was used in the multiplication stage. Two-step rooting procedure was applied in the rooting stage; full MS medium and 0.5xMS medium supplemented with IBA (3 mg/l). Both of them were prepared solid and liquid and all the cultures were kept in the dark during one week. At the end of the week, they were transferred new rooting media which are include 1 mg/l IBA (solid and liquid).Damil showed best root development in B1 solid medium (0.5xMS medium + 1 mg/l IBA). Edabriz showed best root development in A1 solid medium (MS medium + 1 mg/l IBA). More hairy root formation was obtained from these media.

2. Introduction

The usage of clonal rootstocks have not been common in Turkey, yet. Various attempts were performed on clonal rootstocks several times but not succeded. Only liyyle amounts of vegetative apple rootstocks could be multiplied in last years. Seedling rootstocks still have been used in propagation of many fruit species (Küden, 1995).

Cherries is one of the important plant grown in Turkey. Seedling rootstocks is mostly used in cherry propagation, too; because of the propagation problem. But, clonal rootstocks have big advantages for the modern fruit growing. Rapid multiplication of clonal rootstocks takes long time by the classical propagation techniques. But rapid propagation of disease-free plants can be achieved using meristem culture. Therefore, in vitro propagation of clonal dwarf rootstocks of cherries (Damil and Edabriz) was aimed in this study.

3. Materials And Methods

Clonal rootstocks of cherries called "Damil" and "Edabriz" wereused as plant material. Shoot-tips were taken from the rootstocks in spring. After surface sterilization in the lab., meristems were isolated from shoots in the laminar air flow cabinet. Culture tubes were incubated at 25±2 ^{0}C under the 16 hours photoperiod for the development.

Two different media were tested for the initiation stage;

1-Murashige & Skoog medium+Sucrose (30 g/l)+BAP (0.5 mg/l)+Agar(7 g/l)

2-Murashige & Skoog medium + Sucrose (30 g/l)+BAP(1mg/l) + IBA (0.1mg/l) + GA_3 (0.1mg/l) + Agar(7 g/l)

Meristems which were developed in the initiation media, were transfered to the multiplication medium given below.

Murashige & Skoog medium+Sucrose (30 g/l) + BAP (2 mg/l)+Agar (7 g/l)

Two different rooting media were tested for the rooting stage and two-step rooting procedure was applied (Thimann, 1977). In the first step of the rooting stage, plants were transfered two media given below.

A- Murashige & Skoog medium+Sucrose (30 g/l) + IBA (3 mg/l)

B- 0.5xMurashige & Skoog medium+Sucrose (30 g/l) + IBA (3 mg/l)

Both of the medium were prepared solid and liquid. Solid media were solidified with 7 g/l agar (Sigma Co.). Paper bridge system was used in liquid media. All the cultures were kept in dark during one week. At the end of the week, some parts of the plants were transferred new rooting media given below (solid and liquid) for the second step of the rooting stage.

A1→ M & S medium+Sucrose (30 g/l)+IBA (1 mg/l)

A2→ M & S medium+Sucrose (30 g/l)

B1→ 0.5xM & S medium+Sucrose(30 g/l)+IBA(1 mg/l)

B2→ M & S medium+Sucrose (30 g/l)

4. Result And Discussion

This study was carried out to determine the most suitable conditions for in vitro propagation of clonal rootstocks of cherries Damil and Edabriz. Meristems showed better development when cultured in medium 2; MS medium supplemented with GA_3 (Table 1). Both of the rootstocks showed 80 % development in media number 2.

Four subculture were applied in the multiplication stage. Multiplication rate increased step by step until 4th. Step of the subcultures. In the last step of the subcultures, multiplication rate decreased from 5 to 3,2 in Damil; from 9 to 4,5 in Edabriz (Table 2). This is related with the age of cultures and concentration of the cytokinin.

Last step of multiplication (4th. Step), BAP concentration was reduced to 1 mg/l. This treatment was decreased the multiplication rate, but increased the quality of the plants before rooting stage. Generally, multiplication rate of Edabriz was determined higher than Damil.

Two-step rooting procedure showed better root development than the continued presence of 3 mg/l IBA for the both of the rootstocks. Callus formation was seen at the base of the shoot in the medium, continued presence of high auxine. However, some researchers reported that some cultivars of apple root readily in the continued presence of 0,2 mg/l IBA, and certain cherry rootstock cultivar Colt, gave 100 % rooting in the continued presence of 3 mg/l IBA, with little callusing (James and Thurbon, 1979).

Damil rootstock showed best root development in B1 solid medium (0.5xMS medium+1 mg/l IBA) (Table 3). 100 % rooting rate was determined in this medium. Also, average root lenght was the highest in this medium (9.2 cm).
Edabriz rootstock showed best root development in A1 solid medium (MS medium+1 mg/l IBA) (Table 4). 83 % rooting rate and 15 root Per plant were obtained in this medium. In addition the highest root lenght (10 cm.) was determined in this medium.

Table 1. Survival rate of the meristems on the different initiation media

Rootstock	Media	Survival rate (%)
Damil	1	46
	2	80
Edabriz	1	54
	2	80

Table 2. Effect of multiplication stages on the multiplication rate of Damil and Edabriz

Rootstock	Average Multiplication Rate per stage			
	1st.	2nd.	3rd.	4th.
Damil	1.6	4.8	5	3.2
Edabriz	4.4	7.2	9	4.5

Table 3. Effect of different rooting media on rooting of the cultured Damil rootstock

Media	Rooting rate (%)	Average root number/ plant	Average root lenght (cm)
A solid	85	7	7.0
liquid	80	1	1.o
B solid	75	8	4.0
liquid	-	-	-
A1 solid	100	8	4.5
liquid	66	6	2.2
A2 solid	83	4	6.0
liquid	16	1	1.0
B1 solid	100	9	9.2
liquid	33	3	2.2
B2 solid	100	7	6.5
liquid	83	1	3.5

Table 4. Effects of different rooting media on rooting of the cultured Edabriz plants

Media		Rooting rate (%)	Average root number/ plant	Average root length (cm)
A	solid	33	4	5.6
	liquid	-	-	-
B	solid	-	-	-
	liquid	30	3	8.0
A1	solid	83	15	10.0
	liquid	25	10	3.0
A2	solid	50	7	6
	liquid	60	6	6.5
B1	solid	50	12	4.1
	liquid	50	6	6.2
B2	solid	50	11	3.6
	liquid	50	8	4

5. References

James,D.J. & I.J. Thurbon, 1981. Shoot and root initiation in vitro in the apple rootstock M) and the promotive effects of phloroglucinol. J. Hort. Sci, 56(1), 15-20.

Küden, A., 1995. Propagation of fruit trees by grafted cuttings. 2nd. National Hort. Cong., 3-6 Oct. 1995, Adana, Turkey, p.25-29

Murashige, T. & F. Skoog, 1962. A revised medium for rapid growth and bioassay with tobacco tissue cultures. Physiol. Plant., 15, 473-497.

Thimann, K.V., 1977. The initiation of roots on stems in hormone action in the whole life of the plant. University of Massachusetts Press, Amherst, p. 190-205.

DETERMINATION OF CONDITIONS FOR GENE TRANSFER INTO BARLEY (*Hordeum vulgare* L.) BY TISSUE ELECTROPORATION AND BIOLISTIC

F. GÜREL, [1] AND N.GÖZÜKIRMIZI [1,2]

1. *University of Istanbul, Department of Biology, Molecular Biology Section, Vezneciler-34459 Istanbul/TURKEY*
2. *TÜBİTAK-MRC, Research Institute of Genetic Engineering and Biotechnology Gebze-Kocaeli/TURKEY*

1. Introduction

Gene transfer to plants is one of the most popular topics in plant biotechnology. Cereals have some difficulties in transformation such as transfer and expression of foreign genes. In barley, particle gun (biolistic) has recently been used as an effective system [1,2]. Electroporation has also been used in several cereal protoplasts and tissues as an economic transformation system compared to biolistic. However, there are problems in both methods in obtaining transgenic offsprings and stable expression of genes. Biological conditions are important as much as the physical parameters in the procedure to be applied. In barley, a direct regeneration system which may be less affected from undesirable mutations (e.g. somaclonal variations) is needed.

In this work, we examined mostly the biological factors such as timing of transformation and selection parameters.

2.Material and Methods

2.1. PLANT MATERIAL.

Mature seeds of barley (*Hordeum vulgare* L.cv. Zafer 160) were obtained from İzmir-Menemen Agriculture Research Institute.

2.2. PLASMIDS.

pBI 121 (Clontech Inc.) and p35SGUSINT plasmids were used. pBI 121 contains GUS (*β-glucuronidase*) and *npt II* (*neomycin phophotransferase II*) genes. p35SGUSINT [3] contains GUS-INT and *npt II* genes.

2.3. CULTURE CONDITIONS.

Embryos were cultured on Murashige and Skoog(MS) [4] medium supplemented 2mg/l 2,4-Dichlorophenoxyacetic acid and kept in a growth chamber with 26°C temperature, 16 hours dark/8 hour light period. Explants electroporated at 1st, 2nd and 3rd days of culture.

2.4. GENE TRANSFER

2.4.1. Electroporation.

Mature embryos transferred to a petri dish which included *bufferA* (20mM $CaCl_2$, 10mM hepes, 0.4M mannitol, pH 5.6) for 1 hour in room temperature. Explants were transferred to another petri dish which included *buffer A* with 0.3% macerase and 0.5% cellulase for 10 minutes. Twenty embryos were washed twice and transferred to cold electroporation cuvettes (0.4cm gap distance) includes *buffer A* plus 50mM KCl. 30μg circular plasmid DNA was added to each cuvette mixed and put on ice for 30 min. Combination of several voltages (0.2, 0.3 and 0.4kV) with two capacitance (500, 960μFd) values were applied by using an electroporation apparatus (BIORAD, Gene Pulser).

2.4.2. Biolistics.

BIOLISTIC (PDS1000/HE) version was used in gene transfer. A mixture containing 20μg plasmid DNA, 50μl 2.5M $CaCl_2$, 20μl 0.1M spermidin was incubated at room temperature for 30 min. This mixture washed with ethanol (99%) and 10μl was loaded onto each macrocarriers. 1100psi helium pressure and two shots were applied for each petri dish including 30 embryos. 1.0 and 1.6μm gold particles and several microcarrier flight distances (6, 9 and 12cm) were tested for optimum gene transfer efficiency.

2.5. HISTOLOGICAL GUS EXPRESSION AND SELECTION.

GUS expression was assayed histochemically by the method of Jefferson [5]. Electroporated embryos were transferred to MS and selective medium (MS+geneticin sulfate). G418 (geneticin sulfate) was applied in three different concentrations as 40mg/l, 80mg/l and 100mg/l . In addition, genomic DNA was extracted [6] from mature leaves of 20 days plantlets grown on 40mg/l G418 medium and amplified by primers for GUS and *npt II* genes according to a method described previously [7].

3. Results and Discussion

3.1.ELECTROPORATION CONDITIONS.

The germination frequency of non-electroporated mature embryos was nearly 100% on MS medium. The germination frequencies of embryos under non-selective MS conditions were determined in different voltage-capacitance values . Those embryos were electroporated on 2nd day of culture. The voltage of 0.2kV (500V/cm) and the capacitance of 500μFd were found to be optimum to obtain higher germination frequency as 75%. It was previously reported that using 500μFd capacitance instead of 960μFd caused a higher regeneration frequency in wheat embryoids [8]. In order to detect the most appropriate biological period for DNA transfer with minimal damage, culture days before gene transfer were tested. Electroporation caused a decrease in germination capacity on embryos electroporated on 1st(20%), 2nd (24%) and 3rd (95%) days of culture. Electroporation applied on 3rd day of culture gave higher germination than first and second days (Table 1). Embryos electroporated on 1st day of culture did not germinate on selective medium. Electroporation has been used in barley protoplasts in several studies. However, our study demonstrates that the mature embryos of barley can also be used directly in electroporation. Embryos from mature seeds are available commercially and they were highly responsive to regeneration compared to callus and suspension cultures. Physical conditions (capacitance, voltage) used in electroporation directly effect the tissue viability. In our work, one more day period increased the germination rate on MS approximately 3 fold compared to first and second day electroporation (Table 1). The results can be attributed to the potential germination capacity of embryos cultured on both MS and selective media depends on the number of cells under division of those include. When higher voltage was applied, there was no detectable GUS expression and germination. However, histochemical GUS expression was observed on the surface and transversal sections of embryos (Fig. 1A) after 7 days applying in several voltage-capacitance combinations tested. GUS gene expression was also detected in primer root tips, leaf primordia and mature leaf segments from 20 days regenerants developed on selective and MS media (Fig. 1B). On the leaf of regenerants growed in selective medium (40mg/l G418), GUS expression was chimeric after 3 months. The presence of GUS expression on leaf segments 3 months after gene transfer is an indication of stable integration of GUS gene into genome. It was previously pointed out that the amount of selective agent is critical in eliminating untransformed cells after transformation [9]. 11% of embryos were germinated on selective medium contained 40mg/l G418 (Table 1). 80-100mg/l G418 concentration was found to be inhibitory for viability and also caused albinism. In 8 plantlets grown in 40mg/l G418, 1 GUS-*npt II* and 3 *npt II* positive individual were detected by PCR . Three individual plantlets in which *npt II* genes were detected, GUS gene was not appeared even though both genes were present on the same plasmid. A similiar result was reported in a transformation study of barley protoplasts with *npt II* and GUS genes [10].

3.2. MICROPROJECTILE BOMBARDMENT OF BARLEY EMBRYOS.

Transient GUS expression was detected on the embryos after 48 hours. Mature embryos bombarded with p35SGUSINT regenerated and showed root formation basal MS medium. PCR analyses on the 8 putative transgenic leaves showed the presence of only *npt II* genes in 4 individual . This is a preliminary study to determine several conditions for barley transformation. Our study continue with the analyses of stability of introducing genes in transformants and their progeny.

4. Acknowledgement. This study was supported partly by TÜBİTAK (TBAG-1404) and Istanbul University Research Fund (T/93) as research projects.

TABLE 1 . Effects of different preculture periods before electroporation and selective agent concentrations on germination.

No.of Embryos	Selection	Culture days before *Electroporation*	No. of germinated embryos
30	40mg/l Geneticin	2. day	3
40	80mg/l Geneticin	2. day	2
42	100mg/l Geneticin	2. day	-
54	-	2. day	13
44	40mg/l Geneticin	3. day	5
36	80mg/l Geneticin	3. day	2
30	100mg/l Geneticin	3. day	1
40	-	3. day	38

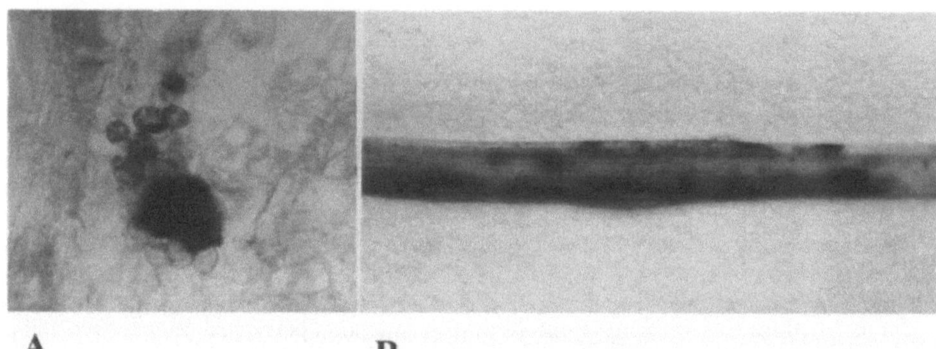

Figure 1(A, B). GUS expressed cells on electroporated embryo with pBI 121(A), histochemical GUS expression on mature leaf segment (B)

5. References

[1] Mendel, R.R., Müller, B., Schulze, J., Kolesnikov, V., Zelenin, A.(1989). Delivery of Foreign Genes to Intact Barley Cells by High-Velocity Microprojectiles., *Theor.Appl.Genet.*, **78**, 31-34.

[2] Stiff, C.M., Kilian, A., Zhou, H., Kudrna, D.A., Kleinhof, A. (1995). Stable Transformation of Barley Callus Using Biolistic[R] Particle Bombardment and the Phosphinothricin Acetyltransferase (*bar*) Gene. *Plant Cell, Tissue and Organ Culture*, **40**, 243-248.

[3] Vancanneyt,G., Schmidt, R., O'connor-sanchez, A., Willmitzer, L., Rocha-Sosa, M. (1990). Construction of an Intron-Containing Marker Gene: Splicing of the Intron in Transgenic Plants and its Use in Monitoring Early Events in *Agrobacterium*-Mediated Plant Transformation, *Mol.Gen.Genet.*, **220**,245-250.

[4] Murashige, T., Skoog, F. (1962). A Revised Medium for Rapid Growth and Bioassay with Tobacco Tissue Cultures, ***Physiol.Plantarum***, **15**, 473-497.

[5] Jefferson, R.A. (1987). Assaying Chimeric Genes in Plants: The GUS Gene Fusion System, *Plant Mol. Biol. Rep.*, **5(4)**, 387-405.

[6] Walbot, V. (1988). Preparation of DNA From Single Rice Seedlings. **Rice Genet** *Newslett.*, **5**,149-151.

[7] Arı Ş., Bayroviç K., Gürel F., Gözükırmızı N.(1995). Determination Transgenic Plants with Polymerase Chain Reaction, *Doğa Turkish Journal of Biology*, **19**,1-4.

[8] Gustafson, V.,D., Baenziger, P., S., Mitra, A., Kaeppler, H.,F., Papa, C.,M., Kaeppler ,S., M. (1995). Electroporation of Wheat Anther Culture-Derived Embryoids, *Cereal Research Communications*. **23 (3)**, 207-213

[9] Christou, P., Ford, T.L. (1995). The Impact of Selection Parameters on the Phenotype and Genotype of Transgenic Rice Callus and Plants, *Transgenic Research*, **4**, 44-51.

[10] Lazzeri, P.A., Brettschneider R., Lührs, R., Lörz, H.(1991). Stable Transformation of Barley via PEG-Induced Direct DNA Uptake into Protoplasts, *Theor.Appl.Genet.*, **81**, 437-444.

MICROPROPAGATION OF *ORIGANUM VULGARE* SUBSP. *HIRTUM* (MT. TAYGETOS)

G. N. ICONOMOU-PETROVICH[1] AND I. NIANIOU-OBEIDAT[2]
Vitro Hellas S.A., Nisseli, Alexandria, HELLAS
Present address: [1] 64, J. Kennedy str., 555 35 Pylea, Thessaloniki, HELLAS
[2] Dept. of Genetics & Plant Breeding, Aristotelian University of Thessaloniki, HELLAS 540 06

1. Introduction

Origanum vulgare is member of the family Lamiaceae, the genus *Origanum*, and possesses the widest distribution among *Origanum* species [1][2]. Under the commercial name "oregano", a large number of plant taxa are found, belonging to different species, genera or even families, resulting in much confusion [3][4]. *O. vulgare* subsp. *hirtum*, known with the vernacular name "Greek oregano", is generally accepted to have the best quality [3].

In contrast to conventional methods of propagation, *in vitro* vegetative propagation, *i.e.* micropropagation, using nodal segments, allows the production of a large homogeneous population of plants in a relatively short period of time, irrespective of the season.

In vitro culture methods are available for callus induction, regeneration and rooting of *O. vulgare* plants [5][6]. So far, no methods have been reported for *in vitro* culture of *O. vulgare* subsp. *hirtum*, which possesses significant morphological as well as chemical differences in comparison to other *O. vulgare* subspecies [2][3]. The aim of this work was to develop methodology for commercial *in vitro* vegetative propagation of *O. vulgare* ssp. *hirtum* [7].

2. Materials & Methods

2.1. PLANT MATERIAL & DISINFECTION

Plants of *O. vulgare* ssp. *hirtum* were collected from the wild from the location Arna, altitude 700 m, of mount Taygetos, in Peloponnese, south-western Hellas. The plants were maintained under greenhouse conditions in large pots in a mixture of peat moss and perlite (3:1), and were sprayed every 20-30 days against soil-borne fungi.

Young and rapidly growing shoots carrying axillary buds were obtained from the source plants, defoliated, and cut into small nodal segments of about 10-15 mm in length. After disinfection, pieces 1-2 mm long which were killed by the disinfectant, were excised from the original explants. The explants 7-12 mm in length carrying only one node were placed vertically on the medium. For rooting, contaminated, deformed, vitrified, or brown shoots and leaves were removed and apical shoots of average height 20 ± 3 mm carrying three nodes were placed in the rooting medium. Plantlets selected for acclimatisation were longer than 15 mm, and rinsed under tap water to remove any adhering medium.

The nodal segments were initially washed for 20-30 seconds in running tap water to free from the dust. Following, using full sterile techniques, they were either surface sterilised in 70 % ethanol for 30 or 60 seconds followed by 3-4 rinses in sterile distilled water, or initially disinfected by rinsing in 70 % ethanol for 10-15 seconds followed by surface sterilisation in an aqueous solution of 0.5 % (w/v) sodium hypochlorite (NaOCl) for 3 or 4 minutes, 1 % (w/v) for 10 or 12 minutes, 1.5 % (w/v) for 10 or 12 minutes, and 2 % (w/v) for 10 minutes with one drop of Tween 20, followed by 3-4 rinses in sterile distilled water.

2.2. CULTURE MEDIA

Growth & shoot proliferation. The explants were transferred individually to glass tubes containing the nutrient medium. For large-scale micropropagation, 500 ml jars were used. Full strength Gamborg's B5 macronutrients, micronutrients, vitamins, with 20 g/l sucrose, and 6 g/l agar were used (Table 2), supplemented with factorial combinations of BAP at 0, 0.25, and 0.50 mg/l, and NAA at 0, and 0.01 mg/l. The experiment was a two-factor randomised (BAP concentration, and NAA concentration). The pH of the media was adjusted to 5.5 prior to autoclaving. For all experiments a minimum of 17 replicates were used. Data were recorded after 1, 2, 3, and 4 weeks and analysed by mean ± standard error.
Rooting. For the rooting of the shoots 330 ml jars were used containing nutrient medium. The experiment for this phase was a three-factor (NAA concentration, macronutrient concentration, and sucrose concentration; Table 3) randomised, with three jars per treatment, and five shoots per jar (total of 15 shoots per treatment).

2.3. CULTURE CONDITIONS & ACCLIMATISATION

The total treatment period was 8 weeks with one transfer from the growth and proliferation phase to the rooting phase at the end of the fourth week. All cultures were incubated in a growth chamber, maintained at temperature 20 ± 2 °C, relative humidity 35-50 %, under standard fluorescent white light, illuminance 2300 lx, and 16 hours photoperiod. *In vitro* plantlets were transplanted in a mixture of peat moss and perlite (3:1) in the greenhouse (Table 1).

Table 1: Acclimatisation conditions of *in vitro* propagated plants

Treatment	Illuminance (klx)	Temperature (°C)	Humidity (%)	Duration (days)
I	5-10	20-29	>90	10
II	10-12	20-29	80	3
III	12-15	20-29	70	2
IV	15-25	>33	50	1
V*	25-75	day: 28-30/ night 18-20	30-40	30
VI*	75-225	outdoor conditions		

* non-fog conditions

3. Results

3.1. GROWTH & SHOOT PROLIFERATION

The best results for disinfection were obtained by submersion of nodal segments for 10-15 seconds in 70 % ethanol, rinsed with tap water, and then for 10 minutes in an aqueous solution of 1 % (w/v) NaOCl. Then the segments were rinsed 4 times in sterile distilled water. Under such conditions more than 93 % (up to 100 %) of the explants obtained were free of fungal or exogenous bacterial contamination.

Optimum shoot growth, proliferation, and overall appearance among the six combinations of growth regulators tested, was observed in Gamborg's B5 medium supplemented with 0.01 mg/l NAA in the absence of BAP (Table 2; Fig. 1).

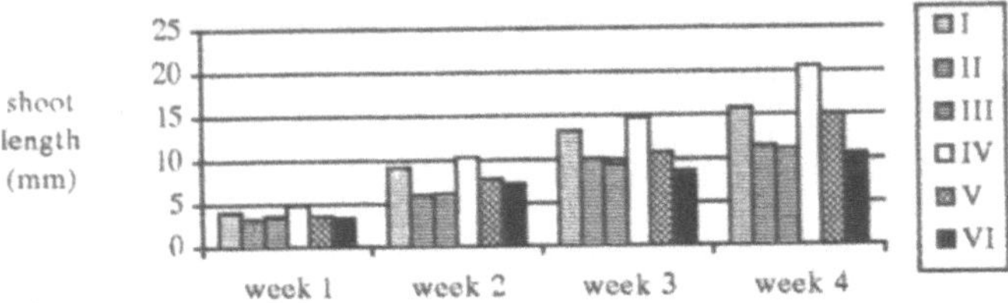

Fig. 1: Effect of BAP and NAA concentrations (treatments I to VI, table 2) on shoot length of *O. vulgare* ssp. *hirtum* after 1, 2, 3, and 4 weeks *in vitro*.

Table 2: Effect of BAP and NAA concentrations on shoot production from nodal segments of *O. vulgare* ssp. *hirtum* after 4 weeks in culture.

Treatment	I	II	III	IV	V	VI
BAP (mg/l)	0.00	0.25	0.50	0.00	0.25	0.50
NAA (mg/l)	0.00	0.00	0.00	0.01	0.01	0.01
Total explants	18	19	17	20	19	18
Mean shoot length (mm) ± S.E.	15.8 ±1.7	11.4±1.4	11.1 ±1.1	20.6 ± 1.0	15.2 ± 1.1	10.5 ± 0.8
Average no. of nodes shoot^{-1}	5.5	5.8	5.6	6.5	6.3	6.3
Vitrification (%)	7	5	< 5	0	0	18
Light green leaves (%)	50	100	100	21	100	100
Browning (%)	58	79	78	< 8	26	68
Pale leaves (%)	16	68	68	0	17	25
Callus formation (%)	5	63	35	< 5	11	16
Rooting (%)	73	0	0	100	0	0

3.2. ROOTING & ACCLIMATISATION

The best root response and overall appearance induced from 20 ± 3 mm long shoots was observed in full strength Gamborg's B5 medium with 20 g/l sucrose, supplemented with 0.01 mg/l NAA (Table 3).

Table 3: Effect of NAA, macronutrient, and sucrose concentration on the rooting of *O. vulgare* ssp. *hirtum*.

Treatment	1	2	3	4	5	6	7	8
Macronutrients*	1/1	1/1	1/1	1/1	1/2	1/2	1/2	1/2
Sucrose (g/l)	10	10	20	20	10	10	20	20
NAA (mg/l)	0.01	0.02	0.01	0.02	0.01	0.02	0.01	0.02
Rooting (%)	100	65	100	100	100	100	100	100
Root length (mm)	1 to 2	c. 2	4 to 10	2 to 4	2 to 3	< 2	6 to 12	4 to 6
Root browning (%)	60	70	10	-	38	40	-	-
Shoot length	36.5	44.2	50.1	54.8	36.6	41.9	48.0	52.1
± S.E. (mm)	± 4.2	± 5.7	± 3.6	± 5.0	± 2.2	± 2.8	± 3.2	± 1.7
Average nodes/explant	7.40	6.13	7.40	6.67	6.93	7.00	7.20	7.13
Vitrification (%)	33	67	-	[40 ap.]	-	-	-	[70ap.]
Leaf colour	dgr	gr	dgr	dgr	dgr	dgr	dgr	dgr

ap., apices; dgr, dark green; gr, green; * 1/1, full strength; 1/2 half-strength

During preliminary analyses it was found that the essential oils of *in vitro* plantlets were characterised by the same main components *i.e.* thymol and carvacrol, as the source plants and all oregano oils.

4. Discussion

4.1. GROWTH & SHOOT PROLIFERATION

In vitro propagation in the medium with 0.01 mg/l NAA is combining optimal overall appearance, and rooting, with minimal negative effects *i.e.* vitrification, light green leaves, browning, and pale leaves. In addition, swelling of the lower cut end of the explants, followed by callus formation is also minimal, thus eliminating the chance of genetic variable plants. It is a good practice however, to cut and discard the lower cut end of the explants, in order to maintain the population genetically stable.

4.2. ROOTING & ACCLIMATISATION

The *in vitro* rooting stage does not serve for the rooting of the shoots only, but also for elongation and improvement of the overall appearance of the shoots. Otherwise in practice, *in vitro* rooting may prove unnecessary, and *ex vitro* methods should be used instead [8].

The concentration of sucrose was found to be the most critical factor in determining the rooting studies. Growth of the shoots was negatively influenced by reducing sucrose concentration in the medium. Also shoot growth (elongation) decreased with sucrose reduction, the number of nodes per shoot was lowered, and shoots showed vitrification symptoms.

The results obtained suggest that multiplication, elongation and root induction, can be carried out in one medium (treatment IV for growth and shoot proliferation, and treatment 3 for rooting, *viz.* Gamborg's B5 medium supplemented with 0.01 mg/l NAA), thus significantly minimise labour cost during micropropagation.

The plantlets transferred to the greenhouse for acclimatisation, developed a new rooting system or elongation of the one induced *in vitro* occurred, without any serious loss (< 3 %) suggesting that *ex vitro* rooting is practicable.

Abbreviations: BAP, 6-benzylaminopurine; B5, Gamborg's B5 culture medium [9]; NAA, 1-naphthaleneacetic acid.

Acknowledgements. Financial support from the European Union through the ECLAIR (AGRE 0060) research programme is gratefully acknowledged.

5. References

1. Ietswaart, J.H. (1980). *A taxonomic revision of the genus Origanum (Labiatae).* London: Leiden University Press.
2. Kokkini, S., Vokou, D., Karousou, R. (1991). Morphological and chemical variation of *Origanum vulgare* L. in Greece. *Botanika Chronika*, 10, 337-346.
3. Bruneton, J. (1995). *Pharmacognosy, phytochemistry, medicinal plants.* Intercept Ltd., Hampshire, England.
4. Lawrence, B.M., Reynolds, R.J. (1984). The botanical and chemical aspects of oregano. *Perfumer and Flavourist,* 9, 41-51.
5. Kumari, N., Saradhi, P.P. (1992). Regeneration of plants from callus of *Origanum vulgare* L. *Plant Cell Reports,* 11, 476-479.
6. Shetty, K., Curtis, O.F., Levin, R.E., Witkiwsky, R., Ang, W. (1995). Prevention of vitrification associated with *in vitro* shoot culture of oregano (*Origanum vulgare*) by *Pseudomonas* spp. *Journal of Plant Physiology*, 147, 447-451.
7. Iconomou-Petrovich, G.N. (1997). *Micropropagation of Origanum vulgare subsp. hirtum.* Extra-Mural Dissertation. King's College, University of London, UK.
8. George, E.F. (1993). *Plant propagation by tissue culture.* Second edition. Part 1: The Technology. Exegetics Ltd., Edington, Wilts, England.
9. Gamborg, O.L., Miller, R.A., Ojima, K. (1968). Nutrient requirements of suspension cultures of soybean root cells. *Experimental Cell Res.*, 50, 151-158.

IN VITRO PROPAGATION OF BETULA PENDULA 'DALECARLICA'

I. ILIEV[1], V. BESENDORFER[2], T. PESKAN[3]
[1]Forest University, 10 "Kliment Ohridski" blvd., Sofia 1756, Bulgaria, [2]University of Zagreb, Department of Molecular Biology, Rooseveltov trg. 6, 10000 Zagreb, Croatia, [3]Tatjana Peskan, Botanisches Institut der Ludwig-Maximilians-University, Menzinger str. 67, D-80638 Munchen 19, Germany

1. Introduction

Preserving and multiplying of 'dalecarlica' form through generative reproduction is difficult, because its male sterility in Bulgaria and the heterozygocity of silver birch.

On the other hand propagation coefficient of heterovegetative hybrids is very little and the production of grafts is related to the loss of time.

It has demonstrated that, limiting factors for in vivo cloning of different silver birch genotypes are the age, physiological stage i.e. the season of cuttings collection and the genotype peculiarity of the donor plant (Ondrushka and Schmidt, 1984; Kajba, 1996 etc.).

Therefore, the main purpose of our work was to develop a method for in vitro cloning of 'dalecarlica' form.

2. Materials and Methods

2.1 PLANT MATERIAL AND CULTURE CONDITIONS

As a donor plant 25 year old individual was used. For callus induction and adventitious bud formation aseptic leaf segments obtained in vitro were used. They were cultivated on the following media:

- medium C: MS (Murashige and Skoog, 1962)+ zeatin riboside of 0, 2, 5, 10 and 15 mg/l;
- medium E: WPM (McCown and Lloyd, 1981) + zeatin riboside of 0, 2, 5, 10 and 15 mg/l;
- medium F: macroelements and chelate after Simola (1985) + trace elements and vitamins after MS + 10 mg/l zeatin riboside.

Each media used for adventitious bud induction were enriched with 30 mg/l adenine. Every variant and medium contained 15 leaf segments.

The regenerants were cultivated on half-strength MS medium without hormones (RK) and with addition of: 0.3 mg/l NAA (RB1), 0.5 mg/l NAA (RB2), 0.5 mg/l

NAA + 0.5 mg/l IBA (RB3), 0.3 mg/l IBA (RB4) and 0.5 mg/l IBA (RB) for inducing of rhizogenesis. Every medium contained 28 shoots obtained in vitro. The pH value before autoclaving was adjusted to 5.6. All the media and their variants were developed in 3 replications.

2.2 CHROMOSOME TECHNIQUE

The karyotipe of in vitro obtained plants was compared with the karyotipe of in vivo cloned plants. The radicles of rooted plants were pretreated with the 0.002 M solution of 8-oxyquinoline for 4 hours at 16°C, and fixed in ethanol-acetic acid (3:1) for 24 hours. The root-tips were hydrolyzed in 1N HCl, for 15 min at a temperature of 60°C and stained with Feulgen for 1.5 hours. The slides were prepared by "squash" technique in 45% acetic acid, dehydrated by ice-dry method and mounted in Euparal.

2.3 PROTEIN EXTRACTION AND PROTEIN ELECTROPHORESIS

Green shoots setting for rooting in vitro were taken for extraction on 3rd, 6th and 9th day after inoculation on the RK and RB rooting media. The experimental conditions describing protein extraction and isoelectric focusing were already reported (Peskan *et al.*, 1996).

The SDS-polyacrylamide gel electrophoresis was carried out according to Laemmli (1970), using 12% separating gel. The gels were stained with Coomassie Brilliant Blue R (Sigma).

Peroxidase (EC 1.11.1.7) isozymes were visualised by incubating the gel for 30 min in the 0.2 M phosphate buffer (pH 5.8) that contained 0.5 M guaicol and 0.5 mM H_2O_2.

3. Results and Discussion

3.1 IN VITRO CULTURE

Differentiation of adventitious buds on leaf explants was induced about 4-5 weeks after the experiment had begun and one week later stem morphogenesis was noticed. In the controle group cultivated on hormone-free medium (R) leaf segments gradually reached complete necrosis.

The strongest stimulating effect was obtained of nutrition media containing 5 mg/l and 10 mg/l ribosid zeatin. However, Medium F appeared to be the most suitable where the average number of the regenerants developed from explant was biggest. The callus developed on F_3 medium could be distinguished from others due to its constant regeneration capacity (Table 1).

In rooting experiments conducted on media containing NAA and/or IBA, the roots were successfully developed but always followed by callus formation.

Adventitious roots were developed after 6-7 days on rooting media and after 9-10 days the process ended. The largest amount of callus was formed on media with NAA (RB1 and RB2), and the formed roots were the thickest, with well

TABLE 1. Effect of different modified media for induction of callus and adventitious bud formation on aseptic leaf segments.

Nitritive media and their variants		Leaf segments which have formed callus - %		Average number of regenerants obtained from the total number of cultivated leaf segments.	
		M ± SE	SD	M ± SE	SD
Meduim C	R	0.00	0.00	0.00	0.00
	1	33.33 ± 3.85	6.67	3.07 ± 0.30	0.52
	2	71.11 ± 4.44	7.70	8.18 ± 0.35	0.60
	3	77.78 ± 2.22	3.85	9.36 ± 0.29	0.51
	4	64.44 ± 5.88	10.18	6.95 ± 0.72	1.24
Medium E	R	0.00	0.00	0.00	0.00
	1	44.45 ± 2.22	3.85	3.76 ± 0.23	0.39
	2	84.44 ± 4.44	7.70	11.33 ± 0.70	1.22
	3	88.89 ± 2.22	3.85	14.69 ± 0.69	0.80
	4	73.33 ± 7.70	13.34	8.29 0.68	1.18
Medium F	R	0.00	0.00	0.00	0.00
	1	53.33 ± 3.85	6.67	4.42 ± 0.27	0.47
	2	97.78 ± 2.22	3.85	37.18 ± 0.87	1.50
	3	**100.00 ± 0.00**	0.00	**41.18 ± 0.22**	0.38
	4	80.00 ± 3.85	6.67	15.76 ± 1.07	1.85

Note: **R** - reference; **1)** + 2 mg/l zeatin riboside; **2)** + 5 mg/l zeatin riboside; **3)** + 10 mg/l zeatin riboside; **4)** + 15 mg/l zeatin riboside.

developed second-order roots and root hairs. The rhizogenesis in this case ended a few days later than on the media with IBA, as the duration of this process was in an inverse ratio with the NAA concentration used. In the experiments with IBA the results were opposed of those obtained with NAA.

On the medium ontaining the combination of NAA and IBA (RB3) small callus tissue were developed and the roots had more lateral roots and root hairs than on the media with IBA.

On the hormone-free medium (RK) formation of callus was not noticed but the rhizogenesis period were longer. The roots were the thinnest, without second-order roots and without root hairs.

On all tested media the percentage of rooted shoots reached high value -from 89.29 ± 2.06% (RK) up to 98.81 ± 1.19% (RB and RB3). These results provide us to consider that rejuvenilization of 'dalecarlica' silver birch could be obtained through the method of adventitious bud formation in vitro.

3.2 CHROMOSOME ANALYSIS

Karyological analysis of in vitro propagated plants of 'dalecarlica', obtained through callus culture, showed stable diploid chromosome complements (2n = 2x = 28) in comparison with some others species (Gajdosova and Vookova, 1991; Libiakova and Gajdosova, 1993 etc.).

The results obtained showed that the mitotic activity of this form is lowest than other silver birch genotypes (Besendorfer *et al.*, 1989). The highest mitotic index (M.I. = 5.6) was observed in the root-tip cells of the in vitro obtained plants, and could be connected with the rapid growth of the regenerants in the multiplication phase as a result of the induced rejuvenilization. The lowest mitotic index of in vivo cloning plants (M.I. = 3.4) showed that deviation of cells decrease with the increasing of ontogenetical age which is in correlation with lowest percentage of the development and absence of nodal segments elongation in vitro.

3.3 PROTEIN EXTRACTION AND PROTEIN ELECTROPHORESIS

The pattern of peroxidase isozymes was not changed during the 9-day period inoculation to RB or RK medium although first roots appeared on 7th day.
An appearance of new peroxidase fractions was revealed in biochemical investigations of metabolic changes in the rhizogenesis process of some conifer species (Berthon *et al.*, 1987; Tzvetkova *et al.*, 1996 etc.). As a result of this it was indicated that they could be used as a marker of inductive and expressive phases of in vitro rooting.
The absence of new peroxidase isozymes in our study outlines differences between conoferous and broadleaved trees, regarding the rhizogenesis peculiarities.

4. References

1. Berthon, J., Boyer, N. and Gaspar, Th. (1987). Sequential rooting media and rooting capacity of Sequoiadendron giganteum in vitro. Peroxidase activity as a marker, *Plant Cell Reports*, **6**, 341-344.
2. Besendorfer, V., Kolevska-Pleticapic, B., Papes, D. and Krsnik-Rasol, M. (1989). Plant regeneration in the root callus culture of Betula pendula Roth., *Biol. Vestn.*, **37**, 13-22.
3. Gajdusova, A. and Vookova, B. 1991. Karyological study of Abies sp. callus culture, *Biologia (Bratislava)* **46**, 211-217.
4. Kajba, D. (1996). Possibilities of cloning for silver birch (Betula pendula Roth.) and black alder (Alnus glutinosa (L.) Gaertn.) adult trees, in Iv. Iliev, P.Zhelev and Pl. Aleksandrov (Eds.), *Propagation of decorative plants*, Second Scientific Conference October 5-7, Sofia, pp. 55-61.
5. Laemmli, U. K. (1970). Cleavage of structural proteins during assembly of the head of bacteriophage T4, *Nature*, **227**, 680-685.
6. Libiakova, G. and Gajdosova, A. (1993). Karyological analysis of the long-term cultivated Abies sp.calli, *Biologia (Bratislava)*, **48** (1), 93-94.
7. McCown, B. and Lloyd, G. (1981). Woody Plant Medium (WPM) - a mineral nutrient formation for microculture of woody plant species, *Hort. Science*, No **16**, 453.
8. Murashige, T. and Skoog, F. (1962). Arevised medium for rapid growth and bioassey with tobacco tissue cultures, *Physiol. Plant.*, **15**, 473-497.
9. Ondruska, R. and Schmidt, G. (1984). A nyirfa - (Betula pendula) fajtak szaporitasa zolddugvanyrol. *Kulonlenyomat a Kerteszeti Egyetem 1984. evi Kozlemenyeibol*, 139-144.
10. Peskan, T., Curkovic-Perica, M. and Krsnik-Rasol, M. (1996). Biochemical characterisation of house-radish tumour and teratoma tissues, *Plant Physiol. Biochem.*, **34** (3), 385-391.
11. Simola, L. (1985). Propagation of plantlets from leaf callus of Betula pendula f.purpurea, *Scientia Hortic.*, vol. **26**, 77-85.
12. Tzvetkova, N., Iliev, Iv. and Zhelev, P. (1995). Morphological and biochemical investigations of in vitro rhizogenesis of Sequoia sempervirens Endl., in *Recent advances in plant biotechnology*, Inst of Plant Genetics, SAS, Nitra, Slovak Republic, October 2-6, pp. 40-46.

AGROBACTERIUM-MEDIATED TRANSFORMATION OF PEROXIDASE GENE TO TISSUE CULTURE OF *POPULUS TREMULA*

Z. İPEKÇİ[1], K.BAJROVIC[1], T. OĞRAŞ[1], N. GÖZÜKIRMIZI[1], M. BOYDAK[2], T. TANK[2], T. AKALP[2], Ö. ÖZDEN[2], M. ÇALIKOĞLU[2], K. TUNÇTANIR[3], M. TULUKÇU[3], H. BALKAN[4] AND H. TANRIYAR[4]

[1]TÜBITAK, MRC, Research Institute for Genetic Engineering and Biotechnology, Gebze-Kocaeli, Turkey, [2]University of Istanbul, Faculty of Forestry, Istanbul, Turkey, [3]Poplar Research Institute, İzmit, Turkey, [4]Turkish Paper and Pulp Mills, İzmit, Turkey

1. Abstract

Tissue culture of *Populus tremula* Dursunbey 2 clone from stem explants was established. peroxidase gene isolated from *Stylosanthes humilis* (Shpx6), in antisense orientation was transferred to poplar via *Agrobacterium tumefaciens.* The effects of Shpx6 gene leads to modification of total peroxidase activity and total lignin amount. A reduction of total peroxidase activity for 50% and total lignin amont for 20% was observed in transgenic plants compared with control plants. The results obtained here, could be useful for utilization of poplar in paper and pulp industry.

2. Introduction

Poplars have a number of characteristics that make them well suited as a model system for biotechnologies due to their small genome size, fast growth rate, short rotation cycle and ability to be propagated vegetatively. In addition, poplar woods are used for many processed wood products such as paper, plywood and matchstick (Ahuja, 1984). Investigations on molecular biology on woody plants have a potential with its application to the breeding and improvement of trees, especially on wood biotechnology perspective (Boudet et al., 1996). Lignins are a major component of the cell of woody plants. They have an important role in adaptation of plants to terrestrial life by providing mechanical support to plant tissue, water-carrying xylem elements and allowing for novel defense strategies against pathogen attack (Imberty et al., 1985). However, lignin may have a negative impact on the utilization of the biomass. During paper pulping, for example, lignin must be eliminated from wood by chemical trearments that are expensive and polluting. Since almost all of the genes of the classical lignin biosynthesis pathway and some genes in alternative pathway have now been characterized, the modulation of lignin content and composition has become a realistic objective by genetic engineering approaches by antisense and sense suppression of gene expression (Whetten and Sederoff, 1991). Many enzymes are involved in lignification process. Peroxidases (EC 1.11.1.7) are known to act on the last step of lignin biosynthesis, that is the polymerization of substituted cinnamyl alcohols in the presence of hydrogen peroxide. They play an important role in several metabolic plant function of potential importance in plant-pathogen interaction, cross-linking of cell wall component, wound healing and auxin oxidation (Sato et al., 1995). Genetic manipulation of the peroxidase isoenzymes thought

to catalyze the oxidative polymerization of precursors, into lignin could result in changes in the rate of lignin deposition. An anionic peroxidase supposed to be involved in lignification was overproduced and down-regulated in tobacco (Lagrimini et al., 1987). Moreover, a few cationic isoenzymes of peroxidase bound ionically to cell walls were shown to be involved in lignification in Zinnia cells in contrast to the suggestion that anionic peroxidase catalyzes lignin synthesis in tobacco (Mäder et al., 1980).
Here we report the molecular effects of *Stylosanthes humilis* cationic antisense Shpx6 gene, that shown high homology with poplar perxidase, on the expression of peroxidase activity and effects on lignin content.

3. Materials and Methods

3.1. PLANT TİSSUE CULTURE AND TRANSFORMATION

A tissue culture system for regeneration of *Populus tremula,* Dursunbey 2 clone obtained from Poplar Research Institude, Turkey, from stem explants were established. Internodal segments from short branches were cultured on Woody Plant Medium (WPM) containing 1 mg/l zeatin. After two weeks shoots were rooted on WPM containing o.5 mg/l indol butyric acid (IBA) (Han et al., 1994). Two weeks after root formation plantlets with well developed roots transferred into sterile soil mixture of humus and sand.
For transformation stem explants were co-cultivated with *Agrobacterium tumefaciens* LBA 4404 strain carrying pGA643 plasmid according to Horsch et al. (1985). This plasmid contains Shpx6 gene under the control of CaMV35S promoter and NPTII gene under the NOS promoter. After co-cultivation explants were transferred to WPM medium containing 1 mg/l zeatin, 50 mg/l kanamycine, 500 mg/l cefotaxime. Cıltures were subcultured with 15 days intervals. Molecular anaysis were performed on two-month old transgenics.

3.2. DETERMINATION OF TRANSGENİC PLANTS

To detect the presence of transfer DNA in putative transformed plants Southern blot of NPTII gene were performed on plantlets grown on selective media.

3.3. DETERMINATION OF PEROXIDASE ACTIVITY AND LIGNIN CONTENT

Total peroxidase activity was screened spectrophotometrically at 470 nm by using guaiacol as subtrate. In addition gel electrophoresis was applied quantitative determination of peroxidases (Krsnik-Rasol, 1991). Lignin content was determined usin TGA method (Campbell and Ellis, 1992).

4. Results

Using *Populus tremula* as a model system we have generated antisense transgenic plants with altered peroxidase activity and lignin content. Transformation efficiency on selective media was 78% and 5-6 shoots were obtained from each explant. In another study, transformation of poplar leaf segments with a sense peroxidase gene was performed and transformation efficiency was 60%, but involving a callus phase and 3-4 shoots were obtained from each explant (Kajita et al., 1994). Southern analysis indicated the integration of transferred genes in the genome of poplar (Fig. 1).

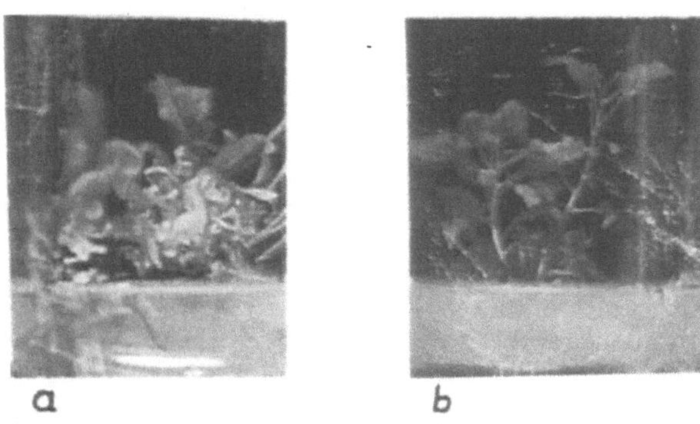

FIGURE 1. Transgenic plant containing antisense Shpx6 (a) and control plant (b)

Red to yellow color changes were observed in Shpx6 transgenics one month duration after transformation and the color disappeared and plants turn into normal development (Fig. 2).

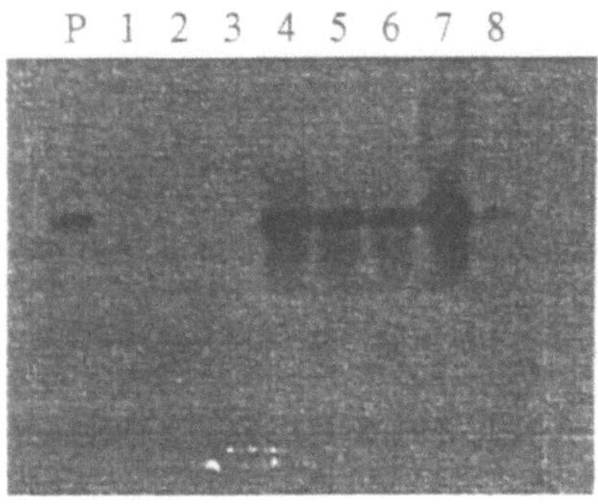

FIGURE 2. Southern analysis of Shpx6 transgenics and control plants; 1-3 controls; 4-8 transgenics

Red coloration of xylem is the typical clue of transgenic plants with reduced cinnamyl alcohol dehydrogenase (CAD) (Baucher et al., 1996). These color changes would be related with similar mechanisms in CAD activity, which plays a role in lignin biosynthesis. In transgenic plants with antisense Shpx6, peroxidase activity was down-regulated for 50% compared with controls. Peroxidase patterns screened with native gel electrophoresis, also confirmed this result. Lignin content was decrease for 20% in transgenic plants compared with controls without changes in lignin composition. Although transgenic tobacco lines have been obtained with significant overexpression (Kajita et al., 1994) and downexpression (Lagrimini et al., 1987) with anionic peroxidase, in our knowledge for the first time we report the down regulation of cationic peroxidase and decrease in lignin content without alteration in lignin composition. Similarly, Ni et al. (1995) obtained transgenic tobacco plants with reduced lignin content, but no changes in lignin composition. The results presented here indicated a significant reduction in peroxidases have a great importance for industrial application in paper and pulp industry.

5. Acknowledgements

This study was a part of the project No. 62-02-001 TÜBİTAK-MAM-GMBAE named *Investigation on biotechnologically developed poplar clones as a raw material for pulp and paper industry* anf financially supported by General Management ot Turkish Pulp and Paper Mills.

6. References

1. Ahuja, M.R. (1984) A Commercially Feasible Micropropagation Methods for Aspen, Silvae Genetica **33**, 174-176
2. Baucher, M., Chabbert, B., Pilate, G., Van Doorsselaere, J., Tollier, M.T., Petit-Conil, M., Cornu, D., Mantiea, B., Van Mantagu, M., Inze, D., Jouanin, L., Boerjan, W. (1996) Red Xylem and Higher Lignin Extractability by Down-Regulating a Cinnamyl Alcohol Dehydrogenase in Poplar, Plant Physiol. **112**, 1479-1490.
3. Boudet, A.M., Grima-Pettenati, J. (1996) Lignin Genetic Engineering. *Molecular Breeding* **2**, 25-39.
4. Campbell, M.M., Ellis, B.E. (1992) Elicited Phenylpropanoid Metabolism in Pine Cultures, Planta **186**, 409-417.
5. Han, K.H., Bradshaw, H.D., Gordon, M.D. (1994) Adventitious Root and Shoot Regeneration *In Vitro* is Under Major Gene Control in an F2 Family of Hybrid Poplar (*Populus trichocarpa x P. deltoides*). *Forest Genetics* **1** (3), 139-146.
6. Horsch, R.B., Fry, J.E., Hoffman, N.L., Eicholtz, D., Rogers, S.G., Fraley, T.A. (1985) A Simple and General Method for Transferring Genes into Plants, Science **227**, 1229-1231.
7. Imberty, A., Goldberg, R., Catesson, A.M. (1985). Isolation and Characterization of *Populus* Isoperoxidases Involved in the Last Step of Lignin Formation, Planta **164**, 221-226.
8. Kajita, S., Osakabe, K., Katayama, Y., Kawai, S., Matsumoto, Y., Hata, K., Morohoshi, N. (1994) *Agrobacterium*-Mediated Transformation of Poplar Using a Disarmed Binary Vector and The Overexpression of a Specific Member of a Family of Poplar Peroxidase Genes in Transgenic Poplar Cell. *Plant Science* **103**, 231-239.
9. Krsnik-Rasol, M. (1991) Peroxidase as a Developmental Marker in Plant Tissue Culture. *Int. J. Dev. Biol.* **35**, 259-263.
10. Lagrimini, L.M., Burkhary, W., Moyer, M., Rothstein,S. (1987) Molecular Cloning of Complementary DNA Encoding the Lignin-Forming Peroidase From Tobacco: Molecular Analysis and Tissue Specific Expression, Proc. Natl. Acad. Sci., **84**, 7542-7546.
11. Mäder, M., Ungemach, J., Schloss, P. (1980) The Role of Peroxidase Groups of *Nicotiana taboccum* in hydrogen Peroxide Formation, Planta **17**, 467-470.
12. Ni, W., Paiva, N.L., Dixon, R.A. (1994) Reduced Lignin in Transgenic Plants Containing a Caffeic Acid O-Methyltransferase Antisense Gene, Transgenic Research **3**, 120-126.
13. Sato, Y., Sugiyama, M., Komamine, A., Fukuda, H. (1995) Seperation and Characterization of the Isoenzymes of Wall-Bound Peroxidase from Cultured *Zinnia* Cells During Tracheary Element Differentiation, Planta **196**, 141-147.
14. Whetten, R.W., Sederoff, R.R. (1995) Lignin Biosynthesis. *The Plant Cell* **7**, 1001-1013.

TAILORING THE ONION CROP FOR THE 21ST CENTURY: DEVELOPMENT OF HIGH QUALITY FRESH AND PROCESSED ONIONS

E. KARAVOKIROU[1], D. TZANOUDAKIS[1] & C.KIK [2]
[1]*University of Patras, Department of Biology, Division of Plant Biology Patras, Greece.*
[2]*Centre for Plant Breeding and Reproduction Research (CPRO-DLO), Wageningen, the Netherlands.*

During the recent years there is a continuous increase in the demand of agricultural products of high quality, grown and processed with the minimum reliance on synthetic chemical inputs and according to methods harmless to the environment.

Onion, *Allium cepa*, can be classified among the horticultural crops with the largest use all over the world. Consequently, an amelioration of the quality of its fresh and dried forms, as a result of a deeper knowledge of the corresponding factors, is of crucial importance.

Therefore eleven research groups on different disciplines, from 5 European countries (see list 1), joined their efforts in a 5 year precompetitive project, granted by European Commission. The objectives of this project, regarding the analysis of the principal determinants of onion quality components (carbohydrates and flavours) are the following:

a) to evaluate the *Allium* germplasm to reveal those parts of the gene pool that harbour maximum diversity for flavour and carbohydrate content, b) to analyse flavour and carbohydrate pathways from a molecular, genetical, biochemical and a physiological point of view, c) to apply a new processing technology to onions, d) to apply new molecular markers and develop a high density genetic linkage map for onions to speed up breeding of better varieties (fig.1).

Among the main components, which are closely related to several parameters regulating onion quality (bulb development, storage ability and sprouting) in fresh and processed form, are carbohydrates. Bulb formation is characterised by an accumulation of carbohydrates (storing fructans) and is influenced by environmental factors (length of photoperiod and light spectrum). In addition to this, sucrose and other carbohydrates regulate gene expression during bulb initiation. Storage ability is related to a prolonged dormancy period. Its end is signaled by sprouting, which is characterised by changes in fructans (the main storage carbohydrates) sucrose and reducing sugars. The current use of maleic hydrazide (which has carcinogenic effects) to prolong dormancy has to be substituted by better methods based on the carbohydrate metabolism. In addition to this enzymes associated with carbohydrates synthesis are going to be analyzed as regards to their enzymatic activity (sucrose synthase, sucrose phosphate synthase, invertase, sucrose- sucrose fructosyltransferase [SST], fructan-fructan fructosyltransferase [FFT]). Further more, molecular marker linkage analysis of the genes involved in the carbohydrate pathway will be carried out using a specific cross between onion cultivars which differ in the percentage of dry matter and the amount of reducing sugars.

A better quality of onion processed forms depends on carbohydrates too. More specificly, it is based on the diminution of the unwanted browning effect (Maillard reaction of onion reduced sugars) which could be achieved by a production of cultivars with a high dry matter content (higher amount of fructans and lower amount of reducing sugars).

The principal characteristics of *Allium* species and the main reason for their vast therapeutic and dietary use are their special odour and flavour. It is due to emission of volatile sulphur compounds, released from their non volatile precursors (S-alk(en)yl - L- cysteine sulphoxides) through an enzyme mediated degradation (the key enzymes being γ-glutamyl transpeptidase and alliinase) when their tissue is disrupted. Among the four different S-alkenyl cysteine sulphoxides (methyl-, propyl-, propenyl-, and allyl-) existing in genus *Allium,* propenyl cystein sulfoxide is the principal component of onion *(Allium cepa).* It is also the source of the lacrymatory factor, which is directly related to onion pungency, a crucial determinant of its quality. High pungency favours onion processing, while a low one is important for the fresh market. In this project all the factors related to pungency (alkenyl cysteine sulfoxides and their derivatives thiosulfinates and pyrouvate) will be studied according to several techniques. Also genetic material responsible for high and low flavour will be established from which segregating populations for mapping studies will be set up. In addition to this, further studies will take place related to the enzymology of flavour generation, and to the genes coding for the key enzymes (allinase and γ -glutamyl transpeptidase) which will be cloned and compared to the molecular markers for pungency characteristics.

A deeper knowledge regarding the genetical characteristics of *Allium cepa* and especially those dealing with its quality is still lacking. The main cause for this is the long generation time and the strong inbreeding depression they suffer. Molecular markers have proved to be more helpful in plant breeding programs compared to morphological, cytogenetical and biochemical ones. In this project two different sets of molecular markers will be studied: the AFLP (Amplified Fragment Length Polymorphism) and the STMS (Sequence Tagged Microsatellite). A high density linkage map of onion will be produced on the basis of these two marker types and key genes involved in dry matter content, reducing sugar composition and pungency mapped. Furthermore the linkage map will be physically linked to individual chromosomes using monosomic addition lines.

Referring to the processing, drying (dehydration) of crops is one of the most common ways of preserving the product from enzymatic and microbial degradation. In order to overcome the disadvantages of the standard procedures which imply thermal treatment (loss of aroma and nutritional value, non enzymatic browning, modification of the product texture, high energy consumption), new methods based on combinations of microwaves and electromagnetic energy will be studied. The principal objectives will be the homogeneous distribution of heat and humidity in the interior of the product and the deeper knowledge of the dielectric properties (permittivity, loss factor) and the thermal ones (thermal capacity and thermal conductivity) of onions and consequently the optimization of onion processed forms.

During the first year, preliminary studies and optimization of techniques were carried out. A core collection was established, after an initial flavour and carbohydrate screening. A general screening for dry matter content and non-structural carbohydrates was performed on 103 accessions. It was found the correlation between dry matter content and the relative amount of fructans is a positive one, while its correlation with the relative amounts of fructose and reducing sugars is negative one. Also a screening for genes involved in the carbohydrate metabolism (invertase, sucrose synthase, sucrose-sucrose fructosyltransferase [SST], fructan-fructan fructosyltransferase [FFT]) was carried out. In addition to this the accessions to be used as parents in crossing for gene mapping were chosen. After a refinement of the analytical methods, to be used in the flavour analysis, the pyruvate content, S-alk(en)yl cysteine sulphoxides composition, and the primary volatiles (HPLC and UV detection) of 90 onion accessions and 10 wild species were screened. Apart from the pungency (pyruvate content, lacrymatory factor and its degradation products), their «aroma index» consisting in the relationship between pyrouvate and aroma was analysed.

In the field of molecular markers and linkage map the AFLP (Amplified Fragment Length Polymorphism) method was optimised for *Allium* and the STMS (Sequence Tagged Microsatellite) markers were developed. A complete set of monosomic addition lines, based on crosses between *A.cepa* and *A.fistulosum*, has started to be developed. Concerning the cp DNA and crossability project, variable DNA sequences have been determined and tested. Three universal primer sets (2 for amplification of two intergenic spacers and one for amplification of an exon of the gene family *trn*) for analysis were tested.

As regards to the processing research, four categories of physical scales (structure and composition, thermal, scales featuring the balance between the product water and its environment balance) were measured. Although it is still the beginning of the research it can be stated that the results obtained up to the moment favour drying of onion by microwaves.

As a conclusion, taking under consideration that the objectives and millestones of the first year were achieved and that the communication and corporation between all participants has been successful, we are able to predict that, by the end of the 5 years the aims of this project will be completed successfully. The deeper knowledge regarding flavour and carbohydrate metabolism together with the genome organisation will lead to the production new cultivars of better quality. In addition to this, new technologies are going to be developed having as a consequence wider scientific, technical, environmental and economical benefits.

ACKNOWLEDGEMENTS

We gratefully acknowledge the financial support of the EU FAIR program to this project (CT95-465).

NOTE

For more information (http://www.symmetrix.ch/Public/mdp/onion.html)

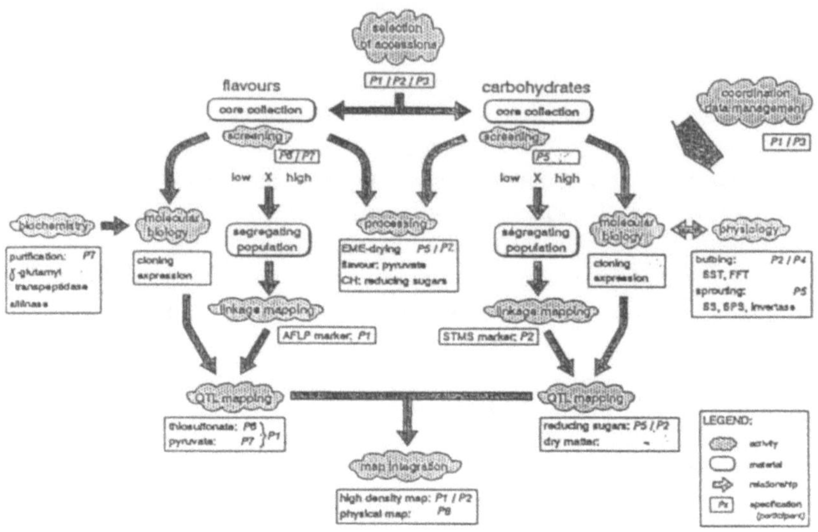

Fig 1. Activities and relationships of the different parts of the project

LIST OF PARTICIPANTS OF THE "ONION QUALITY" PROJECT

- Centre for Plant Breeding and Reproduction Research (CPRO-DLO), Wageningen, the Netherlands:
  dr. C.Kik, coordinator, dr. L.W.D. van Raamsdonk, ir. I.W.Boukema, dr. A.W.van Heusden, dr. J.W.van Ooijen.
- Institute of Plant Genetics and Crop Plant Research (IPK), Gatersleben, Germany:
  prof.dr. K.Bachmann, dr. D. Fischer, dr. J.Keller.
- Horticultural Research International (HRI), Wellesbourne, UK:
  dr. B.Thomas, dr. L. Trueman, dr. D.Astley.
- COOPD'OR - INRA, Dijon France:
  dr.R.Kahane.
- Institute Superieur de Agriculture Beauvais (ISAB), Beauvais and Cergy, France:
  dr. J.C.Laguerre, ir.E.Vialle-Guerin.
- University of Patras, Patras, Greece:
  dr. D.Tzanoudakis, dr. E.Karavokirou .
- Universite Paris XI, Paris and Universite F.Rabelais (IBEAS), Tours; France
  dr J.Boscher, dr. J.Auger, dr. A.Vannereau.
- Universite Paris Sud, Paris, France:
  dr A.Ricroch
- Van der Have, the Netherlands
  ir J Hopmans
- Nickerson Zwaan, the Netherlands:
  ir J L Harrewijn
- S&G Seeds, the Netherlands:
  ir P.Boorsma.

IDENTIFICATION OF GENETIC POLYMORPHISM OF POPLAR *(Populus sp.)* BY DNA MARKER TECHNOLOGY

N. KÖRÜKMEZ[1] , A. ZEHİR[2] and N. GÖZÜKIRMIZI[3,4]

[1]Department of Biology, Gebze Institute of Technology, Gebze-Kocaeli, TURKEY
[2]University of Marmara, Faculty of Science and Art, Department of Biology, Göztepe-İstanbul, TURKEY
[3]University of İstanbul, Faculty of Science, Vezneciler-İstanbul, TURKEY
[4]TÜBİTAK, MRC, Research Institute for Genetic Engineering and Biotechnology, Gebze-Kocaeli, TURKEY

1. Abstract

In order to identify new Populus lines, STS (Sequence Tagged Site) and RAPD (Random Amplified Polymorphic DNA) marker analysis were performed on eight *Populus deltoides x Populus deltoides* hybrids and *Populus deltoides (77/51)* which was used as control. For RAPD analysis STS primers were used in single forms. Four of the five STS primers and seven of ten RAPD markers produced amplified PCR (Polymerase Chain Reaction) bands. Polymorphism values were found to be between 7-70%. As a result, it is concluded that STS marker analysis, just like RAPD can also be utilized in selecting and identifying the high-quality poplar lines for paper and pulp industry.

2. Introduction

Populus is the paradigm of genetic linkage studies in forest trees. The ease of vegetative propagation of individual *Populus* in green houses has contributed to simpler genetic manipulations (Burr *et al.*, 1988; Reiter *et al.*, 1992). The chromosome number of all Populus specials are identical (2n=38), and the nucler genome is relatively small (Dhillon 1987; Bradshaw and Stettler 1993). In this study, in addition to RAPDs, STS markers which are combination of PCR, RFLP and RAPDs were also used to decected Polymorphism in poplar trees. (Bradshaw *et al.*, 1994). Recentely, DNA based marker systems have gained wide application for genetic linkage mapping, identification of polymorphism, pedigree analysis, and the determination of family relationship, in plant genetics (Villar *et al.*, 1996). RAPD marker has several advantages of most importantly being rapid and inexpensive (Williams *et al.*, 1990; Keselli *et al.*, 1992). In this article the polymorphism between eight hybrids and the control of *P. deltoides* using RAPD and STS markers are reported.

3. Materials and Methods

3.1. PLANT MATERIAL

Leaves of hybrids (*P.deltoidesx P.deltoides*) and control *P.deltoides* (77/51) were obtained from İzmit Poplar Research Institute, TURKEY.

3.2 AMPLIFICATION CONDITIONS

The genomic DNAs were extracted from 8 hybrids and from *P.deltoides* as a control according to methods used by Walbot (1988). Single forms of primers for RAPD and double forms of primers for STS markers were given in table 2. Amplification reaction were performed in a total volume of 25 μl reaction containing 100 ng genomic DNA, 100 μM dATP, dTTP, dCTP, dGTP mixture, 50 ng forward primer, 50 ng reverse primer, and 1 unit Taq DNA polymerase in 1xPCR buffer (10 mM Tris-HCl, pH 8.3, 50 mM KCl, 2 mM $MgCl_2$, gelatine mg/ml). In the program used for STS marker, amplification was performed in a thermocycler, which was programmed for the first 4 cycles of 1.5 min 94°C, 1.5 at 55°C, and 2.5 min at 72°C, for 30 cycles of 1 min at 94°C, 1 min at 55°C, and 1.5 min at 72°C. Amplification followed a 10 min extention period at 72°C. The amplification for RAPD was performed as followed. For first 4 cycles of 1.5 min at 94°C, 1.5 min at 35°C, and 2 min at 70°C, the next for 30 cycles of 1 min at 94°C, 1 min at 35°C, and 1.5 min at 70°C. A 10 min extention period at 70°C amplification followed. The amplification products were analysed by on 1.2 % agarose gel, photographed.

TABLE 1. Primers used in this sudy.

Locus	Forward primer	Reverse primer
P781	TGCAGGACTAGCATTGGAACTC	CTGGGAGGCAACTATCCAAAG
P856	GAAAAAAGTTTGCCCAGCAT	CATGCTTCATTTGATGCAATTG
P1010	GATCTCCTCCAACACCAGAAAA	CTGCAGGTTCTGTTGTCCCG
P1054	CTACTGCCGCTTCCGAATTAAG	TGCAGTTGGAAACACTTACTCG
P1079	CTGCAGTGGCTTTCACCTTC	TGCAGATTGTACAGGGCTGG

4. Results and Discussion

After the DNA isolation, and analysis of DNA was performed according to STS and RAPD techniques and the data from these two techniques were compared.The amplifications were obtained with five primers specific to P781, P856, P1010, P1054 and P1079 loci. However, there was no amplification in the P1010 locus. This would indicate the absence of corrosponding loci in our tested material for the respective primers used (Waugh *et al.,* 1992). The polymorphic bands were obtained from the primers for P781 and P856 loci. The dense and homolog bands were obtained with the primers for P1054 (*Figure 1*) and P1079 loci. Bradshaw *et al.* (1994) suggested that these bands could be the mixture of similar DNA sequences.

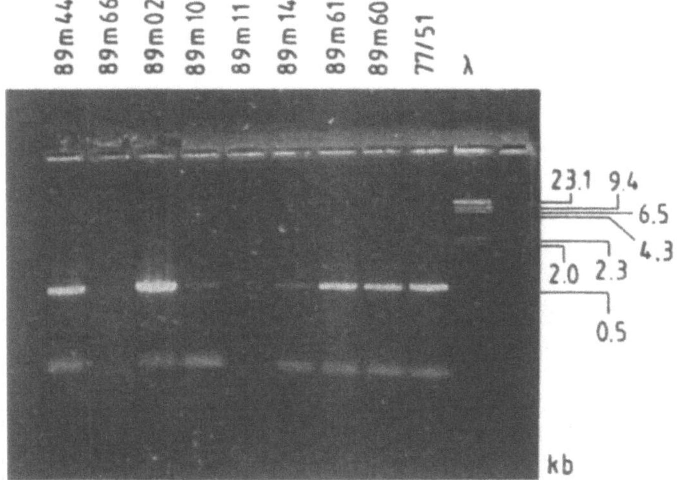

Figure 1. STS analysis using (5' CTACTGCCGCTTCCGAATTAAG 3') and (3' TGCACTTGGAAACACTTACTCG 5') primers with DNA samples from 8 hybrids *P.deltoides* and 1 control λ DNA marker digested with Hind III.

In the RAPD techniques, ten primers were used for the amplification of the genomic DNA. With the use of the same primers in the technique, more bands were obtained. This could result from the specific priming of the STS primer pairs to the DNA sequences, contarary to non-specific priming of RAPD single primers to template DNA. In addition seven of RAPD some monomorphic and somewhat less number of polymorphic bands were obtained between 2.3 kb and 0.5 kb base pairs (*Figure 2*). Amongst the varieties studied, the polymorphism value was shown to be between 7-70% (*Table 2*). In conclusion more definitive evolutionary studies would require the use of larger numbers of specific primers and samples.

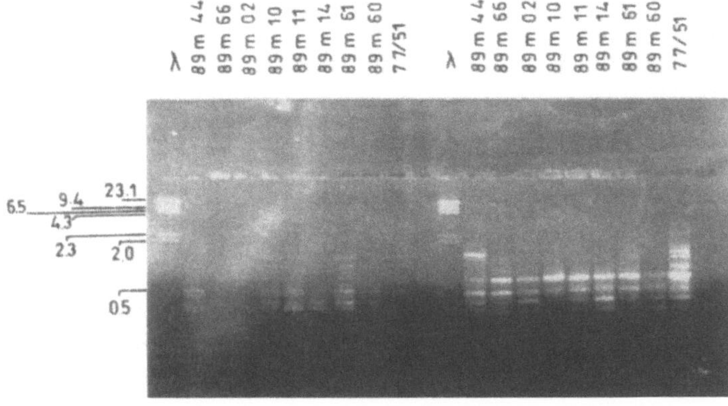

Figure 2. RAPD analysis using (5'CTACTGCCGCTTCCGAATTAAG 3') (on left) and (3' TGCACTTGGAAACACTTACTCG 5') (on rigth) primers with DNA samples from 8 hybrids *P.deltoides* and 1 control λ DNA marker digested with Hind III.

TABLE 2. Polymorphism were estimated by using RAPD markers among 8 *(P.deltoidesx P.deltoides*) hybrids and one control *Populus deltoides* (%).

	Materials	1	2	3	4	5	6	7	8	9
1	K(77/51)	-								
2	89M60	70	-							
3	89M61	70	70	-						
4	89M14	46	57	59	-					
5	89M02	29	33	26	24	-				
6	89M11	36	31	46	34	27	-			
7	89M10	31	42	39	34	32	24	-		
8	89M66	34	46	47	36	19	16	27	-	
9	89M44	29	36	37	32	19	36	21	7	-

5. References

Bradshaw Jr, H.D., Villar, M., Watson, B.D., Otto, K.G., Stewart, S., and Stettler, R.F. (1994) Molecular genetics of growt and development in P*opulus*. II. Agenetic linkage map of a, hybrid poplar composed of RFLP, STS and markers *Theor. Appl. Genet.* **89,** 167-178.

Bradshaw, H.D., Stettler, R.F. (1993) Molecular genetics of growth and development in *Populus*. I.Triploidy in hybrid poplars. *Theor Appl Genet* **86,** 310-307.

Burr, B., Burr, F.A., Thompson, K.H., Albertson, M.C., and Stuber, C.W. (1988) Gene mapping with recombinant inbreds in maize, *Genetics* **118,** 519-526.

Dhillon, S.S., DNA in tree species. In: Bonga, J.M., Durzan, D., (eds) (1997) Cell and tissue culturel in foresty, Martinus Nijhoff Pub, Dordrecht, *The Netherlands*, **1,** 298-313.

Kesselli, R. V., Paran, I., Michelmore, R. W. (1992) Efficient mapping of specially targeted and genomic regions and the tagging of these regions with reliable PCR-based genetic markers, *Crop Sci. Soc. Amer.-Amer. Soc. Hort. Sci.-Amer. Genet. Assn. Joint Plant Breeding Symp. Ser., Minneapolis. Crop. Sci. Soc. Amer., Madison, Wis* p. 31-36.

Reiter, R.S., Williams, J.G.K., Feldmann, K.A., and Rafalski, J.A . (1992) Global and local genome mapping in *Arabidopsis thaliana* by using recombinant inbred lines and random amplified polymorphic DNAs *Proc. Natl. Acad. Sci USA* **89,** 1477-1481.

Villar, M., Lefevre, F., Bradshaw, H.D. Jr. and Cros, E. (1996) Molecular genetics of rust resistance in poplars, *Genetics* **143 (1),** 531-536.

Walbot, V. (1988) Preparation of DNA single rice seedlings, *Rice Genet.* **5,** 149-151.

Waugh, R., and Powell, W. (1992) Using RAPD markers for Crop Improvement, *Tibtech* **10,** 186-191.

Williams, J.G.K., Kubelik, A. R., Kenneth, J.L., Rafalski, J.A. and Tinger, S. V. (1990) DNA Polymorphisms amplified by arbitrary primers are useful as genetic markers. *Nucleic Acids Research,* **18(229),** 6531-6535.

IN VITRO CULTURES OF LEGUMINOUS SPECIES

G.P. KOSTURKOVA, A.D. MEHANDJIEV
Institute of Genetics, Bulgarian Academy of Sciences,
1113 Sofia, Bulgaria

1. Introduction

Leguminous species are the third largest plant group displaying various possibilities of use, particularly in human diet and animal feeding. Grain legumes were research object of many famous scientists in genetics, breeding and biometrics [1]. These plants offer a great number of benefits but they also cause a lot of problems. Grain legumes are not outstanding yielders and are easily attacked by pests and weeds. These were some of the reasons for their decreased production in Europe for the last 40 years [2]. However, new technologies and efficiency in research could contribute to overcome many of the problems handling legumes.

The establishment and application of biotechnology and genetic engineering methods are still up-to-date approaches in modern biology. However, their success varies for the species. *In vitro* protocols are routine for many plants, especially those belonging to *Solanaceae* but for others as cereals and grain legumes still difficulties exist [3]. Although tissue, cell and protoplast cultures have been established for a great number of leguminous species (alfalfa, clover, pea, soybean, lentils etc.) it is still essential to elaborate *in vitro* protocols for new genotypes especially those of particular interest.

For Bulgaria the most important legumes are common beans, pea, soybean and alfalfa. The last three will be object of this report revealing the potentials of *in vitro* cultures of these species.

2. Results and Discussions

2.1. PROTOPLAST CULTURES OF ALFALFA

Medicago species have become target for *in vitro* development almost from the beginning of these techniques' establishment. Alfalfa became a suitable model both, to study forage legumes and to elaborate new protocols in genetic manipulations *in vitro*. Reliable and easy to handle procedures are prerequisites for more sophisticated experiments and protoplast cultures of alfalfa could be a good example for that. They meet many requirements like high yield, viability, division and plating efficiency.

Protoplasts of *M. varia* and *M. sativa* were isolated from cell suspension, mesophyll tissue and cotyledons [4]. Adapting the source material and optimizing enzyme digestion of the cell wall, it was possible to release millions ($1\text{-}2.10^6$ prts/g suspension cells, $2\text{-}6.10^4$ prts/cotyledon pair and 100% release of mesophyll protoplasts) of viable (90%-95%) protoplasts, necessary conditions for performance of large scale experiments. Alfalfa protoplasts divide (60%-80%) and develop in richer or poorer liquid, semisolid or agarose bead culture media [4, 5] However, the medium viscosity, its composition, phytoregulators' supplement and protoplast density are in close relation and altering one of these parameters cell division and formation of colonies can be influenced.

Alfalfa is interesting because regeneration of plants is possible via both ways: somatic embryogenesis or/and organogenesis. From another hand, in *Medicago* protoplast cultures direct embryogenesis and indirect embryogenesis are observed. Alternating culture conditions these processes can be regulated and directed [6] what enables us to choose the appropriate culture and gives possibility to study the mechanisms of these processes and the factors influencing them. For example, electroporation [5] was a factor that caused changes in cell division pattern of cotyledon protoplasts. Formation of unequally sized daughter cells during mitosis was common, especially when higher voltage (750 V/cm and 1000 V/cm) and longer pulses (1.5-3 ms) were applied. Small cells (about 10 μm in size) appeared as a result of unequal division, giving rise to compact colonies and further indirect embryogenesis was observed. When 500 V/cm were applied for 0.75 ms there was 15 % increase of production of embryogenic calli and the number of embryos formed per callus. Electrotreatment not only stimulated but sustained embryogenesis, too, keeping production of embryos up to 30 days after it has ceased in the control cultures.

Having appropriate protocols it was possible to study γ-rays irradiation, selection for resistance to antibiotics and fuzaric acid [4] and genetic transformation by direct DNA uptake [5, 7]. It is interesting to notice that an excellent procedure allowed extremely high transient gene expression (99% acetylation of chloramphenicol) following protoplast electroporation [5].

2.2. *IN VITRO* CULTURES OF PEA

Pea is an excellent and well-developed model plant for investigations in many areas of plant science. However, from *in vitro* aspect it is not so elaborated as some other model plants like tobacco, tomato, petunia and maize are. Induced mutagenesis and hybridization were the main methods used in grain legumes breeding in Bulgaria. A series of pea genotypes with valuable agronomic traits (such as higher protein and lysine content, higher yield, shorter vegetative growth period, morphological differences in grain color, leaf shape, plant height etc) have been created through experimental mutagenesis. However, traditional plant breeding techniques have limitations concerning selection for resistance to some biotic and abiotic factors. These problems can be overcome using *in vitro* approach when efficient methods for plant regeneration are available.

Ten genotypes of *P. sativum* and *P. arvense* were tested for their regeneration abilities *in vitro*. To allow more possibilities to the genotypes to express their morphogenetic potentials embryonic axes from immature embryos were cultured applying two different schemes for induction of organogenesis - through direct bud formation or indirectly through callus stage [8]. Most of the genotypes had very good organogenesis response through both pathways (Table 1) indicating that they can be used *in vitro* for their further improvement and for genetic studies, too.

Table 1. *In vitro* response of embryonic axes cultured on callus induction medium and transferred to organogenesis induction media MSPB or cultured directly for adventitious bud formation on MSPE medium

Genotype/ variety	Organogenesis on MSPB[1] [% ± SE]	Mean number of shoots on MSPB[1] ± SE	Shoot formation on MSPE[2] medium [% ± SE]	Mean number of shoots on MSPE[2] medium ± SE
192 cv Gena	59 ± 6.7	2.5 ± 0.1	100 ± 1.6	3.0 ± 0.1
268 Sofia 135	86 ± 4.8	3.3 ± 0.2	100 ± 1.2	2.7 ± 0.2
281 Kubrat-3	87 ± 3.8	3.1 ± 0.2	45 ± 6.5	1.0 ± 0.1
275 Kubrat-3	78 ± 5.7	3.0 ± 0.2	100 ± 1.4	3.2 ± 0.1
120 Kubrat-3	92 ± 2.3	3.7 ± 03'	100 ± 1.7	3.7 ± 0.3'
65 cv Auralia	100 ± 2.0	2.9 ± 0.1	97 ± 1.8	2.0 ± 0.1
G-19	55 ± 7.1	3.8 ± 0.2	100 ± 1.8	4.1 ± 0.2
G-14	57 ± 7.2	3.3 ± 0.2	58 ± 7.8	1.9 ± 0.1
2-594 cv Ina	100 ± 2.1	2.6 ± 0.1	n.r	n.r
cv Ran-1	50 ± 5.5	3.0 ± 0.3	100 ± 1.1	3.0 ± 0.3

[1] MSP medium containing 5μM BA; [2] MSP +10 μM BA + 1 μM NAA

In vitro cultures of pea are successfully used to study the effect of a novel substances. Various concentrations (from 10 μM to 100 μM) of o-allylthioureidobenzoic acid (ATB) stimulated specifically adventitious bud formation, shoot growth and rhyzogenesis. ATB radioprotective role was proved by increasing up to 100 % viability of γ-rays irradiated embryos. Higher number of adventitious buds and enhancement of the growth was observed in irradiated explants after ATB supplement to the culture medium (data unpublished).

Pea *in vitro* protocols are used for selection for herbicide and disease resistance. Plants were regenerated after including in the medium the herbicide sencor in sublethal and lethal concentrations of 100 mg/l and 200 mg/l. Fungal culture filtrates of *Ascochyta pisi* (in concentrations of 10%-40% v/v) were inoculated in nutritive medium for pea development to select for resistance to ascochyta blight. Several regenerants developed at lethal concentrations of 30% and 40% v/v (data unpublished).

2.3. CALLUS CULTURES OF SOYBEAN

Soybean could be considered a reculcitrant species because of the great difficulties of its development *in vitro*. Although somatic embryogenesis and organogenesis were reported more than ten years ago success in genetic manipulations was achieved recently [9].

Establishment of *in vitro* protocols for Bulgarian genotypes is essential for plant breeding programs because the production of soybean in Bulgaria has decreased nearly ten times for the last ten years. Poor disease and drought resistance of local varieties are great problem which could be overcome with the help of *in vitro* techniquesG.P.

In vitro cultures of three genotypes (Boryana, Biser and Merit) were initiated from immature embryos following the protocol of Barwalle et al [10]. Within six weeks 50% to 75% of the explants formed callus on EB medium. Transferred to MSR medium organogenesis occur in 76% to 92% of the calli. Plant development was stimulated in 15% to 50% on R5 regeneration medium and roots were formed by 10-15 % of the plants. These preliminary results indicate that some Bulgarian genotypes can be included in *in vitro* programs for soybean improvement and genetic studies.

3. Conclusions

Establishment of efficient *in vitro* protocols is multiple factor dependent, and the more factors are checked the more closer is the success. Once reliable procedures are available, they can be applied to various kinds of investigations giving possibility for broader and deeper research. However, it seems that *in vitro* techniques have been applied, so far, in agriculture mainly, but they have tremendous potentials in protection of rare and endangerous species and in environmental protection concerning problems like pollution, ozone layer effects, nitrogen fixation, etc. Probably the next century *in vitro* techniques will focus there.

Acknowledgements: Research was supported by Project SS-540 of the Bulg. Ministry of Science and Education, an IAEA grant, and "CABINET" INCO-COPERNICUS project.

4. References

1. Cubero, J.I. Grain legume breeding: research prospects. *Proceedings of First European Conference on Grain Legumes. Angers-France* (1992), 25-28
2. Carrouee, B. An encreasing EU deficit in protein rich materials. *Grain Legumes* **15** (1997), 22-23.
3. Jacobsen, H.J. Biotechnology applied to grain legumes - current state and approache. *Proceedings of First European Conference on Grain Legumes. Angers-France* (1992), 99-103.
4. Kosturkova, G.P. Application of alfalfa protoplast cultures in genetic studies. *Biotechnol.&Biotechnol. Eq.* **7**, 2 (1993), 40-42.
5. Kosturkova, G.P. Expression of foreign genes following electroporation of Medicago protoplasts. *Biotechnol.&Biotechnol. Eq.* **7**, 2 (1993), 43-46.
6. Antonova-Kosturkova G.P. Embryogenesis in mesophyll protoplast cultures of alfalfa. *Proceedings of IV Int. Conference of Genetics*, Varna, Sept (1989), 376-380.
7. Jones, B., Antonova-Kosturkova G.P., Viera M.L.C., Rech E., Power J.B. and Davey M.R. High transient gene expression , with reserved viability, in electropotated protoplasts of *Glycine*, *Medicago* and *Stylosanthese* species. *Plant Tissue Culture*, **3**, 2 (1993), 59-65.
8. Kosturkova, G.P., Mehandjiev, A., Dobreva, I., and Tzvetkova, V. Regeneration systems from immature embryos of Bulgarian pea genotypes. *Plant Cell, Tissue and Organ Cult.*, **48** (1997), 139-142.
9. Trick H.N., Dinkins R.D., Santarem, E.R., Rong Di, Walker D.R., Parrot W.A., Finer J.J., Collins G.B. Recent Advances in soybean transformation. *Plant Tissue Culture and Biotechnology*, 3,1 (1997), 9-26.
10. Barwale, U.B., Kerns H.R., Widholm, J.M. Plant regeneration from callus cultures of several soybean genotypes via embryogenesis or organogenesis. *Planta* **167** (1986), 473-481.

PROPAGATION BY CUTTINGS OF THE SPECIES *ILEX AQUIFOLIUM* L.[1]

D.E. MINOGLOU and K.P. PANETSOS
Laboratory of Forest Genetics and Plant Breeding, School of Forestry and Natural Environment, Aristotle University of Thessaloniki, 54006 Thessaloniki, Greece.

1. Introduction

Ilex aquifolium L. grows in most of Europe and is native to Greece. Recently, it has a high economic value as decorative plant at Christmas time. Propagation by seeds is problematic because of the delayed germination, the slow subsequent growth of the new seedlings and their late fruition. Contrary, vegetative propagation remains the most efficient method of reproduction, though without the adequate knowledge for its operational use. In this investigation the purpose was to propagate the species by cuttings and to determine factors that affect rooting.

2. Materials and Methods

2.1. WINTER EXPERIMENT

Branches were collected from plants growing in natural stands on Mt. Chortiatis, (1100m, Thessaloniki). Five plants were the donor plants, (Table 1). Collection was done in winter (December 6), during morning from the middle of the crown. Cuttings were put in a greenhouse on a rooting bench where the ambient (day and night) temperatures were between 15° and 20°C, and around 20°C, respectively. Additional light was supplied from 6.00pm to 6.00am with an interruption from 12.00 to 3.00am and mist system, was working from sunrise to sunset, every 10-12 minutes for 2-3 seconds.

Table 1. Sex and growing environment of the genotypes.

G_2	unidentified	Northern exposure – Shade
G_3	male	North eastern exposure – Semi-shade
G_4	female	Eastern exposure – Open space
G_5	male	Eastern exposure – Open space
G_6	female	North eastern exposure – Semi-shade

Mean temperatures were: in the medium 22°C, under the polyfilm 22.2°C with maximum 24.3°C and minimum 14°C. The rooting medium was sphagnum peat and perlite 1:1 per vol. Hormone, K-IBA, 2500ppm dissolved in water, was used. Wounding downwards to the basal 1-2 cm proceeded the hormone treatment. The basal 0.5-1 cm of each cutting was dipped in for 5sec and swirled around in a fungicidal powder. Final preparation included the stripping of some leaves (3 or 4 leaves were left) and of flowering buds existing on the genotypes G_3, G_5 and G_6. In this experiment it was examined how factors such as: genotype, hormone (2500ppm), wound and the number of internodes (one or two for all genotypes and only two for G_3) affect rooting. On this respect the percentages of rooted

[1] This is a part of the work submitted by D. Minoglou as partial fulfillment of the MSc degree, to the Department of Renewable Natural Resources, Mediterranean Agronomic Institute, Chania, Greece.

cuttings, the number and the length of roots were recorded. Any cutting that had at least one root being 2 mm was considered rooted. Afterwards, the height of new plants was also recorded. The experiment was run as a 3x2x2x2 factorial experiment involving genotype, internodes included, hormone and wound treatments, each one in two levels except for genotypes that were in three levels. Here the treatments were the combinations of the factors. For each treatment eight cuttings were included.

2.2. SPRING EXPERIMENT

To assess the influence of season to the rooting, cuttings of genotypes G_2, G_5 and G_6 with one and two internodes, treated with 2500ppm, were used in spring (May 7). Hormone and wound were applied to all cuttings since their behavior to non-hormone and non-wound treatments was known from the winter experiment. Ten cuttings from each genotype were included. The same greenhouse was painted with lime for shading and for temperature control the roof was opened and closed automatically according to the ambient thermostats. Mean temperatures were as follows: in the medium 21.75°C the ambient 25.5°C with mean maximum 33.9°C and mean minimum 20°C. Mist system was working from 6.00am to 8pm spraying every 5 min for 1.5 sec. The rooting percentage, number and length of roots were taken down.

In both experiments split-plot design was applied. Analysis of variance was performed for the rooting percentage, number of roots, length of roots and height of new plants (in the winter experiment). Analysis of variance for root number and root length was based on the number of rooted cuttings in the different treatments. Significant differences (p=0.05), obtained by the analysis of variance were subjected to Duncan's test.

3. Results

3.1. WINTER EXPERIMENT

Rooting: 2 months after insertion the results on rooting, number and length of roots were based on genotypes G_3, G_5 and G_6 since the other two genotypes failed to root.

Table 2. Rooting %, and height of new plants in relation to genotype, number of internodes and treatments.

		TREATMENTS OF CUTTINGS									
		NW-NH (Control)		NW-H		W-NH		W-H		TOTAL	
	IN	%	MH	%	MH	%	MH	%	MH	%	MH
G_3	2	12.5	10.5	37.5	14.5	12.5	-	100	11.37	40.6a	12.05
G_5	1	0	7	0	9.5	0	-	75	24.16	20.3b	15.25
	2	0	-	37.5	23	0	7	50	15.75		
G_6	1	0	-	62.5	16.3	0	10	87.5	22.85	43.75a	21.74
	2	37.5	18.5	37.5	-	37.5	22.66	87.5	24.14		
TOTAL		10	11.9	10	13.2	35	15.83	80	19.65		

W: Wound, H: Hormone (2500ppm), NW: No Wound (control), NH: No Hormone (control)
IN: Number of internodes, %: Rooting percentage, MH: Mean Height (cm)

Genotype, hormone and number of internodes, proved to be highly significant for the appearance of roots. Thus, genotype G_6 in average had the highest rooting percentage. Cuttings with two internodes were significantly better than cuttings with one internode (Table 2). It is noticeable that all the cuttings of genotype G_3 with two internodes gave roots after hormone and wound treatments.

Height of new plants: The height of new plants was measured in August, six months after transplanting. It was found that genotype and number of internodes of cuttings were highly significant for the growth of new plants and genotype G_6 had the tallest plants.

Number of roots: It was clear that the factors genotype, hormone and wound significantly influenced the number of roots while the number of internodes did not. In Table 3 appears the number of roots per cutting as it is affected by the genotype of the donor plant, the number of internodes and the treatments.

Length of roots: The length of roots along with number of roots consist the quality of rooting system of the rooted cuttings. For the length of roots genotype proved to be statistically significant where hormone and wound were highly significant. At the same time the number of internodes proved to be insignificant for the length of roots however, cuttings with one internode had longer roots (Table 3). It is evident that the ranking of genotypes in root characteristics is not the same with that in rooting.

Table 3. Root characteristics in relation to genotype, number of internodes and treatments.

		TREATMENTS OF CUTTINGS									
		NW-NH (Control)		NW-H		W-NH		W-H		TOTAL	
	IN	MN	ML	MN	ML	MN	ML	MN	ML	MN	ML
G_3	2	1	3	8	3	5	3	8.87	4.31	5.7b	3.32ab
G_5	1	-	-	-	-	-	-	9.3	5.66	7.1a	3.54a
	2	-	-	5	2.23	-	-	7	2.75		
G_6	1	-	-	7.8	1.58	-	-	9	3.14	7.79ab	3.12b
	2	2	2.73	12	3.8	2.66	2.33	13.28	5.14		
TOTAL		1.3	2.86	3.83	2.66	8.2	2.65	9.49	4.2		

W: Wound, H: Hormone (2500ppm), NW: No Wound (control), NH: No Hormone (Control)
IN: Number of internodes, MN: Mean Number of roots, ML: Mean Length of roots (cm)

Assessment at the end of four months period: Two months later, after a second observation, it was found that there were more cuttings rooted. Each genotype had an increase as follows: G_3: 65.5% from 40.65%, G_5: 32.8% from 20.3%, G_6: 50% from 43.75%.

3.2. SPRING EXPERIMENT

Results concerning rooting, number and length of roots can be seen on Table 4.

Rooting: Genotype, hormone and the number of internodes, strongly affected rooting as they did in the winter experiment too. Genotype G_6 achieved the highest total rooting success, (70%) and G_2 gave acceptable results regardless of its failure in winter. Cuttings with two internodes had better rooting than cuttings with one internode.

Number of roots: The factors of genotype and the number of internodes significantly affected the number of roots. Hormone treatment alone as well as when combined with genotype was highly significant for the amount of roots formed. Genotype G_6 had more roots not significantly differing from G_5.

Table 4. Rooting % and root characteristics of cuttings in relation to genotype, number of internodes, and hormone treatment.

	TREATMENT: 2500ppm.						TOTAL		
	IN	R	%	C	MN	ML	%	MN	ML
G_2	1	2	20	4	7	5	45b	8.75b	6ab
	2	7	70	2	10.5	7			
G_5	1	-	0	4	-	-	20c	11a	4.5b
	2	4	40	2	11	4.5			
G_6	1	6	60	1	7.5	7.5	70a	11.76a	7.5a
	2	8	80	1	13.6	7.5			
Total			45		9.92	6.3			

IN: Number of internodes, R: Rooted, C: Callus, MN: Mean Number of roots, ML: Mean Length of roots (cm)

Length of roots: The root length was significantly affected by the genotype and the hormone treatment. Genotype G_6 had longer roots than the other two genotypes, eventhough the difference with G_2 was not significant.

Cuttings with two internodes had better root number and root length than cuttings with one internode.

3.3. COMPARISON BETWEEN WINTER AND SPRING EXPERIMENT

The comparison can be based on genotypes G_5 and G_6 because they were included in both experiments. All cuttings were wounded and treated with 2500 ppm, K-IBA. Season proved to be critical for the rooting of *I. aquifolium.* Rooting percentage decreased from 75% in winter to 45% in spring, while the number and the length of roots increased in the spring experiment (Table 5). It is also clear that the two genotypes were affected by season in a different degree.

Table 5. Rooting % and root characteristics affected by the two different seasons.

	Winter	Spring
Root. %	75	45
MN	10	11
ML	4.29	6.83

		Winter			Spring		
	IN	%	MN	ML	%	MN	ML
G_5	1	75	9.3	5.6	0	-	-
	2	50	7	2.75	40	11	4.5
G_6	1	87.5	9	3.14	60	7.5	7.5
	2	87.5	13.3	5.14	80	13.6	7.5

IN: Number of internodes, MN: Mean Number of roots, ML: Mean Length of roots (cm)

4. Discussion-Conclusions

Rooting of *I. aquifolium* was realized in controlled conditions, and was 10% for the control (Table 2), in accordance with the 8% referred by Kralik & Psota (1985). When hormone (IBA) and wounding were applied the rooting percentage in winter fluctuated from 50% to 100%. In difficult to root genotypes (G_5), the above factors are critical. The factor of genotype proved significant for the rooting of the species in both seasons, thing also found for other forest species (Panetsos *et al* 1994a, 1994b).

The duration of rooting time proved to be also important along with the environmental conditions of the donor plants. Burns, (1985) states that cuttings from donor plants growing under semi-shade, (here G_6), root better than others under full sunlight (G_5). The degree of response to rooting depends also on the number of internodes. Cuttings with two nodes root generally better than that with one. Nevertheless, several authors (Merchant, 1985, Bonaminio, 1983, Sanders, 1987) refer the utilization of terminal cuttings without any other distinction. Rooting ability and root quality differ among genotypes of the same population and are also related to the growing conditions. Jones (1985), supports the efficiency of selection based one the above factors and the application of breeding programs for production of clones combining rooting ability, hardiness and desirable ornamental traits.

5. References

Bonaminio, V.P. (1983) Comparison of IBA quick – dips with talk for rooting cuttings. *Combined Proceedings of International Plant Propagators Society* **33**, 565 – 568.

Burns, S.V. (1985) Propagation of hollies. *Holly society journal* **3(3)**, 13 – 14.

Jones, R.L. (1985) The *Aquifoliaceae* of Kentucky. *Transactions of the Kentucky Academy of Science* **10(3I4)**, 104 – 107.

Kralik, J. and Psota, V. (1985) Importance of auxin regulators in rooting some decorative woody species as related to endogenous gibberellins. *Acta Universitatis Agriculturae* **33(3)**, 225-231.

Merchant, K.K. (1985) Propagation of *Ilex opaca* by cuttings. *Combined proceedings – International Plant Propagators' Society* **34**, 536 – 537.

Panetsos, K., Scaltsoyiannis, A., and Alizoti, P. (1994a) Effect of genotype and cutting type on the vegetative propagation of the pine hybrid (*Pinus brutia* (Ten) x *Pinus halepensis* (Mill)). *Ann. Sci. For.* **51**, 447-454.

_ (1994b) Vegetative propagation of *Platanus orientalis* x *P. occidentalis* F1 hybrids by stem cuttings. *Forest genetics* **1(3)**, 125-130.

Sanders, C.R. (1987) Some aspects of the Propagation of *Rhododendron, Mahonia* and *Ilex* by cuttings. *Combined proceedings. Intern. Plant Propagators Society* **28**, 228 – 232.

COMMERCIAL MICROPROPAGATION OF *SIDERITIS CLANDESTINA* (MT. TAYGETOS), *ORIGANUM VULGARE* SUBSP. *HIRTUM* (MT. OLYMPOS), *MELISSA OFFICINALIS* (MT. PINDOS), AND *MENTHA LONGIFOLIA* (MT. PINDOS)

I. NIANIOU-OBEIDAT[1] AND G. N. ICONOMOU-PETROVICH[2]
Vitro Hellas S.A., Nisseli, Alexandria, HELLAS
Present address: [1] Dept. of Genetics and Plant Breeding, Aristotelian University of Thessaloniki, HELLAS 540 06
[2] 64 J. Kennedy str., 555 35 Pylea, Thessaloniki, HELLAS

1. Introduction

A large number of aromatic plants are economically important, and serve as irreplaceable sources of industrial oils—both volatile and fixed—,flavours, fragrances, resins, gums, and other specialty products [1].

S. clandestina, known in Hellas with the common name "mountain tea", is widely distributed all over the mountains of the country and is used as a herbal tea [2]. *O. vulgare* ssp. *hirtum*, known as the "Greek oregano", is occurring almost all over the country, grows wild from sea level up to 1500 m, and is generally accepted to have the best quality [3][4][5]. It is widely used as raw for culinary purposes but is also used in the pharmaceutical and other industries [6]. *M. officinalis* which is known with the common name "balm" or "lemon balm", is used as a seasoning [1]. *M. longifolia*, is a valuable plant which is widely used as a flavour, and has carminative properties [1].

Micropropagation, as a method of propagation, has many advantages over conventional methods, in addition it provides a tool for preserving commercially or ecologically important plant species. The aim of this work was to develop methodology for commercial micropropagation and maintenance of stock plants *via in vitro* vegetative propagation for the four species described.

2. Materials and Methods

2.1. PLANT MATERIAL

The plants used to provide the explant source for *in vitro* culture, were collected from the wild. *S. clandestina* was collected from the location Arna of Mt. Taygetos, *O. vulgare* ssp. *hirtum* from the location Ai-Giannis of lower Mt. Olympos, and *M. officinalis* and *M. longifolia* from the location Aristi of Mt. Pindos.

All plants were individually maintained under greenhouse conditions in large pots in a mixture of peat moss and perlite (3:1), and were sprayed every 20-30 days against soil-borne fungi.

2.2. DISINFECTION

Using full sterile techniques, nodal segments, making cuts approximately midway between the nodes, were surface sterilised in an aqueous solution of 1 % sodium hypochlorite (NaOCl) with one drop of Tween 20 for 10-15 minutes after pre-treatment in 70 % alcohol for 15-20 seconds. The explants were then rinsed in sterile distilled water for 3-4 times.

2.3. CULTURE MEDIA

For the initiation of *in vitro* culture nodal segments were individually transferred to glass tubes containing *c.* 15 ml of nutrient medium. For proliferation and rooting 500 ml jars containing 125 ml nutrient medium were used.

Full strength MS macronutrients, micronutrients, and vitamins with 20 g/l sucrose, and 3.2 g/l agar supplemented with BAP, NAA, and GA_3, were used for *S. clandestina* (table 1), *M. officinalis* (table 3), and *M. longifolia* (table 3). Full strength Gamborg's B5 with 20 g/l sucrose, and 6 g/l agar supplemented with BAP, and NAA, were used for the initiation phase, and full strength WPM with 20g/l sucrose, and 5 g/l agar for the proliferation and the rooting phase of *O. vulgare* ssp. *hirtum* (table 2).

2.4. CULTURE CONDITIONS

The culture period for each *in vitro* stage was four weeks. Cultures were maintained at 20 ± 2° C, relative humidity 35-50 %, under 16 hours photoperiod provided by standard fluorescent white light.

3. Results

S. clandestina

The best initiation was observed in the medium supplemented with 0.7 mg/l BAP, 0.01mg/l NAA, and 0.1 mg/l GA_3 (Table 1; Fig. 1(a)). After two subcultures on the same medium the shoots were growing less vigorously than during the initial culture. Rapid shoot proliferation was achieved on the MS medium supplemented with 0.5 mg/l BAP, or in the hormone-free medium. The proliferation rate was 2.7-3.0. Two weeks after the transfer of the shoots on the rooting medium, only long explants rooted (74 %). The plantlets were successfully acclimatised (80 %).

Table 1. Optimal concentrations of growth regulators supplemented to the MS nutrient medium for commercial micropropagation of *S. clandestina*

Stage		Growth regulators (mg/l)	
Initiation		0.7 mg/l BAP 0.01 mg/l NAA 0.1 mg/l GA_3	
Proliferation	1st Phase	0.7 mg/l BAP 0.01 mg/l NAA 0.1 mg/l GA_3	
	2nd Phase	0.5 mg/l BAP 0.01 mg/l NAA 0.1 mg/l GA_3	or MS (hormone-free)
Rooting		0.5 mg/l IBA 0.1 mg/l GA_3	

O. vulgare ssp. *hirtum*

After initiation in the B5 medium supplemented with 0.01 mg/l NAA, optimum proliferation and rooting were observed in the WPM medium supplemented with 0.2 mg/l BAP and 0.05 mg/l NAA for the proliferation stage and 0.01 mg/l NAA for the rooting stage (Table 2; Fig. 1(b)).

Table 2. Optimal BAP and NAA concentration supplemented to the B5 or WPM nutrient medium for commercial micropropagation of *O. vulgare* ssp. *hirtum*.

Stage	Growth regulators (mg/l)	
Initiation	B5 +	0.01 mg/l NAA
Proliferation and Rooting	WPM +	0.2 mg/l BAP 0.05 mg/l NAA

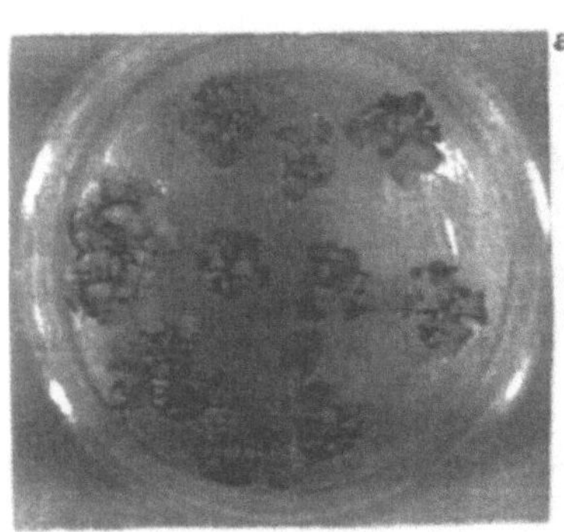

Fig. 1: Micropropagation of *S. clandestina* (a) and *O. vulgare* ssp. *hirtum*.(b).

M. officinalis & *M. longifolia*

Optimum shoot growth, proliferation, and rooting was observed in the MS medium supplemented with BAP, NAA, and GA_3 (Table 3; Fig. 2).

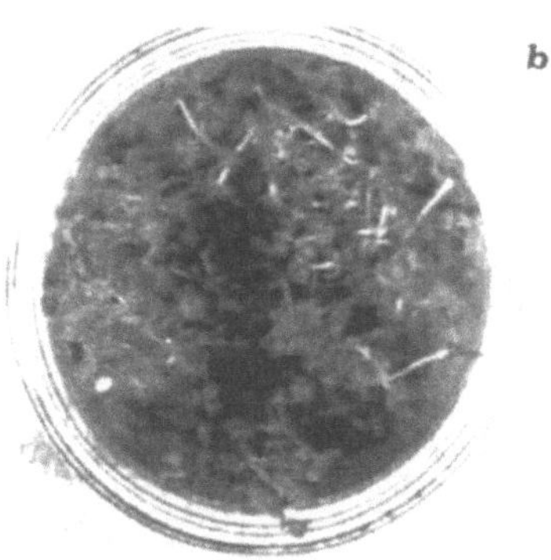

Fig. 2: *In vitro* propagation of *M. officinalis* (a) and *M. longifolia* (b).

Table 3. Optimal concentration of growth regulators supplemented to the MS nutrient medium for commercial micropropagation of *M. officinalis* and *M. longifolia*.

Stage	Growth regulators (mg/l)
Initiation,	0.7 mg/l BAP
Proliferation,	0.01 mg/l NAA
Rooting	0.1 mg/l GA_3

4. Discussion

For all four species after a long period *in vitro, i.e.* after 16 weeks, a period of "rest" in a hormone-free medium, was required as limited growth and proliferation was observed. This was pronounced more for *S. clandestina*, and transfer to a hormone-free medium was necessary after the 10th week. A number of *in vitro* plantlets for all four species, died due to vitrification. Initiation of *O. vulgare* ssp. *hirtum* was also achieved in the proliferation and rooting medium however it was less vigorous.

The highest cost during micropropagation, corresponding to more than 65 % of total costs is labour. It is therefore desirable to minimise the number of transfers during *in vitro* culture and in general handling of the plant material. This may be achieved by performing two *in vitro* stages in one medium (*e.g.* proliferation with elongation and rooting), and/or use alternative methods of *ex vitro* propagation.

Abbreviations: BAP, 6-benzylaminopurine; B5, Gamborg's B5 culture medium [7]; GA, gibberelic acid; IBA, indole-3-butyric acid; MS, Murashige and Skoog medium [8]; NAA, 1-naphthaleneacetic acid; WPM, Lloyd and McCown medium [9] Woody Plant Medium.

Acknowledgements
Financial support from the European Union through the ECLAIR (AGRE 0060) research programme is gratefully acknowledged.

5. References

1. Balandrin, M.F., Klocke, J.A. (1988). Medicinal, aromatic, and industrial materials from plants. In: *Biotechnology in agriculture and forestry*, vol. 4, Medicinal and aromatic plants I, 3-36, Bajaj, Y.P.S. (ed.). Springer-Verlag, Berlin.
2. European Collaborative Linkage of Agriculture and Industry through Research (ECLAIR), research project, contract no. AGRE 0060 (1996). *Biotechnology applications for plant regeneration, identification, and certification aiming at the production of healthy propagation material and the industrial utilisation of certain aromatic plants.* Dissemination report for the fourth period. Directorate General VI, Commission of the European Communities.
3. Kokkini, S., Vokou, D. (1989). Carvacrol-rich plants in Greece. *Flavour and Fragrance Journal*, 4, 1-7.
4. Ietswaart, J.H. (1980). *A taxonomic revision of the genus Origanum (Labiatae).* London: Leiden University Press.
5. Lawrence, B.M., Reynolds, R.J. (1984). The botanical and chemical aspects of oregano. *Perfumer and Flavourist*, 9, 41-51.
6. Bruneton, J. (1995). *Pharmacognosy, phytochemistry, medicinal plants.* Intercept Ltd., Hampshire, England.
7. Gamborg, O.L., Miller, R.A., Ojima, K. (1968). Nutrient requirements of suspension cultures of soybean root cells. *Experimental Cell Research*, 50, 151-158.
8. Murashige, T., Skoog, F. (1962). A revised medium for rapid growth and bio-assays with tobacco tissue cultures. *Physiologia Plantarum*, 15, 473-497.
9. Lloyd, G., McCown, B. (1981). Commercially-feasible micropropagation of mountain laurel, *Kalmia latifolia*, by use of shoot tip culture. *Proceedings of the International Society for Plant Propagation*, 30, 421-427.

LIGNIN PATHWAY MODIFICATIONS BY ANTISENSE O-METHYL-TRANSFERASE AND PEROXIDASE GENE CONSTRUCT IN TOBACCO

T. TALAS OGRAS[1], A. KURU[2] AND N. GÖZÜKIRMIZI[1,2]
[1]TUBITAK, MRC, Research Institute for Genetic Engineering and Biotechnology, Gebze, Kocaeli, TURKEY [2]University of Istanbul, Faculty of Science, Vezneciler, Istanbul, TURKEY

1. Abstract

In order to shuttle down lignin biosynthesis, antisense constructs of O-methyltransferase pOMT8) and peroxidase (Shpx6) from *Stylosanthes humilis* were transferred into leaf dicsc of tobacco by *Agrobacterium tumefaciens*. The analysis of transgenics showed that T-DNA genes were permanently integrated into tobacco genomes. Tobacco plants that contain antisense peroxidase gene, peroxidase enzyme activity was decreased up to 75%, however lignin amount was decreased up to 20% compared with non-transgenic controls. In transgenic tobacco plants were transferred with antisense OMT gene constructs decreased up to 68% in OMT activity and the lignin amount decreased up to 62%.

2. Introduction

Lignin is an important aromatic polymer of phenylpropanoid pathway and is one of the major component of plant cell wall. It also provides mechanical strength to plant, provides solute transport in xylem and important in plant defence. Lignin is synthesised by the polymerisation of three monolignols, such as coumaryl, coniferyl and sinapyl alcohol. The monomeric compositions of lignin are variable in different species and tissues related with developmental stages (Lewis and Yamamoto 1990). O-methyltransferase (OMT, EC 2.1.1.6) is an earlier enzyme in lignin monomer synthesis and plays an important role in plant growth. OMT is responsible for catalysing methoxylation of lignin precursors to form ferulic and sinapic acids. *Stylosanthes humilis* is important annual forage legume in Australia, Asia and other tropical regions of the world (Harrison *et al.*, 1995). In dicotyledonous plant species OMT sequences are highly conserved and *Stylosanthes humilis* OMT clone shows 74% homology with tobacco OMT sequences (McIntyre *et al.*, 1995). Peroxidases (EC 1.11.1.7) are found to be effective in the final step of lignin biosynthesis where catalyse polymerisation of cinnamyl alcohols in the presence of hydrogen peroxide. Genetic manipulations are in progress with the lignin biosynthetic enzymes that lignin amount and monomeric composition changed transgenic plants have already been reported (Ni *et al.*, 1994, Dwivedi *et al.*, 1994). High lignin content in plant materials cause monetary and environmental costs during removing lignin from cellulose in paper and pulp industries. In addition to this, lignin amount limits the digestibility of forage crops. In this study, we used tobacco as a model system to modify lignin content and amount with transferring antisense heterologous constructs of OMT and peroxidase genes.

3. Material and Methods

cDNA clone of *S.humilis* belongs to OMT is 1371 nucleotides in length and is named pOMT8. cDNA clone of peroxidase gene from *S. humilis* is named Shpx6. The antisense constructs of pOMT8 and Shpx6 is located in pGA643 under the 35S CaMV promotor. The constructs were obtained from Dr. Kemal Kazan, Tropical Plant Pathology, The University of Quensland, Australia. The constructs were mobilized into LBA 4404 strain of *Agrobacterium tumefaciens* and tobacco (*Nicotiana tabacum* L.cv. Samsun) leaf discs were infected (Horsch *et al.*, 1985). The plantlets were transferred into soil when they reached 4 cm long. Six peroxidase and eight OMT transgenics were analysed. The screening of transgenics were performed with the eight months old plantlets. Total peroxidase activity was identified spectrophotometrically at 470 nm by mixing stem extract with 5mM guaiacol and 5mM hydrogen peroxide. Peroxidases were also screened on 10% nondenaturing polyacrylamide gels with guaiacol /H_2O_2 solution (Muraja *et al.*, 1994). OMT activity was determined at 280 nm by using caffeic acid as substrate and *S*-adenosyl-L-(*methyl*-^{14}C)methionine as methyl donor (Pellegrini *et al.*, 1993). Determination of lignin amount was done with TGA (thioglycollic acid) technique (Campbell and Ellis 1992) and total soluble lignin were determined in stem and leaf extracts of transgenics.

4. Results and Discussion

The aim of this study is to modify lignin into easy soluble forms in tobacco model system. Transgenic tobacco plants expressing OMT and peroxidase genes in antisense orientation were screened. Peroxidase activities were determined in stem extracts at 470nm for 2min. with 15 sec. intervals. (Figure 1). The peroxidase activity decreased in all transformants but in P4 75% reduction was observed. OMT activity was determined by ^{14}C labelled ferulic acid in transformant stem extracts. O5 is one of the eight transformants and OMT activity reduced 68% (Figure 2). Among 8 antisense OMT gene containing transformants, lignin amount reduced 50% in O5 comparing with controls(Figure 3). Lignin amount in antisense peroxidase gene containing transformants were also determined (Figure 4) and lignin amount reduced 20% in P4 transformant. Lignin compositions were also determined by infrared spectroscopy in transgenics but obvious peaks were not observed on related wavelenghts. Reduced lignin amount in tranformants related with antisense constructs have given better result with antisense OMT containing transformant comparing with antisense peroxidase containing transformant. In this study, antisense technology applications were used as a tool to modify lignin in tobacco model.

Our research is continuing on the older stages of transgenics and the next generations. Genetically modified lignin in plant materials can be used in pulp industry and as forage on trial.

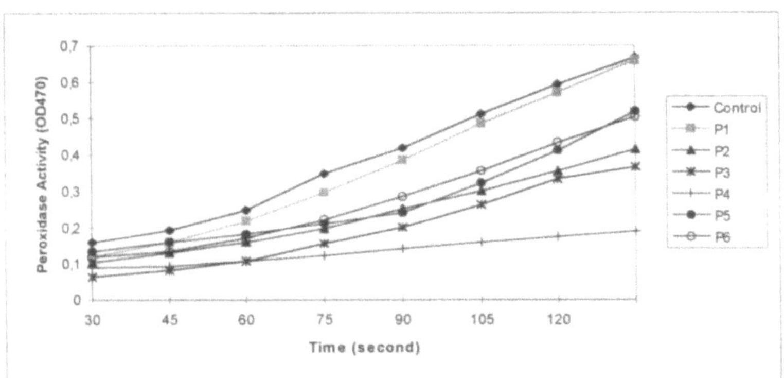

Figure 1. Peroxidase activities in 6 antisense peroxidase gene transferred plants (P).

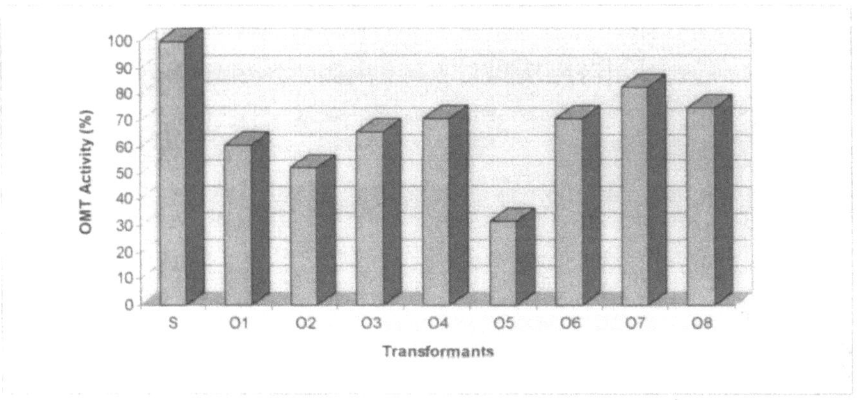

Figure 2. OMT activities in antisense OMT gene containing transformant stem extracts (O). S, control.

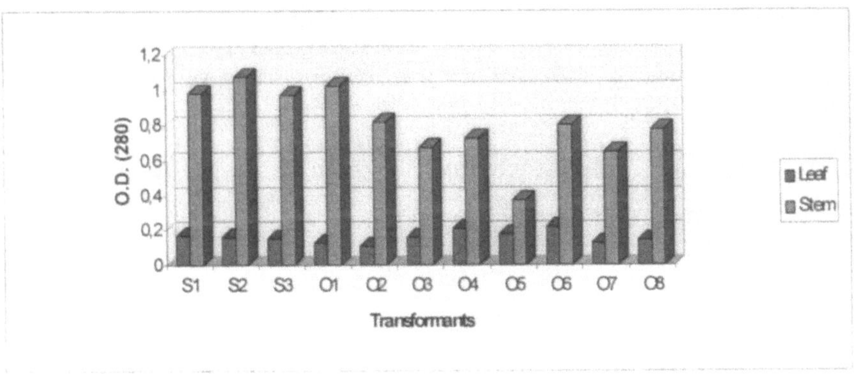

Figure 3. Lignin amounts of antisense OMT gene transferred plants (O). S1,S2,S3, controls.

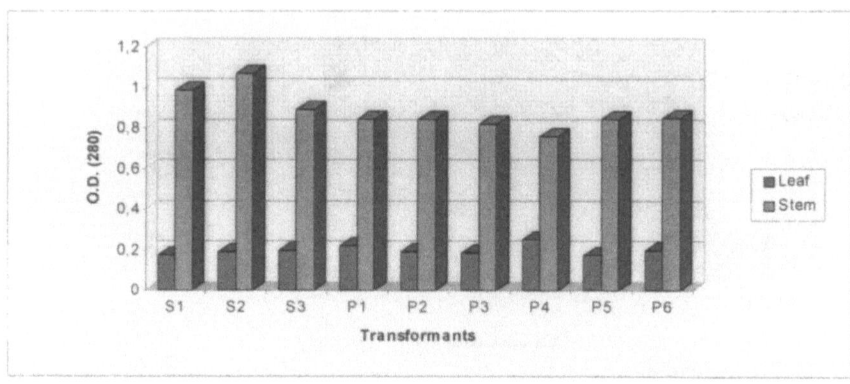

Figure 4. Lignin amounts of antisense peroxidase gene transferred plants (P).

5. Acknowledgement

This work was supported by TÜBİTAK, MRC, Plant Biotech Project Grant No. 62-1-001.

6. References

Campbell, M.M., Ellis, B.E. (1992) Elicited phenylpropanoid metabolism in pine cultures, *Planta* **186**, 409-417.

Dwivedi, U.N., Capbell, W.H., Yu, J., Datla, R.S.S, Bugos, R.C., Chiang, V.L., Dodila, G.K. (1994) Modification of

lignin biosynthesis in transgenic *Nicotiana t*hrough expression of an antisense O-methyltransferase gene from *populus*. *Plant Molecular Biology* **26**, 61-71.

Harrison, S.J., Curtis, M.D., McIntyre, L., Maclean, D.J., Manners, J.M. (1995) Differential expression of peroxidase isogenes during the early stages of infection of the tropical forage legume *Stylosanthes humilis* by *colletotrichum glooesporioides, Molecular Plant Microbe Interactions* **3**, 398-406.

Horsch, R.B., Fry, J.E., Hoffmann, N.L., Eicholtz, D., Rogers, S.G., Fraley, T.A. (1985) A simple and general method for transferring genes into plants, *Science* **227**, 1229-1231.

Lewis,N.G., Yamamoto,E. (1990)Lignin: occurance, biogenesis and biodegradation, *Annu. Rev. Plant Physiol.Plant Mol. Biol.* **41**, 455-496.

McIntyre, C.L., Rae, A.L., Curtis, M.D., Manners, J.M. (1995) Sequence and expression of a caffeic acid O-methyltransferase cDNA homologue in the tropical forage legume s*tylosanthes humilis, Aust. J. Plant Physiol.* **22**, 471-478.

Mutaja,J., Peskan,T., Lorkovic,Z., Krsnik-Rasol,M. (1994) Ultrastructure and protein pattern of horse-radish (*Armoracia lapathifolia gilib*) crown-gall tissue, *Acta Bot. Croat.* **53**, 1-6.

Ni, W., Paiva, N.L., Dixon, R.A. (1994) Reduced lignin in transgenic plants containing a caffeic acid O-methyltransferase antisense gene, *Transgenic Research* **3**, 120-126.

Pellegrini,L., Geoffroy,R., Fritig,B., Legrand,M. (1993) Molecular cloning and expression of a new class of ortho-diphenol-*O*-methyltransferase induced in tobacco (*Nicotiana tabacum L.*) leaves by infection or elicitor treatment, *Plant Physiol.* **103**, 509-517.

MULTILOCUS STUDY OF SIX BALKAN NATURAL POPULATIONS OF *Pinus sylvestris*(L.)

G. PASSAYIANNES[1*], D. MOULALIS[2], A. SCALTSOYIANNES[3] AND M. TSAKTSIRA[1]
1 Researcher, 2 Professor, 3 Assistant Professor
Aristotle University of Thessaloniki, Department of Forestry & Natural Environment, Laboratory of Forest Genetics and Plant Breeding, 54006 Thessaloniki

1. Abstract

Polymorphism of 9 enzyme systems (EST-b, GDH, PGI, LAP, MR, 6PGD, PGM, IDH and ACP) were analyzed by horizontal starch gel electrophoresis technique, on 6 natural Balkan populations of *Pinus sylvestris L.* A total of 37 alleles were detected at 11 loci. Approximately, the same level of diversity (He=0.209) was observed among the populations studied. Interpopulation genetic diversity (Gst) was very low (2.9% of the total diversity), indicating that the above populations are due to fairly recent colonization after the glaciation and they also share a common gene pool. Presumably, the extensive gene flow that exists among populations should be responsible for the lack of differentiation at allelic frequencies observed.

2. Introduction

Pinus sylvestris (L.) is characterized by the most extensive continuous range in Europe and Northern Asia with natural distribution from Scotland to Pacific Ocean and from Scandinavia to Greece, South Spain, Northern and Central Turkey. In Balkan peninsula it occurs in Croatia, Western Yugoslavia, Western and Central Bulgaria and Northern Greece (Pieria mountains is the Southest part in Europe).

Studies of genetic structure of *Pinus sylvestris (L.)* were mainly based on observation of traits belonging in two categories: morphological - anatomical and biochemicals (terpenoids and isozymes). Most of the above studies based on morphological - anatomical traits have shown great differentiation among populations which end up to divide the species into various races. On the other hand, studies based on biochemical traits have shown very low interpopulation genetic diversity although they exhibited high amount of genetic variation within the populations.

* This work is part of the Doctorate thesis of the senior author

3. Materials and methods

Seeds were collected from six native stands distributed over large parts of Northern Greece, FYROM, and Southern Bulgaria. They were germinated at room temperature on moist filter paper. Seeds from 180 single trees were analyzed (30 trees from each population).

The electrophoretic methods, gel and tank buffer, stain recipes and the scoring of alleles and the genetic analysis of the isozymes that we followed in this study, were according to Cheliak and Pitel (1985), Scaltsoyiannes *et al.* (1994) and Moulalis *et al.* (1996).

4. Results

A total of 37 alleles were detected at 11 loci belonging to 9 enzyme systems (Table 1). In the same table the frequency allozyme distributions are also shown. Among the populations studied, no specific pattern of allozyme frequency distributions were observed, except the most southern population (Pieria) which exhibited a slight differentiation from the rest populations.

In Table 2 the values of genetic variability parameters (He, A/L, P) are presented. According to this table all the populations exhibited high values of variation and comparable with the findings of other workers for populations of *P. sylvestris* from Central and North Europe (Prus-Glowacki and Bernard 1994; Prus-Glowacki and Stephan, 1994). Unexpectedly the higher level of heterozygosity was observed in the Southern populations.

The low level of the coefficient of diversity (Gst=0.029) (data are not presented in this work) means that low interpopulation variability exists among the population. Low interpopulation value in *P. sylvestris* was also found by other workers (Goncharenko *et al.* 1994, Prus-Glowacki and Bernard 1994, Prus-Glowacki and Stephan 1994).

In dendrogram that was constructed by Nei's genetic distances two clusters were formed. One major cluster, consisted of Xanthi, Lailias, Mala Kruza populations and one minor, that includes the populations from the mountain Rodopi (Jundola and Elatia). The most Southern population (Pieria) seems to be differentiated from both clusters.

5. Conclusions

The six Balkan populations tested exhibited high amount of genetic variation, comparable to other populations from Central and Northern Europe and other European conifer species.

The low value of Gst (2.9%) indicates the lack of any differentiation among the populations due to the extensive gene flow and to fairly recent colonization of the species after the glaciation period. On the other hand the higher amount of genetic variation exists within the populations.

The estimation of the amount and the pattern of genetic variability of *Pinus sylvestris (L.)* in the Balkan peninsula, gave us new information concerning the breeding of the species in this area.

TABLE 1. Allele frequencies of 11 loci in 6 populations of *Pinus sylvestris (L.)*

population / alleles	Pieria	Lailias	Elatia	Xanthi	Mala Kruza	Jundola
EST -A1	0.89	1.00	0.96	0.96	0.96	0.87
EST -A2	0.11	0.00	0.04	0.04	0.04	0.13
GDH-A1	0.42	0.33	0.29	0.18	0.32	0.24
GDH-A2	0.58	0.63	0.71	0.82	0.68	0.75
GDH-A3	0.00	0.04	0.00	0.00	0.00	0.01
PGI-B1	0.09	0.02	0.06	0.04	0.00	0.04
PGI-B2	0.88	0.98	0.91	0.95	1.00	0.91
PGI-B3	0.03	0.00	0.03	0.01	0.00	0.05
LAP-A1	1.00	0.94	0.98	0.94	1.00	0.99
LAP-An	0.00	0.06	0.02	0.06	0.00	0.01
LAP-B1	0.00	0.07	0.02	0.05	0.00	0.01
LAP-B2	0.93	0.92	0.98	0.95	0.95	0.99
LAP-B3	0.07	0.01	0.00	0.00	0.05	0.00
DIA-A1	0.01	0.00	0.00	0.00	0.00	0.00
DIA-A2	0.75	0.57	0.64	0.53	0.71	0.60
DIA-A3	0.00	0.03	0.01	0.04	0.03	0.04
DIA-A4	0.01	0.14	0.04	0.18	0.00	0.10
DIA-A5	0.20	0.26	0.31	0.24	0.26	0.26
DIA-A6	0.03	0.00	0.00	0.01	0.00	0.00
6PGD-A1	0.00	0.05	0.00	0.04	0.01	0.00
6PGD-A2	0.67	0.59	0.74	0.49	0.50	0.62
6PGD-A3	0.04	0.09	0.04	0.11	0.05	0.04
6PGD-A4	0.25	0.27	0.22	0.36	0.44	0.34
6PGD-A5	0.04	0.00	0.00	0.00	0.00	0.00
6PGD-B1	0.89	0.89	0.98	0.85	0.89	0.95
6PGD-B2	0.11	0.09	0.02	0.15	0.11	0.05
6PGD-B3	0.00	0.02	0.00	0.00	0.00	0.00
PGM-A1	0.00	0.00	0.06	0.02	0.00	0.10
PGM-A2	1.00	1.00	0.89	0.90	0.91	0.86
PGM-A3	0.00	0.00	0.05	0.08	0.09	0.04
IDH-A1	1.00	1.00	1.00	1.00	1.00	1.00
ACP-A1	0.88	0.93	0.90	0.89	0.96	0.93
ACP-A2	0.06	0.00	0.03	0.04	0.00	0.02
ACP-A3	0.06	0.07	0.07	0.07	0.04	0.05

TABLE 2. Genetic variability of 11 loci in 6 populations of *Pinus sylvestris (L.)*

Populations	Mean number of alleles per locus A/L	Percentage of polymorphic loci P (%)	Expected Heterozygosity He
Pieria	2.36	72.73	0.226
Lailias	2.36	72.73	0.208
Elatia	2.45	90.91	0.188
Xanthi	2.64	90.91	0.234
Mala Kruza	2.00	72.73	0.185
Jundola	2.55	90.91	0.215

Figure 1. Dendrogram based on genetic distances

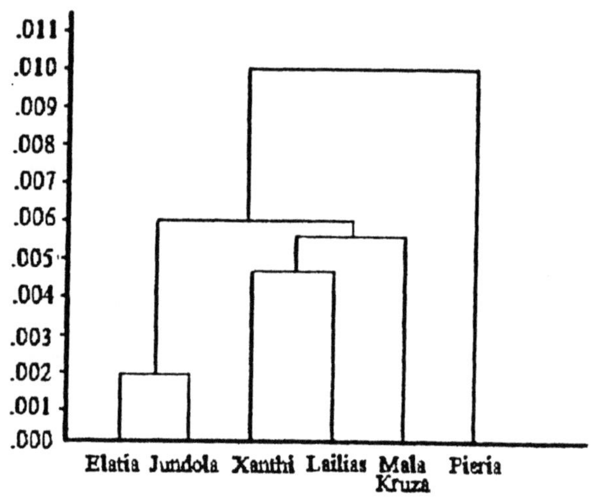

6. References

Cheliak, W.M. and Pitel, J.A. (1985) Techniques for starch gel electrophoresis of enzymes from forest tree species, Information report PI-X-42, Petawawa Nat.For.Serv., Agric. Canada.

Goncharenko, G.G., Silin, A.E. and Padutov, V.E. (1994) Allozyme variation in natural populations of Eurasian Pines. 3. Population structure, diversity, differentiation and gene flow in central and isolated populations of *Pinus sylvestris* L. in Eastern Europe and Siberia, Silv.Genet., Vol. 43, 2-3: 119-132.

Moulalis, D., Scaltsoyiannes, A. and Tsaktsira, M. (1996) Isozyme analysis of Scots pine (*Pinus sylvestris* L.) of the most Southern part of its distribution in South-East Europe, Proceedings of the 6th Conference of the Plant Breeding Hellenic Society, 206-216.

Prus-Glowacki, W. and Bernard, E. (1994) Allozyme variation in populations of *Pinus sylvestris* L. from a 1912provenance trial in Pulawy (Poland), Silv.Genet., Vol. 43, 2-3: 132-138.

Prus-Glowacki, W. and Stephan, B.R. (1994) Genetic variation of *Pinus sylvestris* from Spain in relation to the European populations. Silv.Genet., Vol. 43, 1: 7-14.

Scaltsoyiannes, A., Rohr, R., Panetsos, K.P. and Tsaktsira, M. (1994) Allozyme frequency distributions in five European populations of black pine (*Pinus nigra* Arn.). I. Estimation of genetic variation within and among populations. II. Contribution of isozyme analysis to the taxonomic status of the species. Silvae Genet. 43, 1: 20-30.

CHARACTERIZATION OF THE MITOCHONDRIAL GENOME OF CYTOPLASMIC MALE STERILE HYBRID BETWEEN *LYCOPERSICON PERUVIANUM MILL.* AND *LYCOPERSICON PENNELLII CORR.*

M. PETROVA[1], Z. VULKOVA[2], A. ATANASSOV[1*], P. STOEVA[1*]
[1]*Institute of Genetic Engineering, 2232 Kostinbrod-2, Bulgaria*; [1]**De Montfort University, The Norman Borlaug Centre for Plant Science Research, Institute of Genetic Engineering, 2232 Kostinbrod-2, Bulgaria*; [2]*Institute of Genetics 1113 Sofia, Bulgaria*

1. Introduction

Tomato is one of the major vegetable crops grown in diverse climatic conditions. To exploit hybrid vigour lines with high combing ability have been developed and successfully used to produce hybrid varieties. Tomato has a tremendous potential of heterosis for earliness, uniformity, total yield and resistance attributes and the main constraint for a large scale hybrid seed production is the lack of cytoplasmic male sterility (CMS) system.

The CMS phenotype is inherited through the female parent and results from the interaction between the nuclear and cytoplasmic genomes. It can arise by intraspecific, interspecific or intergeneric crosses, spontaneously or after mutagenic treatment (Hanson, Conde 1985; Kaul 1988; Hanson 1991). CMS arising through crosses between different species led to the concept that it is a result of the incompatibility between the nuclear genome of one species and the cytoplasm of another. CMS is caused by mutation in the mitochondrial (mt) DNA of the male-fertile progenitor as a result of aberrant recombination events (Leaver et al. 1988).

A description of male sterile phenotype in genus *Lycopersicon* as a result of incompatibility between the cytoplasm of one species and the nuclear genome of another has been communicated in the studies of Andersen (1963, 1964) in crosses between *L. esculentum, L. esculentum v. minor, L. minutum and L. hirsutum f. glabratum* used as pistillate parents and *L. pennellii* as pollinating recurrent parents and by Vulkova - Achkova (1980) in crosses between *L. peruvianum* as pistillate parent and *L. pennellii* and *L. hirsutum typicum* as recurrent pollinating parents.

The aim of this investigation is to study the mitochondrial genome of an alloplasmic CMS line carrying the cytoplasm of *L. peruvianum* and the nuclear genome *L. pennellii* and to compare it the mitochondrial genome of the donor of the cytoplasm *L. peruvianum.* This data will allow to distinguish the changes which have taken place as a result of the interaction between the alien cytoplasm of *L. peruvianum* and the genome of *L. pennellii.*

2. Material and methods

Plants from CMS-*pennellii*, *L. peruvianum*, *L. pennellii* (LA 716) and *L. esculentum* have been grown in the greenhouse and leaf material was used for isolation of total DNA. The CMS-*pennellii* line used in the study represents the BC_6-P_2 (*L. peruvianum* x *L. pennellii* LA716*)*. Total DNA was isolated according the protocol of Fulton et. al., (1995). Bulk samples of DNA were prepared from leaf material gathered from 5 to 10 individual plants from each genotype. Ten μg of DNA were digested with *HindIII, EcoRI, BamHI, DraI, EcoRI* + *HindIII* and *BamHI* + *HindIII* restriction enzymes (Boehnringer Mannheim and Stratagene) and fractionated by electrophoresis on 0.8% agarose gels. The fractionated DNA was blotted and hybridised to DIG-labelled heterologous mt gene probes according to the manufacturer's protocols for the Gene Screen Plus membranes (Du Pont) and for the Non-radioactive Labelling and Detection of Nucleic Acids System (Boehringer Mannheim).

The following probes were used: *atp A* (1.5 kb a *HindIII/EcoRI* fragment, containing the coding region of subunit A of the ATPase gene of *Pisum sativum* provided by Dr. Toro Terachi, Kyoto Sangyo University, Japan), *atp 6* (2.7 kb *HindIII* fragment containing subunit 6 of ATPase gene and flanking sequences from *Zea mays* - Dewey et al. 1985), *18S+5S* ribosomal gene region from mt DNA of *Zea diploperennis* (6.0 kb *BamHI* fragment - Gwynn et al. 1987), *nad 3* (0.8 kb *BamHI/HindIII* fragment containing NADH dehydrogenase subunit 3 from *Petunia hybrida* - Yesodi et al 1995) and clone 2 from *Petunia hybrida* containing the CMS associated *S-Pcf* locus (2.9 kb *SalI* fragment including part of the *cox II* gene - cytochrome oxidase subunit 2, *urf-s* - unidentified open reading frame, *nad 3* gene and *rps 12* - ribosomal protein - Yesodi et al. 1995).

3. Results and discussion

Phenotypically the CMS-*pennellii* is identical to the recurrent parent *L. pennellii*, the only established differences concern the characteristics of the flower: the corolla size is smaller, pale yellow in colour and the petals are not well opened, the anthers are reduced in size with mean length 2.3 mm, pale green and are not coalesced laterally to form a normal staminal cone. The anthers do not shed pollen. When pollinated with pollen from *L. pennellii*, the fruit set, the seed set and the size of the formed fruits show that the female fertility of CMS-*pennellii* is normal.

The molecular studies of the organisation of the mitochondrial DNA were performed both on bulk DNA and on DNA isolated from individual plants. The analysis with buck DNA samples was carried out with the CMS *pennellii* line, *L. peruvianum*, *L. pennellii* and *L. esculentum* with *atp A, atp 6, nad 3, 18S+5S* rRNA heterologous gene probes while individual plant DNA from CMS *pennellii* and *L. peruvianum.* was screened with: *atp A* and clone No 2. (Table 1.).

A difference were established in the hybridisation pattern with *atp A* gene after double digestion with *EcoRI* and *HindIII* restriction enzymes. One 3.6 kb fragment hybridised on the CMS line RFLP profile in comparison to *L. peruvianum* where a second 5.0 kb fragment hybridised to the probe (Fig.1. B.). A difference was found with the same probe also after *Dra I* digestion. The RFLP profile of CMS line revealed four fragments: 10.0 kb, 5.3 kb, 3.4 kb and 2.5 kb while the hybridisation pattern of the donor of the cytoplasm *L. peruvianum* have two additional fragments: 6.3 kb and 4.4 kb (Fig. 1. A.). With *nad 3* gene a polymorphism between CMS-*pennellii* and *L. peruvianum* was detected on the blots with *EcoRI/HindIII* digested DNA. The CMS- *pennellii* line hybridised to a 3.6 kb fragment, while *L. peruvianum* - to two additional fragments with lower intensity - 2.9 kb and 5.0 kb (Fig.1. C.). These results demonstrated the presence of *atp A* and *nad 3* gene on the same *EcoRI/HindIII* 3.6 kb fragment that had been detected in both *L. peruvianum* and *CMS-pennellii* mitochondial genome. No differences between CMS-*pennellii* and *L. peruvianum* were revealed with the rest of the listed (Table 1.) probe/enzyme combinations. The restriction fragment length polymorphism data with *nad 3*, with *atp A* clearly demonstrated the differences between the organisation of the mt DNA of the different *Lycopersicon* species (Fig.1).

Table 1. Mt gene probes and restriction enzymes applied in the Southern blot analysis of mt DNA

MITOCHONDRIAL GENE PROBES	RESTRICTION ENDONUCLEASES
18S+ *5S* rRNA	*DraI, BamHI, EcoRI, HindIII*
atp 6	*BamHI, EcoRI, HindIII*
atp A	*DraI, BamHI + HindIII*
	EcoRI + HindIII
nad 3	*BamHI, EcoRI, BamHI + HindIII,*
	EcoRI + HindIII
clone No2 (*cox II gene* , *urf-s* , *nad 3* and *rps 12*)	*HindIII*

To verify the results obtained with the bulk DNA Southern blots, the hybridisation analysis was repeated with *atpA* with *Dra I* digested DNA samples from five individual CMS and *L. peruvianum* plants (data not shown). The detected on bulk DNA differences between the CMS line and *L. peruvianum* (two additional fragments) were present in the pattern of only one of the individual *L. peruvianum* plants. This result suggests that specific subgenomic molecule(s) is (are) present with different stoichiometry in the mitochondrial genome of *L. peruvianum*. These fragments are absent from the RFLP profile of all studied CMS-*pennellii* plants.

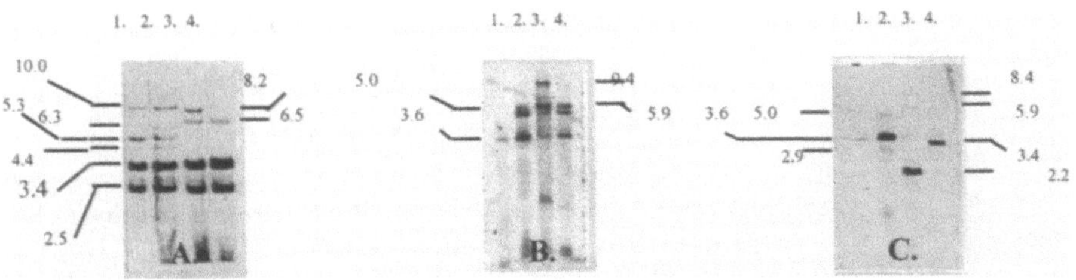

Fig.1. Hybridisation pattern with *atp A* of DNA digested with *DraI* (A.) and *EcoRI/HindIII* (B.) and with *nad 3* of *EcoRI/HindIII* digested DNA (C.). 1. CMS-*pennellii*; 2. *L. peruvianum*; 3. *L. pennellii*; 4. *L. esculentum*.

On individual CMS and *L. peruvianum* plants was studied the hybridisation pattern with *S-Pcf* locus from *Petunia hybrida* (data not shown). This complex locus carries part of *cox II* gene, *urf-s*, *nad 3* and *rsp 12* and a complex pattern of hybridisation was expected to appear. Digestion with *HindIII* produced six fragments with different intensity, with highest intensity in all individual CMS plants is the 5.2 kb fragment. In addition to these six fragments a 7.7 kb fragment with low intensity is present in four of the studied individual plants of *L. peruvianum* which points at the existence in the last of a recombination mt DNA molecule.

4. Conclusion

The characterisation of the mitochondrial genome of CMS-*pennellii* line derived from *L. peruvianum* x *L. pennellii* with different heterologous plant mt DNA genes has shown that two of the studied genes: *atp A* and *nad 3* are proximal and situated on a 3.6 kb *EcoRI/HindIII* fragment. The additional fragments revealed in the hybridisation profile of individual plants of *L. peruvianum* with *atp A* and clone 2 in comparison with CMS-*pennellii* plants can be explained by the presence of recombinant mt DNA molecule(s) with low stoichiometry in the cytoplasm of the progenitor. The absence of these fragments from the RFLP pattern of the CMS-*pennellii* might be discussed as an evidence for the generation of diversity of the mitochondrial genome due the interaction of the cytoplasm of *L. peruvianum* with the nuclear genome of *L. pennellii.*

5. References

Andersen, W. (1963) Cytoplasmic sterility in hybrids of *Lycopersicon esculentum* and *Solanum pennellii. Report of Tomato Genet. Coop.*, **13**, 7.

Andersen, W. (1964) Evidence of plasmon differentiaon in *Lycopersicon. Rep. Tomato Genet. Coop.*, **14**, 4-6.

Dewey, R.E., Levings III Charles S., and Timothy, D. H. (1985) Nucleotide sequence of ATPase subunit 6 gene of maize mitochondria, *Plant Phisiol.***79**, 914-919.

Fulton, T. M., Chunwongse, J., and Tankesley, S. D. (1995) Microprep protocol for extraction of DNA from tomato and other herbaceous plants, *Plant Molecular Biology Reporter* **13**, 3, 207-209.

Gwynn B., Dewey R.E., Sederoff, R.R., Timothy, and D.H. Levings III, C.S. (1987) Sequence of the 18S+5S ribosomal gene region and the cytochrome oxidase II gene from mtDNA of *Zea diploperennis, Theor. Appl. Genet.* **74**, 781-788.

Hanson, M. (1991) Plant mitochondrial mutations and cytoplasmic male sterility. *Annu. Rev. Genet.*, **25**, 461-486.

Hanson, M., and Conde M. (1985) Function and variation of cytoplasmic genomes: lessons from cytoplasmic nuclear interactions affecting male fertility in plants. *Int. Rev. Cytol.*, **94**, 213-267.

Kaul, M.L.H., (1988) Male sterility in higher plants. *Monographs on Theor.Appl.Genet.* **10**. Springer Verlag, Berlin.

Leaver, C., Isaac, P., Small, I., Bailey-Serres, J., Liddell, A., and Hawkesford, M. (1988) Mitochondrial genome diversity and cytoplasmic nale sterility in higher plants, *Phil. Trans. R.Soc. Lond.*, **B 319**, 165-176.

Vulkova- Achkova Z. (1980) *L. peruvianum* a source of CMS. *Report of Tomato Genet.Coop.*, **30**, 36.

Yesodi, V., Izhar, S., Gidoni, D., Tabib, Y., and Firon, N. (1995) Involvement of the different *urf-s* related mitochondrial sequences in the molecular evolution of the CMS-specific *S-Pcf* locus on petunia. *Mol. Gen. Genet.* **248**, 540-546.

GENETIC ANALYSIS ON THE T-DNA INSERTION MUTANTS OF TOBACCO (*NICOTIANA TABACUM*)

S. ŞAHİN, S.D. GÜRLE AND Ş. ARI
University of İstanbul, Department of Biology, Section of Molecular Biology, 34459-Vezneciler, İstanbul-TÜRKİYE

1. Abstract

Two leaf color mutants were obtained from transformation of *Nicotiana tabacum* L. cv. Maden leaf discs with Agl-1 strain of *Agrobacterium tumefaciens* containing pBI-121 plasmid. R_1 and R_2 generations of these mutants were analysed phenotypically. Segregation of mutant phenotype in progenies was found in accordance with Mendelian inheritance. Excistence of GUS and NPT II genes in mutant plants and in their progenies were shown with PCR and DIG labelling techniques. All plants were tested on kanamycin containing selective media which is the indicator of NPT II gene. Histochemical GUS analysis indicated the active expression of GUS gene.

2. Introduction

Desirable foreign genes can be stably introduced into the genome of many plants by manipulation of the naturally occurring transformation mechanism of the T-DNA in *Agrobacterium tumefaciens* (Gasser et al.,1989). However, sometimes the T-DNA inactivates or changes of the plant structural genes integrating randomly or more than one copy into the plant genome. In this way, it occurs some metabolic and developmental mutations which are called insertion mutations. Therefor the structural genes that are releated this mutations can be isolated and characterized using T-DNA. In order that T-DNA insertion mutagenesis can be used as an alternative method of tagging gene in plants besides the transposable elements (Feldmann, 1991). In this study, leaf discs of *Nicotiana tabacum* L. cv. Maden were transformed with Agl-1 strain of *Agrobacterium tumefaciens*. Two leaf color mutants were selected after transformation. These mutants and their generations were analyzed genetically, histochemically and phenotypically.

3. Materials and Methods

3.1. PLANT MATERIAL

Leaf discs of *Nicotiana tabacum* L. cv. Maden were transformed with Agl-1 strain of *Agrobacterium tumefaciens* carrying pBI-121 plasmid that includes GUS reporter and NPT II marker genes. After transformation two leaf color mutant (green-white) were selected among various mutants. They were called Transgenic-1 (TR-1) and Transgenic-2 (TR-2). These mutants and their R_1 and R_2 progenies were used as plant materials for genetic, histochemical and phenotypic analysis.

3.2. GENETIC ANALYSIS

3.2.1. Polymerase Chain Reaction (PCR)

Transferred GUS and NPT II genes were amplified from genomic DNA of TR-1, TR-2 and their R_1 and R_2 progenies according to Saiki et al. (1988).

3.2.2. Digoxigenin Labelling (DIG)

Amplification products were transferred to nitrocellulose membrane and hybridized with DIG labelled GUS and NPT II probes. After hybridization immunological determination was performed.

3.3. HISTOCHEMICAL ANALYSIS

Expression of GUS gene in the plants was confirmed with histochemical X-Gluc analysis according to Jefferson et al., (1987).

3.4. PHENOTYPICAL ANALYSIS (SEGREGATION ANALYSIS)

After germination approximately 784 seeds from TR-1 mutant and 800 seeds from TR-2 mutant in progenies were scored to determine segregation of mutant trait.

4. Results and Discussion

Integration and expression of the GUS and NPT II genes in transgenic plants were confirmed by PCR amplification, DIG labelling (Figure 1) and histochemical GUS assays into primary leaf color mutant and their R_1, R_2 progenies. However, in segregation analysis of TR-1 (R_1 progeny), only normal green leaves plants and a few albino plants were accounted. Since the albino plants died during developing, R_2 progeny was only grown up from normal green leaves plants. All plants in R_2

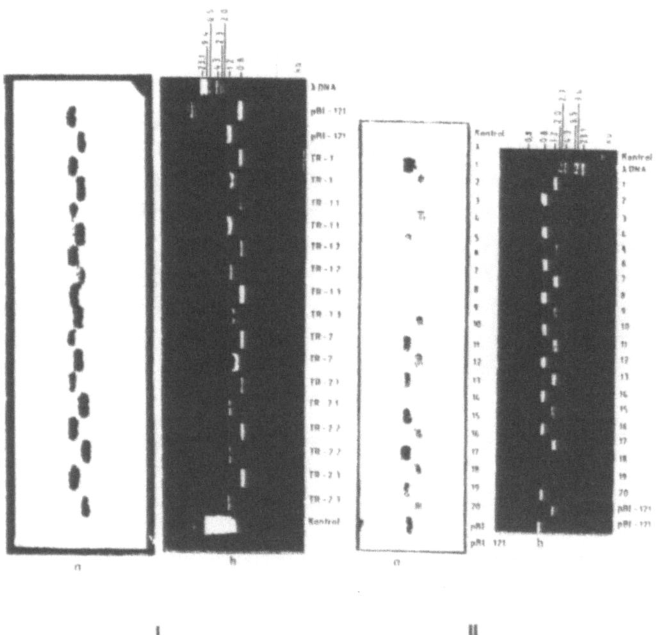

Figure1. Existence of GUS and NPT II genes in Transgenic-1, Transgenic-2 and their generations. **(+)control:** pBI-121 plasmid, **(-) control:**non-transgenic *Nicotiana tabacum,* **1,2 kb band:**GUS gene, **0,8 kb band:**NPT II gene, **a:**DIG labelling, **b:**PCR products, **I:**From TR-1.1 to TR-1.3 are R_1 generation of TR-1, From TR-2.1 to TR-2.3 are R_1 gercneration of TR-2, **II:** R_2 generation of TR-1 and TR-2.

Figure 2. Four different phenotype in TR-2 generations. **(A):**Normal-green leaves, **(B):**Light-green leaves, **(C):**Green-white leaves, **(D):**Albino plants.

pogeny had normal green leaves. All these experiments showed the loss of mutant phenotype in progenies even though the precence of transferred genes in plant genome. It was suggested that the insertion mutation in TR-1 occurred chimerically in part of leaf tissue cells which differentiated as the somatic tissue during plant regeneration. Therefore, the mutant trait has been lost in next progenies.

In segregation analysis of TR-2 (R_1 progeny), four different phenotypes were obtained whose traits were (1)normal-green leaves, (2)green-white leaves, (3)light-green leaves and (4)albino leaves (Figure 2). Since light-green leaves plants grew very slow and albino plants died during developing, R_2 progenies was only grown up from normal-green leaves plants and green-white plants. The segregation of R_1 and R_2 progenies is showed in Table 1. In both of progenies, the segregation ratio for four phenotype was 9:3:3:1 and this result was confirmed with chi-square test. Moreover this ratio was in according to Mendelian Inheritance for two unlinked genetic loci. Similar to our findings Katavic et al. (1994) observed integrated T-DNA inherited according to Mendelian patterns.

TABLE 1. Segregation analysis of TR-2 (NG: normal green leaves plants, GW: green-white leaves plants, LG: light-green leaves plants and A: albino plants)

	GW (R_1)	Segregation ratio (R_1)	GW (R_2)	Segregation ratio (R_2)	NG (R_2)
NG	105	9	234	9	213
GW	23	3	79	3	-
LG	39	3	73	3	-
A	9	1	25	1	-

In addition, localization of T-DNA in genome of TR-2 mutants and their generations was suggested that it can be effected on a related way of chlorophyll syntase, differentiation of plastid or photosynthesis. In the recent years, the plastome mutants that are related with changing leaf color like yellow, white, cream, green-white and light-green are researched importantly. In most cases the plastome mutants are kept and propagated in the form of green-white variegated plants (Hagemann et al.,1996). Besides one of the important inducing factor for plastome mutations is nuclear genes. In conclusion, all our experiments and this knowledge were suggested that TR-2 and their mutant progenies may be plastome mutants.

5. References

1. Feldmann K.A. (1991) T-DNA insertion mutagenesis in Arabidopsis:mutational spectrum, *Plant Journal* **1**, 71-82.
2. Gasser C.S. and Fraley R.T.(1989) Genetically engineering plants for crop improvement, *Science*, **244**, 1293-1299.
3. Hagemann R., Bock R. and Hagemann M.M. (1996) Extranuclear inheritance:Plastid Genetics, *Progress in Botany*, **57**, 197-215.
4. Jefferson R.A., Kavanag T.A. and Bevan M.W. (1987) GUS fusion: β–glucuronidase as a sensitive and versatile gene fusion marker in higher plants, *The EMBO Journal*, **6(13)**, 3901-3907.
5. Katavic V., Haughn G.W., Martin M. and Kunts L. (1994) In plant transformation of Arabidopsis thaliana, *Molecular and General Genetics*, **245(3)**, 363-370.
6. Saiki R.K., Gelfald D.H., Stoffel S., Scharf S.J., Higuchi R.,Ham G.T., Mullis K.B. and Erlich H.A. (1988) Primer directed enzymatic amplification of a thermostable DNA polymerase, *Science*, **239**, 487-491.

ISOZYME POLYMORPHISM AT THE 6-PHOSPHOGLUCOSE DEHYDROGENASE (6PGD) AND LEUCINE AMINOPEPTIDASE (LAP) LOCI IN 18 POPULATIONS OF CEDAR SPECIES

A. SCALTSOYIANNES
Assistant Professor
Aristotle University of Thessaloniki, Department of Forestry & Natural Environment, Laboratory of Forest Genetics and Plant Breeding, 54006 Thessaloniki

1. Abstract

Isozyme variation pattern in 18 Cedar species populations (4 from Turkey, 6 from Lebanon, 1 from Himalayas, 1 from Afghanistan, 1 from Cyprus, 4 from Algeria and 1 from Morocco) belonging to the 4 Cedar species (*Cedrus libani, C. atlantica, C. deodara, C. brevifolia*) was studied at two loci (6PGD-B, LAP-A) by the horizontal starch and polyacrylamide gel electrophoresis techniques. Species specific alleles and alleles with geographic variation pattern were detected in both loci. *Cedrus deodara* and *C. atlantica* have diagnostic alleles in both loci (*C. deodara* for 6PGD-B4 and LAP-A1; *C. atlantica* for 6PGD-B5 and LAP-A2). LAP-A locus is nearly monomorphic at different alleles in the Lebanese (LAP-A3) and the Turkish (LAP-A2) populations of C. libani, while in the same species 6PGD-B locus exhibited geographic variation pattern discriminating the Turkish populations from the Lebanese ones. In *C. brevifolia* variation was observed in both loci.

2. Introduction

The genus *Cedrus* includes coniferous evergreen trees and it is distributed in four geographically separated regions: a) Algeria and Morocco, b) Cyprus, c) Lebanon, Syria and Turkey and d) Afghanistan and the Himalayas.

Due to its exceptional adaptation in a variety of environmental conditions, *Cedrus* has been successfully introduced in many countries outside of its natural distribution, first as a ornamental and then as a reafforestation species.

Today, in spite of its considerable ecological and economic importance, little information concerning the amount and pattern of its genetic variability is available (FAO, 1989). Besides this, according to a number of provenance trials and taxonomic studies based on morphological and anatomical characteristics there is no general consensus on its taxonomy, especially for the species of the Mediterranean basin.

Cedrus was first described as a genus by Trew in 1757 and the most taxonomists now agree on 4 species classification: *Cedrus atlantica* in Morocco and Algeria, *C. brevifolia* in Cyprus, *C. libani* in Lebanon, Syria and Turkey and *C. deodara* in Afghanistan and India.

Recent works (Panetsos *et al.*, 1993), based on isozyme variation among different populations showed that considerable inter-specific allozyme variation exists.

The purpose of this work was to investigate the isozyme variation pattern of two loci (6PGD-B, LAP-A) in 18 Cedar species populations from the whole natural distribution of the genus.

3. Materials and Methods

Electrophoresis was carried out on haploid endosperms of germinated seeds of *Cedrus* species.

The seed material was taken from 18 bulk stand collections (4 from Turkey, 6 from Lebanon, 1 from Afghanistan, 1 from Himalayas, 1 from Cyprus, 4 from Algeria and 1 from Morocco provided by INRA and the National Forest Services.

6PGD-B locus was resolved by means of horizontal starch gel (10.5% w/v) electrophoresis technique (Cheliak and Pitel, 1985), while LAP-A in horizontal polyacrylamide gel (7.5%) electrophoresis technique (Pasteur *et al.*, 1988). The 6PGD and LAP enzyme systems was run in Histidine buffer system pH 8.1 and in Morpholine-Citrate buffer system pH 6.1, respectively.

Isozyme patterns of enzyme systems were examined in random samples consisting of 75-100 endosperms in each population.

4. Results

Leucine Aminopeptidase (LAP). Gels stained for LAP displayed two zones of activity corresponding to two loci LAP-A and LAP-B and in the present work only LAP-A was scored.

Although in the previous works (Panetsos *et al.*, 1993) LAP-A was appeared to be monomorphic in all Cedar species, under our electrophoretic conditions (Morpholine buffer system, polyacrylamide gel) we observed variation among species and populations due to the better resolution among the enzyme variants.

Among 18 Cedar populations tested, three distinct single banded variants were detected at the LAP-A locus. High allozyme frequencies variation was observed among the species and the populations of *C. libani* (Table 1).

In particular, LAP-A locus was fixed at LAP-A1 allele in C. deodara and at LAP-A2 allele in *C. atlantica* while the same locus was nearly monomorphic in the Lebanese and the Turkish populations of *C. libani* at different alleles, LAP-A3 and LAP-A2 respectively.

Phosphoglucose Dehydrogenase (6PGD). Under our electrophoretic conditions, gels stained for 6PGD displayed two zones of activity sometimes overlapped, with 6PGD-A so poorly resolved that only 6PGD-B could be reliable scored.

Among 18 Cedar populations tested, six distinct single-banded variants were detected at the 6PGD-B locus. Allozyme frequencies varied by species and populations and imply that considerable genetic differentiation exists among the species and populations.

6PGD-B locus was fixed at 6PGD-B4 allele in *C. deodara* and at 6PGD-B5 allele in *C. atlantica* except the Moroccan population of *C. atlantica* in which a rare allele (6PGD-B6) was present.

C. libani can be discriminated from the above species by the presence of the 6PGD-B2, 6PGD-B3, 6PGD-B4, 6PGD-B5 alleles in a wide range of frequencies. The allele frequency data (Table 1) shown a latitudinal cline pattern variation (from North to South) dividing the whole species into two discreet geographical races (Turkey and Lebanon).

Concerning *C. brevifolia*, it seems to be closer to Turkish populations (Taurus populations) and it is discriminated from them by the higher frequency of 6PGD-B allele.

Table 1. Allele frequencies at LAP-A, 6PGD-B loci in 18 populations of Cedar sp.

Species & Provenances	Loci and Alleles								
	LAP-A			6PGD-B					
	a1	a2	a3	b1	b2	b3	b4	b5	b6
C. libani									
1. Catalan	-	1.00	-	-	1.00	-	-	-	-
2. Sutlegen	-	0.90	0.10	-	0.65	-	0.35	-	-
3. Pozanti	-	0.90	0.10	0.02	0.42	-	0.56	-	-
4. Arslankoy	-	0.94	0.06	0.02	0.50	-	0.48	-	-
5. Akkar East	-	0.06	0.94	-	0.18	0.04	0.56	0.22	-
6. Akkar Ouest	-	-	1.00	-	0.08	0.03	0.52	0.37	-
7. Zhalta	-	0.06	0.94	-	0.17	0.17	0.63	0.03	-
8. Masser Schouff	-	0.09	0.91	-	0.11	-	0.89	-	-
9. Niha	-	0.05	0.95	-	0.12	-	0.88	-	-
10. Barouk	-	0.05	0.95	-	0.04	0.04	0.92	-	-
11. C. brevifolia	-	0.87	0.13	-	0.80	-	0.20	-	-
C. deodara									
12. Afghanistan	1.00	-	-	-	-	-	1.00	-	-
13. Himalayas	1.00	-	-	-	-	-	1.00	-	-
C. atlantica									
14. Babors	-	1.00	-	-	-	-	-	1.00	-
15. Ouarsenis	-	1.00	-	-	-	-	-	1.00	-
16. Djurdjura	-	1.00	-	-	-	-	-	1.00	-
17. Aures	-	1.00	-	-	-	-	-	1.00	-
18. Morocco	-	1.00	-	-	-	-	-	0.94	0.06

Figure 1. Frequency distribution of LAP-A and 6PGD-B loci in 18 populations of Cedar species.

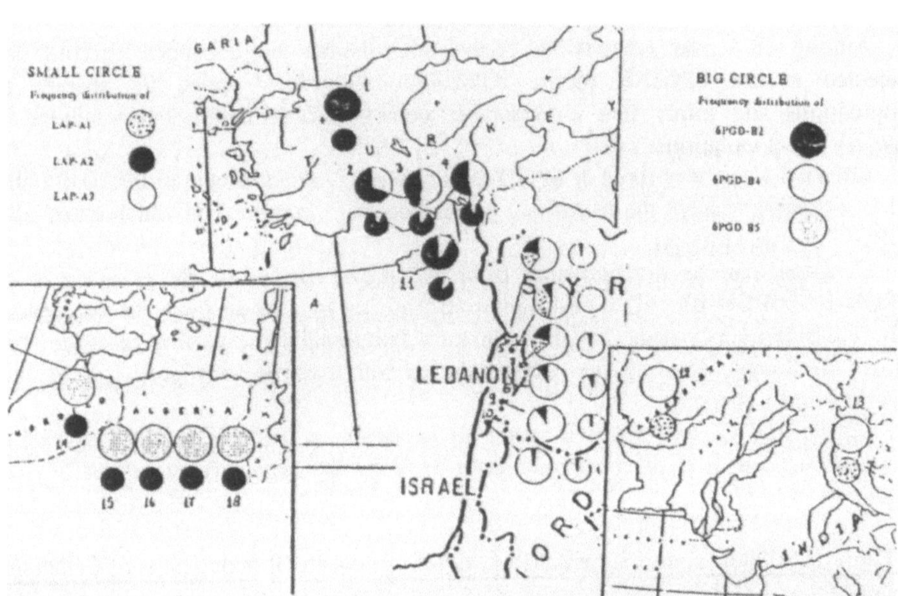

5. Conclusions

Both loci (6PGD-B and LAP-A) possess alleles with diagnostic value discriminating the Cedar species into four distinct species and also distinguishing the Lebanese populations of *C. libani* from the Turkish ones, corresponding to *C. libani* ssp. *libani* and *C. libani* ssp. *stenocoma* according to Davis (1965), Greuter *et al.* (1984) and Arbez (1978).

6. References

Arbez, M. (1987) Les ressources genetiques forestieres en France. Les coniferes 1. INRA, Paris.

Cheliak, W. M. and Pitel, J. A. (1985) Techniques for starch gel electrophoresis of enzymes from orest tree species, Information report PI-X-42, Petawawa National Forestry Institute, etawawa, Canada.

Davis, P. H. (1965) Flora of Turkey and East Aegean Islands, Vol. I. 71-72, Univ. Press, dinburgh.

FAO (1989) Sylviculture of Species: *Cedrus* spp. "Silva Mediterranea", FO: SCM/89/6.

Greuter, W., Burdet, H. M. and Long, G. (1984) Med-Checklist. 1. Pteridophyta (ed. 2) ymnospermae Dicotylidonae (Acanthaceae - Gneoraceae), Edition de Conservatoire et ardin botaniquesde la ville de Geneve.

Panetsos, K. P., Scaltsoyiannes, A. and Tsaktsira, M. (1993) Genetic variation in allozymes of *edrus libani* A.Rich and *Cedrus atlantica* Mannetti. Seminaire International sur le Cedre de 'Atlas, Ifrane 7-11 Juin 1993.

Pasteur, N., Pasteur, G., Bonhomme, F., Catalan, J. and Britton-Davidian J. (1988) Practical Isozyme enetics, 215 pp., Ellis Horwood Limited, Chichester, England.

IDENTIFICATION OF PINUS GENOTYPES WITH RAPD MARKERS

F. ŞENTÜRK [1], A. AKÇİN [1] AND N. GÖZÜKIRMIZI [2,3]
[1]*GIT, Gebze Institute of Technology, Gebze/ Kocaeli-TURKEY*
[2]*TÜBİTAK, MRC, Research Institute for Genetic Engineering and Biotechnology Gebze/Kocaeli-TURKEY*
[3]*University of İstanbul, Faculty of Science Vezneciler/İstanbul-TURKEY*

1. Abstract

Genetic polymorphism was studied five different pine species; *P. pinaster, P. nigra, P. sylvestris, P. brutia,* and *P. pinea*. Genomic DNA isolated from needles was used to determine genetic polymorphism among different pine species by using RAPD (Random Amplified Polymorphic DNA) molecular marker analysis. Among tested thirty-one primers, twelve primers produced amplified bands on the electrophoretic gels. Primer GYTE 9 produced homomorphic bands for all the tested genotypes. However, primer TUB 365 (5' GGACTGGAGT 3') produced polymorphic RAPD bands for these genotypes that could be used species identification. According to results of polymorphism calculations, the highest polymorphism rate was 52.75% between *P. nigra* and *P. pinaster.* The lowest polymorphism was 28.25% between *P. nigra* and *P sylvestris.*

2. Introduction

DNA markers are essential for many types of genetic analysis in forest genetics and tree improvement research such as individual identity, family relationship and linkage mapping (Carlson *et al.*, 1991). In between the available molecular markers types random amplified polymorphic DNAs (RAPDs) (Welsh and McClelland, 1990; Williams *et al.*, 1990) have a wide application since they are rapid, easy, and economical when compared to other markers e.g. RFLPs, STS etc (Staub *et al.*, 1996). Pinus is an important forest plant in Turkey both environmental and various industrial aspects such as paper and pulp industry. In addition, Turkey has a rich germplasm on *Pinus sp.* (Uslu, 1988). Genetic polymorphism was studied on five different pine species; *P. pinaster, P. nigra, P. sylvestris, P. brutia,* and *P. pinea* by using RAPD markers. Genetic polymorphisms in between the species were calculated and evolutionary implications derived from RAPD analysis were presented.

3. Materials and Methods

3.1. PLANT MATERIAL

Needles of *P. pinaster, P. nigra, P. sylvestris, P. brutia,* and *P. pinea* were obtained from plantations around Gebze-TURKEY.

3.2. DNA ISOLATION

A modified method was developed using the main step of Wagner *et al.* (1987) and Saghai-Maroof *et al.* (1984) avoid of the contamination effects of phenolic compaunds on DNA isolated from pine needles. Needle tissue (0.5 g, fresh weight) was powdered with liquid nitrogene and was homogenized a mortar with pestle in 12.5 ml extraction buffer (50 mM Tris pH 8, 5 mM EDTA, 0.35 M sorbitol, 10% polyethylene glycol (PEG), 0.5% spermidine, 0.1% 2-mercaptoethanol). The homogenate was filtered through four layers of cheescloth. A pellet was collected from the homogenate by centrifugation (13000 rpm, 15 min). The pellet was suspended in 100 µl of wash buffer (50 mM Tris pH 8, 5 mM EDTA, 0.35 M sorbitol, 0.5% spermidine, 0.1% 2-mercaptoethanol) and brought to room temperature. N-laurylsarcosine was added to a concentration of 1% (w/v). After 15 min at the room temperature, the suspension was brought to final concentration of 1% (w/v) Hexadecyltrimethylammonium bromide and 0.7 M NaCl and incubated at 60 °C for 10 min. 1 ml of chloroform/octanol (24:1) was added, and the solution was mixed by inversion to form an emulsion that was centrifuged at 5000 rpm for 10 min at room temperature. The aqueous phase was removed, and 2/3 volume of isopropanol was added and mixed by two-to-four quick, gentle inversions. The precipated DNA was lifted out with a glass hook, transferred to a tube containing 1 ml of 70% ethanol, 10 mM NH_4Ac for 20 min, and then dissolved in 150 µl of 10 mM NH_4Ac and 0.25 mM EDTA.

3.3. AMPLIFICATION CONDITIONS FOR RAPD

Various base long oligonucleotides (*Table*) were used in the Polymerase Chain Reaction (PCR). Amplifications were performed in volumes of 25 µl containing 100 µg genomic DNA, 0.2 mM each dATP, dTTP, dGTP, dCTP mixture, 100 ng primer, 1 unit Taq DNA polymerase, 10 mM Tris-HCl pH 8.3, 50 mM KCl, 0.01% gelatine, 2 mM $MgCl_2$. Amplification was performed in the thermalcycler (MJ Reseach PTC100) programmed for first 4 cycles of 1.5 min at 94 °C, 1.5 min at 35 °C, and 2 min at 72 °C, for 26 cycles of 1 min at at 94 °C, 1 min at 35 °C, and 1.5 min at 72 °C. A 10 min extension period at 72 °C followed the amplification. Gel containing 1.2% agarose, ethidiumbromide, at 0.4 $mg.L^{-1}$ and TBE (90 mM Tris-borate, 90 mM boric acid, and 2 mM EDTA pH 8.2) were used to separate amplification products. After pipetting 22-24 µl from under the mineral oil and loading it onto the gel, electrophoresis was conducted for 1.5 hours. Amplification products were detected under UV and photographed.

3.4. CALCULATION OF POLYMORPHISM

Polymorphisms at DNA level were calculated according to the method of Nei and Li (1979) and dendrogram was developed using Statgraphics Plus Ver 2.1.

Table: Primers used in this study

Primer Name	Sequence (5'....3')	Primer Name	Sequence (5'....3')
TUB 341	TGAGTGGGTG	GYTE1	TGCAGGACTAGCATTGGAACTC
TUB 352	GTAGACCCGT	GYTE2	CTGGGAGGCAACTATCCAAAAG
TUB 354	TTCCCCCGCT	GYTE3	GAAAAAAGTTTGCCCAGCAT
TUB 357	TGCTCTGCCC	GYTE4	CATGCTTCATTTGATGCAATTG
TUB 359	GTCCACACGG	GYTE5	GATCTCCTCCAACACCAGAAAA
TUB 364	CATCCCCCTG	GYTE6	CTGCAGGTTCTGTTGTCCCG
TUB 365	GGACTGGAGT	GYTE7	CTACTGCCGCTTCCGAATTAAG
TUB 370	CAAACGTCGG	GYTE8	TGCAGTTGGAACACTTACTCG
TUB 374	CAATCGCCGT	GYTE9	CTGCAGTGGCTTTCACCTTC
TUB 376	CAGCACCCAC	GYTE10	TGCAGATTGTACAGGGCTGG
TUB 379	GGTCCCTGAC	MAM 1	CAATCGCCGT
TUB 380	GAAACGGGTG	MAM 2	TCGGCGGATAG
TUB 385	GTGAGGCGTC	MAM 3	CCTTGACGCA
TUB 386	GGGGGTCTTT	MAM 4	TGGACCGGTG
TUB 388	GATGACCGCC	TUB 396	AAGCCTCGTC
TUB 392	CTCACCGTCC	TUB 397	TGCGTGCTTG
TUB 393	TGTCTGGGTG		

4. Results and Discussion

DNAs from five different pinus species were randomly amplified using RAPD technique for genotype identification and polymorphism rates were calculated by pairwaise comparition of amplified fragments. Isolation of high quality plant DNA from woody plants is known to be difficult because of the presence of contaminating polysaccharides and polyphenolic compounds which can irreversibly react with nucleic acids (Rowland and Nguyen, 1993). We also faced to same difficulties because of highly polyphenolic compounds, but it became possible to isolate high purity DNA using especially polyethylene glycol step. RAPD patterns show differences among tested 31 primers, only 11 primers showed heteromorphic patterns of amplification. One primer, GYTE 9 produced monomorphic bands, and TUB 365 (5' GGACTGGAGT 3') produced polymorphic RAPD bands for all genotypes which can be used species identification (*Figure 1*). However, no amplification was obtained with tested 19 primers. Kaya and Neale (1993) was used RAPD analysis to detect genetic polymorphism in *P.nigra subsp* Pallasiana and *P.brutia.*

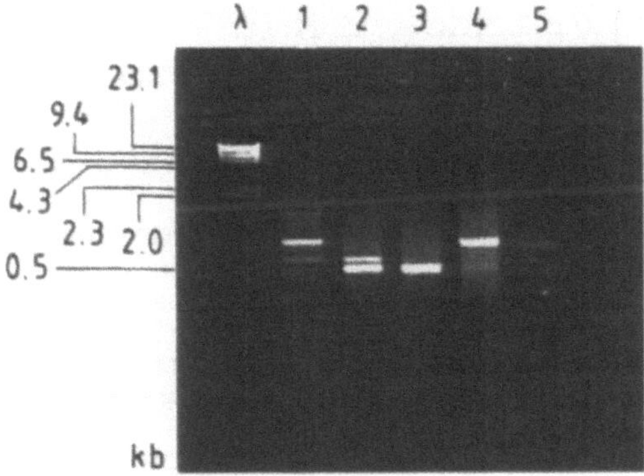

Figure 1: RAPD analysis using TUB 365 primer with DNA samples from 1. *P. pinaster,* 2. *P. nigra,* 3. *P. sylvestris,* 4. *P. brutia,* and 5. *P. pinea* λ, DNA marker digested with Hind III.

The highest polymorphism was 52.75% between *P.nigra* and *P.pinaster*. The lowest polymorphism was 28.25% between *P.nigra* and *P.slyvestris*. The result were demonstrated that there is a close genetic relationship between *P.nigra* and *P.slyvestris* (*Figure 2*). Devoloping techniques for DNA isolation from highly phenolic pine needles and in particular obtaining proper primers which will be used in the determination of polymorphism will provide and essential contribution to the genetic research in the field of pinus genotypes.

Average distance between clusters

0 20 40 60 80 100

P.pinaster

P.nigra

P.sylvestris

P. brutia

P. pinea

Figure 2: Dendogram of genetic relatedness of five different pine genotypes based on 59 RAPD markers by using statistical program Statgraphics Plus 2.1.

5. Acknowledgements

This work was performed at TÜBİTAK Marmara Research Center Plant Biotechnology Unit Labaratories and was supported by the project (96-A-01-04-02) of Gebze Institute of Technology.

6. References

Carlson, J. E., Tulsieram, L. K., Glaubitz, J. C., Luk, V. W. K., Kauffeldt, C., and Rutledge, R. (1991) Segregation of random amplified DNA markers in F_1 progeny of conifers. *Theor. Appl. Genet.* **3**, 194-200.

Kaya, Z., and Neale, D. B. (1993) Random Amplified Polymorphic DNA (RAPD) Polymorphisms in *Pinus nigra subsp. pallasiana* and *Pinus brutia*. *Tr. J. of Agricultural and Forestry*, **12**, 295-306.

Nei, M., and Li, W. H. (1979) Mathematical model for studying genetic variation in terms of restriction endonucleases. *Proc. Natl. Acad. Sci. USA*, **76**, 5269-5273.

Rowland, L. and Nyugen, B. (1993) Use of polyethylene glycol for purification of DNA from leaf tissue of woody plants, *Biotechniques*, **14 (5)**, 734-736.

Saghaı-Maroof, M. A., Soliman, K. M., Jorgensen, R. A., and Allard, R. W. (1984) Ribosomal DNA spacer-length polimorphisms in barley: Mendelian inheritance, choromosomal location, and population dynamics. *Proc. Natl. Acad. Sci. USA*, **81**, 8014-8018.

Staub, J. E., Serquen, F. C. and Gupta, M. (1996) Genetic markers, map construction, and their application in plant breeding, *Hortscience*, **31 (5)**, 729-741.

Uslu, S. (1988) Türkiye Ormancılığına Genel Bir Bakış. *Kavak ve Hızlı Gelişen Yabancı Tür Orman Ağaçları Araştırma Enstitüsü Dergisi*, **1**, 1-34.

Wagner, D. B., Furnier, G. R., Saghai-Maroof, M. A., Williams, S. M., Dancık, B. P., Allardi, R. V. (1987). Chloroplast DNA polymorphisms in lodgepole and jack pines and their hybrids, *Proc. Natl. Acad. Sci. USA*, **84**, 2097-2100.

Welsh, J., McClelland, M. (1990) Fingerprinting genomes using PCR with arbitrary primers, *Nucleic Acids Research*, **18: (24)**, 7213-7218.

Williams, J. K. G., Kubelik, A. R., Livak, K. J., Rafalski, J. A., and Tingey, S.V. (1990) DNA polymophisms amplified by arbitrary primers are useful as genetic markers, *Nucleic Acids Research*, **18: (22)**, 6531-6535.

GENETIC DETERMINATION AND APPLICATION OF PHOTOSYNTHETIC INDEXES AS A REAL WAY IN CROP PRODUCTION IMPROVEMENT

O.I. SIDORENKO and E.V. SHPAK
Dr., Head of Lab. for Photosynthesis and Ph. D. student
Institute of Plant Physiology, Genetics and Bioengineering, National Center of Biotechnology, Timiryasev 45, Almaty, 480090, Kazakstan

1. Abstract

A real approach to plant production process' regulation is the creation of physiology - genetic active photosynthetic apparatus by crop breeding. Increasing of photosynthetic apparatus activity can be created by a way of photosynthetic tests' elaboration and application and its genetic determination in limiting environments. It has been shown that wheat crop production based on functional activity and structure all of the levels of photosynthetic apparatus organization: from photosystems - to whole plant. The more suitable indexes for crop breeding in field conditions are the morpho-physiological ones and CO2 -gas - components in leaf and whole plant. We have characterized 30 photosynthetic indexes for tests on productivity and drought adaptability elaboration and 11 photosynthetic test-systems with different economic-valuable meaning . Based on these indexes with using of correlation, combine ability and polygene analysis, we have elaborated the ways of wheat productivity and ecological adaptability determination. The new lines of wheat - donors of photosynthetic indexes and first productive sort of winter wheat with high photosynthetic activity were produced.

2. Introduction

Agriculture need new sorts with optimal combination of economical - valuable properties and resistance to biotic and abiotic limiting factors of environments. For creation of such sorts crop breeding need new ideas, approaches and methods - new ways of Biotechnology of adjacent biological sciences: plant physiology, biochemistry, genetic. The crop breeding enters into a new stage, which may be called «synthetic crop breeding» based on physiology-genetic management of photosynthesis, growth, development processes and is characterized by passage from extensive to intensive way of crop production improvement. A real approach to plant production process' regulation is the creation of physiology - genetic active photosynthetic apparatus by crop breeding. Increasing of photosynthetic apparatus activity can be created by a way of photosynthetic tests' elaboration and application and its genetic determination in favorable and limiting environmental conditions. In these connection its necessary new ways' stages: a) systematic study of photosynthetic indexes on different levels of photosynthetic apparatus organization, b) differentiation of selective collections by photosynthetic tests based on its genetic determination, c) revealing of forms - donors of high photosynthetic activity, d) creation of a model of Optimal photosynthetic type wheat plant with maximum productivity and adaptability to environments, e) creation of new productive and adaptive sorts with optimal combination of economic-valuable properties.

3. Methods

It have been used more 100 contrast productive and environmental adaptive wheat objects during 1980-1996 years. Morpho-physiological indexes have been determined by measuring of leaf area, distribution and redistribution of biomass by organs in ontogenesis, elements of yield structure. Co_2-gas-changes have been measured in open and closed system with using of IRGA Infralyt, Junkalor in special chambers. Polygene, combine ability and correlation analysis were conducted by Hayman, Griffing and Dospehov methods.

4. Results and discussion

It has been shown that wheat crop production based on functional activity and structure on all of the levels of photosynthetic apparatus organization: from photosystems - to whole plant. The more suitable indexes for crop breeding in field conditions are the morpho-physiological ones and CO2 -gas - components in leaf and whole plant. We have characterized 30 photosynthetic indexes for tests on productivity and drought adaptability elaboration. More informative among them are: the shoot biomass, ear and leaf biomass interaction in flowering, organ biomass distribution from flowering to wax ripeness, net photosynthesis, laf area (table 1). Biological and economic-valuable meanings of photosynthetic tests permit to unite them into photosynthetic test-system: I - Extensive indexes of Photosynthetic apparatus' sizes. II - Intensive indexes of functional activity of Photosynthetic apparatus. III - ^{Inq} - «inquiry» of ear in assimilates. IY - ^{Real} - realization of ear's inquiry in assimilates in ontogenesis. Y - ^{Incr} - increase of ear mass, YI - ^{LInq} - leaves' function in grain mass forming. YII - Rate of biomass accumulation in reproductive and vegetative period. YIII - Distribution and redistribution of assimilates by source and sink organs in ontogenesis. IX - Part of harvesting-valuable grain mass in common plant biomass. X - Real organ's biomass creation. XI - Real crop yield mass creation.

Based on these indexes with using of polygene, combine ability and correlation analysis (table 2), we have elaborated the ways of wheat productivity and ecological adaptability determination. It have been shown photochemical activity (PCA) of chloroplasts is controlled by gene-systems with additive and non-additive effects. Effects of sub-dominance are typical, ones stipulates for positive and negative heterosis in wheat hybrids and can be inherit it. High PCA predominates over low ones in these group of forms, low PCA is inherited by recessive type, that facilitates selection by this index from F_2. Many of heteroseous hybrids is characterized by low number of chloroplasts in cell but high photosynthetic activity of one chloroplast. It corresponds to idea about possibility of existence of genetics «strong» chloroplasts and their using as a test in crop breeding on productivity. Genetic variation of Net Photosynthesis is operated by polygene nuclear system with non-additive and less additive effects. High rate of Net Photosynthesis is controlled by recessive genes as rule, independently of gene-type. Leaf area is characterized by polygene type of inheritance with genetics polymorphism. For nuclear genes which controlled the development of assimilation apparatus, inherent to sub-domination effects. Revealing

TABLE 1. Photosynthetic indexes as the morpho-physiological tests on productivity and adaptation in wheat

Type of test-system	NN	Indexes	Standard for zone	Maximum	Minimum
I.	1.	*Leaf area max, cm^2/plant	304 ±0,36	406 ± 0,47	215 ± 0,29
	2.	Photosynthetic potential, cm^2 days /plant	1209 ± 3,4	1507±2,0	1052±12,0
	3.	Relative leaf area's density, g/ dm^2	0,45±0,029	0,69±0,017	0,36 ± 0,023
II.	4.	*Net Photosynthesis, mg CO_2 / dm 2 . h	20,20 ± 3,00	24,30	7,40
	5.	Gross Photosynthesis, mg CO_2 / dm 2 . h	22,60 ± 1,62	27,40	10,50
	6.	*Net Photosynthetic Productivity, g/m 2.day	12,20 ±1,7	15,13 ± 1,5	11,48± 1,0
III.	7.	*^{Inq}Mass ear in flowering/ Mass leaf in flowering	0,92 ±0,022	1,10 ±0,077	0,51± 0,036
IY.	8.	^{Inq}Mass ear in flowering/ Mass stem in flowering	0,30 ± 0,012	0,49±0,037	0,22 ±0,010
	9.	Mass ear in flowering/ Mass shoot in flowering	0,19± 0,001	0,25±0,01	0,13 ± 0,004
Y.	10.	^{Real}Mass ear in milky ripeness/ Mass leaf in flowering	4,83± 0,21	6,57 ±00,61	3,47± 0,088
	11.	Mass ear in milky ripeness/ Mass shoot in flowering	0,98±0,043	1,23 + 0,032	0,61± 0,012
	12.	Mass ear in wax ripeness/ *Mass leaf in flowering	5.19 ± 0,31	7,00 ±0,64	3,92±00,07
	13.	Mass ear in wax ripeness/ Mass shoot in flowering	1,06 ±0,041	1,46±0,091	0,80± 0,054
YI.	14.	^{Incr}Mass ear in wax ripeness/ Mass ear in flowering	5,67± 0,29	7,69 ±0,17	4,56 ± 0,41
	15.	Mass ear in wax ripeness/ Mass ear in start grain forming	5,33 ±0,32	7,24±0,17	3,90 ±0,19
YII.	16.	^{LInq}Mass 1000grains/ Mass leaf in flowering	29,30 ±1,86	30,80 ±3,43	15,13 ±1,23
	17.	Mass1000grains/ Leaf area max	13,05 ±1,42	18,12 ±2,16	8,43 ±0,33
YIII.	18.	*Daily Increase of Biomass max, g/plant. day	0,57 ±0,18	0,71 ±0,054	0,40 ±0,021
	19.	Increae of ear mass during reproductive period (Ie), g/plant	6,77 ±0,37	8,90 ±0,49	4,49 ±0,098
	20.	Ie/ Mass leaf in flowering	4,19 ±0,20	5,90 ±0,56	3,24 ±0,16
	21.	Daily Increase of ear mass, g/plant.day	0,15±0,010	0,23±0,037	0,12±0,007
	22.	Shoot mass decrease from wax ripeness to flowering, g	2,01 ±0,44	2,01 ±0,44	0,82 ±0,29
	23.	Relation of plant grain productivity to Leaf area max, g/dm 2	2,29 ±0,10	4,46 ±0,47	1,50 ±0,20
IX.	24.	Relation of grain mass to plant mass (Harvest Indexes), %	47,41±1,27	54,44 ±2,18	38,84 ± 0,54
X.	25.	*Shoot mass , g	2,39 ±0,61	3,63 ±0,27	1,68 ±0,08
	26.	*Shoot mass in whole ripeness/ shoot mass in flowering	2,23 ±0,47	3,40 ±0,62	1,85 ±0,33
	27.	*Plant mass (Biological Yield), g/plant	14,75 ±1,00	22,80 ±0,53	12,08 ±0,52
	28.	*Main shoot's grain mass, g	1,72 ±0,032	2,33 ±0,035	1,61±0,024
	29.	*Lateral shoot's grain mass,g	1,35 ±0,028	1,99 ±0,030	1,21±0,026
XI.	30.	*Grain Yield, g/plant	7,00 ±0,61	10,37 ±0,46	5,51±0,21

*marks the more informative indexes for crop breeding.

TABLE 2 . Genetic determination of photosynthetic indexes: correlation analysis

Indexes: / Wheat Forms	Hill reaction	Non-cyclic electron transport	Chloroplast number in cell/leaf area	Net Photo-synthesis	Leaf area
Progress	epistaz	-20,25	0,83/-0,23	-1,58	-2,93
Gostianum-88	-31,80	-24,20	0,63/-3,80	-0,70	2,91
Eritrospermum-350	50,12	-50,40	-0,38/12,80	3,11	1,93
Omskaya-iso	-2,61	94,90	-1,08/	-0,85	-1,89

forms were sufficiently contrast and hybrids showed intermediate inheritance of leaf area. As the result of wheat differentiation by photosynthetic tests, four types of winter wheat production processes in watered conditions of South - East of Kazakstan were determined. For each type inhered in concrete photosynthetic indexes interaction for production forming during vegetation (table 3).

TABLE 3. Types of photosynthetic productivity processes in wheat

Wheat forms	LA*	NP*	Inq.	Real.	BY	GMS	HI	GY
			I. Intensive type					
Almaty's	2,15	24,3	1,10	6,58	17,47	1,96	51,5	8,82
			II. Semi-intensive type					
Progress	3,04	20,2	0,92	5,67	14,75	1,72	50,2	7,38
			III. Semi-extensive type					
Bezostaya-1	3,79	15,1	0,83	5,11	14,40	1,69	38,8	5,59
			IY. Extensive type					
PhnPr-10	3,92	9,4	0,56	7,70	18,39	1,79	46,2	8,49

*LA - leaf area max, cm 2/plant; NP - Net Photosynthesis, mg CO_2 /dm 2 . h; Inq. - ear inquiry in assimilates; Real. - realization of ear inquiry in ontogenesis; BY - biological yield, g; GMS - grain mass of shoot, g; HI - harvesting indexes; GY - grain yield. Wheat forms are typical for each concrete type of photosynthetic productivity process.

I - Intensive - adaptive and productive type of production process is characterized by high photosynthetic activity and not so large size of leaf area. It is a perspective type for new future of wheat crop breeding with grain yield more 100 c/ha.. II - Semi- intensive, but relatively adaptive type of production process is produced relatively more high demands to environments because of higher leaf are. It is a good type for mod ern wheat agriculture. III - Semi-extensive - relatively productive in watered conditions but don't adaptive type of production process, which needs favorable environments and intensive agrotechnology for growth of large leaf area. IY - Extensive type - crop and biomass production in which is possible only on the base of high extensive parameters - sizes of photosynthetic apparatus, large leaf area. It is a traditional morpho-physiological type of crop breeding, result of green revolution with productivity not more 55-60 c/ha.

With using of revealing ways is conducted search of wheat genotypes with high photosynthetic activity. The new lines of wheat - donors of high photochemical activity, high net photosynthesis and low leaf area were revealed and gathered in own core collection. Forms-donors of high photochemical activity of chloroplasts are: sort Omskaya , heteroseous hybrids Omskaya x Progress with heterosis and epystaz phenomena. The highest heterosis on number chloroplasts in cell is observed on hybrids Progress x Omskaya. Rate of Net Photosynthesis were highest in hybrids of Gostianum-88, Progress and Omskaya. Eritrospermum-350 had the best combine ability on Photosynthetic CO_2-gas-changes parameters. Reciproc gene of leaf area is prevailed in Gostianum-88 and Eritrospermum-350. Their hybrids had low value of these indexes. The contrast productive and drought resistant wheat lines were created. The first productive sort of winter wheat with high photosynthetic activity was produced and given to State examination.

5. References

Sidorenko O.I., Bedenko V.P., Urazaliev R.A. (1990) Photosynthesis of Heteroseous Hybrids in Winter Wheat, «Gilim» Publishers, Almaty.

Sidorenko O. (1996) Primary photochemical and photophysical processes in optimal photosynthetic type of wheat plant, Plant Physiology and Biochemistry, Special Issue, 10th FESPP Congress, Florence, Italy, Sept. 9-13, p. 87.

Sidorenko O., Shpak E. (1997) Photosynthetic Indexes and its Genetic Determination as a New Way of Biotechnology for Adaptive and Productive Crop Breeding in Wheat. XYIII Congress of SPPS, The Scandinavian Society for Plant Physiology Publishers, Uppsala, Sweden, p. 171.

GENETIC AND PHENOTYPIC CORRELATIONS FOR SOME SUGAR BEET ROOT CHARACTERISTICS

P. SKLENAR, L. KOVAČEV, N. ČAČIĆ, S. MEZEI AND N. NAGL
Institute of Field and Vegetable Crops, M Gorkog 30, 21000 Novi Sad, Yugoslavia

1. Abstract

Sugar beet taproot weight; length; volume and crown width; height and weight of thirty families were measured. Calculation of coefficients of genetic and phenotypic variation were based on genetic and phenotypic correlation. Phenotypic correlation between crown width and taproot weight, taproot weight and volume, crown width and weight, taproot volume and crown width, were highly significant, while between crown and taproot weight and taproot weight and volume were significant. Heritability varied from h^2=38.50% for taproot weight to h^2=70.23% for crown width. Coefficients of genetic variation were CVg=7.87% for crown width, up to CVg=20.07% for crown weight, while coefficients of phenotipyc variation were CVp=9.85% for crown width to CVp=24.91% for crown weight.

2. Introduction

Numerous attempts have been made to identify physiological or anatomical traits that contribute to sugar beet yield and quality (Campbel & Cole, 1986,Stojaković, 1990). Root weight is with number of plants main component of sugar beet yield per unit area and have significant influence on crystal sugar yield (Čačić et al, 1995).

Individual selection of sugar beet is principally based on root characteristics and sugar or dry matter content. Number of root's traits could cause looses in yield and technological traits of sugar beet (Mesken & Dieleman, 1988, Coe & Theurer, 1987). Therefore is valuable to have information of relationship between some notable traits because, according to Vančetović & Drinić (1996) modification of one trait under selection will result in changes of all other traits to degree in which they are correlated with "principal trait".

Core and canonical correlations between some root characters conducted by Campbell and Cole (1986) has shown that some root traits are significantly interrelated. Using path-coefficient analysis, Čačić et al. (1995) and Sklenar et al. (1997a) have also found that the root traits they investigated have significant direct effect on sugar yield. The main shortcoming of these papers was that they did not distingiish between genetic and environment effects on traits under investigation, since the genotype and environment have consiredable influence on metric traits in sugar beet (Sklenar, 1997b).

Genetic and phenotipyc correlations among the studied traits can be due to pleiotropy, on the one hand, and to the environment, on the other.

3. Materials and methods

Thirty families of the 1102-5-8 population of a monogerm self fertile maintainer of nuclear-citoplasmatic male sterility of sugar beet with the recessive nuclear sterility gene "a" were sown in randomized block design in four replications at Rimski Šančevi. A randomized block design wasw used. The row to row and plant to plant distances were 45 cm and 20 cm, respectively. The following traits were studied: taproot weight (g), length (cm) and volume (cm^3) and crown width (cm), height (cm) and weight (g). The calculations of genetic and phenotipic correlations, broad sense heritability and coefficients for genetic and phenotipic variation were done as per standard procedure according to Falconer (1981).

4. Results and discussion

Broad sense heritability ranged from h^2=0.39 for root weight to h^2=0.70 for root length. The coefficients of genetic variation ranged between Vg=7.87 for crown width and Vg=17.63 for taproot volume. The coefficients of phenotypic variation were generaly higher than the coefficients of genetic variation, ranging from Vp=9.85 for crown width to Vp=23.36 for taproot volume (Tab.1).

The genetic correlations between the traits under investigation were not significant. The highest positive correlations were found between taproot volume and crown width (r=051) and taproot volume and weight (r=0.47), while negative genetic correlations were detected between taproot volume and crown height (r=-0.29) and crown height and root weight (r=-0.09) (Tab.2).

The phenotypic coefficients of correlation between taproot weight and crown width, crown weight and taproot volume, and crown width and crown weight were highly significant, while those between crown weight and root weight and taproot volume and crown weight were significant (Tab.2).

Absence of genetic correlations between measured traits allows us increase genetic pressure on root weight without fear of increasig of the other undesirable root traits.

Fig. 1.Heritability in broad sense, coefficients of genetic and phenotypic variation and mean of measured traits

	h^2	Vg	Vf	$\bar{X}$
root weight	0.39	14.43	23.26	395.02
root lenght	0.70	9.62	11.47	18.18
crown width	0.63	7.87	9.85	7.73
crown height	0.46	9.17	13.60	6.45
crown weight	0.65	20.07	24.91	150.20
taproot volume	0.57	17.63	23.36	354.22

Tab. 2. Genetic (above) and phenotipic (below) correlations for root traits

TRAITS	**root length**	**crown width**	**crown height**	**crown weight**	**taproot volume**
root weight	0.26 0.86	0.45 1.16**	-0.09 0.60	0.34 0.95*	0.47 1.36**
root lenght		0.12 0.55	0.23 0.68	0.31 0.70	0.29 0.84
crown width			0.12 0.64	0.45 1.03**	0.51 1.20**
crown height				0.25 0.90	-0.29 -0.02
crown weight					0.40 0.95*

*, **: significant for probability 5% and 1% respectively

5. References

1. Campbell, L. G. and Cole, D. F. (1986) Relationships between taproot and crown characteristics and yield and quality traits in sugarbeets; *Agron. J.*, Vol.78, 971-973.
2. Coe, G. E. and Theurer, C. (1986): Progress in the development of soil-free sugarbeets; *Journal of the J.A.S.S.B.T.*, Vol.24, No. 1, 49-56.
3. Čačić, N., Kovačev, L., Mezei, S., Mrkovački, N. & Sklenar, P. (1995): Path coefficient analysis for some sugar beet characteristics, 58-th IIRB Congress, Beune 19-25 June, 139-142.
4. Falconer, D.S. (1981): *Introduction to quantitative genetics*, 2nd ed., Longman Group Ltd., p.p. 281-300.
5. Mesken, M. and Dielman, J. (1988): Breeding sugar beets with globe-shaped roots: selection and agronomical performance, *Euphytica S*, 37-44.
6. Sklenar, P., Kovačev, L. i Čačić, N. (1997a): Karakteristike korena S3 populacija monogermnih održivača citoplazmatsko nuklearne muške sterilnosti kod šećerne repe,*Selekcija i semenarstvo* (in press), Dnevnik, Novi Sad.
7. Sklenar, P. (1997b*): Adaptability and stability of sugar beet varieties in different agroecological conditions* Master of sci degree thesis, Faculty of Agriculture, Novi Sad.
8. Stojaković, M. (1990): *Ocena kombinacionih sposobnosti na osnovu morfo-fizioloških osobina ponika kod roditelja i F1 generacije šećerne repe*, Doktorska disertacija, Poljoprivredni fakultet Novi Sad.
9. Vančetović, J. and Drinić, G. (1996): Correlations among more important agronomic traits in two maize composites, *Genetics*, Vol. 28, No.1, 23-29.

ALLOZYME VARIATION IN NINE (9) BALKAN BLACK PINE (*Pinus nigra* ARNOLD) POPULATIONS

M. TSAKTSIRA[1*], A. SCALTSOYIANNES[2] AND
N. ATHANASSIADIS[3]
1 Researcher, 2 Assistant Professor, 3 Professor
Aristotle University of Thessaloniki, Department of Forestry & Natural Environment, Laboratory of Forest Genetics and Plant Breeding, 54006 Thessaloniki

1. Abstract

Allelic frequencies at 6 enzyme loci (MDH-D, DIA-A, LAP-A, AAT-B, PGI-B, ACO-A) were determined in nine black pine populations of south-eastern Europe (4 from Greece, 1 from Austria, 1 from Cyprus, 1 from Yugoslavia, 1 from Bulgaria and 1 from Turkey) by the technique of the horizontal starch gel electrophoresis. A total of 25 alleles were detected. Great variation at the heterozygosity (from 0.279 to 0.438) was observed among the populations. Unexpectedly, the highest value of the heterozygosity was observed in the island of Samos while the lowest one was observed in Cyprus. Useful allozyme markers were detected, discriminating some populations (Austria, Cyprus etc.). The dendrogram of the Nei's genetic distances partially supports the classification system of Fukarek (1958).

2. Introduction

European black pine (*Pinus nigra* ARNOLD) is distributed in a discontinuous area which includes Southern Europe, Minor Asia, Cyprus and north-western Africa (Critchfield and Little, 1966). Black pine is well adapted to a large range of environmental conditions and it is widely used for reforestation. Thus, there is considerable interest in knowledge on geographic patterns of genetic variation.

According to many authors, black pine possesses a lot of taxonomic problems and there is no general consensus on its taxonomy because of the widespread distribution, disjunct area of distribution, prolonged isolation and the existence of transitional forms between subspecies and geographic races (ecotypes) (Vidakovic, 1974; Bassiotis, 1967). In the Balkan peninsula, for example, the variability and taxonomy

*This work is part of the Doctorate thesis of the senior author

of *P. nigra* are not yet completely clarified (Fukarek, 1958a). In Yugoslavia only, European black pine was classified into five subspecies, while in Greece, Wheeler *et al.* (1976), found such a high variation, in most of the traits examined, as the one found in the whole Eastern Mediterranean region.

The above seems to be a good explanation of the fact that so many researchers resulted in different classifications and more than a hundred different Latin names have been given to populations of the species.

Up to date, most of the taxonomic studies of black pine were based on morphological and anatomical characteristics (Tutin *et al.*, 1964; Vidakovic, 1974), except of very few studies that were based on biochemical markers, namely: a) monoterpenes and b) isozymes.

The purpose of our study was to determine the genetic variation pattern of *P. nigra* in south-east Europe and to clarify the taxonomic status of the species in this region by means of isozyme technique.

3. Materials and Methods

Electrophoresis was conducted on haploid endosperms of germinated black pine seeds originated from Greece (Peloponnese, Rodopi, Samos, Kalambaka), Austria, Bulgaria, Yugoslavia, Cyprus, Turkey. The seed material was bulk and was provided by the National Forest Services of Greece, French (INRA), Turkey and Morroco. Information on the locations of the examined populations of *P. nigra* will be given in another work (Tsaktsira *et al.*, in preparation).

The electrophoretic techniques, the scoring of alleles and the genetic interpretation of the isozymes that we followed in this study were the same as those referred in a previous work (Scaltsoyiannes *et al.*, 1994)

4. Results and Discussion

Six loci belonging to 6 enzyme systems, were resolved with sufficient consistency and clarity in the nine European black pine populations (Table 1). From the above loci, PGI-A, AAT-B, were polymorphic over all European black pine populations assayed, where as, MDH-D and LAP-A were nearly monomorphic in the most populations. A total of 25 scorable alleles (allozymes), were detected in the 6 readable loci. The allele frequencies at each locus and population are shown in Table 1. No special variation pattern was observed among the population studied, but some alleles (e.g. PGI-A4, DIA-An for Peloponnese, PGI-A2 for Cyprus, AAT-B3 for Austria etc.) seem to have a diagnostic value.

Table 2, lists genetic variation values calculated for the populations of *P. nigra*. It is shown the proportion of polymorphic loci (P), the number of alleles detected per locus (A/L) and the average amount of genetic variability (expected heterozygosity) exhibited a considerable variation among the populations. The highest value of heterozygosity was, unexpectedly, was found in the population from island Samos.

TABLE 1. Allele frequencies at 6 loci in 9 populations of *Pinus nigra*.

proven. alleles	Austr ia	Yugos lavia	Bulg aria	Rod opi	Tur key	Kalam baka	Samos	Pelopo nnese	Cyprus
MDH-D1	0.94	1.00	1.00	1.00	0.96	0.83	0.84	0.96	1.00
MDH-D2	0.05	0.00	0.00	0.00	0.04	0.17	0.16	0.04	0.00
MDH-D3	0.01	0.00	0.00	0.00	0.00	0.00	0.00	0.00	0.00
DIA-A1	0.02	0.02	0.00	0.00	0.03	0.00	0.00	0.00	0.00
DIA-A2	0.48	0.42	0.40	0.53	0.26	0.66	0.45	0.26	0.05
DIA-A3	0.44	0.51	0.55	0.47	0.70	0.30	0.49	0.52	0.92
DIA-An	0.06	0.05	0.05	0.00	0.01	0.04	0.06	0.22	0.03
PGI-A1	0.00	0.00	0.00	0.00	0.00	0.00	0.00	0.00	0.00
PGI-A2	0.02	0.04	0.00	0.05	0.02	0.00	0.00	0.01	0.24
PGI-A3	0.02	0.00	0.00	0.01	0.00	0.00	0.00	0.02	0.00
PGI-A4	0.33	0.24	0.33	0.25	0.21	0.37	0.20	0.09	0.53
PGI-A5	0.50	0.39	0.36	0.41	0.29	0.38	0.47	0.53	0.19
PGI-A6	0.00	0.00	0.05	0.00	0.00	0.04	0.02	0.00	0.00
PGI-A7	0.13	0.33	0.26	0.28	0.48	0.21	0.31	0.35	0.04
LAP-A1	0.96	0.99	0.99	0.98	1.00	0.97	1.00	1.00	0.96
LAP-An	0.04	0.01	0.01	0.02	0.00	0.03	0.00	0.00	0.04
AAT-B1	0.03	0.00	0.03	0.01	0.00	0.05	0.01	0.00	0.00
AAT-B2	0.81	0.58	0.61	0.64	0.62	0.45	0.61	0.69	0.45
AAT-B3	0.16	0.42	0.36	0.35	0.38	0.50	0.38	0.31	0.55
ACO-A1	0.00	0.00	0.01	0.00	0.00	0.01	0.00	0.00	0.05
ACO-A2	0.00	0.02	0.04	0.00	0.00	0.00	0.09	0.00	0.00
ACO-A3	0.76	0.70	0.66	0.62	0.80	0.84	0.49	0.75	0.82
ACO-A4	0.16	0.15	0.10	0.11	0.03	0.11	0.21	0.06	0.05
ACO-A5	0.05	0.13	0.14	0.27	0.15	0.03	0.21	0.19	0.08
ACO-A6	0.03	0.00	0.05	0.00	0.02	0.01	0.00	0.00	0.00

TABLE 2. Genetic variability parameters at 6 loci in 9 populations of *Pinus nigra*.

Populations	Mean number of alleles per locus A/L	Percentage of polymorphic loci P (%)	Expected Heterozygosity He
Austria	3.50	100.00	0.352
Yugoslavia	2.83	83.33	0.372
Bulgaria	3.17	83.33	0.382
Rodopi	2.67	83.33	0.373
Turkey	2.83	83.33	0.330
Kalambaka	3.00	100.00	0.405
Samos	2.83	83.33	0.438
Peloponnese	2.67	83.33	0.353
Cyprus	2.67	83.33	0.279
Mean	2.91	87.03	0.365

The coefficient of diversity (Gst) was 5.5% (data was not presented in this work), that means the 94.5% of the total variation exist within the populations.

In the dendrogram (Figure 1) that was constructed from the Nei's genetic distances, a distinct cluster consisted of the populations from Bulgaria, Yugoslavia, Rodopi, Samos, Peloponnese and Turkey was observed. On the other hand, populations from Austria, Kalambaka and Cyprus are highly differentiated from the above mentioned populations. The above findings are close to classification of Fukarek (1958a) on the Balkan peninsula. Concerning the population from Cyprus, our results verifing once again Bonnet-Masimbert and Bikay-Bikay (1978) and Wheeler *et al.* (1976) findings, that is, the Cyprian populations belong to a distinct variety of *Pinus nigra*.

Figure 1. Dendrogram based on genetic distances

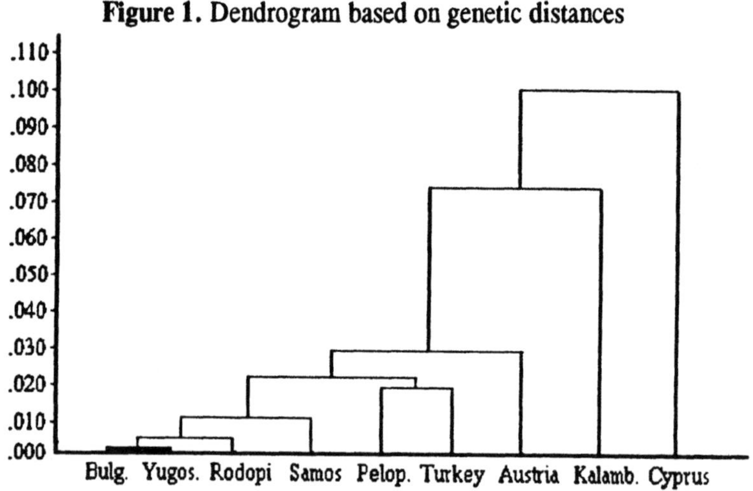

5. References

Bassiotis, C. (1967) Variation in characteristics of black pine in Greece, Scientific Annals of Agriculture and Forestry Faculty, Thessaloniki **11**, 79-133.

Bonnet-Masimbert, M. et Bikay-Bikay, V. (1978) Variabilite intraspecifique des isozymes de la glutamate-oxalacetate-transaminase chez *Pinus nigra* ARNOLD. Interet pour la taxonomie de sous especes, Silvae Genet. **27**, 71-79.

Critchfield, W.B. and Little, L., JR. (1966) Geographic distribution of the pines of the world, USDA Misc. pub. 991, 97 pp.

Fukarek, P. (1958a) Prilog poznavanju crnog bora (*Pinus nigra* Arn.S.Lat.) Beitrag zur Kenntnis der systematischen Stellung. Gliederung und der rezenten Verbeitung der Schwarzkiefer.Radovi Poljoprivredno-sumarskog falkuteta III, 3,B. Sumarstvo, Sarajevo: 3-92.

Scaltsoyiannes, A., Rohr, R., Panetsos, K.P. and Tsaktsira, M. (1994) Allozyme frequency distributions in five European populations of Black pine (*Pinus nigra* Arnold). I. estimation of genetic variation within and among populations. II. Contribution of isozyme analysis to the taxonomic status of the species. Silvae Genet. **43**, **1**: 20-30.

Tutin, T. G., Heywood, V. H., Burges, N. A., Valentine, D. H., Walters, S. M. and Webb, D. A. (Eds.) (1964) Flora Europaea Vol. 1. Cambridge University Press, 464 p.

Vidakovic, M. (1974) Genetics of European black pine (*Pinus nigra* ARN.). Annales Forest. Acad. Sci. et Artium Slavorum Meridionalium **6/3**. 86 pp.

Wheeler, N. C., Kriebel, H. B., Lee, C. H., Read, R. A. and Wright, J. W. (1976) 15-Year performance of European black pine in provenance tests in North Central United States. Silvae Genetica **25**, 1-7.

VARIABILITY OF GROWTH AND PRODUCTIVITY OF HYBRID PLANE PROGENY ON MEADOW BLACK SOIL IN SERBIA

A. TUCOVIĆ, V. ISAJEV, M. ŠIJAČIĆ-NIKOLIĆ
Faculty of Forestry,
1 Kneza Višeslava, 11030 Belgrade, Yugoslavia

1. Abstract

General and group analysis of 29-year old progeny of European plane indicates that, along with individual variability, the progeny is also characterised by a highly expressed polymorphism of a series of properties. This study deals with the polymorphism of crown form, types of bark, straightness of the stem, and anatomic structure. A low share of parent phenotypes (2.70% of the type *occidentalis* and 3.71% of the type *orientalis*) has been explained by postsyngamic lethality and by differential reproduction in nursery conditions

2. Introduction

The progeny of interspecific hybrids is distinguished by high variability, often higher than that of parental species. In forest trees, the first generation is usually raised, e.g. *Larix x eurolepis*, while hybrid progeny is rarely cultivated or it is improved, owing to expressed variability. As an exception, a great number of cultivars of significant economic value were cultivated from black poplar hybrids by back crossing with parent trees or hybrid trees (Jovanović *et al.*, 1971). In homogamete species, meiosis occurs in the more or less normal way, and the variability in successive generations is mainly conditioned by hybrid combinations. In a more numerous progeny, without an expressed competition, there are more chances for sharply deviating combinations, designed as notomorphic or as cultivars, such as *Platanus x acerifolia n.m.hispanica* Muench. etc. *Platanus x acerifolia* (A.) Willd. is a sub-spontaneous interspecific hybrid of *Platanus occidentalis* L. and *Platanus orientalis* L. (Jovanović, 1967, Tucović, 1971, Vukićević, 1987, Đukić, 1994, etc.).

3. Object and Method

The analysed four-line 29-year old population of European plane was raised on the soil classified as meliorated meadow black soil of ancient origin (Košćel et Menković, 1994). All the trees were measured, i.e. total height, diameters at breast height, stem circumference at 1.30 m, crown diameter, crown circumference, and yield. Yield phenophase was observed in the field and evaluated ocularly from 0 - 5 (0 = 0%; 5 = 100%). The justification of differences between mean values for the analysed properties was shown by Student's t-test. The correlations were recorded between the heights of all trees and diameters and circumferences at 1.30 m. A relative parameter of variability is tree height expressed as the percentage of tree crown circumferences.

4. Results and Discussion

The main criteria of ontogenetic adaptation were qualitative and quantitative parameters of potential productivity and ecological resistance. Table 1 presents the statistical parameters of 7 analysed parameters. The analysed multi-line population is characterised by a prominent group variability (polymorphism) in a series of properties. Only 7 properties of group intrapopulation polymorphism were analysed: crown form, bark type, stem straightness, anatomic structure, and the share of parent and hybrid phenotypes according to differential characters of leaves and flower clusters (Tables 2, 3). As for crown characteristics, four phenogroups can be distinguished in the population: trees with wide crowns, wide conical, conical and narrow conical crowns and the ratios are 67 : 77 : 123 : 29. Three phenogroups can be distinguished by the characteristics of dead bark: smooth bark, shallow fissured and fissured bark, the ratio being 133 : 95 : 68. The population has two phenotypes of the straightness of stem: trees with monopodial, slightly tapered and straight stems and trees more or less forked (low, medium and high), their ratio being 186 : 110. As for anatomical structure of the stem, two phenogroups can be distinguished: without and with anatomical defects in the ratio 279 : 17. Trees with anatomical defects in wood can be classified into two groups: trees with outgrowths along the stem, due to accumulation of dormant buds or short shoots (type *mazerica*), and trees with stem deformations due to nonuniform cambial function (type *lignotuber*). During the morphophysiological analysis of stems, a very low share of parent phenotypes was observed (2.7% *occidentalis,* and 3.7% *orientalis*), according to leaf differential properties (length of middle lobe, dentated leaves, appearance of leaf base, etc.) and fruit balls (number and morphology). The analysed group polymorphism of eight properties of hybrid plane 29-year old progeny was also recorded in the younger populations aged 4 to 6, i.e. crown types, bark types, stem straightness, etc., which indicates that these characters are genetically controlled. This hypothesis is confirmed by the values of variation coefficients for all analysed characters (Table 2), as well as Student's t-test (Table 3).

5. Conclusions

Based on the general and group analysis of 296 trees of 29-year old hybrid plane progeny, it was evaluated that the analysed progeny, in addition to individual one, is characterised by a very expressed polymorphism of a series of properties. Hybrid plane is characterised by a homogamete complex (parents and hybrid species have the same numbers of chromosomes in the gametes), which can be progressively developed, taking into account the characteristics of the chromosome complex, wind pollination, abundance of flowering, yield, tree longevity, and potential cultivation on heterogeneous sites. In the successive generations of hybrid plane, phenotypes of parent species can disappear due to postsyngamic lethality and differential reproduction in nursery conditions, without endangering the evolution potential of the species *Platanus x acerifolia* (A.) Willd. The process of hybrid plane cultivation in urban coenoses and in forest plantations is related to the characteristics of the so-called "syndrome of domestication". Namely, hybrid plane is characterised by (1) significant properties, (2) an expressed variability of properties, and (3) its cultivation is cost-effective, both regarding biomass production and by its so-called indirect beneficial functions.

TABLE 1. Statistic parameters of seven characters of 296 hybrid plane trees (*Platanus x acerifolia* Willd.) grown on meadow black soil

min. – max.	$\bar{x} \pm S_{\bar{x}}$	$S \pm S_S$	$V \pm S_V$
Tree height (m)			
8 – 24	16.50 ± 0.15	2.54 ± 0.10	15.39 ± 0.63
Tree diameter at 1.30 m (cm)			
22 – 68	43.18 ± 0.44	7.58 ± 0.31	17.55 ± 0.72
Tree circumference at 1.30 m (cm)			
70 – 220	135.78 ± 1.44	24.80 ± 1.02	18.23 ± 0.75
Crown diameter (m)			
6 – 21	12.26 ± 0.11	1.96 ± 0.08	15.66 ± 0.64
Crown circumference (m)			
18 – 64	38.94 ± 0.37	6.34 ± 0.26	16.30 ± 0.26
Height in % of crown circumference			
10 – 100	43.05 ± 0.65	11.20 ± 0.46	26.05 ± 1.07
Yield with grades 0 (0%) to 5 (100%)			
0 – 5	2.94 ± 0.08	1.46 ± 0.06	49.66 ± 2.04

TABLE 2. Coefficients of variation of the analysed properties and phenogroups of hybrid plane

	Crown				Bark			Straightness		Anatom. str	
Property	wide	wide conical	conical	narrow conical	smooth	shallow fissured	fissured	monopodial	forked	without defects	w
Frequency	67 :			29 :	133 :	95 :	68 :	186 :	110 :	279 :	1
Tree height (m)	17.89	12.78	14.21	12.98	12.39	17.18	19.50	16.08	14.58	14.31	17.
Tree diameter at 1.30m	8.12	8.36	7.84	7.92	7.16	7.44	6.42	8.42	7.98	16.29	8.
Circumference at 1.30	20.25	19.47	17.88	20.31	20.81	18.98	15.85	27.29	22.58	18.67	17.
Crown diameter (m)	11.70	11.54	13.16	15.76	18.90	17.10	16.96	19.33	14.92	15.64	16.
Crown circumfer. (m)	13.47	12.05	13.35	16.17	18.69	16.70	17.15	17.86	15.70	18.32	13.
Height in % crown circ.	17.02	12.29	17.31	19.71	30.07	23.48	34.99	43.56	22.60	22.39	25.
Yield	42.10	29.60	46.85	96.49	52.59	46.30	42.48	47.21	51.80	49.49	38.

TABLE 3. Justification of differences between mean values for analysed properties and 11 phenogroups of hybrid plane (t ≥ 2)

Property	Crown form				Bark			Straightness		Anat. struct.	
Phenogroup	w	wc	c	nc	sm	sf	f	m	fo	without defects	with defects
Height in m (right) Diameter at 1.30 m (left)											
w	-	**4.1**	**4.5**	**3.5**	**3.1**	**3.7**	**2.4**	**3.6**	**3.1**	**4.0**	**2.3**
wc	0.2	-	0.6	0.3	0.1	0.7	1.4	0.9	1.1	0.7	**4.9**
c	1.3	1.0	-	0.1	**2.0**	1.4	**2.4**	**2.0**	1.8	1.7	**6.3**
nc	**3.2**	**3.0**	**2.5**	-	1.2	0.9	1.2	1.0	0.9	0.9	**4.9**
sm	**2.3**	**2.7**	1.9	**2.2**	-	0.6	0.3	0.4	0.2	0.7	**4.8**
sf	0.6	0.3	0.8	1.6	**2.8**	-	**2.3**	0.7	1.2	0.4	**5.4**
f	1.0	1.3	**2.5**	1.7	**4.2**	1.7	-	**2.1**	1.3	**2.6**	**4.1**
m	1.8	1.6	0.6	1.5	1.5	1.5	**3.1**	-	0.3	0.2	**3.9**
fo	0.8	0.6	0.5	1.6	**2.4**	0.3	**2.0**	1.1	-	0.5	**3.6**
without defects	1.1	0.8	0.5	1.5	**2.8**	0.7	**3.1**	1.2	0.2	-	**5.8**
with defects	**5.7**	**8.2**	**3.1**	**2.1**	**4.1**	**5.8**	**6.6**	**4.5**	**5.6**	**5.9**	-
Height in % of crown circumference											
w	-	**9.7**	**4.4**	**8.9**	**11.0**	**7.6**	**5.1**	**7.9**	**6.8**	**11.7**	**2.9**
Crown diameter in m (right) Crown circumference in m (left)											
w	-	**6.4**	**12.4**	**16.8**	**10.4**	**25.8**	**3.3**	**10.6**	**4.7**	**10.6**	**7.1**
wc	**4.6**	-	**5.9**	**12.1**	**4.8**	**2.3**	1.5	**4.5**	1.3	**3.3**	**3.0**
c	**12.2**	**6.2**	-	**8.6**	0.2	**2.5**	**5.6**	1.0	**6.9**	**3.7**	1.2
nc	**4.6**	**11.5**	**8.6**	-	7.7	**9.2**	**11.2**	**8.7**	**12.7**	**11.4**	1.3
sm	**9.0**	**5.5**	**3.4**	1.2	-	**2.0**	**5.3**	0.6	**5.7**	**2.6**	1.3
sf	**6.9**	**3.1**	**5.8**	1.3	**2.4**	-	**3.2**	1.5	**3.6**	**4.2**	**2.4**
f	**2.6**	0.9	**7.2**	1.3	**3.8**	**3.0**	-	**4.8**	**6.5**	**7.1**	**8.2**
m	**10.5**	**7.0**	**2.3**	1.0	0.6	**2.0**	**4.9**	-	**2.4**	**2.9**	1.6
fo	**3.9**	1.0	**16.3**	1.2	**8.0**	**3.3**	0.2	**5.7**	-	**4.9**	0.5
	9.2	**2.6**	**16.0**	0.8	**4.9**	0.0	**3.4**	**2.4**	**4.0**	-	**2.5**
with defects	**5.8**	**3.9**	0.9	**4.1**	0.9	**3.1**	**3.6**	1.9	**5.4**	**3.3**	-
Yield (right) Diameter at 1.30 m (left)											
w	-	1.6	0.8	**7.7**	0.7	1.0	1.9	0.1	1.3	0.4	1.6
wc	0.2	-	0.9	0.8	**2.5**	0.4	**3.8**	1.9	**3.1**	**2.4**	**2.8**
c	0.9	0.8	-	**8.1**	1.4	0.3	**2.6**	0.8	**2.0**	1.2	**2.1**
nc	**2.3**	**2.2**	1.8	-	7.3	**8.7**	**6.3**	**8.7**	**6.4**	7.7	**5.2**
sm	**2.4**	**8.0**	1.9	0.6	-	1.7	1.1	0.9	0.5	0.5	1.1
sf	0.4	0.2	0.6	**2.1**	**2.3**	-	**2.9**	1.1	**2.3**	1.6	**2.3**
f	1.4	1.7	**2.8**	**3.5**	**4.4**	**2.0**	-	**2.3**	0.6	1.8	0.2
m	1.2	1.1	0.3	0.9	0.7	1.2	**3.2**	-	1.5	0.6	0.2
fo	0.0	0.1	1.2	**2.5**	**3.0**	0.4	1.8	1.5	-	1.1	0.7
without defects	0.3	0.1	1.0	**2.4**	**3.1**	0.1	0.9	**5.4**	**3.4**	-	1.5
with defects	**5.2**	**5.3**	**5.1**	2.5	**3.9**	**5.3**	**6.7**	**5.0**	**5.8**	**8.0**	-

6. References

Djukić, M., (1994): Istraživanja ekofizioloških svojstava potomstva polusrodnika hibridnog platana. Dok. disertacija odbranjena na Šumarskom fakultetu u Beogradu, pp 176.

Jovanović, B., (1971): Dendrologija sa osnovama fitocenologije. Izd. Naučna knjiga, Beograd

Jovanović, B., Tucović, A., (1971): Fam. *Salicaceae* Lindl. In: Flora SR S. (Jasifović, M., ed.). SANU. Bg., III.

Košćel, M., Menković, LJ., (1994): Prirodno stanje Deliblatske peščare i mogućnosti korišćenja izvorišta za vodosnabdevanje. Zbornik radova "Deliblatski pesak", Pančevo, 1, pp 127-138.

Tucović, A., (1970): Fam. *Platanaceae* Dum. In: Flora SR S. (M. Josifović, ed.). SANU. Beograd, II, pp 36-40.

Vukićević, E., (1987): Dekorativna dendrologija. Naučna knjiga, Beograd, pp 185-189.

Zhuchenko, A.A.,(1988): Adaptive potential of cultivated plants (Genetic and ecological bases). Publisches "Shitiisa". Kichinev, pp 135-158.

APPLICATION OF INDUCED ANDROGENESIS TO OBTAIN TOMATO PLANTS RESISTANT TO *CLAVIBACTER MICHIGANENSE* SUBSP. *MICHIGANENSE*

N. ZAGORSKA, V. SOTIROVA, L. SHTEREVA, B. DIMITROV AND N. BOGATSEVSKA
Institute of Genetics, Sofia 1113, Bulgaria

1. Introduction

Bacterial canker caused by *Clavibacter michiganense* subsp. *michiganense* occurs in many tomato growing regions all over the world and leads to considerable yield losses. As chemical control of the pathogen is difficult, development of resistant tomato cultivars would be the most effective and reliable. Some authors (Laterrot et al., 1978; Sotirova, 1991) reported resistance in wild tomato species. However they possess a different resistance level and often resistant genes are linked with genes determining undesirable agronomic traits. For this reason, besides the screening of wild *Lycopersicon* species, genetic variation induced by tissue cultures has been explored in many crops (Scowcroft et al., 1987; Karp, 1991). A wide range of plant characteristics can be altered as a result of regeneration from cell and tissue culture, including agronomically important traits such as disease resistance (Evans et al., 1984; Van den Bulk, 1997). The application of somaclonal variation in tomato made it possible to obtain resistance to several pathogens such as TMV (Smith et al., 1986), *Fusarium oxysporum* f. sp. *lycopersici* (Shahin and Spivey, 1987).

The purpose of the present work was to screen plants regenerated from tomato anther cultures and their progenies of different genotypes for resistance to *Clavibacter michiganense* subsp. *michiganense*.

2. Material and Methods

The regenerants obtained from anther cultures of different genotypes (cultivars and hybrids) Roma *ms*, Bella, Roma *ms* x L. 31 and Roma *ms* x UC 82A (Zagorska et al., 1997) were used in this investigation. The bacterial strain *Cmm* 7 of *Clavibacter michiganense* subsp. *michiganense* (*Cmm*) used in this study was obtained from the collection of the Institute of Plant Protection, Kostinbrod. Bacterial suspension was prepared from 36h culture of bacteria grown on Potato Dextrose agar (PDA) at 28^{0} C in concentration 10^{8} cfu/ml.

The R_1 , R_2 and R_3 regenerant progenies obtained from anther cultures of different genotypes were tested for resistance to *Cmm* 7. The inoculation was carried out in phase 3-4 true leaves by the method of De Jong and Homna (1976). The infected plants were grown under greenhouse conditions at 25° C. The testing was performed on ten pots in 5 replications (50 plants). The disease symptoms appeared 8-12 days after inoculation. The evaluation for disease resistance was made by Laterrot's scale (1974). The experimental data were statistically analyzed using Student's *t*-criterion (Lidansky, 1988).

3. Results and discussion

The data concerning the resistance of regenerants (R_1 - R_3) of the above mentioned (investigated) cultivars and hybrids obtained via anther cultures are presented in Table 1. The regenerants differed in their resistance to *Cmm* 7. The average rate of infection in Roma *ms* x UC 82A hybrid was similar or higher than that in the donor plants. One of the regenerated plants (93) of cultivar Bella in R_1 progeny was symptomless at the end of the vegetation. Some regenerants of the same material were characterised by slightly manifested symptoms of the disease (rate 3). Regenerants and their progenies (R_2 and R_3) obtained from the hybrid Roma *ms* x L. 31 showed the lowest average rate of infection compared to the donor material as well as to the other cultivars and hybrids.

The variation observed in the resistance to *Cmm* 7 in the tomato regenerants

TABLE 1. Resistance of regenerants to *Cmm* 7 obtained from anther cultures

Genotypes	No of regenerants	No of infected plants with rate 1	3	5	7	9	Average rate	CV
Roma *ms*	K	0	0	12	16	25	7..35±2.40	0.17
R_1	50	0	0	15	18	24	7.42±2.20	0.18
	51	0	0	8	25	17	7.30±2.10	0.19
	52	0	0	10	17	21	7.23±2.00	0.20
Bella	K	2	15	25	11	13	5.63±2.91	0.30
R_1	90	0	13	15	10	12	6.30±2.97	0.27
	91	0	21	10	17	8	5.87±2.63	0.25
	92	0	18	20	13	10	5.68±2.90	0.30
	93	1	15	17	12	12	5.41±2.62	0.25
	94	0	7	23	15	10	6.62±2.70	0.23
	93-1	2	20	16	10	8	5.52±2.15	0.29
R_2	93-2	0	18	21	12	9	5.70±2.30	0.26

continued

	93-3	2	23	15	10	3	5.40±2.26	0.26
Roma *ms* x L 31	K	3	15	13	12	8	5.25±2.12	0.28***
R_1	101	0	23	17	10	10	6.50±2.73	0.26
	102	0	18	5	17	0	5.83±2.90	0.30***
	103-2	3	23	14	12	8	4.53±0.39	0.39***
	103-3	0	18	10	12	0	4.85±2.25	0.40
	103-4	3	22	15	7	5	4.57±2.15	0.26***
	104-1	6	18	10	17	0	4.53±2.65	0.36
	104-2	2	21	17	10	10	4.73±2.95	0.35***
	105-1	0	17	10	13	0	4.80±2.05	0.31
R_3	103-1-2	1	25	14	10	0	4.32±1.98	0.43***
	103-1-3	0	18	15	19	3	5.40±2.90	0.33
	103-1-4	3	23	10	14	0	4.40±2.00	0.41***
	103-2-1	2	19	13	12	4	5.01±2.05	0.31***
	103-2-3	1	23	14	12	0	4.68±2.17	0.43
	103-4-2	1	20	12	10	2	4.60±2.15	0.43
	105-4-2	2	23	10	10	0	4.51±2.12	0.45
	104-1-1	0	21	10	15	2	4.90±2.92	0.33
	104-1-3	1	17	12	17	0	5.25±2.98	0.37
	104-1-5	2	23	8	10	0	4.30±2.13	0.41
	104-2-1	0	22	10	10	5	5.15±2.50	0.30
Roma *ms* x UC 82 A	K	0	0	3	30	10	7.80±2.46	0.16
R_1	80-1	2	17	12	16	10	6.52±3.15	0.26***
	80-2	0	0	25	15	20	6.90±2.45	0.20***
	82-1	0	0	8	24	18	7.53±2.20	0.19
	82-2	0	3	15	11	15	7.11±2.85	0.20***
	83-1	0	0	15	15	30	7.40±2.43	0.21***
	83-2	0	0	20	10	20	7.60±3.00	0.21
	84-1	0	0	10	10	20	7.25±2.15	0.20
R_2	80-1-1	0	18	10	10	8	5.93±2.20	0.30***

Scale: **1** - no symptoms, **3** - several wilted leaflets, **5** - several wilted leaves, **7** - 50% to 80% wilted leaves, **9** - 90% wilted leaves. P = 0.1%

obtained by anther cultures from different genotypes was proved by the statistically performed comparison with the donor material. The changes that were observed lead to raised resistance or influence the susceptibility. Regenerants characterized by statistically significant variation of resistance in comparison with the donor plants have been found in Roma *ms* x L. 31 hybrid while no such variation was found in the regenerants from other genotypes, which may be due to the initial material.

The mechanisms of gametoclonal and somaclonal variation is not completely known but it is thought to be due to a combination of factors. Genetic background of plants can affect the amount of variability that occurs (Lörz, 1984). In their study on disease resistance in tobacco somaclones Daub and Jenes (!989) concluded that the observed variability depend on the disease as well as the plant genotype. Van den Bulk (1997) tested about 3000 plant progenies from 300 regenerants for resistance to bacterial canker and found a few plants showing only a delay in disease development.

Our results suggest the possibility to induce the resistance to *Cmm* 7 in regenerants obtained from tomato anther cultures. The major problem to obtain resistnant lines is to assess progenies and stabilize the resistance which will be our task in the future.

4. References

Duab, M. F. and A. E. Jenns (1989) Field and greenhouse analysis of variation for disease resistance in tobacco somaclones. *Phytopathology*, **79**, 600-605.

De Jong, R. and S. Homna (1976) Evaluation of screening technique and determination of criteria for assessing resistance to *Corynebacterium michiganense* in tomato. *Euphytica* **25**, 405-414.

Evans, D. A. (1984) Somaclonal and gametoclonal variation. *Amer. J. Bot.* **71**, 759-774.

Karp, A. (1991) On the current understanding of somaclonal variation. In B. G. Miflin (ed.), *Oxford Surveys of Plant Molecular and Cell Biology*, Oxford Univ. Press, 7, 1-8.

Laterrot, H. (1974) Value of the resistance of the tomato Bulgarian 8/12 to *Corynebacterium michiganense* (Smith) Jensen. Eucarpia of Tomato working group, Barry, Italy, 144-149.

Laterrot, H., R. Brand and M.C. Duanay (1978) La resistance a *Corynebacterium michiganense* chez la tomat. *Ann. Amelior. Plants*, **28**, 579-591.

Lidansky, T. (1988) *Statistical methods in biology and agriculture.*, Zemizdat, Sofia, p. 276.

Lörz, H. (1984) Variability in tissue culture derived plants. In W. Arber (ed.) *Genetic Manipulation Impact on Man and Society*, Cambridge University Press, Cambridge, pp. 103-114.

Shahin, E. A. and Spiviey R. (1987) In vitro breeding for disease resistance in tomato. In D. J. Nevins and R. A. Jones (eds). *Tomato Biotechnology*, Alan R. Liss Inc., New York, pp. 89-97.

Smith, S. S. and H. H. Murakishi (1986) Nature of resistance of tomato somaclons to tomato mosaic virus. VI Intern. Congress on Plant Tissue and Cell Culture, August 3-8, University of Minnesota.

Scowcroft, W. R., R. Brettell, S. Ryan, P. Davies and M. Palota (1987) Somaclonal variation and genomic flux. In: C. E. Green, D. A. Lamers, W. P. Hackett and D. D. Biesboer (eds.). *Plant Tissue and Cell Culture*, Alan R. Liss Inc., New York, pp. 275-286.

Sotirova, V. (1991) Resistance of tomato (*Lycopersicon esculentum* Mill) to bacterial disease, D. Sc. Thesis, Sofia.

Van den Bulk, R. W., J. Jansen, W. H. Lindhout and Loffer H. J. (1997) Screening of tomato somaclons for resistance to bacterial canker (*Clavibacter michiganense* subsp. *michiganense*). *Plant Breeding* (in press).

Zagorska, N., L. Shtereva, B. Dimitrov and Kruleva M. (1997) Induced androgenesis in tomatoes (L. *esculentum* Mill) I. Influence of genotype on androgenetic ability. *Plant Cell Reports*, (In press).

Growth, Development and Differentiation

BLUE-LIGHT RECEPTION IN PLANTS AND FUNGI

P. GALLAND

Fachbereich Biologie/Botanik
Philipps-Universität
Lahnberge
D-35032 Marburg
Germany

1. Introduction

Probably all plants and fungi and numerous protists are responsive to light of wavelengths ranging from 300 to 500 nm. Known photoreactions, which include those of photodifferentiation and photoorientation, are mediated by a photoreceptor, the near-UV/blue-light receptor (for short blue-light receptor), which is sometimes also referred to as cryptochrome. More than 40 years ago it had been proposed by Galston (1950) that flavins would be ideal photoreceptor contenders for the investigated spectral range. At the time the proposal was meant as an alternative to the earlier proposition of a carotenoid receptor. During the past decades the flavin hypothesis has largely replaced the carotenoid hypothesis and enjoys nowadays universal acceptance (Ahmad and Cashmore, 1996; Galland and Senger, 1991). A recent revival of the carotenoid hypothesis came from the suggestion that zeaxanthin may play a photoreceptor role in coleoptile phototropism (Quiñones and Zeiger, 1994). In the eighties pterins were proposed as co-photoreceptors in addition to the flavins (Siefermann-Harms *et al.*, 1985; Galland and Senger, 1988; Galland, 1992).

2. Photophysiology

Blue-light physiology has relied greatly on the generation of action spectra for responses such as pigment induction, phototropism, regulation of intermediary metabolism and various photomorphogenic events. Typical action spectra display maxima near 370 and 450 nm as well as a sharp decline of the effectiveness above 500 nm. These features represent useful though not unambiguous criteria for the flavin nature of the blue-light receptor. Because most pterins absorb in the near-UV these action spectra are also equally well explained on the basis of flavins in combination with pterins (Galland and Senger, 1988, 1991; Galland 1992).

Some action spectra, particularly those for photodifferentiation of several fungi, have only a single peak either in the UV-A or in the UV-B region. Such action spectra are reasonably well explained on the basis of a pterin or pteridin receptor (Galland and Senger, 1988). Near-UV photoreceptors are found also in higher plants. An example is anthocyanine induction in broom sorgum, which is under the control of phytochrome, a blue-light receptor and, in addition, a separate UV-B receptor (Yatsuhashi *et al.*, 1982). Evidence for the existence of separate UV-A and blue-light receptors for the control of hypocotyl lengthening was obtained for *Arabidopsis* (Young *et al.*, 1992).

In cases of photoreceptor dichroism one can obtain further clues about their identity from the orientation of the transition dipole moments at different wavelengths. In riboflavin for example the transition dipole moments in the far UV and the blue region are about 45 degrees apart (Kurtin and Song, 1968). A similar relative orientation in the far UV and blue regions were deduced for *Phycomyces* on the basis of polarized light experiments supporting thus the flavin hypothesis (Jesaitis, 1974).

Flavin analogs with altered absorption spectra may cause a corresponding shift of the action spectrum as was observed with roseoflavin for *Phycomyces* (Otto *et al.*, 1981) and with 1-deazariboflavin for *Neurospora* (Paietta and Sargent, 1983). The alteration of the action spectrum suggests photoreceptor displacement, a fact from which the flavin nature of the receptor may be inferred.

3. Physicochemical properties of Flavins and Pterins

The most ubiquitous flavins are FAD, FMN and riboflavin from which the former two are derived. Riboflavin is a tricyclic heteroaromatic isoalloxazine ring with methyl groups in positions 7 and 8 and a ribityl group in position 10. The absorption spectra of riboflavin, FMN and FAD have in their oxidized states major bands near 450, 370 and 265 nm, which are caused by π-π* transitions and correspond to the electronic transitions from the S_0 ground state to the excited states S_1, S_2 and S_3. Flavins are highly fluorescent (emission at 525 nm) and have efficient intersystem crossing from the S_1 singlet to the T_1 triplet state and rather long-lived phosphorescence at 605 nm (Kurtin and Song, 1968). Essentially all the excitation energy of riboflavin and FMN is accounted for by fluorescence (lifetime 5.6 ns) and intersystem crossing to the relatively long-lived triplet state (lifetime 19 μs), indicating that little energy is dissipated by non-radiative decay of excited states. This means that excitation energy is potentially available as required for a photoreceptor.

Oxidized free flavins can be photoreduced in the presence of an electron donor. The amount of flavosemiquinone (1 e^- equivalent) and hydroquinone (2 e^- equivalent) is strongly pH-dependent (Ehrenberg and Hemmerich, 1968). The absorption spectra of the semiquinones can vary greatly with the environmental conditions. The blue semiquinone type of certain flavoproteins has a broad absorption peak that can extend into the far-red, while the red semiquinone type has a relatively sharp peak near 470 nm. It appears likely that blue-light receptors can exist in such semiquinone states *in vivo*. Various physiological studies do indeed indicate that such semiquinones are of paramount importance (Galland *et al.*, 1995; Ahmad and Cashmore 1996, see below).

Like flavins also pterins are ubiquitously present in nature. Pterins are derived from pteridines through introduction of an amino group in position 2 and an oxo-group (a hydroxyl-group as the tautomeric form) in position 4 (Pfleiderer, 1984). The optical properties of the pterins are determined by their tautomerism, which in turn depends on pH and redox state. Pterins exist in three different redox states, the fully oxidized form, the dihydro form and the tetrahydro form. The redox reactions which pterins undergo occur between the dihydro and the tetrahydro forms. This is in contrast to most of the flavin biochemistry which usually operates from the oxidized state.

Relevant to the near-UV/blue light problem are the pterins which absorb between 300 and 500 nm. These include metabolically important compounds such as folates, neopterin and biopterin. The oxidized neutral forms of these pterins absorb generally in the near-UV between about 350 to 380 nm while the 7,8-dihydro and the 5,6,7,8-tetrahydro forms absorb at shorter wavelengths. Beside pterins (2-amino, 4-oxo pteridines) also the closely related lumazines (2-oxo, 4-oxo pteridines) which generally absorb below 400 nm, represent potential contenders for near-UV photoreceptors.

4. Redox reactions and possible spectroscopic assays for the blue-light receptor

If flavins and pterins cooperate in the reception of near-UV and blue light, they could do so *via* redox reactions and/or *via* internal energy transfer. An example for a sequence of redox reactions is nitrate reductase from various eucaryotic organisms. Very probably the following sequence of a two-electron transfer takes place:
NAD(P)H → FAD → cyt *b557* → pterin-Mo-complex → NO_3. In *Neurospora* nitrate reductase appears to be necessary for the photoinitiation of conidia and replacement of the molybdenum, which is complexed by molybdopterin, by tungstate or vanadate renders the system light insensitive (Ninnemann, 1990).

The characterization of a near-UV/blue-light receptor without an spectroscopic assay is hard to achieve. It is indeed the lack of such an assay which has hampered for so long progress in this field. Early attempts to develop a spectroscopic photoreceptor assay were centered around the idea that a flavin-type photoreceptor could possibly induce a redox reaction which could then be measured as light-induced absorption change (LIAC). A first attempt along these lines was undertaken by Berns and Vaughn (1971) with *Phycomyces* material that showed LIACs which were later interpreted as photoreduction of flavins and pterins (Siefermann-Harms *et al.*, 1985). Other LIACs indicated the photoreduction of *b*-type cytochromes, *e.g.* in *Phycomyces*, *Neurospora*, *Dictyostelium*, corn coleoptiles (Leong *et al.*, 1981) and various higher plant material (review: Galland and Senger, 1991). Action spectra for these LIACs indicated a flavin-type photoreceptor. The LIAC approach has been neglegted somewhat during the past decade because of several objections with which this approach has met. For example, the irradiances necessary to achieve such LIACs are many orders of magnitude above the *in vivo* light responses which they are supposed to mediate.

5. DNA Photolyases: Models for Blue-light Receptors

Far-UV generates in DNA pyrimidine dimers (mainly *cys-syn* thymine-thymine dimers) that would be lethal if they were not removed by DNA photolyases. Upon absorption of light of wavelengths between 300 and 600 nm, procaryotic and eucaryotic photolyases split the dimers so that the two original pyrimidines are regenerated. The chromophores of the DNA photolyases of *E. coli* and yeast are $FADH^-$ and 5,10-methenyl-tetrahydrofolyl-polyglutamate (= MTHF) (review: Sancar, 1994). Other photolyases possess beside $FADH^-$ a deazaflavin as second chromophore. The MTHF functions as a light collector which effectively transfers the absorbed energy to the $FADH^-$ which in turn donates reversibly one electron to the pyrimidine dimer causing this way its monomerization. The outlined mode of action has model character for blue-light photoreceptors. For the blue-light receptor system of *Euglena* for example there exists preliminary evidence for internal energy transfer between pterins and flavins (Sineshchekov *et al.*, 1994).

Several photolyases have been identified for higher plants, among them for *Sinapis* (Batschauer, 1993) and *Arabidopsis* (Pang and Hays, 1991). The aminoacid sequence derived from the cloned gene has great similarity to microbial photolyases and a close relationship to enzymes with a deazaflavin chromophor. The expression of photolyases in these systems as well as in certain fungi, for example *Phycomyces*, is itself under blue-light control (*e.g.* Galland, 1996).

6. The CRY1-Photoreceptor of *Arabidopsis thaliana*

The first blue-light receptor which has been identified is the so-called CRY1 photoreceptor (CRY for cryptochrome) of *Arabidopsis thaliana* (Ahmad and Cashmore, 1993). CRY1 mediates the stimulation of anthocyanine biosynthesis and inhibition of hypocotyl growth. Mutants with defects in the *hy-4* gene, which code for the CRY1 receptor, are insensitive to blue light in that they display a long hypocotyl under blue light. The blue-light mediated phototropism is nevertheless normal in these mutants. This fact indicates the existance of several blue-light photoreceptor for different functions.

The cloned *hy-4* gene showed a great amino acid sequence homology to microbial DNA photolyases. It displayed, however, a greater relation to the enzymes of the deazaflavin type especially in the NH_2-terminal half. The *hy-4* gene and a related gene from *Sinapis* (called SA-phr1) were overexpressed in *E. coli* and the respective photoreceptor proteins were shown to contain as chromophores FAD and methenyl-tetrahydrofolate (Lin *et al.*, 1995a; Malhotra *et al.*, 1995). These overexpressed proteins were, however, fusion proteins (containing maltose-binding protein) and thus do not fully represent the *in-vivo* situation. Indeed, when CRY1 photoreceptor was expressed in a baculovirus system, *i.e.* without fusion protein, it was found to bind FAD non-covalently but no second chromophore (Lin *et al.*, 1995). A physiologically significant property of the CRY1 protein is its stability in the semiquinone state which is absorbing green light (Lin *et al.*, 1995). This green-light absorbing form of the blue-light photoreceptor could be responsible for several long-wavelength effects associated with blue-light reactions (Galland and Senger, 1991; Galland 1992).

Consistent with the photoreceptor function of CRY1 was the finding that overexpression of the CRY1 protein in transgenic tobacco and *Arabidopsis* conferred a short-hypocotyl phenotype and also a hypersensitivity to near-UV and blue light (Lin *et al.*, 1995a,b). The CRY1 photoreceptor is functionally dependent on phytochrome A and B; in severely affected phytochrome mutants blue light is ineffective despite the presence of CRY1 receptor (Ahmad and Cashmore, 1997). The CRY1 protein has no predicted hydrophobic domains and does thus not constitute a membrane-bound protein. It appears to be expressed in all tissues and is found in the soluble fraction (Ahmad and Cashmore,1996).

7. Sensory transduction: blue light-induced protein phosphorylation

In various plants blue light induces a rapid phosphorylation of a protein (116 - 124 kDa depending on the species) (Short and Briggs, 1994). In oat coleoptiles for example a 116 kDa protein is phosphorylated. Several lines of evidence suggest a direct involvement of this phosphorylation in the transuduction chain: (1) the action spectrum for phosphorylation in coleoptiles resembles the one for phototropism of coleoptiles (Palmer *et al.*, 1993); (2) in pea there is good agreement for the blue-light sensitivities of phosphorylation and phototropism (Short and Briggs, 1990; Short *et al.*, 1993); (3) the highest activity resides in the coleoptile tip, the place where also phototropic sensitivity is highest (Short *et al.*, 1992), and (4) the level of phosphorylation is higher at the side close to the unilateral light than at the shaded side (Salomon and Rüdiger, 1997).

The phosphorylation can be obtained even *in vitro* with plasma membrane preparations; these show highest activity and ATP consumption when they are obtained from dark-grown plant material (Palmer *et al.*, 1993). The reaction represents the fastest blue-light reaction described so far. Most significant is the conclusion that the phosphorylation occurs at the level of the blue-light photoreceptor for phototropism. This is shown by the observation that *Arabidopsis* mutants with defective phototropism (genotype *nph*) also lack phosphorylation; indeed the 120 kDa protein itself is missing (Liscum and Briggs, 1995). It is presently hypothesized that the 120 kDa proteins represent blue-light receptors and efforts are made to clone the requisite genes.

Blue light induces a rapid depolarization at the plasma membrane of cucumber and *Arabidopsis* (Spalding and Cosgrove, 1992) an event which apparently depends on the inactivation of an plasma membrane proton ATPase and opening of anion channels. In a *hy-4* mutant of *Arabidopsis* this response is largely reduced indicating thus that the CRY1 photoreceptor is involved (Cho and Spalding 1995). Probably the blue-light receptor acts *via* a GTP-binding protein and a proteinkinase. Membrane fractions of dark-grown pea buds contain a G protein which apparently mediates a blue-light enhanced binding of GTP (Warpeha *et al.*, 1991). The light-induced binding could be mediated by a flavin type of photoreceptor, because fluorescence emission at 550 nm could be detected which was quenched by quenchers of flavin excited states.

References

Ahmad, M. and Cashmore A.E. (1993) HY4 gene of *Arabidopsis thaliana* encodes a protein with characteristics of a blue-light photoreceptor, *Nature* **366**, 162-166.

Ahmad, M. and Cashmore A.E. (1996) Seeing blue: the discovery of cryptochrome, *Plant Mol. Biol.* **30**, 851-861.

Ahmad, M. and Cashmore A.E. (1997) The blue-light receptor cryptochrome 1 shows functional dependence on phytochrome A or phytochrome B in *Arabidopsis thaliana*, *The Plant J.* **11**, 421-427.

Batschauer, A. (1993) A plant gene for photolyase: an enzyme catalyzing the repair of UV-light-induced DNA-damage, *Plant J.* **4**, 705-709.

Berns, D.S. and Vaughn, J.R. (1970) Studies of the photopigment system in *Phycomyces*, *Biochem. Biophys. Res. Comm.* **39**, 1094-1103.

Cho, M.H. and Spalding, E.P. (1995) An anion channel mediating growth-inhibition by blue light in *Arabidopsis* seedlings, *Plant Physiol.* **108**, 86.

Ehrenberg, A. and Hemmerich, P. (1968) Flavoenzymes: chemistry and molecular biology, in T.P. Singer (ed.), *Biological Oxidations*, Interscience Publishers, New York, pp. 239-262.

Galland, P. (1992) 40 years of blue-light research and no anniversary, *Photochem Photobiol* **56**, 847-853.

Galland, P. (1996) Ultraviolet killing and photoreactivation of *Phycomyces* spores. *Microbiol. Res.* **151**, 1-9.

Galland, P. and Senger, H. (1988) The role of pterins in the photoreception and metabolism of plants, *Photochem. Photobiol.* **48**, 811-820.

Galland, P. and Senger, H. (1991) Flavins as possible blue-light photoreceptors, in G.H. Holmes (ed.), *Photoreceptor Evolution and Function*, Academic Press, London New York, pp. 65-124.

Galland, P., Amon, S, Senger, H. and Russo, V.E.A. (1995) Blue light reception in *Phycomyces*: Red light sensitization in *madC* mutants, *Bot. Acta* **108**, 344-350.

Galston, A.W. (1950) Riboflavin, light and the growth of plants, *Science* **111**, 619-624.

Jesaitis, A.J. (1974) Linear dichroism and and orientation of the *Phycomyces* photopigment, *J. Gen. Physiol.* **63**, 1-21.

Kurtin, W.E. and Song, P.-S. (1968) Photochemistry of the model phototropic system involving flavins and indoles - I. Fluorescence polarization and MO calculations of the direction of the electronic transition moments in flavins, *Photochem. Photobiol.* **7**, 263-273.

Leong, T.-Y., Vierstra, R.D. and Briggs, W.R. (1981) A blue light-sensitive cytochrome-flavin complex from corn coleoptiles - Further characterization, *Photochem. Photobiol.* **34**, 697-703.

Lin, C., Robertson, D.E., Ahmad, M., Raibekas, A.A., Jorns, M.S., Dutton, P.L. and Cashmore, A.R. (1995a) Association of flavin adenine dinucleotide with the *Arabidopsis* blue-light receptor CRY1, *Science* **269**, 968-970.

Lin, C., Ahmad, M., Gordon, D. and Cashmore, A.R. (1995b) Expression of an *Arabidopsis* cryptochrome gene in transgenic tobacco results in hypersensitivity of blue, UV-A, and green light, *Proc. Natl. Acad. Sci. USA* **92**, 8423-8427.

Liscum, E. and Briggs, W.R. (1995) Mutations in the *nph1* locus of *Arabidopsis* disrupt the perception of phototropic stimuli, *Plant Cell* **7**, 473-485.

Malhotra, K., Kim, S.T. Batschauer, A., Dawut, L. and Sancar, A. (1995) Putative blue-light photoreceptors from *Arabidopsis thaliana* and *Sinapis alba* with a high-degree of sequence homology to DNA photolyase contain the 2 photolyase cofactors but lack DNA repair activity, *Biochemistry* **34**, 6892-6899.

Ninnemann, H. (1990) Participation of pterin-molybdenum cofactor in the regulation of nitrate reductase activity and signal transduction during light-stimulated conidiation of *Neurospora crassa*, in W.R. Ullrich, C. Rigano, A. Fuggi and P.J. Aparicio (eds.), *Inorganic Nitrogen in Plants and Microorganisms*, Springer, Berlin, pp. 190-195.

Otto, M.K., Jayaram, M., Hamilton, R.M. and Delbrück, M. (1981) Replacement of riboflavin by an analogue in the blue-light photoreceptor of *Phycomyces. Proc. Natl. Acad. Sci. USA* **78**, 266-269.

Paietta, J. and Sargent, M.L. (1983) Modification of blue light photoresponses by riboflavin analogs in *Neurospora crassa.*, *Plant Physiol.* **72**, 764-766.

Palmer, J.M., Short, T.W. and Briggs, W.R. (1993) Correlation of blue light-induced phosphorylation to phototropism in *Zea mays L.*, *Plant Physiol.* **102**, 1219-1225.

Pang, Q. and Hays, J.B. (1991) UV-B-inducible and temperature-sensitive photoreactivation of cyclobutane pyrimidine dimers in *Arabidopsis thaliana*, *Plant Physiol.* **95**, 536-543.

Pfleiderer, W. (1984) Pteridines, in A.R. Katritzky and C.W. Rees (eds.), *Comprehensive Heterocyclic Chemistry*, vol. 3, Pergamon Press, Oxford, pp. 263-327.

Plaut, G.W.E., Smith, C.M. and Alworth, W.L. (1974) Biosynthesis of water soluble vitamins, *Ann. Rev. Biochem.* **43**, 899-922.

Quiñones, M.A. and Zeiger, E. (1994) A putative role of the xanthopyll, zeaxanthin, in blue-light perception of corn coleoptiles, *Science* **264**, 558-561.

Salomon, M. and Rüdiger, W. (1997) Phototropism and protein phosphorylation in higher plants; unilateral blue light irradiation generates a directional gradient of protein phosphorylation across the oat coleoptile., *Bot. Acta* **110**, 214-216.

Sancar, A. (1994) Structure and function of DNA photolyase, *Biochemistry* **33**, 2-9.

Short, T.W. and Briggs, W.R. (1990) Characterization of a rapid, blue light-mediated change in detectable phosphorylation of a plasma membrane protein from etiolated pea (*Pisum sativum* L.) seedlings, *Plant Physiol.* **92**, 179-185.

Short, T.W. and Briggs, W.R. (1994) The transduction of blue-light signals in higher-plants, *Annu. Rev. Plant Physiol. Plant Mol. Biol.* **45**, 143-171.

Short, T.W., Porst, M. and Briggs, W.R. (1992) A photoreceptor system regulating *in vivo* and *in vitro* phosphorylation of a pea plasma membrane protein, *Photochem. Photobiol.* **55**, 773-781.

Siefermann-Harms, D., Fritz, B. and Ninnemann, H. (1985) Evidence for a pterin-derivative associated with the molybdenum cofactor of *Neurospora crassa* nitrate reductase, *Photochem. Photobiol.* **42**, 771-778.

Sineshchekov, V.A., Geiß, D., Sineshchekov, O.A., Galland, P. and Senger, H. (1995) Fluorometric characterization of pigments associated with isolated flagella of *Euglena gracilis*: evidence for energy migration, *J. Photochem. Photobiol. B: Biol.* **23**, 225-237.

Spalding, E.P. and Cosgrove, D.J. (1992) Mechanism of blue-light induced plasma-membrane depolarization in etiolated cucumber hypocotyls, *Planta* **188**, 199-205.

Warpeha, K.M.F., Hamm, H.E., Rasenick, M.M. and Kaufman, L.S. (1991) A blue-light-activated GTP-binding protein in the plasma membranes of etiolated peas, *Proc. Natl. Acad. Sci. USA* **88**, 8925-8929

Yatsuhashi, H., Hashimoto T. and Shimizu, S. (1982) Ultraviolet action spectrum for anthocyanin formation in broom sorghum first internodes, *Plant Physiol.* **70**, 735-741.

Young, J.C., Liscum, E., Hangarter, R.P. (1992) Spectral-dependence of light-inhibited hypocotyl elongation in photomorphogenetic mutants of *Arabidopsis*: evidence for a UV-A photosensor, *Planta* **188**, 106-114.

MORPHOLOGICAL-ANATOMICAL CHARACTERISTICS OF TURKEY OAK (*Quercus cerris* L.) OFFSPRING DEPENDING ON LIGHT INTENSITY IN REGENERATION AREAS

M. BOBINAC and D. VILOTIĆ
Faculty of Forestry, Belgrade University,
Kneza Višeslava 1, Belgrade, Yugoslavia

Abstract

This paper deals with morphological-anatomical characteristics of Turkey oak seedlings with regular – monophase and polyphase height growth, formed on regeneration areas in stand conditions. In closed stands, during the vegetation period, Turkey oak height growth is monophase, and on felling areas, under full light, polyphase growth was observed along with monophase growth. Namely, during the vegetation period, some individuals activated their embryonic terminal buds and new morphological characters were formed. In this way, on regeneration areas with different light intensities, we recorded induced variability of height growth and the formation of new morphological characters. The results of morphological-anatomical analysis of cross sections of primary axes point to a different degree of development of Turkey-oak one-year old seedlings, both in the conditions of regeneration areas with different light intensities, and in different types of growth on felling areas under full light intensity.

1. Introduction

From the aspect of silviculture, the definition of the state of seedlings of major edificator species in forest communities, it is important to know their main morphological and growth characteristics in the particular environmental conditions.

The significance of the study of Turkey oak ecology and development is based on plant geographical distribution of the species in the oak belt in southeast Europe, and from the aspect of silviculture, the generally unfavourable state of forests with this edificator, as well as the need of reclamation of vast areas resulting from regressive succession of more valuable edificators.

In Serbia, great variability of Turkey oak morphological characters was observed (Janković, 1956, 1970, Erdeši, Gajić, 1977). The form of leaves, bark and a rather thin crown indicate that Turkey oak is a rather xerophilic and thermophilic species. According to Jovanović, (1954), Turkey oak is a step further in oak philogeny in their invasion (ecological and spatial) of the terrain adjacent to the Mediterranean.

The aim of this paper is to point to some specificities of the development of Turkey oak seedlings in different environmental conditions, and their main morphological-anatomical characteristics. Seedling growth can be modified depending on environmental conditions. The knowledge of morphological-anatomical structure and the scope of modifications caused by different impacts, represents the knowledge of the norms of tree species reactions.

2. Material and Method

Turkey oak seedlings were researched on two types of regeneration areas, in stands with closed canopy and in fully illuminated felling areas. The stands belong to some ecological varieties of the community *Tilio-Carpino-Quercetum robori-cerris* Jov.79 and *Orno-Quercetum cerris-virgiliane* Jov. et Vuk. 77 (Tomić, 1991) on a series of soil over loess.

The morphometric was performed on the seedlings with undisturbed growth, and with different types of shoot axis formation: A-monophase under closed canopy, B_1-monophase, B_2- twophase and B_3- threephase in the conditions of full light. Seedling heights and lengths of individual phase shoots, root collar thickness and the number of leaves on the primary axis and on the respective phase shoots. Anatomic structure of stems of overwintered seedlings was analysed on permanent preparations. Cross sections were taken in the zone of root collar of the seedlings with different growths, to measure the following: thickness of primary bark and primary xylem, pith diameter, vessel diameter in tangential and radial directions, as well as the number of vessels per mm^2 on five randomly selected view fields of each sample.

3. Results and Discussion

In closed canopy stands, Turkey oak seedlings growth was the monophase formation of shoot axis, and in felling areas, under full light, along with monophase growth, numerous individuals had the so-called (Bobinac, 1994, Bobinac, Vilotić, 1994, 1996). During the polyphase growth in height, seedling axes grew during the vegetation period, with a conspicuous seeming period of growth stagnation between the previous and the following phase of height growth. (plumural prolepsis, Gruber, (1987, 1992). In fully illuminated areas, till the middle of September, primary axes of some individuals had a polyphase growth in the form of maximum two phases of growth. As the result of such growth, the seedlings attained considerable height, diameter of root collar and a numerous assimilation apparatus. The characteristics of Turkey oak shoot axis formation in different environmental conditions have been shown in Table 1 and Figure 1.

Table 1. Characteristics of Turkey oak growth (*Quercus cerris* L.) in the first vegetation period under different light

Morphological	Growth type			
	A (n=75)	B_1 (n=25)	B_2 (n=25)	B_3 (n=25)
Character	Average ± Standard Error (*cm*)			
Height: h_t	22.34±0.56	20.83±1.28	23.29±0.76	32.88±1.55
Height: h_1	22.34±0.56	20.83±1.28	12.84±0.59	12.83±0.73
Height: h_2			10.45±0.75	9.05±0.74
Height: h_3				11.00±1.04
Diameter	2.25±0.05	3.99±0.13	4.89±0.16	6.15±0.23
Leaf number	4.34±0.15	7.44±0.48	11.56±0.63	22.64±0.91

Growth type: **A** Closed canopy - monophase growth
B_1 Full light - monophase growth
B_2 Full light – twophase growth
B_3 Full light – threephase growth

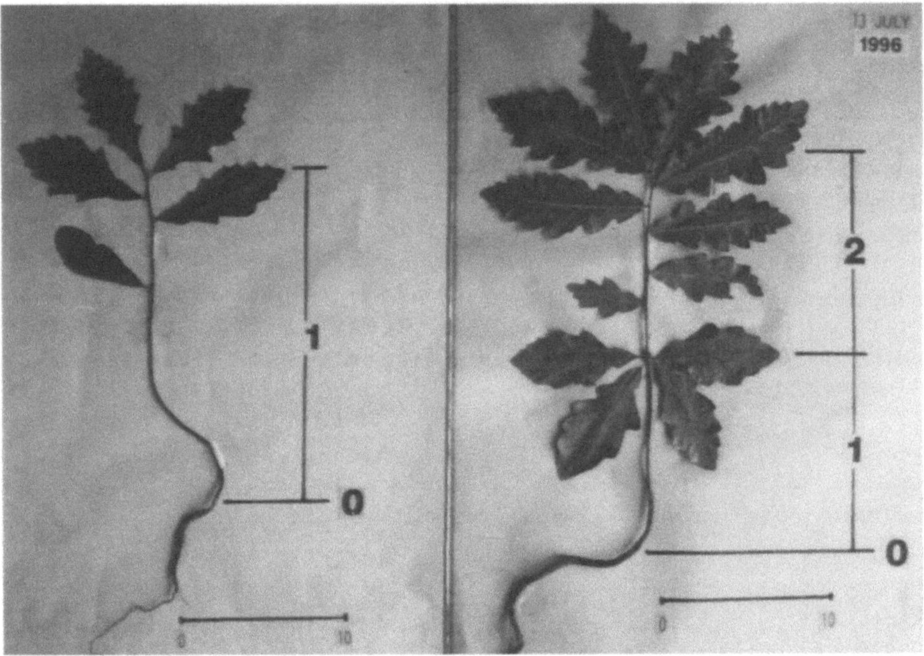

Figure 1. Seedlings (*Q. cerris L.*) with monophase (type A) and polyphase growth (type B_2)

Based on the morphological-anatomical analysis of cross sections, the seedlings with monophase and polyphase growth do not show any differences in morphological structure. Based on the comparison of elements of microscopic structure (Table 2), with the morphological characteristics of seedlings, it can be concluded that the induced polyphase height growth of Turkey oak seedlings is characterised by accelerated growth.

TABLE 2. Microscopic structure of one-year old seedlings (*Q. cerris* L.) in the zone of root collar

Growth type	Bark thickness	Width of primary xylem	Pith diameter	Vessel diameter µm			Number of vessels
	µm	µm	µm	Max	Avg.	Min	pcs/ m^2
A	375.0	2700.0	1350.0	31.5	19.1	7.5	27
B_1	255.0	3037.5	1650.0	31.5	25.5	15.0	26
B_2	435.0	6150.0	2625.0	50.0	35.7	16.5	39
B_3	420.0	4912.5	2700.0	67.5	47.4	30.0	31

4. Conclusions

1. Induced variability of height growth of Turkey oak seedlings was observed on differently illuminated regeneration areas. In the conditions of closed stand canopy, Turkey oak seedling growth was in the form of monophase shoot axis, and on felling areas, under full light, along with monophase growth, polyphase height growth was also recorded.
2. The comparison of elements of microscopic structure with the morphological characteristics of seedlings shows that induced polyphase height growth of Turkey oak seedlings is characterised by accelerated growth.

5. References

Bobinac, M. (1994) Višefazni rast u visinu jednogodišnjih biljaka lužnjaka *(Q. robur L.)* i neki aspekti značajni za semenu obnovu. *Šumarstvo* 1-2, Beograd. 47-57.

Bobinac, M., Vilotić, D. (1994) Višefazni rast jednogodišnjih jedinki hrasta lužnjaka *(Q. robur L.)* sa aspekta morfološke analize poprečnih preseka ose izdanka. *Deliblatski pesak-Zbornik radova* VI, Pančevo. 399-406.

Bobinac, M., Vilotić, D. (1996) Contribution to the study of morphological-anatomical characteristics of beech *(Fagus moesiaca/Domin, Maly/Czeczott.)* seedlings. *Zemljište i biljka*, Vol. 45, No. 1, Beograd. 57-65.

Erdeši, J., Gajić, M. (1977) *Nove forme i nalazišta cera (Quercus cerris L.) u SR Srbiji i SFR Jugoslaviji.* Glasnik Šumarskog fakulteta, br. 52, Jubilarni broj. Beograd. 91-97.

Gruber, F. (1987) *Über die sylleptische Verzweigung der Johannistriebe von Rotbuche und Stieleiche*, AFZ 49.

Gruber, F. (1992) *Dynamik und Regeneration der Gehölze.* Berichte des Forschungszentrums Walökosysteme, Reihe A, Bd.86, Teil I. Göttingen.

Janković, M.M. (1956) *Polimorfizam listova cera (Quercus cerris L.) na Fruškoj Gori i njgov ekološki i taksonomski značaj.* Zbornik Matice Srpske za prirodne nauke, 11. Novo sad. 136-149.

Janković, M.M. (1970) Rod *Quercus* L., Red *Fagales*; "In: Flora SR Srbije (M.Josifović)", tom II, Beograd. 77-86.

Jovanović, B. (1954) *Fitocenoza Quercetum confertae-cerris kao biološki indikator.* Glasnik Šumarskog fakulteta br. 8. Beograd. 207-219.

Tomić, Z. (1991) Zajednica *Orno-Quercetum cerris-virgilianae* Jov. et Vuk. 77. na južnom obodu Panonije. *Glasnik Šumarskog fakulteta*, 73, Beograd. 23-32.

EFFECTS OF IAA ON MESOPHYLL CELL MORPHOGENESIS OF *Triticum turgidum*

G. KOMIS, P. APOSTOLAKOS AND B. GALATIS
University of Athens, Department of Biology, Division of Botany
Greece, 157 84

1. Introduction

Plant hormones affect cell morphogenesis via direct hormone-induced rearrangements of the cortical microtubules [3, 4]. Our aim is to study the effects of indolyl-3-acetic acid (IAA) on mesophyll cell (MC) morphogenesis of wheat (*T. turgidum*). Mature MCs exhibit a lobed outline (see Fig. 1), forming transverse constrictions, due to localized cellulose microfibril (MF) deposition, that determines hoop-like wall reinforcements. Localized MF deposition is correlated with corresponding patterns in the organization of the cortical microtubules (Mts) since the latter form transverse Mt bands, that strikingly coalign with the adjacent wall reinforcements [2, 5]. For the description of our results we have divided the primary leaf in linear regions (2-5 mm, 5-8 mm, 8-11 mm), starting from the leaf base (meristem) to the tip (fully differentiated MCs) [2, 6]. IAA-treatment was either continuous (5 hours) or followed by recovery (post-incubation, 19 hours). Studies included tubulin immunolocalization (method described in [2]) and light and electron microscopy (method described in [1]).

2. Results

In the 2-5 mm leaf region normal MCs undergo longitudinal (bilateral) expansion. At the basal end of this region cortical Mts are obliquely or transversely arranged. At the apical end of this region cortical Mts begin to form transverse bands and single Mts remain between the Mt bands. As cited in [2] the number of Mt bands is somehow related to MC length but with no distinct interdependence. When IAA was applied, an increase in the extent of cell elongation was observed. The increase in cell length defined as a mean relative difference between normal and IAA-treated MCs was ~65% (see also Fig. 6). During continuous IAA-treatment, Mt-bundling was inhibited (data not shown) but when IAA-treatment was subsequently followed by post-incubation, Mt bands were formed.

In the 5-8 mm leaf region, Mt band formation in normal MCs continues and Mt-bands become conspicuous (Fig. 2). MCs continue to elongate but expansion is restricted by the already deposited local wall reinforcements [2]. Hoop-like MF depositions affect the cell shape and a slightly lobed outline is visible. In IAA-treated seedlings the mean MC length increases to ~80% compared to normal MCs (see also Fig. 7). In IAA-treated MCs (during continuous treatment) the transverse Mt configuration is unchanged and MCs remain unlobed. In IAA-treated and post-incubated seedlings, MCs regain the ability to display the morphogenetical pattern via Mt band formation. In this case the number of Mt bands formed is increased compared to normal MCs (Fig. 3).

In the 8-11 mm leaf region, MC morphogenesis is completed during the transition from the basal to the apical end. Mt bands progressively disintegrate and lobe formation is gradually completed (Fig. 1). IAA-treated MCs that failed to organize Mt bands at the previous leaf regions obtained a smooth outline (Fig. 4) whereas MCs from post-incubated seedlings developed outnumbered constrictions resulting in a multilobed shape (Fig. 5). The mean cell length difference falls to ~35% (see also fig. 8) indicating a reduced growth potential of IAA-treated MCs. Another effect of IAA on MC morphogenesis is the increase in the number of the lobes formed observed in the post-incubated MCs. The latter display in average 5 lobes while normal MCs, exhibit only 3 (see also Fig. 9). In addition, the distance between constrictions in multilobed, IAA-treated MCs is decreased (data not shown).

3. Discussion

Our results indicate that *T. turgidum* MCs are susceptible to IAA but only in a limited leaf area. IAA promotes cell elongation to a further extent than that observed in normal MCs by increasing the magnitude of MC bipolar longitudinal expansion. The latter seems to be the most probable cause of the observed inhibition of Mt band assembly and lobe formation during continuous IAA-treatment. When IAA is depleted during post-incubation, primary effects are reversed and MCs recover the ability to perceive and respond to a hypothetical morphogenetical signal that normally acts by determining the positions of Mt band formation at the cell cortex. It appears that IAA-induced elongation secondarily results in the increase of the cell surface, exposed to the morphogenetical field therefor in, multiplying the sites of signal perception. This probably explains the increased lobe number and lobe frequency displayed by IAA-treated and post-incubated MCs.

Figure 1. Longitudinal section of normal mesophyll from the 8-11 mm leaf region, where MC lobe formation is completed (arrows). ×650

Figure 2. Normal MC from the 5-8 mm leaf region after tubulin immunolocalization. Three well organized Mt-bands appear at the cortical cytoplasm. ×650

Figure 3. IAA-affected MC (after IAA-treatment and subsequent recovery) after tubulin immunolocalization. Five Mt-bands are organized. ×650

Figure 4. Longitudinal section of IAA-affected (after continuous treatment) mesophyll where MC lobe formation is inhibited (asterisks). ×250

Figure 5. Longitudinal section of IAA-affected mesophyll (after treatment and subsequent recovery) showing MCs (asterisks) exhibiting outnumbered lobes (compare with Fig. 5). ×650

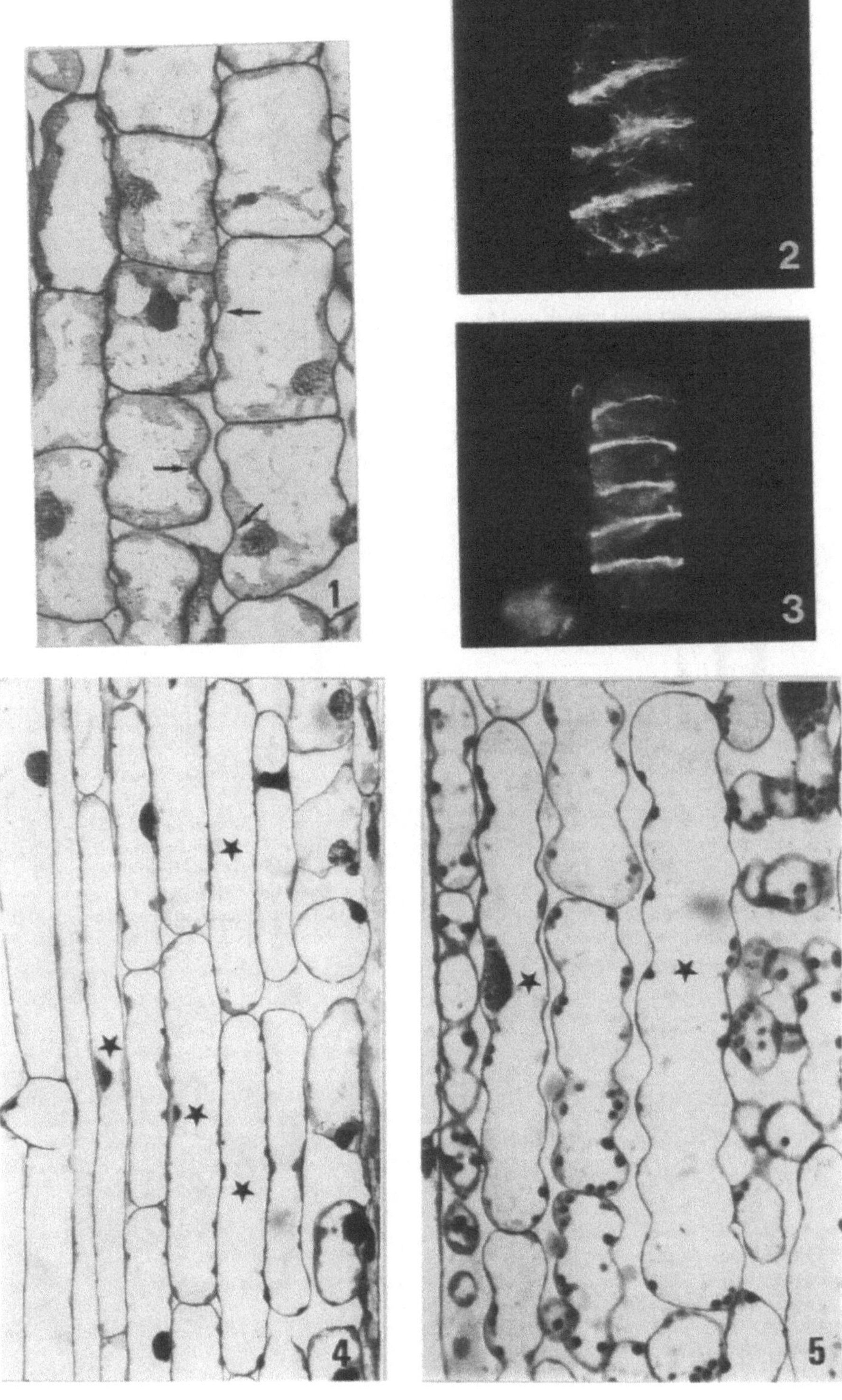
1
2
3
4
5

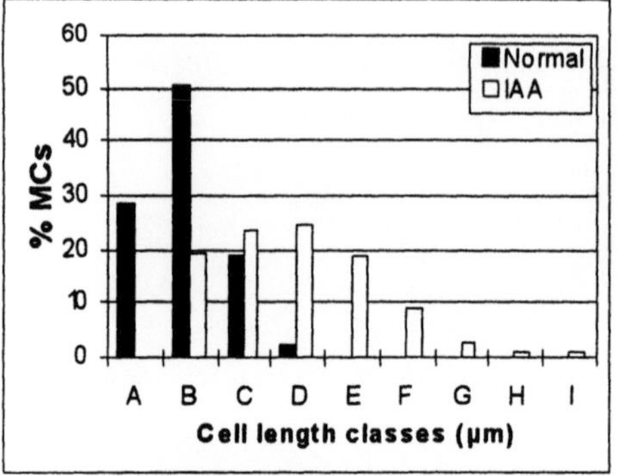

Figure 6. Distribution pattern of normal and IAA-treated MCs from the 2-5 mm leaf region vs. cell length classes (μm) [A: 16-23, B: 24-31, C: 32-39, D: 40-47, E: 48-53, F: 54-61, G: 62-69, H: 70-77, I: 78-85]. 230 MCs were accounted for each case.

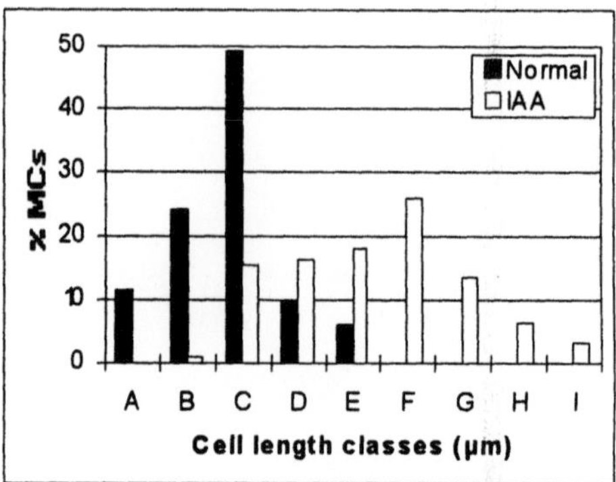

Figure 7. Histogram presenting the distribution of normal and IAA-treated MCs from the 5-8 mm leaf region in cell length classes (μm) [A: 20-40, B: 41-60, C: 61-80, D: 81-100, E: 101-120, F: 121-140, G: 141-160, H: 161-180, I: more than 180]. 230 MCs were accounted for each case.

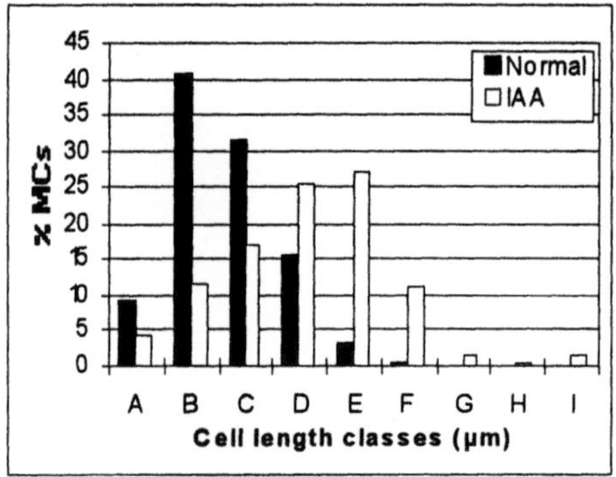

Figure 8. Distribution of normal and IAA-treated MCs from the 8-11 mm leaf region vs. cell length classes (μm) [A: 42-63, B: 64-84, C: 85-105, D: 106-126, E: 127-147, F: 148-168, G: 169-189, H: 190-210, I: more than 210]. 230 MCs were accounted for each case.

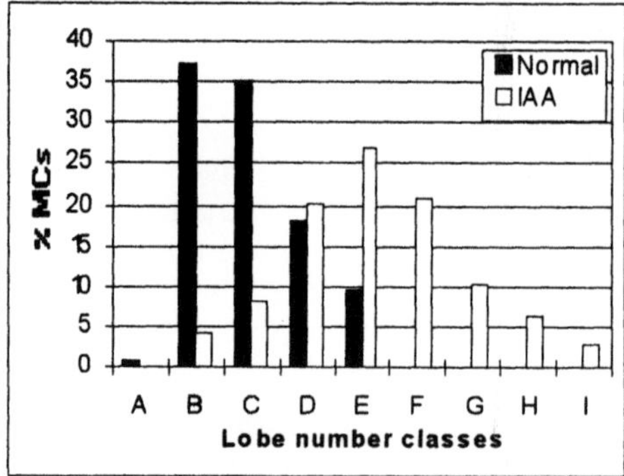

Figure 9. Distribution of normal and IAA-treated MCs from the 8-11 mm leaf region in lobe number classes (lobe number per MC) [A: 1, B: 2, C: 3, D: 4, E: 5, F: 6, G: 7, H: 8, I: more than 8]. 230 MCs were accounted for each case.

4. References

1. Apostolakos P., Galatis B., Panteris E. (1991). Microtubules in cell morphogenesis and intercellular space formation in *Zea mays* mesophyll and *Pilea cadierei* epithem. *Journal of Plant Physiology* **137**, 591-601
2. Jung G., Wernicke W. (1990). Cell shaping and microtubules in developing mesophyll of wheat (*Triticum aestivum* L.). *Protoplasma* **153**, 141-148
3. Shibaoka H. (1991). Microtubules and the regulation of cell morphogenesis by plant hormones. In Lloyd CW (ed), *The cytoskeletal basis of plant growth and form*. Academic Press, London pp. 159-168
4. Shibaoka H. (1994). Plant hormone-induced changes in the orientation of cortical microtubules: Alterations in the cross-linking between microtubules and the plasma membrane. *Annual Review of Plant Physiology Plant Molecular Biology* **45**, 527-544
5. Wernicke W., Jung G. (1992). Role of cytoskeleton in cell shaping of developing mesophyll of wheat (*Triticum aestivum* L.). *European Journal of Cell Biology* **57**, 88-94
6. Wernicke W., Milkovits L. (1984). Developmental gradients in wheat leaves-response of leave segments in different genotypes cultured in vitro. *Journal of Plant Physiology* **115**, 49-58

MORPHOLOGICAL AND PHYSIOLOGICAL CHANGES DURING THE ONTOGENETIC DEVELOPMENT OF PEAS SEEDS

N. PANAYOTOV AND N. STOEVA
Higher Institute of Agriculture, 4 000 Plovdiv, Bulgaria

1. Introduction

Denif (1989) investigated the possibilities of germination of peas seeds during the different periods of their development. Investigating the linear sizes of seeds of pea, Dution (1991) point out, that the last stage in which abortion can occur is in seed with diameter - 8.5 - 6.0 mm. Gorecki *et al.* (1993), and Ross (1990) studied the physiological and biochemical changes in the growth and sprouting of the peas seeds. The main purpose of this study was to investigate the changes in the morphological and physiological condition of the seeds of pea, during the different stages of the development.

2. Materials and Methods

The investigations were carried out with cultivars of green pea - Ran 1 and Prometej. After preliminary marking of the flowers, seeds for analysis were taken on the 10, 20, 25,30 and 40 day from the beginning of flowering,.The weight of one seed (a), weight(b) and diameter (D) of the cotyledons, weight (c) and thickness (T) of spermoderm and length of embryo (LE) were measured. The germinative energy (GE) and germinability (G) were determinated by Bulgarian State Standard (BBS 601 - 85). The fresh matter of sprouts per seed (FM) and length of embryonic root (LER) were established. The physiological behaviours include absolutely dry matter (ADM) (weghtly), intensity of respiration (IR), activity of the enzymes: peroxidase (PA), (Bojarkin, 1951),catalase (CA) (Ermakov, 1972) and nitrtatreductase (NRA) (Jaworsky, 1971), total content of protein (TP) (Lowry, 1951). The seeds for analysis were obtained from 25 plants. Average values of the results are obtained during three years of investigation.Thestatistic analysis followed the method of Fisher (1958).

3. Results and Discussion

The weight of one seed (Fig 1.) in cultivar Ran 1 increased to the 30 day after flowering - from 137.9 mg to 313.3 mg - the insrease was 2.27 times. In cultivar

Prometej the increase of weight of one seed was observed up to the 20 day after flowering (with 1.39 time). The decrease of seed weight from 25 or 30 day most probably was due to the decrease of water content. The similar tendency was observed in the establishment of the weight of the cotyledons. In cultivar Ran 1 it reached from 106.6 mg on the 20 day to 246.6 mg on the 25 day, which was the biggest increase (2.31 times). This characteristic for the seeds of cultivar Prometej increased to the 20 day after flowering (with 1.6 time) and then the weight of the cotyledons was lower with 1.38 time. The weight of the spermoderm in both of cultivars increased to the 20 day after flowering with 1.85 and 1.6 time, respectively. The proportion of cotyledons in the total weight of the seed from Ran 1 was largest on the 25 day - 93.65 % and from Prometej on the 20 day - 92.96 %. In the different periods of the development of the seeds the spermoderm constituted - 5.29 % - 36.60 % of the total weight for Ran 1 and - 10.76 % - 15.16 % of Prometej.

Diameter of cotyledons (Table 1.) in Ran 1 varied from 4.66 mm (10 day) to 7.96 (25 day). In Promotej the increase was up to the 20 day with 1.81 time. The thickness of spermoderm increased to the 25 day and till the end of the investigation did not change essentially. The biggest change in length of embryo of peas seeds of Ran 1 was in the period between 20 - 25 days and of Prometej between 10 - 20 days. The germinative energy (Table 2) in Ran 1 varied from 9.0 % for the seeds from the 20 day after flowering to 33.8 % for the seeds from the 40 day. The germinability in this cultivar increased from 10.2 % (20 day) to 97.5 % (30 day) . It must be pointed out that on the 10 day germinated seeds were not found. On the contrary, in Prometej on the 10 day after flowering germinated seeds were found. The germinative energy from 5.96 % on the 10 day reached 27.3 % on the 40 day. The highest germinability was in the seeds from the 25 day - 96.0 %. In this cultivar decrease in the germinabilty on the 40 day was establiched too. The decrease in this characteristic

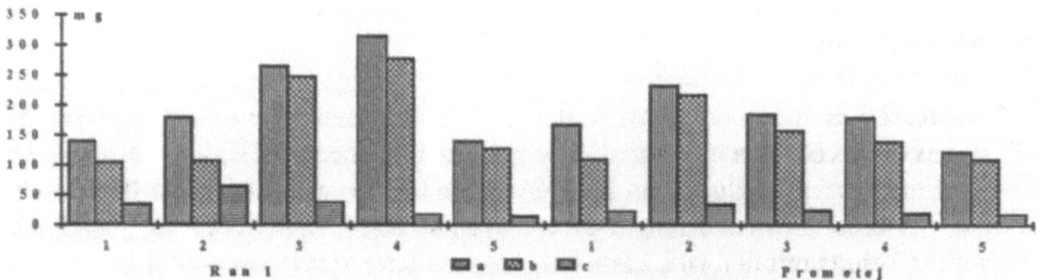

Figura 1. Weight of seed (a),weight of cotiledons (b),weight of spermoderm (c) 1-10 -, 2-20 - ,3-25 - , 4-30 - , 5-40 days from flowering.

Table 1. Morphologo - anatomical characteristics of peas seeds during the ontogenetic dev1elopment.

N	Days	Ran 1			Prometej		
		D - mm	T- µm	LE - mm	D - mm	T- µm	LE - mm
1	10	4.66	182.1	1.45	4.14	200.1	2.84
2	20	6.15	204.3	2.02	7.53	238.3	3.21
3	25	7.96	212.2	3.52	6.50	247.0	3.58
4	30	7.58	184.0	3.58	6.34	206.1	3.64
5	40	7.38	180.1	3.58	6.30	206.1	3.72
	5.0%	0.93	1.92	0.80	0.68	0.67	0.91
	GD1.0%	1.35	2.74	1.16	0.99	1.19	1.32
	0.1%	2.03	4.11	1.74	1.48	2.74	1.99

Table 2. Sowing quality of peas seeds during the ontogenetc development.

		Ran 1				Prometej			
N	Days	GE-%	G- %	LER-cm	FM- mg	GE-%	G-%	LER-cm	FM-mg
1	10	0.0	0.0	0.0	0.0	5.96	26.3	2.83	70.0
2	20	9.0	10.2	0.98	21.9	12.0	49.3	7.14	98.4
3	25	13.7	92.8	4.78	110.3	12.6	96.0	8.54	90.6
4	30	25.0	97.5	3.76	132.5	11.3	94.6	6.99	75.2
5	40	33.8	86.3	3.74	137.1	27.3	85.3	6.72	97.5
	5.0%	4.1	18.0	1.82	9.30	6.23	20.1	2.30	8.02
GD	1.0%	9.4	26.1	2.64	23.35	12.48	29.1	3.31	15.07
	0.1 %	13.6	39.1	3.96	35.03	18.72	43.6	5.00	22.6

in the seeds in advanced stage of development probably was due to the morphologo-anatomical and biochemical changes gathering in the seeds during the development (Ovcarov. 1976) and also could be due to the passing of the seeds in dormancy period (Rahimov *et al.*, 1979). The greatest length of embryonic root was established in the seeds from the 25 day after flowering - 4.78 cm and 8.54 cm for Ran 1 and Prometej, respectively. In Ran 1 with the advance of ontogenetic development the fresh matter of sprout increasedand with greatest matter were for Prometej from 20 and 40 days after flowering - 98.4 mg and 97.5 mg. Very strong positive correlation (Table 3.) was established between the weight of seeds and weight of cotyledons. Positive, but with lower quotient, was the correlation with diameter of cotyledons too. Correlation between weight of the seeds and germinability varied from +0.53 to + 0.98 and from + 0.37 to + 0.97 for Ran 1 and Prometej, respectively. The quotient of correlation was not highest in the heaviest and with biggest germinability seeds. Correlation dependence between length of embryo and germinability was positive too. The quotient was highest for the seed with highest germinability. There is a strong positive correlation between fenologcal age of seeds (days from flowering) and germinability - r =+ 0.81 r =+0.86 for Ran 1 and Prometej, respectively. The content of absolutely dry matter of peas seeds (Table 4.) was highest in the 40 - days seeds. The intensity of respiration in Ran 1 was lowest on the 25 day and in Prometej on the 30 day. The results of the activity of the enzyme peroxidase were not essentially different in both of cultivars. The catalasis activity increased to the 25 day. In the activity of enzyme nitratreductase tendency for decrease was observed. The content of total protein decreased too. This could be due in consequence of partial denaturation and proteoletic and oxygenic changes in nitrogen-containing combinations (Krasnook *et al.*,1978) or from the aggregation of the protein molecules is a results of the formation of disulphide connection (Saio *et al.*,1982).

4. Conclusion

The weight of one seed, of cotyledons and of spermoderm of pea increased to the 20 - 30 day from flowering and diameter of cotyledons and the thickness of spermoderm - to the 20 - 25 day . The growth was most intensive between the 10 - 20 and the 20 - 25 day from flowering. The length of embryo followed gradual growth. Variety differences were registered at the beginning of germination - it was on the 10 day for Prometej and on the 20 day for Ran 1. The highest germinability was on the 25 day and on the 30 day. Positive correlation was established between weight of the seed: and weight and diameter of cotyledons and germinability, and between fenological age of seeds and germinability. The content of absolutely dry matter in the seeds of pea

Table 3. Correlation deppendings of peas seeds during the ontogenetik development.

N	Days	a /b	a/D	a/G	LE/G
		Ran 1 r* = +0.81 +++			
1	10	0.97 +++	0.74 +++	-	-
2	20	0.89 +++	0.41	0.92 +++	0.67
3	25	0.99 ++	0.51 ++	0.53	0.46
4	30	0.99 +++	0.55	0.66 +	0.89 +++
5	40	0.99 +++	0.67 ++	0.98 +++	0.61
		Prometej r* = +0.86 +++			
1	10	0.59 +	0.38 +	0.37 +	0.29
2	20	0.99 +++	0.43 +	0.41 +	0.28
3	25	0.50	0.22	0.41 +	0.70 ++
4	30	0.64 ++	0.67 ++	0.61 ++	0.59 +
5	40	0.60 +++	0.45 +	0.97 ++	0.69 ++

* -r -correlatin quotient, between fenological age of seeds (in days from flowering) and germinability.

Table 4. Physiological characteristics of reas seeds during the ontogenetic development

N	Days	ADM - %	IR- CO_2/g	AP-E/g fr.m.	AC-cm^3O_2	NRA-mg/ NO_3	TP-γ /ml
				Ran 1			
1	10	20.89	0.528	240.0	11.42	26.8	22.50
2	20	20.39	0.532	215.0	11.85	23.1	21.35
3	25	32.03	.0285	218.6	13.40	20.4	18.40
4	30	40.48	0.358	208.8	10.35	18.6	19.60
5	40	42.67	0.400	225.0	10.60	18.2	18.58
				Prometej			
1	10	19.82	0.473	228.6	12.60	22.18	19.48
2	20	21.15	0.496	226.0	13.10	24.50	20.80
3	25	28.18	0.450	248.2	13.21	21.35	22.20
4	30	37.50	0.368	217.9	10.18	19.20	17.08
5	40	39.75	0.384	208.6	10.68	17.62	17.00

gradually increased up to the 40 day after flowering and the intensity of respiration, enzyme activity and content of total protein decreased. This results may by used in seed production with aim to improve the growth, formation and quality of the seeds.

5. References

Bojarkin A. N. (1951) Biochimia, Tehnika, Moskva.

Denif Y.(1989) Pea seed devel.and eval.of its abil.to germ.,IVInt,Sym.on seed res.inHort.,Wageningen,73-81.

Duthion C.(1991)Seed lenghts corresp.to the final stage in seed abort. of legumin.,Cr.Scien.,USA31,179-183.

Ermacov A.I. (1972) Metodi biochimiceskogo izsledovanii rastenii, Kolos, Moskva.

Fiser R. E. (1958) Statisticeskie metodi dlia izsledovatelej, Nauka, Moskva.

Gorecki R.and Esashi Y.(1993) Ultr. biolum.of pea seeds of dif. vigior., An. Acad. Agr. (Poland), v53, 69-79.

Jaworsky E. G.(1971) Method of derminate of NRA activity, Biochim.,Biophys.,Res.Comm.,43, 6, 274-280.

Krasnook N. P. and Visnnakova A. (1978) Sodergan i raspredel belkov v goroha,Buletin VNNIR,77,26- 30.

Lowry R.N. (1951) Protein measurement with Foliu-phenol reagent,J. Biolog. chem., 2, 193, 265 -267.

Ovcarov E.K.(1976) Fiziologia formirovania i prorastania semian, Kolos, Moskva.

Rahimov A.S.i T. G. Muminov(1979)Fiziolog. osn. povis. produkt. semian ovos. i bohc. kultur,Fan,Taskent.

Ross G.S.(1990) The distribution of abscis acid in P.sativum pl. during seed devel., J pl.phys.,136,137 - 142.

Saio K., Kobajokama K. and Cito M.(1982) Content of protein in pea plant,.Coreal chem., 59 (5), 408 - 412.

THE EFFECTS OF COLD TREATMENTS AND ACTIVATED CHARCOAL ON ABA CONTENTS OF ANTHERS AND *in vitro* ANDROGENESIS IN EGGPLANT (*Solanum melongena* L.)

R. TIPIRDAMAZ[1] AND Ş. ELLİALTIOĞLU[2]
[1]Hacettepe University, Faculty of Science, Department of Biology Beytepe Ankara - TURKEY, [2]Ankara University, Faculty of Agriculture Department of Horticulture, Ankara- TURKEY

1. Abstract

The effects of cold pretreatments on eggplant buds and additions of activated charcoal to the nutrient medium on the abscisic acid contents of the anthers in eggplant anther culture were investigated. As cold shock treatments 80 hours at +4°C and 9 days at +9°C were used. Anthers with no cold shock treatment cultured in DDV[1] medium without any activated charcoal led to embryo development. These results indicated that ABA in anthers is not the only factor in obtaining embryos from anther cultures and that eggplant ABA levels of anthers were unlikely to be at the level to hinder embryogenesis.

2. Introduction

One of the problems in obtaining haploids from anther cultures is obtaining very limited number of embryos. To overcome this problem, many techniques have been proposed. It has been shown that pretreatment of whole buds or anthers can increase embryo production. It has been suggested that a thermal shock administered to excised flower buds before the removal of the anthers is essential for the culture of anthers [2]. The shock is proposed to alter the course of the highly polarized first pollen division. Formation of generative cell, a key feature of gametophytic development, is by-passed and the microspores are diverted directly into embryogenesis by a series of equal cell divisions [3]. The addition of activated charcoal to the nutrient medium has been the subject of some interest during recent years. Activated charcoal plays an extremely important role; it adsorbs not only toxic substances (which result in a brown color) from degenerated anthers, but also ABA which prevents embryogenesis [4].
In this paper the effects of activated charcoal and some cold pretreatments to flower buds of eggplant on the ABA content of anthers and embryo formation in anther culture were investigated.

3. Material and Methods

Anthers from *Solanum melongena* L. cv. Kemer were used. Flower buds with anthers containing microspores at the first mitotic stage were harvested [5]. Flower buds were treated at +4°C for 80 h and at +9°C for a 9-day period. The buds in the control group were not subjected to cold conditions. The basal medium used for anther culture was as described by Dumas de Vaulx and Chambonnet [1]. The C medium containing 12% sucrose, and 5 mg/l 2,4-D and 5 mg/l kinetin was used for the first 12 days of culture. The R medium used during the remainder of the culture period contained 3%sucrose and 0.1 mgl^{-1} kinetin. The C and R media were solidified with 0.8% agar (Difco-Bacto agar). % 1 activated charcoal was added to one of the media. The buds were surface sterilized in 20% commercial bleach for 15 minutes and then rinsed 3 in sterile distilled water. Anthers were dissected and cultured on medium C. Cultures were incubated at +35°C in dark for eight days. After this period anthers were transferred to C medium and were subjected to 25°C in the 16/8 h photoperiod illumination for an additional 4 days. After a total twelwe days on medium C, anthers were transferred to medium R. ABA analyses were made at three different stages with the buds obtained. After the buds were separated from the plant and after cold pretreatment was made they were used for the analyses at the first stage. The control anther samples, after separating from the plant, were taken from the buds on which no pretreatment was made. ABA analyses were made in the second stage, after putting the anthers in the culture and after the termal shock made at +35°C, and in the third stage after the incubation for 50 days. ABA content of anthers was analysed as its methyl ester (ABA -Me) by following the method of Johansson et al. [6].

4. Results

The eggplant anthers obtained from the buds on which cold treatment was made at different low temperatures and different periods of time, produced lower ABA values was compared with the control samples. The decrease appearing from the viewpoint of the ABA content in the anthers kept in the cold was statistically at the probability level of 0.01 and was not found to be significant. At the progressing stages of the culture (Stages 1 and 2), a significant difference was not observed from the aspect of the ABA content between the anthers in which cold treatment was applied or not applied. In spite of this, the ABA amount in the anthers that had cold treatment always remained lower than in the control anthers (Table 1).

TABLE 1. The ABA amounts (ng/g fresh weight) in the samples taken at different stages in the anthers of the Kemer eggplant variety in which cold treatments were applied or not applied

	Stage 1	Stage 2	Stage 3
Control	0.649378	0.711460	0.239471
+4°C for 80 h	0.537624	0.700455	0.375190
+9°C for 9 days	0.487134	0.626598	0.187250
Means*	0.558045 a	0.679504 a	0.267303 b

* Means with the same letters in each column do not differ significantly at the 1% level.

In the experiment, an interaction was not found between the cold treatment and the sample taking stages, but the sample taking stages displayed a statistically significant difference among themselves. Even if an increase in the ABA content was observed after a +35°C thermal shock treatment, this increase was found at the probability level of 0.01. On the 50th day of the culture, the ABA content of anthers decreased in all the applications in the experiment. The ABA amount in the anthers decreased to levels lower than those at the beginnig of the culture (Figure 1). The results obtained from the anther culture have been shown in Table 2. In media, where activated charcoal was added, the vigor of the eggplant anthers was lost, particularly after +35°C. In spite of this, the anthers transferred to an R media (containing activated charcoal) dried up and even no development was observed. For this reason, Table 2 shows only the results obtained from media to which no activated charcoal was added.

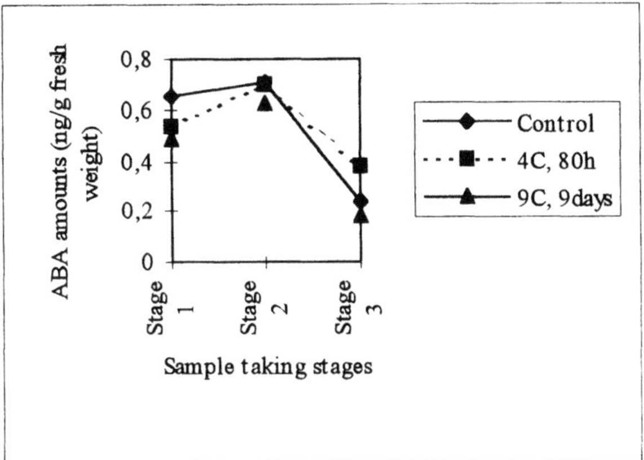

Figure 1. The influences on the ABA amount (ng/g fresh weight) in the anther samples taken at different times from the cold shock treatment that was applied to the buds of Kemer eggplant variety.

TABLE 2. The responses given to the anther culture in regeneration media of the anthers that have undergone different cold shocks in the Kemer eggplant variety.

Cold treatments	No. of cultured anthers	Developed anthers (%)	No. of anthers for ABA analyses	Embryo (n)	formation (%)	Plantlet (n)	formation (%)
Control	500	74.9	100	20	7.75	18	6.99
+4°C,80h	500	67.5	100	0	0	0	0
+9°C,9days	500	86.7	100	0	0	0	0

The formation of embryos was obtained directly from the anthers put in the culture without applying cold pretreatment (7.75%). The anthers on which cold treatment was applied exhibited a lighter brown color and a more lively appearance compared to the anthers in the control group. However, embryos were not obtained from the anthers of buds that underwent +4°C for 80 hours or +9°C for 9 days.

5. Discussion

Along with cold treatment applications in the eggplant anthers making a decreasing influence on the ABA amount, the difference remained at the probability level of 0.01. Whereas, in a study that we made on peppers, the cold shock we applied at the same temperatures and periods determined that the ABA content of anthers decreased in a very definitemanner [7]. Eventhough statistically, there was no difference that would draw attention, in the eggplant held in particular at +9°C for 9 days, a rather low level of ABA amount was found in the anthers of the buds when it is compared with controls.This situation shows that in general, the cold treatment applications makes a decreasing influence on the ABA content of anthers.

In the ABA analysis results made in the anthers after transferring them to a C medium and incubating them +35°C for a period of 8 days, in general, the ABA content was higher than the beginning values. However, this increase was not as great as those appearing in peppers and this make a peak in the graphs [7], in fact it was found to be insignificant at the probability level of 0.01. However, high temperature shock also causes an increase in the ABA content of eggplant anthers.

The ABA amount in the anthers on the 50th day of the culture also decreased below the amount at the beginning. This situation probably stemms from the loss of vigor in the anther and the conclusion of the synthesis of ABA as well as the other metabolites.

The fact of only being able to obtain embryos from the control groups in the anther cultures reveals the conclusion that the cold treatment applications made to the buds, and addition of the activated charcoal to the culture medium creats a negative effect in the eggplant anther cultures. It is thought that the existing ABA content of the eggplant anthers is not at a level that will prevent embryogenesis. In just the same way, it has been put forward by Johansson et al. [6] that the preventive effect of ABA on the pollen embryogenesis, in any case, only appears in case of being above a specific concentration. In conclusion, it has been determined that the ABA in the anthers cannot be the only factor that will influence embryogenesis in the eggplant anther culture.

6. References

1. Dumas de Vaulx, R. and Chambonnet, D. (1982) Culture in vitro d'antheres d'aubergine (*S.melongena* L.); Stimulation de la production de plantes qu moyen de traitements a +35°C associes a de faibles teneurs en substances de croissance. *Agronomie* **2** (10): 983-988.
2. Nitsch, J.P. (1977) in: Plant Cell, Tissue and Organ Culture (eds: J.Reinert and Y.P.S. Bajaj) Springer Verlag, New York, p: 268-278.
3. Sunderland, N., and Roberts, M. (1979) Cold-pretreatment of excised flower buds in float culture of tobacco anther. *Ann. Bot.* **43:** 405-414.
4. Johansson, L., and Eriksson, T. (1977) Induced embryo information in anther cultures of several *Anemone* species. *Physiol. Plant* **40**: 172-174.
5. Karakullukçu, Ş. (1991) Değişik Patlıcan Genotiplerinde *in vitro* androgenesis ve haploid bitki oluşumunu uyarıcı bazı etmenler üzerinde araştırmalar. A.Ü. Fen Bilimleri Enst. Doktora Tezi, Ankara, Paper No.135.
6. Johansson, L., Andersson, B., and Eriksson, T. (1982) Improvement of anther culture technique: Activated charcoal bound in agar medium in combination with liquid medium and elevated CO_2 concertration. *Physiol Plant.* **54**: 24-30.
7. Ellialtıoğlu, Ş., and Tıpırdamaz, R. (1997) Soğuk uygulamaları ve aktif kömürün patlıcan ve biberde *in vitro* androgenesis üzerine etkileri. TÜBİTAK TOGTAG 87 No'lu proje sonuç raporu, Temmuz 1997, Ankara, Paper No.72.

PROTOCOLS OF ORGANOGENESIS AND SOMATIC EMBRYOGENESIS ON *Juglans regia*

P. TSOULPHA[1]. and A. SCALTSOYIANNES[2]
1 Researcher 2 Assistant Professor
Laboratory of Forest Genetics and Plant Breeding,
Department of Forestry and Natural Environment,
Aristotle University of Thessaloniki, 54006 Thessaloniki.

1. Abstract

Organogenesis. Mature embryos of *Juglans regia*, selected from elite trees were cultured on Driver and Kuniyuki Walnut medium (DKW) supplemented with 4.44 μM 6-benzylaminopurine (BA). The obtained adventitious shoots were proliferated on DKW supplemented with 4.44 μM BA and 0.005 μM indole-3-buturic acid (IBA) and tested, successfully, for rooting on a vermiculite medium gelified with DKW. Great variation among *J. regia* clones was observed in, both, proliferation and rooting phases.

Somatic embryogenesis. Immature embryos and cotyledons of *J. regia* collected, weekly, during their developmental phase, were tested for their embryogenetic ability on DKW supplemented with: 1) no hormones 2) 0,1 μM 2,4-Dichlorophenoxyacetic acid (2,4-D) and 5.0 μM thidiazuron (TDZ), 3) 0.44 μM BA, 0.93 μM kinetin and 5.0 μM IBA and 4) 4.4 μM BA, 9.3 μM kinetin and 0.05 μM IBA. Cultures reacted best when cotyledons were gelified and the embryos were, completely, formed, 7-10 weeks after pollination.

2. Introduction

Walnut is a cultivated forest species which is widely recognised for its economic importance both for wood and nut production. To date, traditional vegetative propagation techniques of elite walnuts have proved, either inefficient (cuttings) or costly (grafting). Thus the majority of the research on the field, recently, was focused on the application of tissue culture techniques (micropropagation, somatic embryogenesis).

This work is part of the Doctorate thesis of the senior author.

Micropropagation. Among the first researchers that reported the establishment of walnut cultures *in vitro* were Rodriguez and Sachez-Tames (1981). Later, other workers
presented their work on cultures of *J. nigra, J. regia* embryos (Jay-Allemand and Cornu, 1986) and shoot tips, axillary buds and apical meristem cultures (Gruselle *et al.*, 1987).

Driver and Kuniyuki (1984) presented the successful application of a new medium for the *in vitro* culture of *Juglans* spp. while Jay-Allemand *et al.*, (1992) reported the successful rooting *in vitro* of the hybrid (*J. nigra* 23 x *J. regia*) microshoots in a mixture of vermiculite and DKW medium.

Somatic embryogenesis. one of the most efficient and economically attractive technique for the rapid propagation of plants is that of somatic embryogenesis which renders possible the production of large numbers of plantlets from immature tissues, mainly. Somatic embryos are similar to the zygotic both in their morphology and development (Tautorus *et al.*, 1991).

Somatic embryogenesis has been successfully applied in *Juglans* spp. by Tulecke and McGranahan (1985) and Cornu (1988) who reported the production of somatic embryos from immature cotyledons.

The present work describes the application of protocols of a) organogenesis on mature embryos and b) somatic embryogenesis on immature tissues of walnut individuals selected for their wood quality.

Abbreviations. BA: 6-benzylaminopurine, DKW: Driver and Kuniyuki Walnut medium (1984), 2,4-D: 2,4-Dichlorophenoxyacetic acid, IBA: indol-3-butyric acid, SE: Standard error, TDZ: Thidiazuron, LP: Long and Preece medium (1995).

3. Materials and Methods

a) *Organogenesis.* Nuts were collected from walnut trees, selected for their wood quality.

The procedures of surface sterilisation,, establishment multiplication and rooting were as described in previous work (Tsoulpha *et al.*, 1996).

b) *Somatic embryogenesis.* 20 immature nuts from selected walnuts were collected every week from the end of May till the third week of July. Immature embryos and/or parts of cotyledons were established for three weeks on DKW supplemented with: 1) no hormones (M1) 2) 0,1 μM 2,4-Dichlorophenoxyacetic acid (2,4-D) and 5.0 μM thidiazuron (TDZ) (M2), 3) 0.44 μM BA, 0.93 μM kinetin and 5.0 μM IBA (M3) and 4) 4.4 μM BA, 9.3 μM kinetin and 0.05 μM IBA (M4). The cultures were maintained at 25±1°C with 16 hr photoperiod. After three weeks the cultures were transferred half on DKW and half on LP medium free of hormones.

4. Results and Discussion

Organogenesis. After three weeks in darkness the disinfected embryos (41%) germinated, the shoot was swollen and adventitious shoots appeared. Subsequently, the elongated shoots were isolated and proliferated. The multiplication rates of eight *J regia* clones are shown in Table 1.

Table 1. Multiplication rates of eight *J. regia* clones after the fourth subculture.

Clone	Number of shoots	Multiplication rate
Π_2	127	2.8
Π_3	125	2.8
Π_4	15	1.5
Π_6	300	8.7
Π_7	81	3.5
Π_8	94	4.2
Π_{10}	118	4.1
Π_{11}	23	1.9

In Table 2 the results on rooting of the above *J. regia* clones are represented. These results are in agreement with the findings of Jay-Allemand *et al.*, (1992) who reported rooting percentages 80-100% for the walnut hybrid (*J. regia* x *J. nigra*).

Table 2. Rooting results of eight *J. regia* clones.

Clone	Rooting (%)	Total root length (cm)	Secondary roots
Π_2	40±10.95	5.72±0.87	many
Π_3	93.3±6.45	17.6±2.71	many
Π_4	5±4.47	3.0	none
Π_6	64.3±12.8	19.5±2.57	many
Π_7	67±12.14	7.41±1.99	few
Π_8	40±10.95	8.5±11.98	many
Π_{10}	33.4±12.2	7.3±1.28	many
Π_{11}	25±9.7	22.2±7.0	few

Mean number ± SE. Sample size 20.

Table 3. Effect of immature embryo (e) and cotyledon (c) developmental stage on culture reaction (formation of callus, organogenesis, embryogenesis).

Collection date		Type of reaction		
		callus	embryogenesis	organogenesis
29/5	e	-	-	M4
	c	-	-	-
3/6	e	-	-	M4
	c	M2	-	M2
9/6	e	-	-	M4
	c	-	-	-
16/6	e	-	-	-
	c	-	-	-
20/6	e	-	M4	M4
	c	-	-	-
29/6	e	M2	M3	M1,M3
	c	M1,M2,M4	-	-
7/7	e	M1,M2	M2	M1,M3
	c	M1,M2,M4	M2	M1
13/7	e	M2	M2	M1,M2,M3
	c	M1,2,3,4	M2	M1,M2,M3
20/7	e	-	-	M1
	c	-	-	M1

M1: DKW, **M2**: DKW+0.1 μM 2,4-D and 5.0 μM (TDZ), **M3**: DKW+0.44 μM BA, 0.93 μM kinetin and 5.0 μM IBA and **M4:** DKW+4.4 μM BA, 9.3 μM kinetin and 0.05 μM IBA

Somatic embryogenesis It was observed (table 3) that the cultures of immature walnut tissues did not show any reaction until the mid of June. From the end of the same month somatic embryogenesis appeared both on zygotic embryos (direct) and on callus formed on cotyledon parts (indirect). The media used were M2, M3 and M4. Organogenesis was, also, observed in this stage. Germination of the zygotic embryo occurred in later stages (July) on M1, M2 and M3. Cotyledon pieces in these late stages showed no reaction.

5. References

Cornu D. 1988. Somatic embryogenesis in tissue cultures of walnut (*Juglans nigra, J. major* and hybrids *J. nigra* x *J. regia*).

Driver J., and Kuniyuki A.H. 1984. *In vitro* propagation of Paradox walnut rootstock. HortScience 19(4) 507-509.

Gruselle R., Badia N., Boxus Ph. 1987. Walnut micropropagation: First results. Acta Horticulturae 212.

Jay-Allemand and D. Cornu. 1986. Culture *in vitro* d'embryons isoles de noyer commun (*Juglans regia* L.). Ann Sci For, 43:189-198.

Jay-Allemand C., Capelli P. and Cornu D. 1992. Root development of *in vitro* hybrid walnut microcuttings in a vermiculite containing gelrite medium. Scientia Horticulturae 51: 335-342.

Long L.M., Preece J.E., and Van Sambeek J. W. 1995. Adventitious regeneration of *Juglans nigra* L. (eastern black walnut). Plant Cell Rep 14: 799-803.

Rodriguez R., R. Sanchez-Tames 1981. Cultivo detejitos y differentiation en nogal. Rev Fac Cienc. Univ. Oviedo (Ser Biologia) 22: 21-28.

Tautorus T.E., Fowke L.C. and Dunstan D.I. (1991). Somatic embryogenesisin conifers. Can. J. Bot. 69: 1873- 1899.

Tsoulpha P., A. Scaltsoyiannes, K.P. Panetsos , and .D. Moulalis. 1996 Application of Biotechnology in the

MORPHOGENESIS OF "FLOATING" STOMATA IN THE FERN *Anemia mandioccana* Raddi. STOMATAL PORE FORMATION

M. ZACHARIADIS, P. APOSTOLAKOS AND B. GALATIS
University of Athens, Department of Biology, Division of Botany
Athens 157-84, Hellas

1. Introduction

The stomatal pore opening is one of the most important processes accompanying guard cell (GC) morphogenesis. In angiosperms this process is carried out by the gradual detachment of the median thickened regions of the partner ventral walls (VWs), which starts from the external and/or the internal periclinal walls (PWs) and proceeds inwards [1], [2], and [6]. A different mode of stomatal pore formation operates in the ferns *Polypodium vulgare* [7] and *Adiantum capillus-veneris* [3]. In the latter plant the adjacent plasmalemmata at the mid-region of the post-cytokinetic VW of the GC pair move apart from each other forming the central pore aperture, while the PWs remain intact. At a later stage of differentiation the PW regions covering the central aperture are disrupted to complete the stomatal pore. The detailed mechanism by which this type of stomatal pore opens was examined in the fern *Anemia mandioccana* Raddi. In this study we used light and electron microscope techniques, as described in [1] and [2].

2. Results

The post-cytokinetic VW in *A. mandioccana* stomata is very thin and wavy, contains some loose material and is lined by anticlinal microtubules (Mts) along its whole length. Stomatal pore formation commences at the post-cytokinetic stage at which the anticlinal Mt bundles lining the mid-region of the VW start being organized (Figure 1). It starts at the center of the VW by the movement of the adjacent plasmalemmata apart from each other. In this way an internal rudimentary stomatal pore opens (Figures 1, 2), which is bordered by plasmalemmata lined by numerous Mts; the latter are associated with the former by electron dense bridges (Figure 1). The further growth of this intercellular space up to the PWs keeps pace with the organization of the anticlinal Mt bundles along the mid-region of the VW. Gradually, wall materials are deposited on the whole surface of the VW (Figure 3). Furthermore, unequal triangular wall thickenings are deposited at the junctions of the PWs with the mid-region of the VW. The outer thickening (Figure 4) is more intense than the inner one and is progressively impregnated by amorphous, lipophilous material. At this stage, the rear-

and fore- pore chambers are formed (Figure 3). Gradually, the stomatal pore, especially the pore chambers are lined by a thin layer of cuticle (Figure 4). It still remain closed, being covered by thin layers of PWs (Figure 4). The Mts lining the VW maintain their organization. They are arranged in one to three units deep into the cytoplasm (Figure 3). Under the periclinal walls they radiate away from the VW thickenings. At an advanced stage of GC differentiation the PWs are disrupted. Thus, the stomatal pore is completed and the GC ledges are formed.

Stomatal pore formation was also examined in stomata developed in the presence of colchicine, which induces an almost complete Mt disorganization in the differentiating GCs. Mt disorganization in young GC pairs inhibits central pore aperture formation, i.e. the mutual separation of the plasmalemmata at the middle of the VW does not proceed. In contrast the fore- and rear- chambers of the stomatal pore are formed as usually (Figure 5).

3. Discussion

The formation of stomatal pore in *Anemia mandioccana* proceeds as in the fern *Adiantum* [3]. The entire process is accomplished in three successive steps: i) The formation of the central aperture of the stomatal pore by the local separation of the adjacent plasmalemmata at the mid-depth of the nascent VW, before the deposition of a distinct wall layer on it. ii) The formation of the fore- and rear- pore chambers, which gradually are lined by cuticular materials. iii) The stomatal pore completion by disruption of the PWs, covering its external and internal ends. In colchicine-treated stomata the first step is inhibited while the next two proceed normally. The above support that Mts play an essential role in stomatal pore formation. Somehow, through Mt-membrane interactions, the Mts pull locally inwards the plasmalemmata forming the central pore aperture. In other vascular plants, pore formation seems to depend completely on mechanical forces, generated by an increase in turgor [4], [5] and [6]. However, in these plants the middle lamella of the VW becomes structurally differentiated appearing electron transparent. It lacks pectic polysaccharides [1], [2].

Figure 1. The mid region of the VW of a very young stoma in paradermal view. The plasmalemmata have moved apart from each other forming the stomatal pore. Mts (arrows) line the plasmalemmata. The former are attached to the later via electron dense bridges (arrowheads). *X52000*

Figure 2. Young GC pair in paradermal view. The VW is still thin. The stomatal pore (arrow) has been broadened. *X4000*

Figure 3. High magnification of the upper portion of the VW region of a GC pair in transverse section. Triangular wall thickenings are deposited at the junction of the PWs with the mid-region of the VW. The fore- pore chamber (large arrow) is formed. The small arrows point to Mts. *X36000*

Figure 4. Transverse section of an almost completely differentiated GC. The outer VW thickening is much larger than the inner one and is impregnated by amorphous material. Cuticle line the stomatal pore, which is still closed, being covered by thin layers of PWs (arrows). *X10000*

Figure 5. GC pair affected with colchicine. The central pore aperture is not formed, while the fore- and rear- pore chambers are (arrows). The VW exhibits an irregular thickening (arrowhead). Compare with Figure 4. *X8000*

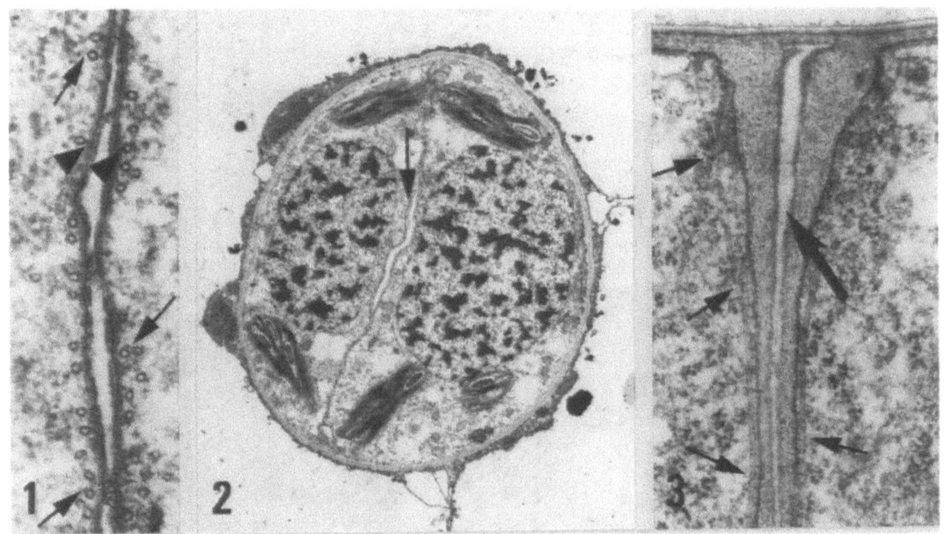
1
2
3

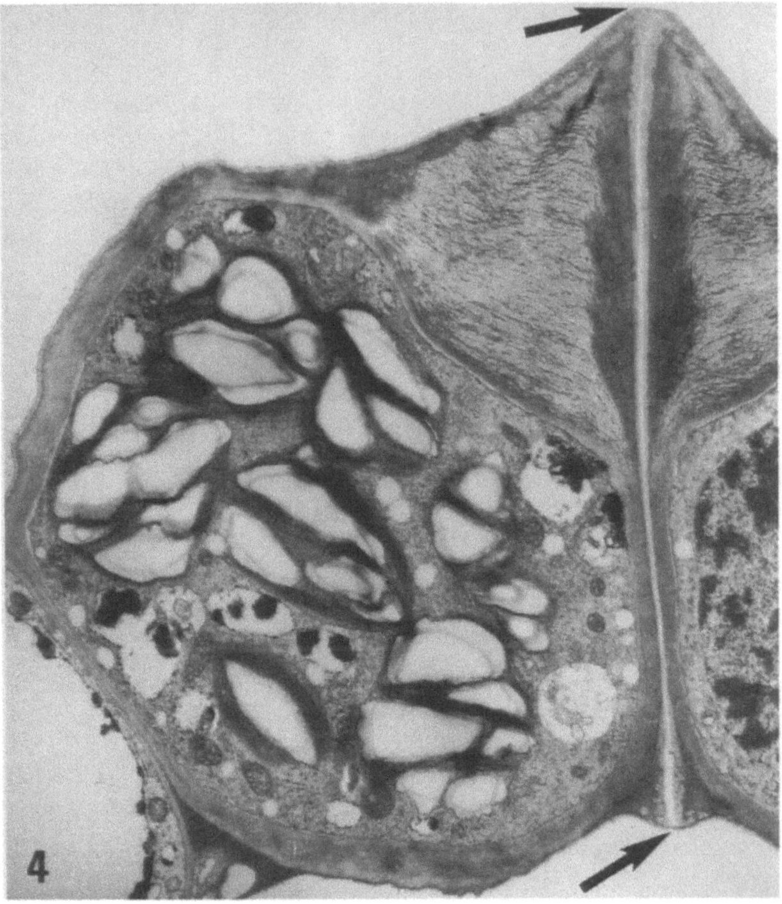

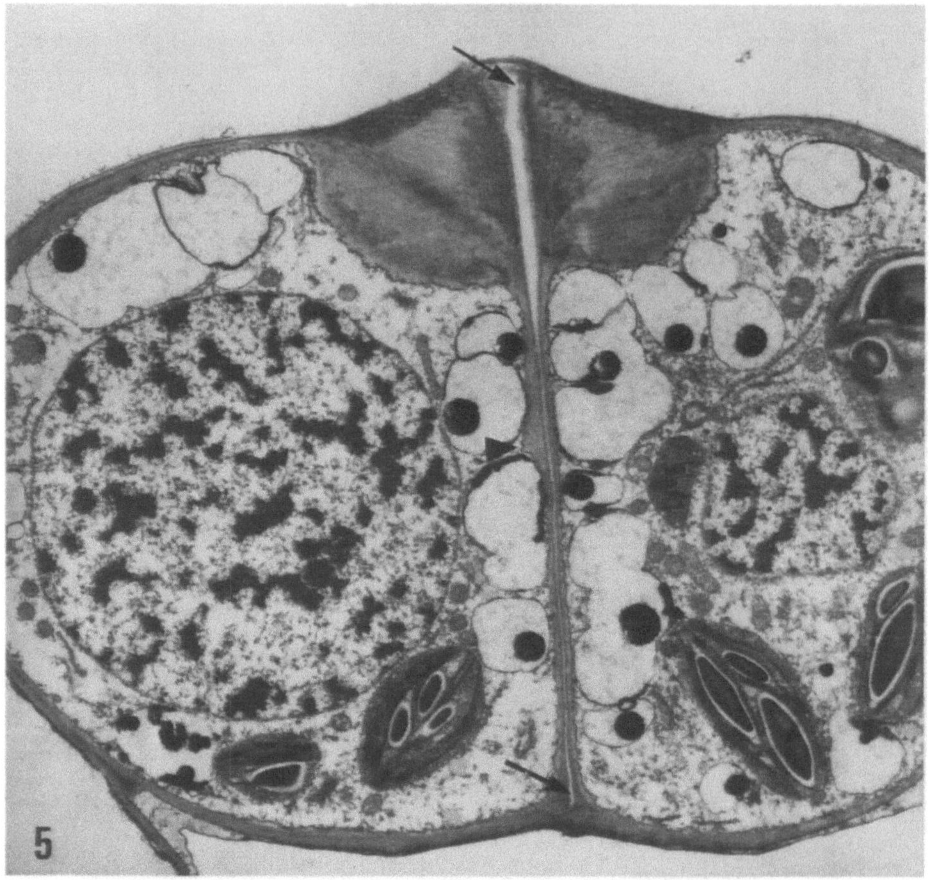

4. References

1. Galatis B. (1980). Microtubules and guard-cell morphogenesis in *Zea mays* L. *Journal of Cell Science* **45**, 211-244
2. Galatis B., Mitrakos K. (1980). The ultrastructural cytology of the differentiating guard cells of *Vigna sinensis. American Journal of Botany* **67**, 1243-1261
3. Galatis B., Apostolakos P., and Katsaros Chr. (1983). Microtubules and their organizing centers in differentiating guard cells of *Adiantum capillus veneris. Protoplasma* **115**, 176-192
4. Palevitz B. A. (1981). The structure and development of stomatal cells. In Jarvis P. G., Mansfield T. A. (eds.), *Stomatal physiology.* Cambridge University Press, Cambridge pp. 1-23
5. Palevitz B. A. (1982). The stomatal complex as a model of cytoskeletal participation in cell differentiation. In Lloyd C. W. (ed.), *The cytoskeleton in plant growth and development.* Academic Press, London pp. 345-376
6. Sack F. D. (1987). The development and structure of stomata. In Zeiger E., Farquhar G. D. Cowan I. R. (eds.), *Stomatal function.* Stanford University Press, Stanford pp. 59-89
7. Stevens R. A. and Martin E. S. (1978). Structural and functional aspects of stomata. I. Developmental studies in *Polypodium vulgare. Planta* **142**, 305-316.

INDEX OF AUTHORS

M

N

O

P

R

S

SUBJECT INDEX

T

U

MIX
Papier aus verantwortungsvollen Quellen
Paper from responsible sources
FSC® C105338

If you have any concerns about our products, you can contact us on
ProductSafety@springernature.com

In case Publisher is established outside the EU, the EU authorized representative is:
Springer Nature Customer Service Center GmbH
Europaplatz 3, 69115 Heidelberg, Germany

Printed by Libri Plureos GmbH
in Hamburg, Germany